OPTICAL BIOSENSORS: TODAY AND TOMORROW

OPTICAL BIOSENSORS: TODAY AND TOMORROW

Edited by

FRANCES S. LIGLER

and

CHRIS ROWE TAITT

Center for Bio/Molecular Science & Engineering
US Naval Research Laboratory
Washington DC, USA

ELSEVIER

Amsterdam • Boston • Heidelberg • London • New York • Oxford
Paris • San Diego • San Francisco • Singapore • Sydney • Tokyo

Elsevier
Radarweg 29, PO Box 211, 1000 AE Amsterdam, The Netherlands
Linacre House, Jordan Hill, Oxford OX2 8DP, UK

1st Edition, *Optical Biosensors: Present and Future*, published 2002.
2nd Edition, *Optical Biosensors: Today and Tomorrow*, published 2008.

Library of Congress Cataloging-in-Publication Data
A catalog record for this book is available from the Library of Congress

British Library Cataloguing-in-Publication Data
A catalogue record for this book is available from the British Library

ISBN: 978-0-444-53125-4

For information on all Elsevier publications visit our website at books.elsevier.com

Printed and bound in the United Kingdom
Transferred to Digital Print 2010

Dedication

The authors thank Kristyn Thiel for her capable assistance in the compilation of these chapters. We acknowledge financial support from NRL Work Unit 62-6006. We dedicate this book to our husbands, George and John, whose support gave us the courage to even contemplate undertaking a second edition of this book.

Contents

Preface

When we prepared the first edition of *Optical Biosensors*, titled *Optical Biosensors: Present and Future*, the field of optical biosensors was rapidly evolving on the coattails of advances in optical telecommunications. Since that time, there has been an explosion of new and exciting science and engineering that has direct bearing on the ability of optical biosensors to have increased and widespread application. For instance, advances in the "nano" realm – nanoelectronics, nanomaterials, nanofluidics – have paved the way for development and integration of new methods and materials for signal generation and transduction, as well as for miniaturization and automation of existing systems. The inherently cross-disciplinary nature of biosensing creates fertile ground for new perspectives for addressing critical challenges. In our experience, the best team for creating biosensors includes experts not only in biochemistry, chemistry, and optical physics, but also mechanical, chemical, electrical and bio-engineering. For this reason, *Optical Biosensors: Today and Tomorrow* includes a number of chapters describing future relevant technologies, in addition to chapters describing state-of-the-art optical biosensor systems.

This book is divided into two parts. "Optical Biosensors: Today" includes comprehensive discussions of technologies that have proven utility and, in many cases, are commercially available now. Seven of these chapters have been updated from the previous edition of *Optical Biosensors*, with descriptions of highly noteworthy developments over the last 6 years in the areas of biosensors based on fiber optics and planar waveguides, flow immunosensors, electrochemiluminescence, and surface plasmon resonance. Due to the increasing popularity of surface-enhanced Raman sensors, a chapter on this technology has been added. "Optical Biosensors: Tomorrow" consists of discussions of science and technologies that the editors consider exciting in terms of their potential to revolutionize future biosensor systems. Chapters with new material on aptamers, molecularly imprinted polymers, intracellular sensing, and microfluidics are again featured, and new chapters on cavity ring down spectroscopy, optical cantilevers, microarrays, single-domain antibodies, and nanoparticles have been added.

Contributors were tasked with describing the underlying principles behind each technology, detailing the demonstrated and potential applications, and providing a (hopefully!) objective view of the advantages and disadvantages of the technology. We also asked them to provide a short history of their system or science, not only to provide a context for the current state of the art, but also to recognize the pioneering work on which the present and the future of optical biosensing is based. Finally, the authors gaze into their crystal balls to envision the place of their technology in the world of tomorrow. These last observations are meant to be both candid and thought-provoking.

The successful marriage of biomolecules, cells, and tissues with opto-electronic detection platforms, not to mention the transformation of such tools into user-friendly systems, requires broad understanding of the possible scientific and technical options. This task becomes less daunting with the ability to reference factual summaries and informed opinions provided by the leaders in the field. We therefore thank our contributing authors for the exposition of their insights into this rapidly expanding field. We are also very grateful to both the pioneering spirits who have historically led the field and the hardy souls who continue to make breakthroughs in biological sensing. We thank them all for their hard work, willingness to share their ideas and perspicacity, and their devotion to a field whose challenges continue to change and inspire.

And finally, thank you for opening this book. It is for you, the curious reader, that *Optical Biosensors: Today and Tomorrow* has been created.

Fran and Chris

List of Contributors

Mar Alvarez
Centro Nacional de Microelectronica (CNM)
Biosensors Group
IMM-CNM-CSIC
Madrid
Spain
Email: malvarez@imm.cnm.csic.es

George P. Anderson
Center for Bio/Molecular Science & Engineering
Naval Research Lab
Code 6900
4555 Overlook Avenue, SW
Washington, DC 20375-5348
USA
Email: george.anderson@nrl.navy.mil

Jutta Bachmann
Bachmann Consulting
Nokkefaret 12
1450 Nesoddtangen
Norway
Email: info@jutta-bachmann.com

Israel Biran
Applied CleanTech Inc.
Email: ibiran01@yahoo.com.

Diane A. Blake
Department of Biochemistry
Tulane University School of Medicine

New Orleans, LA 70112-2699
USA
Email: blake@tulane.edu

Y. Charles Cao
Department of Chemistry
University of Florida
Leigh Hall
Gainesville, FL 32611
USA
Email: cao@chem.ufl.edu

Eun Jeong Cho
Department of Chemistry and Biochemistry
University of Texas at Austin
Austin, TX 78712
USA
Email: euncho@mail.utexas.edu

James B. Delehanty
Center for Bio/Molecular Science & Engineering
Naval Research Lab
Code 6900
4555 Overlook Avenue, SW
Washington, DC 20375-5348
USA
Email: james.delehanty@nrl.navy.mil

Andrew D. Ellington
Department of Chemistry and Biochemistry
University of Texas at Austin
Austin, TX 78712
USA
Email: andy.ellington@mail.utexas.edu

Ellen R. Goldman
Center for Bio/Molecular Science & Engineering
Naval Research Lab

Code 6900
4555 Overlook Avenue, SW
Washington, DC 20375-5348
USA
Email: ellen.goldman@nrl.navy.mil

Andrew Hayhurst
Southwest Foundation for Biomedical Research
Department of Virology and Immunology
San Antonio, TX 78227
USA
Email: ahayhurst@sfbr.org

Jiří Homola
Academy of Sciences of the Czech Republic
Institute of Photonics and Electronics
182 51 Prague
Czech Republic
Email: homola@ufe.cz

James W. Jacobson
Luminex Coporation
12212 Technology Blvd.
Austin, TX 78727
USA
Email: jwjacobs@luminexcorp.com

Thomas O. Joos
NMI Natural and Medical Sciences Institute
University of Tübingen
Markwiesenstr. 55
72770 Reutlingen
Germany
Email: joos@nmi.de

Huaizhi Kang
Department of Chemistry
University of Florida

Leigh Hall
Gainesville, FL 32611
USA
Email: hkang@chem.ufl.edu

Anne W. Kusterbeck
Center for Bio/Molecular Science & Engineering
Naval Research Lab
Code 6900
4555 Overlook Avenue, SW
Washington, DC 20375-5348
USA
Email: anne.kusterbeck@nrl.navy.mil

Laura M. Lechuga
Centro Nacional de Microelectronica (CNM)
Biosensors Group
IMM-CNM-CSIC
Madrid
Spain
Email: laura@imm.cnm.csic.es

Joo-Woon Lee
Division of Liberal Arts and Sciences
Chungju National University
Chungju, Chungbuk, 380-702
Korea
Email: jwoonlee@chungju.ac.kr

Kevin K. Lehmann
Department of Chemistry
University of Virginia
Charlottesville, VA 22904
USA
Email: lehmann@virginia.edu

Frances S. Ligler
Center for Bio/Molecular Science & Engineering
Naval Research Laboratory
4555 Overlook Avenue, SW
Washington, DC 20375-5320
USA
Email: frances.ligler@nrl.navy.mil

Jinny L. Liu
Center for Bio/Molecular Science & Engineering
Naval Research Lab
Code 6900
4555 Overlook Avenue, SW
Washington, DC 20375-5348
USA
Email: jinny.liu@nrl.navy.mil

Igor L. Medintz
Center for Bio/Molecular Science & Engineering
Naval Research Lab
Code 6900
4555 Overlook Avenue, SW
Washington, DC 20375-5348
USA
Email: igor.medintz@nrl.navy.mil

Miguel Moreno
Centro Nacional de Microelectronica (CNM)
Biosensors Group
IMM-CNM-CSIC
Madrid
Spain
Email: mmoreno@imm.cnm.csic.es

David Myszka
Center for Biomolecular Interaction Analysis
University of Utah
Salt Lake City, UT 84132
USA
Email: dmyszka@cores.utah.edu

Meghan O'Donoghue
Department of Chemistry
University of Florida
Leigh Hall
Gainesville, FL 32611
USA
Email: mod@chem.ufl.edu

Sergey A. Piletsky
Cranfield Health
Cranfield University
Silsoe, Beds, MK45 4DT
UK
Email: s.piletsky@cranfield.ac.uk

Manjula Rajendran
Althea Technologies, Inc.,
11040 Roselle Street,
San Diego, CA 92121
USA
Email: mrajendran@altheatech.com

Mark M. Richter
Department of Chemistry
Missouri State University
Springfield, MO 65897
USA
Email: MarkRichter@missouristate.edu

Kim Sapsford
US Food and Drug Administration
CDRH/OSEL/DB
Silver Spring, MD 20993
USA
Email: kim.sapsford@fda.hhs.gov

Abraham D. Stroock
School of Chemical and Biomolecular Engineering
Cornell University
Ithaca, NY 14850
USA
Email: ads10@cornell.edu

Chris Rowe Taitt
Center for Bio/Molecular Science & Engineering
Naval Research Lab
Code 6900
4555 Overlook Avenue, SW
Washington, DC 20375-5348
USA
Email: chris.taitt@nrl.navy.mil

Weihong Tan
Department of Chemistry
University of Florida
Leigh Hall
Gainesville, FL 32611
USA
Email: tan@chem.ufl.edu

Peter B. Tarsa
Massachusetts Institute of Technology
Cambridge, MA 02139
USA
Email: ptarsa@mit.edu

Richard B. Thompson
Department of Biochemistry and Molecular Biology
University of Maryland
Baltimore, MD 21201
USA
Email: rthompso@umaryland.edu

Anthony P.F. Turner
Cranfield Health
Cranfield University
Silsoe, Beds, MK45 4DT
UK
Email: a.p.turner@cranfield.ac.uk

Tuan Vo-Dinh
Fitzpatrick Institute for Photonics
Duke University
Durham, NC 27708-0281
USA
Email: tuan.vodinh@duke.edu

David R. Walt
Department of Chemistry
Tufts University
Medford, MA 02155
USA
Email: david.walt@tufts.edu

Lin Wang
Department of Chemistry
University of Florida
Leigh Hall
Gainesville, FL 32611
USA
Email: lwang@chem.ufl.edu

Sinclair S. Yee
Department of Electrical Engineering
University of Washington
Seattle, WA 98105
USA
Email: yee@ee.washington.edu

Xin Yu
Department of Chemistry
Tufts University
Medford, MA 02155
USA
Email: Xin.Yu@tufts.edu

Kirill Zinoviev
Centro Nacional de Microelectronica (CNM)
Biosensors Group
IMM-CNM-CSIC
Madrid
Spain
Email: Kirill.Zinoviev@cnm.es

Part I

Optical Biosensors: Today

Optical Biosensors: Today and Tomorrow (2nd Edition)
Frances S. Ligler and Chris Rowe Taitt (Editors)

Chapter 1

OPTRODE-BASED FIBER OPTIC BIOSENSORS (BIO-OPTRODE)

Israel Biran[a], Ph.D., Xin Yu[b], Ph.D., and David R. Walt[b], Ph.D.
[a] Applied CleanTech Inc.
[b] Department of Chemistry, Tufts University, Medford, MA 02155, USA

Optrode-based fiber optic biosensors (bio-optrodes) are analytical devices incorporating optical fibers and biological recognition molecules. Optical fibers are small and flexible "wires" made out of glass or plastic that can transmit light signals, with minimal loss, over long distances. The light signals are generated by a sensing layer, which is usually composed of biorecognition molecules and dyes, coupled to the fiber end. Light is transmitted through the optical fibers to the sensing layer where different optical phenomena such as absorption or luminescence are used to measure the interactions between the analyte and the sensing layer. Bio-optrodes can be used for remote analytical applications including clinical, environmental, and industrial process monitoring. In the last decade, due to the rapidly growing use of fiber optics for telecommunication applications, new fiber optic technologies have been developed resulting in high-quality and inexpensive optical fibers that can be used for bio-optrode applications. Recent advancements in bio-optrode technologies include the development of nanoscale bio-optrodes, enabling measurements inside single living cells, and the development of multianalyte and reagentless bio-optrodes. Although currently there are only a very limited number of bio-optrodes commercially available, it is expected that the development of advanced bio-optrode technologies will lead to many commercially available devices for various analytical applications.

1.1. Principle of operation

The word "optrode" is a combination of the words "optical" and "electrode" and refers to a fiber optic-based analytical device that can measure the concentration of a specific chemical or a group of chemicals in a sample of interest. The basic design of an optrode system is shown in Figure 1.1. The main components of an optrode are (a) a light source; (b) an optical fiber to both transmit the light and act as the substrate; (c) the sensing material, which is usually immobilized to the surface of the end face of the fiber; and (d) a detector to measure the output light signal. Computers or microprocessors are used to control the optrode instrumentation and are employed to analyze the output signals.

The "heart" of the optrode is the sensing element. When the sensing element interacts with the analyte, it undergoes physicochemical

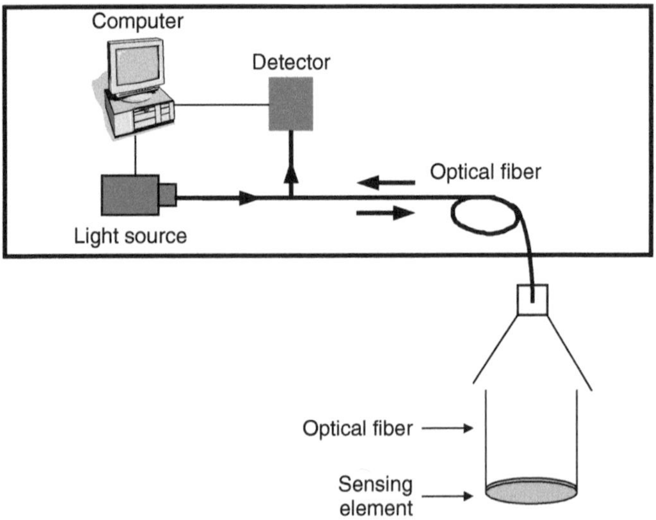

Figure 1.1 Schematic diagram of optrode system.

transformations that change its optical properties. This transduction mechanism generates optical signals that can be correlated to the analyte concentration. The optical signals are measured by launching light from the light source through the optical fiber to the fiber end, where the sensing element is immobilized. The same fiber (Figure 1.1), or a different fiber, is used to guide the output light to the detector (e.g., spectrophotometer, fluorometer) where the reflected, emitted, or absorbed light is measured. Optrode biosensors or bio-optrodes are optrodes in which the sensing elements are of biological origin. Biological sensing elements, such as enzymes, nucleic acids, antibodies, and cells, are immobilized on optical fibers and used for specific recognition of many different analytes (Cunningham, 1998; Mehrvar *et al.*, 2000; Wolfbeis, 2000a, 2000b, 2000c, 2000d, 2002, 2004, 2006; Kuswandi *et al.*, 2001; Lopez-Higuera, 2002; Narayanaswamy *et al.*, 2003; Monk and Walt, 2004a, 2004b; Brogan and Walt, 2005). Since most biological sensing elements and most analytes do not possess intrinsic spectral properties, the biorecognition events are transduced to optical signals (e.g., changes in fluorescence or absorbance) by coupling optically responsive reagents to the sensing elements. For example, fluorescent dyes are used to label nucleic acids and convert the biorecognition interaction between two complementary DNA strands into a fluorescence signal. In another example, an indicator dye, which is optically sensitive to changes in H^+ concentrations, is used to transduce enzymatic activity that consumes or releases H^+ into an optical signal. The signals are generated on the fiber optic face and transmitted by the optical fiber to a remote measurement device. The small dimensions of bio-optrodes allow measurement in very small sample volumes, which make them suitable for various clinical applications (Meadows, 1996; Vo-Dinh and Cullum, 2000; Vo-Dinh *et al.*, 2005). Bio-optrodes are useful for different sensing applications in the industrial, clinical, and environmental fields (Mulchandani and Bassi, 1995; Rogers and Poziomek, 1996; Scheper *et al.*, 1996; Rogers and Mascini, 1998; Marose *et al.*, 1999; Merhrvar, *et al.*, 2000; Kumagai and Kajioka, 2002; Berthold and Lopushansky, 2004).

In this section, the basic characteristics of optical fibers and the optical methods used to transduce a biorecognition event to an optical signal are described. The instrumentation employed in optrode biosensors, the biological sensing elements, and the methods to immobilize them on the fiber optic surfaces are summarized.

1.1.1. Optical fiber characteristics and use in bio-optrodes

Optical fibers are small and flexible "wires" made out of glass or plastic that can transmit light signals, with minimal loss, for long distances. Optical fibers are remarkably strong, flexible, and durable and therefore can be used in harsh and hazardous environments. Optical fibers are non-electrical, which make them highly suitable for applications where the presence of electric current is detrimental (e.g., *in vivo* monitoring inside a patient body). In the last decade, due to the rapidly growing use of fiber optics for telecommunication applications, new fiber optic technologies have been developed, resulting in high-quality and inexpensive optical fibers that can also be used for sensing applications. Optical fibers can transmit multiple optical signals simultaneously, thereby offering multiplexing capabilities for sensing.

Optical fibers consist of a core with a refractive index, n_1, surrounded by a cladding with a lower refractive index, n_2 (Figure 1.2). The difference in the refractive indices between the core and the cladding enables the core–clad interface to effectively act as a mirror such that a series of internal reflections transmits the light from one end of the fiber to the other as shown in Figure 1.3a. Light undergoes total internal reflection (TIR) at the core–clad interface if two basic conditions are fulfilled: (a) The light strikes the cladding at an angle greater than the *critical*

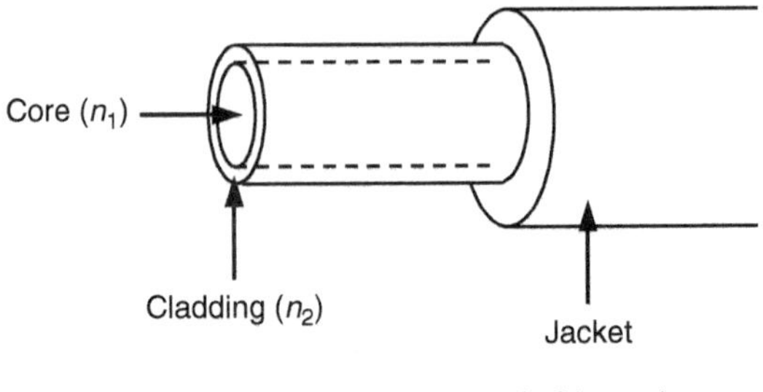

Figure 1.2 Schematic diagram of an optical fiber shows core and clad structure.

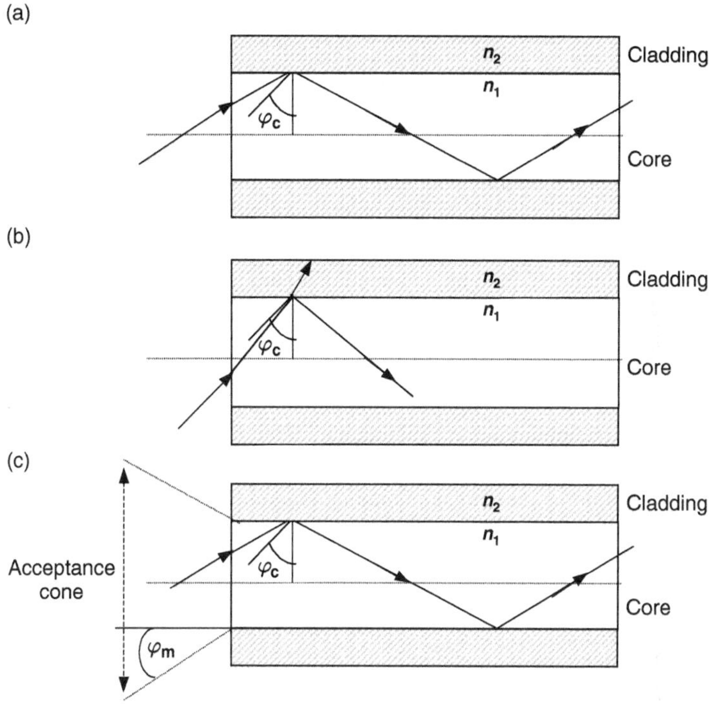

Figure 1.3 Propagation of light through the optical fiber occurs when the TIR condition exists at the interface between the core, (n_1), and clad, (n_2), such that $n_1 > n_2$. (a) Light entering the fiber will experience TIR, if the light angle is greater than the critical angle, φ_c. (b) Light will be partially reflected and partially refracted, if the light angle is less than the critical angle, φ_c. (c) Light will propagate by TIR only when the entering light angle is within the acceptance cone angle (φ_m) range.

angle, φ_c, (Figure 1.3a and b). The critical angle is defined by the ratio between the clad and the core refractive indices, as shown in Eqn (1.1):

$$\sin \varphi_c = \frac{n_2}{n_1} \tag{1.1}$$

(b) The angles of the light entering the fiber should be within the *acceptance cone* as shown in Figure 1.3c. The acceptance cone angle, φ_m, depends on the refractive indices of the core and the clad and also on the refractive index of the medium from which the light enters the fiber, n_o:

$$\sin \varphi_m = \frac{\sqrt{(n_1^2 - n_2^2)}}{n_0} \tag{1.2}$$

Another important parameter that defines the fiber's light collection efficiency is the *numerical aperture* (NA). This parameter is related to the acceptance cone's angle and is given by:

$$\mathrm{NA} = n_o \sin \varphi_m \tag{1.3}$$

A high NA indicates a wide acceptance cone and better light-gathering capabilities of the fiber. A typical NA value for a high-quality glass fiber is 0.55, but fiber NAs as high as 0.66 or as low as 0.12 (Masson *et al.*, 2006) have been used for sensing.

Optical fibers are usually made out of plastic and glass and have many different configurations, formats, shapes, and sizes. Glass fibers are the most commonly used fibers in optrode biosensors. Glass optical fibers can transmit light in the visible and near-infrared regions of the optical spectrum ($400\,\mathrm{nm} < \lambda < 700\,\mathrm{nm}$) and are therefore suitable for measuring fluorescence signals generated by most fluorescent dyes. For applications in which light in the UV region is required, quartz (pure silica) is used as the fiber's core material and doped silica (with a lower refractive index) is used as the cladding material. For most fiber optic-based biosensors, optical fibers with diameters ranging from 50 to 500 μm are employed.

Recently, fiber optic bundles (Figure 1.4a) comprising thousands of identical single fibers, each with a diameter of a few micrometers, were employed for bio-optrodes. The fibers can be bundled in a coherent or random fashion. In coherent fiber bundles, the position of each fiber on one end is identical to its position on the other end. These fibers were originally designed for imaging applications as shown in Figure 1.4b and are also often called "optical imaging fibers". Imaging fibers are suitable for multianalyte optrode biosensor design (Healey and Walt,

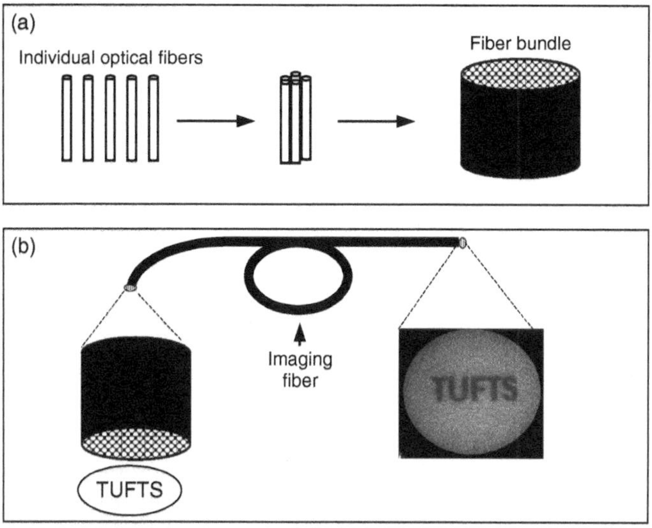

Figure 1.4 Optical fiber bundle fabrication and its use for imaging. (a) Fiber bundles are constructed from thousands of individual single fibers that are fused together. (b) Coherent bundles can be used for imaging (Pantano and Walt, 1995). Reprinted with permission from the American Chemical Society.

1995; Healey *et al.*, 1997a; Michael *et al.*, 1998; Steemers and Walt, 1999; Walt, 2000, 2002; Epstein and Walt, 2003; Wygladacz and Bakker, 2005), since each small individual fiber in the bundle can carry its own light signal from one end of the bundle to the other. Moreover, optical imaging fiber-based biosensors can be used for sensing and imaging simultaneously, providing remote spatial sensing capabilities (Walt, 1998; Issberner *et al.*, 2002).

1.1.2. Optical phenomena employed for biosensing in bio-optrodes

In bio-optrodes, dyes are coupled to the biological sensing element and transduce the biorecognition events to an optically detectable signal. Different optical phenomena—including fluorescence, luminescence, absorption, evanescent wave, and surface plasmon resonance (SPR)—are employed for monitoring these optical changes. In this section, the basic principles of these phenomena and their use in bio-optrodes are described.

Fluorescence is commonly used in bio-optrodes. Fluorescence occurs when molecules are excited at a specific wavelength and re-emit radiation at a lower energy, i.e., a longer wavelength. The absorption of the excitation light promotes the molecule's energy from its ground state to a higher energy state. The molecule emits fluorescent light when it returns to the ground state. Each fluorescent molecule has a unique fluorescence spectrum since the excitation and emission occur only at distinct energy levels corresponding to particular wavelengths. The characteristic fluorescence spectrum of particular molecules allows multiple fluorescent dyes to be used simultaneously in a single analytical assay. In fluorescence-based bio-optrodes, the fluorescence signals are measured by transmitting the excitation light through an optical fiber and measuring the light emission using a detector. Usually, the increase or decrease in fluorescence intensity is measured and then correlated to the analyte concentration. For example, when a fluorescent-labeled antibody is used as the sensing element, the fluorescence intensity is proportional to the amount of antigen (analyte) bound to the optical fiber. Both time- and frequency-domain methods could be used to measure fluorescence lifetime (Chandler *et al.*, 2006). In the time-domain approach, a short pulse of light (the impulse) is used to excite the fluorophore and the statistical decay curve (the impulse response) at a sequence of time steps is observed. In the frequency-domain method, sinusoidally modulated light is used to excite the fluorescent molecule. The resulting emission light also oscillates at the same frequency. The emission light is phase shifted (delayed) and demodulated with respect to the excitation light because of the finite lifetime of fluorescence. The phase shift is expressed as a phase angle from which the lifetime can be determined using simple relationships between the modulation frequency and the degree of demodulation. The concentration of analyte that induces changes in the molecule's fluorescence lifetime can be determined by measuring phase angle values (Thompson *et al.*, 1996).

A decrease in fluorescence intensity due to quenching can also be used for sensing. In this case, the biorecognition event causes a decrease in fluorescence (quenching) of the fluorescent reporter molecule. The fluorescence decrease is related to the analyte concentration. For example, a dye that undergoes fluorescence quenching when the pH decreases can be coupled to an enzymatic reaction that converts a substrate into an acidic product and results in a pH drop. Thus, the decrease in fluorescence can be correlated to the analyte concentration (see also Section 1.1.4.1). Fluorescence quenching is also a manifestation of another fluorescence phenomenon used for sensing in bio-optrodes – *fluorescence resonance energy transfer* (FRET). This phenomenon occurs when two distinct fluorophores are present. If the emission spectrum of one fluorophore overlaps with the excitation spectrum of a second fluorophore, and the two fluorophores are in proximity (<100 Å), then the excited fluorophore (donor) can transfer energy non-radiatively to the second fluorophore (acceptor). There are two types of acceptors. Quenchers are acceptors that are not fluorescent and therefore cause the donor simply to decrease its fluorescence emission intensity. Acceptors can also be fluorescent dyes that accept the energy non-radiatively from the donor, and then re-emit the energy at specific emission wavelength. This energy transfer results in an increase in light emission by the acceptor and a decrease in light emission from the donor. When an energy transfer pair of fluorophores is used to label two interacting molecules (e.g., antibody–antigen, enzyme–substrate), they can be used for sensing. Recently, both the donor and the acceptor molecules have been incorporated into single biological molecules such as proteins (Hellinga and Marvin, 1998; Grant *et al.*, 2005; Komatsu *et al.*, 2006) and nucleic acids (e.g., molecular beacons) (Tyagi and Kramer, 1996; Tyagi *et al.*, 2000; Wabuyele *et al.*, 2003). When these sensing molecules are in their native conformation, the donor and the acceptor are in proximity and therefore low-fluorescence signals from the donor are obtained. When the molecule interacts with the analyte, conformational changes occur, which separate the donor and the acceptor molecules and cause an increase in the fluorescence from the donor (see Section 1.3.3).

The most commonly used fluorescent molecules in bio-optrodes are organic dyes. Recently, self-fluorescent proteins have also been used. The sources of these proteins are marine organisms such as the jellyfish *Aequorea victoria* that produce the green fluorescent protein (GFP)

(Chalfie *et al.*, 1994). When GFP is excited, it emits light at a lower energy and therefore at a longer wavelength. GFP is highly fluorescent, with a quantum efficiency of approximately 80% and is very stable to heat and pH (5.5–12). GFP has been expressed in different cell types (bacteria, yeast, mammalian, and plant) and used as a reporter gene for different cellular events (Naylor, 1999; Serganova and Blasberg, 2005). In order to allow monitoring of several cellular events simultaneously, several GFP mutants have been developed, each with unique excitation and emission wavelengths. Cells expressing fluorescent proteins and purified proteins have been used to construct different bio-optrodes (see Sections 1.1.4.1 and 1.3.3).

Time-resolved fluorescence is another phenomenon used in bio-optrodes. This method is based on the fluorescent molecule's excited state lifetime. The light intensity emitted from a molecule excited by a short pulse of light decays exponentially with time. The decay time pattern is unique for each molecule and can be used for analytical purposes. Barker *et al.* (1999) used this method to improve the performance of a bio-optrode for nitric oxide detection. Fang *et al.* (2004) recently reported the design and development of a compact optical fiber-based apparatus for *in situ* time-resolved, laser-induced, fluorescence spectroscopy of biological systems.

Chemiluminescence is another light emission phenomenon used in bio-optrodes. In contrast to fluorescence, chemiluminescence is produced when a chemical reaction yields an excited species that emits light as it returns to its ground state. The use of chemiluminescence in biosensors, including fiber optic-based biosensors, was recently reviewed (Aboul-Enein *et al.*, 2000; Gubitz *et al.*, 2001; Zhang *et al.*, 2005; Marquette and Blum, 2006). In many bio-optrodes, the chemiluminescence of luminol is used to generate the light signal. The reaction between luminol and H_2O_2 produces a luminescence signal and is also catalyzed by certain ions or molecules (e.g., MnO_4^{2-}, I_2, Cu^{2+}). This reaction can be used, e.g., in enzyme-based bio-optrodes in which the enzymatic reaction generates H_2O_2 (see Section 1.3.3). Enzymes such as horseradish peroxidase can also catalyze or induce a chemiluminescence reaction by producing H_2O_2. In addition, alkaline phosphatase (AP) and β-galactosidase can be used to label biological sensing elements such as antibodies or nucleic acids. In the presence of a 1,2-dioxetane substrate (Bronstein *et al.*, 1996), these enzymes catalyze light formation proportional to the analyte concentration. Most recently, a fiber optic array biosensor was developed by Magrisso *et al.* (2006), in which chemiluminescence generated from both the complex cellular activity of granulocytes and the oxidative species involved in the bactericidal activity of granulocytes was used to monitor circulating phagocyte activity from multiple samples simultaneously. Chemiluminescence-based bio-optrodes are usually used in conjunction with flow cells. An optical fiber with an immobilized sensing element is placed inside the flow cell and transmits the light signals to the detector (Figure 1.5).

Bioluminescence is a biological chemiluminescent reaction. Many organisms produce bioluminescence for cell–cell signaling, self-protection, mating, attracting prey, and finding food (Campbell and Sala-Newby, 1993). The bioluminescence reaction is catalyzed by the

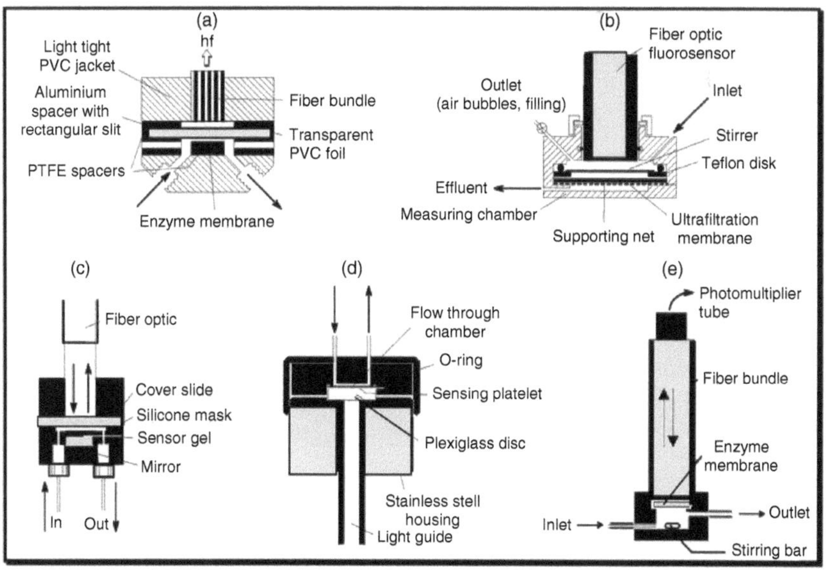

Figure 1.5 Design of flow cells incorporating bio-optrodes (Kuswandi *et al.*, 2001). Reproduced with permission of the Royal Society of Chemistry.

enzyme luciferase and requires the presence of oxygen. The bioluminescent substrate used in this reaction is called luciferin. Different luciferin molecules are used by different organisms. For example, aldehydes and flavins are used by bacteria and imidazolopyrazines are employed by some fish and squid. Bioluminescence can be applied for analytical measurements in two ways: (1) One can detect cellular events inside living cells by fusing the luciferase gene (e.g., the *luc* gene coding for firefly luciferase or the *lux* gene coding for *Vibrio fischeri* luciferase) to the gene of interest. The *in vivo* activity of the selected gene can be detected by monitoring the luminescence signal (LaRossa, 1998). (2) Alternatively, one can use purified recombinant luciferase and synthetic luciferin substrates for *ex vivo* detection assays for analytes such as ATP, NADH, and FMN (Blum *et al.*, 1993). In bio-optrodes, the cells or the purified enzymes are immobilized on the fiber tip and the luminescence signals are transmitted through the fiber to the detector (Polyak *et al.*, 2001; Hakkila *et al.*, 2004; Fine *et al.*, 2006).

Electrochemiluminescence (ECL) refers to the light emitted from species generated at electrode surfaces that undergo electron-transfer reactions to form excited states. ECL involves the production of reactive intermediates from stable precursors, such as $Ru(bpy)_3^{2+}$, at the surface of an electrode through application of a voltage. These intermediates then react under a variety of conditions to form excited states that emit light (Bard, 2004). ECL-active species have been used as labels on biological molecules and have found various applications in both immunoassays and DNA analysis with detection limits as low as 10^{-11} M. Commercial systems have also been developed that use ECL to detect many clinically important analytes such as R-fetoprotein and steroid hormones with high sensitivity and selectivity (Leland and Powell, 1991; Bard *et al.*, 2000). The principle of ECL and its biosensing applications are reviewed in Chapter 7 and elsewhere (Knight, 1999; Kuboka, *et al.*, 2000; Richter, 2004).

Using ECL detection in bio-optrodes is a relatively new approach and several such devices have recently been described (Jin *et al.*, 2001; Szunerits *et al.*, 2003; Szunerits and Walt, 2003; Monk and Walt, 2004a, 2004b). In one report (Szunerits *et al.*, 2003), an *optoelectrochemical* sensor was fabricated by chemically etching an optical fiber bundle using NH_4F and HF solution. Since the core and cladding etched at different rates, a nanotip array was generated; the array was then coated with Au (Figure 1.6a), followed by resin insulation at its base (Figure 1.6b).

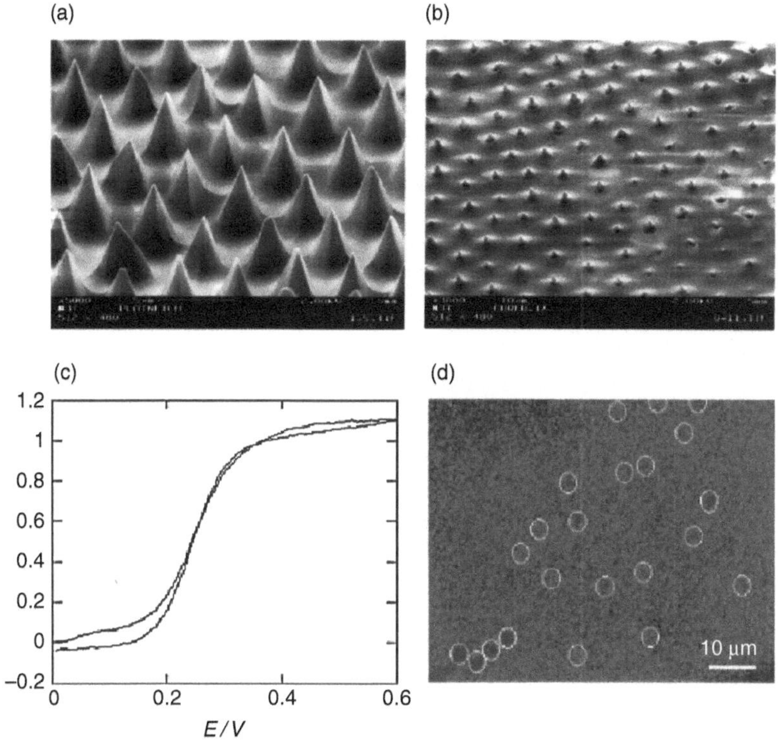

Figure 1.6 (a) SEM images of an etched and gold-coated fiber array and (b) an etched, gold-coated, and resin-insulated electrode array; (c) Cyclic voltammogram of the resin-insulated electrode array in 10 mM $Fe(CN)_6$ at $0.1\,V\,s^{-1}$; (d) ECL image monitored with an applied potential at selected areas of an etched, gold-deposited, and partially polymer-insulated electrode array. White circles indicate identified ECL spots. 1.2 V versus Ag/AgCl applied for 60 s, acquisition time 2 s, in 2 mM $Ru(bpy)_3^{2+}$/100 mM TPrA/pH 7 phosphate buffer. The red color corresponds to the most intense ECL signal; green and blue corresponds to little or no ECL signal. Reprinted with permission from the American Chemical Society. (see Plate 1)

Near-ideal microelectrode behavior was observed by measuring a cyclic voltammogram of potassium ferrocyanide (Figure 1.6c). ECL generated from the tips of the electrode array in an aqueous solution containing $Ru(bpy)_3^{2+}$ and tri-*n*-propylamine (TPrA) was detected with a high-resolution CCD camera (charge-coupled device; Figure 1.6d). Optoelectrochemical bio-optrodes are uniquely advantageous since they combine all the benefits of microelectrodes, such as fast response time and high signal-to-noise ratio, with the imaging properties offered by optical fiber bundles. This combination enables the simultaneous acquisition of both spatially resolved electrochemical and optical information from analytical samples.

Absorption is a simpler process than fluorescence and has also been used in bio-optrodes. Absorption is a process in which light energy is absorbed by an atom or a molecule, promoting the molecule from the ground energy state to a higher energy excited state. The resulting energy is dissipated non-radiatively (i.e., thermally) to the medium when the excited state relaxes to the ground state. The absorbance changes are related to the concentration $[C]$ via the Beer–Lambert relationship:

$$A = \log\left(\frac{I_0}{I}\right) = \varepsilon \cdot [C] \cdot l \tag{1.4}$$

where A is the optical absorbance, I_0 and I are the intensities of transmitted light in the absence and presence of the absorbing species, respectively, l is the effective path length, and ε is the molar absorption coefficient. In practice, optical fibers are used to measure absorbance by transmitting light through the fiber to the sensing layer and measuring changes in the scattered light. Alternatively, light is transmitted through one arm of bifurcated optical fiber to the sensing region, and reflected light signals are measured through the other arm of the fiber (Figure 1.7b). In a different configuration, two fibers are placed with one fiber facing the other, creating an optical cell in which the distance between excitation and collection fiber is the path length.

Surface plasmon resonance is another light phenomenon that has found wide use in bio-optrodes during the past decade. SPR refers to the

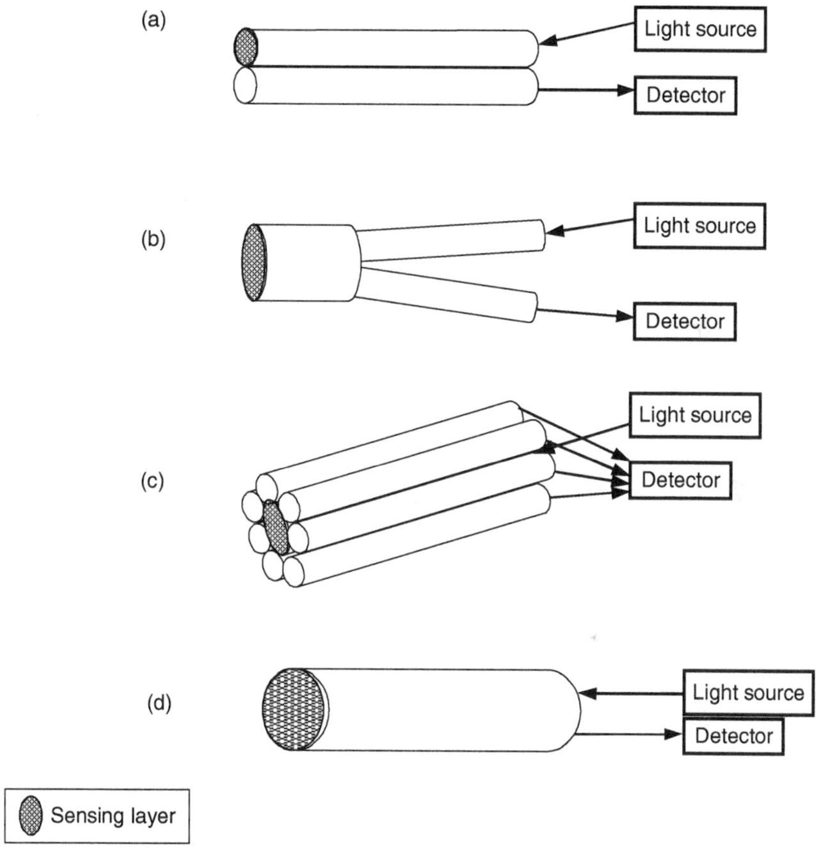

Figure 1.7 Design principle of a bio-optrode. (a) Two fibers: one carries light to the sensing layer and one carries the signal to the detector. (b) Bifurcated fiber: the biosensing layer is placed on the fused end of the fiber. (c) The biosensing layer is placed on the central fiber and the surrounding fibers are used to collect the light signals. (d) Single imaging fiber: the biosensing layer is placed on the distal end of the fiber, excitation and emission light are carried through the same fiber, and light signals are collected at the proximal end.

optical excitation of surface plasmons, which are collective oscillations of free electrons in a metallic film at the interface between a conductor (Au or Ag) and a dielectric (gas, liquid, or solid). The excitation of surface plasmons is accompanied by the transfer of optical energy into the metal layer, which then results in a narrow dip in the spectrum of the reflected light. Since surface plasmon waves are extremely sensitive to small changes in the refractive index near the sensor surface, and these changes are proportional to the sample mass on the surface, SPR can be used in *label-free* biosensing (Liu *et al.*, 2005; Ince and Narayanaswamy, 2006; see also Chapter 4). Many optical coupling schemes has been used for SPR biosensors, such as the most widely used Kretschmann prism configuration (Otto, 1968), grating (Cowan and Arakawa, 1970), channel waveguide (Lambeck, 1991), light pipe (Homola and Yee, 1996), and optical fiber (Jorgensen and Yee, 1993) schemes.

The idea of the SPR fiber optic sensor was first proposed by Jorgensen and Yee (1993) to overcome the drawbacks associated with the traditional Kretschmann prism configuration, such as its high cost, large size, and incompatability for remote sensing. The SPR sensor developed by Jorgensen and Yee was fabricated by first removing part of the cladding of a conventional polymer–clad silica fiber and then symmetrically depositing a thin (300 nm) metal (Ag) layer around the exposed section of the fiber core. White light was used as the source of a range of optical wavelengths guided into the optical fiber and the detection limit of the sensor was as low as 5×10^{-5} refractive index units (RIU). Many fiber optic SPR sensors have been developed since then for various biosensing purposes, such as monitoring red blood cell–ligand interactions (Quinn *et al.*, 2000), detecting myoglobin and cardiac troponin I (Masson *et al.*, 2004a), and quantifying survival motor neuron protein (Masson *et al.*, 2004b).

1.1.3. Optrode biosensor (bio-optrode) design and instrumentation

The design and use of different bio-optrode systems have been recently reviewed (Mehrvar *et al.*, 2000; Wolfbeis, 2000a, 2000b, 2000c, 2000d, 2002, 2004, 2006; Kuswandi *et al.*, 2001). The design of bio-optrodes is similar to chemical optrode design and two basic configurations are used: (a) a single fiber is used to transmit the light from the light source to the sample region and back to the detector, as shown in Figure 1.1, or

(b) multiple fibers are used in which one fiber is employed to transmit the light to the sample region and the other fiber or fibers are used to transmit light from the sample region to the detector, as shown in Figure 1.7a. For the second configuration, the most common format is a bifurcated fiber. Bifurcated fibers are fabricated by fusing two fibers on one end leaving the other ends free. The sensing elements are immobilized on the fused side and the other ends of the bifurcated fiber are connected to the light source and to the detector as shown in Figure 1.7b. In a different configuration, multiple fibers comprising one central fiber surrounded by several fibers are employed. The central fiber carries the immobilized sensing elements and is connected to the light source; the surrounding fibers collect the output light signals and transmit them to the detector (Figure 1.7c). Another configuration made use of a single imaging fiber. The biosensing layer is placed on the distal end of the fiber, both the excitation and the emission light are carried through the same fiber, and light signals are collected at the proximal end (Figure 1.7d).

The light sources used for bio-optrodes should provide sufficient light intensity within the sensor wavelength operating range. In addition, the light output should be stable over long time periods since light fluctuations may add noise to the measurement and reduce the sensor sensitivity. The different light sources used in bio-optrodes and their characteristics are summarized in Table 1.1.

In most fiber optic biosensor systems, the light transmitted from the sensing element (output light) is measured by using photon detection devices, which absorb photons and convert them into electrical signals. Several photon detectors are available as shown in Table 1.2.

1.1.4. Biological sensing elements

Bio-optrodes are constructed by immobilizing biological recognition components, such as enzymes, antibodies, nucleic acids, or cells to optical fibers. In nature, interactions between biological molecules, such as receptor–ligand, antibody–antigen, or two complementary DNA strands, are highly specific. Some of these recognition molecules can be purified

Table 1.1 Light sources.

Type	Wavelength (nm)	Characteristics
Tungsten lamp	IR/NIR, visible	High power output, bulky, expensive, used together with wavelength selection device
Deuterium lamp	200–300	
Xenon lamp	200–1000	
LEDs	470–1300	Low power output, high stability, long life, robust, compact size, inexpensive
Laser (N_2, Ar^+, He-Ne)	377, 488–568, 633	Monochromatic, very high power output, directional, bulky, expensive
Laser Diodes	800–904	High power output, long life, narrow spectral band, inexpensive, compact size

Table 1.2 Light detectors.

Detector type	Advantages	Limitations
Photomultipliers (PMT)	Sensitive, fast, low noise, internal amplification, compact	Need for high power voltage supply, destruction by over exposure
Photodiodes (PD)	Fast, robust, compact, inexpensive	High noise, no internal amplifier
Charge-coupled devices (CCD)	Very sensitive, can be used for imaging	Slow, expensive, need for a cooling system
Avalanche PDs	Lower noise than PD, fast, sensitive, can tolerate intense illumination	More expensive than PD
Complementary-metal-oxide semiconductor (CMOS) imaging sensors	Lower power consumption, lower cost, and higher system integration than CCD	Lower S/N than CCD

and used in fiber optic biosensors. Moreover, through genetic engineering, the original recognition element's structure can be modified and designed for a specific analytical application (Hellinga and Marvin, 1998; Looger, 2003; Kasper *et al.*, 2005). Biological sensing compounds can be divided into two main categories based on their bioactivity: biocatalysts (enzymes and cells) and bioaffinity molecules (antibodies, receptors, and nucleic acids).

1.1.4.1. Biocatalyst-based optrodes

Enzymes are proteins that selectively bind and catalyze the conversion of a substrate to product. Enzymes are used as sensing elements in bio-optrodes based on their ability to bind specific substrates (e.g., the analyte) and catalyze their conversion into an optically detectable product. Many novel enzyme-based bio-optrodes have been developed (Kulp *et al.*, 1987; Luo and Walt, 1989; Healey and Walt, 1995; Li and Walt, 1995; Zhang *et al.*, 1997; Michel *et al.*, 1998a, 1998b; Marquette *et al.*, 2000; Choi *et al.*, 2001; Wu *et al.*, 2004) since the first one was reported by Arnold in 1985. This approach has been reviewed in detail by Kuswandi *et al.* (2001) and in other recent reviews (Marazuela and Moreno-Bondi, 2002; Monk and Walt, 2004a, 2004b; Wolfbeis, 2004). The optical signal obtained – absorbance, fluorescence, reflectance, or chemiluminescence – is proportional to the product concentration and, consequently, to the analyte concentration.

Products that possess intrinsic optical properties can be measured directly, but the most common enzymatic reactions products, such as H^+, ammonia, oxygen, carbon dioxide, and hydrogen peroxide, do not possess intrinsic optical properties and are therefore measured indirectly using indicators (Wolfbeis, 1997). Indicators change their optical properties when interacting with these products. For example, fluorescein is a pH indicator and its emission intensity can be correlated to changes in H^+ concentration. Other indicators employed in enzyme bio-optrodes were recently reviewed by Kuswandi *et al.* (2001).

An interesting example that demonstrates the simple fabrication and function of enzyme-optrodes is the one used for glucose detection based on the enzyme glucose oxidase. Glucose oxidase catalyzes the oxidation of glucose with oxygen to produce gluconolactone and H_2O_2.

$$\text{Glucose} + O_2 \xrightarrow{\text{Glucose oxidase}} \text{Gluconic acid} + H_2O_2$$

Several approaches have been employed to determine glucose concentration with glucose oxidase: (1) monitoring of the oxygen consumed in the enzymatic reaction using a ruthenium complex as an indicator (Rosenzweig and Kopelman, 1996a, 1996b; Endo *et al.*, 2006); (2) measurement of H_2O_2 produced using luminol chemiluminescence (Marquette *et al.*, 2000) or ECL (Marquette *et al.*, 2001; Zhu *et al.*, 2002); and (3) direct measurement of the amount of H_2O_2 using the fluorescent probe europium (III) tetracycline (EuTc). Weakly fluorescent by itself, EuTc and the enzymatically generated H_2O_2 form a strongly fluorescent complex that can be detected in the time-resolved mode (Wu *et al.*, 2004).

In many cases, a sequence of enzymatic reactions is required to detect a specific analyte. In order to fabricate bio-optrodes for detection of such analytes, two or three enzymes are co-immobilized on the optical fiber in such a way that sequential reactions can occur. The first enzyme catalyzes the conversion of the analyte to a product that serves as a substrate for subsequent enzymatic reactions that eventually convert the initial analyte to an optically detectable product (Michel *et al.*, 1998a, 1998b). Using this methodology, analytes that could not be detected in a single reaction step can be detected. In addition, co-immobilizing two enzymes can achieve signal amplification through recycling systems, as shown in Figure 1.8 (Zhang *et al.*, 1997).

Inhibition of enzymatic reactions can also be used as a sensing mechanism in bio-optrodes (Freeman and Bachas, 1992). In this approach, the inhibitor is the analyte, and the measured signal is the decrease in enzymatic activity. One example is the detection of organophosphate and carbamate pesticides using an enzyme inhibition-based optrode. The bio-optrode is based on the inhibition of acetylcholinesterase by organophosphate pesticides. The enzyme is co-immobilized together

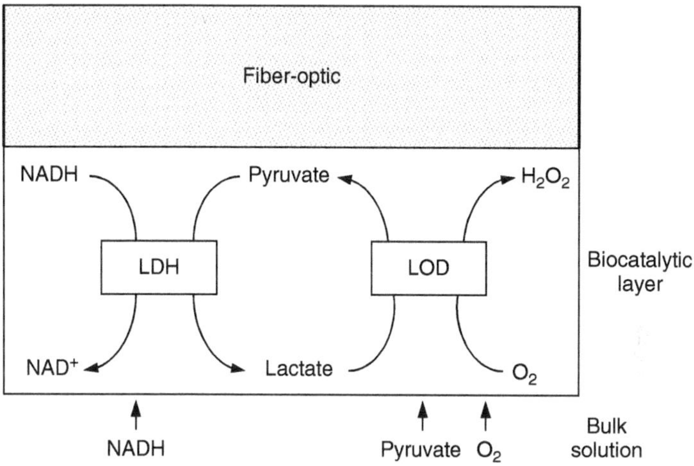

Figure 1.8 Schematic diagram of signal amplification using a dual-enzyme bio-optrode. Pyruvate is detected using lactate dehydrogenase (LDH) and lactate oxidase (LDO), which are co-immobilized on a fiber optic tip. Pyruvate concentration is determined by measuring NADH fluorescence. Pyruvate and NADH diffuse from the bulk solution into the enzyme layer, LDH catalyzes the formation of lactate and NAD^+ during the reduction of pyruvate. LDO then catalyzes the regeneration of pyruvate causing additional consumption of NADH by the LDH-catalyzed reaction. Thus, the signal obtained using a dual-enzyme system is higher than when a single enzyme is used (Zhang *et al.*, 1997). Reprinted with permission from Elsevier Science.

with a pH-sensitive dye at the fiber's distal end. The substrate acetylcholine is hydrolyzed by acetylcholinesterase, causing a change in the local pH and thereby the fluorescence signal. The inhibition of this reaction can be correlated to the pesticide concentration in the sample (Hobel and Polster, 1992; Choi *et al.*, 2001; Doong and Tsai, 2001).

In *living cells*, cellular functions are carried out by enzymes that simultaneously catalyze numerous biochemical reactions. Some enzymatic activities that occur in cells have been applied for bio-optrode fabrication. Although enzymes can be isolated and purified, their activity outside the cells is usually reduced compared to their activity within the cells where they function in an optimum environment containing all the necessary cofactors. Whole cell biocatalysts are particularly advantageous when the detection is based on a sequence of multiple enzymatic reactions. These enzymatic cascade reactions are very difficult and complicated to accomplish *ex vivo* by co-immobilizing the enzymes, but are relatively straightforward when employing whole cells. In practice, whole cells possessing unique or enhanced enzymatic activities that transform the analyte into detectable products, or cells with specific cellular responses such as oxygen consumption or production, are immobilized on optical fibers (Preininger *et al.*, 1994). The methods for detecting the products in cell-based fiber optic biosensors are similar to those employed in enzyme optrodes.

In a more recent approach, cells are genetically engineered to overexpress specific enzymes involved in the analytical measurement. An example of this approach is the use of *Escherichia coli* cells engineered to overexpress the enzyme organophosphorus hydrolase (OPH) on their outer cell membrane (Mulchandani *et al.*, 1998). This enzyme catalyzes the hydrolysis of organophosphorus pesticides to form a chromophoric product that can absorb light at a specific wavelength. The cell optrode is fabricated by immobilizing the cells on a bifurcated fiber optic tip and using a photomultiplier detection system to measure the light signals. A fiber optic biosensor was recently reported by Durrieu and Tran-Minh (2002) for the detection of heavy metals by measuring the inhibition of AP present on the external membrane of *Chlorella vulgaris* microalgae cells. In general, although the specificity of whole cell optrodes is reduced compared to enzyme optrodes, cells are very simple to use and obtain (e.g., growing the cells for a few hours), and there is no need for purification steps, which makes cell bio-optrodes inexpensive to assemble (Campbell *et al.*, 2006; Kumar *et al.*, 2006).

A different approach for sensing with whole cells that does not directly involve biocatalysis is based on utilizing genetic responses and signal transduction mechanisms in living cells (Naylor, 1999; Daunert *et al.*, 2000; Kohler *et al.*, 2000; Leth *et al.*, 2002; Brian *et al.*, 2003; Kuang *et al.*, 2004a, 2004b). Cells may express a specific gene or set of genes when a specific molecule (e.g., analyte) is present in the cell's environment. By fusing reporter genes, coding for optically detectable enzymes or proteins (e.g., luciferase, β-galactosidase, GFP) to the responsive gene, the genetic response is measured and correlated to the analyte concentration.

Another interesting setup combining living cells and fiber optics was recently reported by DiCesare *et al.* (2005), in which an optical fiber-based technology was used to analyze whole cell migration; the ability of cells to migrate can predict the invasiveness and/or metastatic potential of tumor cells. Cells were labeled with a membrane-bound fluorescent dye, Vybrant DiO, and distributed onto a polished optical fiber bundle. When a cell passed over one of the individual fibers in the bundle, the membrane-bound dye caused a large intensity increase, which remained high for a given "residence time" until the cell moved to an adjacent fiber. A significant decrease in cell migration was observed when cells were exposed to an antimigratory drug.

1.1.4.2. Bioaffinity-based optrodes

The natural high selectivity of antibodies, receptors, and nucleic acids make them very powerful sensing elements for recognizing their binding partners. Such bioaffinity optrodes are used as probes because the recognition reaction is essentially irreversible. The bio-optrode sensing elements must be regenerated or recharged before the probe can be used to make another measurement. In many cases, a probe-based bio-optrode configuration involves the use of a permanent fiber optic and a disposable sensing layer that can be placed on the fiber optic's distal end (Figure 1.9).

Immuno/receptor optrodes are a major group of bioaffinity fiber optic biosensors based on transducing antibody–antigen (analyte) interactions into an optical signal that is proportional to the antigen concentration. Monoclonal antibodies that can recognize a specific antigenic epitope region (i.e., a specific spatial structure on the antigen molecule) or polyclonal antibodies that recognize different antigenic epitopes are used in immuno-optrodes. Several detection schemes are employed. The simplest scheme involves the detection of intrinsically fluorescent analytes such as polynuclear aromatic hydrocarbons(PAHs) (Vo-Dinh *et al.*, 2000). Antibodies are immobilized on the fiber surface and a

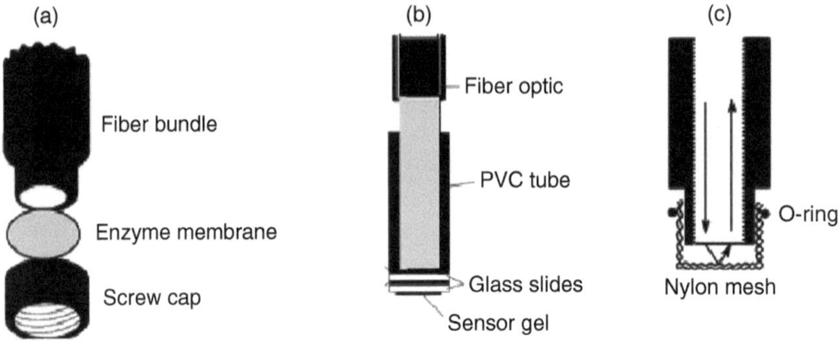

Figure 1.9 Configuration of probe-based bio-optrodes with disposable biosensing elements. (a) Biorecognition sensing molecules immobilized on a membrane, which is held by a screw cap on the optical fiber tip. (b) Disposable glass slide with gel-entrapped enzyme. (c) Nylon membrane with immobilized sensing molecules attached to the fiber using an O-ring (Kuswandi *et al.*, 2001). Reproduced with permission of the Royal Society of Chemistry.

fluorescence signal is obtained when the analyte (antigen) binds to the optrode's surface as shown in Figure 1.10a. One example of a direct assay where analytes are not intrinsically fluorescent was reported by Battaglia *et al.* (2005). Their fiber optic SPR biosensor quantified cytokines involved in wound healing by measuring the shift of SPR wavelength when analytes bound to antibodies immobilized on the fiber. The detection limits for interleukin-1 (IL-1), IL-6, and tumor necrosis factor-α (TNF-α) were reported to be 1 ng/ml in buffered saline solution and in spiked cell culture medium.

A competition assay is a more generalized detection scheme that can be applied to any antibody–antigen pair. The detection is based on competition for the antibody-binding site between the antigen present in the sample (analyte) and an externally added fluorescent-labeled antigen as shown in Figure 1.10b. A known concentration of fluorescent-labeled antigen is added and captured by an antibody, which is immobilized on the optical fiber surface. The fluorescence signal obtained is measured and set as the initial signal. To perform an analysis, the same fluorescent-labeled antigen concentration is mixed with a sample containing an unknown antigen concentration. When this

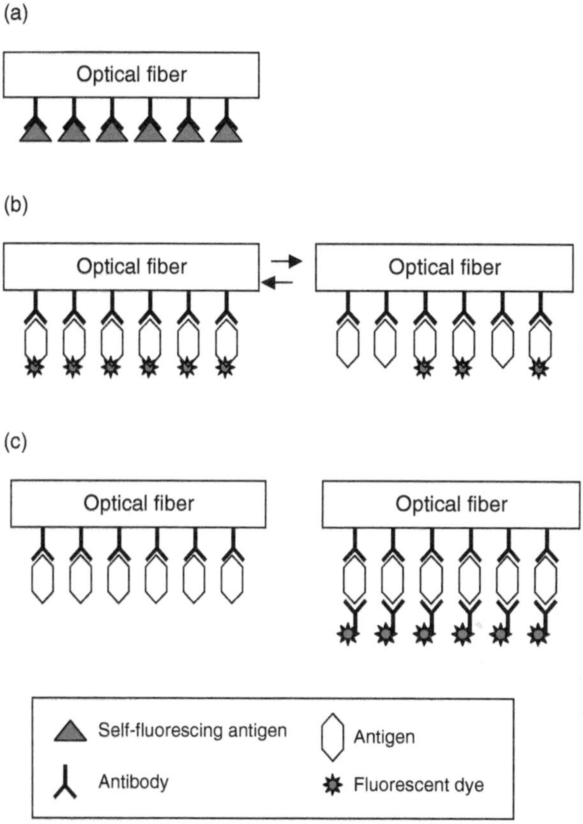

Figure 1.10 Schematic principle of immuno bio-optrodes. (a) Detection of intrinsically fluorescent molecules using immobilized antibodies. (b) Competition assay using a fluorescent-labeled antigen. (c) Sandwich immunoassay using an immobilized antibody and a fluorescent-labeled antibody.

mixture is analyzed using the bio-optrode, the resulting fluorescence signal obtained is lower than the initial signal because of competition with the labeled antigen in the sample. The relative decrease in the initial signal is proportional to the analyte concentration in the sample. Using this detection scheme, bio-optrodes for the detection of different analytes have been developed (Zhao *et al.*, 1995; Wittmann *et al.*, 1996; Maragos and Thompson, 1999).

The preferred detection scheme is the sandwich immunoassay, which involves the use of two antibodies. The first antibody is immobilized to the fiber and used to capture the antigen while the second antibody, which is labeled by a fluorescent dye or enzyme, is used to generate the signal (Figure 1.10c) (Szurdoki *et al.*, 2001; Balcer *et al.* 2002).

Competition and sandwich assays require using labeled antigens or antibodies. Fluorescent molecules and enzymes are employed for labeling using different chemistries (Wortberg, 1997). Although the majority of published reports used fluorescent molecules as labels, enzyme labels are useful and powerful since they catalyze the production of many readily detectable product molecules. Some enzymes used for labeling, such as AP, catalyze the conversion of a non-fluorescent substrate to a fluorescent product and can be detected by monitoring the fluorescent signal generated (Michael *et al.*, 1998). Other enzymes, such as horseradish peroxidase, can catalyze chemiluminescence reactions and are detected by monitoring the emitted light signals (Spohn *et al.*, 1995; Diaz *et al.*, 1998; Aboul-Enein *et al.*, 2000; Xing *et al.*, 2000; Gubitz *et al.*, 2001; Leshem *et al.*, 2004). Enzyme labeling is more sensitive than fluorescent dye labeling since the signal is amplified by the enzymatic reaction. Another new technology to increase the labeling efficiency of biological molecules using quantum dots was proposed at the time of writing and is discussed in Chapter 15.

Nucleic acid-based optrodes are the second major group of bioaffinity-based optrodes. Nucleic acid base pairing is used as the sensing mechanism in bio-optrodes for nucleic acid detection. The presence of a specific DNA sequence, the "target," among millions of other different sequences is detected by hybridization to its complementary DNA sequence, the "probe," which is immobilized on the optical fiber. Hybridization with complementary target DNA can be detected through the use of fluorescently labeled target sequences or by intercalating dyes that exhibit enhanced fluorescence intensity upon hybridization (Ferguson *et al.*, 1996; Healey *et al.*, 1997; Monk and Walt, 2004a; Brogan and Walt, 2005; Massey *et al.*, 2005; Wang and Krull, 2005a, 2005b; Wolfbeis 2006). In a typical assay (Figure 1.11), the target DNA is first amplified and fluorescently labeled using fluorescent primers

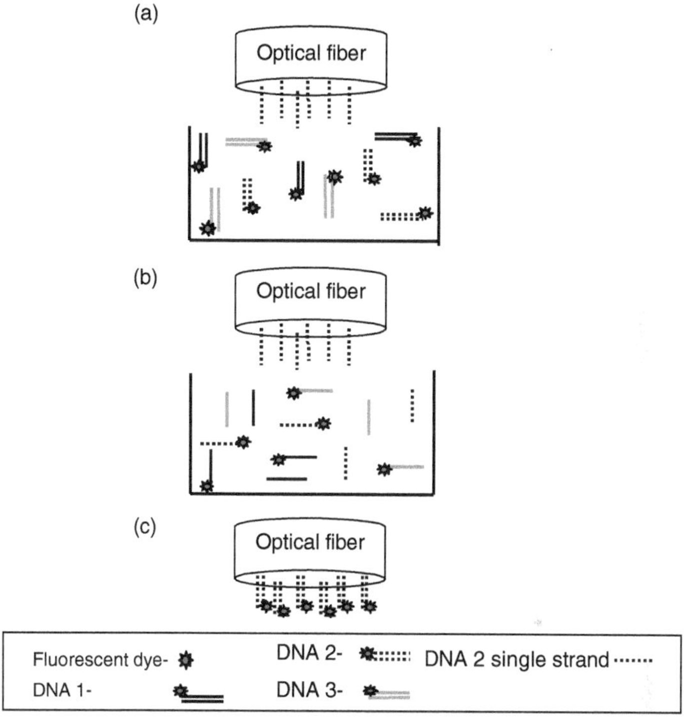

Figure 1.11 Principle of DNA fiber optic biosensors. (a) Single strand DNA probe molecules, with a sequence complementary to the target DNA sequence, are immobilized onto the fiber. (b) The fluorescent-labeled sample DNA molecules are first dehybridized and the fiber is dipped into the sample solution. (c) After hybridization, the complementary strands of the target DNA are attached to the probe DNA on the fiber and a fluorescence signal is obtained.

and the polymerase chain reaction (PCR). The resulting fluorescent double-stranded DNA molecules are dehybridized (usually by heating) (Figure 1.11b) and then allowed to rehybridize (by cooling) to the single-strand DNA probe molecules immobilized on the fiber surface (Figure 1.11c). The excess DNA molecules are washed away, and if the complementary target DNA sequence is present in the sample, a fluorescence signal is detected (Ferguson *et al.*, 1996).

Nucleic acid bio-optrodes have attracted considerable attention in the last few years, partially due to the tremendous amount of genetic information generated from various sequencing efforts but more importantly due to the unique characteristics of DNA oligonucleotides as sensing elements. These characteristics include high selectivity, the predictability and similarity in binding of all the different probe molecules, chemical stability, easy amplification through PCR, and readiness to be functionalized and immobilized on the surface of optical sensing substrates (Watterson *et al.*, 2001).

Nucleic acid bio-optrodes have been used in a variety of applications including the detection of pathogenic bacteria. The target pathogen DNA can be easily extracted from water, wastewater, or clinical samples, and the presence of pathogenic microorganisms can be determined by the bio-optrode (Pilevar *et al.*, 1998; Iqbal *et al.*, 2000; Chang *et al.*, 2001; Ahn and Walt, 2005). Such bio-optrodes have also been used to detect biological warfare agents (Song *et al.*, 2006).

Nucleic acid bio-optrodes also enable genetic analysis for detecting single nucleotide polymorphisms (SNPs) (Watteson *et al.*, 2003), high-throughput genotyping with SNP interrogation (Gunderson *et al.*, 2005), and potentially whole genome analysis. Epstein *et al.* (2003a) developed a novel optical fiber-based DNA microarray by randomly distributing oligonucleotide-functionalized microspheres on the face of an etched imaging fiber bundle. Batches of microspheres were modified with different oligonucleotide probes and then pooled to form bead libraries containing multiple sensor types. Beads with different probe types were encoded using different fluorescent dyes at various concentrations to create a unique signature for each bead type, enabling the positional registration of every bead in the array. In a continuation of this research, the same authors (Epstein *et al.*, 2003b) developed a novel method for decoding random arrays of every possible *n*-mer by performing consecutive hybridizations with fluorescently labeled combinatorial oligonucleotide decoding libraries.

New categories of nucleic acid molecules, such as aptamers (Lee and Walt, 2000) and molecular beacons (Liu *et al.*, 2000; Steemers *et al.*,

2000; Spiridonova and Kopylov, 2002), have been incorporated as sensing molecules into bio-optrodes. These different DNA-sensing schemes can be multiplexed by fabricating an array of hundreds to thousands of probes as will be described later in Section 1.3.2.

1.1.5. Sensing element immobilization

Immobilization of sensing biomolecules to the optical fiber is a key step in bio-optrode development. A good immobilization method should not only be simple, fast, and durable but, more importantly, gentle so that the biological molecule being immobilized can retain its biochemical activity. In addition, biological recognition elements are often co-immobilized with indicator dyes, so that ideally the immobilization method should be suitable for both molecules. In some cases, the recognition compounds are immobilized directly to the optical fiber surface. Alternatively, the molecules are first immobilized on membranes, such as cellulose acetate or polycarbonate, which are later physically attached to the optical fiber (Figure 1.9). There are three main methods for immobilizing a biological sensing compound: adsorption/electrostatic interaction, entrapment, and covalent attachment. A schematic representation of these methods is shown in Figure 1.12.

Adsorption immobilization methods involve adsorbing the sensing material onto a solid surface or polymer matrix through the formation of van der Waals and hydrogen bonds. Sensing materials can be adsorbed directly on the fiber optic end. This immobilization method is very simple; however, the adsorbed molecules tend to gradually leach from the solid support, decreasing sensing performance and/or lifetime. In order to overcome leaching problems, the solid support surface may first be modified with complementary functional groups. For example, a hydrophobic surface can be prepared to immobilize a hydrophobic species. Electrostatic interaction can also be employed for immobilization. This immobilization scheme is based on interaction between oppositely charged molecules. For example, an optical fiber surface can be coated with a positively charged layer (i.e., using poly-L-lysine) that interacts electrostatically with negatively charged recognition molecules (Figure 1.12a). The electrostatic immobilization method is very easy

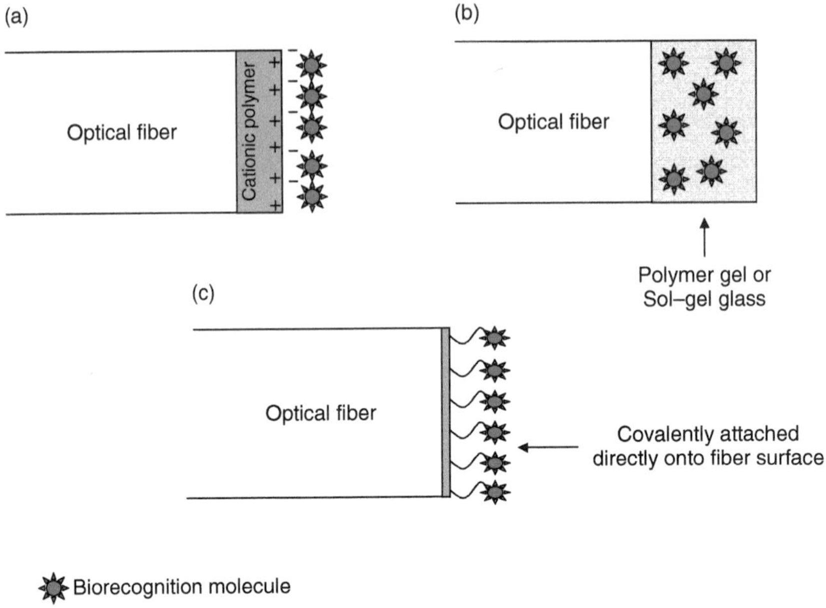

Figure 1.12 Schematic diagram of three different immobilization techniques employed in bio-optrodes. (a) Absorption/electrostatic. (b) Entrapment. (c) Covalent immobilization.

and highly reproducible but may be affected by changes in the medium pH or by changes in other ion concentrations.

Entrapment immobilization involves physical entrapment of sensing biomolecules within a porous matrix (Figure 1.12b). The biomolecules are suspended in a monomer solution, which is then polymerized to a gel causing the molecules to be entrapped. Such polymers can be either thermally or photochemically initiated and attached to the fiber surface by dip-coating procedures (Healey *et al.*, 1995). The immobilized molecules usually do not leach out of the matrix and can retain their biochemical activity. Polyacrylamide gels are most commonly used for entrapment immobilization, although agarose and calcium alginate gels have also been used (Polyak *et al.*, 2001). Fine *et al.* (2006) recently described a novel entrapment method using PVA-based hydrogels; the mild entrapment conditions allowed immobilization of living luminescent yeast cells for biodetection of estrogenic endocrine-disrupting chemicals. One important limitation of this approach is the slow diffusion rates of the analytes and products through the immobilization matrix, which increases the bio-optrode response time.

Optically transparent sol-gel glasses are also used for biological sensing molecule entrapment (Jordan *et al.*, 1996; Dunn *et al.*, 1998, 2001; Kishen *et al.*, 2003; Kwok *et al.*, 2005). Sol-gel glasses are produced by hydrolysis and polycondensation of organometallic compounds, such as tetraethyl orthosilicate ($Si(OCH_3)_4$). The sensing biomolecules are added to the reaction mixture during the formation of the sol or gel. Sol-gel glasses prepared by this method contain interconnected pores formed by a three-dimensional SiO_2 network. As a result, the biomolecules and dyes are trapped but small analytes can readily diffuse in and out of the pores. The main advantages of the sol-gel glass immobilization method are the chemical, photochemical, and mechanical stability of the immobilized layer. Disadvantages of sol-gel glass immobilization are the slow response times in aqueous media and the fragility of thin sol-gel glass films compared with polymer films.

Functional groups in the sensing biomolecules can be covalently bound to reactive groups on the surface of optical fibers, allowing robust immobilization and hence giving the bio-optrodes long lifetimes in storage of anywhere from 4 to 14 months (Eggins, 1996) (Figure 1.12c). The fiber surface can be chemically modified using silanization reactions (Weetall, 1993). For example, the fiber surface can be aminosilanized to form amine functional groups on the fiber surface followed by reaction with –COOH groups on the enzyme or antibody. Amine-modified surfaces can also covalently bind to the biomolecule's amine groups using bifunctional cross-linkers such as glutaraldehyde. Covalent immobilization methods are usually more complicated and time-consuming compared with the other immobilization techniques, but are very reliable since the biomolecules and dyes are not likely to leach out. It should be noted that covalent immobilization might change the biomolecule activity. If the binding occurs at crucial sites (e.g., an enzyme-active site or an antibody-binding site), activity can be lost completely. To avoid such inactivation, substrate, inhibitors, and other effectors are often included in the immobilization medium to protect the active or binding site of the biomolecules. In recent years, new techniques such as using Protein A for optimized antibody orientation (Anderson *et al.*, 1997) have been developed, which controls the immobilized molecule's orientation on the sensing surface and may result in an increase in the immobilization efficiency (Sackmann, 1996).

A more generalized and widely used binding method involves the use of avidin–biotin chemistry. The fiber surface can be modified with biotin groups and bind avidin-modified biomolecular conjugates or vice versa. This method is very attractive since many biotin- or avidin-labeled enzymes, antibodies, and nucleic acids are commercially available. Luo and Walt (1989) first reported using avidin–biotin coupling to fabricate enzyme-based bio-optrodes for penicillin, ethyl butyrate, and urea. This method has hence been widely used by other research groups (Wilchek and Bayer, 1990; Konry *et al.*, 2003; Viveros *et al.*, 2006).

A novel photon-initiated covalent-binding method was recently reported by Konry *et al.* (2005). An electropolymerizable pyrrole–benzophenone film was deposited on optical fiber tips, and a biological receptor was then covalently bound to the film through photoreaction. The photoreaction process included triplet-state excitation, hydrogen abstraction, and radical recombination, resulting in covalent immobilization of nucleic acids or proteins bearing amino acids with sterically accessible C–H bonds. This immobilization technique is easily applicable to a wide variety of biomolecules and combines the advantages of photolithography with electrochemical addressing of polymer films.

1.2. History

Optical fiber-based biosensors evolved from chemical optrodes. The first optical fiber-based chemical sensor was developed by Lubbers and Opitz (1975). Their device was designed to measure CO_2 and O_2 and was used in biological fluids. A few years later, biological molecules were coupled to the optical fiber-based chemical sensors and bio-optrodes were formed. One of the first bio-optrodes included a glucose biosensor (Arnold, 1985) fabricated by coupling the enzyme glucose oxidase to an O_2 optrode. A penicillin biosensor was also fabricated by immobilizing the enzyme penicillinase and a pH sensitive fluorescent dye on a polymer membrane covalently attached to the tip of a glass optical fiber (Kulp *et al.*, 1987). In the following years, many bio-optrodes with different recognition molecules have been developed and are described in several books (Wolfbeis, 1991; Blum *et al.*, 1994; Lopez, 2002) and reviews (Rabbany *et al.*, 1994; Fraser, 1995; Aboul-Enein *et al.*, 2000; Mehrvar *et al.*, 2000; Wolfbeis, 2000a, 2000b, 2000c, 2000d, 2002, 2004, 2006; Epstein and Walt, 2003; Monk and Walt, 2004a; Brogan and Walt, 2005). Although the bio-optrode basic configuration has not changed much from the one proposed by Lubbers and Opitz (1975), new types of optical fibers, optical instruments, biorecognition molecules, and indicators have been integrated into bio-optrodes. These materials, combined with new immobilization techniques and advanced optical approaches, have led to the development of more sophisticated, selective, and sensitive bio-optrodes. Advances in two fields influenced bio-optrode development in the last decade. First was the development of new fiber optic technologies developed for telecommunication applications. Second, advances in molecular biology techniques have allowed specific biorecognition molecules to be designed. Integration of technologies from these two fields has led to the development of advanced bio-optrode technologies such as multianalyte bio-optrodes, reagentless bio-optrodes, and nano bio-optrodes.

1.3. State of the art for bio-optrodes

In this section, a few examples of new bio-optrode technologies and applications will be described. Although many novel and interesting papers related to bio-optrode developments have been published in recent years, we focus here on several examples that emphasize the diversity of existing bio-optrode technologies. In addition, a few examples of bio-optrode applications in the industrial, environmental, and clinical fields will be described.

1.3.1. Nano bio-optrodes

One of the most exciting advances in bio-optrode development is the miniaturization of sensors to submicron dimensions. Nanotechnology facilitates research in this field and leads to the development of new nano bio-optrodes (Cullum and Vo-Dinh, 2000; Kasili *et al.*, 2004; Kasili and Vo-Dinh, 2005; Vo-Dinh and Kasili, 2005). The main importance of such biosensors is their ability to monitor physiological and biological processes inside a single living cell and thereby expand our knowledge about complex intracellular process.

In order to prepare nano bio-optrodes, optical fibers a few nanometers in diameter are fabricated. The fabrication process involves pulling optical fibers with an initial diameter of a few microns, using a modified micropipette puller optimized for optical fiber pulling. After pulling, tapered fibers are formed with typical distal end (tip) diameters of 20–80 nm.

This technique was used by Kopelman and coworkers to make a nanofiber optic chemical sensor for monitoring intracellular pH inside living cells (Tan *et al.*, 1992). Changes in pH were measured by immobilizing a pH-sensitive dye to the fiber tip. The same design was used to prepare an enzyme-based nano bio-optrode for nitric oxide detection (Barker *et al.*, 1998). Fluorescently labeled cytochrome c', which undergoes conformational changes in the presence of NO, was immobilized to the fiber tip. Changes in NO concentrations were correlated to changes in the energy transfer between cytochrome c' and the fluorescent dye.

An antibody-based nano bio-optrode for the fluorescent analyte benzo[a]pyrene tetrol (BPT) was also fabricated for detection inside a single living cell (Vo-Dinh *et al.*, 2000). The nano bio-optrode was prepared by first fabricating nanofiber tips by pulling them from a larger diameter (600 μm) silica optical fiber using a special fiber-pulling device (Sutter Instruments P-2000). This method was based on local heating of a glass fiber using a CO_2 laser while pulling the fiber apart and then coating the tapered fiber's outside walls with a thin silver, gold, or aluminum layer (100–200 nm) using a vacuum evaporator as shown in Figure 1.13a. In this system, the fiber was held at an angle relative to the

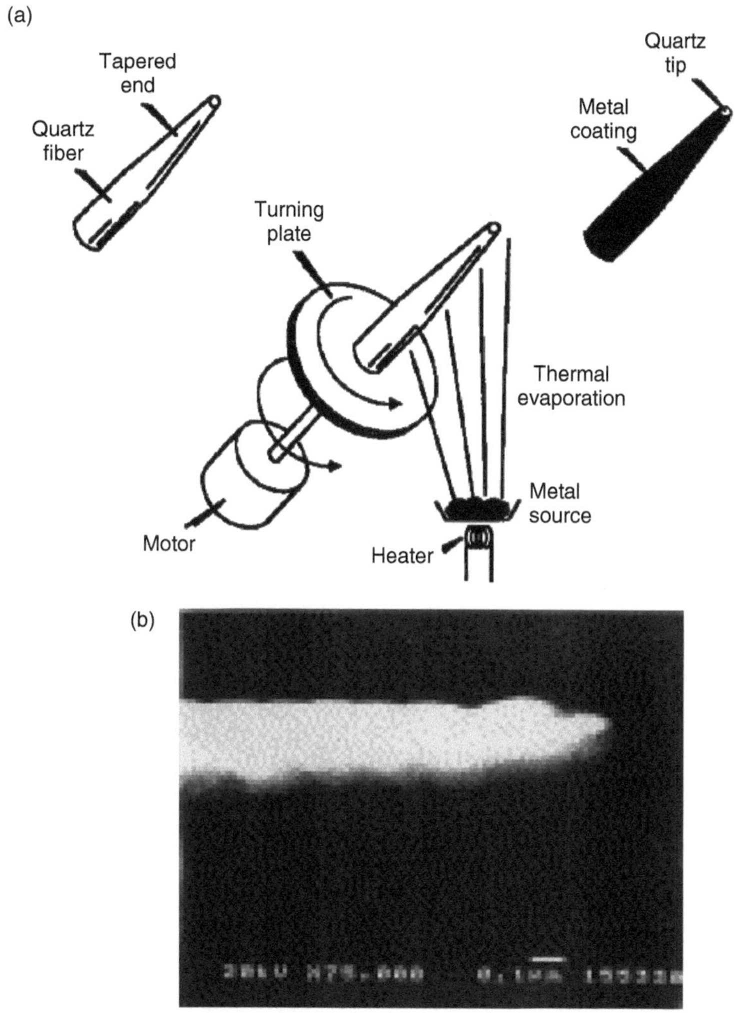

Figure 1.13 Nano bio-optrodes. (a) Fabricating a nanofiber optic tip. An optical fiber is heated and pulled and a tapered end with submicron diameter is formed. The tapered fiber side walls are then coated with a thin metal layer, using thermal evaporation, in order to prevent excitation light leakage. Biorecognition molecules can be immobilized on the fiber tip (Vo-Dinh *et al.*, 2000). Reprinted with permission from *Nat. Biotechnol.* (b) Scanning force micrograph (SFM) of nanofiber (Vo-Dinh *et al.*, 2001). Reprinted with permission from Elsevier Science.

metal vapor, resulting in a coating on the side of the fiber, leaving the tip uncoated. This coating prevented light leakage from the fiber's walls and enabled propagation of the excitation energy down the tapered sides of the nanofiber, bringing maximum light intensity to the fiber tip. The fiber's uncoated tip surface was then silanized in order to covalently attach anti-BPT antibodies. The final nano bio-optrode tip diameter was 200–300 nm. Bio-optrodes of this size have several advantages over larger bio-optrodes including fast response time and higher sensitivity. Using BPT nano bio-optrodes, BPT concentrations as low as ~300 zeptomoles were detected (Vo-Dinh *et al.*, 2001).

The optical measurement system used with the nano bio-optrode is shown in Figure 1.14a. Laser light was transmitted through the fiber and used to excite the captured BPT molecules. Changes in fluorescence

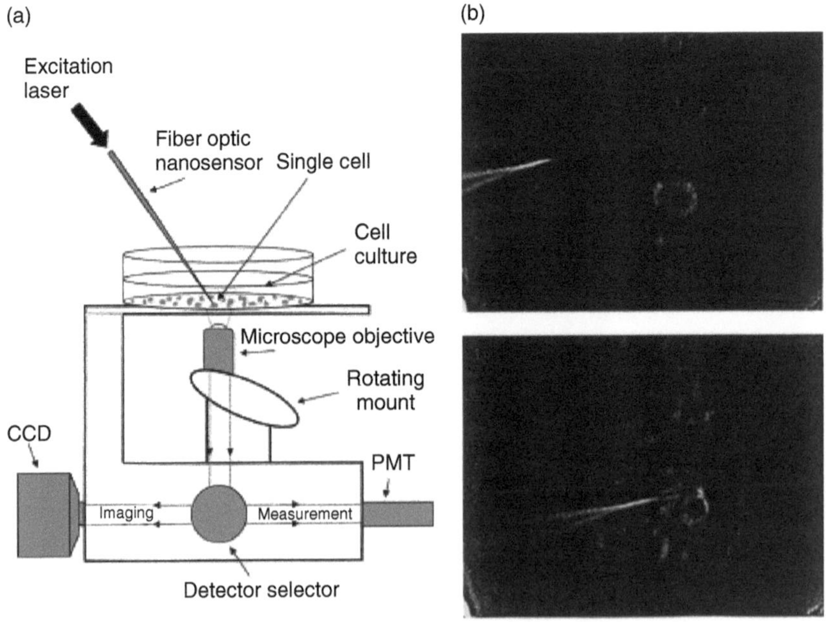

Figure 1.14 Measurements inside a single live cell using a nano bio-optrode. (a) The optical measurement system. (b) A nano bio-optrode inside a single cell (Vo-Dinh *et al.*, 2001). Reprinted with permission from Elsevier Science.

signals due to the presence of bound BPT molecules were transmitted through the microscope objective and measured using a photomultiplier tube (PMT). Using this experimental setup, BPT molecules inside single living cells were measured. The fiber's tip was inserted into the cell (Figure 1.14b) and incubated for 5 min inside the cells to allow the antibodies to bind the antigen (BPT). The fiber was then removed from the cell and the fluorescence signal obtained from the bound BPT was immediately measured. Concentrations as low as 9.6×10^{-11} M were measured inside the cells. The ability to measure concentrations of specific analytes inside single living cells with nano bio-optrodes can lead to a better understanding of many cellular processes such as transport mechanisms through cellular membranes, signal transduction pathways, complex enzymatic reactions, and even gene expression.

1.3.2. Multianalyte sensing

One of the main challenges of any sensor device is to detect several analytes simultaneously. Multianalyte sensing is important for clinical, environmental, and industrial analysis. For example, measuring the presence of proteins, antibodies, DNA sequences, antibiotics, viruses, and bacteria in single blood samples can provide physicians with rapid and comprehensive information about a patient's medical condition. Several approaches have been described for multianalyte bio-optrode fabrication (Li and Walt, 1995; Healey *et al.*, 1997a; Michael *et al.*, 1998; Anderson *et al.*, 2000; Walt, 2000; Chovin *et al.* 2004).

The conventional approach to preparing multianalyte sensors is to simply bundle multiple individual optical sensors. In this approach to multianalyte sensing, several optical fibers are assembled, each containing a different immobilized biorecognition molecule on a single fiber bundle. This approach was used to fabricate multianalyte biosensors for detecting different DNA target sequences simultaneously (Ferguson *et al.*, 1996). Eight optical fibers, each with a different immobilized DNA probe, were bundled together as shown in Figure 1.15a. The bundled fiber's distal end was inserted into the sample solution containing a fluorescein isothiocyanate–labeled oligonucleotide with a sequence complementary to one of the probe sequences. The fluorescence signals

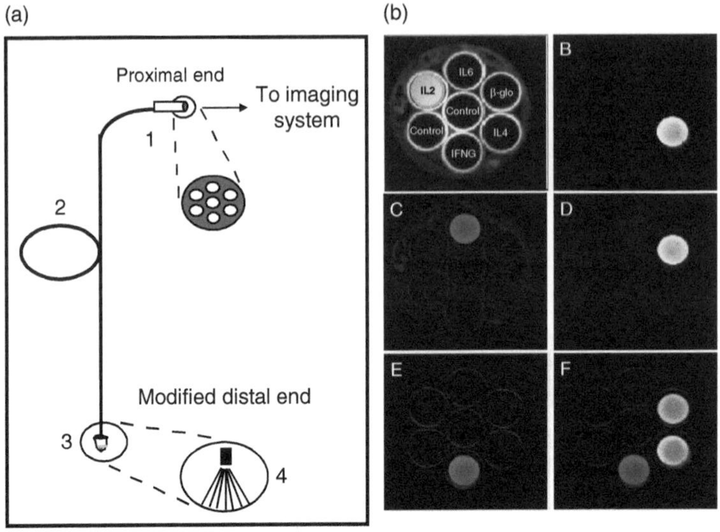

Figure 1.15 Multianalyte bio-optrode for oligonucleotide detection. (a) Schematic diagram of bio-optrode setup. Individual optical fibers, each with a specific immobilized oligonucleotide probe sequence are bundled together. The fiber's distal end is incubated with the sample and the signals obtained at the proximal end are measured using a CCD detector. (b) Fluorescence images acquired after incubating the multianalyte bio-optrode in solutions containing different target sequences. Image F show the bio-optrode response to the presence of three different targets in the sample (Ferguson *et al.*, 1996). Reprinted with permission from *Nature Biotechnol.*

were measured from the fiber's proximal end. Figure 1.15b shows that, when only one target sequence is present, a signal is obtained only from the fiber (bright signal) that contains the complementary probe sequence; the rest of the fibers in the bundle do not respond. When several target sequences were present, signals from several fibers carrying the complementary probes were observed (Figure 1.15b).

In different work, the specificity of this approach was demonstrated (Healey *et al.*, 1997b). Two probes were prepared, one that was complementary to the H-Ras oncogene sequence and a second probe containing a similar sequence but with a single base pair mismatch. When the hybridization reaction was performed at low temperature, both sequences hybridized to the probe, but at high temperature only the wild-type sequence hybridized. This experiment shows that these sensors can be used for point mutation detection.

The same sensor configuration can be applied for different sensing elements such as antibodies, enzymes, or whole cells. This approach, theoretically, is not limited in the number of individual fibers (each with a different sensing chemistry) that can be used simultaneously; however, the array size grows as more sensing elements are added.

An alternative approach involves fabrication of discrete sensing regions, each containing different biosensing elements, at precise spatial locations on an imaging fiber's distal end (Figure 1.16a). The sensing regions can be formed using photopolymerization techniques (Pantano and Walt, 1995). The imaging fiber's proximal end is first prefunctionalized with a polymerizable silane. The fiber is then dipped into a solution containing monomer, cross-linker, indicators, photoinitiator, and the sensing biomolecules. Using a pinhole, light is focused onto a small area ($\sim$30 μm in diameter) on the imaging fiber's proximal end. Light travels through the imaging fiber, from the illuminated area at the proximal end to the distal end. At the distal end, the light activates a photoinitiator and the polymer layer is formed only on the illuminated area (Figure 1.16b). For the formation of the next sensing polymer, light is focused on a different area at the proximal end and the fiber's distal end is dipped into a polymerization solution containing different sensing biomolecules.

Initially, this approach was used to fabricate a multianalyte sensor for pH, CO_2, and O_2 by forming sensing regions with different fluorescent dyes on a single optical imaging fiber face (Ferguson *et al.*, 1997). Based on this initial work, multianalyte biosensors for detecting penicillin and pH were developed (Healey and Walt, 1995; Healey *et al.*, 1997a). This sensor incorporated two sensing regions: in one region the enzyme penicillinase was immobilized together with a pH indicator and in the second region only the pH indicator was immobilized. In the presence of penicillin, the penicillinase activity resulted in the formation of H^+,

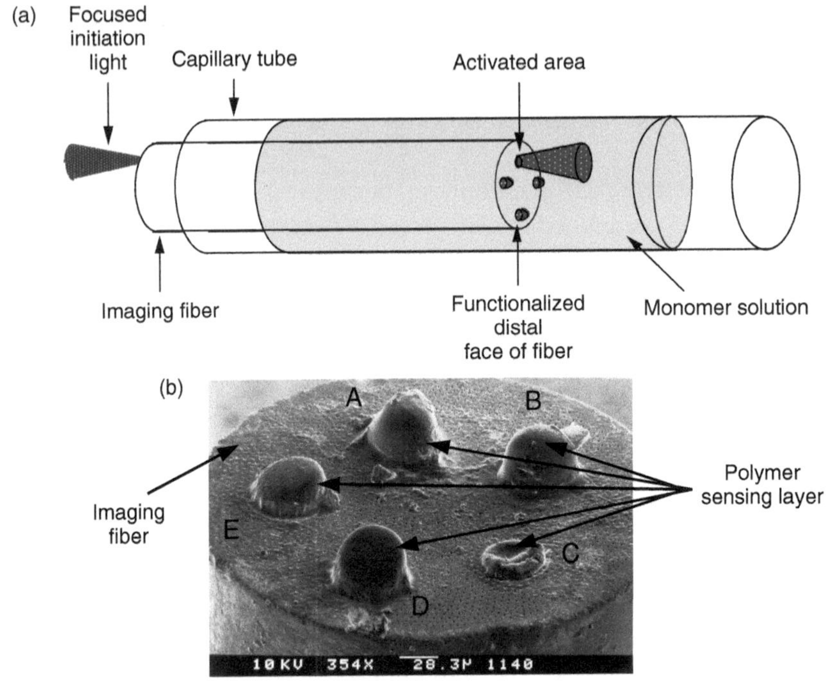

Figure 1.16 Multianalyte bio-optrode with different biosensing elements immobilized in polymers attached to an imaging fiber. (a) Setup of photopolymerization procedure used to fabricate the bio-optrode. Reprinted from Pantano and Walt (1995) with permission from the American Chemical Society. (b) Scanning electron micrograph of immobilized sensing polymer on an imaging fiber (Ferguson *et al.*, 1997). Reprinted with permission from Elsevier Science.

causing a decrease in the local pH in the polymer's microenvironment. By simultaneously monitoring pH changes in both the sensing regions (with and without the enzyme), the changes related to the enzymatic activity could be discriminated from pH changes in the bulk solution. Thus, this dual sensor was able to detect penicillin and could account for changes in the solution pH (Figure 1.17). A similar approach was used to fabricate glucose and O_2 biosensors. The enzyme glucose oxidase was used and the depletion of O_2 in the presence of glucose was monitored (Li and Walt, 1995). A separate sensor for O_2 was also

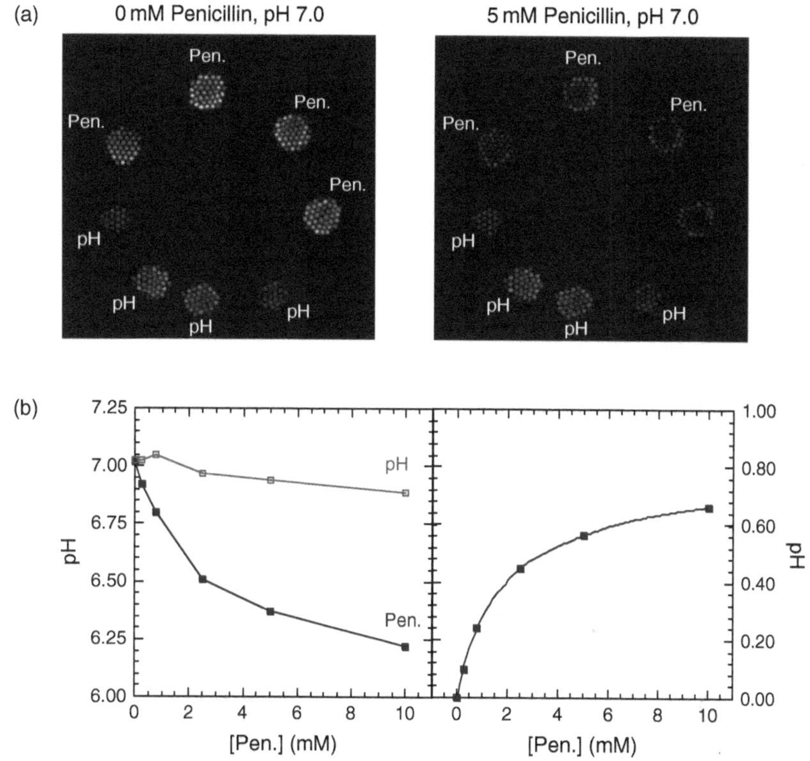

Figure 1.17 Imaging fiber-based penicillin and pH bio-optrode. (a) Response of bio-optrode, similar to the one described in Figure 1.15, with penicillin-sensitive polymer regions (containing the enzyme penicillinase) and pH-sensitive polymer regions. When the penicillin concentration is increased, only the fluorescence intensity from the penicillin-sensitive polymer regions increases. (b) Bio-optrode response to penicillin (solid squares) and pH (empty squares) are shown in the left plot. The difference between the buffer pH and the microenvironmental pH at the penicillin-sensitive polymer is shown in the right plot (Healey and Walt, 1995). Reprinted with permission from the American Chemical Society.

prepared on the same imaging fiber. When the glucose sensor signals were compared with the signals obtained from the sensing region that contained only the O_2 indicator, the concentration of glucose could be quantitatively determined. Both biosensors can be used to determine the analyte concentrations in different environments. In addition, they can provide information about both the biochemical analytes and the pH or O_2 concentrations using a single imaging optical fiber. A possible future application for such biosensors may be for *in vivo* multianalyte sensing, where early changes in drug levels, glucose, O_2, and pH are important.

In both of these approaches (sensor bundling or photopolymerization), when more than 20 optical fibers or polymer regions are required, the bundle of fibers may become too big or the photopolymerization protocol may become complicated. A new approach to overcome this limitation based on the unique characteristics of optical imaging fibers (see Section 1.1.1) has been proposed (Michael *et al.*, 1998; Walt, 2000) and explored extensively (Ferguson *et al.*, 2000; Steemers *et al.*, 2000, 2005; Biran and Walt, 2002; Epstein *et al.*, 2002, 2003a, 2003b, 2003c; Biran *et al.*, 2003; Szunerits *et al.*, 2003; Kuang *et al.*, 2004a, 2004b; Ahn and Walt, 2005; Bowden *et al.*, 2005; DiCesare *et al.*, 2005; Gunderson *et al.*, 2005, 2006; Kuang and Walt, 2005; Song *et al.*, 2005, 2006; Ahn *et al.*, 2006; Blicharz and Walt, 2006; Fan *et al.*, 2006; Rissin and Walt, 2006a, 2006b; Dragoy Whitaker and Walt, 2007). Imaging fibers consist of thousands of optical fibers coherently bundled together, with each individual fiber maintaining its ability to carry its own light signal from one end of the fiber to the other. Thus, by attaching a sensing material to each individual fiber's distal end, an array of thousands of sensing elements can be constructed on the tip of a single imaging fiber array. In practice, microwells are fabricated on the end of each individual fiber by selectively etching the fiber cores. This process results in the formation of a high-density microwell array on the imaging fiber tip as shown in Figure 1.18a. The sensing elements are prepared by immobilizing fluorescent indicators and/or biorecognition molecules to the microsphere surfaces. The microspheres and microwells are matched in size such that the microspheres can be distributed into the microwells to form an array of sensing elements (Figure 1.18b). When different biorecognition molecules are immobilized on different microspheres, the array can be used to detect multiple analytes. A CCD detector is used to monitor and spatially resolve the fluorescence signals obtained from each microsphere (Figure 1.18c and d). Imaging and data analysis software are used to calculate the analyte concentrations.

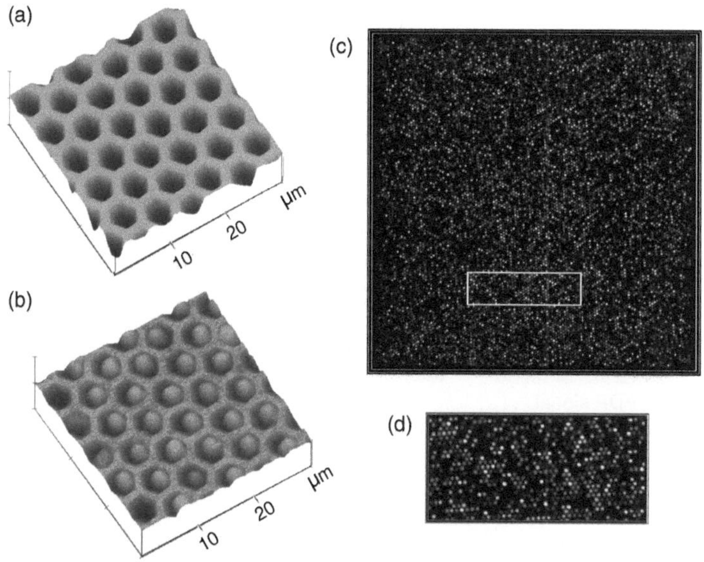

Figure 1.18 High-density multianalyte bio-optrode composed of microsphere array on an imaging fiber. (a) SFM of microwell array fabricated by selectively etching the cores of the individual fibers composing the imaging fiber. (b) The sensing microspheres are distributed in the microwell. (c) Fluorescence image of a DNA sensor array with ~13 000 DNA probe microspheres. (d) Small region of the array showing the different fluorescence responses obtained from the different sensing microspheres (Walt, 2000). Reprinted with permission from the American Association for the Advancement of Science. (see Plate 2)

These sensor arrays are prepared by randomly distributing the microspheres into the wells. In order to allow multianalyte sensing, the location of each sensing microsphere must be determined. The microsphere registration process involves using one of several encoding/decoding schemes. When the microspheres are prepared, each type of microsphere is modified such that it carries a unique optical marker in addition to the biorecognition element. This marker can be a fluorescent dye or a combination of several different fluorescent dyes. Different markers are used for the different microsphere types, allowing each of the microspheres carrying a certain type of biomolecule to be encoded with a unique optical signature. For example, as shown in Figure 1.19, three

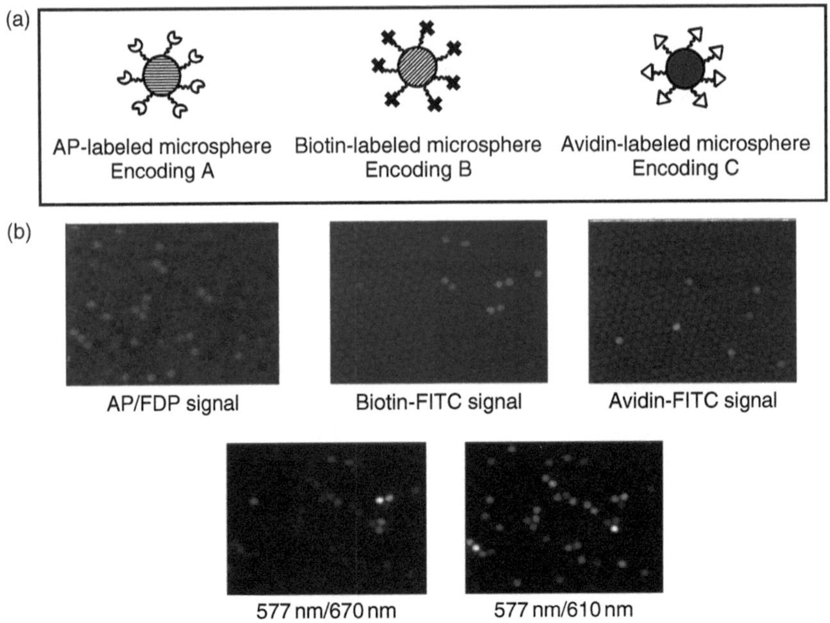

Figure 1.19 Randomly ordered array bio-optrode. (a) Schematic representation of the biorecognition elements immobilized on different sets of encoded microspheres (AP). The microspheres were encoded using three different ratios of two fluorescence dyes: indodicarbocyanine (DiIC) and Texas red cadaverine (TRC), both dyes are excited at 577 nm and emit at 670 and 610 nm, respectively. (b) The three microspheres types are mixed and randomly distributed into the microwell array. Fluorescence responses in the presence of the AP fluorogenic substrate, avidin–FITC, and biotin–FITC are shown on the top images. The identity of each microsphere was determined by calculating the emission ratio 670 nm/610 nm obtained using 577 nm excitation light (bottom images) (Michael *et al.*, 1998). Reprinted with permission from the American Chemical Society.

types of microspheres were prepared by immobilizing the enzyme AP to one group of microspheres, avidin to the second group, and biotin to the third group (Figure 1.19a). Each type was encoded with different concentrations of two fluorescent dyes. When the three microsphere types were mixed and randomly distributed into the microwell array, their location could be determined by applying the appropriate excitation and emission wavelengths to establish the different fluorescent markers on each bead. This biosensor was used for multianalyte detection of fluorescein diphosphate, biotin–FITC, and avidin–FITC, as shown in Figure 1.19b. For each analyte, several different microspheres produced fluorescence emission signals, indicated by the bright spots on the array images.

These images demonstrate two main advantages of this technology. First, the presence of replicates of each microsphere type provides statistically significant results and reduces the probability of both false negatives and false positives. Second, averaging signals from many identical individual sensing elements results in higher signal-to-noise ratios. This multianalyte biosensor design was also used to develop a DNA biosensor with the ability to detect 25 different fluorescently labeled DNA sequences simultaneously (Ferguson *et al.*, 2000). Using a similar approach, Epstein *et al.* developed a microsensor array for simultaneous detection of multiple DNA sequences with a detection limit of 10^{-21} mol (approximately 600 DNA molecules). Recently, this methodology was applied to the detection of pathogenic bacteria (Ahn and Walt, 2005), biological warfare agents (Song *et al.*, 2006), and fabrication of an advanced platform for genomic analysis (Epstein *et al.*, 2003). Another biosensor, comprising microspheres with different immobilized molecular beacons, was used to detect three different unlabeled DNA sequences (Steemers *et al.*, 2000). This microsphere array platform has been commercialized by Illumina Inc., and many novel applications, such as high-throughput genotyping, have been successfully carried out using this system (Gunderson *et al.*, 2005, 2006; Steemers *et al.*, 2005; Fan *et al.*, 2006).

Microspheres with immobilized antibodies were used for simultaneous detection of the clinically important drugs digoxin and theophylline (Szurdoki *et al.*, 2001). Most recently, microsphere immunoassay arrays were used for simultaneous detection of five inflammatory cytokines (Blicharz and Walt, 2006). This multiplexed array is presently being evaluated for its potential for using saliva as a noninvasive diagnostic fluid for pulmonary inflammatory diseases such as asthma.

Alternatively, living cells could be directly immobilized on the surface of the microwells. By utilizing the genetic responses and signal transduction mechanisms in living cells, Walt and co-workers (Kuang *et al.*, 2004a, 2004b, 2005) developed cell assays for genotoxin monitoring and simultaneous monitoring of "promiscuous" drug effects using single cells of multiple cell types.

Multianalyte bio-optrodes are still in the early stages of development. Due to their importance for many analytical applications, it is expected that research efforts will continue to advance the capabilities of such sensors.

1.3.3. Single-molecule detection

Single-molecule detection represents the ultimate goal of ultrasensitive chemical analysis. Several approaches for fabricating bio-optrodes for single-molecule detection have been described (Fang and Tan, 1999; Loescher *et al.*, 1999; Rissin and Walt, 2006a, 2006b). Capture and detection of single molecules of β-galactosidase on a femtoliter fiber optic array was recently reported by Rissin and Walt (2006b). Twenty-four thousand reaction vessels, each 46 fl in size, were generated at the distal face of a polished 1 mm fiber optic array using an acid etch. Each well was then functionalized with biotin and was capable of capturing the target, streptavidin β-galactosidase (SβG) (Figure 1.20a).

After array modification, the biotinylated fiber arrays were incubated in PBS buffer containing varying amounts of SβG. The concentration was adjusted so that statistically either one molecule or no molecules bind to each well during incubation. The wells were then incubated with a fluorogenic substrate of β-galactosidase and sealed using a silicone gasket sandwiched between a microscope slide and the fiber array (Figure 1.20b). Single enzyme molecules were detected in the individual reaction vessels after generation of a sufficient number of fluorescent product molecules (Figure 1.20c). A linear relationship was observed between the percentage of reaction vessels that captured an enzyme molecule and the amount of enzyme present in the interrogated sample. This result suggests the viability of using this method for detecting

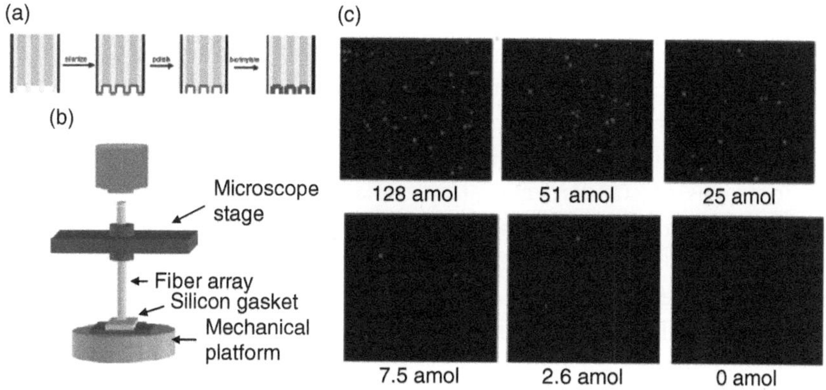

Figure 1.20 (a) A side-view cross-section schematic representing the etched bundle modifications. From left to right: An etched fiber, modification with an amino-functionalized silane, removal of the amine modification from the cladding material via polishing, and functionalization of the fiber bundle with biotin succinimidyl ester (Rissin and Walt, 2006a). (b) A simple schematic of the femtoliter array and microscope platform. The fiber array is locked into the microscope stage, a drop of substrate solution is placed on the silicone gasket, and the mechanical platform applies pressure to the distal end of the fiber, thereby sealing the solution in each femtoliter chamber (Rissin and Walt, 2006b). (c) A portion of the fiber arrays, each incubated with different concentrations of the analyte, SβG. Individual SβG molecules were captured in the biotin-functionalized femtoliter wells at the end of fiber (pseudocolor added using IPlab Software). Reprinted with permission from the American Chemical Society.

ultra-low amounts of target such as DNA or antigens. Such quantitative single-molecule detection techniques could circumvent additional time-consuming and complex steps such as immuno-PCR amplification and will push the limits of ultra-low detection.

1.3.4. Reagentless bio-optrodes for homogeneous assay

One limitation common to many bio-optrode technologies is the requirement for external reagents to be added to the analytical assay. For example, when antibodies are used as recognition molecules in a sandwich assay, there is a need to add a secondary labeled antibody in order to measure the analyte concentration (Figure 1.10c). The same requirement applies to a competition assay where a labeled antigen is used (Figure 1.10b). Most nucleic acid bio-optrodes are based on prelabeling the target sequence with fluorescent dye. The necessity to add reagents complicates the assay procedure and limits the acceptance of bio-optrodes as standard and simple analytical tools. Therefore, many research efforts have concentrated on developing bio-optrodes for "mix and measure" assays where no reagents are added. In this section, several approaches for reagentless (also called homogeneous) bio-optrode fabrication will be described.

One approach for reagentless bio-optrode fabrication is based on monitoring conformational changes in the biorecognition molecule following analyte binding. The conformational changes are usually detected using FRET as the transduction mechanism. In one example, molecular beacons were used to detect unlabeled DNA sequences (Steemers *et al.*, 2000). Molecular beacon structures consist of single-stranded DNA in a hairpin configuration with a fluorophore and quencher attached to opposite termini (Tyagi and Kramer, 1996). The molecule's 3′ and 5′ ends are complementary to one another and form the hairpin structure. The probe sequence, which is complementary to the target sequence, is located in the center (Figure 1.21a). In the absence of target, the fluorophore and quencher are within the requisite energy transfer distance, resulting in fluorescence quenching (Figure 1.21a). Upon target binding, a conformational change occurs, the hairpin separates (denatures), and the fluorescence signal increases (Figure 1.21b). Using an imaging fiber-based, molecular-beacon bio-optrode, three different sequences from mutant genes related to cystic fibrosis have been simultaneously detected (Steemers *et al.*, 2000). The multianalyte, imaging fiber-based bio-optrode was prepared as previously described in Section 1.3.2. Each type of molecular beacon probe was immobilized to beads encoded with unique optical signatures. The resulting three types of beads were randomly distributed into a microwell array and used for the analysis of three different target sequences simultaneously.

In a similar manner, fluorescence donor and acceptor pairs can be incorporated into proteins and used as reporters for substrate-binding

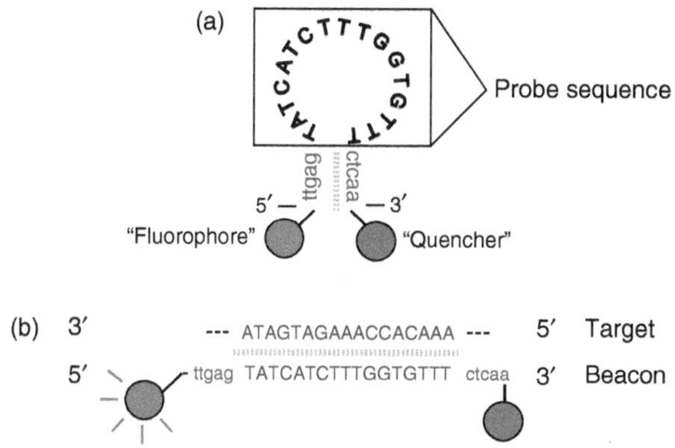

Figure 1.21 Molecular beacon structure. (a) The hairpin structure is formed due to the complementary sequences near the 3′ and 5′ ends. The single strand "loop" contains the probe sequence. In this configuration, the fluorophore and quencher are in proximity and therefore no fluorescence signal is produced. (b) When a target sequence binds, the molecular beacon structure changes causing separation of the fluorophore and quencher resulting in a fluorescence signal change.

events. In one approach, the enzyme carbonic anhydrase, which binds metal ions with high affinity and selectivity, was used to fabricate Zn^{2+}, Co^{2+}, and Cu^{2+} bio-optrodes (Thompson and Jones, 1993; Thompson *et al.*, 1996). Donor molecules, such as Cyt-5 or Cyt-3 dyes, were bound to primary amines in the protein, using *N*-hydroxysuccinimide esters of the dyes as modification reagents. The acceptor molecules in this case were the Co^{2+} and Cu^{2+} analytes themselves, which exhibit weak d–d absorbance bands at long wavelengths. Thus, upon analyte binding, a decrease in the donor fluorescence was observed. The decrease was measured by monitoring the time-dependent phase angle change at a fixed frequency upon binding of the metal ion. Results for two different concentrations of Co^{2+} are shown in Figure 1.22. The fiber configuration included an entrapped enzyme in a polyacrylamide layer immobilized to the tip of an optical fiber.

A related approach for fabricating reagentless enzyme-based biosensors is based on transducing conformational changes occurring upon substrate

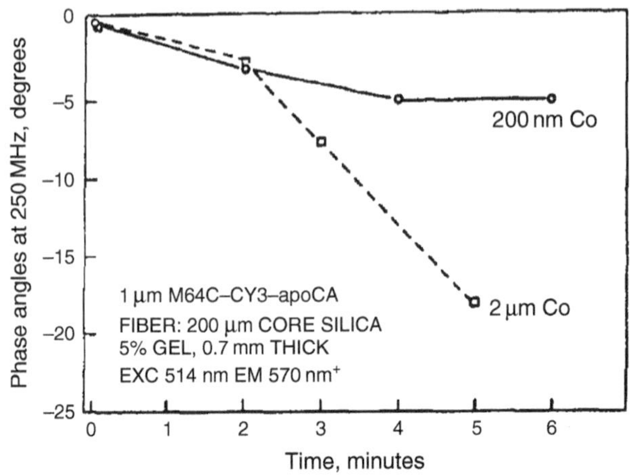

Figure 1.22 Reagentless Co^{2+} bio-optrode. The enzyme carbonic anhydrase was labeled with a fluorescent donor and immobilized to the fiber optic distal end. When Co^{2+} binds to the enzyme, the donor's fluorescence is quenched and the signal decreases (Thompson *et al.*, 1996). Reprinted with permission from Elsevier Science.

binding into FRET signals. Proteins such as calmodulin, maltose-binding protein, and phosphate-binding protein undergo conformational changes upon substrate binding and have been used to prepare such biosensors (Hellinga and Marvin, 1998). Using genetic engineering, two FRET fluorescent groups (acceptor and donor) were incorporated on two different cysteine residues; upon a conformational change, the spatial arrangement of these cysteines was altered, resulting in a FRET signal change.

Another example of a reagentless enzyme-based bio-optrode was recently described (Michel *et al.*, 1998a). The sensor used a controlled-release polymer similar to that used in previous work (Luo and Walt, 1989; Barnard and Walt, 1991) for controlled delivery of substrate in the enzymatic microenvironment. It was designed to detect the three-adenylate nucleotides (ATP, ADP, AMP) using a three-enzyme reaction sequence. Three enzymes were used: adenylate kinase, creatine kinase, and luciferase. The enzymes were compartmentalized in such a way that the product of the first reaction would be accessible to serve as the substrate for the subsequent reactions shown in Figure 1.23a. The final

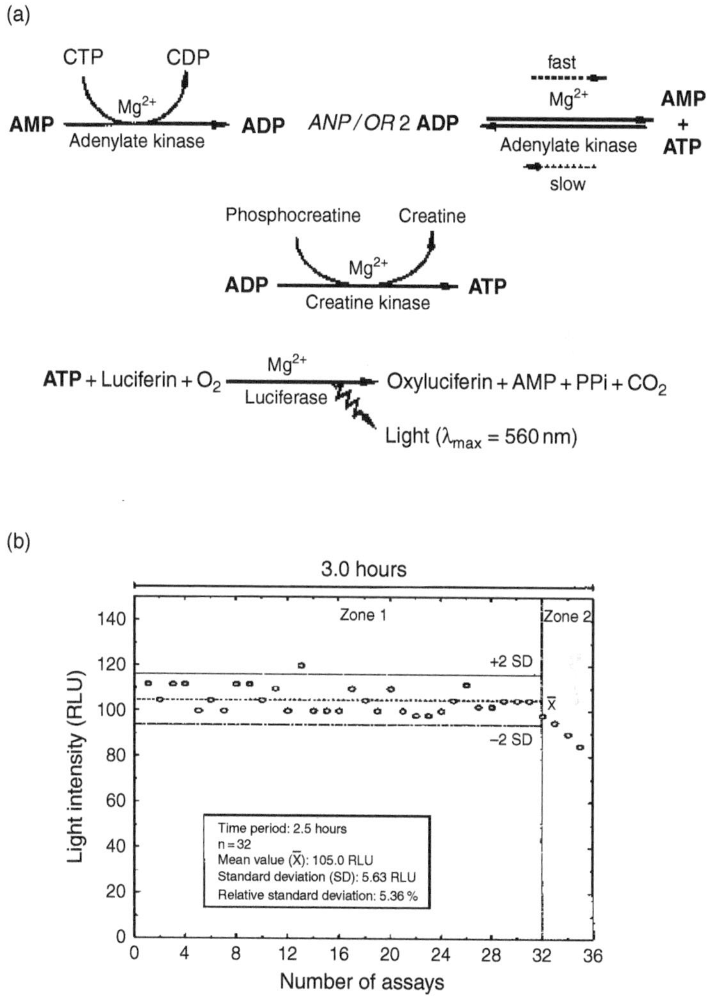

Figure 1.23 Reagentless bio-optrode for AMP, ADP and ATP detection. (a) Schematic of the enzymatic reactions employed for the measurements. (b) Repetitive measurements of ATP using controlled released of luciferin from acrylic microspheres. The light intensity was measured after each injection of 25 pmol ATP. The self-contained bio-optrode reproducibility over three hours (32 repetitive injections) is shown (Michel *et al.*, 1998a). Reprinted with permission from Elsevier Science.

indicator reaction for all three analytes involved the luciferase reaction. In previous designs, the co-substrate for this reaction, luciferin, was externally added to the flow cell. In the new design, luciferin was incorporated into acrylic microspheres. When the microspheres were immobilized together with the enzymes on the fiber surface, they slowly released the luciferin, allowing continuous detection for 3 h (Figure 1.23b). This generic approach is appropriate for the controlled release of cosubstrates or cofactors, which can be used in different enzyme-based bio-optrodes.

In recent years, fiber optic SPR sensors have emerged as a new approach for reagentless biodetection (Battaglia *et al.*, 2005; Chau *et al.*, 2006). In the study by Battaglia *et al.* (2005), the fiber probe tip was coated with 50 nm of Au and then modified with a 16-mercaptohexadecanoic acid to form a self-assembled monolayer. Various anti-cytokine antibodies were immobilized on the self-assembled monolayer via an amine coupling reaction. The shift in the minimum SPR wavelength was proportional to the amount of analyte captured by the immobilized antibodies and was measured using a white-light fiber optic SPR refractometer (Obando and Booksh, 1999). Detection limits at or below 1 ng/ml in both buffered saline solution and spiked cell culture medium were achieved for this label-free assay. The ability to detect multiple markers in complex media with a label-free method offers a promising approach for future clinical applications.

1.3.5. Environmental applications

Many bio-optrodes have been proposed for use in environmental applications (Rogers and Poziomek, 1996; Marty *et al.*, 1998; Rogers and Gerlach, 1999; Schobel *et al.*, 2000; Holst and Mizaikoff, 2002; Wolfbeis, 2002; Vannela and Adriaens, 2006). For remote monitoring, only the fiber tip containing the biorecognition element has to be located at the measurement site (e.g., lakes, rivers, sewage streams), while the optical detection instrumentation can be located in a protected location away from the site. Optical fibers are small in diameter and flexible, and therefore can be located in places inaccessible to other sensing devices. In addition, the optical fiber's durable structure protects it from harsh environmental conditions. At present, although some prototype environmental monitoring fiber optic biosensors are commercially available, such as the Analyte 2000™ and RAPTOR™ developed by Research International and the Naval Research Laboratory, environmental bio-optrodes are still largely in the research and development stage with most of the research focused on detection scheme development and optimization.

Bio-optrodes have been described for detecting pesticides such as terbutryn (Bier *et al.*, 1992), parathion (Eldefrawi *et al.*, 1995), metsulfuron-methyl (Xing *et al.*, 2000), and imazethapyr (Wong *et al.*, 1993). One such example is a bio-optrode for the detection of 2,4-dichlorophenoxyacetic acid (2,4-D) in water (Wittmann *et al.*, 1996). In this system, an optical fiber with an immobilized analyte, 2,4-D, was placed into a flow cell. The assay procedure involved several steps:

(1) The fiber was incubated with fluorescently labeled monoclonal antibody for 2,4-D and the initial (maximum) fluorescence signal was measured. The fiber was then washed with buffer.
(2) The sample was incubated with fluorescently labeled monoclonal antibody for 2,4-D.
(3) The fiber was incubated with a mixture of sample and labeled antibody.
(4) The fiber was washed and the signal was measured.

When a high concentration of analyte (>1000 μg/l) was present in the sample, a low signal was obtained because most of the antibodies were occupied with the sample analyte and could not bind to the 2,4-D immobilized on the fiber. This bio-optrode was used to measure concentrations ranging between 0.2 and 100 μg/l. This concentration range is suitable for environmental applications where the permitted level of 2,4-D in drinking water cannot exceed 0.1 μg/l. The sensing layer could be regenerated by washing with proteinase K. This procedure enabled the bio-optrode to be used for more than eight weeks and in more than 500 successive measurements. Such bio-optrodes have the potential to be useful for online analysis of drinking water and to serve as warning devices for hazardous pesticide contamination.

Fiber optic-based biosensors have also been used to detect toxins and bacteria, such as staphylococcal enterotoxin B, ricin, *Bacillus anthracis*, and *Francisella tularensis* (Anderson and Rowe-Taitt, 2001).

Enzyme-based bio-optrodes for environmental applications have also been developed. The most common approach employs enzyme inhibition as the sensing mechanism. The inhibition of acetylcholinesterase by an organophosphate pesticide has been used in several sensors (Eldefrawi *et al.*, 1995; Xavier *et al.*, 2000; Choi *et al.*, 2001; Doong and Tsai, 2001), and is described in Section 1.1.4.1.

A different enzyme-based bio-optrode that uses a chemiluminescence reaction for detection of phenolic compounds was recently described (Ramos *et al.*, 2001). This bio-optrode is based on the enhancement of the luminol–H_2O_2–horseradish peroxidase chemiluminescence reaction by phenolic compounds. Using this bio-optrode, *p*-iodophenol, *p*-coumaric acid, and 2-naphthol were detected in concentrations as low as 0.83 μM, 15 nm, and 48 nM, respectively. The bio-optrode was fabricated by entrapping the enzyme in a sol-gel layer; the gel was prepared directly on the fiber tip. The assay was performed by inserting the fiber with the immobilized enzyme into a test tube containing the analyte, luminol, and H_2O_2. The chemiluminescence intensity maximum at 5 min was the output signal.

Whole cells have also been used for environmental bio-optrode construction. Recombinant *E. coli* cells overexpressing the enzyme OPH were immobilized to an optical fiber and used to detect organophosphate nerve agents, as described earlier (Section 1.1.4.1) (Mulchandani *et al.*, 1998). The bio-optrode detection limits were 3 μM for paraoxon and parathion and 5 μM for coumaphos. The sensor was stable over a 1-month period and used for over 75 repeated measurements. Recently, a fiber optic biosensor for the measurement of 1,2-dichloroethane in aqueous solution was developed by immobilizing whole cells of *Xanthobacter autotrophicus* GJ10 in calcium alginate on the tip of a fiber optic fluoresceinamine-based pH optrode (Campbell *et al.*, 2006). An enzyme in the *X. autotrophicus* cells, haloalkane dehalogenase DhlA, hydrolytically cleaves a chlorine atom from dichloroethane, generating hydrochloric acid. This enzyme-generated pH change was detected by the fiber optic pH sensor. Dichloroethane as low as 11 mg/l could be reproducibly measured in 8–10 min with a linear response up to at least 65 mg/l.

Using a different approach, in which the cell's genetic response was used as the sensing mechanism, a whole cell bio-optrode was used for detection of naphthalene and salicylate (Heitzer *et al.*, 1994; Ripp *et al.*, 2001). The sensing was performed by *Pseudomonas fluorescens* HK44 cells carrying a plasmid containing a fusion of the *nahG* gene, which is induced by naphthalene and salicylate, and the *luxCDABE* reporter gene, coding for the enzyme luciferase. The cells were immobilized onto the surface of a liquid light guide or an optical fiber by using strontium alginate. The bio-optrode tip was placed in a measurement flow cell that simultaneously received a waste stream solution and a maintenance medium. A rapid increase in bioluminescence was obtained when either naphthalene and salicylate were present in the waste stream. Real environmental samples of pollutant mixtures containing naphthalene were also tested using this system. High bioluminescence was obtained when aqueous solutions saturated with JP-4 jet fuel or aqueous leachates from contaminated soil were tested.

Using a similar approach, Hakkila *et al.* (2004) used recombinant *E. coli* cells carrying genes responsive to the presence of bioavailable heavy metal ions (e.g., Hg^{2+}, Cu^{2+}, Cd^{2+}, and Pb^{2+}) fused to firefly luciferase reporter gene. The cells were immobilized onto the tip of an optical fiber and the device was used to accurately and reproducibly detect bioavailable heavy metals in 17 synthetic and 3 environmental blind samples. In the same manner, Biran *et al.* (2003) fabricated an Hg^{2+} biosensor by immobilizing *E. coli* cells, possessing the *lacZ* reporter gene fused to the heavy metal-responsive gene promoter *zntA*, on the face of an optical imaging fiber containing a high-density array of microwells. A plasmid carrying the gene encoding enhanced cyan fluorescent protein was also introduced into this sensing strain to identify the cell locations in the array. Single cell *lacZ* expression was measured when the array was exposed to mercury and a response to 100 nm Hg^{2+} could be detected after a 1-h incubation time. Another bio-optrode was also reported in which genetically modified *E. coli* cells produced bioluminescence in response to the presence of genotoxic agents (Polyak *et al.*, 2001). This bio-optrode was able to detect mitomycin C as low as 25 μg/l in less than 2 h.

The main importance of genetic response–based bio-optrodes for environmental analysis is the information they provide about the bioavailability of the analytes. This parameter is very important and helps to decide whether and how to treat a polluted site and which remediation strategies to employ.

1.3.6. Clinical applications

The development of bio-optrodes for clinical applications is another promising field and is focused on two types of applications: (a) *in vivo* detection inside a patient, (b) *ex vivo* detection when clinical samples are analyzed at the patient's bedside. The *in vivo* bio-optrodes would enable continuous monitoring of important analyte concentrations and would dramatically improve clinical procedures such as heart bypass surgery and critical care procedures in patients with compromised respiratory conditions. The optical fiber's small diameter, flexibility, nontoxic nature, durability, and lack of direct electrical connections make them highly suitable for *in vivo* applications. Moreover, optical fibers have already proven to be valuable for *in vivo* clinical applications such as endoscopic procedures and laser power transmission for surgical procedures. For example, endoscopes are routinely used in endoscopic surgery for gall bladder removal and for chest and knee surgery. In principle, bio-optrodes can be coupled to such devices and used to provide analytical information during endoscopic surgeries. At present, such bio-optrodes have not been implemented because of blood compatibility problems in which a thrombus (clot) forms around the sensor tip and affects the measurement accuracy. Recently, other *in vivo* chemical optrodes, such as the ones used by Baldini (2003) for *in vivo* monitoring of bile, carbon dioxide, and pH, have been reported.

The second clinical application for bio-optrodes is *ex vivo* diagnostics, mainly in critical care situations. Most diagnostic tests are presently performed in a centralized laboratory. Samples must be collected with the attendant transport, storage, and chain-of-custody issues. The remote location of the laboratory delays the medical diagnosis. In order to provide rapid diagnostic tests, analytical devices, such as bio-optrodes, can be used to bring the laboratory closer to the patient. These point-of-care devices should be sensitive, selective, self-contained (no need to add reagents), and simple to operate. They should also be small in size in order to be conveniently located near the patient. In addition, it is preferable that the sampling unit in contact with the sample (e.g., blood, urine) be disposable. Bio-optrode devices of this type are still not commercially available, but there are similar chemical-based fiber optic sensor devices used routinely in clinics. In these devices, fluorescent dyes are used as indicators for monitoring blood gases (PO_2, PCO_2) and pH. In one device, the immobilized dyes are incorporated into a disposable apparatus that is inserted into an extracorporeal blood circuit on one side and connected to a fiber bundle on the other (Owen, 1996). These sensors are mainly used to monitor blood gases during open-heart surgery. Another device is used for a paracorporeal measurement at the patient's bedside (Martin *et al.*, 1994). The sensors are placed into an external tube connected to an arterial blood line. Blood samples are periodically and automatically pumped into the tube, analyzed by the sensors, and then returned to the blood line. In this way, the blood can be monitored semi-continuously without requiring blood samples to be taken from the patient. It should be possible to incorporate bio-optrodes into such devices and use them to monitor other clinically important analytes.

For both *in vivo* and *ex vivo* applications, the first step in bio-optrode development is to establish sensitive sensing mechanisms that can be used to recognize specific analytes in a complex sample such as blood, urine, or other human fluids. Many examples of bio-optrodes directed for clinical applications have been proposed (Meadows, 1996; Vidal *et al.*, 1996; Vo-Dinh and Cullum, 2000; Knory *et al.*, 2004; Watterson *et al.*, 2004). Several glucose bio-optrodes, based on the enzyme-catalyzed reaction of glucose with glucose oxidase, have been prepared or proposed (Moreno-Bondi *et al.*, 1990; Li and Walt, 1995; Marquette *et al.*, 2000; Zhu *et al.*, 2002). A submicrometer glucose bio-optrode has been prepared (Rosenzweig and Kopelman, 1996a, 1996b). In this bio-optrode, the consumption of molecular oxygen is measured by the fluorescence quenching of ruthenium complexes and serves as a reporter for glucose concentration. The enzyme and the indicator are immobilized in an acrylic polymer support on the fiber tip. The bio-optrode response is very fast (2 s) and concentrations as low as 1×10^{-15} mol were detected. Pasic *et al.* (2006) recently described a microdialysis-based bio-optrode for continuous online glucose monitoring. The bio-optrode was used *in vitro* to monitor glucose-spiked plasma continuously for 3 days. It was further evaluated by online monitoring the glucose level of a healthy volunteer for 24 h. A commercial microdialysis system was used to maintain constant air saturation of the measuring fluid in the cell. A reference oxygen optrode was used to compensate for response changes caused by events such as bacterial growth and temperature fluctuations. The authors suggested that this sensor could potentially be used for continuous glucose monitoring of patients in intensive care units.

Several bio-optrodes were developed for the sensitive detection of important cardiac markers such as myoglobin, cardiac troponin I(cTnI), and brain natriuretic peptide (Hanbury *et al.*, 1997; Masson *et al.*, 2004; Hong *et al.*, 2006). These markers are critical for rapid and accurate heart attack diagnosis and prognosis. The self-contained antibody-based bio-optrode developed by Hanbury *et al.* (1997) is potentially clinically important since it can serve as a method for monitoring the extent of myocardial infarction. Myoglobin was detected by immobilizing a fluorescently labeled monoclonal antibody in polyacrylamide gel on the tip of an optical fiber (Figure 1.24a). Cascade Blue was used both as the fluorescent labeling agent and as a FRET donor molecule. When myoglobin was captured by the antibody, fluorescence energy transfer occurred between Cascade Blue and the myoglobin heme group (acceptor). Fluorescence quenching of Cascade Blue was then correlated to the myoglobin concentration (Figure 1.24b). The polyacrylamide gel was optimized to serve as a size selective filter allowing only low molecular mass molecules to penetrate and interact with the antibodies. Using this approach, myoglobin (16 500 Da) could be discriminated

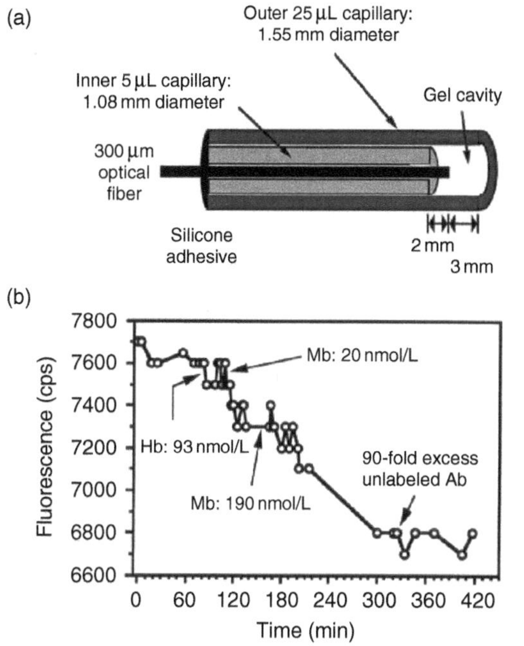

Figure 1.24 Reagentless bio-optrode for myoglobin detection. (a) Bio-optrode setup. (b) Myoglobin bio-optrode responses. The bio-optrode responses in PBS buffer (0–80 min) and after incubations with hemoglobin (Hb), myoglobin (Mb), and unlabeled myoglobin antibody (Ab) are shown (Hanbury *et al.*, 1997). Reprinted with permission from *Clin. Chem.*

from hemoglobin (bigger than 70 kDa). The size selection was necessary since antibodies for myoglobin can also bind hemoglobin. As shown in Figure 1.24b, a significant response was obtained when the bio-optrode was incubated with myoglobin but no response was obtained with hemoglobin (93 nmol/l). The detection limit of this bio-optrode was 5 nmol/l (83 μg/l), which is near the clinical decision limit for myocardial infarction diagnostics. The limitation of using the gel layer was the increased response time due to the low diffusion rate through the gel layer. In addition, when the gel layer was used, the bio-optrode response was irreversible even when the bio-optrode was incubated in a solution containing a high concentration of myoglobin antibodies (Figure 1.22b).

Irreversible sensor responses could limit the use of this bio-optrode for continuous monitoring applications.

Another example of a bio-optrode for heart-related disease diagnostics is the D-dimer antigen bio-optrode (Grant and Glass, 1999). D-Dimer antigen is formed when vascular occlusions are treated with a thrombolytic agent to lyse the clot. This treatment involves inserting a microcatheter at the occlusion site and injecting thrombolytic agents. Although thrombolytic therapy can help in preventing strokes, it suffers from several limitations that bio-optrodes can help to overcome. One limitation of thrombolytic therapy is the difficulty in determining whether the occlusion occurred from an atherosclerotic plaque or from a thrombus. Using the bio-optrode, initiation of D-dimer antigen formation following the injection of a small amount of thrombolytic agent would indicate the presence of a thrombus clot. If D-dimer antigens are not detected after the thrombolytic agent injection, it would be an indication that the occlusion is caused by an atherosclerotic plaque and alternative treatment would be required. In addition, in the case of a thrombus clot, the bio-optrode can be used for online monitoring of the thrombolytic agent dosage needed to dissolve the clot by monitoring the D-dimer antigen formation during the procedure. The principle of operation for this sensor is similar to that of the myoglobin bio-optrode described above. A fluorescently labeled antibody was immobilized in a sol-gel on an optical fiber tip and fluorescence quenching was observed when D-dimer antigen bound. D-Dimer antigen was detected in human plasma and in whole blood at a concentration of 0.56–6 μg/ml, which is within the clinically relevant range. The limitations of this bio-optrode were its poor reversibility and the short storage lifetime of the immobilized antibodies (4 weeks).

In addition to bio-optrodes directed to *in vivo* applications, bio-optrodes for point-of-care medical diagnostics have also been developed. These bio-optrodes offer a miniature design and a fast response time for analytes such as hepatitis C virus (Konry, 2005), bilirubin (Li and Rosenzweig, 1997), cholesterol (Marazuela *et al.*, 1997), and D-amino acids (Zhang *et al.*, 1995).

1.3.7. Industrial applications including bioprocess monitoring

Cell culture-based bioprocesses are very complex to control since they are sensitive to minor changes in the chemical composition of the fermentation medium. Therefore, tools for *online* monitoring of different analyte concentrations during the bioprocess are highly desirable. Bio-optrodes offer several advantages in such applications. The ability to use optical fibers directly inside fermentors (*in situ*) eliminates the need to periodically remove samples for analysis in a remote analytical laboratory. Once inside the fermentor, bio-optrodes can be used for sensitive and selective online monitoring of different analytes. Fermentation substrates and products such as proteins, antibodies, and antibiotics can be monitored. Other parameters related to the biological status of the cells, such as cell viability and activity, can also be measured. The ability to perform this measurement from a remote location (e.g., central control room) without using wires offers another important advantage. The main obstacle, which prevents wide use of *in situ* bio-optrodes (or any type of biosensor), is the need to sterilize the probe, which may damage the sensing biomolecules. For this reason, most bio-optrodes for bioprocess monitoring use a flow system in which a sample of the medium is taken from the fermentor and is delivered to the sensor (Dremel *et al.*, 1992; Mulchandani and Bassi, 1995; Scheper *et al.*, 1996; Marose *et al.*, 1999).

In one example, an FIA-based enzyme bio-optrode system was used for simultaneous detection of five different analytes (glucose, lactate, glutamate, glutamine, and ammonia) in samples removed during animal cell cultivation (Spohn *et al.*, 1995). The system was based on chemiluminescence detection and consisted of five optical fibers, each with a different immobilized enzyme. Each fiber was inserted into a different flow cell and, when the sample was injected, each fiber's response was measured. The results were combined and the concentrations of the different analytes were determined. Figure 1.25 shows results from a 350-h monitoring of an animal cell culture medium. In another example, an optical penicillin/pH biosensor was developed (Healey and Walt, 1995) to simultaneously monitor the pH and the concentration of penicillin produced during *Penicillium chrysogenum* fermentation process. In a

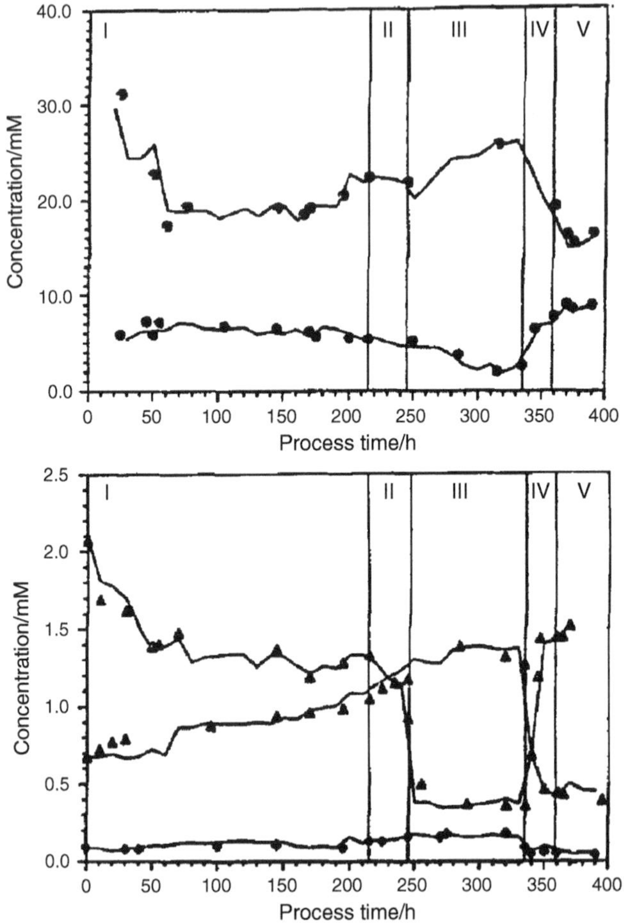

Figure 1.25 Five-channel enzyme-based bio-optrode for continuous monitoring of lactate, glucose, glutamine, ammonia, and glutamate (from top to bottom) during animal cell cultivation. Chemiluminescence measurements were employed to determine the analyte concentrations using five optical fibers, each with a different immobilized enzyme (Spohn *et al.*, 1995). Reprinted with permission from Elsevier Science.

similar way, antibody-based or nucleic acid-based bio-optrodes can be used to monitor different bioprocesses.

1.4. Advantages and limitations of bio-optrode technology

Bio-optrodes offer several advantages over other biosensing technologies based on the unique characteristics of optical fibers. The optical fiber's small dimensions, flexibility, and ability to transmit optical signals for long distances allow them to be used for remote sensing in places where other biosensors cannot be used. In addition, their ability to function without any direct electrical connection to the sample makes them safer than electrochemical biosensors. Optical sensors are also free from electromagnetic interference. Bio-optrodes are intrinsically simpler than electrode-based biosensors since no reference electrodes are needed; if desired, however, reference fibers can be interrogated in parallel to those used to measure analyte. Moreover, the development of new biorecognition molecules, such as those containing FRET-based dyes, enables the fabrication of self-contained bio-optrodes where no additional reagents are needed. Bio-optrodes based on imaging fibers offer additional advantages since they allow multiplexing with multianalyte sensing capabilities.

It is expected that as new optical technologies are developed for telecommunication applications, they will be adopted for bio-optrodes. These technologies include miniaturization of light sources, detectors, and optical fiber components (Kostov and Rao, 2000; Lechuga, 2005). Furthermore, new developments in the area of nanotechnology should eventually enable development of new bio-optrodes at the nanometer scale (Kasili and Vo-Dinh, 2005).

Nevertheless, bio-optrode technologies suffer from several drawbacks. Some of these drawbacks are common to all biosensor devices; the most restrictive of these drawbacks is the poor stability of the biological recognition molecules. Such molecules tend to be sensitive to pH or temperature changes and therefore generally have short lifetimes. Another important limitation is the high cost of some of the purified biological sensing materials. Regeneration of sensing biomolecules is usually problematic and, in most cases, fresh biorecognition molecules are required for each assay. In addition, there are several problems related to the immobilization process including loss in activity, leaching of reagents, and the decreased response time due to slow diffusion of analytes through the immobilized layer.

Several other bio-optrode limitations are related to the nature of optical fibers. Since light signals are the measured parameter, bio-optrodes are sensitive to ambient light interference and precautions must be taken either to exclude light or to employ optical designs with lock-in detection capabilities. In most bio-optrodes, there is a need to use indicator dyes in order to transduce the biorecognition events. The dyes have to be immobilized together with the biomolecules and therefore complicate the bio-optrode fabrication. In addition, the dyes may leach from the immobilization matrix or may lose their characteristics because of photobleaching.

1.5. Potential for improving performance or expanding current capabilities

As with any sensing or monitoring device, the ideal bio-optrode should be specific, sensitive, simple to fabricate and use, well adapted to the measurement environment (e.g., detect specific analytes in a complex sample), reliable, and self-contained. When used as a sensor, it should be operated in a continuous and reversible manner. When used as a probe, it should include a simple and disposable unit that contains the sensing molecules. In addition, for many applications, bio-optrodes should be small, able to detect multiple analytes simultaneously, and enable measurements in remote sites. At present, no bio-optrode device has achieved all these ideal performance capabilities. Nevertheless, based on new bio-optrode technologies currently under development, it is expected that the next generation of bio-optrodes will come closer to achieving these goals.

The development of new bio-optrode technologies and devices is highly dependent on advances in several different fields. Advances in biology, chemistry, materials science, optics, electrical engineering, mechanical engineering, and computer engineering are expected to inspire new bio-optrode technology development. In this section, a few new key technologies and their future impacts on the bio-optrode field are discussed.

1.5.1. New optical fibers and instrumentation

Optical fibers have attracted attention mainly due to their use in telecommunications. New technologies have been developed for fabricating optical fibers with very efficient light transmission capabilities. Fibers can transmit extremely high amounts of data when used in both single or bundle format. These characteristics will advance the development of real-time multianalyte bio-optrodes for various analytical applications.

Improved, smaller, and less-expensive light sources and detectors are driving consumer electronics. Integration of these components into bio-optrodes can lead to miniaturization and commercialization of bio-optrode devices (Kostov and Rao, 2000). Among the different possible light sources, light emitting diodes (LEDs) are very attractive to use in bio-optrodes. LEDs are very small, cover the entire visible spectrum, produce optical power in the range of 0.1–5 mW, and have a very long life (100 000 h) and low cost (~$2). Once LEDs at a particular wavelength have been demonstrated and commercialized, laser diodes are usually available within a few years. Laser diodes have higher power output and are nearly monochromatic, whereas LEDs have a relatively broad spectral emission output. Another interesting new light source is the scintillation light source that can be used as a high-stability light source for the UV and blue region (Potyrailo *et al.*, 1998). These sources are based on long-lived radioisotopes in scintillation crystals, which convert the radioactive emission (typically beta particles) into emitted light. These sources are extremely stable, can be used without external power sources, and have an expected life of 20 years.

In recent years, new generations of miniaturized and improved light detectors, such as photodiodes (PDs) and PMTs, have been developed (Kostov and Rao, 2000). The most sensitive detectors are avalanche

PDs. CCD chips are also rapidly developing; chips with higher signal/noise ratios, wider dynamic ranges, more pixels, and lower dark currents have been developed. In addition, CCD detectors have been miniaturized and integrated into small devices. Image intensifiers have been integrated into CCD cameras to increase light detection sensitivity. Although CCD chip prices have been dramatically reduced, scientific grade CCD cameras are still very expensive (~\$7000–\$20 000). A competing technology to CCD is the complementary metal oxide semiconductor (CMOS) technology. Recent developments in this technology have demonstrated light detection capabilities similar to CCD detectors. The advantages of this technology are lower cost, simpler fabrication process, and the ability to use it for very fast image acquisition (32 000 pictures per second) because frame transfers are not required as all the processing is done on chip. Both CCD and CMOS technologies are being rapidly integrated into future bio-optrode devices.

1.5.2. New biological recognition elements

The "heart" of any bio-optrode is the biological recognition element that initiates the detection process by its interaction with the analyte. Development of new biological recognition elements will increase the number and types of analytes that can be detected by bio-optrodes. New advances in molecular biology techniques allow the design or selection of new recognition molecules. Among new recognition molecules, perhaps the one group attracting the most attention is engineered antibodies, such as small monovalent recombinant antibody fragments (Fab, scFv) and engineered multivalent variants (diabodies, triabodies, tetrabodies, minibodies, and single-domain antibodies) (Holliger and Hudson, 2005; see Chapter 12). For example, using phage display technology it is possible to screen and identify a single-chain antibody (scFv) with specificity for almost any analyte (Hoogenboom *et al.*, 1998). The recombinant scFv molecule is a smaller version of an antibody molecule containing the antigen-binding site. Genetically modified bacteriophages, each presenting a unique scFv molecule on its surface, are used in the screening process. Phage presenting scFvs with higher affinities to the analyte are selected. The system is designed in such a way that the sequence coding for the scFv presented in the selected phage can be readily identified.

Once inserted in an expression vector and transformed into a host, large quantities of the selected scFv molecule can be produced. This process is very powerful since it allows antibodies to be identified and isolated in a very short time (Goldman *et al.*, 2000). Moreover, once the antibody is found, it takes only a few days to produce it in large quantities. In addition, an scFv molecule can be specifically designed to be used in biosensor devices by adding immobilization capabilities to the molecule at the genetic level. For example, scFvs were fused to the cellulose-binding domain) resulting in scFv molecules that can be easily immobilized to a cellulose membrane (Berdichevsky *et al.*, 1999). Engineered proteins (Hellinga and Marvin, 1998; De Lorimier *et al.*, 2002, 2006; Looger *et al.*, 2003) and antibodies (Kramer and Hock, 2004) have already been used widely on biosensing platforms, and they will likely be integrated with bio-optrodes in the near future.

Using molecular biology techniques, several other biorecognition molecules have been designed. Genetic fusion between antibody molecules and the GFP results in self-fluorescent antibodies, also called fluorobodies, and eliminates the need to label the antibody with a fluorescent dye. Biological recognition elements isolated using combinatorial approaches, such as aptamers, are described in detail in Chapter 13. Biomimetic polymer materials for bio-optrode applications are described in Chapter 14.

1.5.3. Imaging and biosensing

The coupling of chemical and biosensing capabilities to optical fiber-based imaging devices (e.g., endoscopes) is expected to attract much attention in the future. Optical imaging fibers are used in endoscopes since they can carry images from one end of the fiber to the other due to the coherent nature of the fibers. This imaging capability can be utilized to simultaneously image and measure local analyte concentrations with micron-scale resolution (Bronk *et al.*, 1995; Michael and Walt, 1999; Issberner *et al.*, 2002). The distal face of an imaging fiber is coated with an analyte-sensitive layer (typically a biorecognition molecule and fluorescent indicator), which produces a microsensor array capable of spatially resolving analyte concentrations. The concept is shown in

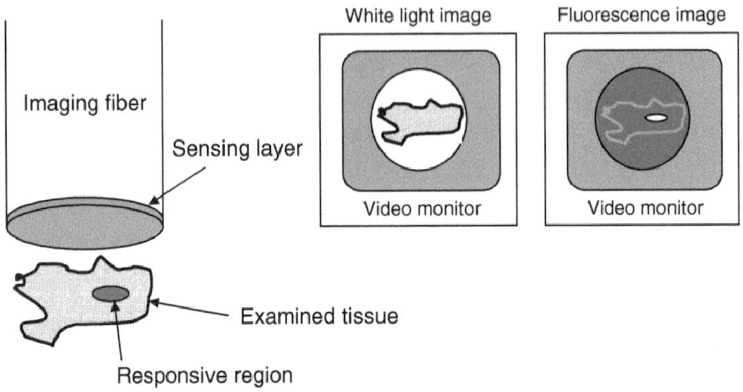

Figure 1.26 Combined imaging and biosensing concept. The technique provides the ability to both view tissue slices or individual cells and measure the release or consumption of different analytes using fluorescence techniques.

Figure 1.26. For example, an L-glutamate imaging fiber bio-optrode was fabricated by coating the imaging fibers with a layer of gel containing L-glutamate oxidase, poly(acrylamide-*co*-*N*-acryloxysuccinimide), and a pH-sensitive fluorescent dye, SNAFL (Issberner *et al.*, 2002). The fiber was used both to image and to chemically detect L-glutamate released from the foregut plexus of the Lepidopteran, *Manduca sexta* with spatial resolution of 3 ~ 4 μm and a detection limit between 10 and 100 μM.

L-Glutamate release was measured *in vivo* following electrical stimulation of the esophageal nerve immediately posterior to the frontal ganglion. First, the fiber was placed over the specimen and a white-light image was taken through the fiber to visualize the specimen's morphology and to observe some of the medium-sized nerve branches (5 ~ 10 μm width) in order to locate and position the fiber tip in a specified area. The imaging fiber bio-optrode was then switched to the fluorescence mode to detect and localize *in vivo* the release of L-glutamate and its subsequent re-uptake or diffusion away from the release site. The ability to observe the location of neurotransmitter release, with microscale spatial resolution, provides a powerful tool for neuroscience researchers. Moreover, this report demonstrated the feasibility of using imaging fiber bio-optrodes to detect neurotransmitter release from nerve endings *in vivo* with physiologically relevant resolution, suggesting the possibility of clinical applications with this technique. For example, a suspected cancer tumor could be examined based both on its morphology and on its response to specific antibodies immobilized on the imaging fiber tip.

1.5.4. Data analysis

New bio-optrode technologies are expected to provide a large amount of analytical data from each measurement. For example, multianalyte bio-optrodes can measure the concentration of many different analytes simultaneously. When performed in continuous fashion (e.g., multiple measurements every second), these measurements generate a high volume of data, especially as bio-optrode arrays are now being developed for genomic analysis. In order to acquire, analyze, and save such high data volumes, sophisticated software should be developed or adapted from other data-intensive applications (Figure 1.27).

The most significant computerized task is the data analysis because it may affect the specificity, sensitivity, and reproducibility of the bio-optrode. Analysis of biosensor measurements may be complicated due to the high variability in the activity of biorecognition molecules and because the measurements are usually performed in a complex sample matrix. It was previously shown that even in a simple FIA biosensor system for measurement of a single analyte in bioprocess samples, there is a need to use advanced computational analysis in order to improve the sensitivity, selectivity, and reproducibility of the measurement (Hitzmann *et al.*, 1998). For example, in typical bioprocess samples, the pH or ion concentrations change during cultivation; these changes may affect the biosensor's enzyme activity or antibody-binding properties. In addition, inhibitors, proteases, and nucleases can be produced during the bioprocess, which will affect protein- or nucleic acid-based biosensors. In order to overcome these problems, multivariate evaluation techniques such as neural networks have been used. Scheper and coworkers (Muller *et al.*, 1997) applied a neural network approach to improve the analysis of signals obtained from a bio-optrode for penicillin. A neural network approach was necessary due to the sensitivity of the measurement to

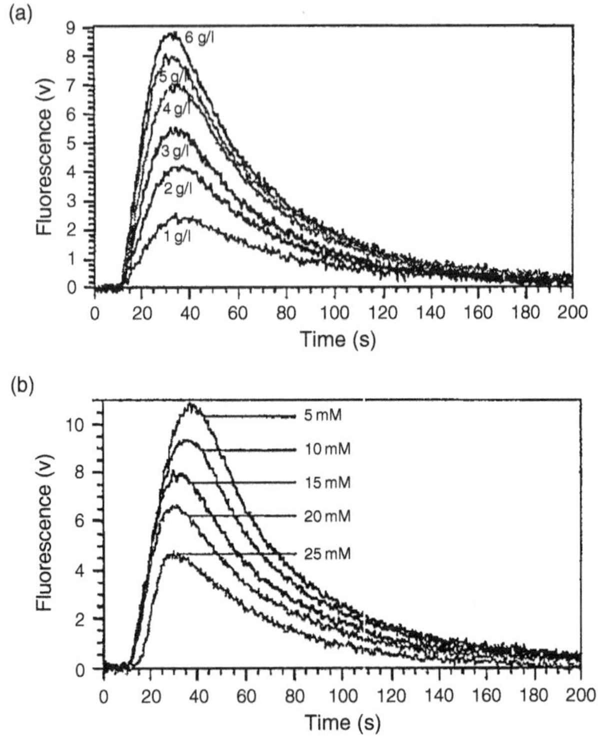

Figure 1.27 Penicillin detection using a FIA-based bio-optrode. (a) Fluorescence signals obtained in response to different penicillin concentrations. (b) Fluorescence signals obtained with different buffer ion concentrations and a fixed penicillin concentration (5 g/l). Neural network analysis was applied to analyze the relationships between these two responses (Muller *et al.*, 1997). Reprinted with permission from Elsevier Science.

changes in buffer ion concentration (Figure 1.25). The neural network was used to simultaneously evaluate the ion and penicillin concentrations from a single measurement based on characteristic signal shape variations. The shape characteristics were thought to be useful because in a preliminary experiment, multiple measurements of the same sample showed reproducible signal shapes. The results from this neural network showed errors of less than 11% for six different penicillin concentrations measured at five different ion concentrations.

To evaluate single-cell reporter-gene transcriptional activation and expression in yeast two-hybrid (Y2H) systems statistically, Walt and coworkers (Whitaker and Walt, 2007) have used a distribution fitting analysis tool, *t*-test, and one-way analysis of variance (ANOVA) model. This statistical analysis allowed elucidation of potential correlations between protein–protein interactions and reporter gene expression levels at varying stringency conditions. Plotting and normal fitting of these data provided a visual illustration of how different yeast strains expressed reporter genes under different conditions. These simple examples demonstrate the power of such computational techniques in the analysis of bio-optrode signals. It is clear that such techniques will be essential for analyzing signals from multianalyte bio-optrodes. Advanced computational methods have recently been used for the analysis of chemical sensor arrays (Jurs *et al.*, 2000), and it is expected that they will be adapted to the analysis of bio-optrode measurements. The generation of high amounts of information from future multianalyte bio-optrodes is expected to shift the emphasis from signal measurement to data analysis.

References

Aboul-Enein, H.Y., Stefan, R.I., van Staden, J.F. *et al.* (2000) *Crit. Rev. Anal. Chem.*, **30**, 271.

Ahn, S. and Walt, D.R. (2005) *Anal. Chem.*, **77**, 5041.

Ahn, S., Kulis, D.M., Erdner, D.L. *et al.* (2006) *App. Environmental Microbiol.*, **72**, 5742.

Anderson, G.P. and Rowe-Taitt, C.A. (2001) *Proc. SPIE – Int. Soc. Opt. Eng.*, **4206**, 58.

Anderson, G.P., Jacoby, M.A., Ligler, F.S., and King, K.D. (1997) *Biosens. Bioelectron.*, **12**, 329.

Anderson, G.P., King, K.D., Gaffney, K.L., and Johnson, L.H. (2000) *Biosens. Bioelectron.*, **14**, 771.

Arnold, M.A. (1985) *Anal. Chem.*, **57**, 565.

Balcer, H.I., Kwon, H.J., and Kang, K.A. (2002) *Ann. Biomed. Eng.*, **30**, 141.

Baldini, F. (2003) *Anal. Bioanal. Chem.*, **375**, 732.

Bard, A.J. (2004) Introduction. In *Electrogenerated Chemiluminescence* (A.J. Bard, ed.) New York: Marcel Dekker, p. 1.

Bard, A.J., Debad, J.D., Leland, J.K. *et al.* (2000) In *Encyclopedia of Analytical Chemistry* (R.A. Meyers, ed.) Chichester, UK: John Wiley & Sons, p. 9842.
Barker, S.L.R., Clark, H.A., Swallen, S.F. *et al.* (1999) *Anal. Chem.*, **71**, 1767.
Barker, S.L.R., Kopelman, R., Meyer, T.E., and Cusanovich, M.A. (1998) *Anal. Chem.*, **70**, 971.
Barnard, S.M. and Walt, D.R. (1991) *Science*, **251**, 927.
Battaglia, T.M., Masson, J.F., Sierks, M.R. *et al.* (2005) *Anal. Chem.*, **77**, 7016.
Berdichevsky, Y., Ben-Zeev, E., Lamed, R., and Benhar, I. (1999) *J. Immunol. Methods*, **228**, 151.
Berthold, J.W. and Lopushansky, R.L. (2004) *Proc. SPIE – Int. Soc. Opt. Eng.*, **5589**, 197.
Bier, F.F., Stocklein, W., Bocher, M. *et al.* (1992) *Sens. Actuators B Chem.*, **7**, 509.
Blicharz, T.M. and Walt, D.R. (2006) *Proc. SPIE – Int. Soc. Optical Eng.*, 6380 (Smart Medical and Biomedical Sensor Technology IV), 638010/1.
Blum, L.J., Gautier, S.M., and Coulet, P.R. (1993) *J. Biotechnol.*, **31**, 357.
Blum, L.J., Gautier, S.M., and Coulet, P.R. (1994) Fiber-optic biosensors based on luminometric detection. In *Food Biosensor Analysis* (G. Wagner and G.G. Guilbault, eds) New York: Marcel Dekker, p. 101.
Biran, I., Rissin, D.M., Ron, E.Z., and Walt, D.R. (2003) *Anal. Biochem.*, **315**, 106.
Biran, I. and Walt, D.R. (2002) *Anal. Chem.*, **74**, 3046.
Bowden, M., Song, L., and Walt, D.R. (2005) *Anal. Chem.*, **77**, 5583.
Brogan, K.L. and Walt, D.R. (2005) *Curr. Opin. Chem. Biol.*, **9**, 494.
Bronk, K.S., Michael, K.L., Pantano, P., and Walt, D.R. (1995) *Anal. Chem.*, **67**, 2750.
Bronstein, I., Martin, C.S., Fortin, J.J. *et al.* (1996) *Clin. Chem.*, 42, 1542.
Campbell, A.K. and Sala-Newby, G. (1993) Bioluminescent and chemluminescent indicators for molecular signaling and function in living cells. In *Fluorescence and Luminescence Probes for Biological Activity* (W.T. Mason, ed.) London: Academic Press, pp. 58–79.
Campbell, D.W., Mueller, C., and Reardon, K.F. (2006) *Biotech. Lett.*, **28**, 883.
Chalfie, M., Tu, Y., Euskirchen, G. *et al.* (1994) *Science*, **263**, 802.
Chandler, D.E., Majumdar, Z.K., Heiss, G.J., and Clegg, R.M. (2006) *J. Fluorescence*, **16**, 793.
Chang, A.C., Gillespie, J.B., and Tabacco, M.B. (2001) *Anal. Chem.*, **73**, 467.
Chau, L.-K., Lin, Y.-F., Cheng, S.-F., and Lin, T.-J. (2006) *Sens. Actuators B Chem.*, **B113**, 100.
Choi, J.-W., Kim, Y.-K., Lee, I.-H. *et al.* (2001) *Biosens. Bioelectron.*, **16**, 937.

Chovin, A., Garrigue, P., Vinatier, P., and Sojic, N. (2004) *Anal. Chem.* **76**, 357.
Cowan, J.J. and Arakawa, E.T. (1970) *Phys. Status Solid*, **1**, 695.
Cullum, B.M. and Vo-Dinh, T. (2000) *Trends Biotechnol.*, **18**, 388.
Cunningham, A.J. (1998) *Introduction to Bioanalytical Sensors*. New York: John Wiley & Sons, Inc., pp. 260–77.
Daunert, S., Barrett, G., Feliciano, J.S. *et al.* (2000) *Chem. Rev.*, **100**, 2705.
De Lorimier, R.M., Smith, J.J., Dwyer, M.A. *et al.* (2002) *Protein Sci.*, **11**, 2655.
De Lorimier, R.M., Tian, Y., and Hellinga, H.W. (2006) *Protein Sci.*, **15**, 1936.
Diaz, A.N., Peinado, M.C.R., and Minguez, M.C.T. (1998) *Anal. Chim. Acta*, **363**, 221.
DiCesare, C., Biran, I., and Walt, D.R. (2005) *Anal. Bioanal. Chem.*, **382**, 37.
Doong, R.A. and Tsai, H.C. (2001) *Anal. Chim. Acta*, **434**, 239.
Dremel, B.A.A., Li, S.Y., and Schmid, R.D. (1992) *Biosens. Bioelectron.*, **7**, 133.
Dunn, B., Cox, J., Lan, E., and Zink, J.I. (2001) *Abstr. Pap. Am. Chem. Soc.*, **221**, 473-COLL.
Dunn, B., Miller, J.M., Dave, B.C. *et al.* (1998) *Acta Mater.*, **46**, 737.
Durrieu, C. and Tran-Minh, C. (2002) *Ecotox. Environ. Safety*, **51**, 206.
Eggins, B.R. (1996) *Biosensors, an Introduction*. New York: John Wiley & Sons, Inc., pp. 1–117
Eldefrawi, M.E., Eldefrawi, A.T., Anis, N.A. *et al.* (1995) *Immunoanalysis of Agrochemicals*. Washington, DC: American Chemical Society, pp. 197–209.
Endo, H., Yonemori, Y., Musiya, K. *et al.* (2006) *Anal. Chim. Acta*, **573–574**, 117.
Epstein, J.R. and Walt, D.R. (2003) *Chem. Soc. Rev.*, **32**, 203.
Epstein, J.R., Ferguson, J.A., Lee, K.H., and Walt, D.R. (2003) *J. Am. Chem. Soc.*, **125**, 13753.
Epstein, J.R., Lee, M., and Walt, D.R. (2002) *Anal. Chem.*, **74**, 1836.
Epstein, J.R., Leung, A.P.K., Lee, K.H., and Walt, D.R. (2003) *Biosens. Bioelectron.*, **18**, 541.
Fan, J.B., Gunderson, K.L., Bibikova, M. *et al.* (2006) *Methods Enzymol.*, **410**, 57.
Fang, Q., Papaioannou, T., Jo, J.A. *et al.* (2004) *Rev. Sci. Instr.*, **75**, 151.
Fang, X. and Tan, W. (1999) *Anal. Chem.*, **71**, 3101.
Ferguson, J.A., Boles, T.C., Adams, C.P., and Walt, D.R. (1996) *Nat. Biotechnol.*, **14**, 1681.
Ferguson, J.A., Healey, B.G., Bronk, K.S. *et al.* (1997) *Anal. Chim. Acta*, **340**, 123.
Ferguson, J.A., Steemers, F.J., and Walt, D.R. (2000) *Anal. Chem.*, **72**, 5618.
Fine, T., Leskinen, P., Isobe, T. *et al.* (2006) *Biosens. Bioelectron.*, **21**, 2263.

Fraser, D. (1995) *Med. Device Technol.*, **6**, 28, 34.
Freeman, M.K. and Bachas, L.G. (1992), *Biosens. Bioelectron.*, **7**, 49.
Goldman, E.R., Pazirandeh, M.P., Mauro, J.M. *et al.* (2000) *J. Mol. Recognition*, **13**, 382.
Grant, S.A. and Glass, R.S. (1999) *IEEE Trans. Biomed. Eng.*, **46**, 1207.
Grant, S.A., Stringer, R.C., Studer, S. *et al.* (2005) *Biosen. Bioelectron.*, **21**, 438.
Gubitz, G., Schmid, M.G., Silviaeh, H., and Aboul-Enein, H.Y. (2001) *Crit. Rev. Anal. Chem.*, **31**, 167.
Gunderson, K.L., Steemers, F.J., Lee, G. *et al.* (2005) *Nat. Genet.* **37**, 549.
Gunderson, K.L., Steemers, F.J., Ren, H. *et al.* (2006) *Methods Enzymol.*, **410**, 359.
Hanbury, C.M., Miller, W.G., and Harris, R.B. (1997) *Clin. Chem.*, **43**, 2128.
Hakkila, K., Green, T., Leskinen, P. *et al.* (2004) *J. Appl. Toxicol.*, **24**, 333.
Harfensteller, M., Schilp, M., Eursch, A., and Zaeh, M.F. (2004) *Proc. SPIE – Int. Soc. Opt. Eng.*, **5590**, 57.
Healey, B.G. and Walt, D.R. (1995) *Anal. Chem.*, **67**, 4471.
Healey, B.G., Foran, S.E., and Walt, D.R. (1995) *Science*, **269**, 1078.
Healey, B.G., Li, L., and Walt, D.R. (1997a) *Biosens. Bioelectron.*, **12**, 521.
Healey, B.G., Matson, R.S., and Walt, D.R. (1997b) *Anal. Biochem.*, **251**, 270.
Heitzer, A., Malachowsky, K., Thonnard, J.E. *et al.* (1994) *Appl. Environ. Micro.*, **60**, 1487.
Hellinga, H.W. and Marvin, J.S. (1998) *Trends Biotechnol.*, **16**, 183.
Hitzmann, B., Ritzka, A., Ulber, R. *et al.* (1998) *J. Biotechnol.*, **65**, 15.
Hobel, W. and Polster, J. (1992) *Fresenius J. Anal. Chem.*, **343**, 101.
Holst, G. and Mizaikoff, B. (2002) Fiber optic sensors for environmental applications. In *Handbook of Optical Fibre Sensing Technology* (J.M. Lopez-Higuera, ed.) Chichester, UK: John Wiley & Sons, Inc, pp. 729–55.
Holliger, P. and Hudson, P.J. (2005) *Nat. Biotech.*, **23**, 1126.
Homola, J. and Yee, S.S. (1996) *Sens. Actuators B*, **37**, 145.
Hong, B. and Kang, K.A. (2006) *Biosens. Bioelectron.*, **21**, 1333.
Hoogenboom, H.R., de Bruine, A.P., Hufton, S.E. *et al.* (1998) *Immunotechnology*, **4**, 1.
Ikariyama, Y., Nishiguchi, S., Koyama, T. *et al.* (1997) *Anal. Chem.*, **69**, 2600.
Ince, R. and Narayanaswamy, R. (2006) *Anal. Chim. Acta*, **569**, 1.
Iqbal, S.S., Mayo, M.W., Bruno, J.G. *et al.* (2000) *Biosens. Bioelectron.*, **15**, 549.
Issberner, J.P., Schauer, C.L., Trimmer, B.A., and Walt, D.R. (2002) *J. Neuro. Met.*, **120**, 1.
Jin E.S., Norris, B.J., and Pantano, P. (2001) *Electroanalysis*, **13**, 1287.
Jordan, J.D., Dunbar, R.A., and Bright, F.V. (1996) *Anal. Chim. Acta*, **332**, 83.

Jorgenson, R.C. and Yee, S.S. (1993) *Sens. Actuators B*, **12**, 213.
Jurs, P.C., Bakken, G.A., and McClelland, H.E. (2000) *Chem. Rev.*, **100**, 2649.
Kasili, P.M., Song, J.M., and Vo-Dinh, T. (2004) *J. Am. Chem. Soc.*, **126**, 2799.
Kasili, P.M. and Vo-Dinh, T. (2005) *J. Nanosci. Nanotech.*, **5**, 2057.
Kaspar, B.H., Amstutz, P., and Plueckthun, A. (2005) *Nat. Bio.*, **23**, 1257.
Kishen, A., John, M.S., Lim, C.S., and Asundi, A. (2003) *Biosens. Bioelectron.*, **18**, 1371.
Knight, A.W. (1999) *Trends Anal. Chem.*, **18**, 47.
Kohler, S., Belkin, S., and Schmid, R.D. (2000) *Fresenius J. Anal. Chem.*, **366**, 769.
Komatsu, T., Kikuchi, K., Takakusa, H. *et al.* (2006) *J. Am. Chem. Soc.*, **128**, 15946.
Konry, T., Novoa, A., Avni, Y.S. *et al.* (2004) *Chem. Sensors*, **20**, 214.
Konry, T., Novoa, A., Cosnier, S., and Marks, R.S. (2003) *Anal. Chem.*, **75**, 2633.
Konry, T., Novoa, A., Shemer-Avni, Y. *et al.* (2005) *Anal. Chem.*, **77**, 1771.
Kostov, Y. and Rao, G. (2000) *Rev. Sci. Instruments*, **71**, 4361.
Kramer, K. and Hock, B. (2004) Antibodies for biosensors. In *Springer Series on Chemical Sensors and Biosensors*. Berlin: Springer GmbH, pp. 3–22.
Kuang, Y. and Walt, D.R. (2005) *Anal. Biochem.*, **345**, 320.
Kuang, Y., Biran, I., and Walt, D.R. (2004a) *Anal. Chem.*, **76**, 2902.
Kuang, Y., Biran, I., and Walt, D.R. (2004b) *Anal. Chem.*, **76**, 6282.
Kuboka, A.V., Bykh, A.T., and Svir, I.B. (2000) *Frsenius J. Anal. Chem.*, **368**, 439.
Kulp, T.J., Camins, I., Angel, S.M. *et al.* (1987) *Anal. Chem.*, **59**, 2849.
Kumagai, T. and Kajioka, H. (2002) Fibre optic gyroscope for industrial applications. In *Handbook of Optical Fibre Sensing Technology* (J.M. Lopez-Higuera, ed.) Chichester, UK: John Wiley & Sons, pp. 619–29.
Kumar, J., Jha, S.K., and D'Souza, S.F. (2006) *Biosens. Bioelectron.*, **21**, 2100.
Kuswandi, B., Andres, R., and Narayanaswamy, R. (2001) *Analyst*, **126**, 1469.
Kwok, N.-Y., Dong, S.J., Lo, W.H., and Wong, K.-Y. (2005) *Sens. Actuators B: Chem.*, **B110**, 289.
LaRossa, R.A. (1998) *Bioluminescence Methods and Protocols*, Vol. 102. Totowa, NJ: Humana Press, pp. 85–299.
Lambeck, P.V. (1991) *Proc. SPIE*, **115**, 100.
Lechuga, L.M. (2005) *Comprehensive Anal. Chem.*, **44**, 209.
Lee, M. and Walt, D.R. (2000) *Anal. Biochem.*, **282**, 142.
Lee, W.-B., Wu, J., Lee, Y.-I., and Sneddon, J. (2004) *Appl. Spectrosc. Rev.*, **39**, 27.

Leland, J.K. and Powell, M.J. (1991) *J. Electroanal. Chem.*, **318**, 91.
Leshem, B., Sarfati, G., Novoa, A. *et al.* (2004) *Luminescence*, **19**, 69.
Leth, S., Maltoni, S., Simkus, R. *et al.* (2002) *Electroanalysis*, **14**, 35.
Li, L. and Walt, D.R. (1995) *Anal. Chem.*, **67**, 3746.
Li, X.P. and Rosenzweig, Z. (1997) *Anal. Chim. Acta*, **353**, 263.
Liu, X.J., Farmerie, W., Schuster, S., and Tan, W.H. (2000) *Anal. Biochem.*, **283**, 56.
Liu, X., Song, D., Zhang, Q. *et al.* (2005) *Trends. Anal. Chem.*, **24**, 887.
Looger, L.L., Dwyer, M.A., Smith, J.J., and Hellinga, H.W. (2003) *Nature*, **423**, 185.
Lopez-Higuera, J.M. (ed.) (2002) *Handbook of Optical Fibre Sensing Technology*. Chichester, UK: John Wiley & Sons.
Loescher, F., Ruckstuhl, T., and Seeger, S. (1998) *Advanced Materials*, **10**, 1005.
Lubbers, D.W. and Opitz, N. (1975) *Pflugers Archiv-Eur. J. Physiol.*, **359**, R145.
Luo, S. and Walt, D.R. (1989) *Anal. Chem.*, **61**, 1069.
Magrisso, M., Etzion, O., Pilch, G. *et al.* (2006) *Biosens. Bioelectron.*, **21**, 1210.
Maragos, C.M. and Thompson, V.S. (1999) *Nat. Tox.*, **7**, 371.
Marazuela, M.D., Cuesta, B., MorenoBondi, M.C., and Quejido, A. (1997) *Biosens. Bioelectron.*, **12**, 233.
Marazuela, M.D. and Moreno-Bondi, M.C. (2002) *Anal. Bioanal. Chem.*, **372**, 664.
Marose, S., Lindemann, C., Ulber, R., and Scheper, T. (1999) *Trends Biotechnol.*, **17**, 30.
Marquette, C.A. and Blum, L.J. (2006) *Anal. Bioanal. Chem.*, **385**, 546.
Marquette, C.A., Degiuli, A., and Blum, L.J. (2000) *Appl. Biochem. Biotechnol.*, **89**, 107.
Marquette, C.A., Leca, B.D., and Blum, L.J. (2001) *Luminescence*, **16**, 159.
Martin, R.C., Malin, S.F., Bartnik, D.J. *et al.* (1994) *Proc. SPIE*, **2131**, 426.
Marty, J.L., Leca, B., and Noguer, T. (1998) *Analysis*, **26**, M144.
Massey, M., Piunno, P.A.E., and Krull, U.J. (2005) In *Springer Series on Chemical Sensors and Biosensors*, 3 (Frontiers in Chemical Sensors) Berlin: Springer GmbH, pp. 227–60.
Masson, J.F., Barnhart, M., Battaglia, T.M. *et al.* (2004) *Analyst*, **129**, 855.
Masson, J.-F., Kim, Y.-C., Obando, L.A. *et al.* (2006) *App. Spec.*, **60**, 1241.
Massona, J.F., Obando, L., Beaudoin, S., and Booksh, K. (2004) *Talanta*, **62**, 865.
Meadows, D. (1996) *Adv. Drug Deliv. Rev.*, **21**, 179.

Mehrvar, M., Bis, C., Scharer, J.M. *et al.* (2000) *Anal. Sci.*, **16**, 677.

Michael, K.L. and Walt, D.R. (1999) *Anal. Biochem.*, **273**, 168.

Michael, K.L., Taylor, L.C., Schultz, S.L., and Walt, D.R. (1998) *Anal. Chem.*, **70**, 1242.

Michel, P.E., Gautier-Sauvigne, S.M., and Blum, L.J. (1998a) *Talanta*, **47**, 169.

Michel, P.E., Gautier-Sauvigne, S.M., and Blum, L.J. (1998b) *Anal. Chim. Acta*, **360**, 89.

Monk, D.J. and Walt, D.R. (2004a) *Anal. Bioanal. Chem.*, **379**, 931.

Monk, D.J. and Walt, D.R. (2004b) *J. Am. Chem. Soc.*, **126**, 11416.

Moreno-Bondi, M.C., Wolfbeis, O.S., Leiner, M.J., and Schaffar, B.P. (1990) *Anal. Chem.*, **62**, 2377.

Mulchandani, A. and Bassi, A.S. (1995) *Crit. Rev. Biotechnol.*, **15**, 105.

Mulchandani, A., Kaneva, I., and Chen, W. (1998) *Anal. Chem.*, **70**, 5042.

Muller, C., Hitzmann, B., Schubert, F., and Scheper, T. (1997) *Sens. Actuators B Chem.*, **40**, 71.

Narayanaswamy, R., and Wolfbeis, O.S. (eds) (2003) *Optical Sensors for Industrial, Environmental and Clinical Applications*. Berlin: Springer-Verlag.

Naylor, L.H. (1999) *Biochem. Pharmacol.*, **58**, 749.

Obando, L.A. and Booksh, K.S. (1999) *Anal. Chem.*, **71**, 5116.

Otto, A. (1968) *J. Phys.*, **216**, 398.

Owen, V.M. (1996) *Biosens. Bioelectron.*, **11**, R5.

Pantano, P. and Walt, D.R. (1995) *Anal. Chem.*, **67**, A481.

Pasic, A., Koehle, H., Schaupp, L. *et al.* (2006) *Anal. Bioanal. Chem.*, **386**, 1293.

Pilevar, S., Davis, C.C., and Portugal, F. (1998) *Anal. Chem.*, **70**, 2031.

Polyak, B., Bassis, E., Novodvorets, A. *et al.* (2001) *Sens. Actuators B Chem.*, **74**, 18.

Potyrailo, R.A., Hobbs, S.E., and Hieftje, G.M. (1998) *Anal. Chim. Acta*, **367**, 153.

Preininger, C., Klimant, I., and Wolfbeis, O.S. (1994) *Anal. Chem.*, **66**, 1841.

Quinn, J.G., Neill, S.O., Doyle, A. *et al.* (2000) *Anal. Biochem.*, **281**, 135.

Rabbany, S.Y., Donner, B.L., and Ligler, F.S. (1994) *Crit. Rev. Biomed. Eng.*, **22**, 307.

Ramos, M.C., Torijas, M.C., and Diaz, A.N. (2001) *Sens. Actuators B Chem.*, **73**, 71.

Richter M.M. (2004) *Chem. Rev.*, **104**, 3003.

Ripp S., Nivens, D.E., Ahn, Y. *et al.* (2001) *Environ. Sci. Technol.*, **34**, 846.

Rissin, D.M. and Walt, D.R. (2006a) *Nano Lett.*, **6**, 520.

Rissin, D.M. and Walt, D.R. (2006b) *J. Am. Chem. Soc.*, **128**, 6286.

Rogers, K.R. and Gerlach, C.L. (1999) *Environ. Sci. Technol.*, **33**, 500A.

Rogers, K.R. and Mascini, M. (1998) *Field Anal. Chem. Technol.*, **2**, 317.
Rogers, K.R. and Poziomek, E.J. (1996) *Chemosphere*, **33**, 1151.
Rosenzweig, Z. and Kopelman, R. (1996a) *Anal. Chem.*, **68**, 1408.
Rosenzweig, Z. and Kopelman, R. (1996b) *Sens. Actuators B Chem.*, **36**, 475.
Sackmann, E. (1996) *Science*, **271**, 43.
Scheper, T.H., Hilmer, J.M., Lammers, F. *et al.* (1996) *J. Chromatogr. A*, **725**, 3.
Schobel, U., Barzen, C., and Gauglitz, G. (2000) *Fresenius J. Anal. Chem.*, **366**, 646.
Serganova, I. and Blasberg, R. (2005) *Nuclear Med. Bio.*, **32**, 763.
Song, L., Ahn, S., and Walt, D.R. (2005) *Emerging Infectious Diseases*, **11**, 1629.
Song, L., Ahn, S., and Walt, D.R. (2006) *Anal. Chem.* **78**, 1023.
Spiridonova, V.A. and Kopylov, A.M. (2002) *Biochem. (Moscow)*, **67**, 706.
Spohn, U., Preuschoff, F., Blankenstein, G. *et al.* (1995) *Anal. Chim. Acta*, **303**, 109.
Steemers, F.J. and Walt, D.R. (1999) *Mikrochim. Acta*, **131**, 99.
Steemers, F.J., Chang, W., Lee, G. *et al.* (2006) *Nat. Methods*, **3**, 31.
Steemers, F.J., Ferguson, J.A., and Walt, D.R. (2000) *Nat. Biotechnol.*, **18**, 91.
Szunerits, S. and Walt, D.R. (2003) *Chemphyschem.*, **4**, 186.
Szunerits, S., Tam, J.M., Thouin, L. *et al.* (2003) *Anal. Chem.*, **75**, 4382.
Szurdoki, F., Michael, K.L., and Walt, D.R. (2001) *Anal. Biochem.*, **291**, 219.
Tan, W.H., Shi, Z.Y., Smith, S. *et al.* (1992) *Science*, **258**, 778.
Tang, L., Kwon, H.J., and Kang, K.A. (2004) *Biotech. Bioeng.*, **88**, 869.
Tang, L., Ren, Y., Hong, B., and Kang, K.A. (2006) *J. Biomed. Optics*, **11**, 021011/1.
Thompson, R.B. and Jones, E.R. (1993) *Anal. Chem.*, **65**, 730.
Thompson, R.B., Ge, Z.F., Patchan, M. *et al.* (1996) *Biosens. Bioelectron.*, **11**, 557.
Tyagi, S. and Kramer, F.R. (1996) *Nat. Biotechnol.*, **14**, 303.
Tyagi, S., Marras, S.A.E., and Kramer, F.R. (2000) *Nat. Biotechnol.*, **18**, 1191.
Vannela, R. and Adriaens, P. (2006) *Crit. Rev. Environ. Sci. Tech.*, **36**, 375.
Vidal, M.M., Delgadillo, I., Gil, M.H., and Alonso-Chamarro, J. (1996) *Biosens. Bioelectron.*, **11**, 347.
Viveros, L., Paliwal, S., McCrae, D. *et al.* (2006) *Sens. Actuators B Chem.*, **B115**, 150.
Vo-Dinh, T. and Cullum, B. (2000) *Fresenius J. Anal. Chem.*, **366**, 540.
Vo-Dinh, T. and Kasili, P. (2005) *Anal. Bioanal. Chem.*, **382**, 918.

Vo-Dinh, T., Alarie, J.P., Cullum, B.M., and Griffin, G.D. (2000) *Nat. Biotechnol.*, **18**, 764.
Vo-Dinh, T., Cullum, B.M., and Stokes, D.L. (2001) *Sens. Actuators B Chem.*, **74**, 2.
Wabuyele, M.B., Farquar, H., Stryjewski, W. *et al.* (2003) *J. Am. Chem. Soc.*, **125**, 6937.
Watterson, J.H., Raha, S., Kotoris, C.C. *et al.* (2004) *Nucleic Acids Res.*, **32**, e18.
Wygladacz, K. and Bakker, E. (2005) *Anal. Chim. Acta*, **532**, 61.
Walt, D.R. (1998) *Accounts Chem. Res.*, **31**, 267.
Walt, D.R. (2000) *Science*, **287**, 451.
Walt, D.R. (2002) *Curr. Opin. Chem. Bio.*, **6**, 689.
Walt, D.R. (2006) *BioTech.*, **41**, 529.
Wang, X. and Krull, U.J. (2005a) *J. Mat. Chem.*, **15**, 2801.
Wang, X. and Krull, U.J. (2005b) *Bioorg. Med. Chem. Lett.*, **15**, 1725.
Watterson, J.H., Piunno, P.A.E., Wust, C.C., and Krull, U.J. (2001) *Sens. Actuators B Chem.*, **B74**, 27.
Watterson, J.H., Raha, S., Kotoris, C.C. *et al.* (2004) *Nucleic Acids Res.*, **32**, e18/1.
Weetall, H.H. (1993) *Appl. Biochem. Biotechnol.*, 41, 157.
Whitaker, D.R. and Walt, D.R. (2007) *Anal. Biochem.*, **360**, 63.
Wilchek, M. and Bayer, E.A. (1990) *Avidin–Biotin Technology*. San Diego, CA: Academic Press.
Wittmann, C., Bier, F.F., Eremin, S.A., and Schmid, R.D. (1996) *J. Agric. Food Chem.*, **44**, 343.
Wu, M., Lin, Z., Duerkop, A., and Wolfbeis, O.S. (2004) *Anal. Bioanal. Chem.*, **380**, 619.
Wolfbeis, O.S. (1991) *Fiber Optic Chemical Sensors And Biosensors*, Vol. 2. Boca Raton, FL: CRC Press, pp. 193–257.
Wolfbeis, O.S. (1997) Chemical sensing using indicator dyes. In *Optical Fiber Sensors – Applications, Analysis, and Future Trends*, Vol. 4 (B. Culshaw and J. Dakin, eds) Norwood: Artech House, pp. 53–107.
Wolfbeis, O.S. (2000a) *Anal. Chem.* **72**, 81R.
Wolfbeis, O.S. (2000b) *Anal. Chem.* **74**, 2663.
Wolfbeis, O.S. (2000c) *Anal. Chem.* **76**, 3269.
Wolfbeis, O.S. (2000d) *Anal. Chem.* **78**, 3859.
Wolfbeis, O.S. (2004) *Springer Series on Chemical Sensors and Biosensors 1 (Optical Sensors)* Berlin: Springer-Verlag, pp. 1–34.
Wong, R.B., Anis, N., and Eldefrawi, M.E. (1993) *Anal. Chim. Acta*, **279**, 141.

Wortberg, M., Orban, M., Renneberg, R., and Cammann, K. (1997) Flurometric immunosensors. In *Handbook of Biosensors and Electronic Noses: Medicine, Food, and the Environment*, (E. Kress-Rogers, ed.) Boca Raton, FL: CRC Press, p. 369.

Xavier, M.P., Vallejo, B., Marazuela, M.D. *et al.* (2000) *Biosens. Bioelectron.*, **14**, 895.

Xing, W.-L., Ma, L.-R., Jiang, Z.-H. *et al.* (2000) *Talanta*, **52**, 879.

Zhang, W., Chang, H., and Rechnitz, G.A. (1997) *Anal. Chim. Acta*, **350**, 59.

Zhang, Z., Zhang, S., and Zhang, X. (2005) *Anal. Chim. Acta*, **541**, 37.

Zhang, Z.J., Gong, Z.L., and Ma, W.B. (1995) *Microchem. J.*, **52**, 131.

Zhao, C.Q., Anis, N.A., Rogers, K.R. *et al.* (1995) *J. Agric. Food Chem.*, **43**, 2308.

Zhu, L., Li, Y., and Zhu, G. (2002) *Sens. Actuators B Chem.*, **B86**, 209.

Optical Biosensors: Today and Tomorrow (2nd Edition)
Frances S. Ligler and Chris Rowe Taitt (Editors)
Published by Elsevier B.V.

Chapter 2

EVANESCENT WAVE FIBER OPTIC BIOSENSORS

George P. Anderson, Ph.D. and Chris Rowe Taitt, Ph.D.
Center for Bio/Molecular Science & Engineering, Naval Research Laboratory, Washington, DC 20375-5348, USA

Evanescent wave fiber optic biosensors are a subset of fiber optic biosensors that perform the sensing function along the fiber's cylindrical length. In all optical fibers, light propagates by means of total internal reflection, wherein the propagating light is launched into waveguide at angles such that upon reaching the cladding–core interface, the energy is reflected and remains in the core of the fiber. Remarkably, however, for light reflecting at angles near the critical angle, a significant portion of the power extends into the cladding or medium which surrounds the core. This phenomenon, known as the evanescent wave, extends only to a short distance from the interface, with power dropping exponentially with distance. The evanescent wave has been exploited to allow for real-time interrogation of surface-specific recognition events.

2.1. Technical concept

Fiber optic biosensors all utilize the concept of total internal reflection (TIR), wherein the light, whether it be source or signal, transits the optical fiber by repeatedly reflecting off the cladding–core interface in a lossless fashion. Traditionally, fiber optics have been utilized as little more than a light pipe, allowing one to bend light and place it at a particular location, potentially extending great distances from the point of origin. However, in the course of TIR, there exists the interesting and useful phenomenon known as the evanescent wave. This power extends beyond the confines of the core into the cladding, which can be utilized for interrogation of a sensing surface. Thus, unlike previously described optrodes, for which the sensing area is the cross-sectional area of the fiber, evanescent wave sensors utilize the cylindrical surface of the core and sense over a much larger area.

2.1.1. The evanescent wave

As described in the earlier chapter on optrode-based fiber optic sensors, TIR is observed at the interface between two dielectric media with different indices of refraction. TIR is described by Snell's law:

$$\frac{n_1}{n_2} \sin \theta_1 = \sin \theta_2 \tag{2.1}$$

where n_1 and n_2 are the refractive indices of the fiber optic core and the surrounding medium, or cladding, respectively; θ_1 is the incident light angle from normal at the cladding–core interface, while θ_2 is the refractive angle either back into the core or out into the surrounding medium. For TIR to occur, n_1 must be greater than n_2 and the angle of incidence must be greater than the critical angle, θ_c, defined as:

$$\theta_c = \sin^{-1}\left(\frac{n_2}{n_1}\right) \tag{2.2}$$

These parameters determine the intrinsic behavior of the optical fiber with respect to the light and the angles at which light must be launched into the core to be totally reflected upon reaching the cladding.

However, Eqns (2.1) and (2.2) do not describe the electromagnetic components of the guided light wave. Of principle interest for our discussion is the evanescent wave which is the electric field that extends beyond the core into the surrounding (lower index) medium. The power of this electric field is highly dependent upon the angle at which the light strikes the interface, with only those rays impacting at angles near critical contributing significantly. Importantly, the power within this field decays exponentially with distance from the core surface. For multimode waveguides, the penetration depth of the evanescent wave, d_p, is the distance at which the strength of the field has decayed to $1/e$:

$$d_p = \frac{\lambda}{4\pi[n_1^2 \sin\theta - n_2^2]^{1/2}} \qquad (2.3)$$

where n_1 and n_2 are the refractive indices of the optical fiber and surrounding medium, as in Eqn (2.1) above, and θ is the angle of incidence (Harrick, 1967). The importance of the evanescent wave in biosensor development is that it provides for the excitation of surface-bound fluorophores, while avoiding generation of a large background signal from the bulk solution (Figure 2.1). This confined range of excitation is one of the key factors that render evanescent wave-based systems relatively immune to interference by matrix components, thereby permitting their operation in the presence of light-absorbing or light-scattering material.

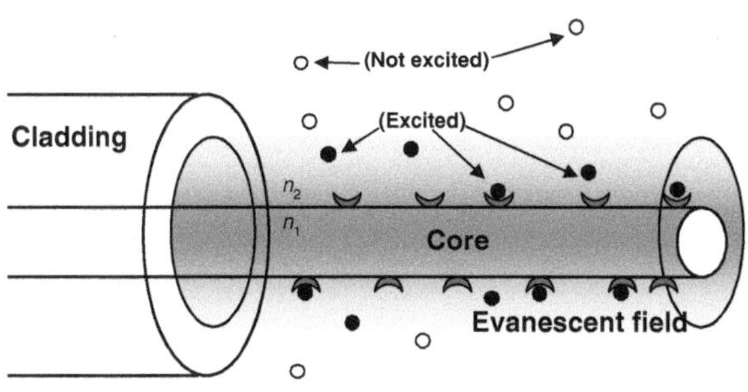

Figure 2.1 Evanescent wave biosensor using partially clad fiber. An evanescent field is set up when the optical fiber core (refractive index, n_1) is placed into a medium of lower refractive index (n_2), under conditions of TIR ($\theta > \theta_c$).

Another crucial factor to consider when designing a fiber optic biosensor is the waveguide parameter of the optical fiber, or V-number. The V-number is a dimensionless factor that determines how many modes a fiber can support and is defined as:

$$V = \frac{2\pi r}{\lambda}\left(n_1^2 - n_2^2\right)^{1/2} \tag{2.4}$$

where r is the radius of the optical fiber, and n_1 and n_2 are defined as above. Low-order modes, or those that propagate at shallow angles with respect to the critical angle, contribute little to the evanescent wave. A fiber's higher order modes propagate at angles near the critical angle and contribute the majority of the power in the evanescent wave (Anderson, 1994). Interestingly, when light couples back into an optical fiber from sources in the cladding, as is the case for evanescent wave biosensors, the process is essentially a tunneling mechanism, where light couples into the fiber's modes in proportion to that mode's potential power carried in the evanescent wave (Figure 2.2).

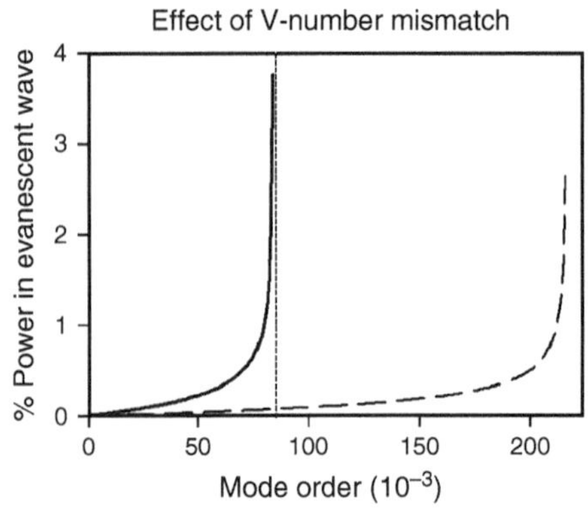

Figure 2.2 Percentage of power in the evanescent wave versus mode order. This figure depicts the loss that would occur because of mismatch in modal capacity between the clad and the sensing regions of the optical fiber. Only 7% of the coupled fluorescence from an untapered sensing region (dashed line) is in lower order modes that propagate in the clad fiber (solid line). Modified from Anderson *et al.* (1996).

The V-number also affects fiber optic probe design, since many probes consist of both a clad portion, which acts merely as a light pipe, and an unclad sensing region. The sensing region is typically immersed in water ($n = 1.33$), whereas the cladding typically has a higher index of refraction ($n \cong 1.4$); this results in an increase in the fiber's mode-carrying capacity at the sensing region. Thus, for a fiber optic probe created by simple removal of the cladding, very little power will be available for excitation; the highest order modes are no longer filled and, even more critically, the fluorescence will couple primarily into modes that fail to propagate in the clad fiber. This is termed a V-number mismatch. In order for the light guided in the unclad region of the core to be propagated in the clad region of the fiber, the following constraint must be satisfied:

$$r_{\text{aq}} < r_{\text{cl}} \left[\frac{n_{\text{co}}^2 - n_{\text{cl}}^2}{n_{\text{co}}^2 - n_{\text{aq}}^2} \right]^{1/2} \tag{2.5}$$

where n_{co} is the refractive index of the fiber core, n_{cl} is the refractive index of the cladded fiber, n_{aq} is the refractive index of the aqueous medium, r_{aq} is the radius of the unclad sensing region, and r_{cl} is the radius of the cladded region.

A significant amount of effort has been expended to minimize the deleterious effects caused by the V-number mismatch. Lackie (1992) patented a mounting system for an unclad fiber where the holder at one end of the fiber had the same refractive index as the fiber itself in order to avoid a V-number mismatch. Glass (1989) patented a similar mounting system for unclad fiber probes with the same goal in mind. Erb and Downward (1998) coated the ends of an unclad fiber with a Teflon having the same refractive index as the aqueous samples ($n = 1.3$) to prevent the V-number mismatch and subsequent loss of light where the fiber probe touched the holder.

For biosensors that utilize partially clad fibers, it was demonstrated that decreasing the fiber radius in the unclad sensor region to eliminate the V-number mismatch increased the light-coupling efficiency from the sensing region into the clad region and subsequent transmission of the

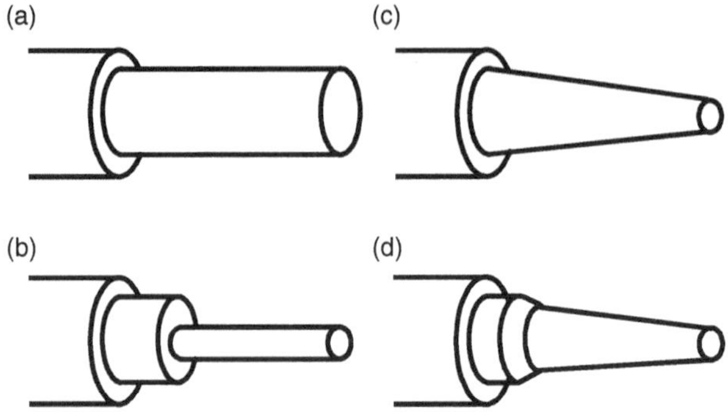

Figure 2.3 Schematic of optical fibers with different tapering geometries. (a) Untapered optical fiber with cladding stripped away from the core. (b) Step-tapered optical fiber to minimize V-number mismatch between sensing and cladded regions. (c) Continuously tapered core designed to maximize excitation at the fiber surface. (d) Combination taper probe designed to reach V-number match quickly, but yet maintain excitation at the surfaces distal to the rapid taper.

fluorescence through the rest of the fiber (Thompson and Kondracki, 1990; Thompson and Villaruel, 1991; Golden *et al.*, 1992, 1994a, 1994b; Anderson *et al.*, 1994a, 1994b; Anderson and Golden, 1995; Gao *et al.*, 1995). Indeed, experiments showed that for partially clad fibers, the ideal geometry of the sensing region was a compound taper (Figure 2.3). This compound taper enhanced both the excitation of fluorescence and its subsequent recovery. As the excitation light moves from the clad region to the unclad sensing region, the lower refractive index of the surrounding medium increases the mode-carrying capacity of the fiber. Thus, even if the fiber's entire numerical aperture (NA; see below) is filled at the distal clad end, an untapered, unclad sensing region lacks equivalent higher order modes and has little power in the evanescent wave. In the region closest to where the cladding ends, an initial rapid taper to a radius that results in matching the V-numbers for the clad and unclad regions will dramatically increase power in the evanescent wave. Then, a gradual taper along the length of the remaining sensing region will replenish the evanescent wave by upconverting light to higher modes at the surface, thereby maintaining excitation power along the probe's length. Conversely, fluorescence that couples back into the fiber and propagates back up the fiber will be converted to lower order modes that can efficiently reach the detector. By altering the geometry of the sensing region, immunoassay sensitivities could also be increased up to 80-fold (Anderson *et al.*, 1993).

2.1.2. Optical fibers

Optical fibers possess a number of physical characteristics that can be used to distinguish them, such as radius, refractive index, V-number (described above), and the NA; the last is a measure of the difference in refractive indices between the fiber core and the cladding and is related to the V-number:

$$\mathrm{NA} = \left(n_1^2 - n_2^2\right)^{1/2} \tag{2.6}$$

Monomode fibers, such as those used in the telecommunications industry, typically have a core size of 5–10 μm and propagate only a single mode at any given wavelength. As they have lower NAs, they have a higher percentage of total power present outside the core; greater than 50% of the power can be in the media surrounding the monomode fibers with low V-numbers (Gloge, 1971) – hence, more power to excite fluorophores via the evanescent field. Single-mode fibers have been used for evanescent sensing by a number of groups (Lew *et al.*, 1986; Villaruel *et al.*, 1987; Carlyon *et al.*, 1992; Hale *et al.*, 1996) and several theoretical studies on the effects of tapering on single-mode fibers have been published (Hale *et al.*, 1996; Ahmad and Hench, 2005; Kharat *et al.*, 2006). While several groups have explored the use of mono-mode fibers for surface plasmon wave-based measurement of refractive index changes (Homola, 1995; Slavik *et al.*, 1999, 2002; Bardin *et al.*, 2002; Monzón-Hernández *et al.*, 2004; Chang *et al.*, 2006), single-mode fibers have not generally received much attention for sensing purposes. Their small core radii render them extremely fragile, difficult to handle, or align with the laser light source. For these reasons, most evanescent wave sensors utilize multimode silica or plastic fibers for light transmission. Multimode fibers possess the advantages of good light transmission over short and medium distances with a wide variety of optical components. While evanescent excitation of surface-bound fluorophores is not as efficient as in the monomode fibers, ease of use and increased coupling efficiency are strong advantages of multi-mode fibers. And most importantly for sensor applications, the larger multimode fibers provide more surface area for the immobilization of the biological recognition molecules.

Fibers composed of fused silica offer excellent optical transmission from the near UV range to the near-IR range of wavelengths and have low intrinsic fluorescence. However, it was found that the amount of intrinsic fluorescence in fused silica fibers was a significant problem for the discrimination of fluorescence below 600 nm (Ligler *et al.*, 1995). Photobleaching the fibers immediately prior to conducting fluorescence assays eliminated the background fluorescence, while operating at wavelengths above 600 nm avoided the problem altogether (Shriver-Lake *et al.*, 1995b). Diameters of silica multimode fiber cores generally range from 50 μm to 1.5 mm, and these fibers are generally resistant to most biological buffer systems. However, methods required to resolve V-number mismatches (i.e., tapering often utilizes hydrofluoric acid) may be both hazardous and time-consuming, and can also be difficult to reproduce from batch to batch.

Plastic fibers, on the other hand, can be injection molded to meet nearly any design requirement (Slovacek *et al.*, 1992; King *et al.*, 1999; Saaski and Jung, 2000). Research International has developed plastic probes molded with a lensing structure on the end proximal to the excitation source. This lens has a dual purpose: to focus the excitation light into the fiber and to collimate the returning fluorescent signal (Jung *et al.*, 2003). The most commonly used plastic fibers are composed of polymethylmethacrylate or polystyrene. While plastic fibers have a very limited range of temperatures at which they can be used in comparison with silica fibers, this range of temperatures (−30 to 80°C) is sufficient for the study of most biological systems. The primary limitation of plastic fibers is the limited spectral range over which they can be used. This limitation is due to the high attenuation in the red and near-IR spectrum from –CH absorption bands. This problem can be partially circumvented by doping the plastics with deuterium; deuterium replaces the –CH absorption band with a –CD band, thereby increasing the range of useful wavelengths. These deuterium-doped plastic fibers, however, tend to lose their optical transmission over time (Boisdé and Harmer, 1996). In practice, an additional problem with the use of plastic fibers is the variation in the starting materials used for molding; plasticizers and other additives which fluoresce at visible wavelengths may be included in company's proprietary formulations. As a result, each batch of starting material must be screened for the resulting optical properties at the specific wavelengths intended for use.

2.1.3. Instrumentation

The majority of the evanescent wave fiber optic biosensors described to date are essentially fluorimeters that monitor fluorescent signals generated at the surface of an optical fiber probe coated with a biological recognition molecule. To automate, the system requires addition of pumps and valves to transfer samples, buffers, and reagents over the fiber probe. The components and configurations for the complete system depend on the geometry of the fiber and the degree of automation of the sensor. A separate discussion of surface plasmon wave-based instrumentation, which does not include a fluorimeter component, is also included.

2.1.3.1. Fluorimeter

The fluorimeter portion of the biosensor can be divided into the excitation and emission collection components. Generally, lenses are used to focus light from the source onto or into the fiber, and line filters are included in the excitation path if necessary. The first fiber optic biosensors used halogen lamps (Block and Hirschfeld, 1987; Glass, 1989; Block *et al.*, 1990) or xenon lamps (Kooyman *et al.*, 1987; Lipson *et al.*, 1992) as light sources; these were soon replaced with lasers with more uniform excitation wavelengths (Thompson and Ligler, 1988; Golden *et al.*, 1992; Hale *et al.*, 1996). The development of hydrophilic, near-infrared (IR) dyes (Shriver-Lake *et al.*, 1995b; Wadkins *et al.*, 1995) and diode lasers (Golden *et al.*, 1994a, 1997; Choa *et al.*, 1996) meant that the advantages of laser excitation could be implemented in small, low-cost devices. An additional advantage of the diode lasers is that the mechanical choppers originally included in order to discriminate fluorescence from ambient light could be eliminated by having the diode lasers pulsed to achieve the same effect.

The ability to discriminate a weak fluorescent signal above the background excitation light is the most critical feature of the fluorimeter. Choppers or pulsed lasers efficiently gate out stray ambient light, minimizing this potential source of noise. Discrimination of the fluorescent signal from stray excitation light depends on the Stokes shift of the fluorophores and the use of high-quality line filters and emission filters. A fluorescent reporter molecule with a large Stokes shift, the separation between the peak excitation and the emission wavelengths, greatly improves the efficiency of filters. Several fluorophores have lower wavelength shoulders in their absorption spectra which permit efficient excitation at wavelengths adequately removed from the emission maximum. Ultimately, filter selection is driven by the properties of the fluorophore intended for use. In addition, to simplify filtering of stray excitation light, the design and physical arrangement of the optical components can aid greatly in this process by separating the excitation and emission paths physically. Figure 2.4 depicts several of these strategies. These strategies rely on coupling the light into the evanescent wave from the center of the core and recovering the fluorescent light from the higher order modes. Appropriate lenses and filters are included with each of these configurations.

While the first fluorescence immunoassays performed on waveguides involved collecting the fluorescent signals perpendicular to the waveguide (Kronick and Little, 1975), nearly all fiber optic biosensors collect the emitted light out of the end of the fiber. This means that the signal is integrated over the active surface of the fiber core and focused on a single detector, usually either a photomultiplier tube or photodiode. In a notable exception to end-face collection, Fang and Tan (1999) determined that they could detect individual fluorescent molecules using evanescent excitation and signal collection normal to the waveguide using a microscope equipped with an intensified charge-coupled device. In this case, there was no intent to integrate the signal from multiple biological recognition events. More recently, Chou *et al.* (2007) used a similar configuration, with a photomultiplier tube mounted above the fiber to quantify fluorescence.

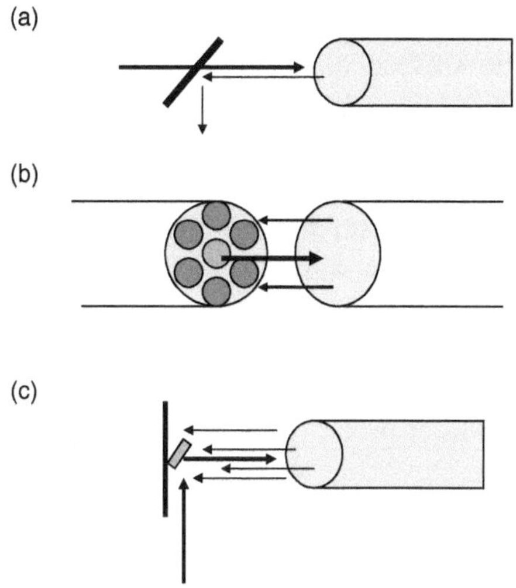

Figure 2.4 Examples of strategies examples that have been utilized to launch excitation light and detect evanescent wave generated fluorescence from the same fiber face. (a) Utilization of a dichroic filter allows the excitation light to pass through the filter, but reflects the fluorescence onto the detector. An alternative is to reflect the excition light off the dichroic filter into the fiber and allow the emitted light to pass straight through. (b) A fiber bundle is used to launch the excitation light using the center fiber, while the surrounding fibers are used to collect the returning fluorescence which is proportioned primarily in the higher order modes (Golden *et al.*, 1997). (c) A small mirror is used to reflect the excitation light into the fiber, while the returning fluorescence goes straight toward the detector. Like (b), this configuration takes advantage of the recovered fluorescence being primarily in higher order modes.

2.1.3.2. The fluidics

In addition to the fluorimeter, the fluidics component for delivering the sample and reagents to the fiber probe is an integral part of the biosensor. Primary considerations include the mechanism for holding the fiber probe in place within the fluid stream, the chamber surrounding the probe with its inlet and outlet ports, and the associated pumps and valves for automated assays in biosensors intended for applications other than research.

Several capillary chambers with individual fluid inlets and outlets to deliver fluid to the surface of the fiber have been described (Block and Hirschfeld, 1987; Glass, 1989; Slovacek and Love, 1992; Anderson *et al.*, 1993; Meserol, 1996; Neel and Lyst, 1997). Where the capillaries are intended for use with unclad fiber probes, significant attention has been given to the mechanism for holding the probe in the center of the capillary to prevent the *V*-number mismatch problems described above. Using a single fiber in a capillary tube, Oroszlan *et al.* (1993) reported the first fully automated assay system based on an optical fiber and computerized the operation of over 200 sequential 15- to 30-min assays over a single-fiber probe. The commercial instrument RAPTOR, manufactured by Research International, also utilizes unclad fibers in a fully automated system. Four fibers are mounted within a single coupon molded to contain four flow channels – one for each fiber. The injection molded plastic fibers possess a lensing portion that helps to focus emitted fluorescence onto the detector and a tab for centering the fiber within each channel and gluing it into the coupon (Figure 2.5)

Figure 2.5 RAPTOR fiber optic biosensor. This biosensor is a fully portable, self-contained unit. It can perform four simultaneous sandwich immunoassays; coupons that contain the polystyrene optical waveguides onto which the capture antibodies are immobilized can be seen on top of the RAPTOR.

(Saaski and Jung, 2000; Jung *et al.*, 2003). Use of partially clad fibers has simplified fluidics connections for many laboratory prototype systems. A simple capillary system (Anderson *et al.*, 1993) can be used for single fibers, but also permits the ganging of multiple probes for simultaneous multianalyte tests on a single sample with minimal cost (Bakaltcheva *et al.*, 1998; Shriver-Lake *et al.*, 1998; Tang *et al.*, 2006).

Movement of fluids through these networks can be accomplished by simple manual manipulations with a syringe or pipette or using automated or semi-automated pumping systems. Automated, timed addition and incubation of assay solutions with multiple fibers was first demonstrated in 1997 using a commercial fiber optic biosensor – the Analyte 2000 (Research International, Woodinville, WA) – and a laboratory prototype fluidics system constructed using commercial off-the-shelf components. This automated prototype was tested for its ability to collect and identify aerosolized bacteria while airborne in a small, unmanned plane (Figure 2.6) (Ligler *et al.*, 1998; Anderson *et al.*, 1999). This was the first demonstration that samples could be both collected and tested without manual operations, with results transmitted to the operator on the ground. Other groups are currently developing similar fluidics systems (Tang *et al.*, 2006).

The demonstration that a small fiber optic biosensor (see Figure 2.6) could be fully automated and operated remotely was the impetus for the commercial development of a self-contained, portable system, with all fluidic, optical, and electronic components fully automated within a single, hardened case. An early prototype developed by Research International was designed to flow fluids through a disposable coupon containing four optical fibers using a pneumatic fluidics system comprising a small air pump, pinch valves, and pressurized fluid reservoirs (King *et al.*, 1999; Saaski and Jung, 2000; Saaski, 2000). The concept behind this design was to avoid clogging and fouling of the pumping components by avoiding contact with the actual sample. However, issues with failure and leakage of the membranes used in the valves required an overhaul of the entire fluidics system, resulting in the evolution of the current version of this system, RAPTOR. The new system utilizes peristaltic pump-driven fluidics system based on small, custom-made

Figure 2.6 Images of the payload, front and back, that was mounted for operation in the UAV (swallow) pictured below. This was the first complete system, payload and airplane, designed specifically for the task of unmanned airborne biodetection. Reprinted from Anderson *et al.* (1999), with permission from John Wiley & Sons, Ltd.

pumps and contains no valves; each fiber can be independently interrogated with a different tracer reagent, allowing use of reagents previously shown to interact with other components of the system (e.g., cross-species cross-reactivity).

2.1.3.3. Systems based on surface plasmon resonance (SPR)

As described in detail in Chapter 4, SPR is a phenomenon encountered under conditions of TIR (see Eqns (2.1–2.4)) and involves excitation of surface electrons of a thin metal film coating the sensor surface. The excited electrons, or plasmons, produce a surface plasma wave that projects into the medium of lower refractive index.

Coupling of light from the waveguide into the metal overlayer and excitation of surface plasmons is measured as a decrease in reflectivity of the light exiting the waveguide or prism. As such, the conditions required for this coupling (coupling angle, wavelength, polarization) are exquisitely sensitive to any change in the optical properties of the bulk medium, as these changes will translate into changes in the interaction of light with the metal layer. Thus, a modification of the refractive index of the bulk solution in close proximity to the probe, such as occurs when a target analyte binds to an immobilized recognition molecule, will result in a measurable change in coupling condition. SPR instruments are therefore considered "reagentless" systems, as binding of target can be directly measured, without addition of exogenous components.

Surface plasmon resonance instruments are commonly classified as angular-, wavelength-, intensity-, or phase-modulated systems, depending on which variables are kept constant and which are monitored for changes in coupling conditions; examples of the different types of SPR-based fiber optic systems have been described (Slavik *et al.*, 2002; Mitsushio and Higo, 2004), although wavelength-modulated systems are most prevalent in the literature. These latter systems commonly employ a polychromatic lamp to supply excitation light, which is then focused into the optical fiber; output of the fiber optic SPR sensor is connected to a spectrograph capable of resolving minute changes in wavelength. Choppers can be installed on the input side of phase-modulated systems, and an optical multimeter installed for measurement of the output light in intensity-modulated systems.

Most of these systems employ side-polished optical fibers overcoated with a thin layer of gold or silver (Figure 2.7). However, an intensity-modulated system has recently been described (Mitsushio and Higo, 2004) using a gold film deposited on half of the core of the optical fiber *without a flattened surface*; this latter configuration, easier to fabricate than side-polished systems, gives rise to an asymmetrical response curve and is somewhat less sensitive than other configurations under most refractivity ranges.

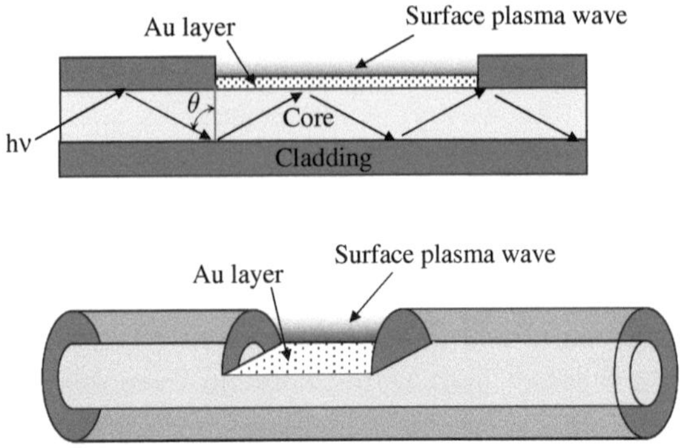

Figure 2.7 Schematic of a side-polished optical fiber used for SPR. The cladding is stripped away in a segment of the optical fiber. After polishing, the exposed core surface is overcoated with a thin layer of gold or silver, which will give rise to a surface plasma wave.

2.1.4. Detection parameters

A number of different optical parameters, such as absorbance, fluorescence emission spectral shifts, fluorescence resonant energy transfer (FRET), and bioluminescence, have been used for detection/quantification with fiber optic optrodes and are reviewed in Chapter 1. However, the number of parameters measured using evanescent wave fiber optic sensors is much more limited. Absorbance has been directly measured in evanescent wave fiber optic chemical sensors (Buerck *et al.*, 2001) and in an evanescent fiber optic immunosensor based on long-period fiber Bragg gratings (DeLisa *et al.*, 2000). The "reagentless" fiber optic biosensors using SPR directly interrogate refractive index changes. However, fluorescence intensity has been the property most often used in evanescent wave biosensors.

For affinity-based biosensors, measurement of fluorescence requires that the analyte of interest or a secondary detection element either possess intrinsic fluorescence that can be distinguished from the background or be labeled with a fluorescent tag. Having suitably fluorescent analytes is rare and fluorescently tagging analytes difficult, leaving most applications to require the use of a fluorescent "tracer" reagent – either a fluorescent derivative of the target of interest or a tagged recognition molecule (e.g., antibody) that binds to the target. Regardless of the assay format, fluorophores with high extinction coefficients and good quantum yields are preferred. Near-IR dyes, in particular, have the advantage of emitting at longer emission wavelengths, where (background) fluorescence from naturally occurring compounds is minimal. The cyanine and AlexaFluor dyes also have high quantum yields, are easy to conjugate under mild conditions, and are significantly more resistant to photobleaching than rhodamine and fluorescein dyes. Many of the newer dyes can also be used to label protein or other molecules at significantly higher levels without self-quenching (Anderson and Nerurkar, 2002). These dyes also have the advantage of exciting in the range of wavelengths emitted by inexpensive photodiode lasers, which are also lightweight and long-lived.

Another method for substantially reducing the background of evanescent wave assays is use of time-resolved fluorescence; a later chapter describes this technology in detail. While not yet in widespread use in the biosensor field due to requirements for high intensity, single-photon, pulsed lasers and sophisticated data analysis, two groups have demonstrated time-resolved fluorescence detection using evanescent wave fiber optic technology. Bock *et al.* (1995) used evanescent excitation to distinguish between binding of analytes labeled with two fluorophores to an optical fiber based only on the difference in their fluorescence lifetimes. Browne *et al.* (1996), using sol-gel-coated fibers, were able to distinguish between encapsulated fluorophores localized in different regions on the optical fiber. While this latter work did not utilize a biologically based system, the authors demonstrated the use of time-resolved detection as a means of resolving the fluorescence kinetics of various fluorophores, as well as the spatial arrangement of the fluorophores on the fiber.

2.1.5. Immobilization of biomolecules on optical fiber probes

As with any biosensor, methods for immobilizing the biomolecules to the fiber surface to maximize the density of the receptors and to maintain their function are critical for success. Two major approaches have been explored extensively: direct attachment to the fiber core and attachment of the receptor to an intermediate biomolecular thin film (i.e., through avidin or a lipid monolayer).

The early work of Hirschfeld (1984) and Hirschfeld and Block (1985) utilized the direct attachment approach, modifying the fiber surface with silanes and using any of a variety of chemistries or cross-linkers to attach proteins. This general methodology has been adapted by numerous groups for attaching a plethora of biomolecules to optical fibers. Inclusion of lengthy hydrophilic spacers between the silane and the immobilized species has improved retention of activity in some instances (Herron *et al.*, 1996).

While non-specific adsorption is effective in attachment of proteins to many surfaces, investigators working with glass or silica fibers found that the density of attached antibodies was highly variable and binding activity was often lost by denaturation and/or leaching from the surface. However, Rogers *et al.* (1989, 1991) found that non-specific adsorption of the large acetylcholine receptor (AChR) was a very satisfactory method to immobilize the protein and maintain its binding function. With the use of plastic fibers instead of glass or fused silica, King *et al.* (2000) found that non-specific adsorption of antibodies to polystyrene was reliable in terms of maintaining antibody density and function, even after long periods of storage or extended use (<40 assays).

In addition to direct immobilization, recognition molecules have been attached to optical fibers through use of intermediate layers. These layers provide a means to immobilize the recognition species non-covalently, using well-characterized biomolecular interactions. The most commonly used interaction is that of avidin with biotin; avidin can be covalently immobilized or adsorbed onto optical fibers, in essence, creating a universal surface onto which any biotinylated recognition species can be subsequently attached. Alternatively, the surface can be coated with biotin, with subsequent (non-covalent) immobilization of avidin and additional biotinylated recognition species (Abel *et al.*, 1996; Liu and Tan, 1999). Significant improvements in sensitivity have been observed with recognition molecules immobilized in this manner, although the improvements may be dependent on the recognition species used (Narang *et al.*, 1997a; Ligler *et al.*, 1998). As avidin possesses four biotin-binding sites, it is believed that inactivation/denaturation of one or two sites by the immobilization procedure – catastrophic for other biomolecules such as antibodies – will still allow a significant portion of the original binding activity to be retained.

A similar approach was also demonstrated using Protein A or Protein G as an intermediate layer (Anderson *et al.*, 1997). Protein A and Protein G are microbially produced IgG-binding proteins with different affinities for the Fc region of IgGs of different species. Although antibodies immobilized via Protein A or Protein G produced brighter signals than the same antibodies covalently attached to the fiber probes, there was no improvement in the limit of detection. However, this method has been demonstrated to be highly regenerable (Quinn *et al.*, 2000). In an interesting twist on the use of an intermediate protein layer, Mauro *et al.* (1996) genetically engineered a DNA-binding protein to include a Protein G terminus and bound this receptor to IgG-coated fiber probes.

In addition to protein layers, organic films fabricated using the Langmuir–Blodgett (LB) method have been deposited on optical fibers. Krull *et al.* (1988) pioneered this approach using lipids both for direct sensing and as membrane receptor supports. They demonstrated the response of an embedded fluorophore to membrane perturbation by phloretin and valinomycin. Zhao and Reichert (1992) doped lipid films with a biotinylated lipid to serve as an anchor for subsequent attachment of avidin; this method allowed very fine control over the density of the biotin and the association of the avidin. Avidin layers attached to immobilized biotin can also serve as the substrate for the attachment of another layer containing biotinylated receptors (Piunno *et al.*, 1995; Abel *et al.*, 1996; Duvenek *et al.*, 1996). While not a biosensor *per se*, Matsuo *et al.* (2000) continued to use LB films on optical fibers as an evanescent sensing layer for albumin, using the response of tetrabromophenol blue to measure pH change. The complexity of using the LB method and problems with reproducibility, however, make it unlikely that this approach will yield a commercial product.

In an excellent study on how various antibody immobilization strategies affect binding activity, Vijayendran and Leckband (2001) found that immobilizing IgGs through their carbohydrate moiety was the most promising surface attachment strategy. Higher biological activity and more homogeneous binding affinities were observed by carbohydrate-immobilized IgGs than in assays using IgGs immobilized using Protein G, avidin–biotin interactions, or through direct covalent attachment. These results may, of course, depend on the recognition molecule used.

2.1.6. Molecular recognition and detection: Affinity sensors

Optical biosensors require a biological recognition element with the following qualities: affinity and specificity for the analyte of interest; stability when incorporated into the biosensor; and ability to cause a detectable change in optical properties upon recognition of analyte. The recognition of analyte may take the form of an enzymatic reaction, such as in catalytic sensors, or may be one or a series of binding events, such as used in immunosensors and other affinity-based sensors. Since the majority of evanescent wave fiber optic sensors are affinity based, we will confine our discussion to these types of assays.

Regardless of the type of immobilized biological recognition molecule, affinity-based assays are formatted so that in the presence of analyte: (1) a fluorescent complex is formed at the surface of the optical biosensor; (2) its formation is blocked; or (3) a pre-formed fluorescent complex disrupted or modified. In general, the assay formats fall into three major categories: direct binding, competitive, and sandwich assays (Figure 2.8).

In direct binding assays (Figure 2.8a), the immobilized recognition biomolecule binds an analyte that is intrinsically fluorescent, carries a label, or changes the fluorescence of the recognition element upon binding. Alternatively, the system monitors the change in refractive index at the surface (i.e., SPR-based systems). Examples in which the analyte is itself fluorescent have been used primarily for demonstrating proof of concept, determining the sensitivity of the system, or measuring the kinetics of binding (Vijayendran *et al.*, 1999). Systems in which the fluorescence of the immobilized recognition molecules changes upon

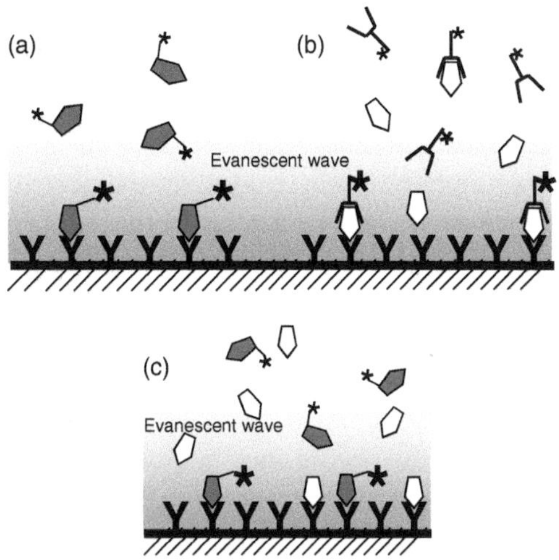

Figure 2.8 Three major classifications of affinity assay formats. (a) Direct binding: binding of fluorescent target molecule to an immobilized recognition molecule is directly measured. (b) Sandwich assays: a fluorescent "sandwich" is formed when target complexed to immobilized recognition molecule is bound by a fluorescently labeled "tracer" species. (c) Competitive assay: fluorescently labeled and unlabeled target compete for binding to immobilized recognition molecules; alternatively, immobilized and free target can compete for binding to labeled antibody in solution.

analyte binding employ molecular beacons (Liu and Tan, 1999), FRET complexes (Ko and Grant, 2006), and fluorescence quenching (Krull and Wang, 2002); these assays do not require a fluorescent target. A number of fiber optic sensors have been described using immobilized or imbedded dyes for detection of proteins and cell products (Matsuo *et al.*, 2000; Kishen *et al.*, 2003; Preejith *et al.*, 2003, 2006). However, as the recognition species in these sensors are not biological, these systems are not categorized as biosensors and are best discussed in other contexts.

Competitive assays are most often used in immunoassays for small molecules containing a single site or epitope for molecular recognition (Oroszlan *et al.*, 1993; Devine *et al.*, 1995; Shriver-Lake *et al.*, 1995a; Thompson and Maragos, 1996; Mosiello *et al.*, 1997; Maragos and Thompson, 1999). In addition, competitive formats have also proven quite useful with other types of receptors and binding molecules exhibiting semi-selective binding characteristics. For example, the antibiotic polymyxin B was used to detect endotoxin using a competitive assay (James *et al.*, 1996; Ligler and James, 1999). In competitive assays using the AChR, a number of neurotoxins were detected by their ability to inhibit binding of labeled α-bungarotoxin to the immobilized receptor (Rogers *et al.*, 1989). Erb *et al.* (2001b, 2001c) have developed a competitive assay for estrogen mimics using a recombinant human estrogen receptor. Kleinjung *et al.* (1998) demonstrated the detection of adenosine using RNA as the binding moiety.

Competition assays can be configured in two ways; in both formats, a decrease in fluorescence is proportional to the amount of unlabeled analyte species in the sample. In the first format, unlabeled and labeled analytes compete for binding to recognition molecules immobilized on the surface of the waveguide (Figure 2.8c). The second format measures the competition between the target analyte and an immobilized target analog for binding to a fluorescent recognition molecule in solution; this assay format has been used to test urine samples for cocaine and its metabolites (Nath *et al.*, 1999) and to screen pond water for estrogen mimics (Erb *et al.*, 2001c). A potential limitation of these formats is the requirement to measure a decrease in fluorescence – i.e., detecting small differences between two large values – rather than an increase in fluorescence. Furthermore, additional controls may be required to exclude other potential sources of inhibition or fluorescence quenching.

In large part due to their versatility, the vast majority of assays described in the literature utilize a sandwich format. Sandwich assays (Figure 2.8b) are commonly used for relatively large targets and require a second (fluorescent) recognition species. This fluorescent "tracer" is used to detect a target that has been "captured" by a recognition element on the fiber surface; most often, the "tracer" is an antibody, although intercalating dyes have been used to measure formation of duplex DNA (Piunno *et al.*, 1994, 1995; Duveneck *et al.*, 1996; Uddin *et al.*, 1997). Sandwich immunoassays have proven effective for the detection of proteins, bacteria, viruses, and protozoa; the majority of antibody-based assays shown in Tables 2.1 and 2.2 were performed using this format.

Examples of displacement assays exist, but are few and far between. Pandey and Weetall (1995a) used double-stranded DNA stained with an intercalating dye as a receptor. When a compound was added that is capable of intercalating into DNA, the dye was displaced and the fluorescence reduced.

In situations where positive samples are encountered only rarely, it is possible to perform many sequential immunoassays on the same optical fibers without additional manipulations (King *et al.*, 2000). However, for many applications, it is useful to have a sensing probe that can be regenerated and reused. When the binding of the analyte to the recognition element is univalent (e.g., competitive immunoassays, assays using monoclonal antibodies), the bound analyte or labeled analog can often be removed without destroying the functionality of the receptor. The simplest method for removing the bound molecule is with mild organic solvents (Oroszlan *et al.*, 1993 (200 assays/fiber probe); Shriver-Lake *et al.*, 1995b). Evanescent wave sensors based on DNA hybridization or that have oligonucleotides as recognition elements can generally also withstand relatively harsh regeneration conditions. Regeneration procedures can involve simply melting the duplex or triplex DNA (Piunno *et al.*, 1995; Abel *et al.*, 1996; Uddin *et al.*, 1997) or may involve chemical treatments with low salt buffers (Rogers *et al.*, 2001), 50% urea (Abel *et al.*, 1996; Duveneck *et al.*, 1996), or 90% formamide (Liu and Tan, 1999; Watterson *et al.*, 2001). Both thermal and urea-based regeneration procedures have been demonstrated to extend fiber optic biosensor uses up to 400 sequential assays (Abel *et al.*, 1996).

Target analytes bound to fibers by multiple interactions are more often difficult to remove without irreversibly denaturing the immobilized recognition species. Oroszlan *et al.* (1992), Wijesuria *et al.* (1994), and Anderson *et al.* (1997) used rather harsh chaotropic agents and pH extremes to accomplish such regeneration; however, in general, the number of times that the biomolecule will tolerate such abuse, if at all, is limited. Moreno-Bondi *et al.* (2000) describe the use of sonication for separating the CA15-3 antigen from immobilized antibody; however, only 65% of the antigen was removed, and the number of binding molecules for which sonication is a useful approach may be limited.

An extremely harsh regeneration protocol involved treatment of an AChR-based biosensor (Rogers *et al.*, 1989, 1991, 1992) with 1% SDS for use in multiple sequential competition assays; in spite of multiple denaturation cycles, the receptor survived the regeneration process for extended periods of use.

2.2. History

The concept of the evanescent wave was first introduced by Hirschfeld (1965) in the mid-1960s. Ten years elapsed before this concept was put into practice for immunoassays by Kronick and Little (1975). These researchers attached haptens to a planar waveguide and detected binding of evanescently excited fluorescent antibodies both in direct binding and in competitive assay formats. At about the same time, Hesse (1974) patented the use of a fiber optic-based sensor for O_2 and iodide. However, the combination of evanescent wave and fiber optic technologies did not occur until 10 years after Hesse's patent. The first evanescent wave fiber optic immunosensor, developed by Hirschfeld (1984), was further optimized by Andrade *et al.* (1985) and Sutherland *et al.* (1984). These latter groups demonstrated the greater efficiency of inline fluorescent detection (versus perpendicular detection), whereby both excitation light and emitted fluorescence could be propagated in the waveguide. In addition to the fluorescence-based sensing described in this chapter, the evanescent wave has been put to use for other surface sensitive analytical approaches. These include attenuated total reflection (ATR) technique in IR spectroscopy (ATR–IR) and evanescent wave dynamic light scattering (EVDLS). ATR–IR permits evaluation of surface chemical composition, while EVDLS permits study of the dynamic behavior of polymer particles near the interface (Gao and Rice, 1989; Matsuoka, 2001).

During the late 1980s, research on evanescent wave systems focused on the development of the immobilization chemistry and the integration of lasers for better discrimination of excitation and emission. In the early 1990s, the advent of near-IR dyes and diode lasers facilitated a movement toward small, portable systems that maintained the sensitivity of the laboratory systems with their heavy lasers. For the first time, the promise of a sensitive, robust system for field use started driving fiber optic biosensors toward applications in environmental monitoring, food safety, and point-of-care clinical applications. Since the early 1990s, the majority of publications in the field of evanescent fiber optic biosensors have described new biochemical assays, targeted toward the detection of analyte in complex, mostly unprocessed samples, with some discussion of further refinements to the optical and fluid delivery systems. The most recent work has been application driven, with food safety and biodefense taking center stage. Although the majority of evanescent wave fiber optic biosensors are still used in laboratory settings, using instruments and protocols under controlled conditions with trained operators, commercial instruments are gaining a foothold. These systems will be discussed in greater detail in the next section.

2.3. State of the art

The majority of recent work in evanescent wave fiber optic biosensors has been in one of the following broad areas: (1) immunoassay development and analyte detection in real-world samples; (2) alternative recognition schemes; (3) detection of nucleic acid sequences; (4) "reagentless" sensing; and (5) extending the capabilities of the instrumentation (miniaturization, automation, multianalyte detection, continuous monitoring.)

2.3.1. Immunoassays for real-world applications

The vast majority of evanescent wave fiber optic biosensors are used for immunoassays. Extensive research has recently expanded the breadth of analytes that can be detected using antibody-based techniques, as well as methods for improving sensitivity of these assays and testing of analytes in non-homogeneous matrices. In the years since 2000, the plethora of analytes for which these biosensor immunoassays have been developed includes the following: viruses (King *et al.*, 2000; DeMarco and Lim, 2002; Donaldson *et al.*, 2004); Gram-negative bacteria (King *et al.*, 2000; Anderson *et al.*, 2000, 2004; Tims *et al.*, 2001; Tims and Lim,

2003; Geng *et al.*, 2004, 2006; Ko and Grant, 2006; Tu *et al.*, 2006); vegetative and spore forms of Gram-positive bacteria (Anderson *et al.*, 2000, 2001; Tims and Lim, 2004); protozoa (Anderson and Rowe-Taitt, 2001); toxins (Anderson *et al.*, 2000, 2001, 2004); physiological markers of health or disease (Anderson *et al.*, 2001; Erb *et al.*, 2001b; Tang and Kang, 2005; Tang *et al.*, 2006; ThreeFold Sensors, 2007); hormones (Erb *et al.*, 2001a); and herbicides (Xing *et al.*, 2000).

Assay development and characterization is, of course, application driven. Several notable areas of recent interest and voluminous publications are food safety, environmental analysis, and biodefense, with large numbers of papers relevant to these applications published by the laboratories of Bhunia at Purdue, Anderson and Ligler at the Naval Research Laboratory, and Lim at the University of South Florida. Sample matrices successfully tested for analyte detection include clinical fluids (blood, serum, plasma, sputum, urine, fecal extracts); soil extracts; food homogenates; aerosolized particles (pollen, clay, smokes); and water from various sources, including seawater (Table 2.1). In most cases, while the complex matrices did have some effect on assay sensitivity or detection limits, the presence of the interfering material did not produce false-positive or false-negative results in terms of the ability to detect the analyte. This general immunity to matrix effects and the ability to measure target analytes with little to no sample preparation is one of the best features of these sensors.

2.3.2. Alternative recognition schemes

A notable number of papers describing detection using recognition elements other than antibodies or nucleic acid probes were published in the early to mid-1990s. Rogers *et al.* (1991, 1992) used AChR to detect various cholinergic agonists and antagonists. In a competitive assay, fluorescein-labeled bungarotoxin in solution competed with unlabeled agonist (or antagonist) to determine inhibition constants for acetylcholine, carbamylcholine, nicotine, *Naja naja* toxin, bungarotoxin, and tubocurarine. Constants for the antagonists (toxins) were similar to K_d values reported with radioligand assays using solubilized preparations of AChR.

Table 2.1 Evanescent fiber optic biosensors for detection in complex matrices

Analyte	Sample matrix	Reference
Cocaine, benzyoylecgonine, and cocaethylene	Urine	Nath *et al.* (1999)
Cocaine	Coca leaf extract	Topporada *et al.* (1997)
TNT	River water, bilge water	Shriver-Lake *et al.* (1995a)
TNT, RDX	Groundwater	Shriver-Lake *et al.* (1997), van Bergen *et al.* (2000)
Bacillus globigii	Air samples	Ligler *et al.* (1998), Anderson *et al.* (1999)
Ricin	River water, urine	Ogert *et al.* (1993), Narang *et al.* (1997b)
Y. pestis F1 antigen	Blood, plasma, serum	Cao *et al.* (1995), Anderson *et al.* (1996)
D-Dimer	Plasma, whole blood	Rowe *et al.* (1998), Rowe (unpublished)
Staphylococcal enterotoxin B	Clay, topsoil, pollen, and smoke extracts	King *et al.* (1999)
	Serum, urine, ham extract	Templeman *et al.* (1996)
Specific antibody	Serum	Nath *et al.* (1997), Anderson *et al.* (1998)
Giardia	Fecal extracts; pond, river and seawater	Anderson and Rowe-Taitt (2001)
Burkholderia cepacia	Ground water	Pease *et al.* (1995)
Lipopolysaccharide	20% plasma	James *et al.* (1996)
β-estradiol, estrone, estriol, diethylstilbestrol, zearalenone, and tamoxiphen	Plasma	Erb *et al.* (2001b)

(Continued)

Table 2.1 Continued

Analyte	Sample matrix	Reference
E. coli O157:H7	Ground beef extract	DeMarco *et al.* (1999), DeMarco and Lim (2002a)
	Unpasteurized apple juice	DeMarco and Lim (2002b)
Vaccinia	Throat swabs	Donaldson *et al.* (2004)
Fumonisin	Maize extract	Thompson and Maragos (1996), Maragos and Thompson (1999)
Listeria monocytogenes	Hotdog, bologna	Geng *et al.* (2004), Nanduri *et al.* (2006)
Bacillus anthracis spores (inactivated)	Cornstarch, sugar, talc, baking soda, and pesticide	Tims and Lim (2004)
Estrogen mimics	Pond water	Erb *et al.* (2001c)
E. coli O157:H7, shiga-like toxin	Ground beef, bacterial culture	Geng *et al.* (2006), Tu *et al.* (2006)
Blood group A antigen	Washed red blood cells	Quinn *et al.* (2000)
STAT3, phosphorylated STAT3	Lysed cell culture supernate	Kapoor *et al.* (2004)
IL-6, IL-12, INF-γ	Serum, urine	Threefold Sensors
S. typhimurium	Homogenized pork	Ko and Grant (2006)
	Irrigation water	Kramer and Lim (2004)
Protein C, Protein S, antithrombin III, plasminogen	Plasma	Tang *et al.* (2006)

Eldefrawi and Rogers (Rogers *et al.*, 1992) also immobilized fluorescein-labeled acetylcholinesterase (AChE) onto optical fibers as a means to detect multiple neurotoxins. This enzyme catalyses the following reaction:

$$\text{Acetylcholine} \rightarrow \text{choline} + \text{acetic acid} \quad (2.7)$$

In the presence of acetylcholine and active enzyme, the reaction above causes the pH to drop, quenching the pH-sensitive fluorescence of fluorescein. In the presence of an AChE-specific neurotoxin, however, the enzymatic reaction was inhibited, causing the pH to remain neutral and fluorescein emission to remain detectable (Figure 2.9). This group was able to detect paraoxon, echotiophate, bendiocarb, and methomyl.

More recently, AChE has been immobilized by means of LB films within a reaction chamber for detection of organophosphorus pesticides; use of LB technique provided a high density of immobilized AChE without loss of activity. Rather than a fluorescent detection mechanism, this system used absorbance of *O*-nitrophenol, a colored enzymatic product, to measure inhibition of enzyme activity by paraoxon in real time (Choi *et al.*, 2001, 2003). To increase the number of potential pesticides detectable by this system, glutathione-S-transferase (GST) was co-immobilized in the same reaction chamber and absorbances read

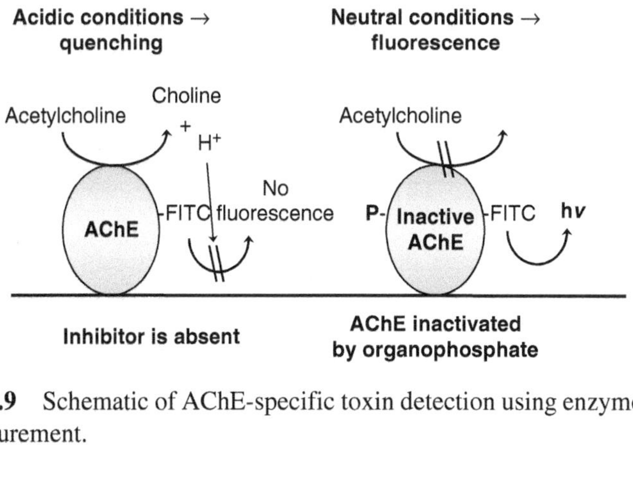

Figure 2.9 Schematic of AChE-specific toxin detection using enzyme activity for measurement.

at two wavelengths to assess and distinguish inhibition by paraoxon (AChE) and captan (GST) (Choi *et al.*, 2003).

Another example of an enzyme-based system used optical fibers coated with fibrinogen to detect thrombin activity in solution (Garden *et al.*, 2004). Thrombin, pre-mixed with fluorescein-labeled fibrinogen in solution, was added to fibrinogen-coated probes and the increase of fluorescence monitored immediately. Thrombin activity – indicated by formation of a fluorescent precipitate on the fiber – was detectable at 0.01 NIH units/ml (3.2 ± 0.7 ng/ml) in less than 60 s. Although this assay may be suitable to measure clotting times in whole blood, the authors present the caveat that blood samples may need dilution due to the high concentrations of unlabeled fibrinogen expected.

James *et al.* (1996) and Ligler and James (1999) developed a biosensor based on the high affinity of the antibiotic polymyxin B for the lipopolysaccharide (LPS) component of Gram-negative bacterial cell walls. After immobilizing polymyxin B onto optical fibers, the authors were able to detect 25 ng/ml LPS spiked into samples containing up to 20% plasma in competitive assays.

More recently, Goldman *et al.* (2000) demonstrated the use of phage-displayed peptides as detector reagents. After several rounds of biopanning to isolate M13 constructs expressing surface-bound peptides with affinity to staphylococcal enterotoxin B (SEB), the phage were labeled with a fluorescent dye and used as detection reagents in an evanescent wave fiber optic biosensor. While these peptides were able to distinguish between optical fibers coated with SEB from those coated with streptavidin, they were unable to detect SEB bound to the fibers in a sandwich assay.

An interesting use of intermolecular interactions was used by Erb *et al.* (2001b, 2001c) to address the effect of various estrogen mimics on the binding between the estrogen receptor and a portion of DNA known as the estrogen response element (ERE). ERE was immobilized onto optical fibers via a fluorescent binding matrix. Binding of estrogen to the immobilized ERE led to a modulation in fluorescence. Upon addition of various estrogenic compounds to estrogen receptor in solution, the conformation of the solution-phase receptor was altered, changing the kinetics of binding of the receptor to immobilized ERE. Competitive assays using estrogen receptor have also been described (Erb *et al.*, 2001c). Using estrone-3-glucuronide-coated fibers, environmental estrogen mimics and synthetic compounds were tested for their ability to inhibit binding of fluorescent estrogen receptor to the surface-immobilized estrone-3-glucuronide. These assays have been used in real matrices (environmental samples, urine) to measure estrogenic compounds; competitive antibody-based assays have also been developed for interleukin-6 (IL-6) and interferon-γ (INF-γ), with detection limits in the low to sub-pg/ml range (ThreeFold Sensors, 2007; Figure 2.10).

2.3.3. Detection of nucleic acid sequences

Since Squirrel and his colleagues (Graham *et al.*, 1992) first described a hybridization assay on the surface of an optical fiber probe, the use of evanescent fiber optic biosensors to detect specific nucleic acid sequences has been steadily increasing. These assays include an intriguing variety of formats and target analytes; the nucleic acid polymer may be the target analyte, the receptor, or both.

Mauro and Coworkers (Mauro *et al.*, 1996; Campbell, 1995) developed a rapid evanescent wave sensor for detection of fluorescent polymerase chain reaction (PCR) products by engineering a chimeric recognition molecule. The chimeric molecule was composed of the DNA-binding domain of a yeast transcriptional regulator protein (N-terminus) with the IgG-binding domain of Protein G (C-terminus). This protein was then immobilized onto IgG-coated optical fibers (via Protein G-IgG interaction) and was used to detect nanomolar concentrations of PCR product encoding the yeast DNA operator sequence in less than 10 min; multiple sequential analyses could be performed with only a 1-min regeneration time. While this biosensor was not designed to reduce the time for PCR amplification, it provided a significant reduction in the time required to detect the PCR products.

A great deal of work has been published on detection of specific nucleic acid sequences using hybridization reactions performed on evanescently

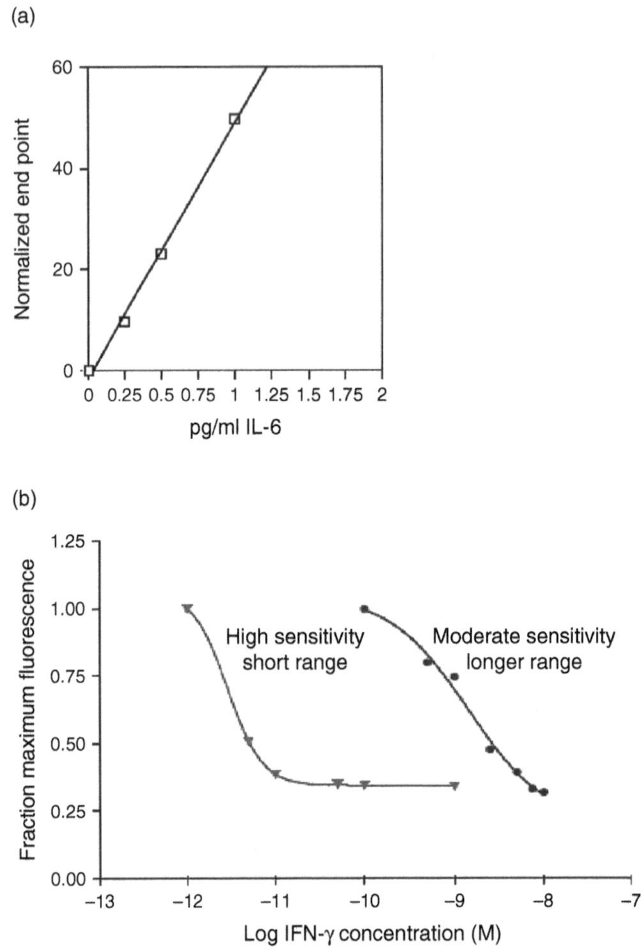

Figure 2.10 Detection of physiologically relevant concentrations of blood markers. (a) IL-6 can be quantified in the low to sub-pg/ml range; dose–response curves have identical slopes when measured in serum and urine. (b) Measurement of IFN-γ using fluorophore-labeled IFN-γ (right-hand curve) or fluorosphere-labeled IFN-γ (left-hand curve) in competitive assays. Reproduced with permission from J. Erb, ThreeFold Sensors.

excited optical fibers. Many of these DNA sensors utilize fluorescently labeled sequences (e.g., labeled PCR products) to demonstrate binding of DNA or RNA samples to immobilized oligonucleotides (Graham *et al.*, 1992; Abel *et al.*, 1996; Pilevar *et al.*, 1998). These "direct" detection schemes require time-consuming labeling steps prior to the biosensor assay, but the labeling can be concurrent with the amplification. In a different labeling approach, multiple groups have utilized intercalating dyes such as ethidium bromide to monitor hybridization of duplex (Piunno *et al.*, 1994, 1995; Duveneck *et al.*, 1996; Uddin *et al.*, 1997) and triplex DNA (Uddin *et al.*, 1997). In one of the few instances of a displacement assay performed on an evanescent wave biosensor, Pandey and Weetall (1995) generated response curves for the displacement of ethidium bromide from DNA duplexes immobilized on optical fibers by other intercalating compounds; this biosensor could therefore be used as a generic sensor for aromatic carcinogens.

Temperature-sensitive DNA-hybridization experiments demonstrate the ability of evanescent wave sensors to characterize DNA/DNA or DNA/RNA interactions. By measuring the effect of temperature on fluorescence intensity of hybrids treated with ethidium bromide, Uddin *et al.* (1997) were able to monitor formation of both parallel and antiparallel triplex DNA complexes. Melting profiles of duplex DNAs were also used by Rogers *et al.* (2001) to assess the extent of radiation-induced damage to an oligonucleotide probe. Real-time measurements have also been utilized by Pilevar *et al.* (1998) in a "sandwich"-type assay to detect *Helicobacter pylori* rRNA. Total RNA prepared from *H. pylori* was hybridized to immobilized complementary oligonucleotides. Subsequent addition of fluorescent detector probe, complementary to a closely related site, allowed detection of the *H. pylori* RNA.

Watterson *et al.* (2001) used an evanescent wave system in a series of elegant experiments to dissect the different processes involved in nucleic acid hybridization versus adsorption. Using a stopped-flow system for injection and mixing, this group followed adsorption of complementary and non-complementary DNA sequences to DNA probes immobilized on optical fibers. They determined that adsorption occurred at a significantly higher rate than hybridization, but that this non-selective adsorption did not inhibit selective interaction between complementary sequences. Surprisingly, their results indicated that the presence of high concentrations of non-complementary DNA reduced the response time of the sensors for hybridization of complementary DNA present in the same sample.

A significant limitation of these assays is the need to purify and/or amplify target sequences prior to analysis. Recently, Lim's group has combined sample preparation with PCR amplification and detection (Tims and Lim, 2004). A fiber optic evanescent wave biosensor was used to capture pathogenic bacteria, which were subsequently enriched by addition of growth medium to the flow cell. After a short incubation, the cells were identified by PCR amplification of the extracted DNA. This approach allowed the presence of *Escherichia coli* O157:H7 to be confirmed by two methods in less than 10 h. The time-to-result has recently been decreased to 2 h by direct PCR of bacteria bound to the fibers, without culture or enrichment (Simpson and Lim, 2005). Other groups have also shown that bacteria bound to optical fibers can be enriched and analyzed subsequently (Bhunia *et al.*, 2004; Geng *et al.*, 2004). We anticipate that lower sensitivity screening followed by high sensitivity/selectivity identification may increasingly become an intrinsic part of automated biosensors.

2.3.4. Reagentless detection

The need for addition of exogenous (fluorescent) reagents to biosensors presents a significant limitation for use outside of the laboratory, especially in environmental extremes where appropriate storage and usage facilities may not be available or practical. For this reason, a large effort has been put forth to develop "reagentless" methods, where target detection is instantaneous and requires no manipulations other than addition of sample.

A subset of evanescent wave fiber optic biosensors ideally suited for reagentless detection is the SPR-based systems. As these sensors directly measure changes in refractive index, no intrinsic fluorescence or fluorescently labeled species are required. Sensors based on monomode and multimode fibers have been reported, typically using gold-coated, side-polished optical fibers. Although first proposed well over 10 years ago (Jorgenson and Yee, 1993), these sensors are still in relative infancy and actual use for measurement of analytes is limited. Most of the papers published in this area have been geared toward demonstrating sensitive discrimination of minute differences in refractive index (Homola 1995; Slavik *et al.*, 1999, 2002; Bardin *et al.*, 2002; Monzón-Hernández *et al.*, 2004; Chang *et al.*, 2006; Peng *et al.*, 2006; Lin *et al.*, 2007). However, one paper describes detection of the toxin SEB in the low ng/ml range in less than 10 min (Slavík *et al.*, 2002). Quinn *et al.* (2000) used a wavelength-modulated fiber optic SPR based on that of Jorgensen and Yee (1993) in one of the few demonstrations of this technology in real-world samples. By using hydrogel (carboxymethyl cellulose)-coated optical fibers functionalized with antibody directed against blood type A antigen, these researchers were able to detect whole red blood cells in various dilutions of packed red cells. Although this system had 30-fold higher noise than the commercial instrument BIACORE 1000, the system was highly specific, since no red cells were bound in the absence in the anti-type A antigen antibody. A key drawback of this method of sensing is the difficulty in depositing a homogeneous coating and achieving good chemical functionalization of the surface.

As an alternative to creating instruments capable of reagentless detection, efforts have been made to modify the biological recognition elements of detection systems, such that the recognition event itself is capable of generating a measurable change in fluorescent signal; a later chapter describes the evolution of these "beacon" systems although, to date, only a limited number have been integrated into biosensors. In 1999, molecular beacons were first demonstrated as reagentless probes in an evanescent wave sensor (Liu and Tan, 1999). These oligonucleotide "molecular beacons" contained a stem-and-loop structure; the loop region was complementary to the target sequence and the two ends were labeled with either a fluorophore (5′) or a quencher (3′). The 3′ end of the molecular beacon (close to the quencher) was modified to posses a biotin, which was used to immobilize it to an avidin-coated optical fiber. In the absence of the target sequence, the stem-and-loop structure of the molecular beacon was conserved, leading to background levels of fluorescence. If target DNA (or RNA) with an appropriate sequence was added to the system, hybridization of the target DNA to the complementary sequence in the loop region caused the stem-and-loop structure to open up. This resulted in movement of the fluorophore and quencher arms away from each other. The increased distance between the ends no longer allowed efficient quenching, and fluorescent emission was observed. Tan's group has extended their studies to include multiplexed assays and direct detection of RNA species without reverse transcription (Liu *et al.*, 2000). With an eye to performing *in situ* detection of transcripts within living cells, this group has also demonstrated sensitive detection using sub-micron optical fibers functionalized with molecular beacons (Liu *et al.*, 2000).

Krull's group at the University of Toronto (Wang and Krull, 2002) took a different approach to reagentless DNA sensing by using single-stranded DNA probes conjugated on the 5′ end to the intercalating dye. Upon hybridization of target DNA to these immobilized probes, the dye – thiazole orange intercalated into the double-stranded DNA, increasing fluorescence. The time to reach maximal fluorescence signal was less than a minute, and non-selective adsorption/hybridization from non-complementary target was largely ameliorated by increased salt concentrations. With appropriate gating of excitation and intermittent, rather than continuous, interrogation of the fibers, the fibers could be used for over 8 h with minimal photobleaching.

Ko and Grant (2006) have recently developed a fiber optic biosensor utilizing FRET to detect *Salmonella typhimurium*. An immunocomplex comprised of Protein G labeled with FRET acceptor fluorophore and anti-*Salmonella* IgG labeled with FRET donor fluorophore was immobilized on the optical fibers. FRET was observed within 5 min of exposure to *Salmonella* cells. Although some sensitivity was lost when spiked pork samples were tested (limit of detection $\sim 10^5$ cfu/g), detection limits in buffer (10^3 cfu/ml) were in the same range as those determined using other rapid testing methods. This work demonstrates the potential for incorporating reagentless assays into rapid testing approaches; however, the effort required to optimize this system was extensive.

2.3.5. Extending the capabilities of the instrumentation

The first evanescent fiber optic biosensor (Hirschfeld, 1984; Hirschfeld and Block, 1985) was made with a halogen lamp and was sold by ORD (Waltham, MA), who packaged the system in a small metal suitcase. The subsequent laser-based breadboard devices were all significantly larger, and it was not until the advent of commercial diode lasers that systems were able to possess the advantages of both collimated, narrow bandwidth excitation and small size. The Analyte 2000 was the first device that was not only portable, but that interrogated more than one fiber probe (Research International, Woodinville, WA; http://www.resrchintl.com). This instrument, which utilized the fiber optic waveguide designs developed at the Naval Research Laboratory (Anderson *et al.*, 1994a), contained four integrated "daughter" cards, each card mounting a 635 nm diode laser and a photodiode. Fiber bundles connected the optics to the fiber probes, as in Figure 2.4b, and the electronic signals from the Analyte 2000 were analyzed on a laptop computer. The configuration of this system allowed four separate assays to be performed simultaneously, either in parallel or in series, depending on the configuration of the fluidics components assembled by the user. This instrument and its subsequent prototypes have been used in many of the early demonstrations of multianalyte detection using evanescent wave technology (Table 2.2). By further expanding this capability to discriminate between closely related targets, Kapoor *et al.* (2004) have recently used ratiometric determinations of unmodified and phosphorylated transcription factor 3 (STAT3) to quantify trophic factor-activated signaling molecules in cell cultures. Table 2.2 also shows a number of "semi-selective" fiber optic sensors that have been used for detection of multiple analytes using a single recognition element. These sensors utilize recognition elements that bind to multiple ligands that are structurally related or have similar modes of action. These latter sensors can be used for detection of multiple analytes, but identification of the specific analyte detected may require additional, more-specific analyses.

Table 2.2 Evanescent wave fiber optic biosensors used for multianalyte detection

Analytes	**Biosensor system**	**Reference**
	Specific identification	
TNT, RDX	Antibody: Analyte 2000	Bakaltcheva *et al.* (1999) van Bergen *et al.* (2000)
Ovalbumin, *B. globigii*	Antibody: Analyte 2000	Anderson *et al.* (1999)
Ovalbumin, *Erwinia herbicola*, *B. globigii*, MS2	Antibody: RAPTOR	King *et al.* (2000)
SEB, ricin, *Francisella tularensis, B. globigii*	Antibody: RAPTOR	Anderson *et al.* (2000)
Ovalbumin, *B. anthracis*, *Giardia*, *F. tularensis*	Antibody: RAPTOR	Anderson and Rowe-Taitt (2001), Anderson *et al.* (2001)
Ricin, SEB, cholera toxin, ovalbumin	Antibody: RAPTOR	Anderson *et al.* (2001)
STAT3, phospho-STAT3	Antibody: laboratory prototype	Kapoor *et al.* (2004)
SEB, *F. tularensis*, *B. anthracis*, ricin, *Y. pestis* F1 antigen, MS2, cholera toxin, *S. typhimurium*	Antibody: RAPTOR	Anderson *et al.* (2004)
Protein C, Protein S, antithrombin III, plasminogen	Antibody: Analyte 2000	Tang and Kang (2005), Tang *et al.* (2006)
Paraoxon, captan	AChE, GST: laboratory prototype	Choi *et al.* (2003)
	Semi-selective analyses	
Remazol brilliant blue, anthraquinone-2,6-disulfonic acid, decacycliene, DAPI	EtBr-duplex DNA (displacement reaction): ORD device	Pandey and Weetall (1995)

Table 2.2 Continued

Analytes	Biosensor system	Reference
Acetylcholine, carbamylcholine, nicotine, *N. naja* toxin, bungarotoxin, tubocurarine	AChE: ORD device	Rogers *et al.* (1991, 1992)
Cocaine, benzyoylecgonine, cocaethylene, cocaine metabolites	Antibody: ORD device Antibody: Analyte 2000	Devine *et al.* (1995) Nath *et al.* (1999)
Acetylcholine, benzoylcholine, benzylacetate	AchR: ORD device	Rogers *et al.* (1989)
Chlordane, heptachlor, other cyclodiene insecticides	Antibody: ORD device	Brummel *et al.* (1997)
PCBs, trichlorophenyl butyrate	Antibody: ORD device	Zhao *et al.* (1995)
Genestein, diethylphthalate	Estrogen receptor: TG01 (Threefold Sensors)	Erb *et al.* (2001c)
β-Estradiol, estrone, estriol, diethylstilbestrol, zearalenone, and tamoxiphen	Estrogen receptor: Endotect (ThreeFold Sensors)	Erb *et al.* (2001a, 2001b)

Certainly, automation is extremely important and thus has been a major thrust in the move toward commercialization of biosensors and has facilitated use of the biosensors in the field. There are presently two commercial producers of evanescent fiber optic biosensors intended for use by non-technically trained individuals: ThreeFold Sensors and Research International.

ThreeFold Sensors (2007; Ann Arbor, MI) manufactures several fully portable fiber optic biosensors. Systems that use modulated visible CW diode laser for evanescent excitation of near-IR fluorophores comprise of the HH series instruments; each utilizes fiber optical cartridges that can be modified as appropriate by the user. The HH01B is fully automated whereas the HH01 can be used with an optional peristaltic pump unit to move fluids through the cartridges (Figure 2.11). These units were designed for kinetic analyses, but have also been used for endocrine disruptor screening (estrogen receptor-binding assays), hormone measurement for fertility testing and hormone replacement therapy, monitoring blood cytokine levels, and screening for lead poisoning (ThreeFold Sensors, 2007; Erb *et al.*, 2001a, 2001b, 2001c; Smith *et al.*, 1999; Figure 2.10). Once the disposable filter cartridge (containing optical fiber) is snapped into the optical coupler, a sample

(a)

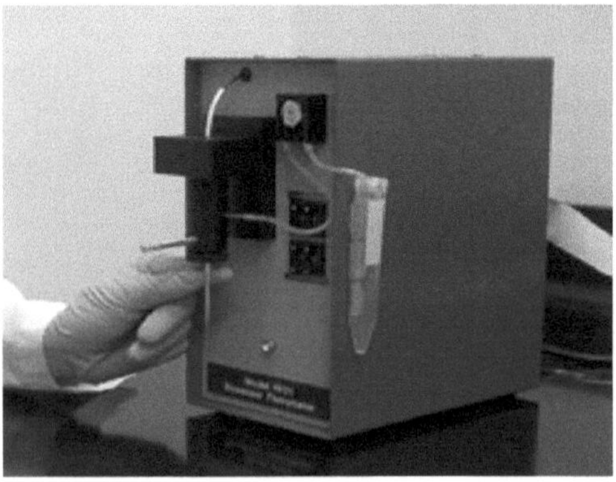

(b)

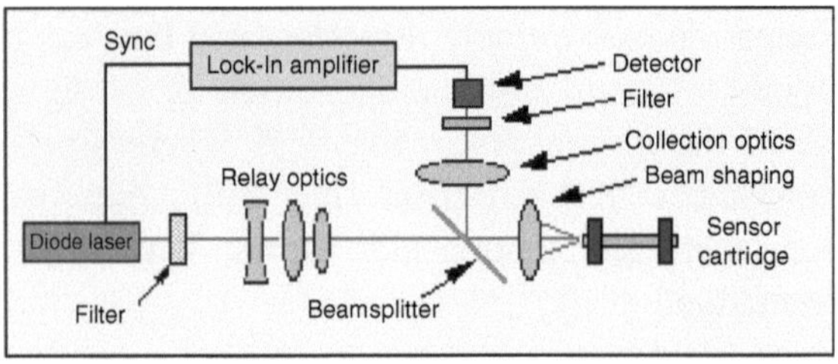

Figure 2.11 (a) ThreeFold Sensors' HH01Bs has a miniature peristaltic pump to move fluids at controlled rates through biosensor cartridges. (b) Schematic of ThreeFold Sensors fluorimeter/biosensor. Reproduced with permission from J. Erb, ThreeFold Sensors.

is injected and the excitation laser is pulsed on and off, while changes in fluorescence are measured. The concentration of target is determined from the slope of the binding curves, rather than from a single time point.

ThreeFold Sensors also produces a time-gated biosensor fluorimeter, the TG01. While not portable, this instrument is well suited to detailed kinetic analyses and utilizes near-IR lanthanide fluorophores having long fluorescence lifetimes. By measuring fluorescence occurring only after any biological and optical component fluorescence has decayed to zero, this instrument is capable of measuring, with high discrimination, fluorescence due to the lanthanide complexes. The TG01 was used to study the effect of estrogen mimics on binding between estrogen receptors and immobilized ERE (Erb *et al.*, 2001c).

Research International produces the RAPTOR. This system weighs about 12 lb, and can be operated independently or controlled using a separate computer. The optics are essentially the same as in the Analyte 2000 (described above), but the RAPTOR contains all the fluidics connections to perform fully automated assays, with user-defined parameters and thresholds for determination of positive results. While most assays require some user interface during operation, the RAPTOR can control a portable air sampler completely unattended (the SASS 2000, Research International). This system was intended to function as a point detector. However, although it can test numerous sequential negative samples, once a positive is encountered, the user must replace the fiber probes before new assays are performed. While this is a simple process, it limits the system's unattended operational capability.

Typically, the RAPTOR is set up to detect four different analytes. However, we have demonstrated that the number of potential targets per four-fiber coupon could be doubled by a simple modification of standard protocols (Anderson *et al.*, 2004). By using different mixtures of capture and tracer antibodies on each fiber, the pattern of binding of each of the eight targets could be distinguished from the others, allowing accurate identification of the target present (Figure 2.12). To achieve this same end, albeit a bit more elegantly, Research International has recently

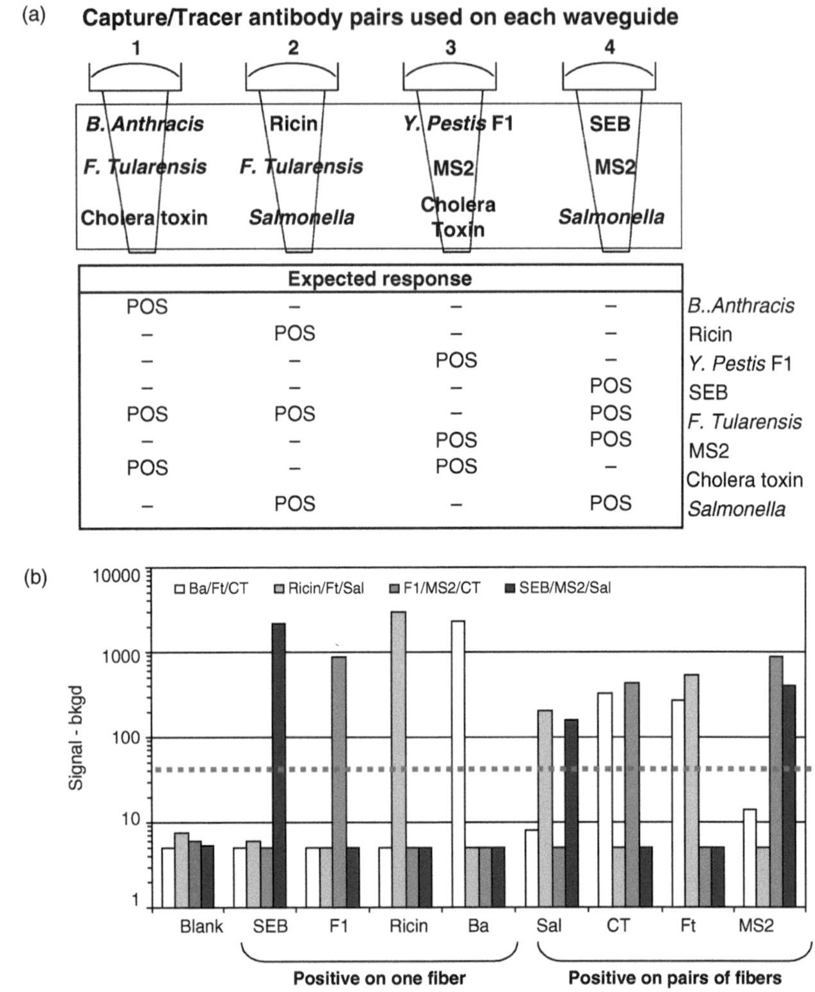

Figure 2.12 Detection of eight targets using the four-fiber optical biosensor, RAPTOR. By using complementary mixes of both capture and tracer antibodies on the four optical fibers, each of the targets can be identified by its pattern of fluorescence. SEB (50 ng/ml), *Yersinia pestis* F1 antigen (50 ng/ml), ricin (Ric, 1000 ng/ml), and *Bacillus anthracis* Sterne (10 μg/ml) were identified by fluorescent signal on a single optical fiber. *S. typhimurium* (Sal, 5×10^5 cfu/ml), cholera toxin (CT, 10 ng/ml), *F. tularensis* LVS (Ft, 10^6 cfu/ml), and MS2 coliphage (10^8 pfu/ml) were identified by fluorescence from pairs of fibers. Figure adapted from Anderson *et al.* (2004).

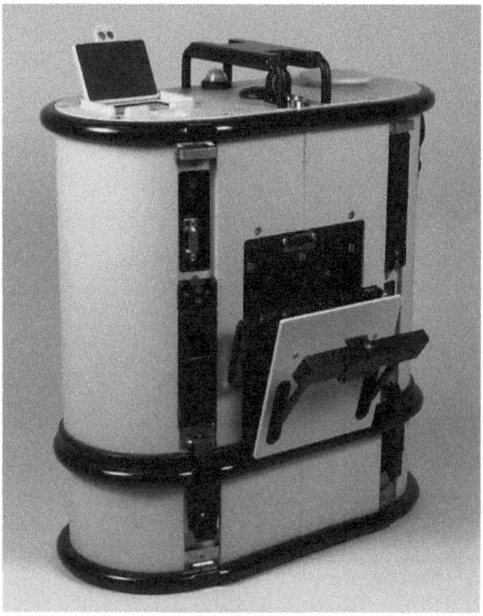

Figure 2.13 Photograph of the BioHawk multianalyte detection system. This is the most recent product by Research International. It can perform eight separate immunoassays, two on each of four waveguides, and the system includes an integrated aerosol sampler, as well as allowing manual sample input. Reproduced with permission from E. Saaski, Research International.

redesigned the fiber optic biosensor into a package called BioHawk that tests for eight analytes simultaneously and includes an integrated aerosol collector (Figure 2.13).

2.4. Advantages and limitations

Like all optical biosensors, fiber optic sensors have several advantages over those that utilize other signal transduction methods. No direct electrical connection to the transduction system is required. Therefore, optical sensors are immune to many interfering electrochemical and electromagnetic effects that plague sensors based on electrochemical transduction, such as electromagnetic radiation, corona discharge, radiofrequency interferences, corrosion, shock/vibrations, and harmonic induction. Furthermore, the lack of requirement for radiolabels or hazardous organic materials for sample preparation or signal generation (such as benzene and other organic solvents) render optical sensors more "user-friendly" than other methods that do. In addition, the current availability and high quality and intensity of miniature light sources continue to decrease the size and price of optical biosensors.

Fiber optic biosensors, in particular, have reaped the benefits of the telecommunications boom that has led to the commercial availability of inexpensive, high-quality optical fibers. The use of optical fibers in biosensors takes advantage of the versatility of these components. System configurations can be changed to suit the user's specifications. Shortened fibers can be used in sensors designed for portability, while fibers of extended length allow assays to be performed in remote or hazardous locations. To date, evanescent fiber probes have not been explored for *in vivo* applications, in large part due to the requirement for additional reagents in sandwich and competitive assays and the inability of many immobilized recognition species to withstand conditions required for sterilization. These two limitations are shared by many sensing systems utilizing taggants or similar biological recognition species. With the development of more heat- and solvent-stable recognition molecules, the latter limitation may become less cumbersome in the near future.

Sensors based on evanescent wave excitation have the advantage of surface-specific detection. The limited penetration depth of the evanescent wave allows spatial separation of surface-bound fluorescent complexes from fluorophores or interfering species present in the bulk solution. Use of the evanescent field thus allows real-time measurement of surface interactions to be performed both in homogeneous and in non-homogeneous samples (Table 2.1). Although this exquisite surface-selectivity should ameliorate the need for washing, in practice, washing steps are often included to remove unbound sample components from the surface prior to interrogating the surface for bound target with the fluorescent tracer reagents. Furthermore, this immunity from optical interferences should be distinguished from matrix effects on the biological recognition event. Such biochemical effects, however, will be platform independent.

In optical sensors, the speed with which results are available is limited by the rate of molecular interaction, not by the rate of signal transduction. This advantage is further improved by the surface-selective nature of the evanescent wave, which allows the capability of real-time measurements. Recent experiments have further demonstrated the potential of such real-time measurements for determination of binding constants and enhanced discrimination between closely related species (Chou *et al.*, 2007). While real-time measurement of surface-binding events is possible with evanescent wave sensors, this capability has not received widespread use; even so, assay times for evanescent wave fiber optic sensors are at least an order of magnitude shorter than those from standard ELISAs, still considered the gold standard for bioassays.

With the advent of smaller, low-cost lasers and detectors, an additional advantage of these sensors is the capability of performing multiple assays in parallel, potentially even outside of a laboratory setting. The ready availability of these smaller optical components means the only limitation to the number of biochemical assays that can be performed simultaneously (in series or in parallel) will depend only on the compatibility of the biochemical reagents and possible limitations inherent in sensors of increased complexity – e.g., increased cost, size, and maintenance. Our recent demonstration that a four-fiber system could be used to detect eight targets in a single assay (Anderson *et al.*, 2004) shows that even minor changes in the assay parameters can improve the multiplexing capabilities of these instruments.

A key disadvantage of evanescent wave systems is the poor coupling efficiency of the generated fluorescence back into guided modes. One cannot simply use a more intense source since photobleaching would be exacerbated; this effect is particularly true for single-mode fibers that have up to 50% of their energy present outside the fiber core. Properly designed fiber optic sensor regions can help maximize fluorescent recovery with appropriate tapering and/or V-number matching; however, coupling efficiency will always be a major limiting factor in achieving good assay sensitivities. Surface area is another limiting factor, especially for monomode fibers, where surface area available for the coupling of recognition elements is low. Multi-mode fibers, on the other hand, have a larger surface area available for immobilization of recognition elements, since surface area is proportional to fiber diameter, and multimode fibers have diameters typically 10 times or more greater than monomode fibers.

As with all sensors using biological recognition elements, stability and regeneration of evanescent wave fiber optic biosensors must be characterized for each recognition system. Retention of activity by the immobilized recognition species is of course highly dependent on the method used for attachment. The widespread use of intermediate layers such as avidin and Proteins G and A has facilitated the generation of "universal" probes, where the user can decide on the immobilized species immediately before use; addition of an intermediate biological layer onto the inorganic surface has also been shown to improve the activity of some, but not all, capture reagents. These generic surfaces, however, may not be appropriate for all uses: Surfaces coated with Protein A or G are useless in non-antibody-based detection schemes, and use of avidin–biotin interactions for signal transduction (e.g., many amplification reactions) precludes assays with avidin-coated surfaces. Long-term storage also depends on the stability of the recognition elements, but storage times of over 12–18 months have been demonstrated for antibodies immobilized on fiber optic probes and lyophilized fluorescent antibody reagents (King *et al.*, 2000). While regeneration of nucleic acid hybridization-based sensors is simply a matter of "melting" the complexes, regeneration of sensors utilizing protein recognition elements may not be as simple. In general, regeneration has been more successful with systems employing monovalent recognition species; disruption of polyvalent binding is generally more difficult, requiring harsh conditions or denaturants to remove an analyte bound by multiple interactions. At this time, long-term monitoring is feasible only if a positive is a rare event (and no regeneration is required) or if the association between the analyte and the recognition molecule is readily reversible.

2.5. Future of evanescent fiber optic biosensors

There are opportunities for improving each part of the evanescent wave fiber optic biosensor while still utilizing the advantages of this technology. We will discuss the areas that we predict will offer such opportunities in terms of the biorecognition molecules, the fluidics, the waveguide materials and optics, and systems integration.

The optical biosensor is only as good as the molecular recognition molecule it uses. As discussed in several other chapters, new types of recognition molecules are being developed using different forms of combinatorial biochemistry. Creation of chimeric recognition elements offers great potential for improved recognition properties, as well as the option of *in situ* signal generation. While not yet used in rapid biosensing systems, peptide nucleic acids – synthetic mimics of DNA in which the sugar backbone is replaced by a peptide polymer – have shown tremendous promise for improving stability in nucleic acid hybridization assays (Nielsen, 2004; Pellestor and Paulasova, 2004); their sensitivity to mismatches may further enhance selectivity and offers potential for detection of single nucleotide polymorphisms (Jensen *et al.*, 1997; Nielsen *et al.*, 1999; Giesen *et al.*, 1998). The chapter by Medintz and Delehanty, for example, describes many new, rationally designed biological constructs capable of both sensing and signal transduction. While these new recognition species have shown tremendous potential for intracellular sensing for research purposes, how they will react when immobilized on biosensor surfaces and used under field conditions remains unknown. In their chapter on nucleic acid-based detection, Cho and co-workers describe additional reagentless recognition molecules that have the benefit of rapid laboratory evolution to new molecules with potentially even greater stability and affinity. Creation of molecularly imprinted polymers (MIPs), on the other hand, has taken the turn from simple templating and polymerization reactions to rational design of more specific and higher affinity reagents, as described by Piletsky and Turner; MIPs too have the capability of reagentless detection, as well as potential catalytic properties. In their chapter on single-domain antibodies, Liu and colleagues discuss the potential for increased stability, storage, and use under environmental extremes of this category of immunoreagents, currently under rapid development. Certainly, technological hurdles must still be jumped before these new receptors begin replacing conventional antibodies. They do, however, show promise for obtaining molecules not only with designer specificity, but also with enhanced stability for extended use or exposure to samples incompatible to most antibodies (e.g., high salt, organic solvent).

The second area in which evanescent wave fiber optic biosensors may be improved is in controlling the transport of sample and reagents to the recognition molecules on the fiber probes. Several approaches have been used to overcome diffusion limitations on analyte transport. Zhou *et al.* (1998) demonstrated that the ultrasonic standing-wave chamber concentrated *S. typhimurium* into parallel layers or in a column along the axis of the testing cell. Sonication had no effect on the antibodies, but forced cells (and cells bound to polystyrene microspheres) to move to the axis of the test cell. Fluorescent signals were an order of magnitude higher in tests utilizing the acoustic standing-wave than those when the ultrasonic treatment was absent. On the other hand, the fluidics systems developed for the Analyte 2000 and RAPTOR use flow to expose the fiber probe to a maximum amount of analyte; Bhunia's group has also extended this concept to the immobilization of capture species on optical fibers, showing a significant improvement in assay performance when flow was used during the immobilization steps (Nanduri *et al.*, 2006). However, the low Reynolds number flow in the capillaries surrounding the fiber probes means that, in general, the effective mixing on movement of fluids over the fiber surface is very limited. Especially appropriate for use with monomode fiber-based systems, passive microfluidic structures to increase mixing at the fiber surface, such as those discussed in Chapter 17, may improve both sensitivity and assay times.

The third area for improvement involves the optical components. The probes could be fabricated from materials that have absolutely no fluorescence and the molds could be made so that the probes are absolutely smooth and the geometry optically perfect. Diode lasers at lower wavelengths would make new labels, such as quantum dots, very useful for single-source excitation, with multiple emissions detected at different wavelengths.

There has also been a recent push in development of microstructured optical fibers (MOFs) for sensing purposes. Whereas the sensing surface of standard solid optical fibers is limited to the exterior surfaces, MOFs possess multiple air holes running the length of the fiber that may also be used for sensing. With careful choice of various parameters (silica versus polymer, pocket diameter, lattice pitch), evanescent fields can be produced within these pockets (Monro *et al.*, 1999; Hoo *et al.*, 2003; Jensen *et al.*, 2004; Emiliyanov *et al.*, 2007). Proof of concept has been demonstrated that these MOFs can be used for biological detection (Ringdorf *et al.*, 2006). However, the challenges inherent in sample delivery to the pockets and effective coupling of excitation and emission light to the sensing portion remain significant and will require great efforts to overcome.

Finally, the importance of system integration cannot be understated. In this chapter, we have discussed the sensing biochemistry, fiber probes, optics, and fluidics as separate elements of the sensor. However, small size and effective automation rely on the ability to make these components work together. Furthermore, the more functions that a single structure can perform – i.e., waveguiding and fluid containment (Ligler *et al.*, 2002) – the more efficient the system might be. Research in the area of microfluidics is already producing sensors on a single substrate that includes a light source and photodetectors. Systems integration has the potential, more than ever before, to produce monolithic structures with multiple functionalities. The biosensor of the future might resemble a pack of cards, with single card constituting a multianalyte probe.

In the end, the issue of total cost ultimately drives the marketplace. For a fiber optic biosensor to succeed, it must provide an enhanced benefit at a reasonable cost. Issues such as sample throughput and sample multiplexing are also serious concerns that impact on the final cost. Certainly, niche markets can be exploited, but often that niche may not support the cost of product development. As alternative detection technologies mature – SPR, flow cytometry, microarray, and electrochemiluminescence, to name but a few – the marketplace becomes even more resistant to marginally better mousetraps. For a new fiber optic sensor system to succeed will take a top-to-bottom integrated effort to transition the next breakthrough to the market.

Acknowledgements

The authors thank the Office of Naval Research (NRL work unit 69-6006-06) for supporting the preparation of this chapter.

References

Abel, A.P., Weller, M.G., Duveneck, G.L. *et al.* (1996) *Anal. Chem.*, **68**, 2905.

Anderson, G.P., Breslin, K.A., and Ligler, F.S. (1996) *ASAIO J.*, **42**, 942.

Anderson, G.P. and Golden, J. (1995) US Patent No. 5,430,813.

Anderson, G.P., Golden, J.P., Cao, L.K. *et al.* (1994a). *IEEE Eng. Med. Biol.*, **13**, 358.

Anderson, G.P., Golden, J.P., and Ligler, F.S. (1993) *Biosens. Bioelectron.*, **8**, 249.

Anderson, G.P., Golden, J.P., and Ligler, F.S. (1994b). *IEEE Trans. Biomed. Eng.*, **41**, 578.

Anderson, G.P., Jacoby, M.A., Ligler, F.S., and King, K.D. (1997) *Biosens. Bioelectron.*, **12**, 329.

Anderson, G.P., King, K.D., Cao, L.K. *et al.* (1998) *Clin. Diagn. Lab. Immunol.*, **5**, 609.

Anderson, G.P., King, K.D., Cuttino, D.S. *et al.* (1999) *Field Anal. Chem. Technol.*, **3**, 307.

Anderson, G.P., King, K.D., Gaffney, K.L., and Johnson, L.H. (2000) *Biosens. Bioelectron.*, **14**, 771.

Anderson, G.P., Lingerfelt, B.M., and Taitt, C.R. (2004) *Sens. Lett.*, **2**, 18.

Anderson, G.P. and Nerurkar, N.L. (2002) *J. Immunol. Meth.*, **271**, 17.

Anderson, G.P. and Rowe-Taitt, C.A. (2001) *Proc. SPIE*, **4206**, 58.

Anderson, G.P., Rowe-Taitt, C.A., and Ligler, F.S. (2001) *Proc. 1st Conf. Point Detect. Chem. Biol. Defense*, October 2000, Williamsburg, VA.

Anderson, G.P., Shriver-Lake, L.C., Golden, J.P., and Ligler, F.S. (1992) *Proc. SPIE*, **1648**, 39.

Andrade, J.D., Van Wagenen, R.A., Gregonis, D.E. *et al.* (1985) *IEEE Trans. Electron. Devices*, **32**, 1175.

Bakaltcheva, I.B., Shriver-Lake, L.C., and Ligler, F.S. (1998) *Sens. Act. B Chem.* **51**, 46.

Bardin, F., Kašík, I., Trouillet, A. *et al.* (2002) *Appl. Optics*, **41**, 2514.

Block, M.J. and Hirschfeld, T.B. (1987) US Patent No. 4,716,121.

Block, M.J., Lackie, S.J., and Glass, T.R. (1990) US Patent No. 4,909,990.

Bock, D., Galla, K., Martin, M. *et al.* (1995) *Sens. Act. B Chem.* **29**, 293.

Boisdé, G. and Harmer, A. (1996) *Chemical and Biochemical Sensing with Optical Fibers and Waveguides.* Boston: Artech House, 389 pp.

Browne, C.A., Tarrant, D.H., Olteanu, M.S. *et al.* (1996) *Anal. Chem.*, **68**, 2289.

Brummel, K.E., Wright, J., and Eldefrawi, M.E. (1997) *J. Agric Food Chem.*, **45**, 3292.

Buerck, J., Roth, S., Kraimer, K. *et al.* (2001) *J. Hazard. Mater.*, **83**, 11.

Campbell, J.R. (1995) US Patent Reg. No. H1398.

Cao, L.K., Anderson, G.P., Ligler, F.S., and Ezzell, J. (1995) *J. Clin. Microbiol.*, **33**, 336.

Carlyon, E.G., Lowe, C.R., Reid, D., and Bennion, I. (1992) *Biosens. Bioelectron.*, **7**, 141.

Chang, Y.-J, Chen, Y.-C., Kuo, H.-L., and Wei, P.-K. (2006) *J. Biomed. Optics*, **11**, 014032.

Choa, F.S., Shih, M.H., Toppozada, A.R. *et al.* (1996) *Anal. Lett.*, **29**, 29.

Choi, J.-W., Kim, Y.-K., Lee, I.-H. *et al.*. (2001) *Biosens. Bioelectron.*, **16**, 937.

Choi, J.-W., Kim, Y.-K., Oh, B.-K. *et al.* (2003) *Biosens. Bioelectron.*, **18**, 591.

Chou, C., Hsu, H.-Y., Wu, H.-T. *et al.* (2007) *J. Biomed. Opt.*, **12**, 02045.

DeLisa, M.P., Zhang, Z., Shiloach, M. *et al.* (2000) *Anal. Chem.*, **72**, 2895.

DeMarco, D.R. and Lim, D.V. (2002a). *J. Food Prot.*, **65**, 596.

DeMarco, D.R. and Lim, D.V. (2002b). *J. Rapid Meth. Automation Microbiol.*, **9**, 241.

DeMarco, D.R., Saaski, E.W., McCrae, D.A., and Lim, D.V. (1999) *J. Food Prot.*, **62**, 711.

Devine, P.J., Anis, N.A., Wright, J., Kim, S. *et al.* (1995) *Anal. Biochem.*, **227**, 216.

Donaldson, K.A., Kramer, M.F., and Lim, D.V. (2004) *Biosens. Bioelectron.*, **20**, 322.

Duveneck, G.L., Oroszlan, P., Abel, A.P. *et al.* (1996) *Proc. SPIE*, **2631**, 14.

Emiliyanov, G., Jensen, J.B., Bang, O. *et al.* (2007) *Opt. Lett.*, **32**, 460.

Erb, J. and Downward IV, J. (1998) US Patent No. 5,854,863.

Erb, J.L., Downward, J.G., Erb-Downward, J.R., and Otho, U. (2001a). US Patent No. 6,251,688.

Erb, J.L., Downward, J.G., Erb-Downward, J.R., and Wittliff, J.L. (2001b). US Patent No. 6,300,082.

Erb, J.L., Garber, E.A.E., Downward IV, J.G. *et al.* (2001c). *Proc. 2nd Intl. Conf. Pharm. Endocrine Disrupt. Chem. Water*, The National Ground Water Assoc., p. 203.

Fang, X.H. and Tan, W.H. (1999) *Anal. Chem.*, **71**, 3101.

Gao, J. and Rice, S.A. (1989) *J. Chem. Phys.*, **90**, 3469.

Gao, H.H., Chen, Z., Kumar, J. *et al.* (1995) *Proc. SPIE*, **34**, 3465.

Garden, S.R., Doellgast, G.J., Killham, K.S., and Strachan, N.J.C. (2004) *Biosens. Bioelectron.*, **19**, 737.

Geng, T., Morgan, M.T., and Bhunia, A.K. (2004) *Appl. Environ. Microbiol.*, **70**, 6138.

Geng, T., Uknalis, J., Tu, S.-I., and Bhunia, A.K. (2006) *Sensors* **6**, 796.

Giesen, U., Kleider, W., Berding, C. *et al.* (1998) *Nucl. Acids Res.*, **26**, 5004.

Glass, T.R. (1989) US Patent 4,844,869.

Gloge, D. (1971) *Appl. Opt.*, **10**, 2252.

Golden, J.P., Anderson, G.P., Rabbany, S.Y., and Ligler, F.S (1994a) *IEEE Trans. Biomed. Eng.*, **41**, 585.

Golden, J.P., Saaski, E.W., Shriver-Lake, L.C. *et al.* (1997) *Opt. Eng.*, **36**, 1008.

Golden, J.P., Shriver-Lake, L.C., Anderson, G.P. *et al.* (1992) *Opt. Eng.* **31**, 1458.

Golden, J.P., Shriver-Lake, L.C., Narayanan, N. *et al.* (1994b) *Proc. SPIE*, **2138**, 241.

Goldman, E.R., Pazirandeh, M.P., Mauro, J.M. *et al.* (2000) *J. Mol. Recog.*, **13**, 382.

Graham, C.R., Leslie, D., and Squirrel, D.J. (1992) *Biosens. Bioelectron.*, **7**, 487.

Hale, Z.M., Payne, F.P., Marks, R.S. *et al.* (1996) *Biosens. Bioelectron.*, **11**, 137.

Harrick, N.J. (1967) *Internal Reflection Spectrosopy*. New York: John Wiley & Sons, 327 pp.

Herron, J.N., Christensen, D.A., Caldwell, K.D. *et al.* (1996) US Patent No. 5,512,492.

Hesse, H.C. (1974) East German Patent 106,086.

Hirschfeld, T. (1965) *Can. J. Spectrosc.*, **10**, 128.

Hirschfeld, T. (1984) US Patent No. 4,447,546.

Hirschfeld, T.E. and Block, M.J. (1984) US Patent No. 4,447,546.

Hirschfeld, T.E. and Block, M.J. (1985) US Patent No. 4,558,014.

Homola, J. (1995) *Sens. Actuators B*, **29**, 401.

Hoo, Y.L., Jin, W., Shi, C. *et al.* (2003) *Appl. Opt.*, **42**, 3509.

James, E.A., Schmeltzer, K., and Ligler, F.S. (1996) *Appl. Biochem. Biotechnol.*, **60**, 189.

Jensen, J.B., Pedersen, L.H., Hoiby, P.E. *et al.* (2004) *Opt. Lett.*, **29**, 1974.

Jorgenson, R.C. and Yee, S.S. (1993) *Sens. Actuators B*, **12**, 213.

Jung, C.C., Saaski, E.W., McCrae, D.A. *et al.* (2003) *IEEE Sens. J.*, **3**, 352.

Kapoor, R., Kaur, N., Nishanth, E.T. *et al.* (2004) *Biosens. Bioelectron.*, **20**, 345.

Kharat, H.J., Kakde, K.P., Shirale, D.J. *et al.* (2006) *Fiber Integ. Opt.*, **25**, 411.

King, K.D., Anderson, G.P., Bullock, K.E. *et al.* (1999) *Biosens. Bioelectron.*, **14**, 163.

King, K.D., Vanniere, J.M., LeBland, J.L. *et al.* (2000) *Environ. Sci. Technol.*, **34**, 2845.

Kishen, A., John, M.S., Lim, C.S., and Asundi, A. (2003) *Biosens. Bioelectron.*, **18**, 1371.

Kleinjung, F., Klussmann, S., Erdmann, V.A. *et al.* (1998) *Anal. Chem.*, **70**, 328.

Ko, S. and Grant, S.A. (2006) *Biosens. Bioelectron.*, **21**, 1283.

Kooyman, R.P.H., de Bruijn, H.E., and Greve, J. (1987) *Proc. SPIE*, **798**, 290.

Kramer, M.F. and Lim, D.V. (2004) *J. Food Prot.*, **67**, 46.

Kronick, M.N. and Little, W.A. (1975) *J. Immunol. Meth.*, **8**, 235.

Krull, U.J., Brown, R.S., DeBono, R.F., and Hougham, B.D. (1988) *Talanta* **35**, 129.

Lackie, S.J. (1992) US Patent 5,152,962.

Lee W.E. and Thompson, H.G. (1996) *Can. J. Chem.*, **74**, 707.

Lew, A., Depeursinge, C., Cochet, F. *et al.* (1986) *Proc. SPIE*, **514**, 71.

Ligler, F.S. and James, E.A. (1999) US Patent Reg. No. H1775.

Ligler, F.S., Anderson, G.P., Davidson, P.T. *et al.* (1998) *Environ. Sci. Technol.*, **32**, 2461.

Ligler, F.S., Breimer, M., Golden, J.P. *et al.* (2002) *Anal. Chem.*, **74**, 713.

Ligler, F.S., Breslin, K.A., Cao, L.K., and Anderson, G.P. (1995) *Proc. SPIE*, **2388**, 16.

Ligler, F.S., Shriver-Lake, L.C., and Wijesuriya, D.C. (1996) US Patent No. 5,496,700.

Lipson, D.G., Loebel, N.G., McLeaster, K.D., and Liu, B. (1992) *IEEE Trans. Biomed. Eng.*, **39**, 886.

Liu, X. and Tan, W. (1999) *Anal. Chem.*, **71**, 5054.

Liu, X., Farmerie W., Schuster, S., and Tan, W. (2000) *Anal. Biochem.*, **283**, 56.

Maragos, C.M. and Thompson, V.S. (1999) *Nat. Toxins*, **7**, 371.

Matsuo, H., Kuniyoshi, S., Kudo, K., and Tanaka, K. (2000) *Synthetic Metals*, **115**, 37.

Matsuoka, H. (2001) *Macromol. Rapid Commun.*, **22**, 51.

Mauro, J.M., Cao, L.K., Kondracki, L.M. *et al.* (1996) *Anal. Biochem.*, **235**, 61.

Medintz, I.L., Anderson, G.P., Lassman, M.E. *et al.* (2004) *Anal. Chem.*, **76**, 5620.
Medintz, I.L., Goldman, E.R., Lassman, M.E. *et al.* (2005) *Anal. Chem.*, **77**, 365.
Meserol, P. (1996) US Patent No. 5,492,674.
Mitsushio, M. and Higo, M. (2004) *Anal. Sci.*, **20**, 689.
Monro, T.M., Richardson, D.J., and Bennett, P.J. (1999) *Electron. Lett.*, **35**, 1188.
Monzón-Hernández, D., Villatoro, J., Talavera, D., and Luna-Moreno, D. (2004) *Appl. Optics*, **43**, 1216.
Moreno-Bondi, M.-C., Mobley, J., Alarie, J.-P., and Vo-Dinh, T. (2000) *J. Biomed. Opt.*, **5**, 350.
Mosiello, L., Nencini, L., Segre, L., and Spanò, M. (1997) *Sens. Actuators B*, **39**, 353.
Nanduri, V., Kim, G., Morgan, M.T. *et al.* (2006) *Sensors*, **6**, 808.
Narang, U., Anderson, G.P., King, K.D. *et al.* (1997a) *Proc. SPIE*, **2980**, 187.
Narang, U., Anderson, G.P., Ligler, F.S., and Burans, J. (1997b) *Biosens. Bioelectron.*, **12**, 937.
Nath, N., Eldefrawi, M., Wright, J. *et al.* (1999) *J. Anal. Toxicol.*, **23**, 460.
Nath, N., Jain, S.R., and Anand, S. (1997) *Biosens. Bioelectron.*, **12**, 491.
Neel, T.G. and Lyst, Jr., J.E. (1997) US Patent No. 5,639,668.
Nielsen, P.E. (2004) *Peptide Nucleic Acids: Protocols and Applications.* London: Horizon Bioscience, 318 pp.
Nielsen, P.E. and Egholm, M. (1999) *Curr. Issues. Mol. Biol.*, **1**, 89.
Ogert, R.A., Shriver-Lake, L.C., and Ligler, F.S. (1993) *Proc. SPIE*, **185**, 11.
Oroszlan, P., Thommen, C., Wehrli, M. *et al.* (1993) *Anal. Meth. Instrumentation*, **1**, 43.
Oroszlan, P., Wicar, S., Teshima, G. *et al.* (1992) *Anal. Chem.*, **64**, 1623.
Orvedahl, D.S., Love, W.F., and Slovacek, R.E. (1991) *Proc. SPIE*, **1487**, 187.
Pandey, P.C. and Weetall, H.H. (1995) *Anal. Biochem.*, **67**, 787.
Pease, M.D., Shriver-Lake, L., and Ligler, F.S. (1995) *ACS Symp. Ser.*, **613**, 33.
Pellestor, F. and Paulasova, P. (2004) *Eur. J. Hum. Genet.*, **12**, 69.
Pilevar, S., Davis, C.C., and Portugal, F. (1998) *Anal. Chem.*, **70**, 2031.
Piunno, P.A.E., Krull, U.J., Hudson, R.H.E. *et al.* (1994) *Anal. Chim. Acta*, **288**, 205.
Piunno, P.A.E., Krull, U.J., Hudson, R.H.E. *et al.* (1995) *Anal. Chem.*, **67**, 2635.
Preejith, P.V., Lim, C.S., and Chia, T.F. (2006) *Meas. Sci. Technol.*, **17**, 3255.
Preejith, P.V., Lim, C.S., Kishen, A. *et al.* (2003) *Biotechnol. Lett.*, **25**, 105.
Quinn, J.G., O'Neill, S., Doyle, A. *et al.* (2000) *Anal. Biochem.*, **281**, 135.

Ringdorf, L., Hoiby, P.E., Jensen, J.B. *et al.* (2006) *Anal. Bioanal. Chem.*, **385**, 1370.

Rogers, K.R., Anis, N.A., Valdes, J.J., and Eldefrawi, M.E. (1992) Fiber-optic biosensors based on total internal reflection fluorescence. In *Biosensor Design and Application* (P.R. Mathewson and J.W. Finley, eds) Washington: American Chemical Society, pp. 165–73.

Rogers, K.R., Apostol, A., Madsen, S.J., and Spencer, C.W. (2001) *Anal. Chim. Acta*, **444**, 51.

Rogers, K.R., Eldefrawi, M.E., Menking, D.E. *et al.* (1991) *Biosens. Bioelectron.*, **6**, 507.

Rogers, K.R., Valdes, J.J. and Eldefrawi, M.E. (1989) *Anal. Biochem.*, **182**, 353.

Rowe, D.A., Bolitho, J.S., Jane, A. *et al.* (1998) *Thromb. Haemost.*, **79**, 94.

Saaski, E.W. (2000) US Patent No. 6,082,185.

Saaski, E.W. and Jung, C.C. (2000) US Patent No. 6,136,611.

Shriver-Lake, L.C., Breslin, K.A., Charles, P.T. *et al.* (1995) *Anal. Chem.*, **34**, 2431.

Shriver-Lake, L.C., Donner, B.L., and Ligler, F.S. (1997) *Environ. Sci. Technol.*, **31**, 837.

Shriver-Lake, L.C., Golden, J.P., Patonay, G. *et al.* (1995) *Sens. Act. B Chem.*, **29**, 25.

Shriver-Lake, L.C., Naz, N.A., and Ligler, F.S. (1998) A fiber optic biosensor for the detection of TNT/RDX in environmental samples. In *Current Protocols in Field Analytical Chemistry* (V. Lopez-Avila, ed.) New York: John Wiley & Sons, pp. 2E1–2E12.

Simpson, J.M. and Lim, D.V. (2005) *Biosens. Bioelectron.*, **21**, 881.

Slavík, R., Homola, J., and Brynda, E. (2002) *Biosens. Bioelectron.*, **17**, 591.

Slavík, R., Homola, J., and Htyroký, J. (1999) *Sens. Actuators B*, **54**, 74.

Slovacek, R.E. and Love, W.F. (1992) US Patent No. 5,156,976.

Smith, R.H., Lemon, W.J., Erb, J.L. *et al.*, 1999, *Clin. Chem.*, **45**, 1683.

Spiker, J.O., Kang, K.A., Drohan, W.N., and Bruley, D.F. (1998) *Adv. Exp. Med. Biol.*, **454**, 681.

Sutherland, R.M., Dahne, C., Place, J.F., and Ringrose, A.S. (1984) *Clin. Chem.*, **30**, 1533.

Tang, L. and Kang, K.A. (2005) *Adv. Exp. Med. Biol.*, **578**, 101.

Tang, L., Ren, Y., Hong, B., Kang, K.A. (2006) *J. Biomed. Opt.*, **11**, 021011.

Templeman, L., King, K.D., Anderson, G.P., and Ligler, F.S. (1996) *Anal. Biochem.*, **233**, 50.

Thompson, R.B. and Kondracki, L. (1990) *Proc. SPIE*, **1204**, 35.

Thompson, R.B. and Ligler, F.S. (1988) *Proc. SPIE*, **904**, 27.

Thompson, V.S. and Maragos, C.M. (1996) *J. Agric. Food Chem.*, **44**, 1041.

Thompson, R.B. and Villaruel, C.A. (1991) US Patent No. 5,061,857.
ThreeFold Sensors (2007), http://mywebpages.combase.net/tfs-jdownward.
Tims, T.B. and Lim, D.V. (2004) *J. Microbiol. Meth.*, **59**, 127.
Tims, T.B., Dickey, S.S., DeMarco, D.R., and Lim, D.V. (2001) *Am. Clin. Lab.*, **20**, 28.
Topporada, A.R., Wright, J., Eldefrawi, A.T. *et al.* (1997) *Biosens. Bioelectron.*, **12**, 113.
Tu, S.-I., Geng, T., Uknalis, J., Bhunia, A. (2006) *Proc. SPIE*, **6381**, 638106.
Uddin, A.J., Piunno, P.A.E., Hudson, R.H.E. *et al.* (1997) *Nucl. Acids Res.*, **25**, 2635.
van Bergen, S.K., Bakaltcheva, I.B., Lundgren, J.S., and Shriver-Lake, L.C. (2000) *Environ. Sci. Technol.*, **34**, 704.
Vijayendran, R.A. and Leckband, D.E. (2001) *Anal. Chem.*, **73**, 471.
Vijayendran, R.A., Ligler, F.S., and Leckband, D.E. (1999) *Anal. Chem.*, **71**, 5405.
Villaruel, C.A., Dominguez, D.D., and Dandridge, A. (1987) *Proc. SPIE*, **798**, 225.
Wadkins, R.M., Golden, J.P., and Ligler, F.S. (1995) *Anal. Biochem.*, **232**, 73.
Wang, X. and Krull, U. (2005) *Bioorg. Med. Chem. Lett.*, **15**, 1725.
Wang, X. and Krull, U.J. (2002) *Anal. Chim. Acta*, **470**, 57.
Watterson, J.H., Piunno, P.A.E., Wust, C.C. *et al.* (2001) *Fresenius J. Anal. Chem.*, **369**, 601.
Wijesuriya, D., Breslin, K., Anderson, G. *et al.* (1994) *Biosens. Bioelectron.*, **9**, 585.
Zhao, C.Q., Anis, N.A., Rogers, K.R. *et al.* (1995) *J. Agric. Food Chem.*, **43**, 2308.
Zhao, C., Pivarnik, P., Rand, A.G., and Letcher, S.V. (1998) *Biosens. Bioelectron.* **13**, 495.
Zhao, S. and Reichert, W.M. (1992) *Langmuir*, **8**, 2785.
Zhou, C., Pivarnik, P., Auger, S. *et al.* (1997) *Sens. Actuators B*, **42**, 169.

Optical Biosensors: Today and Tomorrow (2nd Edition)
Frances S. Ligler and Chris Rowe Taitt (Editors)
Published by Elsevier B.V.

Chapter 3

PLANAR WAVEGUIDES FOR FLUORESCENCE BIOSENSORS

Kim Sapsford[a], Ph.D., Chris Rowe Taitt[b], Ph.D., and Frances S. Ligler[b], D.Phil., D.Sc.
[a]US Food and Drug Administration, CDRH/OSEL/DB, Silver Spring, MD 20993, USA
[b]Center for Bio/Molecular Science and Engineering, Naval Research Laboratory, Washington, DC 20375-5348, USA

While confocal scanners and other microscopy techniques are routinely used for imaging genomic and proteomic microarrays, the leading methods for fluorescence-based, multianalyte detection are based on total internal reflection fluorescence (TIRF). Immobilization of multiple capture biomolecules on planar waveguides provides for multianalyte detection on a single substrate. Planar waveguide TIRF has been used to measure a variety of analytes including hormones, toxins, bacteria, and viruses, leading to applications such as environmental and food safety monitoring, clinical diagnostics, and military defense. Many of the analytes have been measured in both buffer and complex matrices, such as whole blood, nasal secretions, food, beverages, ground water, and soil suspensions. Detection limits in both buffer and complex matrices have been comparable. The continued development and miniaturization of the biosensor instrumentation has led to systems that are fully automated, portable, and highly competitive with laboratory techniques. Such biosensors are now transitioning into the commercial market.

3.1. Technical concept

Biosensors consist of two important features: the molecular recognition element and the signal transduction mechanism (Hall, 1990; Vo-Dinh and Cullum, 2000). The molecular recognition element can take the form of a biomolecule (antibody, enzyme, or nucleic acid), biological system (membranes, tissues, or whole cells), or biomimetic (synthetic bioreceptors) and imparts specificity to the system. The signal transduction mechanism is the process by which the biochemical recognition event is converted into a measurable signal, the intensity of which is proportional to the analyte concentration. In the case of planar waveguide-based biosensors, the molecular recognition event results in the binding of a fluorophore at the surface of the planar waveguide. The planar waveguide provides the localized excitation light to generate a signal from fluorophores at the surface, but is not usually used to collect the light. Biosensor technology has the potential to both address fundamental scientific questions and to develop sensing systems for a variety of practical applications. However, in order to successfully develop commercial biosensors, the final product must offer a number of advantages over existing technology, such as the ability to perform faster, more sensitive, multi-analyte and real-time measurements, preferably with a portable, easy-to-use, stand-alone device.

A number of optical techniques for imaging fluorescent microarrays were reviewed recently by Bally *et al.* (2006). Microscopy techniques that have proven effective include epi-fluorescence microscopy, confocal microscopy, and total internal reflection fluorescence (TIRF) microscopy (Nagl *et al.*, 2005; Schaferling and Nagl, 2006). Apart from the TIRF microscopy, however, the planar surface acts only as a solid support and not a waveguide. The ability to image high density microarrays using these microscopy-based techniques has revolutionized the field of genomics and proteomics. High density DNA and protein microarrays have been used to investigate gene expression, markers for disease and pathogens, drug response and development, and cell function (Brown and Botstein, 1999; Gullans, 2000; Lockhart and Winzeler, 2000; MacBeath and Schreiber, 2000; Haab *et al.*, 2001; Knight, 2001; Kumble *et al.*, 2003; Wulkkuhle *et al.*, 2004; Dawson *et al.*, 2006;

Schafeling and Ngal, 2006). For biosensor applications, TIRF excitation of planar waveguides is the most utilized optical configuration and therefore will represent the bulk of the discussion in this chapter. With a history deeply rooted in the field of fiber optics, planar waveguide TIRF is a means of selectively exciting the fluorescence emission of fluorophores present near the surface of a waveguide and is relatively immune to bulk matrix effects. This technique has found numerous applications in the field of biosensors, in particular immunosensors and sensors for genetic analysis.

3.1.1. Signal transduction and amplification

Based on microscopy, optical imaging technologies have revolutionized the fields of genomics, proteomics, and single-molecule detection and are increasingly being applied to biosensors (Nagl *et al.*, 2005; Tinnefeld and Sauer, 2005; Schaferling and Nagl, 2006). As evanescent wave biosensors using planar waveguides have many facets in common with other methods using planar arrays, it is worthwhile presenting a brief synopsis of these complementary methods. The basic epi-fluorescence microscope combines a number of components including a light source (Xenon or Mercury arc-discharge lamp, lasers, or light emitting diodes), excitation and emission filters, focusing lenses, a dichroic mirror, and a detector, as shown in Figure 3.1 (Nagl *et al.*, 2005). The filters and dichroic mirror, which reflects the excitation light but is transparent to the emission wavelength, are chosen to match the absorption and emission profiles of the particular fluorescent dye used. The excitation light is directed toward the specimen via the dichroic mirror (also referred to as a "beamsplitter"). The resulting fluorescence from the illuminated specimen is separated from the illumination light and measured at the detector. Fluorescence from the planar surface is typically measured using either a charge-coupled device (CCD) or photomultiplier tube (PMT) in a scanning configuration to produce a fluorescent image of the surface. With the incorporation of lasers, more advanced microscope designs, such as confocal and TIRF microscopy, offer a number of advantages over the basic "wide field" microscope design (Lehr *et al.*, 2003a; Semiatin, 2006). Confocal scanning microscopy produces high resolution, 3D images of the specimen by focusing the excitation

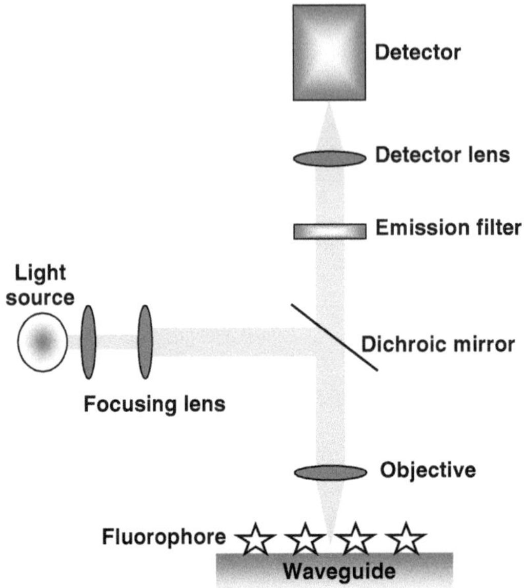

Figure 3.1 A schematic representation of a standard epi-fluorescent microarray imager.

laser beam into a small volume within the fluorescent sample on the planar surface and rejecting out-of-focus emission. TIRF-microscopy (Semiatin, 2006) takes advantage of the waveguiding properties of the planar substrate, is surface-selective in nature, and excites only fluorophores present near the waveguide/sample interface to reduce background signals and improve sensitivity (see below). The majority of the fluorescence microscopy-based techniques have to date been used to study biological processes and interactions rather than for the development of biosensing systems for commercial applications. Also, the instrumentation is often large and bulky and, while perfectly suitable for the laboratory, lacks the portability and stand-alone capability required for many biosensing applications.

In optical biosensor configurations based on the principle of reflectance, light is launched into an optical waveguide at such an angle that it is reflected at the interface of the waveguide and its surrounding matrix, as depicted in Figure 3.2. At the interface of two media with different

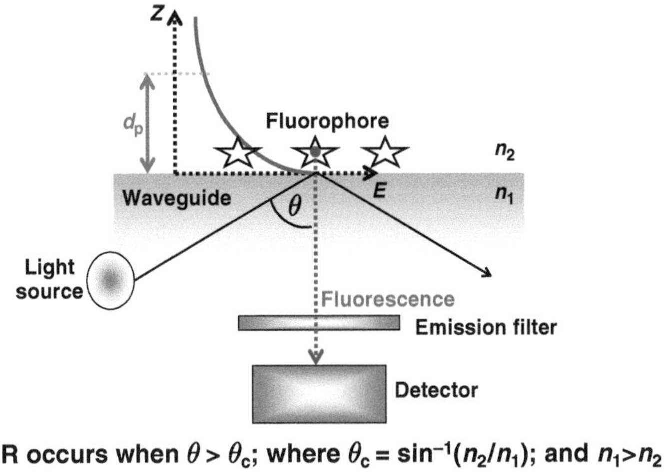

Figure 3.2 The basic experimental arrangement of a system based on the principle of reflectance, with detection of the emitted fluorescence measured at right angles to the waveguide interface.

refractive indices, n_1 and n_2, incident light from the higher refractive index medium will be partly refracted and partly reflected. However, when the angle of incidence is greater than the critical angle (θ_c), the phenomenon of total internal reflection is observed, whereby all the light is reflected and none refracted. Under the condition of total internal reflection, a standing wave of the electromagnetic field is set up at the point of reflection. This standing wave is known as the "evanescent wave" and penetrates into the lower refractive index medium with an exponential decay. The depth at which the intensity of either the electric or magnetic component of the electromagnetic wave drops to $1/e$ of its original value is known as the "depth of penetration" and is described by Equation (3.1), where $n_{21} = n_2/n_1$.

$$d_p = \frac{\lambda_{vac}}{2\pi n_1}\left[\frac{1}{\sin^2\theta - n_{21}^2}\right]^{1/2} \tag{3.1}$$

As shown in Eqn (3.1), d_p is dependent on the incident angle (θ), the refractive indices of the dielectrics (n_1 and n_2), and the wavelength (λ) of the light. Typical penetration depths range between 100 and 300 nm.

It is this evanescent wave, generated at the interface between the two dielectrics, which interacts with the species at the interface.

There are several types of sensors that, under the condition of total internal reflectance, use the interaction of the resulting evanescent wave with the surface-bound species for signal transduction: attenuated total reflectance (ATR) monitors alterations in the IR, visible, and UV regions; surface plasmon resonance (SPR) and interferometric techniques measure variations in refractive index; and TIRF monitors changes in fluorescence (Bradley *et al.*, 1987; Lu *et al.*, 1992; Chittur, 1998; Plowman *et al.*, 1998; Wadkins *et al.*, 1998). The basic arrangement of the TIRF system is shown in Figure 3.2.

Coherent light in the form of lasers is typically used as the excitation source in TIRF studies. The exact choice of the laser is dependent upon the fluorescent label used; the most commonly used lasers are the argon-ion (488 nm) laser for fluorescein and a helium–neon (633 nm) or diode laser (635 nm) for dyes such as the cyanine dye Cy5. A number of devices have been used in the detection of the resulting fluorescence emission, in particular CCD cameras (Silzel *et al.*, 1998; Feldstein *et al.*, 1999; Plowman *et al.*, 1999), PMTs (Schult *et al.*, 1999; Lundgren *et al.*, 2000; Schuderer *et al.*, 2000), photodiodes (Brecht *et al.*, 1998), and a CMOS camera (Askari *et al.*, 2001; Golden and Ligler, 2002).

A variety of options have been described for improving the optics and sensitivity of the type of TIRF instrumentation represented in Figure 3.2. Most of the final systems described consist of a number of similar components such as the light source and detector and also a variety of focusing lenses to improve detector response (Duveneck *et al.*, 1995; Herron *et al.*, 1996, 1997; Golden, 1998; Feldstein *et al.*, 1999). Golden (1998) used a 2D graded index (GRIN) lens to focus the fluorescence from the planar waveguide onto a CCD. The GRIN lens provided a shorter working distance than a standard lens with a concomitant decrease in overall instrument size. The introduction of bandpass and longpass filters was found to improve the rejection of scattered laser light and hence reduce the background fluorescence of the system (Feldstein *et al.*, 1999).

A side effect of using bulk waveguides and collimated light can be the production of sensing "hot spots" along the planar surface which occur where the light beam is reflected, illuminating only discrete regions. These hot spots have been successfully utilized as sensing regions by Brecht and coworkers in the development of an immunofluorescence sensor for water analysis (Brecht *et al.*, 1998; Klotz *et al.*, 1998). Feldstein *et al.* (1999, 2000) overcame this problem by incorporating a line generator and a cylindrical lens to focus the beam into a multi-mode waveguide which included a propagation and distribution region prior to the sensing surface. This resulted in uniform lateral and longitudinal excitation at the sensing region.

Another method of achieving uniform longitudinal excitation of the sensing region is to decrease the waveguide thickness (Herron *et al.*, 1996). When the thickness of the waveguide is much greater than the wavelength of the reflected light, the waveguide is referred to as an "internal reflection element" (IRE). However, when the thickness of the waveguide approaches the wavelength of the incident light, the path-length between the points of total internal reflection become increasingly shorter. At the thickness where the standing waves, created at each point of reflection, overlap and interfere with one another, a continuous streak of light appears across the waveguide, and the IRE becomes known as an "integrated optical waveguide" (IOW) (Plowman *et al.*, 1998).

Integrated optical waveguides, used frequently in TIRF studies, are mono-mode and are prepared by depositing a thin film of high refractive index material onto the surface of a glass substrate. These thin films are typically 80–160 nm in thickness and consist of inorganic metal oxide compounds such as tin oxide (Duveneck *et al.*, 1995), indium tin oxide (Asanov *et al.*, 1998), silicon oxynitride (Plowman *et al.*, 1999), and tantalium pentoxide (Duveneck *et al.*, 1997; Pawlak *et al.*, 1998). The light is coupled into these IOWs via a prism or grating arrangement. Studies comparing IRE- and IOW-based waveguides concluded that the integrated optics improved the sensitivity of the system by a factor of 100 (Brecht *et al.*, 1998; Lehr *et al.*, 2003a).

3.1.2. The molecular recognition element

The different types of binding events that are typically monitored via fluorescence measurements in affinity-based sensors include antibody–antigen interactions, nucleic acid hybridization (DNA/RNA), and receptor–ligand binding (Rogers, 2000). Affinity assays were originally performed using radiolabeled species; however, fluorescent markers are now the more common label of choice due to safety, longer shelf lives, lower costs, and ease of preparation, bioconjugation, and disposal. Although a number of biomolecules contain some intrinsic fluorescence in the form of amino acid residues or cofactors, often this fluorescence has weak intensity and low quantum yields. Therefore, extrinsic fluorescent probes are more often incorporated into one of the binding partners. The introduction of an extrinsic fluorescence probe, such as rhodamine, coumarin, cyanine, or fluorescein dyes, allows both site and spectral selection. These commercial fluorescent dyes are available with a variety of reactive groups that permit simple attachment to the desired species. The Invitrogen handbook, *A Guide to Fluorescent Probes and Labeling Technologies* (http://probes.invitrogen.com), is an excellent resource for choosing appropriate extrinsic fluorescent probes and suitable bioconjugation techniques (Haugland, 2005). The use of fluorescent markers is also favored over enzyme labels due to increased stability and because no additional substrates are required.

3.1.2.1. Immunoassays

Antibodies and antibody fragments, because of their selectivity and sensitivity, have been the most popular choice of selective recognition elements for many biosensor applications (Marquette *et al.*, 2006). Antibodies can be readily generated to a wide variety of target analytes, from whole bacterial cells to small organic species (Moore and Clayton, 2003). Monoclonal and polyclonal antibodies offer the user a range of affinities and specificities. Additional tailoring can be achieved via genetic engineering to produce recombinant antibodies (Kramer *et al.*, 2003). High density proteomic microarrays can employ antibodies as capture molecules or purified enzymes, proteins, and peptides as immobilized target molecules. In a similar vein to the genomic microarrays, such microarrays can be used to address many fundamental scientific questions concerning protein function in cells. They have numerous applications including screening for bio-markers of disease, such as cancer, and drug target identification (Kumble, 2003; Melton, 2004).

To date, immunoassays are the most well-characterized systems employed in biosensors based on TIRF. The assays carried out using antibody–antigen systems can be divided into five main categories: direct, competitive, displacement, sandwich, and ELISA immunoassays, of which competitive and sandwich are the most popular for TIRF biosensing. As can be seen from Figure 3.3, the direct assay (*Panel a*) is the simplest method to perform; however, it requires that the antigen contain some form of intrinsic fluorescence or be labeled prior to the assay. In the absence of a fluorescent antigen, preferred formats are competitive and sandwich assays. In the competitive assay (*Panel b*), two arrangements are possible. The first, Figure 3.3b(i), uses an immobilized antibody, and a fluorescently labeled antigen competes with the non-labeled antigen for binding sites (Brecht *et al.*, 1998; Schult *et al.*, 1999; Sapsford *et al.*, 2002). In the second arrangement, Figure 3.3b(ii), an analog of the antigen is immobilized on the waveguide surface where it competes with the non-labeled antigen for binding sites on the fluorescently labeled antibody (Tschmelak *et al.*, 2004, 2005; Ngundi *et al.*, 2005, 2006a, 2006c; Sapsford *et al.*, 2006d). In the displacement assay (Figure 3.3, *Panel c*), a fluorescently labeled antigen saturates the binding sites of immobilized antibodies prior to the exposure to sample. Upon introduction of the sample containing unlabeled antigen, displacement of the labeled antigen occurs and is then measured (Rabbany *et al.*, 1994; Sapsford *et al.*, 2002). The resulting fluorescence signal from the competitive and displacement formats is inversely proportional to the unlabeled antigen concentration. Competitive formats are especially useful in the detection of molecules not large enough to possess two distinct epitopes, such as explosives, drugs, and mycotoxins (Ngundi *et al.*, 2006c; Sapsford *et al.*, 2006e). Sandwich assays (Figure 3.3, *Panel d*), on the other hand, require relatively large antigens; the antigen is bound to the immobilized capture antibody at one epitope and is detected by a fluorescent tracer antibody bound to a second epitope (Silzel *et al.*, 1998; Plowman *et al.*, 1999; Rowe *et al.*, 1999a, 1999b; Schult *et al.*, 1999). Sandwich and direct assays produce a fluorescence signal that is directly proportional to the amount of bound antigen.

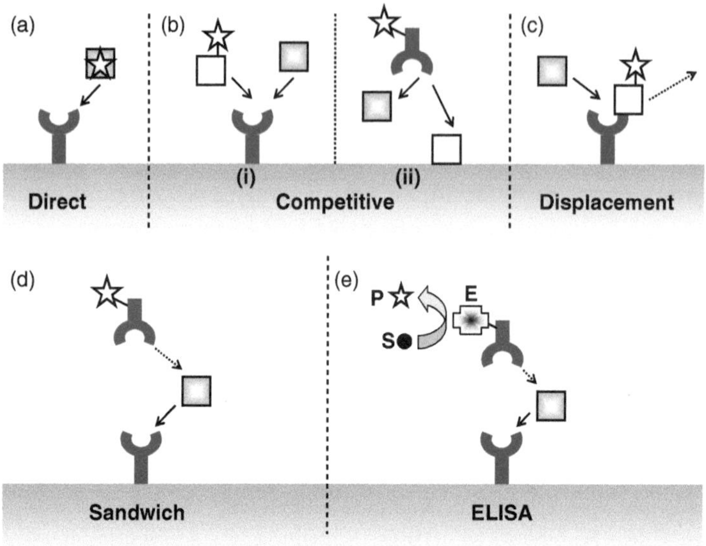

Figure 3.3 Schematic representation of the five main categories of solid-phase immunoassays. (a) Direct assay measuring antigen captured from solution. (b) Competitive assays: (i) competition between a known concentration of the fluorescently labeled antigen and an unknown concentration of unlabeled antigen for binding sites on the immobilized antibody or (ii) competition for binding sites on the fluorescently labeled antibody between an unlabeled antigen in solution and immobilized antigen. (c) Displacement assay: fluorescently labeled antigen is pre-incubated with the immobilized antibody and is then displaced by unlabeled antigen from the solution. (d) Sandwich assay: the amount of immobilized antigen is determined by passing a second fluorescently labeled antibody over the antibody-bound antigen. (e) The ELISA: an assay in which the signal-generating partner is typically labeled with an enzyme. Addition of the appropriate substrate produces an absorbing, fluorescent or chemiluminescent signal.

3.1.2.2. DNA and mRNA analysis

DNA and RNA detection is based on the specific hybridization of oligonucleotides with complementary base pairs. Any target analyte that contains either DNA or RNA, including bacteria and viruses, can be measured using nucleic acid-based biosensors. Such biosensors have been used in biothreat and disease diagnosis applications. These nucleic acid assays can also benefit from PCR amplification of the target DNA/RNA, prior to detection, greatly improving the sensitivity of the measurement. These extra steps, however, come at the price of increasing analysis time and may also require separation of the target oligonucleotides from a complex sample. High density gene microarrays, typically screened using epi-fluorescence microscopy, have revolutionized the study of gene expression, single nucleotide polymorphisms, and disease diagnosis (Debouck and Goodfellow, 1999; Wulfkuhle *et al.*, 2004; Lim *et al.*, 2005; Schaferling and Ngal, 2006). The hybridization between complementary DNA or mRNA fragments and immobilized DNA has become increasingly measured using planar waveguide TIRF (Duveneck *et al.*, 1997; Budach *et al.*, 1999; Schuderer *et al.*, 2000; Lehr *et al.*, 2003b). One advantage of nucleic acid capture agents is the ability to regenerate the sensing surface, as demonstrated by Duveneck (1997) and Budach (1999) using 50% urea or sodium hydroxide (Schuderer *et al.*, 2000). Both systems demonstrated good stability and excellent signal reproducibility during repeated cycles of binding and regeneration.

3.1.2.3. Specific membrane receptor–ligand interactions

Currently, only a limited number of studies describing receptor–ligand binding using planar waveguide TIRF have been published, excluding those utilizing cognate antibody–antigen pairs or nucleic acid hybridization. Schmid *et al.* (1997, 1998) immobilized the green fluorescent protein (GFP) and serotonin receptor; Rowe-Taitt *et al.* (2000a) and Kelly *et al.* (1999) studied gangliosides as receptors; and Pawlak *et al.* (1998) investigated the membrane enzyme Na,K-ATPase in membrane fragments. When successful, receptor–ligand binding studies offer applications in the pharmaceutical industry for drug development and for investigating membrane processes.

3.1.2.4. Semi-selective capture molecules: purified carbohydrates and peptides

Although considered an advantage, the high specificity and selectivity inherent in antibody- and DNA-based molecular recognition results in detection of only pre-selected analytes. An important component of biosensors in the future will be the ability to screen samples both for pre-determined and unknown potential target analytes. Carbohydrates and peptides represent two such classes of semi-selective capture agents that bind to a variety of target analytes. Many bacterial pathogens recognize carbohydrates displayed on the target cells they infect (Nilsson *et al.*, 1994; Park *et al.*, 2004; Ngundi *et al.*, 2006b, 2006d). Thiol-terminated monosaccharides *N*-acetylneuraminic acid (Neu5Ac) and *N*-acetylgalactosamine (GalNAc) were immobilized and screened for their ability to capture Cy5-labeled *Escherichia coli* O157:H7, *Salmonella typhimurium*, *Listeria monocytogenes*, staphylococcal enterotoxin B (SEB), and cholera and tetanus toxins using the TIRF-based NRL Array Biosensor (Ngundi *et al.*, 2006d). Fluorescent signals obtained from the CCD image following the assay demonstrated that only the cholera and tetanus toxins interacted with the immobilized sugars, while none of the tested bacterial targets bound.

Antimicrobial peptides (AMPs) are another class of semi-selective recognition molecules. This category of naturally occurring molecules serves as part of the innate immune system of many organisms and works by binding to and disrupting cell membranes, although other mechanisms of killing have been demonstrated. Kulagina *et al.* (2005, 2006, 2007) took advantage of the natural binding affinities of various AMPs to use them for target recognition in assays for inactivated *E. coli* O157:H7, *S. typhimurium*, *Coxiella burnetti*, *Brucella melitensis*, vaccinia virus, and Venezuelan equine encephalitis virus using a planar waveguide. Detection limits on the highest affinity AMPs were similar to, or in some cases better than, antibody-based assays performed simultaneously on the same waveguide. Significant differences in binding patterns of the different bacterial species were observed; these binding patterns may be used as a means of target identification in future iterations. The immobilized AMPs were significantly more stable than antibodies immobilized on the same waveguides.

3.1.3. Immobilization of the biomolecule to the transducer

Various methods exist in which the biological component of a biosensing system can be immobilized onto the surface of the transducer, including physical adsorption, covalent immobilization, and entrapment in polymer matrices (Hall, 1990). Two important requisites of all immobilization techniques are that the functional integrity of the biomolecule be preserved and that the active site remains accessible to the binding partner. Each immobilization procedure has various advantages and disadvantages; however, physical adsorption and covalent binding to functionalized surfaces are the most commonly used in TIRF measurements. Physical adsorption of a biomolecule to a surface occurs via dipole–dipole interactions, van der Waals forces, or hydrogen bonding, depending on the nature of the substrate surface and the adsorbate. However, physical adsorption in general is not only strongly influenced by changes in the ambient conditions, such as pH and the solvent used, but may also be a reversible process. Furthermore, adsorption may not provide as high density of immobilized biomolecules as covalent immobilization (Ulbrich *et al.*, 1991). Physical adsorption usually occurs in a random molecular orientation, resulting in the inaccessibility of the active binding site in at least a subpopulation of the immobilized molecules. Initial adsorption of a biomolecule can be followed by surface-induced denaturation and tighter attachment, with concomitant decrease in function.

On the other hand, covalent immobilization provides surfaces with reproducibly attached biomolecules at relatively high densities (e.g. 2 ng/mm^2; Bhatia *et al.*, 1989). Most methods of covalent immobilization involve the activation of the surface (e.g., using silane or thiol self-assembled monolayers), followed by covalent linkage of the biomolecule either directly or using a crosslinker. Most of the molecular recognition elements mentioned in Section 3.1.2 either naturally possess specific chemical moieties targeted by crosslinkers or can be synthesized to contain specific groups that aid in conjugation to waveguide surfaces. Typical groups include amines, thiols, and carboxylic acids which are reactive with succinimidyl esters (NHS), maleimides, and carbodiimides (such as EDC), respectively (Hermanson, 1996). Homo- and hetero-bifunctional crosslinkers are commercially available to bind the molecular recognition element covalently to a suitably modified waveguide surface. These covalent attachment chemistries are highly appropriate for attachment of biomolecules to simple bulk waveguides such as glass (Schmid *et al.*, 1998; Schuderer *et al.*, 2000; Ligler *et al.*, 2007), silica (Herron *et al.*, 1993, 1996), polymer surfaces (Silzel *et al.*, 1998; Schult *et al.*, 1999), and the slightly more complicated IOW waveguides (Duveneck *et al.*, 1997; Asanov *et al.*, 1998; Pawlak *et al.*, 1998; Plowman *et al.*, 1999). Additional methods to modify these waveguides for immobilization of biomolecules have been described (Bhatia *et al.*, 1989; Brizzolara *et al.*, 1994; Lee and Saavedra, 1996; Sojka *et al.*, 1999; Pirrung *et al.*, 2000).

Avidin–biotin interaction is extensively used in the immobilization of biotinylated molecular recognition elements. Typically, avidin, streptavidin, or the deglycosylated derivative NeutrAvidin is covalently attached or adsorbed to the waveguide and a biotinylated recognition ("capture") molecule added in a subsequent step. The presence of four biotin–biotin binding sites on each avidin molecule minimizes the potential for complete inactivation of the avidin upon immobilization. Careful biotinylation of the capture molecule and attachment through immobilized avidin does permit careful control of the number of sites though which the biomolecule is attached. While this helps maintain the mobility of a receptor molecule and minimizes steric hindrance, it does not control the location of the attachment site on the molecule, unless site-specific modifications can be made.

Many of the covalent methods mentioned above result in a somewhat random orientation of the recognition element on the surface, resulting in the potential blocking of analyte-binding sites, which can affect the biosensors' reproducibility and sensitivity. This is particularly true for proteins, such as antibodies, where immobilization protocols often take advantage of the many surface amines present on their surface. To address these concerns, an increasing number of researchers working with antibodies have investigated alternative immobilization protocols which impart control over the antibody surface orientation. Several methods involve cleavage of the full antibody into F_{ab} fragments which display thiol groups to aid in site-directed immobilization (Lu *et al.*, 1995; Huang *et al.*, 1996; Vikholm *et al.*, 1998; Weiping *et al.*, 1999). Techniques which target the F_c portion of a full antibody are also popular and include using protein A or G (Vijayendran *et al.*, 2001) or functionalization of the carbohydrate moiety (Weiping *et al.*, 1999;

Vijayendran *et al.*, 2001). The methods for preparing antibodies to be immobilized through either the carbohydrate or the thiol groups result in a significant loss of antibody during the processing procedure but generally produce highly functional surfaces.

An additional method of controlling the attachment site in a biomolecule is site-directed mutagenesis (McLean *et al.*, 1993; Vigmond *et al.*, 1994). Vogel's group successfully immobilized fully functional GFP and serotonin receptors onto planar waveguides in defined orientations (Schmid *et al.*, 1997, 1998). GFP and serotonin were genetically modified to contain poly-histidine tags at the C- and N-terminus, respectively. These tags bound specifically to waveguides coated with thiosilane and nitrilotriacetic acid (NTA)-bound metal ions (Schmid *et al.*, 1998). Kwon *et al.* (2004) developed an unusual method involving protein fusions between the antibody and the enzyme cutinase. The antibody–cutinase fusion was then immobilized using the covalent interaction between cutinase and a suicide substrate.

Attachment of DNA onto gold and glass surfaces has been reviewed by Beaucage (2001). Strategies are similar to protein immobilization and typically involve silanization of the waveguide followed by activation with a heterobifunctional linker to attach either amine- or thiol-terminated nucleic acids (Budach *et al.*, 1999; Charles *et al.*, 2003). Schuderer *et al.* (2000) opted to immobilize biotinylated 18-mer oligonucleotides onto glass surfaces functionalized with adsorbed avidin. Similar attachment methods can readily be applied to the immobilization of purified carbohydrates or peptides to waveguide surfaces. However, surface density and orientation play for a critical role in target recognition by these capture species and must be carefully controlled (Ngundi *et al.*, 2006b, 2006d; Kulagina *et al.*, 2005, 2006).

One major problem with studying receptor–ligand binding has been the immobilization of the membrane-associated receptors and immobilization onto a sensor surface should mimic this environment to ensure that the receptor remains active (Pawlak *et al.*, 1998). Pawlak *et al.* (1998) created multilayers on top of the planar waveguide using a long-chain alkyl phosphate monolayer to make the metal oxide surface hydrophobic. Lipid vesicles were then adsorbed to the monolayer to form a lipid film. Membrane fragments containing the ATPase were finally adsorbed onto the lipid bilayer for use in a fluorescence-quenching assay. Rowe-Taitt *et al.* (2000a) immobilized gangliosides as receptors for toxins by immobilizing a long-chain alkyl silane to create a hydrophobic surface for intercalation of the fatty acid tail of the ganglioside. The immobilized receptor was quite stable to rinsing, dehydration, and rehydration. Both direct binding and sandwich assays using different gangliosides patterned on the waveguide were demonstrated. Groups at Cornell, Sandia National Laboratories, and Los Alamos National Laboratory have utilized supported lipid bilayers (Kelly *et al.*, 1999; Moran-Mirabel *et al.*, 2005) for immobilization of gangliosides for toxin detection.

3.2. History

A number of the researchers currently involved in developing planar waveguide TIRF carried out much of their initial research in the field of fiber optics, including: Herron, Christensen, and Reichert (USA); Duveneck, Ehrat, and Neuschafer (Switzerland); Ligler (USA) and Bilitewski (Germany) and their coworkers. Much of the history of planar waveguide fluorescence-based biosensors has involved development of both the technical components, including the optical design of the instrument used for surface analysis, and the biomolecule immobilization onto waveguide surfaces, both discussed in a previous chapter. The evanescent wave excitation of a surface-bound fluorophore has been studied for a number of years using fiber optic technology.

There are a number of advantages to using a planar waveguide instead of optical fibers, including the relative ease of preparation and integration into fluidic systems. Although early researchers in the field immobilized capture biomolecules uniformly over the planar surface, the most important advantage of using a planar waveguide stems from the possibility of creating patterns of immobilized biomolecules leading to multiple, parallel assays on a single waveguide (Herron *et al.*, 1993; Duveneck *et al.*, 1997; Brecht *et al.*, 1998; Pawlak *et al.*, 1998; Schult *et al.*, 1999).

A number of techniques have been used in the creation of patterned biomolecular assemblies on planar surfaces as reviewed by Blawas and Reichert (1998a). The patterns can be created using photolithography or by depositing the recognition molecules in physically separate locations on the waveguide using techniques such as ink-jet printing, contact and non-contact arrayers, or microfluidic flow cells.

Methods for the photolithographic patterning of proteins on surfaces have been described by a number of researchers (Bhatia *et al.*, 1992, 1993; Liu *et al.*, 2000; Sapsford *et al.*, 2001; Ito *et al.*, 2006). In one of the first examples of protein patterning, ultraviolet light was used to pattern (3-mercaptopropyl) trimethoxysilane on a glass surface (Bhatia *et al.*, 1992, 1993). Exposed regions of the surface became protein-resistant through the conversion of the thiol group to a sulfonate species, while the masked areas went on to bind the biomolecule. This proved to be a convenient method of creating high resolution patterns (less than 10 microns in width) of immobilized capture biomolecules. Unfortunately, this method was effective for patterning of only a single biomolecule.

Subsequently, other investigators developed methods for photopatterning of the surface to immobilize multiple capture molecules. Schwarz *et al.* (1998) photoablated polymer substrates to produce avidin patterns on the exposed regions. Likewise, Wadkins *et al.* (1997) used glass slides coated with a photoactivated optical adhesive and a mask to create wells in the gel layer upon light exposure. A different biomolecule could then be covalently attached to the exposed glass within each well. A similar method was adopted by Mirzabekov's group (Guschin *et al.*, 1997; Arenkov *et al.*, 2000) for the immobilization of oligonucleotides; however, the biomolecules were immobilized within gel pads rather than on the glass; crosslinking of the gel pads was accomplished in the presence of the biomolecules, allowing the recognition element to be fully incorporated into the hydrogel networks. Blawas *et al.* (1998b) used the caged-biotin-bovine serum albumin (BSA) compound, methyl α-nitropiperonyloxy-carbonyl-biotin-BSA, to pattern glass surfaces. First, caged-biotin-BSA was immobilized on the glass slide. Second, the slide was exposed to 353 nm light through a mask, which effectively removed the cage surrounding the biotin in unmasked areas. Third, streptavidin selectively bound to the irradiated regions of the surface. Finally, biotinylated capture antibody was bound to the streptavidin. The disadvantage of this method was the potential for cross-contamination during subsequent cycles of the process. Two additional methods of photolithographically patterning proteins on planar waveguides used polyethylene glycol (PEG) moieties to prevent nonspecific protein adsorption. In the method developed by Conrad *et al.* (1997, 1998), a photochemically active silane, *o*-nitrobenzyl polyethylene glycol trichlorosilane, was attached to a glass waveguide. Photo-oxidation of the PEG-terminated silane through a mask cleaved the carbamate to yield an amino-terminal PEG and a surface-bound *o*-nitrosobenzaldehyde, which could react with the antibody via a Schiff's base reaction. The process was repeated for the addition of different antibodies to additional spots on the waveguide. Similarly, Liu *et al.* (2000) tethered a benzophenone photophore through a PEG spacer to a maleimide group. After exposure to light, antibodies were immobilized and an assay conducted on the polystyrene waveguides. Spacer lengths of five ethylene glycol groups proved optimal for maximizing the signal-to-background ratio.

The use of ink-jet printing is another popular choice for the production of patterned biomolecular surfaces. Silzel *et al.* (1998) ink-jet printed either the capture antibodies or the protein avidin in 200 μm diameter zones on the surface of polystyrene films. Biotinylated antibodies were later immobilized on the avidin spots. A checkerboard pattern of two different oligonucleotides was produced by Budach *et al.* (1999) using the ink-jet printing of capture biomolecules onto a Ta_2O_5 waveguide using (3-glycidoxypropyl) trimethoxysilane. In a similar vein, microcontact printing using elastomeric stamps has also been demonstrated as a means of patterning planar surfaces, producing high resolution sub-micron-sized arrays (Renault *et al.*, 2003).

Recently, contact and non-contact arrayers have become increasingly popular, especially for the production of high density arrays. The users of such arrayers can control the position of spots of capture molecules with dimensions on the order of 50–500 microns. Delehanty and Ligler (2002, 2003) patterned biotin-labeled antibodies onto avidin-functionalized glass slides using a non-contact arrayer and determined the optimal printing conditions for producing bright uniform spots on the waveguide surface. Wu *et al.* (2006) opted for maleimide-activated silicon nitride waveguides for contact printing of thiol-terminated oligonucleotides, while Ito (2006) described the use of photoimmobilization. Wacker *et al.* (2004) used DNA hybridization to direct antibody immobilization at specific locations on a waveguide surface. A non-contact arrayer spotted a number of different short single chain oligonucleotides in different locations on a planar waveguide. Antibodies labeled with specific complementary oligonucleotides were then exposed to the surface and hybridized only to the regions of the slide functionalized with the appropriate capture.

Physically isolated patterning has also been performed using flow cells constructed from a variety of materials, including polydimethylsiloxane (PDMS) (Feldstein *et al.*, 1999; Golden *et al.*, 2005a), a rubber gasket (Plowman *et al.*, 1999), a Teflon block fitted with O-rings (Schuderer *et al.*, 2000), and a microfluidics network made of silicon (Bernard *et al.*, 2001; Delamarche *et al.*, 2005). Typically the flow cell, containing a number of channels, is temporarily attached to the surface of the planar waveguide and each channel filled with a solution of the capture biomolecule, as shown in Figure 3.4a. In this example (Feldstein *et al.*,

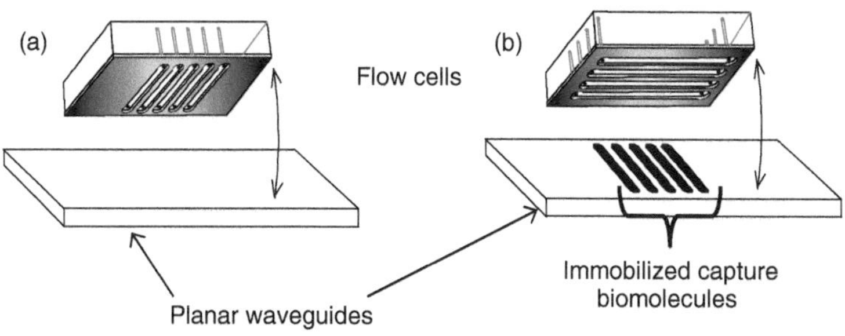

Figure 3.4 The patterning of capture biomolecules using flow cells (adapted from Feldstein *et al.*, 1999). (a) A multi-channel flow cell is pressed onto the planar waveguide and each channel filled with a solution of the capture biomolecule. (b) Sample and fluorescent tracer antibody are passed over the waveguide surface perpendicular to the immobilized capture biomolecule channels using a second flow cell.

1999; Golden *et al.*, 2005a), the resulting waveguide was patterned with stripes of immobilized biomolecules. The sample and fluorescently labeled antibody were then passed over the surface using a second flow cell orientated perpendicular to the immobilized capture biomolecule channels, as shown in Figure 3.4b. The beauty of the physically isolated patterning technique is its ability to immobilize a number of different capture biomolecules onto a single surface with no possibility of cross-contamination. A potential disadvantage of the approach is the difficulty in scaling up for manufacture of large numbers of waveguides; this potential problem may be offset by the ease with which it can be performed by investigators with no special photolithographic or arraying equipment (Golden *et al.*, 2005a).

Processes that may reduce the sensitivity of TIRF measurements include non-specific binding of the fluorescent species to non-sensing regions of the surface, increasing the background signal. Currently, researchers use protein blockers, such as BSA or casein, often coupled with a surfactant, such as polyoxyethylene-sorbitan monolaurate (Tween), to block the surface and minimize non-specific adsorption after immobilization of the capture biomolecule. Addition by 50% cranberry juice has been found to be effective at preventing fouling of bacteria on planar waveguides (Johnson-White *et al.*, 2006). Another possibility is the introduction of monolayers consisting of oligoethylene oxides within the non-sensing regions of the waveguide; these layers are highly protein resistant in SPR (Silin *et al.*, 1997; Lahiri *et al.*, 1999) and TIRF studies (Conrad *et al.*, 1997). This approach, however, may not prevent nonspecific adsorption to the sensing regions of the planar waveguide (Conrad *et al.*, 1997).

3.3. State of the art

As described in the previous sections, much of the initial investigation into the use of planar waveguides in biosensors has centered on instrumentation development and reproducible immobilization and patterning of the capture biomolecules. The driving force behind the development of biosensors is the possibility of quick, cost-effective, user-friendly, field analytical technologies that have sensitivity and specificity comparable to the laboratory measurements. Due to the vast number of biological systems that respond to a variety of analytes, the number of potential applications for biosensors in general and planar waveguide biosensors in particular is huge. Applications include medical diagnostics and healthcare, environmental monitoring, process monitoring in the chemical, food, and beverage industries, and military defense (Kumble *et al.*, 2003; Shriver-Lake *et al.*, 2004; Lim *et al.*, 2005; Taitt *et al.*, 2005; Weller, 2005; Ince *et al.*, 2006; Marquette and Blum, 2006; Rodriguez-Mozaz *et al.*, 2006; Rogers, 2006; Soper *et al.*, 2006).

Microscopy-based technologies, such as confocal scanners, are increasingly used for multi-analyte detection on planar surfaces and, while still very much a laboratory technology, are gaining popularity in clinical applications. Antibody multi-capture arrays are generally produced using arrayer technology with the assay results imaged either using a commercial slide scanner or epi-fluorescence microscopy (Li *et al.*, 2003, 2005; Wolf *et al.*, 2004; Rubina *et al.*, 2005; Rucker *et al.*, 2005). A number of microscopy studies have demonstrated that TIRF-based excitation of the waveguide surface, which is intrinsically less prone to background fluorescence, results in more sensitive detection compared to epi-illumination (Duveneck *et al.*, 2002; Lehr *et al.*, 2003a). TIRF-based microscopy (TIRFM) has been used by a number of researchers to investigate live cell functions and interactions in real time (Vigeant *et al.*, 2002; Mashanov *et al.*, 2003; Jones *et al.*, 2004; Burns *et al.*, 2006; Kellermayer *et al.*, 2006; Webb *et al.*, 2006). TIRFM has been applied to the monitoring of DNA hybridization, in solution, within the microfluidic channels (Hollars *et al.*, 2006), or using molecular beacons immobilized on waveguide surfaces (Yao *et al.*, 2003). The binding of cholera and tetanus toxins to gangliosides immobilized within lipid bilayers has also been investigated (Moran-Mirabal *et al.*, 2005). Diphtheria and tetanus toxins, non-specificially deposited onto planar waveguides, were distinguished using TIRFM using antibodies labeled with different colored quantum dots (QDs) (Hoshino *et al.*, 2005).

An important niche where TIRF biosensors have found use is the study of real-time binding kinetics. The Array Biosensor developed at the US

Naval Research Laboratory (NRL) has demonstrated the capability of performing real-time binding analyses determining kinetic and affinity constants for both specific (Sapsford *et al.*, 2001), semi-selective (Ngundi *et al.*, 2006b), and non-specific (Sapsford *et al.*, 2004a) binding interactions. Real-time TIRF studies have also been used to study the hybridization and melting temperature conditions for DNA interactions (Tolley *et al.*, 2003; Lehr *et al.*, 2003b; Yao *et al.*, 2003).

Many TIRF-based biosensor studies have investigated the use of a single capture biomolecule-analyte assay, as a means to optimize individual assays prior to creating a multiplexed system. Such systems include DNA hybridization assays (Duveneck *et al.*, 1997; Schuderer *et al.*, 2000) and antibody–antigen studies for the human pregnancy hormone human chorionic gonadotropin (Herron *et al.*, 1993; Schult *et al.*, 1999), the pesticide pollutant atrazine (Brecht *et al.*, 1998), the asthma drug theophylline (Schult *et al.*, 1999), various anthrax markers (Martinez *et al.*, 2005), and many of the targets listed in Table 3.1. The majority of these systems were found to have comparable or slightly better detection limits than the conventional methods used to measure the analyte, such as enzyme-linked immunosorbent assay (ELISA). A number of groups have incorporated these individual tests into multi-analyte assays and demonstrated detection of multiple targets on a single substrate.

Silzel *et al.* (1998) demonstrated the feasibility of multi-analyte detection with their TIRF biosensor by measuring four different human IgG subclasses. Zeller *et al.* (2000) developed a unique TIRF system in which the planar waveguide consisted of multiple, single pad, sensing units. Each of these single pads had its own laser light input, background suppression, and coupling of the fluorescence emission to the detector. The authors demonstrated a two-pad sensing device in which one pad was modified with mouse IgG and the other with rabbit IgG. However, only one Cy5-labeled antibody directed against each immobilized antigen was assayed at a time. It was suggested that the laser light could be split into spatially different parts in multi-analyte measurements, using multiple single sensing pads. Such a process would probably involve a number of optical components, and therefore the robustness of the system for use outside the laboratory would be questionable.

Table 3.1 Immunoassay-based limits of detection using NRL Array Biosensor (from Ligler *et al.*, 2007)

Target	**Limit of Detection (LOD)**[a]	**References**
Small Molecules:		
TNT	1–20 ng/ml[b,c]	Sapsford *et al.*, 2002
Deoxynivalenol	0.2 ng/ml[b] (1–10 ng/g)[d]	Ngundi *et al.*, 2006a, 2006c
Ochratoxin A	0.8 ng/ml[b] (3.8–100 ng/g)[d]	Ngundi *et al.*, 2005, 2006a
Aflatoxin B_1	0.3 ng/ml[b] (0.6–5.1 ng/g)[d]	Sapsford *et al.*, 2006e
Protein Targets:		
Toxins:		
SEB	0.1 ng/ml[b] (0.1–0.5 ng/ml)[d] (1 ng/ml)[e] (100 ng/ml)[f]	Sapsford *et al.*, 2005; Shriver-Lake *et al.*, 2003; Rowe *et al.*, 1999a, Rowe-Taitt *et al.*, 2000c
Cholera Toxin	1.6 ng/ml[b] (100 ng/ml)[f]	Rowe-Taitt *et al.*, 2000b, 2000c
Botulinum Toxoid A	20 ng/ml[b] (20–500 ng/ml)[d]	Sapsford *et al.*, 2005
Botulinum Toxoid B	200 ng/ml[b]	Rowe-Taitt *et al.*, 2000b
Ricin	8 ng/ml[b]	Rowe-Taitt *et al.*, 2000b
Allergens:		
Ovalbumin	25 pg/ml[b] (1.3 ng/ml)[d]	Shriver-Lake *et al.*, 2004
Physiological Markers:		
Yersinia pestis F1	25 ng/ml[b,e]	Rowe *et al.*, 1999a
D-dimer	50 ng/ml[b,e]	Rowe *et al.*, 1999a
Antibodies	1:200–1:400 titres	Moreno-Bondi *et al.*, 2006

(Continued)

Table 3.1 Continued

Target	Limit of Detection (LOD)[a]	References
Bacterial Targets:		
B. abortus	3×10^3 cfu/ml[b] (5×10^5 cfu/ml)[f]	Rowe-Taitt *et al.*, 2000b, 2000c
Francisella tularensis	10^5 cfu/ml[b] (7×10^6 cfu/ml)[f]	Rowe-Taitt *et al.*, 2000b, 2000c
S. typhimurium	8×10^4 cfu/ml[b] ($8 \times 10^4 - 4 \times 10^5$ cfu/ml)[d]	Taitt *et al.*, 2004
Shigella dysenteriae	5×10^4 cfu/ml[b] ($5 \times 10^4 - 8 \times 10^5$ cfu/ml)[d]	Sapsford *et al.*, 2004b
Campylobacter jejuni	10^3 cfu/ml[b] (2×10^3–3×10^3 cfu/ml)[d]	Sapsford *et al.*, 2004b
L. monocytogenes	10^4 cfu/ml[b]	Taitt *et al.*, 2003
Bacillus globigii	10^5 cfu/ml[b]	Rowe *et al.*, 1999b
B. anthracis	624 cfu/ml[b] (7×10^4 cfu/ml)[f]	Rowe-Taitt *et al.*, 2000b, 2000c
E. coli O157:H7	5×10^3 cfu/ml[b] ($1 - 5 \times 10^4$ cfu/ml)[d]	Shriver-Lake *et al.*, 2007
Staphylococcus aureus	$\sim 10^6$ cfu/ml[b]	unpublished
Viral Targets:		
MS2 Phage	10^7 pfu/ml[b]	Rowe *et al.*, 1999b
Vaccinia	10^7–10^8 pfu/ml[b]	unpublished

a. cfu/ml (colony forming units/ml); pfu/ml (plaque forming units/ml).
b. Buffer LOD.
c. Buffer LOD depends on immunoassay format.
d. LOD range in food matrices.
e. LOD range in clinical matrices.
f. Fixed concentration measured in environmental matrices (not LOD).

The performance of single and multiple analyte assays was compared by Plowman *et al.* (1999) using IgGs of different species and antibodies specific for cardiac proteins (creatin kinase MB, cardiac troponin I, and myoglobin). Studies also investigated the effects of using polyclonal versus the more specific monoclonal antibodies during assays. Results suggest that polyclonal antibodies are more prone to cross-reactivity in the multi-analyte assay format; therefore, monoclonal antibodies were the capture biomolecule of choice when available. Although some sensitivity was lost in the multiplexed assay, detection limits were within clinically significant ranges. Interestingly, however, Rowe and colleagues (Rowe *et al.*, 1999b; Taitt *et al.*, 2002) demonstrated the use of mixtures of polyclonal antibodies of different species with no cross-species interactions observed; furthermore, the same sensitivity was achieved with mixtures of the fluorescent tracer antibodies as with single antibody assays. The degree of cross-reactivity is most likely to be a function of the antibodies selected.

Gauglitz and coworkers have developed a TIRF optical immunosensor and an associated computer supported system – the River Analyzer (RIANA) and Automated Water Analyser Computer Supported System (AWACSS), respectively – for screening river water for the presence of small organic pollutants including atrazine, isoproturon, estrone, and propanul (Tschmelak *et al.*, 2004, 2005). The TIRF-based biosensor consists of multiple sensing pads coupled with a microfluidic system for automatic sample handling and simultaneous multi-analyte detection. Analyte derivatives are immobilized onto the sensing pads and competitive immunoassays performed to measure the presence of the target analytes in surface water, ground water, industrial waste, and sediment samples. The AWACSS system has been field deployed with remote control and surveillance (Tschmelak *et al.*, 2005). Ehrat, Duvenek, and colleagues in Switzerland have developed the TIRF-based Zeptosens reader with waveguide platforms and flow cells for performing simultaneous multi-analyte detection (Duvenek *et al.*, 2002); it is now marketed by their company Zeptosens (www.zeptosens.com). To date, this latter technology has mostly been applied to nucleic acid detection in a high density array format.

Ligler and coworkers at NRL have produced the greatest number of publications in the field of planar waveguide TIRF in the process of developing the NRL Array Biosensor. The first published papers on TIRF studies in 1998 demonstrated the detection of three analytes: ricin, *Y. pestis* F1 and staphylococcal enterotoxin B (SEB) (Wadkins *et al.*, 1998). The long-term aim for the NRL Array Biosensor has always been the development of a fully automated, portable, low cost instrument as a tool for rapid, onsite screening of biohazards wherever they might be found. While the underlying principles of the NRL Array Biosensor have remained, the design of the biosensing system has continually evolved to take advantage of new and improved technologies (Feldstein *et al.*, 1999, 2000; Rowe-Taitt *et al.*, 2000b; Golden *et al.*, 2005b; Ligler *et al.*, 2007). For example, automation of the biosensor included investigating methods of attaching the flow cell to the waveguide surface, both permanently (Ligler *et al.*, 1998) and reversibly (Leatzow *et al.*, 2002; Johnson-White and Golden, 2005). Also various fluidic designs for storage and introduction of the assay sample and tracer components, including a fluidics cube (Dodson *et al.*, 2001), have been designed and optimized (Golden *et al.*, 2005b).

The current laboratory prototype of the NRL Array Biosensor is shown in Figure 3.5A. All components are mounted within a portable carrying case and the entire system weighs approximately 15 lb. Excitation of the waveguide surface is achieved using a 635 nm diode laser equipped with a line generator, which expands the laser beam and launches it into the edge of the microscope slide waveguide, providing even fluorescence excitation across the viewing area. The cooled CCD camera images the resulting fluorescent array from the microscope slide through 700 ± 35 nm bandpass and 665 nm longpass filters.

The array sensor uses two removable six-chamber reservoir modules (Figure 3.5B), each fitted with a rubber septum on the bottom. The sensor platform is equipped with blunt needles which penetrate pre-punched holes in the rubber septa connecting the reservoirs to the rest of the fluidics system. A vertical slide mount equipped with a spring-loaded plate presses the pre-patterned microscope slide against a six-channel gasket molded in PDMS forming the six-assay flow channels. Each of the six channels is connected at one end to a pump and at the opposite end to a two-way valve; these valves switch between the two sources of fluid (sample reservoirs and tracer reservoirs) in the reservoir modules. The fluidics system was designed with the slide and flow channels in an

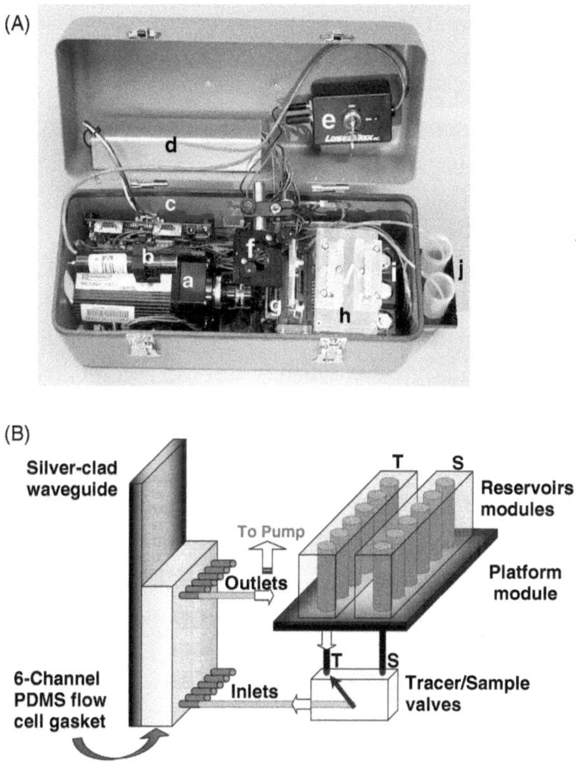

Figure 3.5 The NRL Array Biosensor. (A) Biosensor components: (a) CCD camera (Retiga 1300, Q-Imaging), (b) a 635 nm diode laser (LAS-635-15, Lasermax), (c) RS232 interface board, (d) a 50 W power supply (Sunpower), (e) Laser power switch, (f) mirror, (g) vertical slide mount, (h) removable reservoir modules, (i) the peristaltic pumps (P625/66.143, Instec), and (j) waste and buffer reservoirs. (B) Schematic of the fluidics components showing the valves (Lee Company) located under the reservoir modules. Adapted from Golden *et al.* (2005b).

upright position, separate from the optical and electronic components, reducing the potential for damaging fluid leaks.

The automated NRL Array Biosensor is controlled via a laptop computer using a program written with LabWindows. The software provides the user with a simple interface in which to control the assay conditions, such as the timing and flow of the peristaltic pumps and valves and the CCD measurement. A LabWindows-based image analysis program has also been written to analyze the CCD image of the waveguide following completion of the automated assay.

Because of selectivity and sensitivity, immunoassays have been the first choice for developing rapid identification methods using the fluorescence-based NRL Array Biosensor. Emphasis has been placed on optimization of the instrument design, including automation, and development of the multi-analyte capabilities and as such much of the original assay development was carried out in buffer. Figure 3.6 shows the final CCD image, taken using the NRL Array Biosensor, of a waveguide exposed to simultaneous multi-analyte immunoassays for botulinum toxoid A, SEB, and *C. jejuni* in buffer. Additional assays have been developed using alternative recognition molecules for detection of toxins and biothreats: gangliosides (Rowe-Taitt *et al.*, 2000a), purified mono- and di-saccharides (Ngundi *et al.*, 2006b, 2006d), and AMPs (Kulagina *et al.*, 2005, 2006, 2007). Use of alternative recognition species,

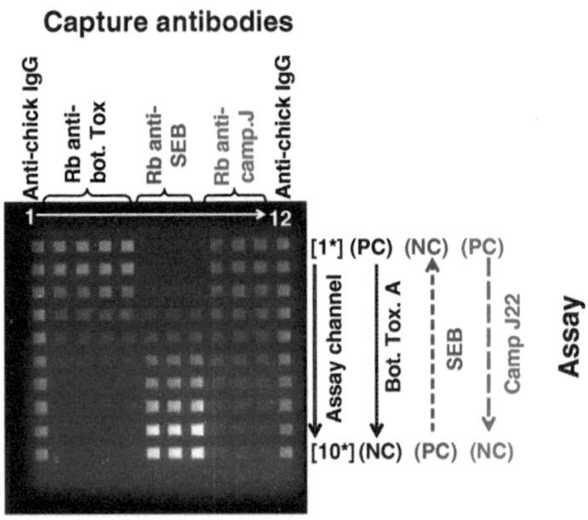

Figure 3.6 The NRL Array Biosensor used for the simultaneous multi-analyte quantification of botulinum toxoid A, SEB, and *C. jejuni* in buffer. The positive (PC) and negative (NC) control assay lanes are important to include when real-world samples are run to ensure the waveguide is performing as expected.

as well as modified immobilization procedures, further emphasizes the versatility of this platform.

Many recent studies have been more application-driven with an essential requirement to evaluate matrix effects on analyte detection. Such studies are essential not only for system development but also for the instrumentation to make the transition into a commercial product. Table 3.1 provides detailed information on the immunoassays developed for various target analytes and includes the limits of detection (LOD) both in buffer and various spiked matrices. The LOD was calculated as the lowest concentration actually tested that produced a signal three standard deviations above the blank.

Simultaneous detection of analytes in clinical fluids has clearly been demonstrated with *Y. pestis* F1, SEB, and D-dimer, all with detection limits suitable for clinical analysis requirements (Rowe *et al.*, 1999a). Multiplexed assays have also been developed to screen human sera for antibodies directed against bacterial and viral antigens, including SEB, tetanus toxin, diphtheria toxin, and hepatitis B (Moreno-Bondi *et al.*, 2006). Such assays could potentially be used in a clinical setting to monitor individuals for exposure to various pathogens or to study the efficacy of vaccination. Efficacy of specific bacterial detection in fecal samples has also been demonstrated (Taitt *et al.*, 2004). The impact of potential environmental interferents has also been addressed (Rowe-Taitt *et al.*, 2000c). Analyte samples of *B. anthracis*, *F. tularensis* LVS, *B. abortus*, SEB, cholera toxin, and ricin were assayed in the presence and absence of interferents such as sand, clay, pollen, and smoke extracts, and results were compared. No false-positive or false-negative responses were caused by the environmental samples; however, in some cases the signal amplitude was affected.

Much of the recent work in assay development has concentrated on the rapid detection of foodborne contaminants. The NRL Array Biosensor has been used for the measurement of protein toxins (SEB and botulinum toxoid A – Shriver-Lake *et al.*, 2003; Sapsford *et al.*, 2005) and bacteria (*Campylobacter* spp., *Shigella* spp., *E. coli*, and

S. typhimurium – Sapsford *et al.*, 2004b; Taitt *et al.*, 2004; Shriver-Lake *et al.*, 2007) in a variety of spiked food matrices. Analyte detection is rapid with immunoassays taking 10–45 min, is simple to perform, and requires minimal sample processing. Sandwich immunoassays work extremely well for the larger protein and bacterial contaminants; however, small analytes such as mycotoxins are not large enough to contain separate recognition sites for both the capture and the tracer antibodies. For this reason, competitive immunoassays have been successfully developed for the detection of the mycotoxins ochratoxin A (Ngundi *et al.*, 2005, 2006a), aflatoxin B_1 (Sapsford *et al.*, 2006e), and deoxynivalenol (Ngundi *et al.*, 2006a, 2006c) in spiked and naturally contaminated foods. The feasibility of running both sandwich and competitive formats on the same waveguide surface was recently demonstrated, allowing the user to measure large and small molecular weight target analytes simultaneously, as well as multiplexing the competitive immunoassays (Ngundi *et al.*, 2006a; Sapsford *et al.*, 2006d).

Johnson-White *et al.* (2007) have recently demonstrated the ability to combine the rapid screening capability of the NRL Array Biosensor with methods for determining cell viability. The waveguide format provides species identification as well as isolated, concentrated antigen samples free of matrix contaminants for further analysis. After fluorescence imaging, the cells captured on the waveguide could be cultured to confirm viability. Genetic analysis of the captured cells using the real-time polymerase chain reaction gave three orders of magnitude improvement in sensitivity when compared to the initial fluorescent immunoassay, while also providing confirmatory identification of the captured microbes.

Three companies have licensed the technology relevant to the NRL Array Biosensor (Ligler *et al.*, 2007). Hanson Technologies (www.hansontechnologies.com) offers the Hanson Leopard™ array biosensor which currently uses immunoassays for pathogen and toxin detection. Constellation Technologies Corp. (www.contech.com) is currently developing the Array BioSensor (ABS) system for biological agent detection. Precision Photonics Corp. (www.precisionphotonics.com)

is adapting the NRL Array Biosensor for the detection of disease markers in clinical samples, primarily for use in third-world countries.

3.4. Advantages and limitations for use in optical biosensing

As with all optical techniques, planar waveguide biosensors based on fluorescent signal generation have associated advantages and limitations (for reviews, see Schobel *et al.*, 2000; Bally *et al.*, 2006). The introduction of a fluorescent probe to a biomolecule has the advantage of allowing both site and spectral selection. Fluorescent labels have longer shelf lives, lower costs, and greater safety than radiolabels. The use of a fluorescent probe is also favored over enzyme labels typically used in ELISA, for stability reasons and because additional substrates are not required.

An inherent advantage of using evanescent wave fluorescence is its surface-specific nature. Only fluorophores within the field of the evanescent wave are excited to give fluorescence emission. This renders the technique relatively immune to matrix effects such as fluorophores in the bulk, particulates, and turbidity, so binding of the fluorophore can be measured in real time and with high sensitivity.

The planar waveguide provides a single surface for the patterning of multiple capture biomolecules and hence provides for the possibility of simultaneous assays for multiple analytes. The number of analytes that can be identified simultaneously is limited only by the reproducibility of the method for making the patterns, the area of the waveguide which can be evenly illuminated and imaged, and the affinity and specificity of available recognition molecules.

While fluorescence-based detection provides sensitivity in terms of signal-to-background discrimination, the requirement for a fluorescent assay component may also be a disadvantage. Labeling with an extrinsic fluorescent dye not only adds extra steps, but it also must be a carefully controlled process, such that the label does not interfere with or influence the binding interactions of the biomolecule. Label-free techniques such as SPR are commonly used in the study of binding events, and the fairly recent development of SPR imaging has greatly advanced the technology (for a review, see Bally *et al.*, 2006). However, sensitivities are still typically lower than corresponding fluorescence assays. Furthermore, SPR measurements are usually performed in buffer rather than clinical or environmental fluids due to signal generation by non-specifically bound sample components.

Planar waveguide TIRF is a sensitive technique with LOD typically between 0.1 and 200 ng/ml for proteins (Duveneck *et al.*, 1997; Plowman *et al.*, 1999; Ligler *et al.*, 2007) and 10^3–10^5 cfu/ml for bacterial targets. Depending on the target, these LODs are often found to be comparable with standard measurement techniques, such as ELISA, and yet the TIRF immunoassay can be carried out in a much shorter time, e.g., 10–45 min (TIRF) versus 2 h (ELISA). However, in some cases these LODs still fall short of LODs desired by many applications. This is particularly true for the bacterial targets where detection of <1 cfu/ml is often required. Under the flow cell conditions used by the NRL Array Biosensor, laminar flow in the microfluidic channels prevents the entire sample from being efficiently interrogated. In most cases, only a small proportion of the sample is actually exposed to the capture molecules immobilized on the sensing surface which may negatively affect the sensitivity of the resulting assay.

3.5. Potential for improving performance or expanding current capabilities

While TIRF-based biosensors offer many benefits for analyte detection, current limitations, mainly regarding sensitivity, are likely to benefit from technologies emerging in the literature. Such technologies include improved optics, system integration, brighter fluorescent labels, and tailored recognition molecules.

Optical components are becoming increasingly sophisticated in terms of miniaturization and capability. MacCraith's group developed waveguides with a geometry that improves collection of the emitted fluorescence as well as providing excitation light (Blue *et al.*, 2005; Holthoff *et al.*, 2005). Neuschafer *et al.* (2003) described an evanescent resonator that confines excitation energy in a thin surface layer and increased fluorescence of bound biomolecules 100-fold. Other IOWs also increased the sensitivity of TIRF-based systems compared to IRE arrangements (Brecht *et al.*, 1998; Lehr *et al.*, 2003a). Duvenek *et al.* (2001) reported the induction of two-photon fluorescence on planar waveguides, potentially a powerful tool for distinguishing very small signals from background. Such waveguides are very promising for clinical and drug screening applications (Duveneck *et al.*, 2002; Martinez *et al.*, 2005) where the instrument can remain in a laboratory. However, as pointed out by Feldstein *et al.* (1999), the alignment constraints for IOW waveguides must be reduced in order to produce a robust system for field deployable applications (Brecht *et al.*, 1998).

Techniques developed by both the optics and microfluidics communities should facilitate automation and system integration. Sol-gels and embossed gratings are being used with planar waveguides to provide multi-wavelength capability (Bradshaw *et al.*, 2002; Ruano *et al.*, 2003). Organic light-emitting devices (OLEDs) are being integrated into biosensors (Choudhury *et al.*, 2004; Burke *et al.*, 2005; Shinar, *et al.*, 2006); at some point OLEDs and organic detectors may both be integrated with planar biosensors. Several approaches to overcoming laminar flow limitations on target delivery to the sensing surface have been reported. In one approach, laminar streams can be used to focus the sample flow near the sensing region (Kamholz and Yager, 2001; Hofmann *et al.*, 2002; Munson *et al.*, 2005). Immiscible fluid vesicles have also been used to generate nanoscopic fluid films of ~200 nm on the sensor surface, improving mass transport limitations (Jennissen and Zumbrink, 2004). In an alternative approach, passive mixing components in the fluidic channels may improve sensitivity by introducing target into the depletion layer at the surface (Vijayendran *et al.*, 2003; Delamarche *et al.*, 2005; Floyd-Smith *et al.*, 2006; Mott *et al.*, 2006; Golden *et al.*, 2007). Microfluidic technology also offers the possibility of introducing upstream sample pretreatment and preconcentration steps prior to exposure to the sensing surface (de Mello and Beard, 2003).

Metal-clad waveguides offer unique refractive index (Skivesen *et al.*, 2007) and fluorescent-based biosensing platforms. Metal-enhanced fluorescence methods to improve sensitivity are a recent development for TIRF-based detection systems. The technology takes advantage of the observed fluorescence enhancement of dyes that occurs in the proximity of silver island films (Matveeva *et al.*, 2004; Stranik *et al.*, 2005; Zhang *et al.*, 2005; Fu *et al.*, 2006) and thin metal titanium films (Zourob *et al.*, 2005). Up to a 40-fold enhancement in signal has been observed, although the extent is dependent upon the dye studied (Matveeva *et al.*, 2004). The origin of this observed enhancement is the result of two simultaneous phenomena: a local field enhancement that increases the efficiency of the dye excitation and an increase in the fluorescent dye's radiative decay rate. A 5-fold fluorescence enhancement was observed in 96-well plates coated with silver colloids and coupled with low power microwave heating, using the biotin–avidin interaction, measured using a TIRF arrangement (Aslan *et al.*, 2006).

Quantum dots can also be used as fluorescent probes to improve sensitivity and multiplexing capabilities. QDs offer a number of advantages over conventional fluorescent dyes, including high photostability, size tunable color, narrow emission profiles, and the ability to excite all sizes (colors) of QDs using UV wavelengths. QD bioconjugates have been used in a variety of imaging, labeling, and sensing applications (Medintz *et al.*, 2005a; Sapsford *et al.*, 2006c). Simultaneous, multiplexed immunoassays using different color QDs as fluorophores have been demonstrated for bacterial protein toxins (Goldman *et al.*, 2004; Hoshino *et al.*, 2005). By simply switching the excitation source and the excitation and emission filters, the NRL Array Biosensor has clearly demonstrated the ability to excite surface-bound QDs using TIRF (Sapsford *et al.*, 2004d; Medintz *et al.*, 2005b, 2006). As shown in Figure 3.7, using novel polypeptides terminated with His_6 functionality, 590 nm QDs were readily imaged using the NRL Array Biosensor (Medintz *et al.*, 2006), demonstrating the possibility of using QDs as fluorophores in a TIRF-based biosensor system.

Other methods of increasing the sensitivity of the assay include increasing the number of dyes attached to the tracer component. Tracer

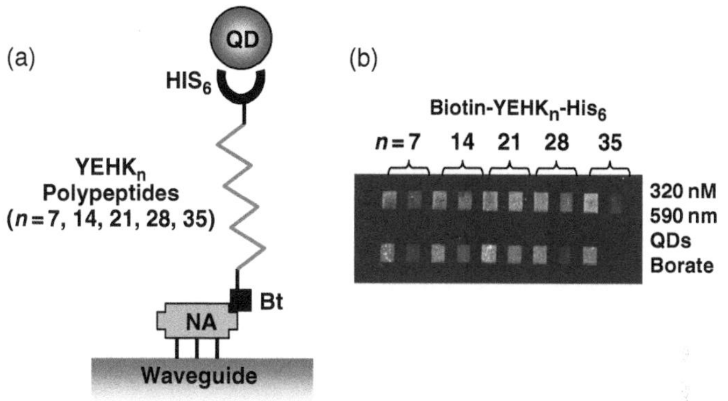

Figure 3.7 An image of immobilized quantum dots obtained using the NRL Array Biosensor.

antibodies offer only a limited number of fluorescent dye attachment sites; overloading with dye can result in self-quenching of the dye and also reduce the interaction of the antibody with the antigen (Anderson *et al.*, 2002). There are a number of companies that offer fluorescent microspheres functionalized to allow bioconjugation; however, they are typically large particles in the 1–2 μm diameter range, lowering the potential for excitation of surfaces distal to the waveguide. Smaller fluorescence nanoparticles can be synthesized (Yao *et al.*, 2006) and are likely to find application in array-based fluorescence biosensors. A cowpea mosaic virus (CPMV) was investigated as a tracer for immunoassays using the NRL Array Biosensor. The ability to label the 30 nm viral capsid with antibody and up to 60 fluorescent dyes (without quenching) resulted in an improved LOD in SEB sandwich immunoassays, when the virus nanoscaffold was used as a tracer, relative to a mole equivalent of dye-labeled antibody (Sapsford *et al.*, 2006b). This same CPMV capsid also demonstrated improved sensitivity when used as a tracer in microarray-based genotyping assays (Soto *et al.*, 2006).

Fluorescence resonance energy transfer (FRET) is a sensing format that could readily be measured using TIRF. There are numerous potential materials, including QDs, that have and could be used in FRET-based assays. The technique has been used for a variety of applications (reviewed in Sapsford *et al.*, 2004c, 2006a), including: nucleic acid assays, (Tan *et al.*, 2004), enzyme kinetic studies, and immunoassays (Sapsford *et al.*, 2004c). One of the benefits of FRET-based detection, when designed correctly, it that is allows direct measurement of binding to the target analyte, without the need for additional tracer reagents. Although FRET combined with planar waveguide-based detection is still in the early stages of development, it promises to be an excellent combination for rapid direct detection biosensors, capable of multiplexed assays for target detection (Kelly *et al.*, 1999). Molecular beacons utilize FRET-based quenching to interrogate hybridization events (Tan *et al.*, 2004; Yao *et al.*, 2004). Liu and Tan (1999) synthesized an oligonucleotide such that it possessed a stem and loop structure with the 5′-labeled quencher end and the fluorophore-labeled 3′ end in close proximity. Addition of the target DNA separated the fluorophore and quencher ends, resulting in increased fluorescence.

The published works using planar waveguides for fluorescence biosensors have used primarily DNA, antibodies, and avidin for analyte recognition. Studies describing the immobilization and use of different kinds of recognition molecules, such as receptors, carbohydrates, and peptides, in planar waveguide TIRF systems are limited to a small number of papers. Phage-display technology may provide interesting capture agents for biosensor applications, providing a dense array of oriented proteins or peptides (Hengerer *et al.*, 1999; Azzazy *et al.*, 2002; Petrenko *et al.*, 2003). The use of nucleic acid aptamers for analyte detection has recently been extended from other fields to use in planar waveguide TIRF biosensors. Potyrailo *et al.* (1998) studied the binding of thrombin to a fluorescently labeled, immobilized DNA aptamer by monitoring changes in the evanescent wave–induced fluorescence anisotropy. DNA and RNA aptamers can be synthesized for a number of target analytes with specificities comparable with the corresponding antibodies (Li and Li, 2000; Wang *et al.*, 2001). These aptamers (Kirby *et al.*, 2004; Collett *et al.*, 2005) as well as catalytic DNAs (Li and Li, 2000) and peptide oligonucleotide conjugates (Venkatesan and Kim, 2006) show great promise for use in planar waveguide biosensors as they can be synthesized.

There are also a number of enzymes, such as oxidases and reductases, which react with a variety of clinically and environmentally important analytes that have been used as biological recognition elements in a wide variety of biosensors (reviewed by Wilson and Hu, 2000). However, they have not as yet been utilized as capture biomolecules in planar waveguide TIRF studies. Though the majority of enzyme biosensors use an electrode for signal transduction, some biosensors utilize enzymes that produce or consume an optically detectable species. Fiber optic systems using a number of luminescence-generating enzymes have been described (as reviewed by Rowe-Taitt and Ligler, 2002). Furthermore, enzymes have been used in a number of optical fiber measurements in which the target analyte is indirectly determined. Here, the enzyme is either labeled or co-immobilized with the fluorescent probe. Typically, upon introduction of the analyte, the enzyme activity produces changes in the concentration of oxygen, ammonia, or pH. These changes can result in either the increase or decrease in the fluorescence intensity of the probe. For example, oxygen quenching of ruthenium fluorescence was the basis of a fiber optic glucose biosensor (Moreno-Bondi *et al.*, 1990). In addition, a fiber optic sensor for acetylcholinesterase (AchE) inhibitors utilizes fluorescein-labeled AchE immobilized onto the surface. Catalysis of its substrate acetylcholine results in the production of protons and acid-mediated fluorescence quenching of fluorescein (Rogers *et al.*, 1991).

Zeptosens (www.zeptosens.com) has recently developed a system using TIRF excitation to measure fluorescence from waveguides coupled with 96-well flow cells for analysis of nucleic acids (Duveneck *et al.*, 2002). Bottomless 96-well cell culture plates are also commercially available that may be readily attached to waveguide surfaces (Bruns *et al.*, 2006). Standard 96-well high throughput screening plates assembled on a total internal reflection stage have also been used in fluorescent studies (Aslan *et al.*, 2006). Such systems could readily be integrated into a multiplexed TIRF system to improve sample throughput, while also taking advantage of the improved sensitivity and real-time detection offered by the TIRF excitation.

This chapter is concerned primarily with the technique of TIRF that utilizes planar waveguides and has discussed its background, history, and current developments. The approach can be used to probe complex samples for targets of clinical, environmental, food, and military relevance. Commercial systems now exist for automated, stand-alone devices that have shown potential in a number of fields (Ligler *et al.*, 2007). However, with the continuing development of smaller electronics and components, such as the CMOS camera currently used with microarrays (Vo-Dinh *et al.*, 1999; Askari *et al.*, 2001), the potential for even smaller, handheld planar waveguide TIRF devices could become reality. The speed of signal transduction and relative immunity to matrix effects and other interfering influences, which are often problematic for other types of transduction methods, are two key advantages of planar waveguide TIRF biosensors. These advantages, and the additional benefit of spatially distinct sensing regions, are enabling these systems to gain advantage over other single-analyte sensing systems. Although the majority of studies have centered on antibody–antigen type systems, expansion of the work concerned with DNA/mRNA and receptor–ligand binding interactions, as well as integration of other types of recognition biomolecules, could lead to a much wider field of applications. In conclusion, the future looks bright for planar waveguide TIRF in both optical detection and exploration of biological processes.

Acknowledgements

This work was supported by NRL 6.2 work unit 6007. The views expressed are those of the authors and do not reflect opinion or policy of the US Navy or Department of Defense.

References

Anderson, G.P. and Nerurkar, N.L. (2002) *J. Immunol. Methods*, **271**, 17.

Arenkov, P., Kukhtin, A., Gemmell, A. *et al.* (2000) *Anal. Biochem.*, **278**, 123.

Asanov, A.N., Wilson, W.W., and Oldham, P.B. (1998) *Anal. Chem.*, **70**, 1156.

Askari, M., Alarie, J.P., Moreno-Bondi, M., and Vo-Dinh, T. (2001) *Biotechnol. Prog.*, **17**, 543.

Aslan, K., Holley, P., and Geddes, C.D. (2006) *J. Immunol. Methods*, **312**, 137.
Azzazy, H.M.E. and Highsmith, Jr., W.E. (2002) *Clin. Biochem.*, **35**, 425.
Bally, M., Halter, M., Voros J., and Grandin, H.M. (2006) *Surf. Interface Anal.*, **38**, 1442.
Beaucage, S.L. (2001) *Curr. Med. Chem.*, **8**, 1213.
Bernard, A., Michel B., and Delamarche, E. (2001) *Anal. Chem.*, **73**, 8.
Bhatia, S.K., Shriver-Lake, L.C., Prior, K.J. *et al.* (1989) *Anal. Biochem.*, **178**, 408.
Bhatia, S.K., Hickman, J.J., and Ligler, F.S. (1992) *J. Am. Chem. Soc.*, **114**, 4432.
Bhatia, S.K., Teixeira, J.L., Anderson, M. *et al.* (1993) *Anal. Biochem.*, **208**, 197.
Birkert, O., Haake, H.-M., Schutz, A. *et al.* (2000) *Anal. Biochem.*, **282**, 200.
Blawas, A.S. and Reichert, W.M. (1998a) *Biomaterials*, **19**, 595.
Blawas, A.S., Oliver, T.F., Pirrung M.C., and Reichert, W.M. (1998b) *Langmuir*, **14**, 4243.
Blue, R., Kent, N., Polerecky, L. *et al.* (2005) *Electron. Lett.*, 41.
Bradley, R.A., Drake, R.A.L., Shanks, I.A. *et al.* (1987) *Phil. Trans. R. Soc. Lond. B*, **315**, 143.
Brecht, A., Klotz, A., Barzen, C. *et al.* (1998) *Anal. Chim. Acta*, **362**, 69.
Brizzolara, R.A. and Beard, B.C. (1994) *J. Vac. Sci. Technol. A.*, **12**, 2981.
Brown, P.O. and Botstein, D. (1999) *Nat. Genet. Suppl.*, **21**, 33.
Bruns, T., Strauss, W.S.L., Sailer, R. *et al.* (2006) *J. Biomed. Opt.*, **11**, 034011.
Budach, W., Abel, A.P., Bruno, A.P., and Neuschafer, D. (1999) *Anal. Chem.*, **71**, 3347.
Burke, C.S., McGaughey, O., Sabattie, J.-M. *et al.* (2005) *Analyst*, **130**, 41.
Charles, P.T., Vora, G.J., Andreadis, J.D. *et al.* (2003) *Langmuir*, **19**, 1586.
Chittur, K.K. (1998) *Biomaterials*, **19**, 357.
Choudhury, B., Shinar R., and Shinar J. (2004) *J. Appl. Phys.*, **96**, 2949.
Collett, J.R., Cho E.J., and Ellington A.D. (2005) *Methods*, **37**, 4.
Conrad, D.W., Davis, A.V., Golightley, S.K. *et al.* (1997) *Proc. SPIE*, **2978**, 12.
Conrad, D.W., Golightley, S.K., and Bart, J.C. (1998) US Patent 5,773,308.
Dawson, E.D., Moore, C.L., Smagala, J.A. *et al.* (2006) *Anal. Chem.*, **78**, 7610.
De Mello, A.J. and Beard, N. (2003) *Lab Chip*, **3**, 11N.
Debouck, C. and Goodfellow, P.N. (1999) *Nat. Genet. Suppl.*, **21**, 48.
Delamarche, E., Juncker, D., and Schmid, H. (2005) *Adv. Mater.*, **17**, 2911.
Delehanty, J.B. and Ligler, F.S. (2002) *Anal. Chem.*, **74**, 5681.
Delehanty, J.B. and Ligler, F.S. (2003) *BioTechniques*, **34**, 380.
Dodson, J.M., Feldstein, M.J., Leatzow, D.M. *et al.* (2001) *Anal. Chem.*, **73**, 3776.

Duveneck, G.L., Abel, A.P., Bopp, M.A. *et al.* (2002) *Anal. Chim. Acta*, **469**, 49.

Duvenek, G.L., Bopp, M.A., Ehrat, M. *et al.* (2001) *Appl. Phys. B*, **73**, 869.

Duveneck, G.L., Neuschafer D., and Ehrat, M. (1995) International Patent GolN 21/77, 21/64.

Duveneck, G.L., Pawlak, M., Neuschafer, D. *et al.* (1997) *Sens. Actuators B*, **38–39**, 88.

Feldstein, M.J., Golden, J.P., Rowe, C.A. *et al.* (1999) *J. Biomed. Microdevices*, **1:2**, 139.

Feldstein, M.J., MacCraith, B.D., and Ligler, F.S. (2000), US Patent 6,137,117.

Floyd-Smith, T.M., Golden, J.P., Howell P.B., and Ligler, F.S. (2006) *Microfluidics Nanofluidics*, **2**, 180.

Fu, Y. and Lakowicz, J.R. (2006) *J. Phys. Chem. B*, **110**, 22557.

Golden, J.P. (1998) US Patent 5,827,748.

Golden, J.P., Floyd-Smith, T.M., Mott D.R., and Ligler, F.S. (2007) *Biosens. Bioelectron.*, **22**, 2763.

Golden, J.P. and Ligler, F.S. (2002) *Biosens. Bioelectron.*, **17**, 719.

Golden, J.P., Shriver-Lake, L., Sapsford K., and Ligler, F. (2005a) *Methods*, **37**, 65.

Golden, J.P., Taitt, C.R., Shriver-Lake, L.C. *et al.* (2005b) *Talanta*, **65**, 1078.

Goldman, E.R., Clapp, A.R., Anderson, G.P. *et al.* (2004) *Anal. Chem.*, **76**, 684.

Gullans, S.R. (2000) *Nat. Genet. Suppl.*, **26**, 4.

Guschin, D., Yershov, G., Zaslavsky, A. *et al.* (1997) *Anal. Biochem.*, **250**, 203.

Haab, B.B., Dunham M.J., and Brown, P.O. (2001) *Genome Biol.*, **2**, research 0004.1.

Hall, E.A.H. (1990) *Biosensors*. Milton Keynes: Open University Press.

Haugland, R.P. (2005) Chapters 1–5. In *The Handbook: A Guide to Fluorescent Probes and Labeling Technologies* (M.T.Z. Spence, ed.) New Brunswick, USA: Carlsbad, CA (for Invitrogen Corp).

Hengerer, A., Kosslinger, C., Decekr, J. *et al.* (1999) *BioTechniques*, **26**, 956.

Hermanson, G.T. (1996) *Bioconjugate Techniques*. San Deigo: Academic Press.

Herron, J.N., Caldwell, K.D., Christensen, D.A. *et al.* (1993) *SPIE – Adv. Fluor. Sen. Tech.*, **1885**, 28.

Herron, J.N., Christensen, D.A., Caldwell, K.D. *et al.* (1996), US Patent 5,512,492.

Herron, J.N., Christensen, D.A., Wang H.-K., and Caldwell, K.D. (1997) US Patent 5,677,196.

Hofmann, O., Voirin, G., Niedermann P., and Manz, A. (2002) *Anal. Chem.*, **74**, 5243.

Hollars, C.W., Puls, J., Bakajin, O. *et al.* (2006) *Anal. Bioanal. Chem.*, **385**, 1384.
Holthoff, W.G., Tehan, E.C., Bukowski, R.M. *et al.* (2005) *Anal. Chem.*, **77**, 718.
Hoshino, A., Fujioka, K., Manabe, N. *et al.* (2005) *Microbiol. Immunol.*, **49**, 461.
Huang, S.-C., Caldwell, K.D., Lin, J.-N. *et al.* (1996) *Langmuir*, **12**, 4292.
Ince, R. and Narayanaswamy, R. (2006) *Anal. Chim. Acta*, **569**, 1.
Ito, Y. (2006) *Biotechnol. Prog.*, **22**, 924.
Jennissen, H.P. and Zumbrink, T. (2004) *Biosens. Bioelectron.*, **19**, 987.
Johnson-White, B. and Golden, J. (2005) *Meas. Sci. Technol.*, **16**, 1.
Johnson-White, B., Zeinali, M., Buquo, L., and Ligler, F.S. (2006) *Anal. Chem.*, **78**, 853.
Johnson-White, B., Lin, B., and Ligler, F.S. (2007) *Anal. Chem.*, **79**, 140.
Jones, J.T., Myers, J.W., Ferrell, Jr., J.E., and Meyer, T. (2004) *Nat. Biotechnol.*, **22**, 306.
Kamholz, A.E. and Yager, P. (2001) *Biophys. J.*, **80**, 155.
Kellermayer, M.S.Z., Karsai, A., Kengyel, A. *et al.* (2006) *Biophys. J.*, **91**, 2665.
Kelly, D., Grace, K.M., Song, X. *et al.* (1999) *Optics Lett.*, **24**, 1723.
Kramer, K. and Hock, B. (2003) *Anal. Bioanal. Chem.*, **377**, 417.
Kriby, R., Cho, E.J., Gehrke, B.T. *et al.* (2004) *Anal. Chem.*, **76**, 4066.
Koltz, A., Brecht, A., Barzen, C. *et al.* (1998) *Sens. Actuators B*, **51**, 181.
Knight, J. (2001) *Nature*, **410**, 860.
Kumble, K.D. (2003) *Anal. Bioanal. Chem.*, **377**, 812.
Kulagina, N.V., Lassman, M.E., Ligler F.S., and Taitt, C.R. (2005) *Anal. Chem.*, **77**, 6504.
Kulagina, N.V., Shaffer, K.M., Anderson, G.P. *et al.* (2006) *Anal. Chim. Acta*, **575**, 9.
Kulagina, N.V., Shaffer, K.M., Ligler, F.S., and Taitt, C.R. (2007) *Sens. Actuators B*, **121**, 150.
Kwon, Y., Han, Z., Karatan, E., Mrksich, M. *et al.* (2004) *Anal. Chem.*, **76**, 5713.
Lahiri, J., Isaacs, L. Tien J., and Whitesides, G.M. 1999, *Anal. Chem.*, 71, 777.
Lang, S., Xu, J., Stuart, F. *et al.* (2000) *Biochem.*, **39**, 15674.
Leatzow, D.M., Dodson, J.M., Golden, J.P., and Ligler, F.S. (2002) *Biosens. Bioelectron.*, **17**, 105.
Lee, J.E. and Saavedra, S.S. (1996) *Langmuir*, **12**, 4025.
Lehr, H.-P., Brandenburg, A., and Sulz, G. (2003a) *Sensors Actuators B*, **92**, 303.

Lehr, H.-P., Reimann, M., Brandenburg, A. *et al.* (2003b) *Anal. Chem.*, **75**, 2414.
Li, J. and Li, Y. (2000) *J. Am. Chem. Soc.*, **122**, 10466.
Li, Y. and Reichert, W.M. (2003) *Langmuir*, **19**, 1557.
Li, Y., Schutte, R.J., Abu-Shakra, A., and Reichert, W.M. (2005) *Biomaterials*, **26**, 1081.
Ligler, F.S., Conrad, D.W., Golden, J.P. *et al.* (1998) *Proc SPIE*, **3258**, 50.
Ligler, F.S., Sapsford, K.E., Golden, J.P. *et al.* (2007) *Anal. Sci.*, **23**, 5.
Lim, D.V., Simpson, J.M., Kearns E.A., and Kramer, M.F. (2005) *Clin. Microbiol. Rev.*, **18**, 583.
Liu, X. and Tan, W. (1999) *Anal. Chem.*, **71**, 5054.
Liu, X.H., Wang, H.K., Herron, J.N., and Prestwich, G.D. (2000) *Bioconj. Chem.*, **11**, 755.
Lockhart, D.J. and Winzeler, E.A. (2000) *Nature*, **405**, 827.
Lu, B., Lu, C., and Wei, Y. (1992) *Anal. Lett.*, **25**, 1.
Lu, E.J., Xie, J., Lu, C. *et al.* (1995) *Anal. Chem.*, **67**, 83.
Lundgren, J.S., Watkins, N., Racz, D., and Ligler, F.S. (2000) *Biosens. Bioelectron.*, **15**, 417.
MacBeath, G. and Schreiber, S.L. (2000) *Science*, **289**, 1760.
Martinez, J.S., Grace, W.K., Grace, K.M. *et al.* (2005) *J. Mater. Chem.*, **15**, 4639.
Marquette, C.A. and Blum, L.J. (2006) *Biosens. Bioelectron.*, **21**, 1424.
Mashanov, G.I., Tacon, D., Knight, A.E. *et al.* (2003) *Methods*, **29**, 142.
Matveeva, E., Gryczynski, Z., Malicka, J. *et al.* (2004) *Anal. Biochem.*, **334**, 303.
Mauro, J.M., Cao, L.K., Kondracki, L.M. *et al.* (1996) *Anal. Biochem.*, **235**, 61.
McLean, M.A., Stayton, P.S., and Sligar, S.G. (1993) *Anal. Chem.*, **65**, 2676.
Medintz, I.L., Sapsford, K.E., Konnert, J.H. *et al.* (2005b) *Langmuir*, **21**, 5501.
Medintz, I.L., Sapsford, K.E., Clapp, A.R. *et al.* (2006) *J. Phys. Chem. B*, **110**, 10683.
Medintz, I.L., Uyeda, H.T., Goldman, E.R., and Mattoussi, H. (2005a) *Nat. Mater.*, **4**, 435.
Melton, L. (2004) *Nature*, **429**, 101.
Moore, P. and Clayton, J. (2003) *Nature*, **426**, 725.
Moran-Mirabal, J.M., Edel, J.B., Meyer, G.D. *et al.* (2005) *Biophys. J.*, **89**, 296.
Moreno-Bondi, M.C., Wolfbeis, O.S., Leiner, M.J.P., and Schaffer, B.H.P. (1990) *Anal. Chem.*, **62**, 2377.
Moreno-Bondi, M.C., Taitt, C.R., Shriver-Lake, L.C., and Ligler, F.S. (2006) *Biosens. Bioelectron.*, **21**, 1880.
Mott, D.R., Howell, P.B., Golden, J.P. *et al.* (2006) *Lab Chip*, **6**, 540.

Munson, M.S., Hawkins, K.R., Hasenbank M.S., and Yager, P. (2005) *Lab Chip*, **5**, 856.

Nagl, S., Schaeferling, M., and Wolfbeis, O.S. (2005) *Microchim. Acta*, **151**, 1.

Neuschafer, D., Budach, W., Wanke, C., and Chibout, S.-D. (2003) *Biosens. Bioelectron.*, **18**, 489.

Nilsson, K.G.I. and Mandenius, C.-F. 1994, *Biotechnol.*, **12**, 1376.

Ngundi, M.M., Shriver-Lake, L.C., Moore, M.H. *et al.* (2005) *Anal. Chem.*, **77**, 148.

Ngundi, M.M., Shriver-Lake, L.C., Moore, M.H. *et al.* (2006a), *J. Food Protect.*, **69**, 3047.

Ngundi, M.M., Taitt C.R., and Ligler, F.S. (2006b) *Biosens. Bioelectron.*, **22**, 124.

Ngundi, M.M., Qadri, S.A., Wallace, E.V. *et al.* (2006c) *Environ. Sci. Technol.*, **40**, 2352.

Ngundi, M.M., Taitt, C.R., McMurry, S.A. *et al.* (2006d) *Biosens. Bioelectron.*, **21**, 1195.

Park, S., Lee, M.-R., Pyo, S.-J., and Shin, I. (2004) *J. Am. Chem. Soc.*, **126**, 4812.

Pawlak, M., Grell, E., Schick, E. *et al.* (1998) *Faraday Discuss.*, **111**, 273.

Petrenko, V.A. and Vodyanoy, V.J. (2003) *J. Microbiol. Methods*, **53**, 253.

Pirrung, M.C., Davis, J.D., and Odenbaugh, A.L. (2000) *Langmuir*, **16**, 2185.

Plowman, T.E., Saavedra, S.S., and Reichert, W.H. (1998) *Biomaterials*, **19**, 341.

Plowman, T.E., Durstchi, J.D., Wang, H.K. *et al.* (1999) *Anal. Chem.*, **71**, 4344.

Potyrailo, R.A., Conrad, R.C., Ellington A.D., and Hieftje, G.M. (1998) *Anal. Chem.*, **70**, 3419.

Rabbany, S.Y., Donner B.L., and Ligler, F.S. (1994) *Crit. Rev. Biomed. Eng.*, **22**, 307.

Renault, J.P., Bernard, A., Bietsch, A. *et al.* (2003) *J. Phys. Chem. B*, **107**, 703.

Rodriguez-Mozaz, S., Lopez de Alda, M.J., and Barcelo, D. (2006) *Anal. Bioanal. Chem.* **386**, 1025.

Rogers, K.R. (2000) *Mol. Biotechnol.*, **14**, 109.

Rogers, K.R., (2006) *Anal. Chim. Acta*, **568**, 222.

Rogers, K.R., Cao, C.J., Valdes, J.J. *et al.* (1991) *Fundam. Appl. Toxicol.*, **16**, 810.

Rowe, C.A., Scruggs, S.B., Feldstein, M.J. *et al.* (1999a) *Anal. Chem.*, **71**, 433.

Rowe, C.A., Tender, L.M., Feldstein, M.J. *et al.* (1999b) *Anal. Chem.*, **71**, 3846.

Rowe-Taitt, C.A., Cras, J.J., Patterson, C.H. *et al.* (2000a) *Anal. Biochem.*, **281**, 123.

Rowe-Taitt, C.A., Golden, J.P., Feldstein, M.J. *et al.* (2000b) *Biosens. Bioelectron.*, **14**, 785.

Rowe-Taitt, C.A., Golden, J.P., Feldstein, M.J. *et al.* (2000c) *Biosens. Bioelectron.*, **15**, 579.

Rowe-Taitt, C.A. and Ligler, F.S. (2002) Fiber optic biosensors. In *Handbook of Fiber Optic Sensing Technology* (J.M. Lopez-Higuera, ed.) New York: John Wiley & Sons, 687–704.

Ruano, J.M., Glidle, A., Cleary, A. *et al.* (2003) *Biosens. Bioelectron.*, **18**, 175.

Rubina, A.Y., Dyukova, V.I., Dementieva, E.I. *et al.* (2005) *Anal. Biochem.*, **340**, 317.

Rucker, V.C., Havenstrite, K.L., and Herr, A.E. (2005) *Anal. Biochem.*, **339**, 262.

Sapsford, K.E., Berti, L., and Medintz, I.L. (2004c) *Minerva Biotec*, **16**, 253.

Sapsford, K.E., Berti, L., and Medintz, I.L. (2006a) *Angew. Chem. Int. Ed.*, **45**, 4562.

Sapsford, K.E., Charles, P.T., Patterson, C.H., and Ligler, F.S. (2002) *Anal. Chem.*, **74**, 1061.

Sapsford, K.E. and Ligler, F.S. (2004a) *Biosens. Bioelectron.*, **19**, 1045.

Sapsford, K.E., Liron, Z., Shubin, Y.S., and Ligler, F.S. (2001) *Anal. Chem.*, **73**, 5518.

Sapsford, K.E., Medintz, I.L., Golden, J.P. *et al.* (2004d) *Langmuir*, **20**, 7720.

Sapsford, K.E., Ngundi, M.M., Moore, M.H. *et al.* (2006d) *Sens. Actuators B*, **113**, 599.

Sapsford, K.E., Pons, T., Medintz, I.L., and Mattoussi, H. (2006c) *Sensors*, **6**, 925.

Sapsford, K.E., Rasooly, A., Taitt, C.R., and Ligler, F.S. (2004b) *Anal. Chem.*, **76**, 433.

Sapsford, K.E., Soto, C.M., Blum, A.S. *et al.* (2006b) *Biosens. Bioelectron.*, **21**, 1668.

Sapsford, K.E., Taitt, C.R., Fertig, S. *et al.* (2006e) *Biosens. Bioelectron.*, **21**, 2298.

Sapsford, K.E., Taitt, C.R., Loo, N., and Ligler, F.S. (2005) *Appl. Environ. Microbiol.*, **71**, 5590.

Semiatin, I.K. (2006) *R&D Magazine*. August, 24.

Schaferling, M. and Nagl, S. (2006) *Anal. Bioanal. Chem.*, **385**, 500.

Schmid, E.L., Keller, T.A., Dienes, Z., and Vogel, H. (1997) *Anal. Chem.*, **69**, 1979.

Schmid, E.L., Tairi, A.-P., Hovius, R., and Vogel, H. (1998) *Anal. Chem.*, **70**, 1331.

Schobel, U., Barzen, C., and Gauglitz, G. (2000) *Fresenius J. Anal. Chem.*, **366**, 646.
Schuderer, J., Akkoyun, A., Brandenburg, A. *et al.* (2000) *Anal. Chem.*, **72**, 3942.
Schult, K., Katerkamp, A., Trau, D. *et al.* (1999) *Anal. Chem.*, **71**, 5430.
Schwarz, A., Rossier, J.S., Roulet, E. *et al.* (1998) *Langmuir*, **14**, 5526.
Shinar, R., Ghosh, D., Choudhury, B. *et al.* (2006), *J. Non-Crystalline Solids*, **352**, 1995.
Shriver-Lake, L.C., Shubin, Y.S., and Ligler, F.S. (2003) *J. Food Protect.*, **66**, 1851.
Shriver-Lake, L.C., Taitt, C.R., and Ligler, F.S. (2004) *J. AOAC Intern.*, **87**, 1498.
Shriver-Lake, L.C., Turnera, S., and Taitt, C.R. (2007) *Anal. Chim. Acta*, **584**, 66.
Silin, V., Weetall, H., and Vanderah, D.J. (1997) *J. Col. Interface Sci.*, **185**, 94.
Silzel, J.W., Cercek, B., Dodson, C. *et al.* (1998) *Clin. Chem.*, **44**, 2036.
Skivesen, N., Horvath, R., Thinggaard, S. *et al.* (2007) *Biosens. Bioelectron.*, **22**, 1282.
Sojka, B., Piunno, P.A.E., Wust, C.C., and Krull, U.J. (1999) *Anal. Chim. Acta*, **395**, 273.
Soper, S.A., Brown, K., Ellington, A. *et al.* (2006) *Biosens. Bioelectron.*, **21**, 1932.
Soto, C.M., Blum, A.S., Vora, G.J. *et al.* (2006) *J. Am. Chem. Soc.*, **128**, 5184.
Stranik, O., McEvoy, H.M., McDonagh, C., and MacCraith, B.D. (2005) **107**, 148.
Taitt, C.R., Anderson, G.P., Lingerfelt, B.M. *et al.* (2002) *Anal. Chem.*, **74**, 6114.
Taitt, C.R., Golden, J.P., Shubin, Y. *et al.* (2003) *ISCAS 2003 Conf. Proc. Int. Symp. Circuits Syst.*
Taitt, C.R., Anderson, G.P., and Ligler, F.S. (2005) *Biosens. Bioelectron.*, **20**, 2470.
Taitt, C.R., Shubin, Y.S., Angel, R., and Ligler, F.S. (2004) *Appl. Environ. Microbiol.*, **70**, 152.
Tan, W.H., Wang, K.M., and Drake, T.J. (2004) *Curr. Opin. Chem. Biol.*, **8**, 547.
Tinnefeld, P. and Sauer, M. (2005) *Angew. Chem. Int. Ed.*, **44**, 2642.
Tolley, S.E., Wang, H.-K., Smith, R.S. *et al.* (2003) *Anal. Biochem.*, **315**, 223.
Tschmelak, J., Proll, G., and Gauglitz, G. (2004) *Biosens. Bioelectron.*, **20**, 743.
Tschmelak, J., Proll, G., Riedt, J., Kaiser, J. *et al.* (2005) *Biosens. Bioelectron.*, **20**, 1509.

Ulbrich, R., Golbik R., and Schellenberger, A. (1991) *Biotechnol. Bioeng.*, **37**, 280.
Venkatesan, N. and Kim, B.H. (2006) *Chem. Rev.*, **106**, 3712.
Vigeant, M.A.S., Ford, R.M., Wagner, M., and Tamm, L.K. (2002) *Appl. Environ. Microbiol.*, **68**, 2794.
Vigmond, S.J., Iwakura, M., Mizutani, F., and Katsura, T. (1994) *Langmuir*, **10**, 2860.
Vijayendran, R.A. and Leckband, D.E. (2001) *Anal. Chem.*, **73**, 471.
Vijayendran, R.A., Motsegood, K.M., Beebe D.J., and Leckband, D.E. (2003) *Langmuir*, **19**, 1824.
Vikholm, I. and Albers, W.M. (1998) *Langmuir*, **14**, 3865.
Vo-Dinh, T., Alarie, J.P., lsola, N. *et al.* (1999) *Anal. Chem.*, **71**, 358.
Vo-Dinh, T. and Cullum, B. (2000) *Fresenius J. Anal. Chem.*, **366**, 540.
Wacker, R. and Neimeyer, C.M. (2004) *Chem. Bio. Chem.*, **5**, 453.
Wadkins, R.M., Golden J.P., and Ligler, F.S. (1997) *J. Biomed. Optics*, **2**, 74.
Wadkins, R.M., Golden, J.P., Pritsiolas, L.M., and Ligler, F.S. (1998) *Biosens. Bioelectron.*, **13**, 407.
Wang, L., Carrasco, C., Kumar, A. *et al.* (2001) *Biochem.*, **40**, 2511.
Webb, S.E.D., Needham, S.R., Roberts, S.K., and Martin-Fernandez, M.L. (2006) *Optics Lett.*, **31**, 2157.
Weiping, Q., Bu, X., Lei, W. *et al.* (1999) *J. Inclusion Phenomena Macrocyclic Chem.*, **35**, 419.
Weller, M.G. (2005) *Anal. Bioanal. Chem.*, **381**, 41.
Wilson, G.S. and Hu, Y. (2000) *Chem. Rev.*, **100**, 2693.
Wolf, M., Juncker, D., Michel, B. *et al.* (2004) *Biosens. Bioelectron.*, **19**, 1193.
Wu, P., Hogrebe, P., and Grainger, D.W. (2006), *Biosens. Bioelectron.*, **21**, 1252.
Wulfkuhle, J., Espina, V., Liotta, L., and Petricoin, E. (2004) *Eur. J. Cancer*, **40**, 2623.
Yao, G., Fang, X., Yokota, H. *et al.* (2003) *Chem. Eur. J.*, **9**, 5686.
Yao, G. and Tan, W.H. (2004) *Anal. Biochem.*, **331**, 216.
Yao, G., Wang, L. Wu, Y. *et al.* (2006) *Anal. Bioanal. Chem.*, **385**, 518.
Zeller, P.N., Voirin, G., and Kunz, R.E. (2000) *Biosens. Bioelectron.*, **15**, 591.
Zhang, J., Matveeva, E. Gryczynski, I. *et al.* (2005) *J. Phys. Chem. B*, **109**, 7969.
Zourob, M., Mohr, S. Brown, B.J.T. *et al.* (2005) *Anal. Chem.*, **77**, 232.

Optical Biosensors: Today and Tomorrow (2nd Edition)
Frances S. Ligler and Chris Rowe Taitt (Editors)

Chapter 4

SURFACE PLASMON RESONANCE BIOSENSORS

Jiří Homola, Ph.D.[a], Sinclair S. Yee, Ph.D.[b], and David Myszka, Ph.D.[c]

[a]Institute of Photonics and Electronics, Academy of Sciences of the Czech Republic, 182 51 Prague, Czech Republic
[b]Department of Electrical Engineering, University of Washington, Seattle, WA 98105, USA
[c]Center for Biomolecular Interaction Analysis, University of Utah, Salt Lake City, UT 84132, USA

Surface plasmon resonance (SPR) biosensors exploit special electromagnetic waves – surface plasmon-polaritons – to probe changes in the refractive index at surfaces of metals. Surface plasmon resonance biosensors can therefore be used to monitor the interaction between an analyte in solution and its biospecific partner immobilized on the metal surface without the use of labels. Major application areas include detection of low levels of biological analytes and study of biomolecular interactions. In the past 15 years, SPR biosensor technology has been commercialized, and SPR biosensors have become a central tool for characterizing and quantifying biomolecular interactions both in life science and pharmaceutical research.

4.1. Technical concept

Surface plasmon resonance (SPR) biosensors use surface plasma waves (SPWs) to probe biomolecular interactions occurring at the surface of a sensor. This section introduces SPWs, presents methods for their excitation and interrogation, and discusses the concept of SPR optical biosensors.

4.1.1. Surface plasma waves

4.1.1.1. Surface plasmon-polaritons at a plane interface between a semi-infinite dielectric and a metal

Surface plasmon-polaritons or SPWs occur at the surfaces of metals, which behave like free-electron plasma. Let us consider a plane interface composed of two semi-infinite homogeneous, isotropic media and choose the coordinate system so that the metal (dielectric function ε_M) occupies the region $z < 0$, while the dielectric medium (dielectric constant ε_D) occupies the region $z > 0$ (Figure 4.1). Electromagnetic field modes supported by this geometry can be found by solving Maxwell's equations in each medium and applying boundary conditions at the interface. We are looking for modes guided by the interface, and thus the magnitudes of all their field vectors have to be decreasing with increasing distance from the interface. In general, stratified media with plane boundaries can support transverse electric (TE) modes (in which the direction of propagation and magnetic and electric intensity vectors form an orthogonal triad and the electric vector is parallel to the interface) and transverse magnetic (TM) modes (where the magnetic vector is perpendicular to the direction of propagation of the wave and parallel to the plane of interface) (Figure 4.1). Transverse electric-polarized surface modes cannot exist in this geometry if the materials involved

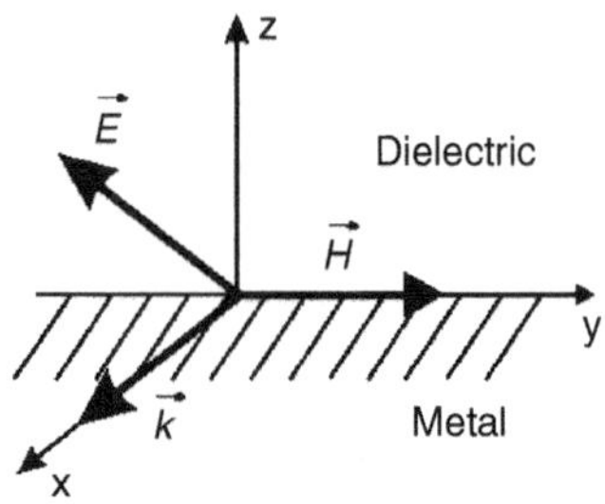

Figure 4.1 Transverse magnetic wave at a metal–dielectric interface.

are non-magnetic (Boardman, 1982). Therefore, only TM modes can be supported by the metal–dielectric interface. In complex notation, electric and magnetic intensity vectors of a TM-polarized mode are:

Dielectric

$$\vec{E} = (E_x, 0, E_z) = \left(1, 0, \frac{i\beta}{\alpha_D}\right) A \exp[-\alpha_D z + i(\beta x - \omega t)],$$

$$\alpha_D^2 = \beta^2 - \left(\frac{\omega}{c}\right)^2 \varepsilon_D$$

$$\vec{H} = (0, H_y, 0) = (0, 1, 0)\frac{-iA\omega\varepsilon_D}{c^2\alpha_D\mu} \exp[-\alpha_D z + i(\beta x - \omega t)] \quad (4.1)$$

Metal

$$\vec{E} = (E_x, 0, E_z) = \left(1, 0, \frac{-i\beta}{\alpha_M}\right) A \exp[\alpha_M z + i(\beta x - \omega t)],$$

$$\alpha_M^2 = \beta^2 - \left(\frac{\omega}{c}\right)^2 \varepsilon_M$$

$$\vec{H} = (0, H_y, 0) = (0, 1, 0)\frac{iA\omega\varepsilon_M}{c^2\alpha_M\mu} \exp[\alpha_M z + i(\beta x - \omega t)] \quad (4.2)$$

where ω is the angular frequency, β is the propagation constant, c is the speed of light in vacuum, μ is the permeability of vacuum, and A is a normalization constant proportional to the energy carried by the mode (Boardman, 1982).

Application of boundary conditions requiring continuity of tangential components of electric and magnetic intensity vectors (H_y and E_x) at the interface yields an equation for the propagation constant (Boardman, 1982):

$$\beta = \frac{\omega}{c}\sqrt{\frac{\varepsilon_M \varepsilon_D}{\varepsilon_M + \varepsilon_D}} \quad (4.3)$$

The propagation constant is generally a complex number because the dielectric function of metal ε_M is a complex function of the angular frequency. Equations (4.1–4.3) represent a true SPW that propagates along the interface and decays exponentially in a direction perpendicular to the direction of propagation, if the real part of ε_M is negative and its absolute value is smaller than ε_D. At optical wavelengths, this condition is fulfilled for several metals of which gold and silver are the most commonly used (Figure 4.2).

The real part of the propagation constant is related to the effective index N – the quantity commonly used in waveguide optics – in the following manner:

$$N = \frac{c}{\omega}\mathrm{Re}\{\beta\} = \mathrm{Re}\left\{\sqrt{\frac{\varepsilon_M \varepsilon_D}{\varepsilon_M + \varepsilon_D}}\right\}, \tag{4.4}$$

where Re{ } denotes the real part of a complex number. The imaginary part of the propagation constant is related to the modal attenuation b (in dB/cm if β is given in 1/m):

$$b = \mathrm{Im}\{\beta\}\frac{0.2}{\ln 10} = \mathrm{Im}\left\{\sqrt{\frac{\varepsilon_M \varepsilon_D}{\varepsilon_M + \varepsilon_D}}\right\}\frac{0.2\omega}{c \ln 10}, \tag{4.5}$$

where Im{ } denotes the imaginary part of a complex number.

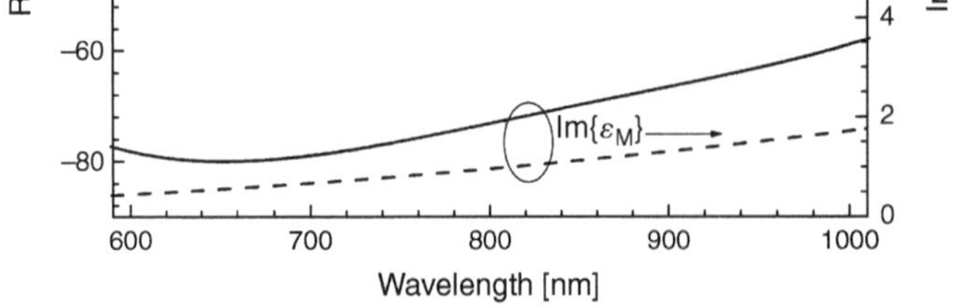

Figure 4.2 Dielectric function of gold and silver. (Data taken from Palik, 1985 and Ordal *et al.*, 1983.)

Spectral dependencies of the effective index and mode attenuation for SPWs supported by silver and gold surfaces are shown in Figure 4.3. As follows from Figure 4.3, the effective index of an SPW increases with decreasing wavelength. The effective index of the SPW supported

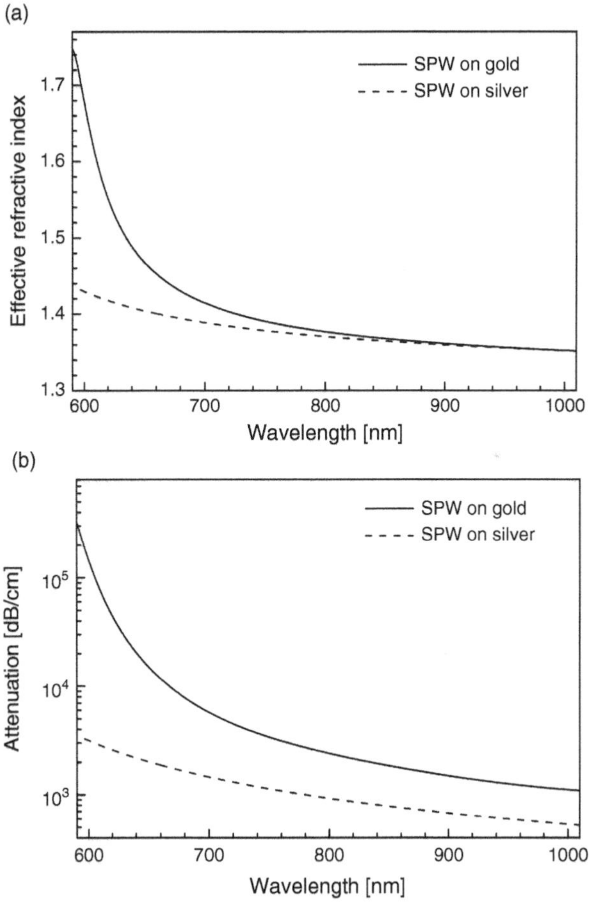

Figure 4.3 Surface plasmon-polaritons: effective index and attenuation. The effective index (a) and attenuation (b) of a surface plasmon-polariton as a function of the wavelength for a surface plasmon-polariton propagating along the interface between a metal (gold or silver) and a non-dispersive dielectric (refractive index = 1.32).

by a gold surface is larger than that of the SPW supported by silver because the real part of the dielectric constant of gold is smaller than that of silver. The mode attenuation of the SPW supported by a gold boundary is larger than that of the SPW supported by a silver surface as

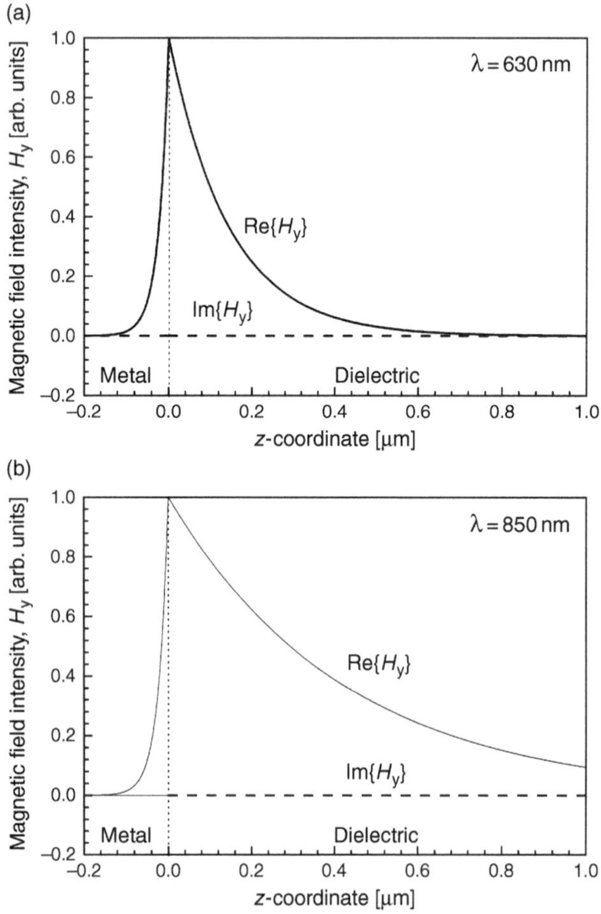

Figure 4.4 Surface plasmon-polariton: field pattern. Spatial distribution of the magnetic intensity for a surface plasmon-polariton at the interface between gold and a non-dispersive dielectric (refractive index = 1.32) in the direction perpendicular to the interface, calculated for the wavelength of (a) 630 nm and (b) 850 nm.

the imaginary part of the dielectric constant of gold is larger than that of silver. The field profile of a surface plasmon-polariton is illustrated in Figure 4.4.

The field decay in the direction perpendicular to the direction of propagation may be characterized by means of the penetration depth, L_p. The penetration depth is defined as the distance from the interface at which the amplitude of the field has fallen to $1/e$ of its value at the surface and is related to field parameters for the dielectric and metal, α_D and α_M, as follows:

$$L_{pD} = \frac{1}{\mathrm{Re}\{\alpha_D\}} \quad L_{pM} = \frac{1}{\mathrm{Re}\{\alpha_M\}}, \tag{4.6}$$

Similarly, the attenuation of an SPW in the direction of propagation can be characterized by means of the propagation length, which is defined as the distance in the direction of propagation at which the energy of the wave decreases by a factor of $1/e$.

$$L = \frac{1}{2\,\mathrm{Im}\{\beta\}} \tag{4.7}$$

Characteristics of SPWs supported by surfaces of gold and silver are given in Table 4.1. As follows from Eqn (4.7), Table 4.1, and Figure 4.3, SPWs propagating along the surface of silver are less attenuated than those propagating along the surface of gold. Surface plasma waves propagate with high attenuation and the attenuation increases with

Table 4.1 Major characteristics of surface plasmon-polaritons at the interface between metal (silver or gold) and a non-dispersive dielectric (refractive index = 1.32) for two different wavelengths

Metal	**Silver**		**Gold**	
Wavelength (nm)	630	850	630	850
Penetration depth into metal (nm)	24	23	29	25
Penetration depth into dielectric (nm)	219	443	162	400
Propagation length (μm)	19	57	2	24

decreasing wavelength. As follows from Figure 4.4, the electromagnetic field of an SPW is distributed in a highly asymmetric fashion and majority of the field (usually more than 90%) is concentrated in the dielectric medium.

4.1.1.2. Surface plasmon-polaritons on a thin metal film surrounded by dielectric media

A thin metal film surrounded by dielectric media may support surface plasmon-polaritons at both the interfaces. For thin metal films, there is coupling between surface plasmon-polaritons associated with each boundary, giving rise to mixed modes – symmetric and antisymmetric surface plasmon-polaritons. These modes are found as solutions of Maxwell's equations in each medium, which satisfy boundary conditions at both the interfaces (Burke *et al.*, 1986). Figure 4.5 shows the effective index and attenuation of these two modes as a function of the thickness of the metal film. Clearly, if the metal film is rather thick ($\sim$100 nm), the propagation constants of the two modes are almost identical. As the metal thickness decreases, the two modes become more different. The symmetric surface plasmon-polariton exhibits an effective index and attenuation, both of which decrease with decreasing metal film thickness, while the effective index and attenuation of the antisymmetric surface plasmon-polariton increase with decreasing thickness of the metal film. Thus, the symmetric surface plasmon-polariton exhibits a lower effective index and attenuation than its antisymmetric counterpart. Therefore, the first mode is often referred as a long-range surface plasmon-polariton, while the other is referred as a short-range surface plasmon-polariton, a reference to their relative attenuation and propagation lengths.

The dispersion properties of symmetric and antisymmetric surface plasmon-polaritons (Figure 4.6) are rather different from those of the traditional SPWs supported by a single interface (Figure 4.3). It should be noted that the effective index of the symmetric mode exhibits much lower "dispersion" (dependence on the wavelength) than that of the antisymmetric mode. This is because the symmetric surface plasmon exhibits a much weaker electromagnetic field in the metal film, which is a strongly dispersive medium (Figure 4.7).

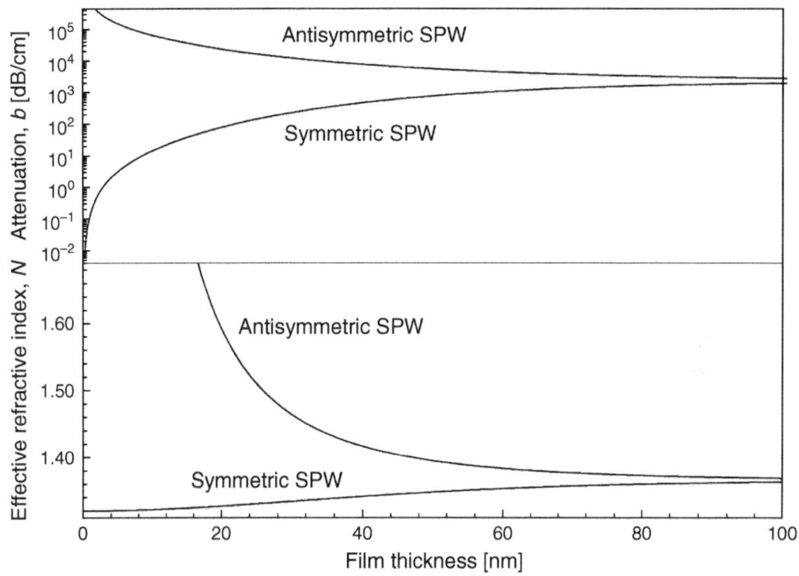

Figure 4.5 Surface plasmon-polaritons on a thin metal film. Effective index and mode attenuation as a function of the metal film thickness for symmetric and antisymmetric surface plasmon-polaritons propagating along a thin gold film embedded between two identical non-dispersive dielectrics (refractive index = 1.32); wavelength = 800 nm. Note the drop in the attenuation for the symmetric bound mode as the metal thickness approaches zero.

As illustrated in Figure 4.7, the distribution of magnetic intensity (and the TE intensity) is symmetric and antisymmetric with respect to the center of the metal film for symmetric and antisymmetric surface plasmon-polaritons, respectively. The field of the symmetric SPW penetrates deeper into the dielectric media than that of the antisymmetric SPW.

4.1.2. Optical excitation of surface waves

A light wave (LW) can couple to an SPW at a metal–dielectric interface if the component of light's wavevector that is parallel to the interface matches that of the SPW. As the propagation constant of an SPW at a metal–dielectric interface (Eqn 4.3) is larger than that which can be provided by the component of the wavevector of light in the dielectric,

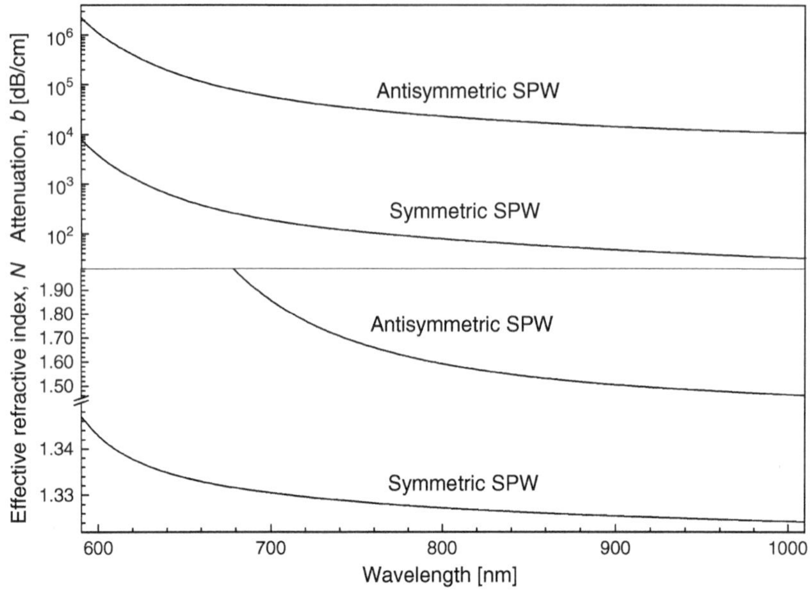

Figure 4.6 Surface plasmon-polaritons on a thin metal film. Effective index and mode attenuation as a function of the wavelength, calculated for symmetric and antisymmetric surface plasmon-polaritons propagating along a thin gold film embedded between two identical non-dispersive dielectrics (refractive index = 1.32), gold film thickness = 20 nm.

SPWs cannot be excited directly by light incident onto a smooth metal surface. In order to allow for excitation of an SPW by a LW, the light's wavevector needs to be enhanced to match that of the SPW. The light's wavevector can be increased by passing the light through a medium with a refractive index higher than that of the dielectric medium at the boundary at which the SPW is to be excited, or by roughness of the metal surface (statistical or regular).

4.1.2.1. Excitation of surface plasmon-polaritons using prism couplers

The enhancement of a LW's wavevector by passing light through an optically denser medium is illustrated in Figure 4.8. The effective index of an SPW at the metal (gold)–dielectric (water) interface ranges from

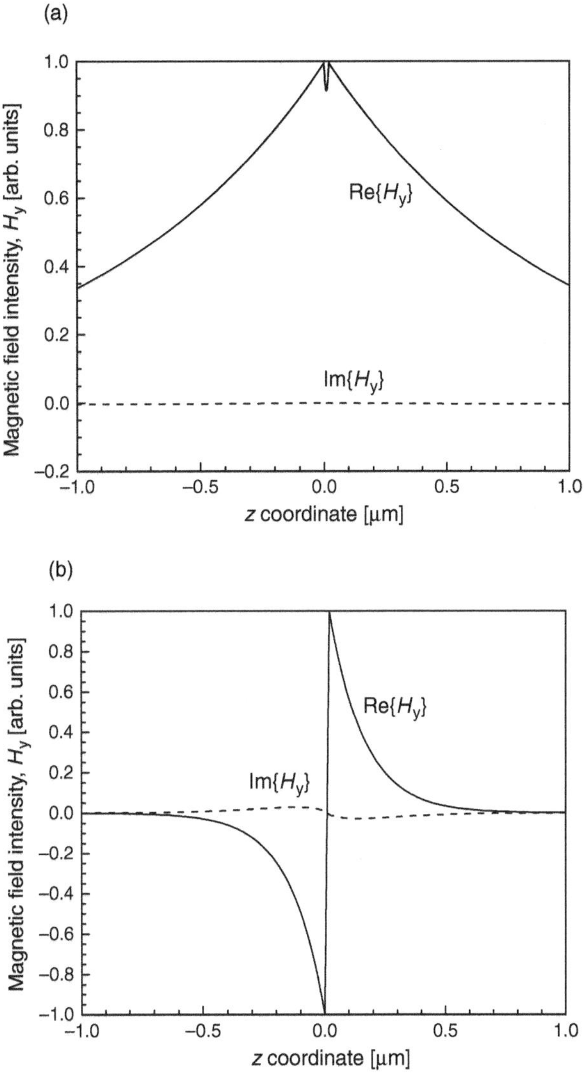

Figure 4.7 Surface plasmon-polaritons on a thin metal film: field patterns. Spatial distribution of the magnetic intensity of (a) symmetric and (b) antisymmetric surface plasmon-polaritons propagating along a thin gold film embedded between two identical non-dispersive dielectrics (refractive index = 1.32), gold film thickness = 20 nm, wavelength = 800 nm.

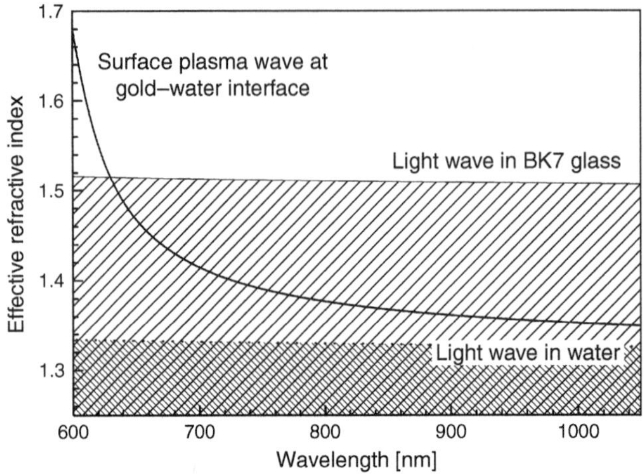

Figure 4.8 Effective index of a surface plasmon-polariton as a function of the wavelength for a surface plasmon-polariton propagating along the interface between gold and water. Normalized wavenumbers provided by LWs in a BK7 glass and water shown for comparison.

1.35 (at a wavelength of 1000 nm) to 1.68 (at a wavelength of 600 nm), while the effective index of light in the dielectric does not exceed 1.335. A LW propagating in the dielectric with a higher refractive index, such as BK7 glass, can provide a component of the normalized wavevector parallel to the metal surface with a value between 0 and about 1.52, depending on the angle between the direction of propagation of the LW and the metal surface. Therefore, the condition for coupling light into the SPW can be fulfilled for wavelengths longer than 630 nm.

The coupling between a LW in the high refractive index dielectric medium and an SPW at the metal–low refractive index dielectric interface can be established by the attenuated total reflection (ATR) method. A LW passes through a high refractive index prism and is totally reflected at the prism base generating an evanescent wave penetrating a metal film in the Kretschmann configuration (Figure 4.9a) or a dielectric layer in the Otto configuration (Figure 4.9b). This evanescent wave propagates along the interface with the propagation constant, which can be adjusted to match that of the SPW by controlling the angle of

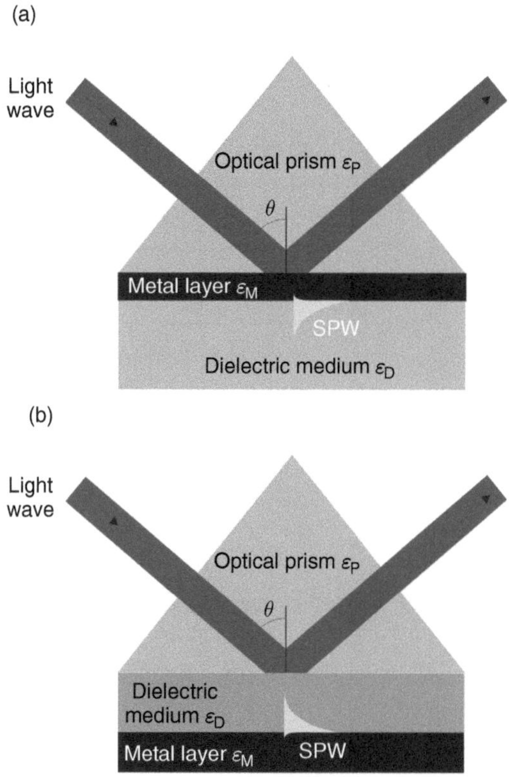

Figure 4.9 Excitation of SPWs on smooth metal surfaces by means of the ATR. (a) Kretschmann configuration and (b) Otto configuration.

incidence. Thus, the matching condition can be fulfilled, allowing the evanescent wave to be coupled into the SPW.

Assuming that the prism has only a minor influence on the propagation constant of the SPW at the interface of a metal and a low refractive index dielectric, the coupling condition may be expressed as:

$$\sqrt{\varepsilon_P}\sin(\theta) = \mathrm{Re}\left\{\sqrt{\frac{\varepsilon_M \varepsilon_D}{\varepsilon_M + \varepsilon_D}}\right\} \tag{4.8}$$

where θ denotes the angle of incidence and ε_P, ε_M, and ε_D denote dielectric functions of the prism, the metal film, and the dielectric medium ($\varepsilon_P > \varepsilon_D$). As discussed in Section 4.1.2.4, this condition is fulfilled for "thick" metal films (50 nm or more for gold and optical wavelengths).

4.1.2.2. Excitation of surface plasmon-polaritons using optical waveguides

Similarly, an SPW can be excited by a LW guided by an optical waveguide. This approach is illustrated in Figure 4.10. Light propagates in a waveguide in the form of guided modes. The electromagnetic field of a guided mode is concentrated in the waveguiding layer. A fraction of the optical energy propagates in the form of an evanescent wave in the low refractive index medium surrounding the waveguiding layer. In the section of the waveguide containing an SPW-active metal film, this evanescent wave can excite an SPW at the outer (in Figure 4.10a) or inner (Figure 4.10b) surface of the metal film. Assuming that the waveguide influences the propagation constant of the SPW only slightly, the coupling condition for a guided mode and an SPW can be expressed as follows:

$$N_{ef} = \mathrm{Re}\left\{\sqrt{\frac{\varepsilon_M \varepsilon_D}{\varepsilon_M + \varepsilon_D}}\right\} \tag{4.9}$$

where N_{ef} denotes the effective index of the waveguide mode, and ε_M and ε_D denote dielectric functions of the metal and the dielectric medium.

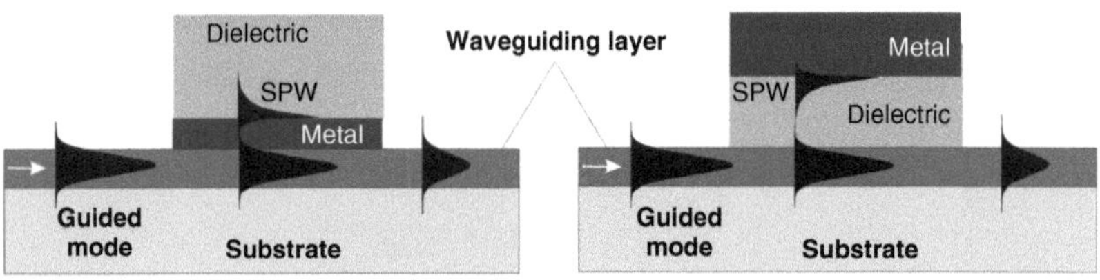

Figure 4.10 Excitation of SPWs by guided modes of optical waveguides. (a) Excitation of SPW on the inner boundary of a thin metal film, and (b) excitation of SPW on the outer metal boundary through a dielectric buffer layer.

4.1.2.3. Excitation of surface plasmon-polaritons using grating couplers

An SPW may also be excited by a LW with its wavevector increased by the wave's interaction with surface roughness of the metal film. Most commonly used configurations for SPW excitation on rough metal surfaces are based on diffraction gratings (Figure 4.11).

If a LW is made incident on a periodically distorted surface of a diffraction grating, a series of waves directed away from the surface at different angles is produced (Hutley, 1982). The components of the wavevector of these diffraction-generated LWs parallel to the interface are:

$$k_x + mG = k_{xm} \tag{4.10}$$

where m is the diffraction order (integer), k_x is the component of the wavevector of the incident light along the grating surface, G is the grating wavevector, and k_{xm} is the wavevector of the diffracted LW. In case of shallow gratings, the coupling condition may be expressed as:

$$\sqrt{\varepsilon_D}\sin(\theta) + m\frac{\lambda}{\Lambda} = \pm\mathrm{Re}\left\{\sqrt{\frac{\varepsilon_M \varepsilon_D}{\varepsilon_M + \varepsilon_D}}\right\}, \tag{4.11}$$

where θ is the angle of incidence of the LW, Λ denotes the pitch of the grating ($\Lambda = 2\pi/G$), ε_M and ε_D denote dielectric functions of the metal and the dielectric medium, and λ denotes the free-space wavelength.

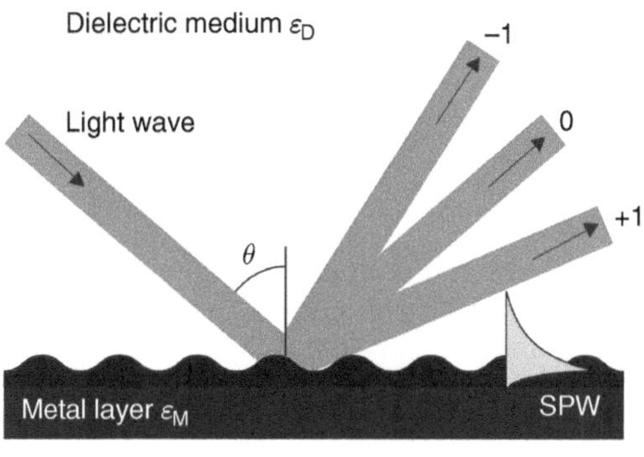

Figure 4.11 Excitation of SPWs at the surface of a diffraction grating.

4.1.2.4. Excitation of surface plasmon-polaritons: energy transfer

In the process of optical excitation of SPWs, a portion of the energy of the LW is transferred into the energy of an SPW and dissipated in the metal film. Assuming $|\mathrm{Re}(\varepsilon_\mathrm{M})| \gg \varepsilon_\mathrm{D}$ and $|\mathrm{Re}(\varepsilon_\mathrm{M})| \gg \mathrm{Im}(\varepsilon_\mathrm{M})$, the reflectivity for the Kretschmann geometry of the ATR method may be expressed as:

$$R(k_x) = 1 - \frac{4\Gamma_\mathrm{i}\Gamma_\mathrm{rad}}{[k_x - \beta^+]^2 + (\Gamma_i + \Gamma_\mathrm{rad})^2}, \tag{4.12}$$

where

$$k_x = \sqrt{\varepsilon_\mathrm{P}}\frac{\omega}{c}\sin(\theta), \quad \beta^+ = \mathrm{Re}\{\beta + \Delta\beta\},$$
$$\Gamma_\mathrm{i} = \mathrm{Im}\{\beta\}, \qquad \Gamma_\mathrm{rad} = \mathrm{Im}\{\Delta\beta\} \tag{4.13}$$

θ is the angle of incidence, ε_P is the dielectric constant of the prism, ω is the angular frequency, c is the velocity of light in vacuum, k_x is the component of the wavevector of the LW that is parallel with the interface, and β is the propagation constant of the SPW as given by Eqn (4.3) (Raether, 1983). Γ_i characterizes the attenuation of the SPW at the metal–dielectric interface due to the dissipation in the metal. The complex quantity $\Delta\beta$ characterizes the influence of the prism on the propagation constant of the SPW. Its imaginary part, Γ_rad, characterizes the attenuation of the SPW due to the coupling of the SPW to optical radiation in the prism. As follows from Eqn (4.12), the reflectivity exhibits a Lorentzian dip located at the angle θ_min, given by the following condition:

$$\sqrt{\varepsilon_\mathrm{P}}\frac{\omega}{c}\sin(\theta_\mathrm{min}) = \beta^+ \tag{4.14}$$

In the approximation of thick metal films, $\mathrm{Re}\{\beta\} \gg \mathrm{Re}\{\Delta\beta\}$ and the right-hand side of Eqn (4.14) is equal to the propagation constant of the surface plasmon at the metal–dielectric interface. This suggests that the reflectivity minimum occurs when the component of the wavevector of the incident LW parallel to the interface matches the propagation constant of the SPW. The depth of the reflectivity dip can theoretically range from 0 to 100%. The reflectivity minimum reaches zero if $\Gamma_i = \Gamma_{rad}$, indicating that all the energy of the incident LW can be lost due to the SPW's excitation if the parameters of the structure and incident LW are chosen properly. It can be shown by the examination of Eqn (4.12) that the width of the dip is proportional to $\Gamma_i + \Gamma_{rad}$ and thus to the total loss the SPW exhibits. Therefore, the excitation of SPWs on a metal with low loss gives rise to a narrow dip in the TM-wave reflectivity and vice versa. For a given metal, the reflectivity minimum and the width of the dip cannot be minimized at the same time, as Γ_i decreases with decreasing metal film thickness while Γ_{rad} exhibits an opposite trend. In order to produce deep and narrow reflectivity dips, one has to use very thin metal films (<20 nm) and reduce the coupling of the SPW to optical radiation in the prism by introducing a buffer layer between the prism and the metal film. Such a structure allows for the excitation of long-range SPWs. Drops in the reflectivity due to the excitation of a surface plasmon in the traditional three-layer Kretschmann configuration and excitation of long-range SPWs in the four-layer configuration are illustrated in Figure 4.12. Besides the change in the amplitude, the LW exciting an SPW undergoes also a change in phase (Figure 4.12).

4.1.3. Optical sensor based on SPWs

4.1.3.1. Surface plasma waves as refractive index probes

When an SPW propagates along the metal–dielectric interface, its field penetrates into the dielectric (with penetration depth given by Eqn (4.6)), probing the dielectric medium. Any change in the dielectric's optical properties results in a change in the characteristics of the SPW, especially its propagation constant. The relationship between variations in optical parameters of the dielectric and the propagation constant of the SPW can be established using the perturbation theory (Snyder and Love, 1983; Homola, 2006).

According to the perturbation theory, a small change in the permittivity distribution, $\varepsilon(z) \rightarrow \bar{\varepsilon}(z) = \varepsilon(z) + \delta\varepsilon(z)$ ($z < 0$ is occupied with a metal,

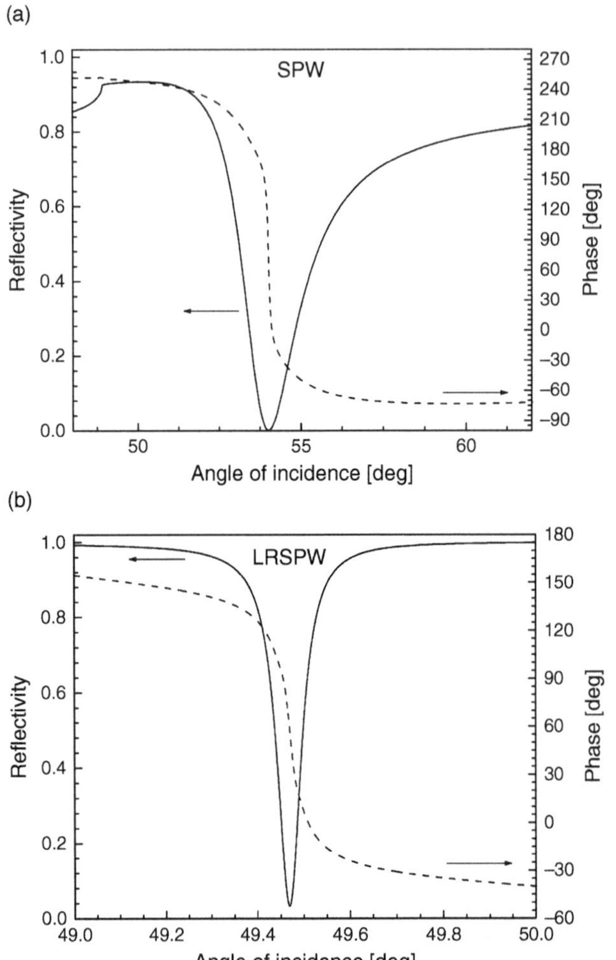

Figure 4.12 Reflectivity for a TM wave in: (a) Kretschmann (three-layer) geometry consisting of an SF14 glass prism (refractive index = 1.65), a gold layer (thickness = 50 nm), and a low-refractive index dielectric medium (refractive index = 1.32), and (b) four-layer geometry consisting of an SF14 glass prism, a low refractive index buffer (refractive index = 1.32, thickness = 1200 nm), a thin gold layer (20 nm), and a low refractive index dielectric medium (refractive index = 1.32), wavelength = 800 nm. Note the difference in the *x*-axis scale.

$z > 0$ with a dielectric), produces a change in the effective index of the SPW, $n_{ef} \rightarrow \bar{n}_{ef} = n_{ef} + \delta n_{ef}$:

$$\delta n_{ef} = \frac{n_{ef} \int\limits_{A_\infty} \frac{\delta\varepsilon}{\varepsilon^2} H_y^2 \, dA + \frac{1}{n_{ef} k^2} \int\limits_{A_\infty} \frac{\delta\varepsilon}{\varepsilon^2} \left(\frac{\partial H_y}{\partial z} \right)^2 dA}{2 \int\limits_{A_\infty} \frac{1}{\varepsilon} H_y^2 \, dA}, \tag{4.15}$$

where H_y denotes the modal field. The integration is performed over the cross section of the waveguide A_∞ (Homola, 2006).

In principle, there are two main types of refractive index perturbations that can occur in SPR sensors (Figure 4.13). The first type is a homogeneous change in the refractive index in the whole dielectric medium (typically referred to as bulk refractive index change). It can be shown that a change in the effective index of a surface plasmon due to a bulk refractive index change can be expressed as:

$$\delta n_{ef} = \frac{n_{ef}^3}{n_D^3} \delta n, \tag{4.16}$$

where the perturbation in the bulk refractive index δn and permittivity $\delta\varepsilon$ are related as $\delta\varepsilon = 2n_D \delta n$. As the effective index of the surface plasmon is always larger than the refractive index of the dielectric, the bulk refractive index sensitivity of the effective index of the surface plasmon $(\delta n_{ef}/\delta n)_B$ is always larger than the sensitivity of a free space plane wave in the infinite dielectric medium (which is equal to one).

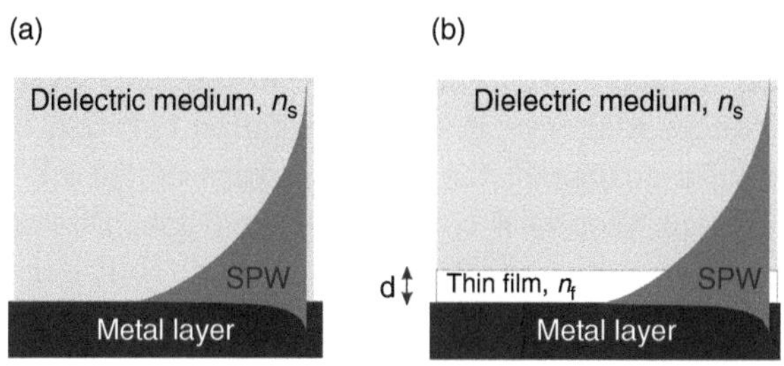

Figure 4.13 Surface plasma waves probing (a) a homogeneous medium and (b) a thin dielectric layer.

A change in the effective index of the surface plasmon induced by a surface refractive index change occurring within a layer with a thickness h can be calculated using

$$\delta n_{\text{ef}} = \frac{\text{Re}\left\{\beta^3\left[1-\exp\left(-2\gamma_{\text{D}}h\right)\right]\right\}}{k^3 n_{\text{D}}^3}\delta n \tag{4.17}$$

The perturbation of the effective index of a surface plasmon depends exponentially on the thickness of the layer within which the refractive index change occurs. For the thicknesses much larger than the penetration depth of the surface plasmon $h \gg L_{\text{pD}} = (1/\text{Re}\{\gamma_{\text{D}}\})$, the exponential term can be neglected and Eqn (4.17) changes into Eqn (4.16). For refractive index changes occurring within a layer much thinner than the penetration depth of the surface plasmon ($h \ll L_{\text{pD}} = 1/\text{Re}\{\gamma_{\text{D}}\}$) and a metal film with the real part of the permittivity of the metal much larger than the imaginary part, the sensitivity of the effective index of a surface plasmon to a surface refractive index change can be reduced to:

$$\left(\frac{\delta n_{\text{ef}}}{\delta n}\right)_{\text{S}} = 2\frac{n_{\text{ef}}^3}{n_{\text{D}}^3}\frac{h}{L_{\text{pD}}} = \left(\frac{\delta n_{\text{ef}}}{\delta n}\right)_{\text{B}}\frac{2h}{L_{\text{pD}}}, \tag{4.18}$$

where L_{pD} is the penetration depth into the dielectric (Homola, 2006).

4.1.3.2. Optical sensors based on surface plasmon-polaritons

Optical sensors based on resonant excitation of SPWs, often referred to as SPR sensors, are optical devices that exploit the sensitivity of the propagation constant of an SPW to refractive index to measure changes in the refractive index or changes in other quantities that can produce changes in the refractive index. A change in the refractive index to be measured produces a change in the propagation constant of the SPW, which results in a change in the characteristics of the LW interacting with the SPW (see Section 4.1.2.4). Based on which characteristic of the LW interacting with the SPW is measured, SPR sensors can be classified as follows:

Surface plasmon resonance sensors with angular modulation. The component of the LW's wavevector parallel to the metal surface matching that of the SPW is determined by measuring the coupling strength at multiple angles of incidence of the LW and determining the angle of incidence yielding the strongest coupling (Figure 4.14a, *upper plot*). The wavelength of the LW used to excite an SPW is fixed.

Surface plasmon resonance sensors with wavelength modulation. The component of the LW's wavevector parallel to the metal surface matching that of the SPW is determined by measuring the coupling strength at multiple wavelengths and determining the wavelength yielding the strongest coupling (Figure 4.14b). The angle at which the LW is incident onto the metal film is kept constant.

Surface plasmon resonance sensors with intensity modulation. The change in the intensity of the LW interacting with the SPW is measured (Figure 4.14b). Both the angle at which the LW is incident onto the metal film and its wavelength are kept constant.

Surface plasmon resonance sensors with phase modulation. The shift in phase of the LW interacting with the SPW is measured (Figure 4.14a, *lower plot*). Both the angle at which the LW is made incident onto the interface and its wavelength are kept constant.

Surface plasmon resonance sensors with polarization modulation. The amplitude and phase of the TM-polarized wave interacting with the SPW change if the propagation constant of the SPW changes. Transverse electric-polarized LW does not interact with SPWs and thus exhibits no resonant amplitude and phase variations. Therefore, the polarization state of the incident LW consisting of both the polarizations would also be sensitive to variations in the propagation constant of the SPW.

4.1.3.3. Surface plasmon resonance biosensors

In principle, any phenomenon that gives rise to a change in the refractive index at the surface of the SPW-active metal film can be observed and quantified by means of an SPR sensor. Surface plasmon resonance biosensors are SPR sensing devices that incorporate biomolecules, which recognize and are able to interact with selected analytes. These biomolecular recognition elements are immobilized on the SPR sensor

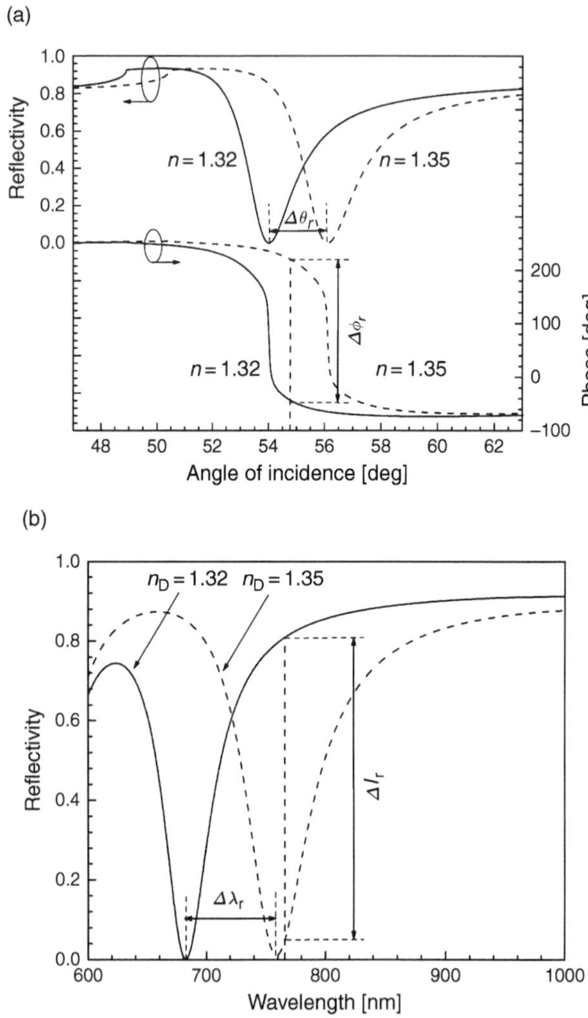

Figure 4.14 (a) Reflectivity and phase for a TM LW exciting SPWs in the Kretschmann geometry (SF14 glass prism – 50-nm-thick gold layer – dielectric) as a function of the angle of incidence for two different refractive indices of the dielectric; wavelength = 682 nm. (b) Reflectivity for a TM wave in the same geometry as a function of the wavelength for two different refractive indices of the dielectric; angle of incidence = 54°.

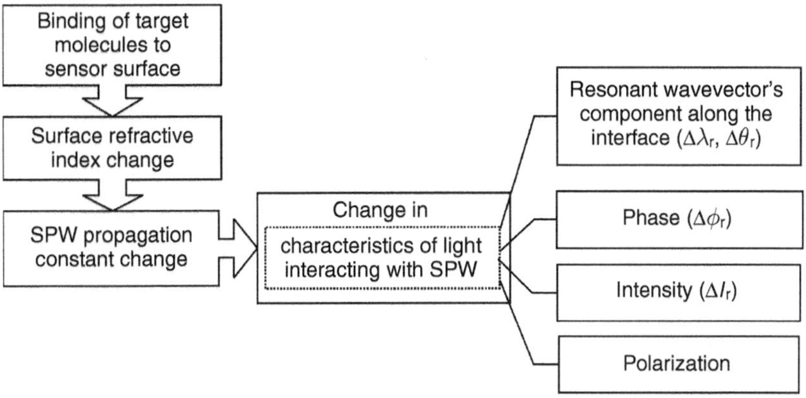

Figure 4.15 Surface plasmon resonance biosensor: principle of operation.

surface. When a solution containing analyte molecules is brought into contact with the SPR sensor, analyte molecules in solution bind to the recognition elements on the sensor surface, producing an increase in the refractive index at the sensor surface. This change produces a change in the propagation constant of the SPW and is determined by measuring a change in one of the characteristics of the LW interacting with the SPW (Figure 4.15).

4.1.3.4. Surface plasmon resonance sensors: main performance characteristics

The main performance characteristics of SPR biosensors include sensitivity, resolution, and the lowest detection limit. Sensor s*ensitivity* is the ratio of the change in sensor output to the change in the value of the measurand. In SPR biosensors, the sensor output is, depending on the modulation method used, the coupling angle of incidence, wavelength, intensity, phase, and polarization of the LW interacting with the SPW, and the measurand is the concentration of analyte. Mathematically, the sensitivity is the slope of the calibration curve Output = Output (Measurand):

$$S = \frac{\partial \text{Output}}{\partial \text{Measurand}} = \frac{\partial P}{\partial c}, \tag{4.19}$$

where P is the output of the sensor and c is the concentration of target analyte. As the output of an SPR biosensor is usually a non-linear function of the concentration of analyte, the sensor's sensitivity is not a constant, but varies with the analyte concentration. The sensitivity of an SPR biosensor can be decomposed into two components – sensitivity to the refractive index change produced by the binding of analyte to a biospecific coating on the sensor's surface, S_{RI}, and the efficiency, E, with which the presence of analyte at a given concentration, c, is converted into the change in the refractive index, n:

$$S = \frac{\partial P}{\partial n}\frac{\partial n}{\partial c} = S_{RI}E. \tag{4.20}$$

The efficiency, E, depends on the ability of the biospecific coating to capture analyte molecules and on the properties of the target analyte, and thus is application dependent. The refractive index sensitivity, S_{RI}, can be expressed as:

$$S_{RI} = \frac{\partial P}{\partial \mathrm{Re}\{\beta\}}\frac{\partial \mathrm{Re}\{\beta\}}{\partial n} = S_1 S_2 \tag{4.21}$$

The first term, S_1, depends on the modulation method and the method of excitation of the SPW and can be determined if the coupling condition is known for the sensor geometry (examples are Eqns (4.8, 4.9, and 4.11)). S_2 is independent of the modulation method and the method of excitation of the SPW and describes the sensitivity of SPW's propagation constant to the refractive index change and can be determined from Eqns (4.16–4.18), depending on whether the refractive index change occurs in the vicinity of the sensor's surface or through the whole depth of the SPW's field. The sensitivity of sensors based on the ATR method and angular modulation decreases with the increasing wavelength and, except for very short wavelengths, depends on the wavelength rather weakly (Homola *et al.*, 1999a). On the contrary, the sensitivity of SPR sensors with wavelength modulation increases rapidly with increasing wavelength. This difference in the spectral properties of refractive index sensitivities is illustrated in Figure 4.16, which shows the coupling between a LW in a prism and an SPW at the metal–dielectric interface for two different refractive indices of analyte.

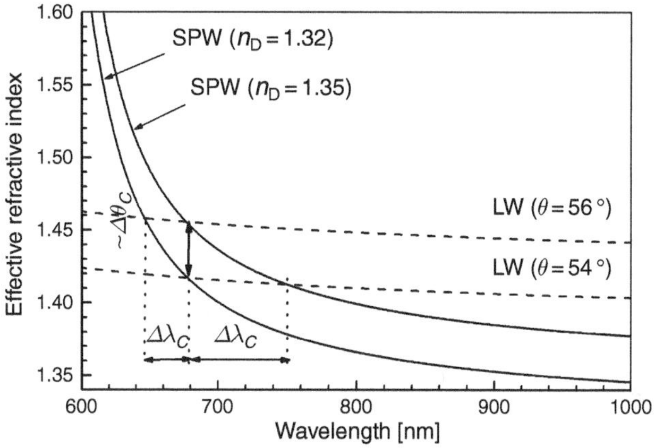

Figure 4.16 Spectral dependence of the effective index of an SPW at the gold–dielectric interface for two different refractive indices of the dielectric medium. Also shown is the spectral dependence of the component of the wavevector of a LW, which is parallel to the SPW-active metal surface, calculated for two different angles of incidence.

For a dielectric refractive index equal to 1.32, the coupling condition between the SPW and LW is fulfilled for an angle of incidence of 54° and a wavelength of 682 nm. A spectrally homogeneous increase in the refractive index of the dielectric by 0.03 produces an increase in the effective index of the SPW of 0.037 (at a wavelength of 650 nm) to 0.032 (at a wavelength of 1000 nm). To reestablish the coupling condition, the angle of incidence has to be increased by about 2°, or the wavelength has to be increased by about 73 nm. The sensitivity of an angle-modulated sensor is proportional to the change in the effective index of the SPW, which does not change much with the operating wavelength. However, the sensitivity of the wavelength-modulated sensor depends on the relative dispersion of the SPW and the LW, which varies substantially with the wavelength. As the dispersions of SPW and LW are more similar at longer wavelengths, a change in the effective index of the SPW produces a larger shift in the coupling wavelength at longer wavelengths. For example, for an angle of incidence of 56°, the coupling condition is fulfilled for a wavelength of 646 nm and the refractive index-induced change in the coupling wavelength is about 32 nm, and thus is smaller by a factor of more than two than for the sensor operating at wavelength of 679 nm and operating at the angle of incidence of 54°. Detailed studies of the sensitivity of SPR sensors have been previously published (Kooyman *et al.*, 1988; de Bruijn *et al.*, 1992; Yeatman, 1996; Homola, 1997, 1999a, 2006; Johansen *et al.*, 2000). *Resolution* is the smallest increment in the measurand that can be resolved by the sensor. Herein, we shall use this term for describing the ability of SPR instrumentation to resolve changes in the refractive index. We shall use the *lowest detection limit* to describe the lowest concentration of analyte that can be measured by the SPR biosensor. Both the resolution and lowest detection limit of an SPR biosensor are ultimately limited by the accuracy with which the SPR sensor's output can be determined, which is limited by the stability of the sensor's baseline (the sensor response stability over a period of time if no analyte is present) and noise of the sensor output. The noise is determined by noise properties of the components of the sensor system (Kolomenskii *et al.*, 1997; Nenninger *et al.*, 2002) and the data processing method used (Chinowsky *et al.*, 1999a; Tobiška and Homola, 2005).

4.2. History

Wood (1902), observing the spectra of a continuous light source using a reflection diffraction grating, noticed narrow dark bands in the spectrum of diffracted light. Theoretical analysis carried out by Fano (1941) led to the conclusion that these anomalies are associated with SPWs supported by a grating. Otto (1968) pointed out that SPWs may be excited using ATR. In the same year, Kretschmann and Raether (1968) reported excitation of SPWs in another configuration of the ATR method. Following the pioneering work of Otto, Kretschmann, and Raether, research in SPWs has gathered momentum and broadened (see, for example, Boardman, 1982; Raether, 1983). The potential of SPWs for characterizing thin films (Pockrand, 1978) and studying processes at metal interfaces (Gordon *et al.*, 1980) was recognized in the late 1970s.

In 1983, Nylander and Liedberg exploited SPWs excited in the Kretschmann geometry of ATR for gas detection and biosensing (Nylander *et al.*, 1982; Liedberg *et al.*, 1983). In the following years,

SPR sensors based on diffraction gratings (Cullen *et al.*, 1987/1988) and planar optical waveguides (Kreuwel *et al.*, 1987) were demonstrated. Reports of the first fiber optic SPR sensors followed several years later (Jorgenson and Yee, 1993; Dessy and Bender, 1994; Homola, 1995). Angular and spectral modulation methods were introduced to SPR sensing in the late 1980s (Matsubara *et al.*, 1988; Zhang and Uttamchandani, 1988). A commercial SPR biosensor based on angular modulation was launched by Biacore International AB in 1990. In the years following, Biacore (2007) has developed a range of laboratory SPR instruments based on the Kretschmann geometry of the ATR method and angular modulation.

In the late 1980s, the potential of the SPR method for spatially resolved measurements was recognized and the first surface plasmon-polariton microscopes were reported (Yeatman and Ash, 1987; Rothenhäusler and Knoll, 1988). In the following decade, the idea of SPR imaging was further advanced, yielding SPR sensing devices with a large number (>100) of sensing channels (Brockman *et al.*, 2000; Brockman and Fernandez, 2001; Fu *et al.*, 2004) and most recently, a high-performance sensor with polarization contrast and spatially patterned multilayer SPR structures (Piliarik *et al.*, 2005). In an effort to further increase the sensitivity of SPR sensors, phase-modulated SPR sensors based on heterodyne phase measurement (Nelson *et al.*, 1996) and interferometry (Nikitin *et al.*, 1999) were introduced in the late 1990s. High-sensitivity SPR sensors exploiting long-range surface plasmons have also been demonstrated (Nenninger *et al.*, 2001; Slavík and Homola, 2007). Recently, localized surface plasmons on metallic nanoparticles and nanostructures have gained a great deal of interest and their potential for development of SPR sensing devices has been evaluated (Stuart *et al.*, 2005).

4.3. State of the art

An SPR biosensor instrument consists of an optical system for excitation and interrogation of SPWs, a biospecific coating incorporating a biomolecular recognition element that interacts with analyte molecules in a liquid sample, a fluidic system comprising one or more flow cells or cuvettes for sample confinement at the sensing surface(s), and systems for sampling and sample delivery. This section reviews the state of the art in the two key components of SPR biosensors – optical instrumentation and biospecific coatings – and discusses the main application areas for SPR biosensors.

4.3.1. Surface plasmon resonance sensor instrumentation

4.3.1.1. Surface plasmon resonance sensors based on prism couplers

Attenuated total reflection in prism couplers has been used widely for the excitation of SPWs in SPR sensors. Prism-based SPR sensors based on the Kretschmann geometry of the ATR method and all the main modulation approaches have been demonstrated. These include SPR sensors based on intensity modulation (Nylander *et al.*, 1982; Vidal *et al.*, 1993), angular modulation (Matsubara *et al.*, 1988; Liedberg *et al.*, 1993), wavelength modulation (Zhang and Uttamchandani, 1988; Homola *et al.*, 1995), phase modulation (Nelson *et al.*, 1996; Nikitin *et al.*, 1999), and polarization modulation (Kruchinin and Vlasov, 1996). Angular and wavelength modulations have been used most frequently; they rely on multipoint measurements, yielding more robust measurements than simple intensity and phase modulation approaches relying on single-point measurement.

Figure 4.17 shows an example of a prism-based SPR sensor with angular modulation. A convergent beam of monochromatic light is launched into a prism coupler and made incident onto a thin metal film. The angular component of light that fulfills the coupling condition (see Eqn (4.8)) excites an SPW at the outer boundary of the metal film. The coupling produces a narrow dip in the angular spectrum of the reflected light. The position of this dip is measured with a position-sensitive photodetector (e.g., a charge-coupled device (CCD) or a photodiode array).

This configuration has been developed into numerous research laboratory SPR sensor systems and several commercial SPR sensor instruments (see Section 4.2). The bulk refractive index sensitivity

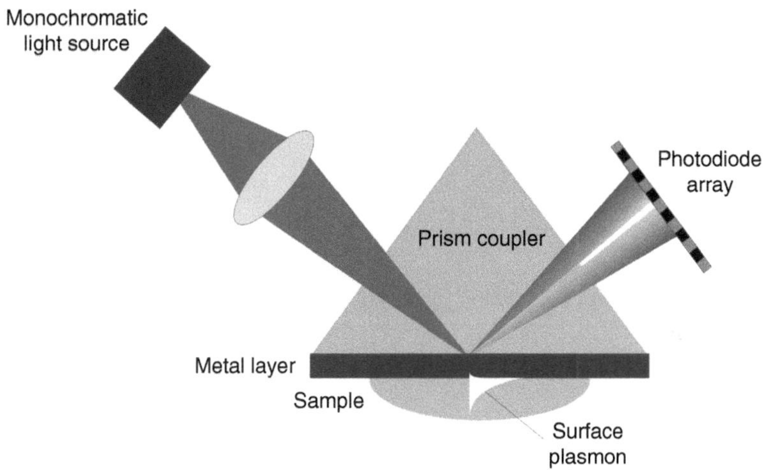

Figure 4.17 Surface plasmon resonance sensor based on angular modulation.

of these devices is typically 50–100° per refractive index unit (RIU), depending on the optical properties of the prism and the SPW-active metal layer, and the wavelength (Homola *et al.*, 1999a). The best laboratory SPR sensors based on prism coupling and angular modulation attain refractive index resolution down to 1×10^{-7} RIU (Biacore, 2007). Several attempts to miniaturize this SPR sensor configuration have been reported. The sensor reported by Foster *et al.* (1994) replaces a conventional prism with a thin rectangular glass substrate with an SPR-active metal film and integrates a source of monochromatic light and a CCD array into a portable system. Another system integrates a light source, a detector, and an optical system for excitation and interrogation of the SPW into a single component (Melendez *et al.*, 1996). This system uses a divergent light beam and attains a refractive index resolution of 5×10^{-6} RIU (Elkind *et al.*, 1999). Recently, a portable/mobile SPR sensor system based on this SPR optics has been reported (Naimushin *et al.*, 2005).

Wavelength modulation-based SPR sensors (Figure 4.18) use a collimated beam of polychromatic light to excite an SPW on a thin metal film. A spectral dip due to the excitation of SPW is observed using

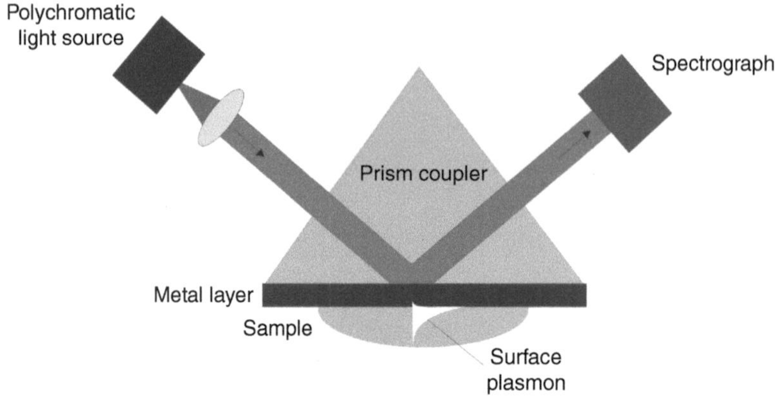

Figure 4.18 Surface plasmon resonance sensor based on spectral modulation.

a wavelength-sensitive photodetector (e.g., spectrograph, monochromator). The refractive index sensitivity of these sensors increases with the coupling wavelength and attains about 7100 nm/RIU around 800 nm. A refractive index resolution as good as 2×10^{-7} has been demonstrated with a wavelength modulation-based SPR sensor (Nenninger *et al.*, 2002).

Miniaturized versions of the wavelength-modulated SPR sensors based on retroreflective designs have been developed in which the reflected light with an encoded SPR dip in its spectrum is collected by the same optics used to launch light into the prism (Figure 4.19) (Cahil *et al.*, 1997; Stemmler *et al.*, 1999). A sensor combining angular and spectral modulations has been proposed by Karlsen *et al.* (1996). In this sensor, SPWs in different locations along a thin glass lightguide are excited by white light incident at different angles. Light beams associated with different angles are separated at the output of the lightguide and spectrally analyzed.

The ATR method is very attractive for the development of multichannel SPR sensing devices. The most straightforward approach to SPR multichannel sensing uses "parallel" sensing channels. In this approach, surface plasmons are excited simultaneously in multiple areas arranged perpendicular to the direction of propagation of SPWs. Light reflected

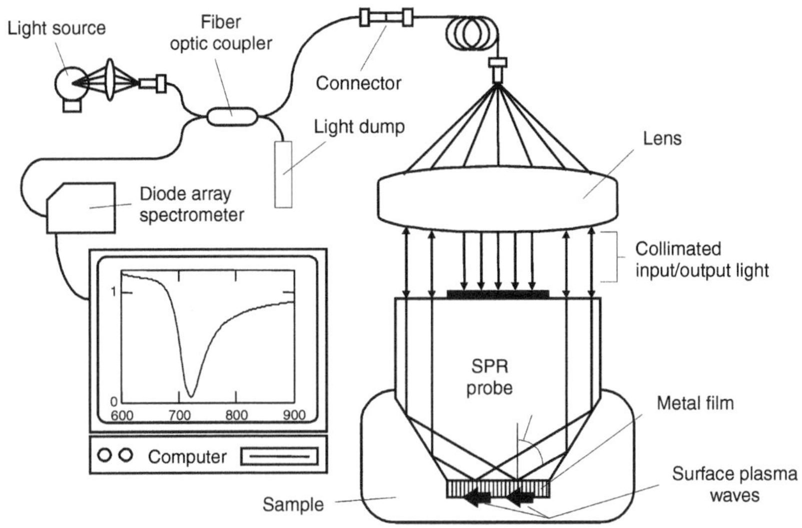

Figure 4.19 Surface plasmon resonance sensor using a miniature retroreflective sensing element. (Reproduced from Cahill *et al.*, 1997, with kind permission from Elsevier Science.)

from each area is separately analyzed to yield information about SPR in every location. Wavelength (Nenninger *et al.*, 1998) and angular (Löfås *et al.*, 1991) modulation-based SPR sensors with up to four parallel sensing channels have been demonstrated (Figure 4.20).

An alternative approach to multichannel sensing uses sensing elements of special designs to excite SPW in different channels by means of different spectral bands of the incident light. This spectral separation of optical signals from different sensing channels may be accomplished by serially exciting SPW in different areas of the sensing element by LWs of different angles of incidence (Figure 4.21a; Homola *et al.*, 2001a). Alternatively, an overlayer can be employed to shift the coupling wavelength for a part of the sensing surface to longer wavelengths (Figure 4.21b) (Homola *et al.*, 1999b).

Another approach to multichannel SPR sensing is based on the same idea as the surface plasmon microscopy (Rothenhäusler and Knoll, 1988).

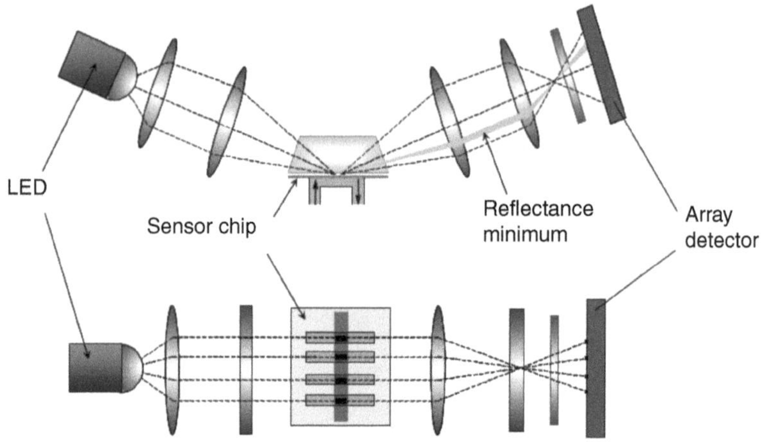

Figure 4.20 Surface plasmon resonance sensor with four parallel sensing channels. (Figure provided by S. Löfås, Biacore AB.)

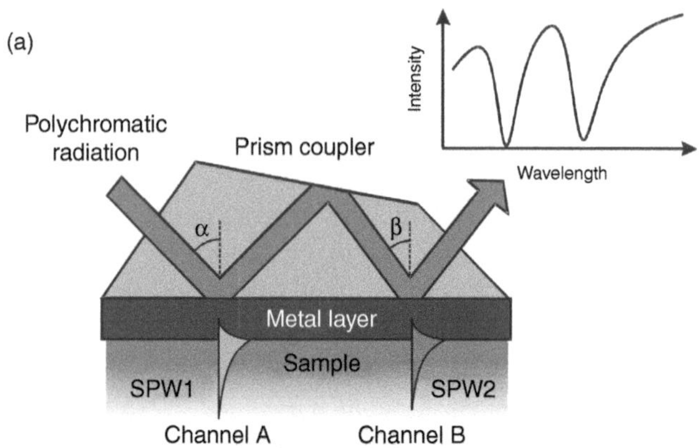

Figure 4.21 Surface plasmon resonance dual-channel sensors based on spectral discrimination of sensing channels. (a) Spectral discrimination of sensing channels by means of altered angles of incidence (Homola *et al.*, 2001a). (b) Spectral discrimination of sensing channels by means of a high refractive index overlayer (Homola *et al.*, 1999b).

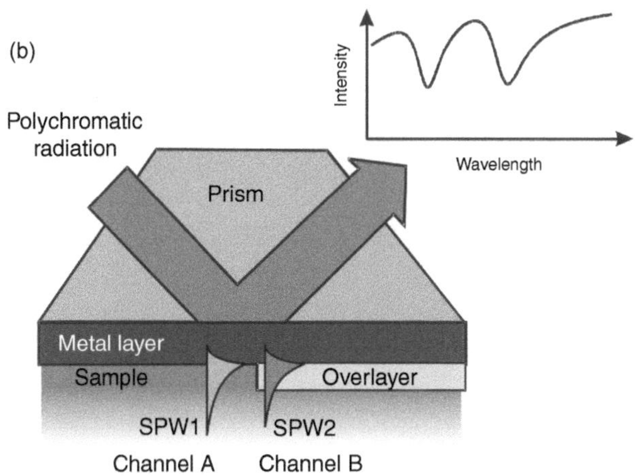

Figure 4.21 Continued

This approach, sometimes referred to as SPR imaging, uses a collimated beam of monochromatic light to illuminate the entire surface to simultaneously couple light to SPWs in multiple areas. The strength of the coupling depends on the refractive index at the sensor surface in each area. When the reflected light beam is imaged onto a two-dimensional detector array, light reflected from each area of the sensing surface is incident onto a different area of the detector (Figure 4.22). Analysis of intensity distribution yields information on refractive index in each area (Berger *et al.*, 1998; Nelson *et al.*, 1999; Brockman and Fernandez, 2001). In order to increase the sensor stability and optimize the contrast of SPR images, Fu *et al.* (2004) introduced an SPR imaging setup employing a white light source and a bandpass interference filter. Their SPR sensor instrument, operating at a wavelength of 853 nm, was demonstrated to provide a refractive index resolution of 3×10^{-5} RIU. The spatial resolution of this system was determined to be better than 50 μm. Recently, Piliarik *et al.* (2005) have reported the combination of SPR imaging with polarization contrast and a spatially patterned multilayer SPR structure. They demonstrated a refractive index resolution better than 2×10^{-6} RIU in more than 100 sensing channels (the size of each sensing spot was $400 \times 800\,\mu m^2$).

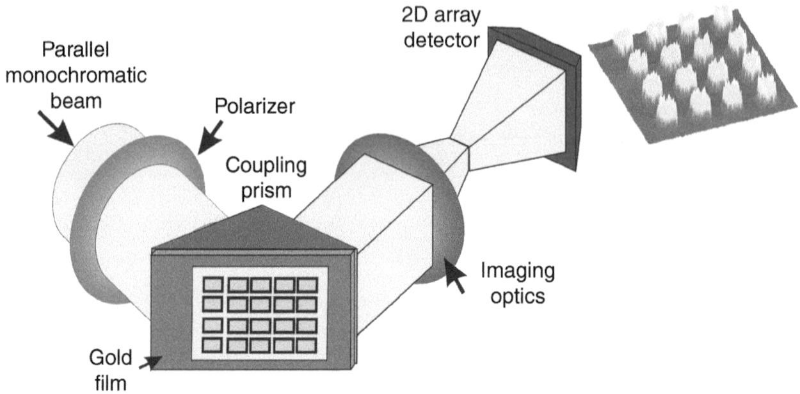

Figure 4.22 Surface plasmon resonance multichannel sensor based on SPR imaging.

In general, the use of prism couplers provides SPR sensors with a number of attractive features such as optical system simplicity, robustness, versatility, and potential for multichannel sensing. Although high-performance SPR sensing devices of this type tend to be rather bulky, laboratory bench-top instruments, high-performance portable SPR sensing devices will likely be developed in the near future.

4.3.1.2. Surface plasmon resonance sensors based on grating couplers

Light diffraction at the surface of a diffraction grating has been used in SPR sensors to a lesser extent than ATR in prism couplers mainly because grating-based SPR sensors require advanced modeling and optimization tools (Chandezon *et al.*, 1982; Hutley, 1982; Moharam and Gaylord, 1986) and complex fabrication procedures. Typical implementations of the grating-based SPR sensors include intensity-modulated (Cullen *et al.*, 1987/1988; Cullen and Lowe, 1990; Lawrence *et al.*, 1996) and wavelength-modulated (Vukusic *et al.*, 1992; Jory *et al.*, 1994) SPR sensing devices. A wavelength-modulated SPR sensor based on a silver-coated grating demonstrated by Jory *et al.* (1994) was shown to exhibit refractive index sensitivity of about 1000 nm/RIU. An intensity-modulated SPR sensor based on a gold-coated grating, described by Cullen (Cullen *et al.*, 1987/1988), showed a refractive index sensitivity of about 900% per RIU.

An interesting modification of the wavelength modulation approach using an acousto-optic modulator has been proposed by Jory *et al.* (1995) (Figure 4.23). In this approach, the acousto-optic modulator is used to control the wavelength of a light beam incident on a diffraction grating. Modulation of the light's wavelength around the minimum reflectivity produces a modulated intensity output that is proportional to the differential of the response with respect to wavelength. By locking to the zero differential corresponding to the SPR reflectivity minimum and monitoring the acousto-optic modulator's frequency, the SPR minimum position is measured with the accuracy of about 5×10^{-4} nm. This leads to a grating-based SPR sensor with a refractive index resolution below 10^{-6} RIU (Jory *et al.*, 1995). An SPR imaging device based on a grating coupler was also reported (Brockman and Fernandez, 2001).

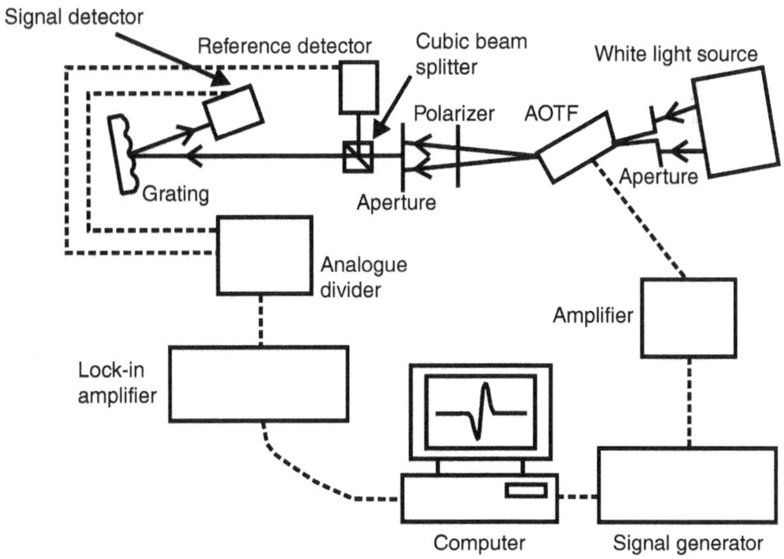

Figure 4.23 Surface plasmon resonance sensor based on a grating coupler and an acousto-optical modulator. (Reproduced from Jory *et al.*, 1995, with kind permission from IOP Publishing Limited.)

In this configuration, a collimated monochromatic light beam (wavelength 860 nm) is made incident onto a plastic chip with a gold-coated diffraction grating supporting an array of 400 sensing channels (each with a diameter of 250 μm). Upon reflection from the chip, the light is projected onto a two-dimensional CCD array. The grating coupler has been also used for the excitation of SPWs in an SPR sensing device based on a Schottky-barrier semiconductor structure, as demonstrated by Nikitin *et al.* (1994, 1997). A refractive index resolution of 1×10^{-5} RIU has been demonstrated with this SPR sensor.

Grating-based sensors offer several advantageous features over other configurations of SPR sensors. Gratings for SPR biosensors can be produced in plastic substrates (Lawrence *et al.*, 1996) by high-volume replication techniques (Gale, 1997). Precise control of the thickness of the SPW-active metal film is not required for grating-based SPR sensors. However, it may be considered a drawback that the light beam needs to pass through a sample and therefore the sample and the flow cell need to be optically transparent.

4.3.1.3. Surface plasmon resonance sensors based on integrated optical waveguides

Research into integrated optical waveguide SPR sensors was pioneered by researchers at the University of Twente in the late 1980s (Lambeck, 1992; additional references cited therein). Since then, SPR sensing devices using slab (Lavers *et al.*, 1994) and channel (Harris and Wilkinson, 1995) single-mode integrated optical waveguides have been reported. These sensing devices exhibit a rather limited operating range. The sensor's operating range can be adjusted by using waveguides fabricated in low refractive index glass (Harris and Wilkinson, 1995), waveguides with a buffer layer (Lavers *et al.*, 1994), a high refractive index overlayer (Čtyroký *et al.*, 1997), or a multilayer (Weiss *et al.*, 1996). Integrated optical SPR sensors exhibit a refractive index sensitivity of about 2000 nm/RIU (Homola *et al.*, 1997). An integrated optical SPR sensor with a reference arm compensating for variations of light levels in the waveguide input has been described by Mouvet *et al.* (1997) (Figure 4.24). An alternative approach, based on a channel-integrated optical waveguide and wavelength modulation, has been demonstrated

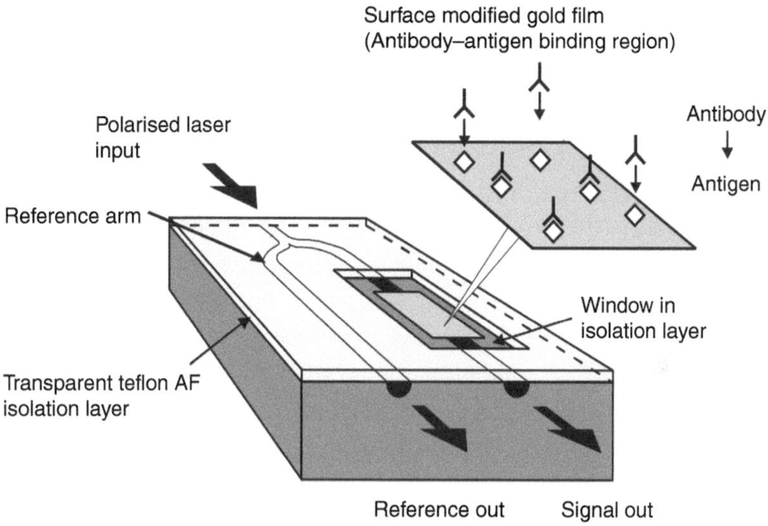

Figure 4.24 Surface plasmon resonance sensor based on integrated optical waveguide. (Reproduced from Harris *et al.*, 1999, with kind permission from Elsevier Science.)

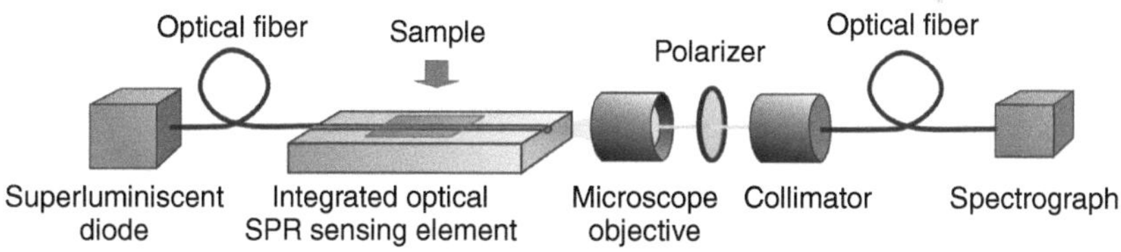

Figure 4.25 Surface plasmon resonance sensor based on an integrated optical waveguide and wavelength modulation. (Reproduced from Dostálek *et al.*, 2001, with kind permission from Elsevier Science.)

by Dostálek *et al.* (2001) (Figure 4.25). This sensor was demonstrated to exhibit a refractive index sensitivity of 2100 nm/RIU and resolution of 1×10^{-6} RIU.

Integrated optical waveguide SPR sensors hold promise for the development of multichannel sensing devices with on-chip referencing and multianalyte detection capability. Challenges for widespread implementation of this SPR sensor configuration lie in the development of efficient and inexpensive methods for coupling light in and out of the integrated optical waveguide.

4.3.1.4. Surface plasmon resonance sensors based on optical fibers
Direct excitation of SPWs by modes of optical fibers presents an interesting approach to the development of miniature SPR sensing devices. The first SPR sensor based on a multimode optical fiber was reported by Jorgenson and Yee (1993). This wavelength modulation-based SPR sensor uses a conventional polymer clad silica fiber with partly removed cladding and a metal film deposited symmetrically around the exposed section of fiber core. This approach yields a miniature optical fiber SPR probe with an interaction area about 10 mm long. This sensor is able to detect refractive index variations with a resolution up to 5×10^{-5} RIU within its operating range of 1.2–1.4 RIU (Jorgenson and Yee, 1993). The operating range of the sensor may be adjusted for sensor applications in the refractive index range 1–1.7 RIU by using a thin high refractive index dielectric overlayer and high refractive index core fibers (Jorgenson and Yee, 1994) (Figure 4.26).

A similar geometry, in which the sensing area is formed not at the tip but in the middle of an optical fiber, has been used for the development of an intensity modulation-based SPR sensor (Ronot-Trioli *et al.*, 1996; Trouillet *et al.*, 1996). In this configuration, a collimated monochromatic light beam is launched into a fiber in such a way that only modes with propagation constants within a narrow range are efficiently excited. Variations in the refractive index of the sample are determined by measuring the transmitted optical power. The sensitivity of this SPR sensor is somewhat reduced because numerous fiber modes, which are incident on the metal surface at slightly different angles, take part in the excitation of the SPW. The reported sensor resolutions are 8×10^{-5} RIU for a gold-based sensor and 5×10^{-5} RIU for a silver-based sensor. These superior resolutions are due to a relatively low system noise level, which allows for accurate measurement of intensity changes as small as 0.2% (Ronot-Trioli *et al.*, 1996; Trouillet *et al.*, 1996).

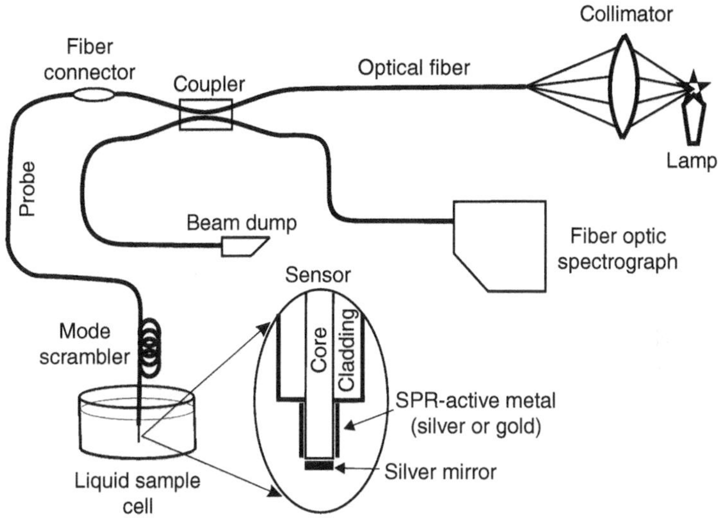

Figure 4.26 Surface plasmon resonance probe based on a multimode optical fiber. (Reproduced from Jorgenson and Yee, 1994, with kind permission from Elsevier Science.)

Surface plasmon resonance sensors based on single-mode optical fibers have also been reported (Dessy and Bender, 1994; Homola, 1995). These sensors employ an optical fiber with a locally removed cladding and an SPW-active metal film. A guided mode propagates in the fiber and excites an SPW at the outer surface of the metal film if the two modes are closely phase matched. This sensor can be designed to operate as a transmissive sensor (Homola, 1995) or a fiber optic probe (Slavík *et al.*, 1998) (Figure 4.27).

An intensity-modulated SPR sensor of this design has been demonstrated to exhibit sensitivity of 3900 dB/RIU in the refractive index range 1.4193–1.4104 and 2300 dB/RIU in the operating range 1.3302–1.3422, which correspond to a resolution better than 2×10^{-5} RIU, if variations in light intensity of 1% can be resolved (Slavík *et al.*, 1999). A wavelength modulated-version of this SPR sensor has been shown to attain a refractive index sensitivity of 3100 nm/RIU and a resolution of 5×10^{-7} RIU (Slavík *et al.*, 2001). Performance of this sensor is,

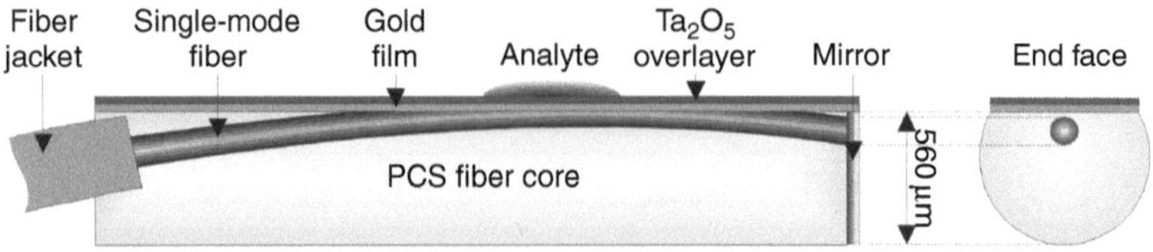

Figure 4.27 Surface plasmon resonance probe based on a single-mode optical fiber. (Reproduced from Slavík *et al.*, 1998, with kind permission from Elsevier Science.)

however, limited by the sensor's sensitivity to the polarization state of the fiber mode in the interaction region of the sensor, which limits its actual resolution to 3×10^{-5} RIU. This sensitivity can be dramatically reduced by employing polarization-maintaining fibers (Homola *et al.*, 2001b; Piliarik *et al.*, 2003), yielding sensor resolutions of 2×10^{-6} RIU.

Surface plasmon resonance fiber optic sensors present the highest degree of miniaturization of SPR sensor technology. The main challenge for SPR fiber optic sensors remains the development of multichannel devices with referencing capabilities.

4.3.2. Biosensor surface chemistries

Various biomolecular interactions have been used in SPR biosensors. These include antigen–antibody, hormone–receptor, protein–protein, DNA–DNA, and DNA–protein interactions. An essential requirement of SPR biosensors is that one of the interacting partners must be immobilized to the sensor surface. The first SPR biosensors were functionalized by simple surface adsorption of antibodies on a gold film (Liedberg *et al.*, 1983). However, attaching biomolecules such as proteins directly to metal surfaces often denatures the molecule, leading to a loss of binding activity. Moreover, this approach provides no control over the orientation of the antibodies.

Several immobilization chemistries that provide desired chemical properties for stable and defined binding of ligands have been developed. One approach is based on covalent attachment of the biological recognition element to the metal film via a linker layer. This can be performed by forming a self-assembled monolayer of thiol molecules (e.g., alkanethiols) with suitable reactive groups on one end of the molecule and a gold-complexing thiol on the other (Duschl *et al.*, 1996). Then the recognition element can be attached to the thiols, forming a monolayer of ligand molecules. Another approach uses a hydrogel matrix composed of carboxyl-methylated dextran chains to yield a two-dimensional matrix for ligand attachment (Löfås *et al.*, 1990). The matrix provides a hydrophilic environment conducive to maintaining the activity of biomolecules. The non-cross-linked dextran is also thought to maintain much of the entropic properties of the immobilized macromolecule and allow for diffusion within the local environment. In addition, the matrix increases the sensitivity of the SPR biosensor by providing an extended volume within the field of the SPW. Carboxyl groups on the dextran are easily modified using standard coupling chemistries allowing proteins to be attached via surface-exposed amine, carboxyl, sulfhydryl, and aldehyde groups. Modified versions of dextran have also been used for different applications, including shorter dextran to characterize larger molecular weight analytes and lower-charge dextran to reduce non-specific binding caused by ionic interactions. Alternatively, metal surfaces may be functionalized by thin polymer films to which ligands may by coupled via amino groups (Nakamura *et al.*, 1997).

A rapidly evolving field of biosensor applications involves the use of membrane or lipid surfaces as the binding substrate. In fact, the solid phase nature of biosensor technology makes it ideal for characterizing the interactions of molecules with membrane surfaces. Self-assembled lipid layers are routinely generated using gold–thiol coupling chemistry, but additional surface designs have been used to create stable hybrid lipid bilayers as well as suspended bilayer systems (Cooper and Williams, 1999). Several groups have used these surfaces to measure specific interactions of proteins with phospholipid head groups (Evans and MacKenzie, 1999; Santagata *et al.*, 2001) as well as to characterize lateral interactions between proteins bound to a lipid interface (Lahiri *et al.*, 1999).

4.3.3. Applications of SPR biosensors

There are two major application areas for SPR biosensing devices – detection of biological analytes and biophysical analysis of biomolecular interactions.

4.3.3.1. Surface plasmon resonance biosensor-based detection of bioanalytes

To date, SPR biosensors have been widely demonstrated for sensitive detection of various analytes. Various detection formats have been employed in SPR biosensing in order to optimize performance of SPR biosensors for specific applications. The main detection formats include direct, sandwich, and competitive assays (Figure 4.28). In the direct detection format, a sample containing target molecules is brought into contact with the sensor surface coated with respective biomolecular recognition elements (e.g., antibodies). Binding of analyte molecules to antibodies produces an increase in the refractive index at the sensor surface. The SPR sensor instrument translates this change into a change in sensor response. A sandwich assay consists of two steps. In the first step, the analyte molecules bind to antibodies immobilized on the sensor surface as in the direct detection format. In the second step, the sensor is incubated with a solution containing secondary antibodies, which bind to the previously captured analyte, enhancing the specific sensor response. Another two-step detection format is based on competitive

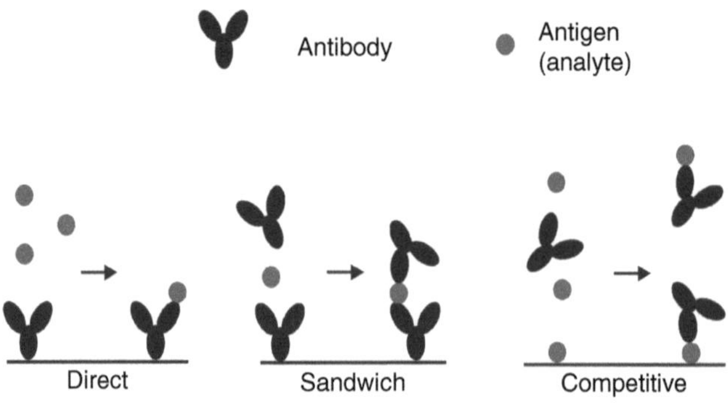

Figure 4.28 Surface plasmon resonance biosensing – main types of detection formats.

reaction of analyte in solution and analyte immobilized on the surface with free antibodies. In the first step of the competitive assay, a sample is mixed with antibodies. Analyte molecules in the sample bind to antibodies and block their binding sites. Then, the sample with antibodies is brought to the sensor surface so that the unbound antibodies can bind to analyte molecules immobilized on the sensor surface. In this type of assay, the SPR biosensor measures the concentration of unbound antibodies, which can be used to calculate the concentration of the target analyte.

In general, the choice of detection format depends on the specifics of the application (size of target analyte molecules, binding characteristics of available biomolecular recognition element, range of analytes concentrations to be measured). Direct detection is usually preferred in applications when binding of analyte at appropriate concentrations produces a sufficient direct response. If necessary, the lowest detection limits of the direct SPR sensors can be improved by using a sandwich assay. The secondary antibodies may also be coupled to large particles such as latex particles (Severs, 1993) and gold beads (Leung, 1994) to further enhance the SPR sensor response. Competitive assays are used in SPR biosensor-based detection of low concentrations of small analytes, whose binding to surface-immobilized recognition molecules does not produce a measurable sensor response.

Small molecules detected by SPR biosensors include the herbicides atrazine (Minunni and Mascini, 1993) and simazine (Mouvet *et al.*, 1997; Harris *et al.*, 1999) and the drugs morphine (Sakai *et al.*, 1998) and methamphetamine (MA) (Sakai *et al.*, 1999). Minunni and Mascini (1993) used an SPR sensor and competitive (binding inhibition) assay to detect atrazine. Monoclonal antibodies against atrazine were mixed with the sample containing atrazine and the free antibody concentration was determined by exposing the sample to the atrazine derivative-coated SPR biosensor. A detection limit of 0.05 ng/ml was achieved. Simazine detection in water samples was demonstrated using an integrated optical SPR sensor and binding inhibition assay (Mouvet *et al.*, 1997; Harris *et al.*, 1999). The lowest detection limit for simazine was determined to be 0.1 ng/ml (Harris *et al.*, 1999). Morphine detection based on a binding

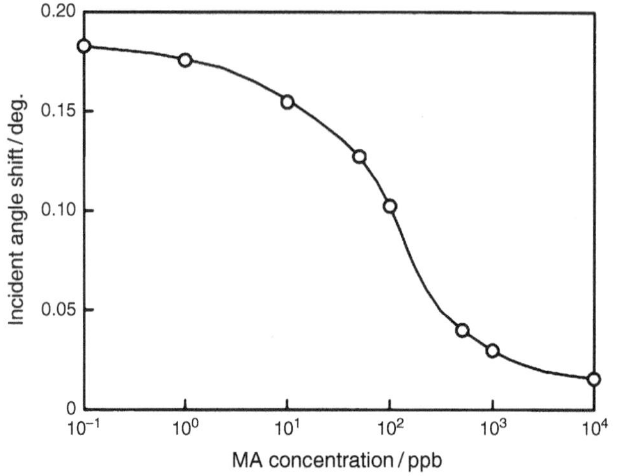

Figure 4.29 Detection of MA using SPR biosensor and binding inhibition assay. Coupling angle change as a function of MA concentration; anti-MA antibody concentration – 20 ppm. (Reproduced from Sakai *et al.*, 1999, with kind permission from Elsevier Science.)

inhibition assay was reported by Miura *et al.* (1997), who detected morphine at concentrations down to 0.1 ng/ml. Sakai *et al.* (1999) developed an SPR biosensor for detection of MA. Figure 4.29 shows their calibration curve for MA concentrations ranging from 0.1 ng/ml to 10 μg/ml. As their SPR biosensor uses a binding inhibition assay and the unbound antibodies rather than the molecules of MA are detected by the sensor, the sensor response is inversely proportional to MA concentration. The lowest detection limit of the SPR biosensor was 0.1 ng/ml (Sakai *et al.*, 1999).

Examples of medium-sized molecules that have been detected by SPR biosensors are staphylococcal enterotoxin B (SEB) (Homola *et al.*, 2001c; Rasooly, 2001) and human choriogonadotropin (Dostálek *et al.*, 2001). Detection of SEB was performed using both angular (Rasooly, 2001) and wavelength (Homola *et al.*, 2001c) modulation-based SPR sensing devices. The lowest achieved detection limits were 10 ng/ml (Rasooly, 2001), and 5 and 0.5 ng/ml for direct detection

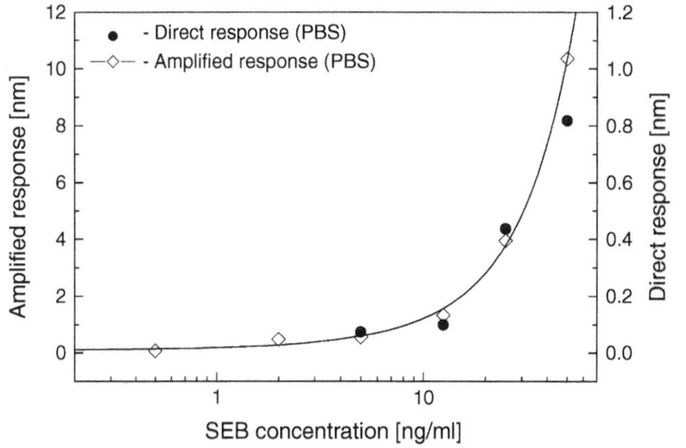

Figure 4.30 Detection of SEB using SPR biosensor. Equilibrium SPR sensor response to different concentrations of SEB in buffer (PBS), using direct and sandwich detection modes (Homola *et al.*, 2001c).

and sandwich assay modes, respectively (Homola *et al.*, 2001c) (Figure 4.30). Surface plasmon resonance biosensor-based detection, followed by matrix-assisted laser desorption/ionization time-of-flight mass spectrometry, made it possible to detect SEB in milk and mushroom samples at levels of 1 ng/ml (Nedelkov *et al.*, 2000). Direct detection of human choriogonadotropin by means of an integrated optical SPR sensor was demonstrated by Dostálek *et al.* (2001). The lowest detection limit was 2 ng/ml. A prism-based SPR sensor with wavelength modulation and a synthetic peptide as a recognition element for detection of antibodies against the Epstein–Barr virus was reported (Vaisocherová *et al.*, 2007). The sensor was capable of detecting concentrations of anti-EBNA as low as 0.2 ng/ml (~1 pM) both in buffer and in 1% human serum (Vaisocherová *et al.*, 2007).

Bacterial pathogens *Escherichia coli* O157:H7, *Salmonella enteritidis*, and *Listeria monocytogenes* are examples of large analytes that have been targeted by SRP biosensor technology. Detection of *E. coli* O157:H7 was performed by Fratamico *et al.* (1998), who used a Biacore SPR sensor and sandwich assay format. They used monoclonal antibodies immobilized on the sensor surface for capturing *E. coli*

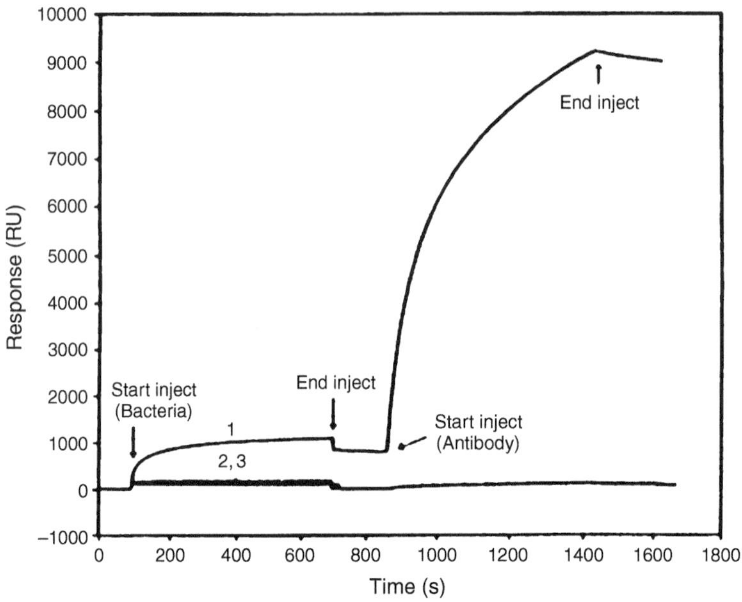

Figure 4.31 Detection of *E. coli* O157:H7 using SPR biosensor and sandwich assay. Curve 1 represents sensor response to *E. coli*; 2, *Salmonella typhimurium*; 3, *Yersinia enterocolitica*; bacterial concentrations are approximately 5×10^9. (Reproduced from Fratamico *et al.*, 1998, with kind permission from Kluwer Academic Publishers.)

and polyclonal secondary antibodies for enhancing the specific sensor response (Figure 4.31). The lowest detection limit for *E. coli* was established at 5×10^7 cfu/ml. Direct detection of *S. enteritidis* and *L. monocytogenes*at concentrations down to 10^6 cfu/ml was demonstrated by Koubová *et al.* (2001). Taylor *et al.* (2006) demonstrated the quantitative and simultaneous detection of *E. coli* O157:H7, *S. choleraesuis*, *L. monocytogenes*, and *Campylobacter jejuni* in buffer and apple juice using an eight-channel SPR sensor with wavelength modulation. The limits of detection were determined to be 1.4×10^4 cfu/ml for *E. coli*, 4.4×10^4 cfu/ml for *S. choleraesuis*, 1.1×10^5 cfu/ml for *C. jejuni*, and 3.5×10^3 cfu/ml for *L. monocytogenes*.

In order to optimize detection limits, SPR biosensors usually use biomolecular recognition elements exhibiting rather high affinities to the target analyte. However, the strong interaction between the biomolecular recognition elements and the target analyte most often does not allow the analyte to dissociate if the concentration of analyte in the sample drops. Thus, such biosensors cannot perform continuous monitoring of concentration of the target. The feasibility of SPR biosensors for continuous monitoring was investigated by Ohlson *et al.* (2000) who demonstrated continuous detection of biopharmaceuticals and antibody titers in patient sera using low-affinity antibodies.

4.3.3.2. Surface plasmon resonance biosensors for biophysical analysis of biomolecular interactions

The key advantages of SPR biosensors are the lack of labeling requirements and the ability to monitor binding reactions in real time. These advantages allow the analysis of interaction of nearly any binding system, and it is therefore no surprise that SPR biosensors have become a mainstay of both life science and pharmaceutical research. Surface plasmon resonance biosensors can be used to measure binding kinetics directly, including association rates (k_a) and dissociation rates (k_d), as well as equilibrium constants (K_D) for complex formation. Surface plasmon resonance instruments that can collect binding data at a variety of temperatures make it possible to extract the temperature dependence of the rate constants, which provides detailed information about the enthalpic (ΔH) and entropic (ΔS) thermodynamic parameters for the transition state and complex (Roos *et al.*, 1999; Myszka, 2000).

The quality of data available from SPR biosensors is high enough to support global data analysis as a method of extracting rate constants (Morton, 1998). In global analysis, the association and dissociation phase data acquired for a set of analyte concentrations over the same ligand surface are fit simultaneously using one set of rate constants. Global fitting results in a more robust evaluation of the shared parameter values, which should be independent of the concentration of the analyte and the surface density of the immobilized ligand. The statistical behavior of the parameter estimates is also improved (Morton *et al.*, 1995). For example, Figure 4.32 shows the binding data collected for the small molecule 4-carboxy-benzenesulfonamide interacting with carbonic anhydrase II immobilized on to the surface of a BIACORE

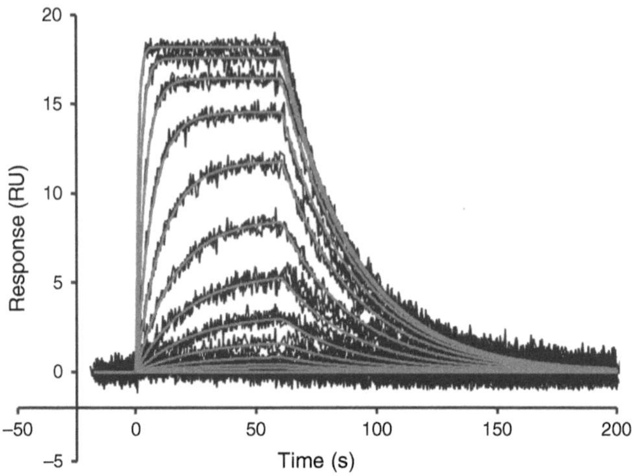

Figure 4.32 Binding of 4-carboxy-benzenesulfonamide to carbonic anhydrase II immobilized on a BIACORE 2000. Triplicate injections of 4-carboxy-benzenesulfonamide were made from a concentration of 20 μM–40 nM in twofold dilutions. The experimental data (black lines) were globally fit to a simple A + B = AB mechanism.

2000 biosensor. The experimental data were globally fit to a simple 1:1 interaction model (A + B = AB) to determine the reaction rate constants ($k_a = 4.8 \pm 0.2 \times 10^4\, M^{-1}\, s^{-1}$ and $k_d = 0.0365 \pm 0.0006\, s^{-1}$). The kinetic rate constants provide more detailed information about a binding interaction than equilibrium constants (K_D). For example, two different systems may have the same affinity, but vastly different rate constants, giving them different biological properties. Association and dissociation rate constants that can be measured with current SPR biosensor instruments (Biacore, 2007) range from 1×10^3 to $1 \times 10^7\ M^{-1}\ s^{-1}$ and from 1×10^{-5} to $0.5\, s^{-1}$, respectively, allowing the determination of equilibrium dissociation constants (K_D) in the range of 1 mM to 1 pM.

From a practical standpoint, commercial biosensors have been most often used to characterize the binding activity of monoclonal antibodies. Biopharmaceutical companies routinely use SPR biosensor technology to determine the kinetics of antibody/antigen interactions during initial screening, as well as to provide quality control during production.

Biosensor analysis has also become an important tool for characterizing engineered proteins and for identifying the roles of side chains within protein interfaces. Examples include mapping ligand/receptor interfaces to identify hot spots of activity (Pearce *et al.*, 1996) and dissecting the roles of residues in complex association and dissociation (Katsamba *et al.*, 2001).

While kinetic analysis has been a major emphasis of biosensor use, there are other important applications. Biosensors are often used in a qualitative format to identify overlapping binding sites for antibodies. Typically, epitope mapping experiments are done by capturing an analyte on a surface using one antibody and testing if other antibodies are capable of recognizing the bound antigen (Daiss and Scalice, 1994). In addition, because reactants can be added sequentially in time, it is possible to characterize molecular assembly processes to identify the order of association of subunits in a large complex. There are also examples of using SPR to monitor the refolding of a protein after acid denaturation. A change in response is observed as an immobilized protein transitions from an unfolded to a folded state due to changes in its dielectric constant (Sota *et al.*, 1998).

4.4. Advantages and limitations of SPR biosensors

Surface plasmon resonance biosensors belong to a family of thin film refractometry-based sensors such as the grating coupler (Nellen and Lukosz, 1990), the resonant mirror (Buckle *et al.*, 1993), and the integrated optical interferometer (Heideman *et al.*, 1993), which all measure refractive index changes produced by biomolecular interactions occurring at the surface of the sensor. These sensors exhibit similar fundamental advantages and limitations. The main advantage of these approaches is their ability to detect molecular interactions directly without the use of radioactive or fluorescent labels; this characteristic makes it possible to observe biomolecular interactions in "real time". Another advantage is the versatility of these techniques for detection and/or analysis of different types of analytes and interactions – analyte molecules do not have to exhibit any special properties such as fluorescence or characteristic absorption and scattering bands.

Currently, a large variety of SPR sensor platforms – from miniature SPR fiber optic probes to robust bench-top laboratory units – are available to meet special requirements of a multitude of biosensing applications. Although the development of parallelized SPR sensor platforms with a large number of sensing channels (>100) has progressed a great deal, their performance characteristics such as resolution and limit of detection are still behind the best spectroscopic SPR sensors by more than an order of magnitude and need to be improved.

The main challenge for the SPR biosensors is in the specificity of detection, which is solely based on the ability of the biomolecular recognition coating to recognize and capture target analyte molecules. Any change in the refractive index of a sample and/or adsorbing of non-target molecules on the sensor surface can produce a false sensor response. Therefore, SPR sensors use reference channels to compensate for these interfering effects. In conventional SPR systems for biophysical analysis of biomolecular interactions, one of the sensing channels (specific channel) is typically treated with a ligand that binds analyte, while the other (reference channel) is treated with the coupling chemistry in the absence of added ligand or coupled with a protein that does not interact with the analyte. Although such sensors deal with pure samples in controlled laboratory environments, changes in the background refractive index introduced by exchanging buffers or by injecting high concentrations of analyte in the study of low-affinity interactions cannot be entirely compensated. The uncompensated non-specific response (typically several percent of the background refractive index change), in particular the transient part following introduction of the background refractive index change, can cause errors in the determination of kinetic constants of biomolecular interactions (Ober and Ward, 1999).

Compensation for non-specific effects is even more crucial in SPR biosensors for detection of bioanalytes in crude samples and out-of-laboratory environments. Development of robust surface chemistries suppressing non-specific adsorption on the sensor surface and referencing methods for discrimination of the specific response (binding of analyte) from the non-specific response (adsorption of non-target molecules on the sensor surface or sample refractive index variations) remains one of the main challenges in development of practical biosensors for detection of bioanalytes in the field.

Typical high-performance SPR sensors exhibit refractive index resolution in the low 10^{-7} RIU range, making it possible to detect small and medium-sized analytes in concentrations well below 1 ng/ml, which is sufficient for a wide range of analyses. For example, minimum lethal doses for most potent proteinaceous toxins are about 100 ng (Gill, 1982). The ability of SPR biosensors to detect larger analytes such are bacteria and cells with detection limits on the order of 10^4–10^5 cfu/ml is, however, less satisfactory (Ryan, 1990). One of the key factors limiting the ability of SPR biosensors to detect bacteria is the size of a bacterium; for a bacterium of typical size ~1 μm, the bulk of the bound cell is situated outside the evanescent field of an SPW. Furthermore, low concentration of the target antigen (relative to total cellular material) and slow diffusion of bacterial cells to the sensor surface also limit detection capabilities (Perkins and Squirrell, 2000). However, in the field of biophysical analysis of biomolecular interactions, SPR biosensors allow for monitoring of both weak and strong interactions (K_D ranging from 1 mM to 1 pM). Compared with traditional biophysical techniques used to measure binding interactions (titration calorimetry, ultracentrifugation, stopped flow, and column chromatography), SPR biosensors typically require less material and have a higher throughput.

4.5. Potential for expanding current capabilities

Improvements in the detection capabilities of SPR biosensors are desired to enable direct detection of biomolecular interactions involving small molecules or extremely low analyte concentrations. In order to accomplish this goal, approaches to improving sensitivity and resolution of SPR sensor instruments are being developed. One promising approach to the development of high-sensitivity SPR sensors is based on the exploitation of phase properties of light coupled to SPW. A phase modulation-based SPR sensor has been developed by Nelson *et al.* (1996), who measured the phase of the LW using the optical heterodyne technique. This approach demonstrated the potential of phase-modulated SPR sensors; however, it required a rather bulky and complex instrument requiring an acousto-optic modulator and sophisticated data processing. Another interesting approach to enhancing sensitivity of SPR sensors is based on the exploitation of long-range SPWs (LRSPWs) (Nenninger *et al.*, 2001; Slavík and Homola, 2007), which exhibit lower dispersion than classical SPWs supported by a metal–dielectric interface (see Figures 4.3 and 4.6). As the long-range SPWs penetrate deeper into the probed sample, the use of these LRSPWs may also potentially enhance detection capabilities of SPR biosensors for large analytes (e.g., bacteria). Besides yielding enhanced sensitivities, long-range surface plasmons produce very narrow angular or spectral dips more than 10-fold narrower than those produced by conventional SPWs (Figure 4.12). This makes it possible to determine the spectral position of the SPR dip with high accuracy, consequently lowering the sensor resolution down to the low 10^{-8} RIU range (Slavík and Homola, 2007).

Referencing approaches are being investigated for improving robustness and specificity of SPR biosensors. One interesting approach to SPR biosensing with compensation for sample refractive index variations was reported by Chinowsky *et al.* (1999b), who combined an SPR sensor with angular modulation with a critical angle refractometer. Another promising approach to separation of specific and non-specific sensor responses is multiple-surface-plasmon spectroscopy, which can be performed in SPR sensors using prism couplers (Homola *et al.*, 2001d) or grating couplers (Adam *et al.*, 2006).

From the standpoint of biological applications, the field of membrane biology could benefit from advances in immobilization methods. There is a tremendous need to develop methods incorporating specific transmembrane receptors, such as G-protein-coupled receptors, onto the biosensor surface. Recently, Salamon *et al.* (2000) have reconstituted transmembrane receptors on synthetic lipid layers over a silver surface. Using a plasmon-coupled waveguide sensor, they were able to detect specific conformational states of the receptor upon the binding of various agonist and antagonist ligands. Routine methods for studying membrane-associated receptors would open up entirely new avenues to study the interactions of these receptors in a more native environment.

Advances in the development of new SPR biosensor hardware will be driven by the needs of the consumer. A large untapped market for biosensors lies in industries that require high sensitivity as well as high throughput. Food and environmental analyses could benefit greatly from the real-time aspects of biosensor analysis, once throughput has been increased. The pharmaceutical industry, which was fast to adopt optical biosensors as tools to characterize biopharmaceutical products, is beginning to implement the technology in small molecule screening, as well as in general bioavailability analysis. Surface plasmon resonance biosensors could also play an important role in defense, where fast, portable, and durable units are needed for early detection and identification of hazardous agents in the field. Development of these systems will require significant advances in miniaturization of SPR detection platforms and their integration with microfluidics as well as special sampling systems. Given their extremely wide capabilities and ever-evolving applications, we envision that the use of SPR biosensor technology will continue to expand as a modern bioanalytical tool.

References

Adam, P., Dostálek, J., and Homola, J. (2006) *Sens. Actuators B*, **113**, 774.

Berger, C.E.H., Baumer, T.A.M., Kooyman, R.P.H., and Greve, J. (1998) *Anal. Chem.*, **70**, 703.

Biacore (2007) Biacore website: http://www.biacore.com

Boardman, A.D. (1982) *Electromagnetic Surface Modes*, Chichester: John Wiley & Sons, pp 1–76.

Brockman, J.M. and Fernandez, S.M. (2001) *Am. Lab.*, **33**, 37.

Brockman, J.M., Nelson, B.P., and Corn, R.M. (2000) *Annu. Rev. Phys. Chem.*, **51**, 41.

Buckle, P.E., Davies, R.J., Kinning, T. *et al.* (1993) *Biosens. Bioelectron.*, **8**, 355.

Burke, J.J., Stegeman, G.I., and Tamir, T. (1986) *Phys. Rev. B*, **33**, 5186.

Cahill, C.P., Johnston, K.S., and Yee, S.S. (1997) *Sens. Actuators B*, **45**, 161.

Chandezon, J., Dupuis, M.T., Cornet, G., and Maystre, D. (1982) *J. Opt. Soc. Am.*, **72**, 839.
Chinowsky, T.M., Jung, L.S., and Yee, S.S. (1999a) *Sens. Actuators B*, **54**, 89.
Chinowsky, T.M., Strong, A., Bartholomew, D.U. *et al.* (1999b) *Proc. SPIE*, **3857**, 104.
Choi, K., Seo, W., Cha, S., and Choi, J. (1998) *J. Biochem. Mol. Biol.*, **31**, 101.
Christensen, L.L. (1997) *Anal. Biochem.*, **249**, 153.
Cooper, M.A. and Williams, D.H. (1999) *Anal. Biochem.*, **276**, 36.
Čtyroký, J., Homola, J., and Skalský, M. (1997) *Electr. Lett.*, **33**, 1246.
Cullen, D.C. and Lowe, C.R. (1990) *Sens. Actuators B*, **1**, 576.
Cullen, D.C., Brown, R.G., and Lowe, C.R. (1987/1988) *Biosens. Bioelectron.*, **3**, 211.
Daiss, J.L. and Scalice, E.R. (1994) *Methods*, **6**, 143.
Danelian, E, Karlén, A., Karlsson, R. *et al.* (2000) *J. Med. Chem.*, **43**, 2083.
de Bruijn, H.E., Altenburg, B.S.F., Kooyman, R.P.H., and Greve, J. (1992) *App. Opt.*, **31**, 440.
Dessy, R.E. and Bender, W.J. (1994) *Anal. Chem.*, **66**, 963.
Dostálek, J., Čtyroký, J., Homola, J. *et al.* (2001) *Sens. Actuators B*, **76**, 8.
Duschl, C., Sevin-Landais, A., and Vogel, H. (1995) *Biophys. J.*, **70**, 1985.
Elkind, J.L., Stimpson, D.I., Strong, A.A. *et al.* (1999) *Sens. Actuators B*, **54**, 182.
Evans, S.V. and MacKenzie, C.R. (1999) *J. Mol. Recogn.*, **12**, 155.
Fano, U. (1941) *J. Opt. Soc. Am.*, **31**, 213.
Foster, M.W., Ferrel, D.J., and Lieberman, R.A. (1994) *Proc. SPIE*, **2293**, 122.
Fratamico, P.M., Strobaugh, T.P., Medina, M.B., and Gehring, A.G. (1998) *Biotechnol. Tech.*, **12**, 571.
Frostell-Karlsson, A., Remaeus, A., Roos, H. *et al.* (2000) *J. Med. Chem.*, **43**, 1986.
Fu, E., Chinowsky, T., Foley, J. *et al.* (2004) *Rev. Sci. Instrum.*, **75**, 2300.
Gale, M.T. (1997) *Microelectron. Eng.*, **34**, 321.
Gaus, K. and Hall, E.A.H. (1998) *Biosens. Bioelectron.*, **13**, 1307.
Gill, D.M. (1982) *Microbiol. Rev.*, **46**, 86.
Gordon II, J.G. and Ernst, S. (1980) *Surf. Sci.*, **101**, 499.
Harris, R.D. and Wilkinson, J.S. (1995) *Sens. Actuators B*, **29**, 261.
Harris, R.D., Luff, B.J., Wilkinson, J.S. *et al.* (1999) *Biosens. Bioelectron.*, **14**, 377.
Heideman R.G., Kooyman, R.P.H., and Greve, J. (1993) *Sens. Actuators B*, **10**, 209.
Homola, J. (1995) *Sens. Actuators B*, **29**, 401.
Homola, J. (1997) *Sens. Actuators B*, **41**, 207.

Homola, J. (ed.) (2006), *Surface Plasmon Resonance Based Sensors*. Springer, pp. 3 and 45.
Homola, J., Čtyroký, J., Skalský, M. *et al.* (1997) *Sens. Actuators B*, **38–39**, 286.
Homola, J., Dostálek, J., and Čtyroký, J. (2001a) *Proc. SPIE*, **4416**, 86.
Homola, J., Dostálek, J., Chen, S. *et al.* (2001c), *Proc. SPIE*, **4416**, 280.
Homola, J., Koudela, I., and Yee, S. (1999a) *Sens. Actuators B*, **54**, 16.
Homola, J., Lu, H.B., Nenninger, G.G. *et al.* (2001d) *Sens. Actuators B*, **76**, 403.
Homola, J., Lu, H.B., and Yee, S.S. (1999b) *Electr. Lett.*, **35**, 1105.
Homola, J., Piliarik, M., Slavík, R., and Čtyroký, J. (2001b) *Proc. SPIE*, **4416**, 82.
Homola, J., Schwotzer, G., Lehmann, H. *et al.* (1995) *Proc. SPIE*, **2508**, 324.
Hutley, M.C. (1982) *Diffraction Gratings*. London: Academic Press.
Johansen, K., Stalberg, R., Lundstrom, I., and Liedberg, B. (2000) *Meas. Sci. Technol.*, **11**, 1630.
Johnston, K.S., Yee, S.S., and Booksh, K.S. (1997) *Anal. Chem.*, **69**, 1844.
Jönsson, U., Fagerstam, L., Ivarsson, B. *et al.* (1991) *Biotechniques*, **11**, 620.
Jorgenson, R.C. and Yee, S.S. (1993) *Sens. Actuators B*, **12**, 213.
Jorgenson, R.C. and Yee, S.S. (1994) *Sens. Actuators A*, **43**, 44.
Jory, M.J., Bradberry, G.W., Cann, P.S., and Sambles, J.R. (1995) *Meas. Sci. Technol.*, **6**, 1193.
Jory, M.J., Vukusic, P.S., and Sambles, J.R. (1994) *Sens. Actuators B*, **17**, 1203.
Karlsen, S.R., Johnston, K.S., Yee, S.S., and Jung, C.C. (1996) *Sens. Actuators B*, **32**, 137.
Karlsson, R. and Ståhleberg, R. (1995) *Anal. Biochem.*, **228**, 274.
Katsamba, P.S., Myszka, D.G., and Laird-Offringa, I.A. (2001) *J. Biol. Chem.*, **276**, 21476.
Kolomenskii, A.A., Gershon, P.D., and Schuessler, H.A. (1997) *Appl. Opt.*, **36**, 6539.
Kooyman, R.P.H., Kolkman, H., van Gent, J., and Greve, J. (1988) *Anal. Chim. Acta*, **213**, 35.
Koubová, V., Brynda, E., Karasová, L. *et al.* (2001), *Sens. Actuators B*, **74**, 100.
Kretschmann, E. and Raether, H. (1968) *Z. Naturforsch.*, **23A**, 2135.
Kreuwel, H., Lambeck, P., Gent, J.V., and Popma, T. (1987) *Proc. SPIE*, **798**, 218.
Kruchinin, A.A. and Vlasov, Y.G. (1996) *Sens. Actuators B*, **30**, 77.
Lahiri, J., Isaacs, L., Tien, J., and Whitesides, G.M. (1999) *Anal. Chem.*, **71**, 777.
Lambeck, P.V. (1992) *Sens. Actuators B*, **8**, 103.
Lavers, C.R. and Wilkinson, J.S. (1994) *Sens. Actuators B*, **22**, 75.

Lawrence, C.R., Geddes, N.J., Furlong, D.N., and Sambles, J.R. (1996) *Biosen. Bioelectron.*, **11**, 389.
Leung, P.T., Pollard-Knight, D., Malan, G.P., and Finlan, M.F. (1994) *Sens. Actuators B*, **22**, 175.
Liedberg, B., Lundstrom, I., and Stenberg, E. (1993) *Sens. Actuators B*, **11**, 63.
Liedberg, B., Nylander, C., and Lundström, I. (1983) *Sens. Actuators*, **4**, 299.
Löfås, S., Malmqvist, M., Rönnberg, I. *et al.* (1991) *Sens. Actuators B*, **5**, 79.
Matsubara, K., Kawata, S., and Minami, S. (1988) *Appl. Opt.*, **27**, 1160.
Melendez, J., Carr, R., Bartholomew, D.U. *et al.* (1996) *Sens. Actuators B*, **35–36**, 212.
Minunni, M. and Mascini, M. (1993) *Anal. Lett.*, **26**, 1441.
Miura, N., Ogata, K., Sakai, G. *et al.* (1997) *Chem. Lett.*, **8**, 713.
Moharam, M.G. and Gaylord, T.K. (1986) *J. Opt. Soc. Am.*, **A3**, 1780.
Morton, T.A. and Myszka, D.G. (1998) *Meth. Enzymol.*, **295**, 268.
Morton, T.A., Myszka, D.G., and Chaiken, I.M. (1995) *Anal. Biochem.*, **227**, 176.
Mouvet, C., Harris, R.D., Maciag, C. *et al.* (1997) *Anal. Chim. Acta*, **338**, 109.
Myszka, D.G. (2000) *Meth. Enzymol.*, **323**, 325.
Nakamura, R., Muguruma, H., Ikebukuro, K. *et al.* (1997) *Anal. Chem.*, **69**, 4649.
Nedelkov, D., Rasooly, A., and Nelson, R.W. (2000) *Int. J. Food Microb.*, **60**, 1.
Naimushin, A., Soelberg, S., Bartholomew, D. *et al.* (2003) *Sens. Actuators B*, **96**, 253.
Nellen, P.M. and Lukosz, W. (1990) *Sens. Actuators B*, **1**, 592.
Nelson, B.P., Frutos, A.G., Brockman, J.M., and Corn, R.M. (1999) *Anal. Chem.*, **71**, 3928.
Nelson, S.G., Johnston, K.S., and Yee, S.S. (1996) *Sens. Actuators B*, **35–36**, 187.
Nenninger, G.G., Clendenning, J.B., Furlong, C.E., and Yee, S.S. (1998) *Sens. Actuators B*, **51**, 38.
Nenninger, G.G., Piliarik, M., and Homola, J. (2002) *Meas. Sci. Technol.*, **13**, 2038.
Nenninger, G.G., Tobiška, P., Homola, J., and Yee, S.S. (2001) *Sens. Actuators B*, **74**, 145.
Nikitin, P.I. and Beloglazov, A.A. (1994) *Sens. Actuators A*, **41–42**, 547.
Nikitin, P.I., Beloglazov, A.A., Kabashin, A.V. *et al.* (1999) *Sens. Actuators B*, **54**, 43.
Nikitin, P.I., Beloglazov, A.A., Valeiko, M.V. *et al.* (1997) *Sens. Actuators B*, **38**, 53.

Nylander, C., Liedberg, B., and Lind, T. (1982) *Sens. Actuators*, **3**, 79.
Ober, R.J. and Ward, E.S. (1999) *Anal. Biochem.*, **271**, 70.
O'Brien II, M.J., Brueck, S.R.J., Perez-Luna, V.H. *et al.* (1999) *Biosen. Bioelectron.*, **14**, 145.
Ohlson, S., Jungar, C., Strandh, M., and Mandenius, C.F. (2000) *Trends Biotechnol.*, **18**, 49.
Ordal, A.M., Long, L.L., Bell, R.J. *et al.* (1983) *Appl. Opt.*, **11**, 1099.
Otto, A. (1968) *Z. Physik*, **216**, 398.
Palik, E.D. (1985) *Handbook of Optical Constants of Solids*. London: Academic Press, 804 pp.
Parriaux, O. and Voirin, G. (1990) *Sens. Actuators A*, **21–23**, 1137.
Pearce Jr., K.H., Ultsch, M.H., Kelley, R.F. *et al.* (1996) *Biochem.*, **35**, 10300.
Perkins, E.A. and Squirrell, D.J. (2000) *Biosen. Bioelectron.*, **14**, 853.
Pfeifer, P., Aldinger, U., Schwotzer, G. *et al.* (1999) *Sens. Actuators B*, **54**, 166.
Piliarik, M., Homola, J., Maníková, Z., and Čtyroký, J. (2003) *Sens. Actuators B*, **90**, 236.
Piliarik, M., Vaisocherová, H., and Homola, J. (2005) *Biosen. Bioelectron.*, **20**, 2104.
Pockrand, I. (1978) *Surf. Sci.*, **74**, 237.
Raether, R. (1988) *Surface Plasmons on Smooth and Rough Surfaces and on Gratings*. Berlin: Springer-Verlag, 136 pp.
Rasooly, A. (2001) *J. Food Protect.*, **64**, 37.
Ronot-Trioli, C., Trouillet, A., Veillas, C., and Gagnaire, H. (1996) *Sens. Actuators A*, **54**, 589.
Roos, H., Karlsson, R., Nilshans, H., and Persson, A. (1999) *J. Mol. Recognit.*, **11**, 204.
Rothenhäusler, B. and Knoll, W. (1988) *Nature*, **332**, 688.
Ryan, K.J. (1990) In *Medical Microbiology: An Introduction to Infectious Diseases* (J.C. Sherris, ed.) New York: Elsevier, p. 357.
Sakai, G., Nakata, S., Uda, T. *et al.* (1999) *Electr. Acta*, **44**, 3849.
Sakai, G., Ogata, K., Uda, T. *et al.* (1998) *Sens. Actuators B*, **49**, 5.
Salamon, Z., Cowell, S., Varga, E. *et al.* (2000) *Biophys. J.*, **79**, 2463.
Santagata, S., Boggon, T.J., Baird, C.L. *et al.* (2001) *Science*, **292**, 2019.
Severs, A.H. and Schasfoort, R.B.M. (1993) *Biosens. Bioelectron.*, **8**, 365.
Shimomura, M., Nomura, Y., Zhang, W. *et al.* (2001) *Anal. Chim. Acta*, **434**, 223.
Slavík, R. and Homola, J. (2007) *Sens. Actuators B*, **123**, 10.
Slavík, R., Homola, J., and Čtyroký, J. (1998) *Sens. Actuators B*, **51**, 311.
Slavík, R., Homola, J., and Čtyroký, J. (1999) *Sens. Actuators B*, **54**, 74.
Slavík, R., Homola, J., and Čtyroký, J. (2001) *Sens. Actuators B*, **74**, 106.

Snyder, A.W. and Love, J.D. (1983) *Optical Waveguide Theory*. London: Chapman and Hall, 734 pp.
Sota, H., Hasegawa, Y., and Iwakura, M. (1998) *Anal. Chem.*, **70**, 2019.
Stemmler, I., Brecht, A., and Gauglitz, G. (1999) *Sens. Actuators B*, **54**, 98.
Stuart, D.A., Haes, A.J., Yonzon, C.R. *et al.* (2005) *IEEE Proc.-Nanobiotechnol.*, **152**, 13.
Taylor, A.D., Ladd, J., Yu, Q. *et al.* (2006) *Biosens. Bioelectron.*, **22**, 752.
Thiel, A.J., Frutos, A.G., Jordan, C.E. *et al.* (1997) *Anal. Chem.*, **69**, 4948.
Tobiška, P. and Homola, J. (2005) *Sens. Actuators B*, **107**, 162.
Trouillet, A., Ronot-Trioli, C., Veillas, C., and Gagnaire, H. (1996) *Pure App. Opt.*, **5**, 227.
Vaisocherová, H., Mrkvová, K., Piliarik, M. *et al.* (2007) *Biosens. Bioelectron.*, **22**, 1020.
Vidal, M.M.B., Lopez, R., Aleggret, S. *et al.* (1993) *Sens. Actuators B*, **11**, 455.
Vukusic, P.S., Bryan-Brown, G.P., and Sambles, J.R. (1992) *Sens. Actuators B*, **8**, 155.
Weiss, M.N., Srivastava, R., Groger, H. *et al.* (1996) *Sens. Actuators A*, **51**, 211.
Wood, R.W. (1902) *Phil. Magm.*, **4**, 396.
Yeatman, E.M. (1996) *Biosens. Bioelectron.*, **11**, 635.
Yeatman, E.M. and Ash, A. (1987) *Electron. Lett.*, **23**, 1091.
Zhang, L.M. and Uttamchandani, D. (1988) *Electr. Lett.*, **23**, 1469.

Optical Biosensors: Today and Tomorrow (2nd Edition)
Frances S. Ligler and Chris Rowe Taitt (Editors)
Published by Elsevier B.V.

Chapter 5

FLOW IMMUNOSENSORS

Anne W. Kusterbeck[a], M.S., and Diane A. Blake[b], Ph.D.
[a]Center for Bio/Molecular Science and Engineering, Naval Research Laboratory, Washington, DC 20375-5348, USA
[b]Department of Biochemistry, Tulane University School of Medicine, New Orleans, LA 70112-2699 USA

Flow immunosensors combine the selectivity and sensitivity of traditional immunoassays with the rapid response of a sensor. Two different flow immunosensors will be described in this chapter: a displacement immunosensor that utilizes a non-equilibrium displacement reaction and a kinetic exclusion immunosensor that measures the amount of free antibody-binding sites in an equilibrium mixture of antibody and antigen. Working assays have been developed for a wide range of molecules including explosives, drugs of abuse, and environmental contaminants. Accurate determinations of analyte concentrations can be made on site, thus providing immediate feedback to field managers and law enforcement personnel. Side-by-side comparisons of these measurements with laboratory instruments (high performance liquid chromatography (HPLC), gas chromatography/mass spectrometry (GC–MS), inductively coupled plasma emission spectroscopy (ICP), etc.) have demonstrated the accuracy and precision of these methods. Commercial versions of the flow immunosensors have been engineered that integrate fluidics, electronics, and computer control into both portable and autonomous instruments. More recently, advanced laboratory prototypes of the displacement immunosensor have been fabricated to improve low-end detection, extend the applications to underwater sensing, enhance field ruggedness, and assist in the manufacturing process. The evolution of this technology from laboratory prototypes to field applications is presented herein.

5.1. Principles of operation

Immunosensors have become a predominant form of biosensor due primarily to the ready availability of many different well-characterized antibodies. Many of the problems of selective ligand binding and molecular recognition have been solved by nature, since an antibody can selectively bind an antigen in the presence of a complex mixture of compounds. The immunoassay format also has a long tradition of use in the medical and pharmaceutical research communities and is accepted as a standard detection platform. One biosensor format that has emerged for applications in a variety of new detection areas is the flow immunosensor. The displacement immunoassay system has been variously described in the literature as a non-conventional immunoassay system (Ghindilis *et al.*, 1998) or as a category of chromatographic immunoassays (Hage and Nelson, 2001). The kinetic exclusion method has also been described previously (Blake *et al.*, 1999; Blake and Blake, 2003; Glass *et al.*, 2004). In any case, the aim in developing these methods has been to provide improved, faster, and more efficient detection. The resulting automated or semiautomated systems provide for more rapid responses without the need for long incubations or extensive sample manipulation.

5.1.1. The displacement immunosensor

In the displacement immunosensor, immunoassays are incorporated into a device that is set up under continuous buffer flow, as shown in Figure 5.1. The essential components of the system are antibodies immobilized on a solid support, antigen analogs that are labeled with a reporter molecule, and the associated hardware needed to establish a controlled flow system. Variations in any one of these basic elements allow the system to be adapted for different testing scenarios. Preparation for an assay first involves covalently immobilizing a monoclonal antibody (Ab) to a prepared surface. In parallel, reporter molecules are chemically coupled to a target antigen/analyte (Ag*). After purification, this labeled molecule is allowed to react with the immobilized Ab until equilibrium is reached, generally 2–15 h. To perform an assay, the solid support coated with the Ab/Ag* complex is placed in a buffer flow. When a sample containing the analyte of interest is injected into the

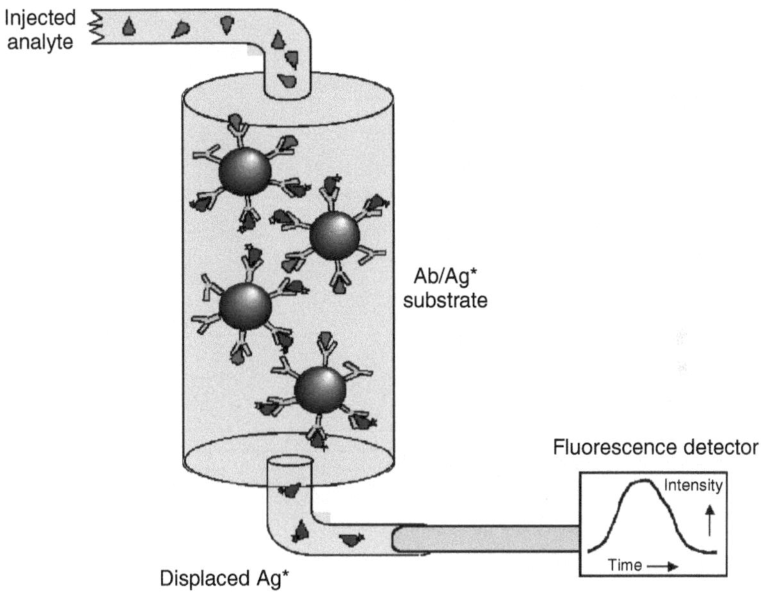

Figure 5.1 Schematic of the displacement immunosensor method. Step 1: monoclonal antibodies are covalently attached to a solid substrate and incubated with labeled antigen analog (Ab/Ag* substrate). Step 2: sample is injected into the flow stream and the labeled antigen is displaced. Step 3: displaced labeled antigen and excess sample move downstream, where a detector measures the label.

flow stream, the labeled antigen molecules are displaced into the buffer and measured downstream using a fluorescence detector. The integrated peak area of the reporter molecules can be calculated and is proportional to the number of analyte molecules injected, within a predetermined linear range for each antibody.

The entire process is complete within 30–90 s, depending on the flow rate and the hardware used. Multiple sample injections can be made on a single prepared surface without the need for additional reagents or regeneration, though high analyte concentrations (>10–20% of surface capacity) will degrade the sensor response more rapidly. Figure 5.2 illustrates a typical dose–response curve observed for increasing concentrations of

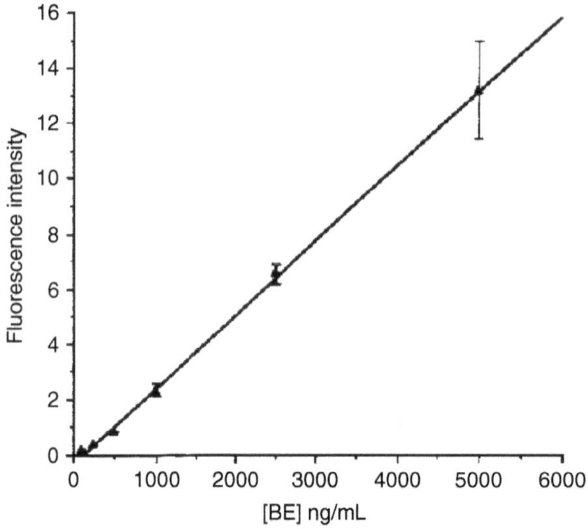

Figure 5.2 Representative displacement immunosensor dose–response curve. Increasing concentrations of the cocaine metabolite, [BE], were introduced to the system. The relative fluorescence intensity was measured for each injection. Values represent the standard error of the mean for three injections (Figure published in Yu *et al.*, 1996, reprinted with permission from Elsevier.).

the cocaine metabolite, benzoylecgonine (BE). As seen in this graph, the response is linear between 250 and 5000 ng/ml. At higher concentrations, the response is not linear, but still indicates that a sample contains a significant level of analyte.

5.1.2. The kinetic exclusion immunosensor

The kinetic exclusion method is designed to measure the concentration of free, uncomplexed antibody in mixtures containing free antibody (Ab^*), antigen (Ag), and Ab^*/Ag complexes. The operating principles of the kinetic exclusion method are shown in Figure 5.3.

The kinetic exclusion immunosensor consists of a capillary flow/observation cell fitted with a microporous screen through which

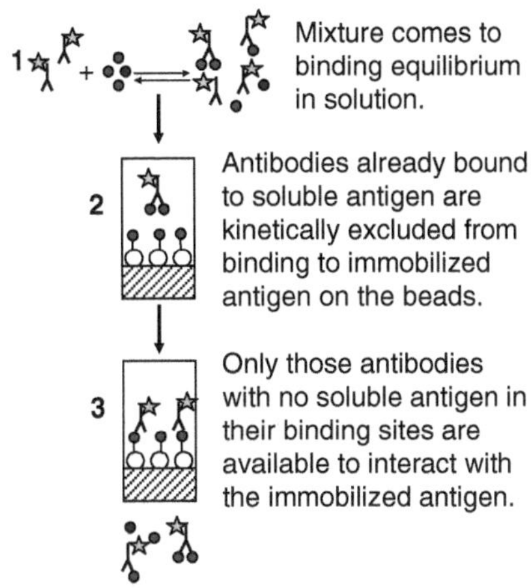

Figure 5.3 Schematic of the kinetic exclusion method. Step 1: sample containing the antigen is mixed with the fluorescently labeled antibody and the reaction mixture is allowed to come to equilibrium. Step 2: the equilibrated antibody–antigen solution is then passed rapidly over the immobilized antigen present on the surface of the microbeads in the capillary flow/observation cell. Step 3: only those antibodies without antigen already in their binding sites are available to bind to the microcolumn. The rest of the equilibrium binding mixture (fluorescently labeled antibody molecules with antigen in their binding sites, antigens, other components in the sample mixture) is removed in an automated wash step.

various solutions are drawn under negative pressure. Antigen molecules are immobilized on particles larger than the average pore size of the screen, and the particles are deposited above the screen to form a microbead column. Preparation for the assay involves mixing a fluorescently labeled antibody (Ab*) with the antigen and allowing the mixture to come to equilibrium (1–5 min, depending upon the binding constant of the antibody). The equilibrium binding mixture is then passed rapidly over the column containing the immobilized antigen, and only those antibody molecules with unoccupied binding sites bind to the microbead column. Exposure of the soluble antibody to the immobilized antigen is sufficiently brief (~480 ms) to ensure that negligible dissociation of the soluble Ab*/Ag complex occurs during the time of exposure to the microbeads. Antibodies with binding sites that are occupied with antigen are thus kinetically excluded from binding to the beads. Soluble reagents are removed from the beads by an immediate buffer wash. The sensor monitors the fluorescent signal in the capillary flow cell upon injection of the equilibrium binding mixture into the instrument.

Representative data traces and a dose–response curve from the field-portable immunosensor are shown in Figure 5.4. The total time for the analysis of each sample is 160 s. The amount of free antibody (i.e., antibody not bound to soluble caffeine) in each equilibrium mixture can be calculated as the difference between the initial and final fluorescence for each trace (shown as Delta in Figure 5.4a). The top trace (labeled "No caffeine") is a negative control that contains fluorescently labeled antibody but no analyte. As increasing concentrations of analyte are

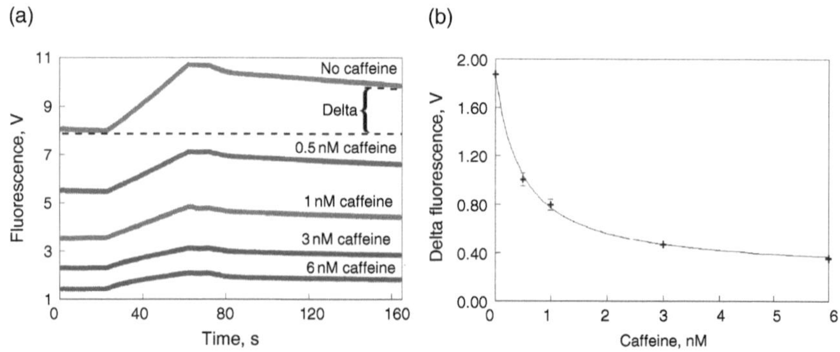

Figure 5.4 Representative data traces obtained with the field-portable immunosensor. (a) Equilibrium reaction mixtures containing fixed quantities of a fluorescently labeled anti-caffeine antibody and decreasing concentrations of soluble caffeine (6–0 nM) were injected through a microcolumn coated with caffeine–BSA. The difference between the baseline and final fluorescence for each individual sample injection (Delta) was calculated automatically by the instrument. (b) Delta fluorescence was replotted versus caffeine concentration. The assay was performed in triplicate and values represent the standard error of the mean for triplicate determinations (in some cases the SE was smaller than the diameter of the plotted points).

added to each equilibrium mixture (0.5–6 nM caffeine in this example), more of the antibody-binding sites are filled with soluble analyte and less antibody is available to bind to the beads; the Delta thus decreases as the concentration of soluble analyte increases. A dose–response curve is generated by replotting these Delta values versus the concentration of soluble antigen (caffeine in this example) as shown in Figure 5.4b.

This chapter will present a brief history of how each sensor format was developed and review pertinent data on sensor performance with a variety of different analytes. The authors will also discuss hardware development and field studies. The chapter will conclude with a discussion about how the user might make decisions about which sensor to use with specific antibodies and for particular applications.

5.2. History of the displacement immunosensor

The displacement immunosensor has evolved over the past 20 years from an idea to working commercial prototype instruments. The many steps taken in this evolution include proof-of-principle experiments, development of a laboratory device, basic studies of antibody behavior, field testing of a laboratory breadboard, engineering of commercial prototypes, and demonstration/validation of the commercial instruments to gain regulatory approval and establish equivalency with accepted analytical methods.

5.2.1. Assay development

Responding to the need of law enforcement agencies in the mid-1980s for a simple and rapid method to detect illicit substances, including explosives and drugs of abuse, the Naval Research Laboratory (NRL) undertook the development of antibody-based biosensors. In that same time frame, Warden *et al.* (1987) demonstrated a "hit and run" assay for T-2 toxin using bound antigen on a substrate and displaced Fab′ antibody fragments after short incubation periods.

Wanting to conserve the expensive antibodies, NRL immobilized the antibody on the substrate and attempted to displace labeled antigen in a real-time reaction. This work led to a new immunoassay system, the displacement immunosensor. Initial proof-of-principle experiments were performed using the well-studied 2,4-dinitrophenol (DNP) molecule as a model antigen for the common explosive 2,4,6-trinitrotoluene (TNT). In these studies, an anti-DNP monoclonal antibody was covalently attached to an agarose gel (Kusterbeck *et al.*, 1990). An I^{125} label was conjugated to DNP via an insulin A-chain spacer molecule developed specifically for coupling a single small molecule to multiple labels (Bredehorst *et al.*, 1991). The antibody/labeled antigen gel was placed in a small column, flow was started, and samples were applied. Detection was accomplished by measuring the level of radioactivity in the eluted fractions (Kusterbeck *et al.*, 1990).

The next transition for the displacement immunosensor came with the demonstration of a working assay replacing the radiolabel with the fluorophore fluorescein isothiocyanate. The sensitivities achieved were comparable to those of the radioassay but did not involve the associated hazards. Immobilization of the antibody has now been accomplished using a variety of covalent chemistries. Tresyl chloride linkages (Kusterbeck *et al.*, 1990; Wemhoff *et al.*, 1992; Whelan *et al.*, 1993) have been used, as well as proprietary commercial chemistries supplied on membranes (Rabbany *et al.*, 1999). Recently, silane chemistries and the immobilization procedure first described by Bhatia *et al.* (1989) have all been found to work exceptionally well for attaching antibodies to glass surfaces (Narang *et al.*, 1997; Charles *et al.*, 2000).

Fluorophores tested in the displacement immunosensor include fluorescein (Judd *et al.*, 1995) and the cyanine-based near-infrared dye Cy5 (Bart *et al.*, 1997a). In all cases, the important factor in synthesizing the fluorescent conjugate has been the inclusion of a spacer molecule of sufficient length to prevent fluorescence quenching by the antigen. One successful strategy was the attachment of multiple fluorophores to an insulin A-chain carrier peptide (Bredehorst *et al.*, 1991). Carbon spacers of varying lengths have also been employed to separate the cyanine dye Cy5 from the target molecule and to modify the affinity of the Ab/Ag interaction.

The solid substrates used for immobilization have included Sepharose gels (Kusterbeck *et al.*, 1990; Ogert *et al.*, 1992; Whelan *et al.*, 1993), glass beads (controlled pore glass or solid) (Yu *et al.*, 1996), bis-acrylamide/azlactone copolymer beads (Judd *et al.*, 1995; Bart *et al.*, 1997b), activated nylon membranes (Charles *et al.*, 2000; Rabbany *et al.*, 2000), fused silica capillaries (Narang *et al.*, 1997; Charles *et al.*, 1999), and poly(methylmethacrylate) (Holt *et al*, 2002). Table 5.1 summarizes the number of ways in which the system has been configured to achieve the appropriate assay parameters and detection thresholds for different analytes.

Table 5.1 Analytes and formats reported for the displacement immunosensor

Analyte	Format used	Sample sources	References
DNP	Beads	Lab standards	Kusterbeck *et al.* (1990)
TNT	Agarose gel	Lab standards	Whelan *et al.* (1993)
	Membranes	Groundwater	Bart *et al.* (1997b)
	Capillary	Air collection	Rabbany *et al.* (1999)
	Sol-gel on poly-methyl methacrylate	Soil	Charles *et al.* (2000)
		Groundwater	Gauger *et al.* (2001)
		Groundwater	Narang *et al.* (1997)
		Lab standards	Holt *et al.* (2002)
		Seawater	Charles *et al.* (2004)
RDX	Agarose gel	Groundwater	Bart *et al.* (1997b)
	Membranes	Groundwater	Charles *et al.* (2000)
	Capillary	Lab standards	Charles *et al.* (1999)
PETN	Copolymer beads	Lab standards	Judd *et al.* (1995)
Benzoyl-ecognine	Agarose gel	Lab standards	Ogert *et al.* (1992)
		Urine	Yu *et al.* (1996)
Cortisol	Agarose gel	Blood, plasma	Kapstein *et al.* (1997)
Alachlor	Agarose gel	Lab standards	Lehotay *et al.* (1994)
PCBs	Agarose gel	Lab standards	Charles *et al.* (1995)
		Oil samples	Charles *et al.* (1995)

After work with the DNP model system, efforts focused on detection of the explosive TNT. Because monoclonal antibodies specific for TNT were not commercially available, custom hybridomas were produced and screened for activity in a displacement immunoassay performed in a 96-well plate. The fluorophore fluorescein isothiocyanate (FITC) was coupled to a TNT analog, trinitrobenzene (TNB). With adjustment in buffer composition and flow rate, detection sensitivities for TNT were achieved at the low ng/ml level (Whelan *et al.*, 1993).

Over the next 10 years, the displacement assay concept was extended to antigens of interest in other real-world scenarios. Assays were demonstrated for the explosives TNT (Whelan *et al.*, 1993), hexahydro-1,3,5-trinitro-1,3,5 triazine (RDX) (Bart *et al.*, 1997a; Charles *et al.*, 2000), and pentaerythritol tetranitrate (PETN) (Judd *et al.*, 1995). For the drugs of abuse, assays were validated for benzoylecognine (Ogert *et al.*, 1992), marijuana, and opiates (unpublished data). Appropriate conditions were also determined for the environmental contaminant polychorinated biphenyls (PCBs) (Charles *et al.*, 1995) and the pesticide alachlor (Lehotay *et al.*, 1994). Finally, efficacy of the displacement flow immunosensor concept was shown with the therapeutic drug cortisol (Kaptein *et al.*, 1997). Sample matrices tested have included air samples, laboratory standards in buffer and solvents, environmental groundwater and soil samples, and body fluids, including urine, saliva, and blood serum (Table 5.1). The samples were generally tested without pretreatment, although filtration was used in several cases as a means of removing large particulates that could impede flow.

The criteria established to validate individual assays were a good linear response over several orders of magnitude, demonstration of a detection level significantly below a regulatory threshold, and accuracy and precision with a statistically significant number of samples. In addition, acceptance by the regulatory communities required good agreement when compared with previously validated standard laboratory analysis by high performance liquid chromatography (HPLC), gas chromatography/mass spectrometry (GC–MS), or gas chromatography/electron capture detector (GC/ECD).

Sample collection, one of the most critical aspects of field analysis, has been accomplished in many ways. Initially, the wetted wall cyclone "Spincon" collector (Midwest Research Institute (MRI), Kansas City, MO) was used to test TNT displacement immunoassay performance with air samples. The TNT vapors/particulates produced by a vapor generator or solid block of TNT were pulled from the air and into a water sample by the Spincon. Though the MRI instrument was relatively inefficient (similar to other air samplers), the prototype flow immunosensor was able to detect TNT (or DNT) when present in the collected samples at levels above 5 ng/ml. These values were independently confirmed using GC/ECD analysis. Later air tests have employed the SCAEP electrostatic air sampler (Team Technologies, Boston, MA) and the SASS™ wetted wall cyclone (Research International, Woodinville, WA).

Other sample collection strategies used include surface swipes with cotton swabs and elution of bound material, solid-phase extraction after extended exposure to contaminated air or water, collection of body fluids (blood, urine, saliva), and simple grab samples of water or process effluents. The last method is by far the easiest and most cost effective for on-site work. Perhaps the most difficult collection approach has been the extraction of analytes from soil samples. Gauger *et al.* (2001) demonstrated that a 3-min solvent extraction using acetone, with subsequent drying of the sample and resuspension in buffer, could be used successfully to measure explosives levels in soil.

5.2.2. Hardware development

In parallel with assay development, the design, fabrication, and engineering of flow immunosensor instrumentation has occurred. In the simplest form, the laboratory prototype device consisted of off-the-shelf parts, including small disposable plastic columns, flexible polypropylene tubing, a standard HPLC injector, a peristaltic pump, and a fluorescence detector with a flow cell (Wemhoff *et al.*, 1992; Whelan *et al.*, 1993). Recording of the signal was accomplished using a simple analog chart recorder. A version of this system was used in a series of field tests conducted in 1995 at two US military bases (Bart *et al.*, 1997). These tests provided the first true demonstration of the ability of the displacement immunosensor to measure environmental contaminants on site.

Though workable, this early system was cumbersome and required lengthy setup and skilled users. Simple instruments with minimal user input were needed. Accordingly, in cooperation with private industry, a flow immunosensor prototype device was configured that incorporated small commercial in-line valves and a dedicated laptop computer with custom-designed software.

After establishing and demonstrating a workable design, engineering of commercial prototypes was undertaken. For both drug detection and explosives detection, years of development were required before production prototypes were fabricated. Early systems were unreliable and gave a number of false positives and/or false negatives. Frequent problems were encountered with the fluidics or computer software. The culmination of the hardware development has been the engineering of second and third generation instruments designed to improve flow immunosensor performance. Shown in Figure 5.5 are the first commercial systems, the IMPACT Test System (Lifepoint, Inc., Rancho Cucamonga, CA)

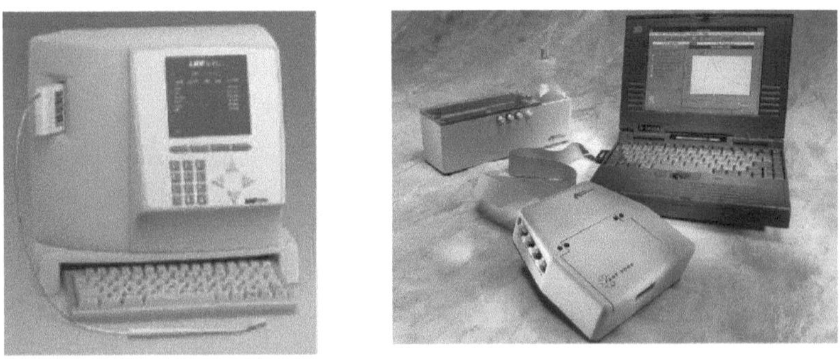

Figure 5.5 Commercial prototypes developed based on displacement immunosensor technology. Shown on the left is the IMPACT Test System manufactured by Lifepoint, Inc., for on-site measurements of drugs of abuse. On the right is the FAST 2000 engineered by Research International for environmental monitoring of explosives (Photos courtesy of Lifepoint, Inc., and Research International, Inc.).

and the FAST 2000 (Research International, Woodinville, WA). These stand-alone instruments are capable of simultaneous analysis of multiple drugs of abuse (IMPACT) or explosives (FAST) within minutes. Though these biosensors are no longer being built, a later version of the FAST instrument, the FAST 6000, with six separate channels for simultaneous analyses, has been developed by Research International and is available.

5.2.3. Basic kinetic studies

Another critical aspect of displacement immunosensor development has been an examination of the basic kinetics of antigen/antibody interactions. Because the displacement event relies on the dissociation rate of the antigen from the antibody, numerous studies have examined the effect of flow rate (Wemhoff *et al.*, 1992), immobilized antibody density (Rabbany *et al.*, 1994), antibody heterogeneity (Rabbany *et al.*, 1997; Selinger and Rabbany, 1997), and more recently, mass transport considerations and fluid dynamics (Holt *et al.*, 2000). In these studies, a number of theoretical equations unique to the displacement immunosensor were derived and experimentally tested. The most important findings are highlighted below.

A starting point for any kinetic study is the well-documented antibody/antigen behavior in solution (Berzofsky and Berkower, 1984). At equilibrium in a solution, the binding constants are defined as:

$$K_a = \frac{k_a}{k_d} = \frac{[\text{AbAg}]}{[\text{Ab}][\text{Ag}]} \tag{5.1}$$

where [Ab] is the free antibody, [Ag] is the free antigen, [AbAg] is the antibody/antigen complex, k_a is the association rate constant, and k_d is the dissociation rate constant.

In contrast to these solution interactions, the properties of immobilized antibodies were not well understood. However, the initial equation for the displacement event following antigen injection can be defined as:

$$\text{AbAg}^* + \text{Ag} \rightleftarrows \text{AbAg} + \text{Ag}^* \tag{5.2}$$

where AbAg* is the immobilized antibody/labeled antigen complex, Ag is the injected antigen, AbAg is the immobilized antibody/unlabeled antigen complex, and Ag* is the displaced labeled antigen. With subsequent injections of unlabeled antigen, more antibody-binding sites are depleted of labeled antigen and the efficiency of the reaction changes significantly. This undissociated fraction, termed θ, represents the difference between the total amount of labeled antigen initially bound to the solid support and the amount displaced after each addition of unlabeled antigen (Wemhoff *et al.*, 1992). A second equation was then derived to measure the displacement efficiency (D_e) for subsequent injections as follows:

$$D_e = \frac{(\text{displaced Ag}^*)}{(\text{loaded Ag})} \times \frac{1}{\theta} \tag{5.3}$$

Finally, looking at the displacement reaction as a first-order rate process led to an equation that could be used to calculate the "apparent" dissociation rate, or k_d as a function of time, t:

$$k_d = \frac{\ln \theta}{t} \tag{5.4}$$

This early analysis, with supporting experimental work, demonstrated that displacement immunosensor assays were dependent on flow rate. Low flow rates allowed a higher displacement efficiency, with more labeled antigen released and lower detection levels achieved. As expected, higher flow rates increased the displacement rate and improved response time, but decreased low-end detection (Wemhoff *et al.*, 1992). These results demonstrated that values for a given Ab/Ag pair could be predicted and outlined successful approaches for adjusting the detection threshold of the displacement immunosensor.

Rabbany *et al.* (1995) extended this work to show how the density of immobilized antibody affected the performance of the displacement immunosensor assays. Yu *et al.* (1996) provided a mathematical solution to account for loss of signal molecules that allowed the system to be recalibrated with subsequent sample injections. Later repetitive displacement experiments also illustrated how multiple samples could be tested on a single Ab/Ag* substrate with extended periods of flow without significant loss of signal intensity (Rabbany *et al.*, 2000).

Finally, Holt *et al.* (2000) investigated the fluid dynamics surrounding the displacement immunoassay. All formats of the assay prior to 2000 (bead, membrane, capillary) were compared based on physical parameters of the system, antibody densities, flow rate, and length of washing. The computational models developed in this study showed that accurate measurements could be made in an immunoassay under flow conditions and clearly defined the direct relationship between assay performance and displacement immunosensor operating parameters.

5.3. Applications/implementation of the displacement immunosensor

5.3.1. Aviation security

As mentioned previously, law enforcement agencies represented the initial impetus for NRL displacement immunosensor efforts. Aviation security programs at the Federal Aviation Administration (FAA) supported early development of the TNT sensor for pre-flight screening of cargo holds, baggage compartments, and passenger cabins of airplanes. In the early field tests, the NRL prototype could detect TNT (or DNT) particles present on the airplane when an MRI Spincon was able to collect the explosives into water at levels above 5 ng/ml. These values were independently confirmed using GC/ECD analysis. Further assay development was pursued for the other explosive molecules of importance in terrorist activities, including RDX and PETN, the primary components of plastic explosives. Reliable assays were demonstrated for all the compounds of interest (Whelan *et al.*, 1993; Judd *et al.*, 1995; Bart *et al.*, 1996). True implementation of the technology, however, has been problematic due to regulatory constraints, instrument costs, and passenger impatience/non-acceptance of increased screening delays. Also, due to the extremely low vapor pressures of explosives, a more critical issue for all detectors is the amount of material available to collect and analyze. Extended air sampling that includes collection of particulates is usually required to detect trace levels (<50 g) of explosives.

5.3.2. Drug detection

A second target of opportunity for the displacement immunosensor has been with the law enforcement community. The US Customs Service and the Drug Enforcement Agency (DEA) often require rapid confirmation of illicit substances at border checkpoints and other field sites. As with explosive compounds, most drugs of abuse are small molecules, making them good candidates for immunosensor applications. Accordingly, displacement immunosensor assays were developed and tested for cocaine and its major metabolite, benzoylecognine (Ogert *et al.*, 1992). The data in this study showed excellent agreement with the "gold standard" analytical method, GS/MS, and led directly to collaboration with private industry. Under a cooperative research and development agreement (CRADA) between NRL and US Drug Testing (later Lifepoint, Inc.), a commercial prototype was built and tested (Yu *et al.*, 1996). From the late 1990s to the present time, the displacement immunosensor developed under this program has demonstrated accurate, quantitative assays for the five National Institutes of Drug Abuse (NIDA-5) drugs: cocaine, opiates (heroin/morphine/codeine), phencyclidine (PCP), amphetamines/methamphetamines, and tetrahydro-cannabinol (THC, marijuana). In addition, the company completed validation of their instrument and received FDA 510K approval of the NIDA assays. Commercially available disposable cassettes accommodated assays for up to 10 analytes simultaneously. Potential markets and customers of this technology include on-site drug screening, law enforcement personnel, and medical testing laboratories. Though the company was successful in marketing the IMPACT Test System instruments for several years, Lifepoint ceased operations in 2005.

A related drug detection application of the displacement immunosensor is the monitoring of patient levels of therapeutic drugs in a continuous mode. Kaptein *et al.* (1997) used displacement immunosensor technology to develop a rapid sensitive assay for continuous monitoring of cortisol levels in patient sera.

5.3.3. Environmental monitoring and field demonstrations

Perhaps the most investigated field of use for displacement immunosensor technology is on-site environmental analysis. In the late 1980s, environmental regulators and field managers at Superfund remediation sites were looking for ways to improve site characterization of contaminated sites and reduce costs involved with long-term testing and monitoring (Rogers *et al.*, 1996). The emergence of immunochemical methods for environmental testing and the acceptance in 1993 of nine immunoassays for inclusion in the US Environmental Protection Agency (EPA) Solid Waste Testing Methods was the first step in this direction (Van Emon and Gerlach, 1996).

Displacement immunosensor assays previously developed for explosives detection in air samples were easily adapted to groundwater and soil analysis. With the support of the EPA, the flow immunosensor was used in an extensive series of tests at US military bases designed to validate on-site instrument performance. Preliminary flow immunosensor data from laboratory tests of explosives standards in uncontaminated groundwater showed a high degree of accuracy in the RDX and TNT assays. Accuracy values for samples in buffer ranged from 93% to 99% when known concentrations were analyzed. Calculations for precision ranged from as low as ±6% to as high as ±15%. An assessment of the displacement flow immunosensor in initial experiments with field samples also suggested that the method was comparable to most immunoassay or HPLC methods when samples were free of other interfering compounds. In fact, the method could very effectively map both the extent of contamination and the concentration of explosives in contaminated aquifers (Bart *et al.*, 1996). Figure 5.6 shows a contour plot generated from the analysis of TNT levels in groundwater monitoring wells at Umatilla Army Depot Activity (UMDA), Umatilla, OR. As seen in this plot, the flow immunosensor field test results were almost identical to those obtained from laboratory HPLC analysis.

Similar agreement with laboratory methods was observed in later field tests using the commercial FAST 2000 flow immunosensor. Figure 5.7 illustrates the agreement between HPLC laboratory data

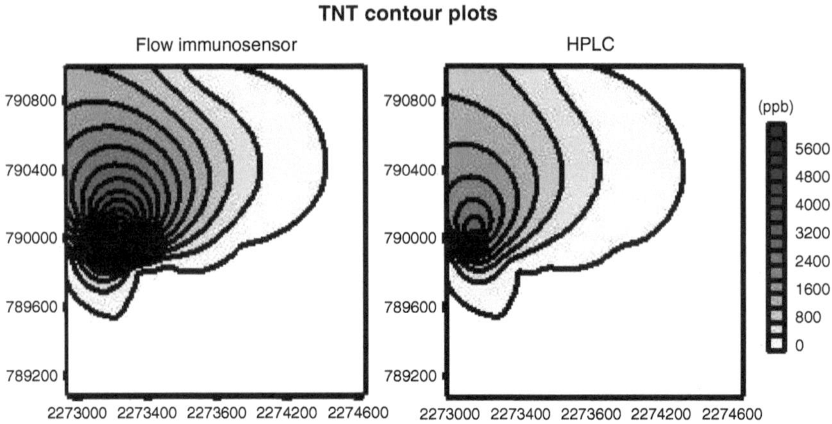

Figure 5.6 Contour plots showing TNT concentrations at UMDA as measured by the displacement immunosensor and HPLC (Bart *et al.*, 1996). (Reprinted with permission from the American Chemical Society).

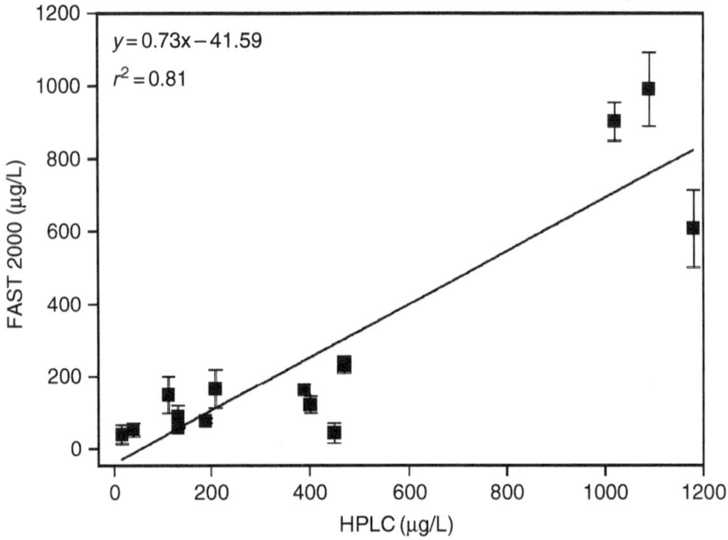

Figure 5.7 Composite of results for groundwater field samples from Umatilla Army Depot. Calculations of RDX concentrations were performed on identical sample splits using the FAST 2000 on-site (replicate analyses, $n = 7$) and an EPA-certified laboratory for HPLC analysis. Note: values below the detection limit have been omitted.

and the FAST 2000 on-site determinations for RDX in groundwater monitoring wells at UMDA. In general, the data showed that FAST 2000 immunosensor measurements for individual explosives were not as affected by matrix interferents at higher target concentrations, although matrix effects were apparent at the low end. Overall, for the five field sites used in field validation studies, the FAST 2000 was able to detect TNT and RDX with reasonable accuracy. The method was also predictive for two of the three field sites selected for detailed statistical analyses using the *t*-test. For most samples, the standard deviations observed were $<10\%$ and the comparison with HPLC results was good. These conclusions echo those of Hennion and Barcelo (1998) who suggest that the most appropriate use of immunoassay methods is primarily as a preliminary field screening technique. If possible, quantitative measurements of field sites can be made for later laboratory confirmation. Van Emon and Gerlach (1998) also determined that immunochemical methods like the displacement immunosensor are well suited to field analysis because they tolerate high sample loading, are sensitive and inexpensive, and require no hazardous materials or solvents.

As mentioned previously, displacement immunosensor assays have been developed for additional environmental contaminants including the pesticide alachlor (Lehotay *et al.*, 1994) and PCBs (Charles *et al.*, 1995). These assays have yet to be validated and field tested.

5.3.4. Underwater chemical sensing

Because samples must be tested in water, a natural extension of environmental monitoring was use of the displacement immunosensor for chemical sensing in the marine environment. In a study sponsored by the Office of Naval Research (ONR), a multi-organizational team conducted controlled underwater experiments at San Clemente, CA, and Duck, NC, in 2000–2003 to examine the feasibility of underwater chemical sensing for small molecules, including TNT. Simulated targets designed to mimic Unexploded Ordnance (UXO), autonomous underwater vehicles (AUV), and multiple sensor approaches were used to study plume dynamics and analyte transport in the ocean. The results for the displacement immunosensor, as detailed by Charles *et al.* (2004)

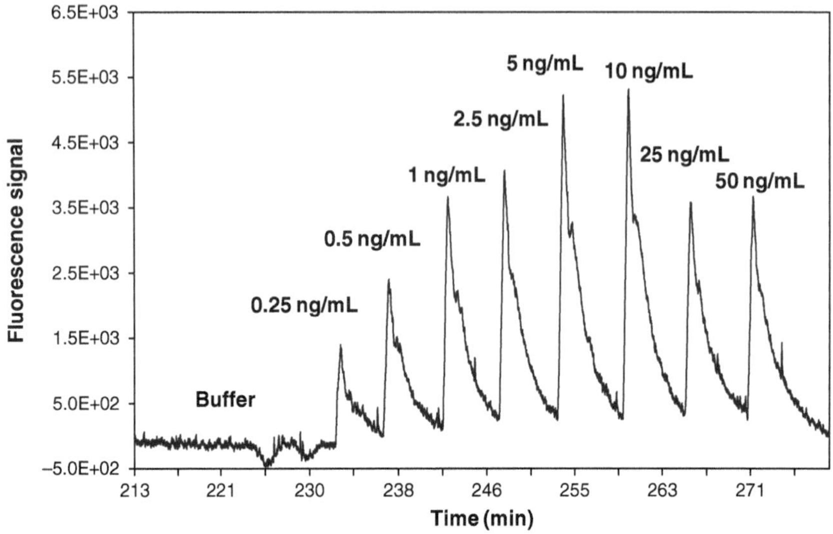

Figure 5.8 Displacement immunosensor detection of TNT in seawater. Increasing concentrations of the TNT in 95% seawater were injected directly into the sensor and the fluorescence intensity was recorded.

and Kusterbeck *et al.* (2005), demonstrated that antibodies could be selected to function well in seawater and that immunosensor assays could detect high parts per trillion levels of TNT in marine samples without preconcentration (Figure 5.8).

5.3.5. Advanced development

A breakthrough in displacement immunosensor technology occurred in the late 1990s when Narang *et al.* (1997) immobilized antibodies inside fused silica capillaries, and with this new surface, performed a displacement immunoassay. By using this format, they were able to lower the detection limit two to three orders of magnitude below that previously observed for displacement immunoassays (1 part per trillion versus 1 part per billion). These additional orders of magnitude in sensitivity opened the way for the detection of trace chemicals in environmental samples (Narang *et al.*, 1998; Charles *et al.*, 2000). This increase in sensitivity also improved the ability of the system to meet regulatory

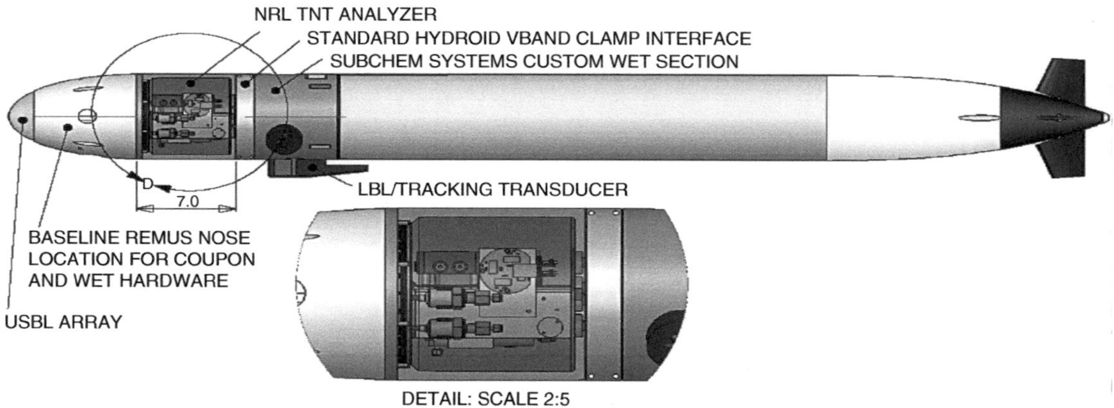

Figure 5.9 Schematic of a displacement immunosensor designed as a payload for the REMUS UUV. Built by Subchem Systems (Newport, RI), the instrument is an integrated system that is powered by the UUV. (Photo courtesy of Subchem Systems, Inc.)

guidelines and decreased matrix effects. The ongoing development of this format into a plastic coupon with machined internal channels (Holt *et al.*, 2001) has improved the ruggedness and extended the utility of the method to other field applications. This initial work to develop plastic coupons is now being extended to microfabricated components designed specifically to improve detection thresholds. As described by Howell *et al.* (2005), these coupons have microfluidic mixers with grooves etched within the channels that alter lateral flow and focus the sample stream, thereby increasing the dye concentration at the optical interface (Figure 5.9).

5.4. History of the kinetic exclusion sensor

Work on the kinetic exclusion sensor was initiated in the 1990s by engineers dissatisfied with the performance of fluorescent evanescent waveguide biosensors for the analysis of environmental samples. The goal was to develop an immunosensor system less susceptible than fluorescent waveguide sensors to variables that were difficult to control in real-world samples (pH, ionic strength, ion composition) (S.J. Lackie, 2007, personal communication). The patent on this technology was issued in 1994 and the first commercial instrument was sold in 1995 (Lackie, 1994). The initial peer-reviewed publication describing the use of the commercially available KinExA™ instrument to quantify antibody–antigen interactions was published in 1996 (Blake *et al.*, 1996). The first applications of this instrument were in the characterization of protein–ligand interactions, and the instrument has become increasingly valuable as a tool in this area of research, as well as in the related field of protein therapeutics (Blake *et al.*, 1996, 1999; Khosraviani *et al.*, 2000; Jones *et al.*, 2002; Nowakowski *et al.*, 2002; Blake and Blake, 2003; Darling and Brault, 2004; Drake *et al.*, 2004; Rathanaswami *et al.*, 2005; Razai *et al.*, 2005; Xie *et al.*, 2005; Luginbühl *et al.*, 2006). As work with the commercial instrument progressed, it became clear that the kinetic exclusion format also had significant applications as a flow immunosensor for low molecular weight contaminants in environmental samples.

5.4.1. Assay development

Because of the early commercial availability of the kinetic exclusion immunosensor, a number of different research groups have simultaneously developed applications using this format, as detailed in Table 5.2. The first published kinetic exclusion assay for an environmental contaminant (the herbicide, 2,4-dicholorophenoxyacetate, 2,4-D) was developed by a group at the EPA (Rogers *et al.*, 1997). Two different assay configurations were described in this early study: the standard kinetic exclusion format described in Figure 5.3, which utilized antigen-coated microbeads, and an alternate format wherein antibody was coated onto the microbeads. In this alternate format, 2,4-D and a fluorescently labeled 2,4-D analog subsequently competed for antibody-binding sites; results in this alternative format were observed as an increase in downstream fluorescence over negative controls. The assay format that used antigen-coated beads showed better precision and a greater dynamic range. In 1999, the EPA group again reported using the KinExA™ in an unconventional format (Rogers *et. al.*, 1999). In this second study, fluorescently labeled organophosphorus hydrolase was immobilized on the microbeads, substrate was added, flow was stopped, and the instrument was used to monitor a decrease in fluorescence that accompanied a pH change concomitant with product formation.

The majority of the applications published to date, however, have used the general format described in Figure 5.3, which employs antigen-coated beads to capture antibody not bound to soluble antigen in equilibrium mixtures. The commercially available KinExA™ instrument (or variations on this instrument, as described in Section 5.4.2, below) have been used to develop assays for aflatoxin B1 in extracts of nuts and seeds (Strachan *et al.*, 1997), estradiol, and other environmental estrogens in reference samples and wastewater (Ohmura *et al.*, 2001; Glass *et al.*, 2004a, 2004b), divalent metal ions (Cd^{2+}, Co^{2+}, Pb^{2+}) in buffer and spiked environmental water samples (Blake *et al.*, 2001), drugs of abuse in urine (O'Connell *et al.*, 1999; Eldefrawi *et al.*, 2000), herbicides in reference materials and plant samples (Chin *et al.*, 2003; Kim *et al.*, 2004, 2006), and UO_2^{2+} in surface and groundwater (Blake *et al.*, 2001, 2006; Yu *et al.*, 2005). In all analyses of real-world samples, the data obtained from the kinetic exclusion

Table 5.2 Assays developed using the kinetic exclusion immunosensor

Analyte	**Sample source**	**References**
2,4-D	Well and river water	Rogers *et al.* (1997)
Aflatoxin B1	Nuts and seeds	Strachan *et al.* (1997)
Metal ions	Reference materials and surface water	Blake *et al.* (2001) Blake *et al.* (2007)
Organophosphate pesticides	Cattle dip wastes	Rogers *et al.* (1999)
BE	Urine	O'Connell *et al.* (1999)
Environmental estrogens	Reference samples and wastewater	Ohmura *et al.* (2001); Glass *et al.* (2004a, 2004b)
Zearalenone	Reference samples	Carter *et al.* (2000)
Opiates	Urine	Eldefrawi *et al.* (2000)
2.4-dinitrophenol	Reference samples	Carter *et al.* (2003)
Imidazolinone herbicides	Reference samples and cucumbers	Chin *et al.* (2003) Kim *et al.* (2004)
Uranium (UO_2^{2+})	Reference samples and groundwater	Blake *et al.* (2001, 2006); Yu *et al.* (2005)
PCBs	Reference samples Transformer oil	Chiu *et al.* (2001); Glass *et al.* (2004b, 2006)
Dioxin F114	Reference samples	Glass *et al.*, 2004b
Fenitrothion	Reference samples	Glass *et al.* (2004b)
Thiamethoxam	Surface and groundwater, apples, potato, cucumber	Kim *et al.* (2006)
Caffeine	Reference sample and surface water	Blake *et al.* (2006)
TNT	Seawater and groundwater	Bromage *et al.* (2006)

sensor showed good agreement with validated methods of analysis (HPLC, GC–MS. Atomic absorption spectroscopy (AAS), ICP/MS).

5.4.2. Hardware development for field instrumentation

As work with the commercially available KinExA progressed, it became clear that the kinetic exclusion format could have significant advantages as a field instrument, especially for the analysis of low molecular weight antigens (metals, pesticides, herbicides, explosives, PCBs). Two research groups (one in Japan and one in the USA) therefore began to work with Sapidyne Instruments, Inc., to develop instruments that could be used for analysis in the field. The Bench Top Immunosensor is shown in Figure 5.10. This sensor, described in Glass *et al.* (2004b), was developed jointly by Sapidyne Instruments, Inc. (Boise, ID), and the Central Research Institute of the Electric Power Industry

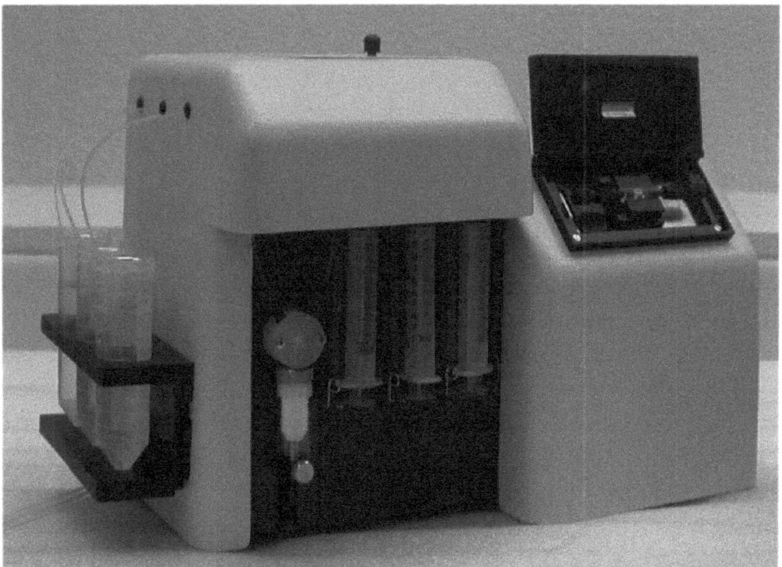

Figure 5.10 Bench Top Immunosensor that uses the kinetic exclusion format, developed jointly by Sapidyne Instruments, Inc. (Boise, ID), and CREIPI (Abiko, Japan). (Reprinted with permission from Elsevier.)

(CRIEPI, Abiko, Japan). This semiautomated instrument had the ability make a consecutive series of unattended measurements but was not field-portable because it required a 110 V power source. The instrument included an easily replaceable disposable flow cell prepacked with antigen-coated beads; each flow cell could be used for the analysis of several samples. The instrument was designed such that the end-user could quickly and easily switch among a panel of tests by changing the flow cell and the fluorescently labeled antibody. The efficacy of this sensor was demonstrated by its ability to measure six analytes of environmental interest: dioxin F114, fenitrothion, three coplanar PCBs, and estradiol (Glass, 2004b). The dynamic range and final sensitivity of each assay depended upon the K_d of the antibody used in the assay rather than upon some limitation in the transduction of the antibody–antigen interaction by the kinetic exclusion format. This prototype is no longer available from either Sapidyne or CRIEPI (S.J. Lackie, 2007, personal communication).

A second fieldable instrument, the Inline Immunosensor, was developed by Sapidyne Instruments, Inc., in collaboration with Tulane University (New Orleans, LA). The operator has the option of using a fresh microbead column for each analysis (Yu *et al.*, 2005) or reusing one beadpack for several analyses (Bromage *et al.*, 2007). Unlike the KinExA 3000™, however, the Inline Immunosensor (Figure 5.11) can collect a sample from a process line, amend the sample with assay reagents, and determine the amount of a specific antigen based on comparisons to an instrument-generated standard curve (Yu *et al.*, 2005). This sensor also requires a 110 V power source. Sensor function was initially verified in two different assay systems: a model assay for biotin that utilized commercially available reagents and an antibody-based analysis of uranium that employed a monoclonal antibody previously isolated and characterized in the Blake laboratories (Blake II, R.C. *et al.*, 2003, 2007). The sensor's multiple sample lines facilitated assay development and the associated software was easily modified to accommodate the requirements of specific antibodies and/or antigens. While the Inline Immunosensor is still considered a prototype, it is available commercially from Sapidyne Instruments, Inc., and several instruments have been purchased by diverse laboratories for additional applications.

Figure 5.11 Inline Immunosensor that uses the kinetic exclusion format, developed by Sapidyne Instruments (Boise, ID) and the Blake laboratory at Tulane University (New Orleans, LA). This instrument is now available commercially from Sapidyne Instruments, Inc.

One such application has been the use of the Inline Immunosensor to monitor TNT concentrations in seawater, river water, and groundwater (Bromage *et al.*, 2006). This assay showed a high degree of sensitivity for TNT in all water sources. The sensor was subsequently used to analyze the TNT content of raw groundwater samples obtained from potentially contaminated groundwater wells at the Naval Weapons Depot (Yorktown, VA). The samples were assessed via traditional methods (GC–MS, HPLC) as well as by the Inline Immunosensor assay. In all instances, the immunosensor demonstrated a close agreement with the traditional methods for the assessment of TNT contamination in the environment (E.S. Bromage, 2007, personal communication). A clear benefit demonstrated in these studies, over traditional methods of assessment, was the speed with which the biosensor could quantify the contaminant (160 s) and the cost per sample ($0.05).

More recently, a smaller, battery-powered, field-portable version of the kinetic exclusion immunosensor was also developed by Sapidyne and the Blake lab at Tulane (Figure 5.12). The field-portable immunosensor contains two modules. The Sample Handling Module contains pumps, valves, motors, and a custom-designed device for syringe manipulation. The Optics Module contains an LED, dichroic mirrors to separate emitted light from that used for excitation, and a photodiode that measures light emitted from the flow cell. The Optics Module also contains the software for data collection/analysis. The instrument is powered by a battery pack from a cordless drill; both the battery and a separate battery charger are available from any well-equipped hardware store. These two modules are sealed into a watertight outer box, designed to make the instrument as resistant as possible to field conditions. The instrument is controlled via a wireless interface (designed to maintain the integrity of the watertight seal) and can be operated by a Palm™, a handheld, or a laptop computer. The field-portable immunosensor employs a disposable flow cell prepacked with antigen-coated beads,

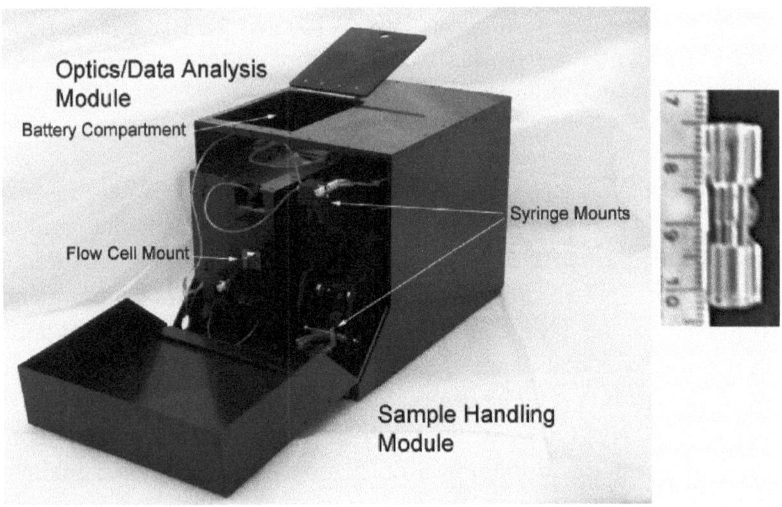

Figure 5.12 (*Left*) Field-portable immunosensor that uses the kinetic exclusion format, developed by Sapidyne Instruments (Boise, ID) and the Blake laboratory at Tulane University (New Orleans, LA). (*Right*) Disposable flow cell used in this instrument (centimeter scale included).

much like that used in the earlier Bench Top model. The field-portable immunosensor, however, has a completely new liquid handling system. The user mixes a fluorescently labeled antibody with an environmental sample containing antigen and loads the assay mixture into a disposable syringe. The instrument then automatically injects the sample over the prepacked flow cell and washes the flow cell with buffer. Fluorescence is monitored once per second from the beginning of the injection and the change in fluorescence (inversely proportional to the concentration of antigen in the assay mixture) is calculated after the buffer wash. Each analysis is complete in approximately 165 s. The disposal flow cell can be used for several analyses and then discarded.

5.4.3. Basic kinetic studies

The generation of an antibody to a low molecular weight molecule (metal, pesticide, herbicide, explosive, PCB) requires the covalent conjugation of the antigen to a carrier protein. The antibodies thus generated almost always bind more tightly to the conjugated antigen than to the antigen being analyzed. In studies where these binding affinities have been quantified and compared, the differences in affinities between the low molecular weight molecule and its conjugated counterpart have been between 150- and 600-fold in favor of the conjugated antigen (Blake *et al.*, 1996, 2004; Khosraviani *et al.*, 2000). When such differences in affinity exist, an assay system that requires a direct competition between the conjugated and unconjugated antigen is likely to yield an assay with low sensitivity (Blake *et al.*, 2001).

The design of the kinetic exclusion system minimizes the time that the antibody is in contact with the immobilized antigen–protein conjugate (usually 240–500 ms). The format thus limits the time in which antibody can dissociate from the soluble, low molecular weight antigen and rebind to the conjugated antigen immobilized on the beads in the capillary column. Other factors that increase efficiency of this kinetic exclusion format include the rapid flow of assay mixtures through the beads and the high surface area of the beads in the capillary column, which minimize the diffusion limitations at the bead surface and maximize capture of free antibody on the bead surfaces (Blake II, R.C. *et al.*,

1999, 2003; Glass, 2004a). Detailed explanations of the basis for the kinetic exclusion method and its theoretical limits of detection have been published (Blake, D.A. *et al.*, 1996; Blake II, R.C. *et al.*, 1999; Ohmura *et al.*, 2001) and only a brief discussion will be included herein, with an emphasis on the variables most likely to effect sensor performance.

As discussed previously in this chapter (see Eqn (5.1)), the solution equilibrium dissociation constant K_d is the reciprocal of K_a, or

$$K_d = \frac{[Ab][Ag]}{[AbAg]} \tag{5.5}$$

If the concentration of antibody in an assay mixture is very much less than the concentration of antigen, then the [Ag] term in Eqn (5.5) will be approximately equal to the concentration of antigen added to the binding mixture. The kinetic exclusion method will then allow the operator to detect the fraction of antibody molecules that do not have antigen in their binding site ([Ab]), and by difference, the fraction that have occupied binding sites ([AbAg]) in a series of equilibrium mixtures made with fixed concentration of antibody and varying concentrations of antigen.

Operationally, the concentration of occupied antibody-binding sites at each antigen concentration (I_{exp}) is determined by integrating the fluorescent signal in the baseline portion at the beginning of the trace and subtracting that value from a similar integral taken at the end of the trace, after a buffer wash. This value can be converted to a fraction of occupied binding sites using the following equation:

$$\text{fraction of occupied binding sites} = \frac{I_0 - I_{exp}}{I_0 - I_\infty} \tag{5.6}$$

where I_0 is difference calculated from an experiment where no antigen was added and I_∞ is the difference calculated from a trace where a saturating concentration of antigen was added. The value for the K_d can then be calculated from a nonlinear regression fit of Eqn (5.7) to the data:

$$\text{fraction of occupied binding sites} = \frac{[Ag]}{[Ag] + K_d} \tag{5.7}$$

where [Ag] represents the concentration of antigen added to each reaction mixture.

Under those experimental conditions where the concentration of antibody is greater than 25% of the lowest free antigen concentration, the value of K_d may be determined from a nonlinear regression fit to the following quadratic equation:

$$\text{fraction of occupied binding sites} = \frac{([\text{Ab}]+[\text{Ag}]+K_d) - \{([\text{Ab}]+[\text{Ag}]+K_d)^2 - 4[\text{Ab}][\text{Ag}]\}^{1/2}}{2} \quad (5.8)$$

where [Ab] and [Ag] are the total concentrations of antibody and antigen in each assay mixture, respectively, and K_d is the equilibrium dissociation constant. Equation (5.8) corresponds to a one-site homogeneous binding model when the binding event causes a significant depletion in the concentration of both the free, unbound Ab and Ag. This binding model is valid when binding to each of the antibody's two sites is separate and independent, and this equation is used for curve fitting in the commercially available KinExA software.

Practically, whether one uses Eqns (5.7) or (5.8) depends upon how much antibody [Ab] is needed in an assay mixture to generate an appropriate the signal/noise response. Signal is generated in the kinetic exclusion method by the binding of free antibody (antibody molecules without antigen in their binding sites) to ligand immobilized on the beads in the flow cell. This binding is in turn controlled by: (i) the on-rate of the binding reaction; (ii) the affinity of the antibody for the immobilized antigen; and (iii) the concentration of immobilized antigen on the microbeads. The easiest of these factors to manipulate is the concentration of immobilized antigen on the beads.

A variety of different solid supports are available for use in the kinetic exclusion immunosensor, and the microbead support best for a specific application must be determined empirically. In general, rigid beads made of poly(methylmethracrylate) or polystyrene, 98 μm in diameter, provide the best flow characteristics. However, the antigen is usually immobilized on these surfaces by simple adsorption and it is sometimes difficult to achieve a concentration on the bead surface sufficient for efficient capture of the free antibody, especially if the affinity of the antibody for the immobilized antigen is not very high. It is possible to achieve higher concentrations of immobilized antigen on the bead surface via a covalent immobilization, using NHS-activated Sepharose (Glass *et al.*, 2004b), or a polyacrylamide/azlactone copolymer (Blake *et al.*, 2005); however, these softer beads do not support the high flow rates required for some applications. Antigens can also be immobilized on poly(methylmethacrylate) beads derivatized as described by Henry *et al.* (2000); such microbeads have both higher concentrations of antigen and superior flow properties (H. Yu and D.A. Blake, 2007, unpublished data), but the derivatization process is tedious and not within the capabilities of most biological laboratories.

At equilibrium, the laws of mass action will determine the relative proportions of bound and free antibody, which will further depend upon the total amounts of antibody and antigen present in the assay. When higher levels of antibody are used to achieve a measurable signal in the kinetic exclusion immunosensor, the sensitivity of the assay is lowered. The greatest sensitivity in the kinetic exclusion format can only be achieved when the concentration of antibody added to the assay is below the antibody K_d (Ohmura *et al.*, 2001). Thus, higher affinity antibodies may require a more efficient immobilized antigen.

5.5. Advantages and limitations

Immunoassays and immunochemical field methods have inherent advantages that have been reviewed extensively (Rogers, 1995; Hennion and Barcelo, 1998; Van Emon and Gerlach, 1998). Hage and Nelson (2001) concluded that the displacement immunosensor provided direct, simple, and rapid detection of the analyte, thus highlighting a unique advantage of this format. In the displacement immunosensor format, sample analysis is continuous, non-reactive compounds are removed from the surface, and subsequent samples can be injected without resetting the sensor. In addition, a single substrate can be used for multiple tests (typically >20), depending on analyte concentration. In the kinetic exclusion immunosensor, flow is not continuous; however, an automated instrument (the Inline Immunosensor) is available that has the ability to sample and analyze samples autonomously. The number of tests that can be performed using the Inline Sensor is usually limited by a reservoir of beads. If new beads are packed for each analysis, the instrument can be used for approximately 55 tests (Yu *et al.*, 2005). Additional sampling capacity is available if the same beadpack is used for multiple analyses (E.S. Bromage, 2007, personal communication; H. Yu and D.A. Blake, 2007, unpublished data).

As highlighted in previous sections, both of the flow immunosensors described in this chapter (displacement and kinetic exclusion) are well suited to the detection of small molecular weight compounds. Assays have been developed for explosives, drugs, pesticides, and metals. The range of compounds that can be analyzed is limited only by the antibodies that are available. Another advantage of both flow immunosensor systems is that they can be made field portable and can be carried on-site for real-time analysis of a variety of analytes. Results are obtained within minutes rather than days or weeks. If there is a question about a particular sample, that sample can be analyzed again immediately.

The ability to alter formats to fit a particular test scenario or application is a major strength of the displacement immunosensor described in this chapter. For example, slowing the flow rate of the displacement immunosensor improved low-end detection (Rabbany *et al.*, 1995) while increased flow rates made the assay less sensitive, but extended the system lifetime. In addition, changes in the solid substrate used to immobilize the antibody, in the reporter molecule, and in the selected Ab/Ag pairs could be also used to modulate assay detection limits in the displacement immunosensor system. An additional advantage of the displacement immunosensor is that signal is seen as a rise from the baseline, not the decrease in signal observed with the kinetic exclusion sensor and competitive flow immunoassay systems (O'Connell *et al.*, 1999; Eldefrawi *et al.*, 2000). A limitation of the displacement immunosensor is the degradation or loss of signal with time and repeated use. Because the labeled antigen is constantly being displaced from the immobilized antibody, there is a measurable loss of assay response (Rabbany *et al.*, 1995). Also, standard deviations begin to increase and low-end detection is lost. This makes it critical that calibration steps be included in any immunosensor protocol. Long-term use of the displacement immunosensor requires a mechanism for re-saturation of the immobilized antibody with the labeled antigen analog. The kinetic exclusion immunosensor uses soluble antibody and a fresh beadpack for each analysis and degradation of signal is usually not a problem, even for long-term use.

Throughout development of both of these technologies, a major obstacle has been incorporation of effective sampling methods. For many relevant applications, air is the preferred medium. Because both flow immunosensors work in a water medium, samples collected by air samplers must be partitioned into an aqueous medium. This can be done in two ways. The antigen may be pulled directly into water using a cyclone sampler or charged particle collector, like the SCAEP and Spincon samplers described earlier. The second method involves solid-phase extraction, in which the sample is trapped on a coated surface such as a selective filter for later extraction. These preparative methods have been extremely effective for air and water samples (Barshick and Griest, 1998; Durrach *et al.*, 1998). Charles *et al.* (2001) have shown that solid-phase extraction can also enhance flow immunosensor performance and improve the accuracy twofold in comparison to direct sample analysis for low concentration analytes. Nevertheless, the process is cumbersome, costly, and time-consuming.

A related problem with operation in the field is the negative effect of environmental matrices on the performance of both flow immunosensors. Routinely, little to no sample preparation is used. Thus, reliable analysis of a sample depends upon a given water sample having minimal effect on the antibody/antigen interaction or the fluorescence intensity. This is decidedly not the case for a non-negligible percentage of field samples, where groundwater samples from selected field sites caused detrimental effects on system accuracy and precision. Similar negative effects on immunoassay performance have been observed with samples high in salt, picric acid, or humic acids (Myers *et al.*, 1994).

In general, the choice of the flow immunosensor to use for a particular application will depend upon the availability of an instrument and reagent kits, the level of sensitivity desired, and the effect of the sample matrix on the assay. The analysis cost per sample and acceptance by regulatory agencies may also be factors in system selection. At this point in time, only the FAST 6000 and the Inline Sensor are available commercially and even these instruments require that the analyst develop his/her own assay reagents.

The level of sensitivity in either flow immunosensor system will be a function of the affinity of the antibody for the analyte to be measured (TNT, pesticides, drug metabolites, etc.). In the displacement immunosensor, a number of variables control the final sensitivity that can be achieved. These variables include flow rate, density of immobilized antibody on the sensor surface, and the relative affinity of the antibody for the fluorescently conjugated and underivatized antigen. Studies have shown that lowering the flow rate will increase the sensitivity of the displacement immunosensor-based assay, while concomitantly decreasing the dynamic range and increasing the time required for analysis (Wemhoff *et al.*, 1992). The density of immobilized antibody on the sensor surface is directly proportional to the sensitivity that can be achieved with the displacement immunosensor; higher antibody densities provide a more sensitive assay (Rabbany *et al.*, 1995). The displacement format requires displacement of a pre-bound, fluorescently conjugated antigen from the immobilized antibody by soluble antigen. Sensitivity of the displacement immunosensor will be compromised if the antibody binds much more tightly to the conjugated than to the unconjugated antigen. In contrast, the conjugated and unconjugated antigen do not directly compete for antibody-binding sites in the kinetic exclusion format, because the soluble antigen and antibody are allowed to come to binding equilibrium in solution, in the absence of the immobilized antigen. The antibody molecules in this equilibrium mixture are in contact with the immobilized antigen for only the 250–750 ms required to pass the equilibrium mixture over the beads in the capillary flow cell. Antibody molecules that already contain soluble antigen in their binding site are kinetically excluded from binding to the immobilized antigen and only those antibody molecules that do not contain soluble antigen in their binding sites are available to interact with the immobilized antigen on the beads (Blake *et al.*, 1999; Blake and Blake, 2003). Thus in the kinetic exclusion format, the immobilized antigen on the beads is merely a tool to capture free antibody molecules from the equilibrium binding mixture. The sensitivity and dynamic range of the kinetic exclusion immunosensor is controlled entirely by the K_d of the antibody used in the assay (Glass, 2004b).

The effects of the sample matrix are specific both to the site being analyzed and to the antibody being used in the immunosensor. When an assay is sufficiently sensitive, sample dilution may be sufficient to overcome any deleterious effects of the sample matrix. In other cases, it may be possible to devise a simple pretreatment strategy to reduce or eliminate the effects of sample matrix. In any case, the best immunosensor to use for a particular measurement at a specific site is best left to the worker performing the analysis.

5.6. Future potential

Future advances in the flow immunosensors described in this chapter will involve improvements in assay components and hardware/software innovations. The ideal flow immunosensor for field use would contain integrated reference/calibration capabilities and could be run for extended periods (days to weeks) with unattended use. To achieve these goals, the system will need to be rugged, reliable, and capable of remote operation. One step in this direction is the engineered version of the displacement flow immunosensor to be deployed on the Hydroid REMUS unmanned underwater vehicle (UUV). Shown in Figure 5.9, the advanced system is a fully integrated payload with the ability to do continuous auto-sampling using appropriate pumps and valves. An autonomous kinetic exclusion sensor is also under development for deployment in AUV.

A promising research effort in the displacement immunosensor system is the development of a "flow immunosensor on a chip" reported by Holt *et al.* (2002). In this format, the substrate can be regenerated so that the system lifetime is extended and low detection limits are maintained. The additional engineering of the microfluidic components of the system demonstrated by Howell *et al.* (2005), and mentioned previously, may also offer significant improvements in mixing, flow properties, and peak resolution.

In the displacement immunosensor the limiting step is not only the kinetics of the binding event, but also the actual engineering of the device. Thus, if the detector is sufficiently distant from the solid substrate, there may be peak broadening as well as a very real delay in response. While not necessary for applications involving environmental cleanup, any lag time may interfere with "real time" monitoring in a displacement immunosensor installed in an autonomous vehicle or in a process control system. Screening of large areas using the displacement immunosensor format would be precluded because the time between sample collection and analysis would be too great. For process monitoring, the instrumental analysis time would need to be matched to the reaction rate. This could be achieved by implementing kinetic determinations that use only the rate of response to calculate analyte concentration. Key factors in this approach would involve significantly improved fluidics and increased signal sampling rates.

To improve detection limits and reduce or eliminate matrix effects on the either immunosensor system, these systems could incorporate on-line solid-phase extraction and preconcentration. Yu *et al.* (2001) have demonstrated such a system and are testing this modification on a microfluidic device that uses monolithic porous polymers. Molecularly imprinted polymers, hydrogels, and mesoporous materials selective for explosives and heavy metals also show promise as a way to increase sensitivity (Pekel *et al.*, 2001; Johnson-White *et al.*, 2007).

With respect to receptor binding and signal transduction, genetic engineering and molecular imprinting are being explored to gain stable molecular recognition elements with appropriate affinities that withstand solvents and work despite the extremes in salt, pH, and temperature found in environmental samples. While conventional monoclonal antibodies have been used for most assay development to date, new materials and improved receptors may provide significant enhancements to all existing sensor formats. Goldman *et al.* (2006) have recently reported single-domain antibodies (sdAb) to proteins and whole viruses that have been engineered using non-immune gene libraries obtained from sharks and llamas. These antibodies have demonstrated exceptional tolerance to extreme conditions and retain almost 90% specific binding after prolonged heating at temperatures between 95°C and 100°C. Though the sdAb affinities are low compared to conventional antibodies, they suggest that affinity maturation could be used to improve binding. Such receptors may provide a way to overcome the limitations of using protein-based sensors for extended periods or in harsh environments. Phage-displayed peptides, single-chain antibodies (scFv), and synthetic aptamers (Lee and Walt, 2000) that have specificity for the target molecules might also play a role in future systems. Alternative reporter molecules that amplify a signal or eliminate the need for fluorescence detector are also possible. Medintz and coworkers have recently demonstrated a self-assembled modular biosensor using single-chain antibody fragments (scFv) for recognition and surface-tethered components that exhibit a concentration dependent change in fluorescence resonance energy transfer (FRET) when an analyte is added (Medintz *et al.*, 2005; see chapter by Delehanty and Medintz). Surface engineering of biomaterials and the use of specific coupling chemistries to improve antibody attachment and retain a higher level of activity on the substrate also have the potential to increase the rate and/or amplitude of signal transduction in the flow displacement immunosensor system.

A possible future area of application is in monitoring food and water quality. There has been an increased interest in maintaining a safe water supply and detecting contaminants in food, including meat and milk. Standard immunoassays with enzyme-linked immunosorbent assays (ELISAs) or kits have been used previously (Aga and Thurman, 1997) and monoclonal and recombinant antibodies have been reported for more than 35 common pesticides (Hennion and Barcelo, 1998; Li *et al.*, 2006); therefore the starting materials for assay development are readily available.

The ultimate challenge for acceptance and broader commercial use of both of the flow immunosensor technologies described in this chapter (and other immunoassay-based methods) is certification by regulatory agencies for on-site use. Before this can occur, the field method must be accurate, must meet all data quality objectives, and must provide consistent analytical performance in a variety of samples (Van Emon and Gerlach, 1998). The need to analyze large numbers of samples may also require high sample throughput.

Despite more than a decade of intense research, including demonstration/validation field studies, prototype fabrication, manufacturing by private industry, and improvements to assay performance, true commercialization of these flow immunosensors has yet to occur. This slow technology transfer mirrors that seen with the rest of the biosensor community (Weetall, 1999). The reality is that transition remains a significant barrier, despite good technology. At the present time, these flow immunosensors can be considered a useful tool for semiquantitative field analysis that can provide valuable information to on-site decision makers.

Acknowledgments

The authors thank Lisa Shriver-Lake, Paul Charles, Elizabeth James, and Scott Melton for technical assistance and advice. This work was supported by the Environmental Security and Technology Certification Program (ESTCP), the ONR NRL Work Unit 8764, by Grant no. N00014-06-1-0307 from the ONR, by Grant no. NA06NOS4260226 from NOAA, by Cooperative Agreement no. 05HQAG0109 from the USGS, and by the Environmental Remediation Sciences Program, Office of Science (BER) US Department of Energy Grant no. DE-FG-98ER62704. The views expressed here are those of the authors and do not reflect those of the US Navy or the Department of Defense.

References

Aga, D.S. and Thurman, E.M. (1997). In *Immunochemical Technology for Environmental Applications* (D. Aga and E.M. Thurman, eds.) Washington, DC: American Chemical Society, ACS Symposium Series 657, p. 1.

Barshick, S.A. and Griest, W.H. (1998). *Anal. Chem.*, **70**, 3015.

Bart, J.C., Judd, L.L., Hoffman, K.E. *et al.* (1996) *Am. Chem. Soc. Symp. Ser.*, **657**, 210.

Bart, J.C., Judd, L.L., Hoffman, K.E. *et al.* (1997b) *Environ. Sci. Technol.* **31**, 1505.

Bart, J.C., Judd, L.L., and Kusterbeck, A.W. (1997a) *Sens. Actuators A Physical*, **38**, 411.

Berzofsky, I.A. and Berkower, I.J. (1984). In *Fundamental Immunology* (W.E. Paul, ed.) New York: Raven Press, p. 595.

Bhatia, S.K., Shriver-Lake, L.C., Prior, K.J. *et al.* (1989) *Anal. Biochem.* **178**, 408.

Blake II, R.C. and Blake, D.A. (2003). In *Methods in Molecular Biology, Vol. 248: Antibody Engineering: Methods and Protocols* (K.C. Lo Benjamin, ed.) Totowa, NJ: Humana Press, p. 417.

Blake, D.A., Blake II, R.C., Abboud, E.R. *et al.* (2007). In *Immunoassay and Other Bioanalytical Techniques* (J.M. Van Emon, ed.) Boca Raton, FL: Taylor and Francis, p. 93.

Blake, D.A., Chakrabarti, P., Khosraviani, M. *et al.* (1996) *J. Biol. Chem.*, **271**, 27677.

Blake II, R.C., Delehanty, J.B., Khosraviani, M. *et al.* (2003). *Biochemistry*, **42**, 497.

Blake, D.A., Jones, R.M., Blake II, R.C. *et al.* (2001) *Biosens. Bioelectron.* **16**, 799.

Blake, D.A., Khosraviani, M., Pavlov, A.R. *et al.* (1997). In *Immunochemical Technology for Environmental Applications* (D.S. Aga and E.M. Thurman, eds) ACS Symposium Series 657, Washington, DC, p. 49.

Blake II, R.C., Li, X., Yu, H., and Blake, D.A. (2007) *Biochemistry*, **46**, 1573.

Blake II, R.C., Pavlov, A.R., and Blake, D.A. (1999) *Anal. Biochem.* **272**, 123.

Blake II, R.C., Pavlov, A.R., Khosraviani, M. *et al.* (2004) *Bioconjugate Chem.*, **15**, 1125.

Blake, D.A., Yu, H., James, E.A. *et al.* (2006). In *Uranium in the Environment: Mining Impact and Consequences*. New York: Springer, p. 87.

Bredehorst, R., Wemhoff, G.A., Kusterbeck, A.W. *et al.* (1991) *Anal. Biochem.*, **193**, 272.

Bromage, E.S., Lackie, T., Unger, M.A. *et al.* (2007) *Biosens. Bioelectron.*, **22**, 2532.

Carter, R.M., Blake II, R.C., Mayer, H.P. *et al.* (2000) *Anal. Lett.*, **33**, 405.

Carter, R.M., Blake II, R.C., Nguyen, T.D. *et al.* (2003) *Biosens. Bioelectron.*, **18**, 69.

Charles, P.T. and Kusterbeck, A.W. (1999) *Biosens. Bioelectron.*, **14**, 387.

Charles, P.T., Conrad, D.W., Jacobs, M.S. *et al.* (1995) *Bioconj. Chem.*, **6**, 691.

Charles, P.T., Dingle, B.M., Bergen, S.V. *et al.* (2001) *Field Anal. Chem. Technol.*, **5**, 1.
Charles, P.T., Gauger, P.R., Patterson Jr., C.H. *et al.* (2000) *Environ. Sci. Technol.* **34**, 4641.
Charles, P.T., Rangasammy, J.G., Anderson, G.P. *et al.* (2004) *Anal. Chim. Acta*, **524**, 199.
Chin, T.E., Wong, R.B., Pont, J.L., and Karu, A.E. (2002) *J. Agric. Food Chem.*, **50**, 3380.
Darling, R.J. and Brault, P.A. (2004) *Assay Drug Devel. Technol.*, **2**, 647.
Drake, A.W., Myszka, D.G., and Klakamp, S.L. (2004) *Anal. Biochem.*, **328**, 35.
Durrach, M.R., Chutjuan, A., and Plett, G.A. (1998) *Environ. Sci. Technol.*, **32**, 1354.
Eldefrawi, M.E., Nath, A.N., Anis, N.A. *et al.* (2000) *Appl. Biochem. Biotech.*, **87**, 25.
Gauger, P.R., Holt, D.B., Patterson, Jr., C.H. *et al.* (2000) *J. Haz. Matl.*, **83**, 234.
Ghindilis, A.L., Atanasov, P., Wilkins, M., and Wilkins, E. (1998) *Biosens. Bioelectron.*, **131**, 113.
Glass, T.R., Ohmura, N., Saiki, H. *et al.* (2004a) *Anal. Chem.*, **76**, 767.
Glass, T.R., Ohmura, N., Saiki, H. *et al.* (2004c) *Anal. Chim. Acta*, **517**, 161.
Glass, T.R., Saiki, H., Joh, T. *et al.* (2004b) *Biosens. Bioelectron.*, **20**, 397.
Goldman, E.R., Anderson, G.P., Liu, J.L. *et al.* (2006) *Anal. Chem.*, **78**, 8245.
Hafner, F.T., Kautz, R.A., Iverson, B.L. *et al.* (2000) *Anal. Chem.*, **72**, 5779.
Hage, D.S. and Nelson, M.A. (2001) *Anal. Chem.*, **73**, 199A.
Hennion, M.C. and Barcelo, D. (1998) *Anal. Chim. Acta*, **362**, 3.
Henry, A.C., Tutt, T.J., Galloway, M. *et al.* (2000) *Anal. Chem.*, **72**, 5331.
Holt, D.B., Gauger, P.R., Kusterbeck, A.W., and Ligler, F.S. (2002) *Biosens. Bioelectron.*, **17**, 95.
Holt, D.B., Kusterbeck, A.W., and Ligler, F.S. (2000) *Anal. Biochem.*, **287**, 234.
Holt, D.B., Rabbany, S.Y., Kusterbeck, A.W., and Ligler, F.S. (1999) *Rev. Anal. Chem.*, **18**, 1999.
Howell Jr., P.B., Mott, D.R., Fertig, S. *et al.* (2005) *Lab Chip*, **5**, 524.
Johnson-White, B., Zeinali, M., Shaffer, K.M. *et al.* (2007) *Biosens. Bioelectron.*, **22**, 1154.
Jones, R.M., Yu, H., Delehanty, J.B., and Blake, D.A. (2002) *Bioconj. Chem.*, **13**, 408.
Judd, L.L., Kusterbeck, A.W., Conrad, D.W. *et al.* (1995) *Proc. SPIE*, **2388**, 198.
Kaptein, W.A., Zwaagstra, J.J., Venema, K. *et al.* (1997) *Sens. Actuators B*, **45**, 63.

Khosraviani, M., Blake II, R.C., Pavlov, A.R. *et al.* (2000) *Bioconj. Chem.*, **11**, 267.
Kim, H.-J., Shelver, W.L., Hwang, E.C. *et al.* (2006) *Anal. Chim. Acta*, **571**, 66.
Kim, H.-J., Shelver, W.L., and Li, Q.X. (2004) *Anal. Chim. Acta*, **509**, 111.
Kusterbeck, A.W., Deschamps, J.R., and Charles, P.T. (2005) *Proc. SPIE*, **5780**, 39.
Kusterbeck, A.W., Wemhoff, G.A., Charles, P.T. *et al.* (1990) *J. Immunol. Meth.*, **135**, 191.
Lackie, S.J. (1994) *Assay system.* U.S. Patent number 5,372,783.
Lee, M. and Walt, D.R. (2000) *Anal. Biochem.*, **282**, 142.
Lehotay, S.J., Wright, S.F., Demilo, A.B., and Stanker, L.H. (1994) *Proc. 21st Annual Fed. Anal. Chem. Spec. Soc. Conference*, St Louis, MO.
Li, T., Zhang, Q., Liu, Y. *et al.* (2006) *J. Agric. Food Chem.*, **54**, 9085.
Luginbühl, B., Kanyo, Z., Jones, R.M. *et al.* (2006) *J. Mol. Biol.*, **363**, 75.
Medintz, I.L., Goldman, E.R., Lassman, M.E. *et al.* (2005) *Anal. Chem.*, **77**, 365.
Myers, K.F., McCormick, F., Strong, B.G. *et al.* (1994) Cold Regions Research Engineering Laboratory Technical Report IRRP-94-4.
Narang, U., Gauger, P.R., Kusterbeck, A.W., and Ligler, F.S. (1998) *Anal. Biochem.*, **255**, 13.
Narang, U., Gauger, P.R., and Ligler, F.S. (1997) *Anal. Chem.*, **69**, 2779.
Nowakowski, A., Wang, C., Powers, D.B. *et al.* (2002) *Proc. Natl Acad. Sci. USA*, **99**, 11346.
O'Connell, K.P., Valdez, J.J., Azer, N.L. *et al.* (1999) *J. Immunol. Meth.*, **225**, 157.
Ogert, R.A., Kusterbeck, A.W., Wemhoff, G.A. *et al.* (1992) *Anal. Lett.*, **25**, 1999.
Ohmura, N., Lackie, S.J., and Saiki, H. (2001) *Anal. Chem.*, **73**, 3392.
Pekel, N., Sahiner, N., and Guven, O. (2001) *J. Appl. Polymer Sci.*, **81**, 2324.
Rabbany, S.Y., Kusterbeck, A.W., Bredehorst, R., and Ligler, F.S. (1994) *J. Immunol. Meth.*, **168**, 227.
Rabbany, S.Y., Kusterbeck, A.W., Bredehorst, R., and Ligler, F.S. (1995) *Sens. Actuators B*, **29**, 72.
Rabbany, S.Y., Lane, W.J., Marganski, W.A. *et al.* (2000) *J. Immunol. Meth.*, **246**, 69.
Rabbany, S.Y., Marganski, W.A., Kusterbeck, A.W., and Ligler, F.S. (1998) *Biosens. Bioelectron.*, **13**, 939.
Rabbany, S.Y. Piervincenzi, R., Judd, L.J. *et al.* (1997) *Anal. Chem.*, **69**, 175.
Rathanaswami, P., Roalstad, S., Roskos, L. *et al.* (2005) *Biochem. Biophys. Res. Commun.*, **334**, 1004.

Razai, A.C., Garcia-Rodriguez, J., Lou, I.N. *et al.* (2005) *J. Mol. Biol.*, **351**, 158.
Rogers, K.R. (1995) *Biosens. Bioelectron.*, **10**, 533.
Rogers, K.R. and Gerlach, C.L. (1996) *Environ. Sci. Technol.*, **30**, 486A.
Rogers, K.R., Kohl, S.D., Riddick, L.A., and Glass, T.R. (1997) *Analyst*, **122**, 1107.
Rogers, K.R., Wang, Y., Mulchandani, A. *et al.* (1999) *Biotechnol. Prog.*, **15**, 517.
Selinger, J.V. and Rabbany, S.Y. (1997) *Anal. Chem.*, **69**, 170.
Strachan, N.J.C., John, P.G., and Millar, I.G. (1997) *Food Agric. Immunol.*, **9**, 177.
Van Emon, J.M. and Gerlach, C.L. (1996). In *Environmental Immunochemical Methods* (J.M. Van Emon, L.L. Gerlach, and J.C. Johnson, eds.) Washington, DC: American Chemical Society, p. 2.
Van Emon, J.M. and Gerlach, C.L. (1998) *Anal. Chim. Acta*, **376**, 55.
Warden, B.A., Allam, K., Sentissi, A. *et al.* (1987) *Anal. Biochem.*, **162**, 363.
Weetall, H.H. (1999) *Biosens. Bioelectron.*, **14**, 237.
Wemhoff, G.A., Rabbany, S.Y., Kusterbeck, A.W. *et al.* (1992) *J. Immunol. Meth.*, **156**, 223.
Whelan, J.P., Kusterbeck, A.W., Wemhoff, G.A. *et al.* (1993) *Anal. Chem.*, **85**, 3561.
Yu, C., Davey, M.H., Svec, F., and Frechet, J.M.J. (2001) *Anal. Chem.*, **73**, 5088.
Yu, H., Jones, R.M., and Blake, D.A. (2005) *Int. J. Env. Anal. Chem.*, **85**, 817.
Yu, K., Kusterbeck, A.W., Whelan, J.P. *et al.* (1996) *Biosens. Bioelectron.*, **11**, 725.
Xie, L., Jones, R.M., Glass, T.R. *et al.* (2005) *J. Immunlog. Meth.*, **304**, 1.

Optical Biosensors: Today and Tomorrow (2nd Edition)
Frances S. Ligler and Chris Rowe Taitt (Editors)

Chapter 6

FLUORESCENCE LIFETIME BIOSENSING: ENTERING THE MAINSTREAM

Richard B. Thompson, Ph.D.
Department of Biochemistry & Molecular Biology, University of Maryland School of Medicine, Baltimore, MD 21201, USA

Fluorescence lifetime biosensing marries the enormous selectivity of biological (or biologically derived) recognition molecules with the sensitivity and accuracy of fluorescence lifetime measurement to determine analytes in complex matrices such as cells, living organisms, growth media, and natural waters. Adaptation of the approach for use with the microscope and with optical fibers makes these sensors uniquely useful for biology and environmental monitoring; of particular importance is their utility for quantitative fluorescence resonance energy transfer (FRET)-based studies in live cells. Examples of the technology are presented.

6.1. Technical concept

The basic concept of fluorescence lifetime-based biosensing is really quite simple, but there are subtleties in its instrumentation and execution that require attention by the practitioner to obtain satisfactory results. Fundamentally, one is measuring the analyte by changes in the fluorescence lifetime of some indicator system. It is important to distinguish these at the outset from the pioneering time-resolved fluorescence immunoassays (TRFIA) introduced by Hemmila, Soini, and their colleagues, which use modest (microsecond) time resolution to suppress (nanosecond) background fluorescence in favor of the (micro- to millisecond lifetime) emission of their lanthanide chelate labels (Lovgren *et al.*, 1985). However, the antigen level in TRFIAs is correlated with the long-lived emission *intensity*, not the *lifetime,* and consequently these assays suffer from most of the usual drawbacks of intensity-based assays (see Section 6.4). By comparison, for true lifetime-based assays (including biosensing) the information carrier is the lifetime itself. For our purposes, a fluorescence biosensor involves the use of a biological (or biologically derived) molecule to serve as a recognition element in transducing the presence or level of the analyte as a change in fluorescence we can measure.

6.1.1. Basics of fluorescence lifetime and its measurement

It is beyond the scope of this chapter to fully treat the theory and instrumentation of fluorescence lifetime measurement, so we shall emphasize the most salient points; for a more extensive introduction the reader is referred to Lakowicz's excellent *Principles of Fluorescence Spectroscopy.* Briefly, when a fluorophore (fluorescent molecule or atom) is raised to an excited state by absorbance of a photon, it spends a very short time (typically a few nanoseconds) in that state before emitting a photon (or decaying by another process). For an individual fluorophore the time interval between absorption and emission varies randomly, but for an ensemble of fluorophores the average amount of time spent in the excited state is a characteristic of the fluorophore called the lifetime, τ. If we excite the ensemble with an instantaneously brief (e.g., δ-function) flash of light, the fluorescence emission will decay exponentially with time (t) from an initial value I_0:

$$I(t) = I_0 e^{-t/\tau} \tag{6.1}$$

On a logarithmic plot (Figure 6.1) the lifetime is inversely proportional to the negative slope of the (straight) line. The process is thus akin

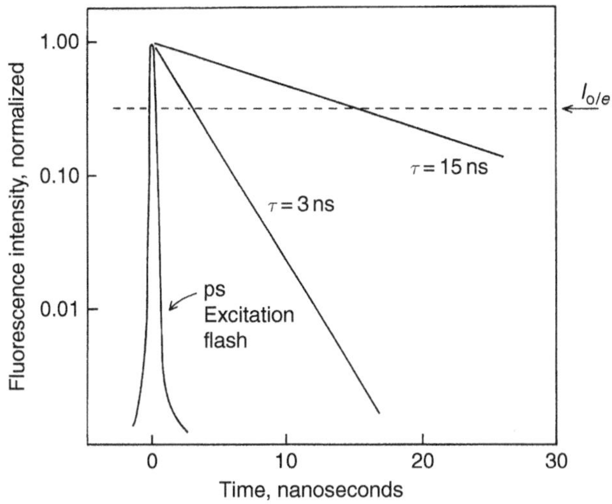

Figure 6.1 Fluorescence emission as a function of time. Immediately following a brief (picosecond) excitation flash, the fluorescence decays in a log-linear fashion. The lifetime is inversely proportional to the slope of the decay and equal to the time for the emission to decay to $1/e$ of the initial value I_0.

to radioactive decay in its kinetics, although for radioactivity the time constant is expressed as a half-life. With an excitation source emitting brief pulses of light and a detector having a sufficiently rapid response, it is possible to measure the time dependence of the emission and calculate the slope (and thus the lifetime) directly. While such so-called monoexponential decays are observed, they are uncommon because some processes can affect the apparent emissive rate, or a fluorophore in a sample may find itself in more than one environment, and/or there may be multiple kinds of fluorophores present. In such cases multiple lifetimes are present, and the time-dependent emission observed is no longer linear on a logarithmic scale. The time-dependent emission is a superposition of the decays of the components (i) of the mixture, and can be described as a multiexponential:

$$I(t) = \sum \alpha_i \mathrm{e}^{-t/\tau_i} \tag{6.2}$$

where α_i and τ_i are the fraction and lifetime of component i, respectively. An example of a multiexponential decay is shown in Figure 6.2. The

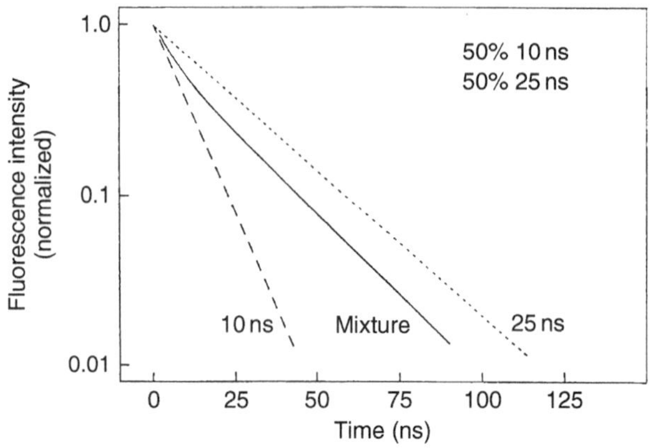

Figure 6.2 Emission of a multiexponential decay. Time-resolved emission is depicted for 10 ns (- - - -) and 25 ns (······) emitters and a 1:1 mixture of the two (—).

curvature of the line corresponding to data from the mixture can be seen by comparing it to the lines from the two components in the mixture. It requires extremely precise knowledge of the time-dependent emission to analytically determine the fractions (or preexponential factors) α and lifetimes τ of the components of even simple mixtures; practically what is done is to fit the observed decay to some assumed set of parameters using an iterative process until the differences between the measured and calculated data are small enough to be satisfactory. Even this process (despite decades of effort) can recover α's and τ's for only a few components with any accuracy or precision, and measuring a change in proportion of two components is the most common type of lifetime-based biosensor. Mixtures more complex than three or four components are often fit to a distribution of components.

6.1.2. Time-domain lifetime measurement

Several means have been described for actually measuring the time-resolved fluorescence intensity depicted schematically in Figure 6.1. As is shown below, it often is necessary to determine multiple components, which in turn requires that the time-dependent intensity be determined with high (picosecond) time resolution for a (typically weak) signal whose intensity may decline three or more orders of magnitude. Consequently, techniques such as boxcar averaging, direct measurement using sampling oscilloscopes, streak cameras, and upconversion techniques are less useful because typically they do not offer an adequate dynamic range in intensity or time, although their time resolution may be very high (Nordlund, 1990; Xu *et al.*, 2003). The most common time-domain technique is time-correlated single-photon counting (TCSPC) (Birch and Imhof, 1991). This technique times the difference in arrival time at the detector between an excitation photon and an emission photon and constructs a histogram of time differences for a large number of excitation flashes (events): for shorter lifetimes the emission photon arrives sooner and thus the histogram of events declines more rapidly from the peak. The technique is sensitive and capable of high accuracy and dynamic range if enough events are collected, and this in turn depends strongly on how frequently the excitation source flashes. For mode-locked lasers and some diode lasers the flash rate is in the megahertz range, and an accurate decay (perhaps a million events) can be collected in less than a second. Pulsed sources such as some flashlamps and nitrogen, excimer, YAG, or ruby lasers with flash durations much longer than a nanosecond, poor flash-to-flash reproducibility, and/or kilohertz flash rates are much less useful. Recently, the commercially available electronics used for TCSPC have become simpler, smaller, and cheaper (e.g., Picoquant, Becker and Hickl), permitting its use in imaging, especially in the microscope (vanderOrd *et al.*, 2001).

6.1.3. Frequency-domain lifetime determination

The other main approach for measuring α's and τ's is the frequency-domain approach, also called phase or phase-modulation fluorometry. This approach is predicated on the Fourier theorem, which states in effect that if you know the time dependence of a system over all time, it is interconvertible mathematically with the frequency response over all frequencies. This is the same basic approach (inverted) used nowadays in Fourier transform nuclear magnetic resonance (NMR) wherein one collects the time-dependent relaxation in response to a radio frequency (RF) pulse and converts it by Fourier transform to the frequency response that is the desired NMR spectrum. Phase fluorometry simply takes the frequency response of the system (see below) over a relevant range, and extracts the α's and τ's by a fitting process exactly akin to that used in the time domain.

In phase fluorometry, the fluorophore is excited with a source amplitude-modulated at a frequency usually in the RF range of 1–500 MHz; the fluorophore is a driven oscillator and emits fluorescence intensity modulated at the same frequency. However, because of the finite delay between absorption and emission (the lifetime), the emission is phase shifted and demodulated with respect to the excitation, as shown in Figure 6.3:

For a single lifetime, the phase shift $\Delta\varphi$ and modulation ratio m between the excitation and emission waveforms are simple functions of the lifetime τ and the circular modulation frequency ω (=2π times the modulation frequency f):

$$\tan \Delta\varphi = \omega\tau \tag{6.3}$$

$$m = (1 + \omega^2\tau^2)^{-1/2} \tag{6.4}$$

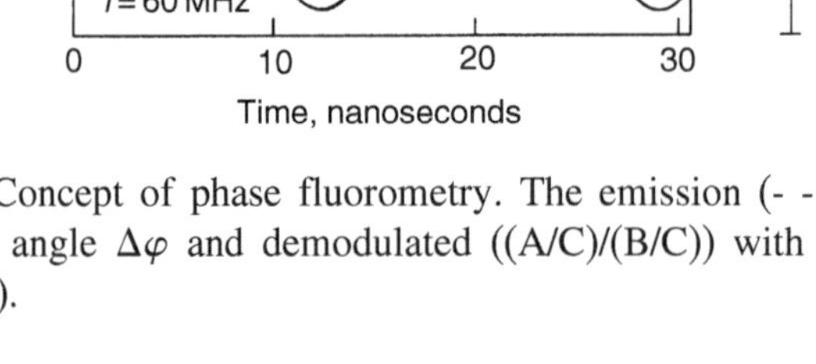

Figure 6.3 Concept of phase fluorometry. The emission (- - - -) is phase-shifted by an angle $\Delta\varphi$ and demodulated ((A/C)/(B/C)) with respect to the excitation (—).

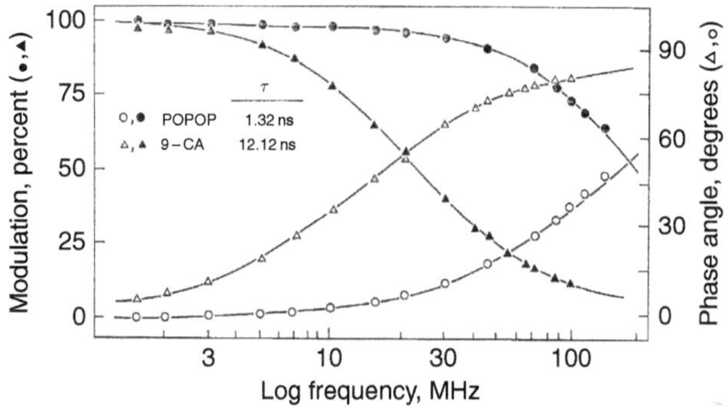

Figure 6.4 Frequency-dependent phase shifts (○,Δ) and modulations (▲,●) for POPOP (circles) and 9-cyanoanthracene (triangles); note that the curves for the two compounds are the same, merely shifted in frequency space.

If we measure the phase and modulation at several different frequencies, we observe that the phase angle increases and the modulation decreases monotonously with frequency, approaching zero and the maxima (90° and 100% modulation) at the extremes of frequency (Figure 6.4). Whereas different lifetimes exhibit straight lines of differing slopes in the time domain, the phase and modulation curves for monoexponentials are always the same; they are just shifted in frequency space.

If multiple lifetimes are present, it is essentially necessary to measure phases and modulations at several frequencies, and use a fitting process similar to that used in the time domain to recover the α's and τ's. An example is shown in Figure 6.5, which depicts measured data for a two-component mixture and the best fits assuming that only one or two lifetimes are present. It is evident that the data cannot be well fit assuming that only a single component is present: the fits differ from the measured data systematically and by amounts much larger than the random errors usually obtained (typically 0.3° in phase angle and 0.3% in modulation). By comparison, the two-lifetime fit approximates the measured data closely and the differences (also called the residuals) are apparently random and comparable to the random errors of the measurements.

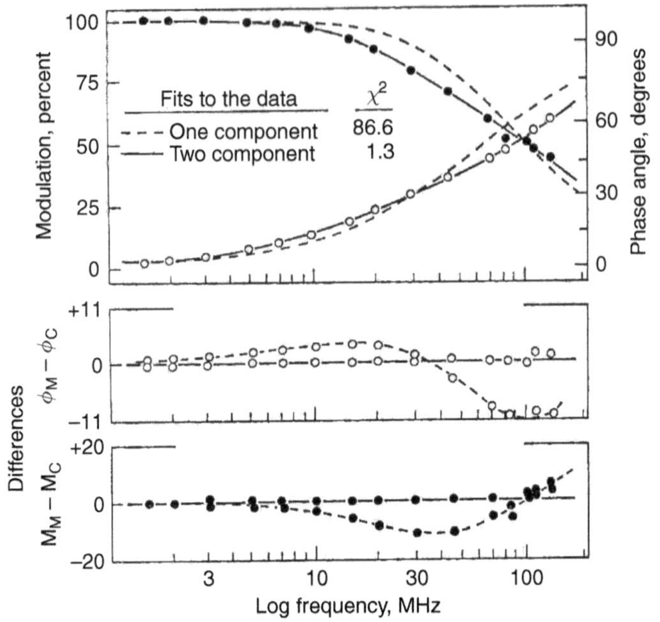

Figure 6.5 Frequency-dependent phase shifts (○) and modulations (●) for a 1:1 mixture of fluorophores, together with the best one- (- - - -) and two-component (—) fits to the data. Note that the differences between measured and fitted phase angles ($\varphi_m - \varphi_c$) and modulations ($M_m - M_c$) are large and vary systematically for the one-component fit, whereas for the two-component fit they vary randomly about zero and are similar to the standard errors of the measurements.

While phase lifetimes were first measured by Gaviola (1926) and primitive phase fluorometers appeared in the Soviet Union in the 1950s, all modern phase fluorometers incorporate the "cross-correlation" approach (essentially the heterodyne method used in radio receivers since the 1930s) proposed by Birks and Dyson (1961) for the instrument built by Spencer and Weber (1969). Briefly, the detector signal at frequency ω (say, 100 MHz) is mixed with another signal whose frequency differs by a small amount ($\omega + \Delta\omega$; say, 100 000 030 Hz in this example). What one observes at the detector are three signals at different frequencies that contain all the phase and modulation information: the fundamental (ω), sum ($2\omega + \Delta\omega$), and difference ($\Delta\omega$, 30 Hz in this example). This provides three absolutely crucial advantages. First, one can measure the phase and modulation using only the low-frequency $\Delta\omega$ signal, which is much easier and cheaper to measure electronically than the sub-nanosecond phase shifts present in the higher-frequency signals. Second, one can use a high-Q filter tuned to 30 Hz (or whatever) to reduce noise in the signal dramatically before amplification and analysis. Finally, the approach is usable over a very broad range of ω, as long as $\Delta\omega$ is maintained constant; thus the electronics used in signal collection, amplification, and phase and modulation meters need not work over a broad frequency range, which is demanding and expensive.

6.1.4. Basic approaches for lifetime-based biosensing

Lifetime-based sensing primarily takes two fundamental approaches. Both take advantage of well-known mechanisms of lifetime perturbation. The first dates back to the early 1980s, which exploits the by then well-known decline in fluorescence lifetime observed when an excited fluorophore is quenched by colliding with (or passing in close proximity to) a suitable quencher. This quenching (termed collisional quenching) reduces the apparent lifetime because in an ensemble of excited fluorophores, those that persist the longest in the excited state are most likely to encounter a quencher and lose their excited state energy, so one preferentially observes emission from fluorophores that emit their photons shortly after excitation. In ideal circumstances, the process follows the Stern–Volmer theory, where the degree of intensity and lifetime reduction (or more precisely the *ratio* of unquenched intensity or lifetime (I_0 or τ_0) to quenched (I, τ)) increases linearly with quencher concentration $[Q]$:

$$\frac{I_0}{I} = \frac{\tau_0}{\tau} = 1 + k_q[Q] \tag{6.5}$$

Figure 6.6 depicts the intensity decline observed as the quencher concentration increases, plotted directly (*left*) and according to the Stern–Volmer equation (*right*). In this formula (Eqn (6.5)), k_q is the quenching

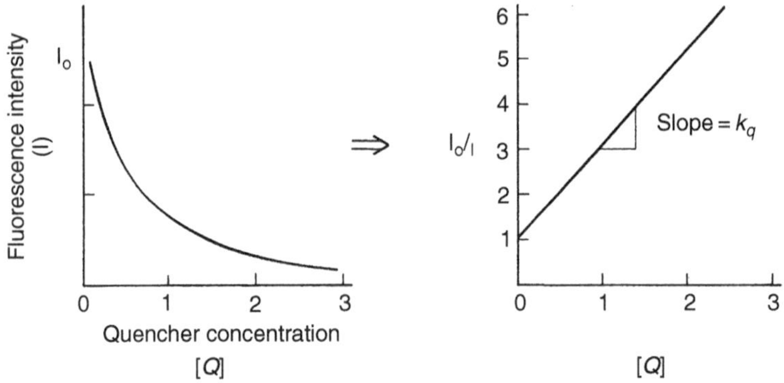

Figure 6.6 Fluorescence quenching. The *left panel* depicts quencher concentration-dependent fluorescence intensity, whereas the *right panel* depicts I_0/I as a function of quencher concentration for the same data: the slope in this Stern–Volmer plot equals k_q, the quenching constant.

constant, and as might be expected for a diffusional process, it is in turn a function of temperature, viscosity, molecular size, and pressure. Factored in is the efficiency of the quenching process itself, which ranges widely.

Thus if the analyte quenches a suitable fluorophore, one can easily determine its concentration by measuring the decline in lifetime. Usually, the story is more complex than the simple Stern–Volmer formulation, but the effect is the same even if the plot is not linear. Two factors combine, however, to make collisional quenching a less important approach in lifetime-based biosensing. The first is that the collisional process is less selective than a binding process, and there seems little hope of directly exploiting the selectivity of biological molecules in a collisional process. Second, with typical fluorophores having nanosecond lifetimes, the concentration of the quencher (analyte) must be fairly high for a significant fraction of fluorophores to encounter a quencher during the excited state. For instance, to quench the sulfopropylacridinium fluorophores developed for halogen sensing (Verkman, 1990), tens of micromolar chloride or bromide are necessary even though the fluorophore lifetime is 26 ns, longer than perhaps 99% of other fluorophores.

The other main method is much better suited to biorecognition and usually involves a proximity-dependent quenching process, most often fluorescence resonance energy transfer (Förster transfer or FRET). Fluorescence resonance energy transfer is a widely used process wherein energy is transferred from an excited fluorophore in a non-photon-mediated manner to a colored acceptor depending primarily on the donor–acceptor distance, relative orientation, and wavelength match between the donor's emission and the acceptor's absorbance (Forster, 1948). From a sensing standpoint, a key feature is that from the donor's perspective energy transfer is a quenching process that reduces its intensity and lifetime. Thus, a binding event that brings an acceptor into close proximity causes a reduction in the donor's lifetime. The acceptor can be the analyte itself or a labeled analog of the analyte for a competitive assay. The fractional saturation of the binding site(s) on the donor can be measured directly as the proportion (α) of the shorter fluorescence lifetime, which is a simple function of the concentration and affinity. An important requirement of this approach is that the bound form should not exhibit complete energy transfer: there must be some emission from the bound as well as the unbound donor to cause a mixture of lifetimes to emit. Otherwise, the bound form does not contribute to the emission, and one observes only the emission from the unbound state.

The difficulty of perturbing the lifetime of most fluorophores accounts for the robustness of lifetime-based sensing. In general, relatively few chemicals are likely to efficiently quench a given fluorophore, and they typically must be present at near-millimolar concentrations. In analysis of typical samples, constituents present at such high concentration are usually known, and it can be assessed whether they will perturb the analysis. For instance, in a sample matrix as complex as seawater, the constituents are all known down to the micromolar range, and their concentrations are often conserved: thus if bromide ion is known to quench the fluorophore, its influence on the lifetime when present at typical oceanic levels (0.84 mmol/kg) may be calculated. Apart from whatever chemical quenchers may be present, some influences may alter lifetimes, most notably temperature. In general, temperature elevated above ambient levels will reduce lifetimes somewhat, and cooling will increase them. At much higher temperatures (e.g., 100°C), many fluorophores will be unstable in solution, whereas at low temperatures (or in the solid state), where diffusion is small or absent, lifetimes will generally plateau as diffusion-based quenching ceases. Thus, it is a good idea to thermostat fluorescence experiments such as these. Pressure can also alter lifetimes, but the effects are generally significant only at pressures of thousands of atmospheres, which is outside the range typically observed in the biosphere (see the work of Weber and Drickamer). Light (or other electromagnetic fields) can perturb lifetimes, but only at the power levels usually used to induce nonlinear phenomena, such as are found in the focused mode-locked laser beams used to induce two-photon fluorescence.

6.2. History

While the importance of fluorescence lifetime as a biophysical measurement (for instance, for interpreting the results of fluorescence anisotropy/polarization measurements) was recognized at least 50 years ago (Weber, 1953), its utility for chemical analysis was really only understood in the 1980s. In part this was due to the rudimentary instrumentation available at the time, which consisted of TCSPC instruments using nanosecond flashlamps and three-frequency phase fluorometers, neither of which provided high-accuracy results without a great deal of time and trouble. Demas and DeGraff (2001) recognized that fluorescence lifetime measurements were potentially a good means of quantifying oxygen by collisional quenching, particularly if long lifetime metal–ligand charge transfer complexes such as ruthenium *tris*(bipyridyl) are used. The long (microsecond) lifetimes of these fluorophores require much less in terms of time resolution than do typical (nanosecond) lifetimes (Lippitsch *et al.*, 1988), and consequently these analyses appeared much more feasible. In the early 1980s, the Lakowicz group began to explore the utility of the two- and three-frequency phase fluorometers then commercially available for essentially analytical applications. In particular, they recognized that coupling a phase-sensitive detector such as a lock-in amplifier to a phase fluorometer provided a DC signal proportional to the sine of the difference between the (arbitrary) phase angle of the detector and the phase of the emission. The importance of this development was that it provided a signal whose intensity related to both the fractional intensity of a particular component *and its lifetime*. Given a mixture of two components with differing lifetimes and unknown proportions, one could determine the fraction of each by adjusting the phase of the phase-sensitive detector to be 90° out of phase with the emission of the component, thus canceling it out and producing a true intensity for the other component (Lakowicz *et al.*, 1992). This can be seen in Figure 6.7, where the phase-sensitive intensity for different fluorophores having lifetimes ranging from 40 ps to 9.9 ns are plotted as a function of the detector phase angle at 49.5 MHz. Note that

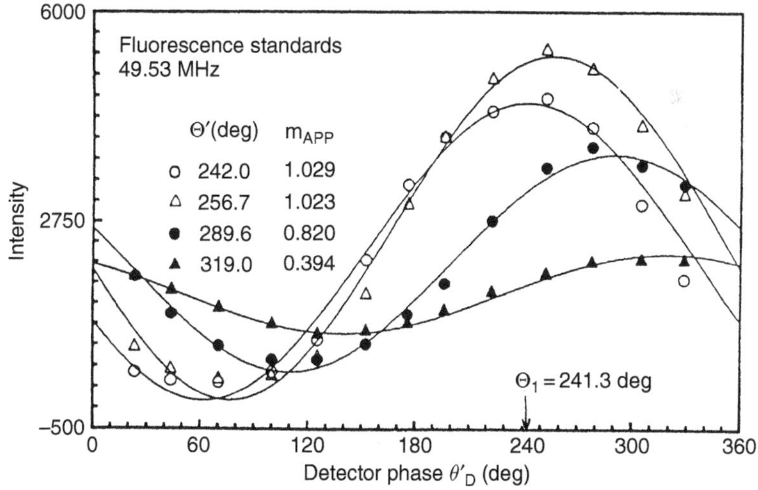

Figure 6.7 Phase-sensitive detection. Phase-sensitive intensities at 49.53 MHz are depicted as a function of detector phase angle (arbitrary offset) for fluorophores having lifetimes of 0.04 (○), 1.1 (Δ), 3.75 (●), and 9.9 (▲) ns. Reproduced from Szmacinski *et al.* (1994), with permission from Academic Press.

while the phase angle where the emission of the 40 ps fluorophore is minimized is about 60°, the minimum for the 9.9 ns lifetime fluorophore is at about 130°.

At about this time, Lakowicz also realized that lifetime-based sensing would have many of the advantages of ratiometric measurements then being introduced for measuring pH (Zhang and Seitz, 1984) and calcium (Grynkiewicz *et al.*, 1985), and that many probes that exhibited only marginally useful intensity changes were excellent lifetime probes (Szmacinski and Lakowicz, 1994). Furthermore, he realized that from the standpoint of the donor, Förster energy transfer is a quenching process resulting in a reduction in lifetime, one, moreover, that could be predicted using Forster's theory. Lakowicz and his colleagues then adapted the energy transfer immunoassay of Ullman and Schwarzberg (1981) to lifetime-based measurement (Ozinskas *et al.*, 1993).

Brand and Stryer and their colleagues (Keating and Wensel, 1991) realized that an important application of lifetime-based analysis would be imaging through the microscope. They appreciated the substantial biological information to be gleaned from images of cells and tissues where the source of contrast in the image was fluorescence lifetime, especially if such images could be used to create quantitative maps of the concentrations of analytes of interest, such as calcium. Constructing an image of time-resolved decays pixel by pixel (versus measuring the decay of a region of interest within the image) would be an extremely tedious process in view of the low (5%) duty cycle of TCSPC and the low repetition rates of the pulsed light sources readily available at that time. By comparison, use of a gated microchannel plate image intensifier to image in the microscope together with an amplitude-modulated light source permitted incorporation of phase lifetime information into the image with relative ease (Szmacinski *et al.*, 1994). The fluorescence lifetime imaging (FLIM) approach of Lakowicz employed phase-sensitive detection to produce a series of images over a range of phase angles, which together could be used to calculate lifetime maps and the like (Figure 6.8).

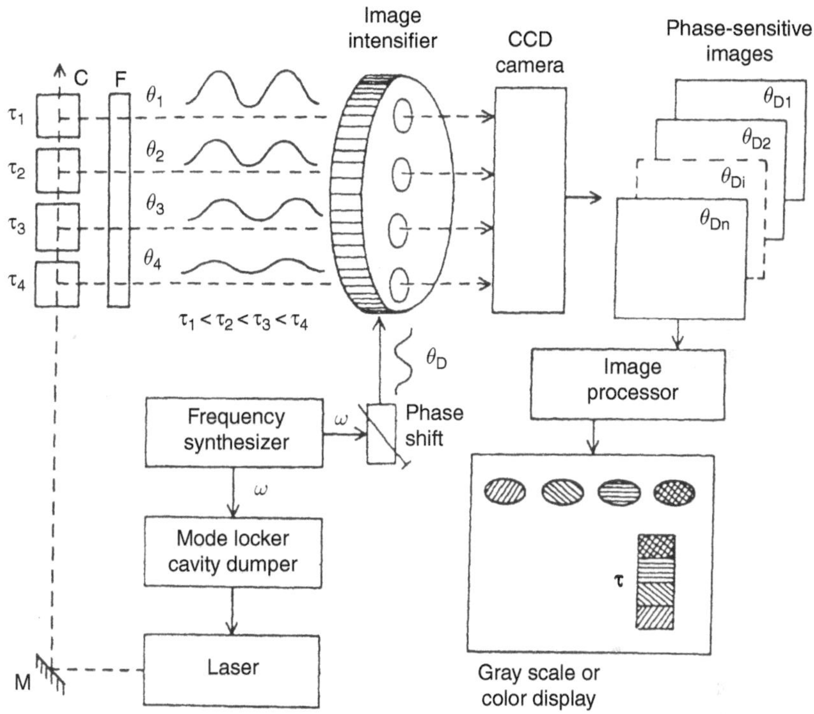

Figure 6.8 Fluorescence lifetime imaging in the frequency domain. Different areas of the specimen having different lifetimes τ_n (or mixtures of lifetimes) emit modulated fluorescence having characteristic phase angles θ_n; the gated image intensifier gain is modulated at the same frequency at some arbitrary phase angle θ_D, such that the CCD camera detects an intensity for that pixel dependent upon the difference between the emission and detector phase angles. A series of CCD images are collected at different detector phase angles and the images processed to yield an image whose contrast is based on differing lifetimes, not intensities. Reproduced from Szmacinski *et al.* (1994), with permission from Academic Press.

Most recently, the commercial availability of very high-speed, multiplexed TCSPC instrumentation adaptable to existing fluorescence microscopes has made FLIM a reality for laboratories that do not specialize in fluorescence spectroscopy. In a word, fluorescence lifetime-based sensors have entered the mainstream.

6.3. State of the art

6.3.1. Instrumentation

Until recently, fluorescence lifetime-based sensors of all types had been relatively exotic devices constructed for use by specialists in time- or frequency-domain fluorometry. While time- and frequency-domain instruments have been available commercially, they (as well as the mode-locked lasers often used with them) were not operable by scientists outside the field without significant training; the situation was analogous to that for other sophisticated instrumentation such as multiphoton fluorescence microscopes, NMR, and mass spectrometers. However, the availability of essentially turnkey instrumentation has brought lifetime-based sensing much more into the mainstream.

In the frequency domain, directly modulatable diode lasers and light-emitting diodes (LEDs) are vastly cheaper and easier to operate than gas lasers and lamps that require external (usually Pockels cell) modulators. The use of the cross-correlation approach introduced by Birks and Dyson (1961) means that the electronics used for amplifying the signal and measuring the phase and modulation can be very simple and inexpensive because only the low-offset-frequency ($\Delta\omega$) signal need be processed; nowadays, these electronics comprise a single printed circuit board. At one point, single-frequency phase fluorometers with diode laser excitation (Levy *et al.*, 1997) were offered for sale by Oriel for as little as \$20 000. Broadband multifrequency instruments have been available commercially for about 20 years and are now highly automated; a virtue of frequency-domain instruments is that artifacts in the data are for the most part overt and easily recognized. From the standpoint of instrumentation for acquiring lifetime images (instead of just the lifetime of selected region(s) of interest in the microscope), the frequency domain offers simpler instrumentation.

By comparison, the most popular time-domain method (TCSPC) requires that weak, high-frequency (short pulse) signals be amplified substantially, and this can introduce significant artifacts that can be rather subtle. However, the growth in FLIM has been fueled by the commercial availability of much cheaper and faster multichannel TCSPC electronics, together with picosecond pulsed diode lasers having high repetition rates, and nanosecond pulsed LEDs.

6.3.2. Applications

Some of the clearest examples of the performance achieved by combining the biologically based recognition chemistry with fluorescence transduction come from our own work in fluorescence-based biosensors. For some time we have exploited the unmatched affinity and selectivity of human apo-carbonic anhydrase II for metal ions such as Zn and Cu to make fluorescence-based biosensors. A key advantage of such biological recognition molecules is that – as our collaborators in the Fierke lab have demonstrated (Fierke and Thompson, 2001) – their affinity and selectivity for certain metal ions may be usefully modified: for instance, while the wild-type protein has affinity for Cu(II) roughly 40-fold higher than for Zn(II), other variants have much higher affinity for Cu(II) *and* higher selectivity over zinc ion (e.g., T93S95V97, Figure 6.9), whereas others have reduced affinity for Cu(II) compared with the wild type, but also much reduced affinity for Zn(II) (H119N). Thus, the former is well suited for open ocean analyses where free Cu(II) is expected to be extremely low, whereas the latter is better suited for coastal waters where Cu(II) is likely to be in the picomolar range, but the weak affinity for Zn(II) assures it will not interfere even at nanomolar levels. Conversely, Dr Fierke's lab has developed a variant that inverts the Irving–Williams order affinity, in that it has higher affinity for zinc ion than for copper (McCall and Fierke, 2004). We know of no small molecule indicators for which this has been shown. Similarly, the (rather slow) zinc binding kinetics of the wild-type protein have been improved 800-fold while only decreasing affinity 10-fold; evidently, it is difficult to improve the speed of response of small molecule indicators without also reducing the affinity (Nolan *et al.*, 2005).

Among other transduction approaches (reviewed in Fierke and Thompson, 2001), we have developed three different means of transducing metal ion binding to carbonic anhydrase as a change in fluorescence

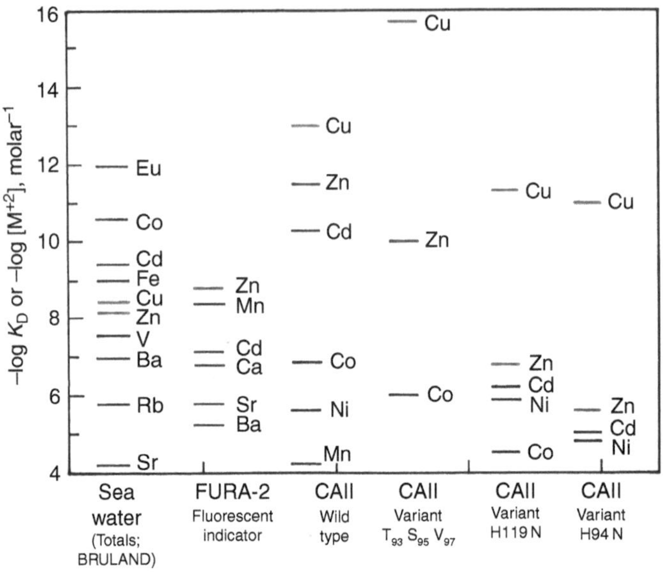

Figure 6.9 Relative affinities of carbonic anhydrase II variants for selected metal ions compared with the affinity of Fura-2 and total concentrations present in seawater (data of Bruland *et al.*). Reproduced from Thompson (2002), with permission from Elsevier.

lifetime. The first is suited for metal ion analytes such as Cu(II), Co(II), Ni(II), and Cd(II), where the metal ion binding to the fluorescent-labeled apoprotein quenches the fluorescence and reduces the lifetime (Thompson *et al.*, 1996b). The second two work almost exclusively with zinc, and take advantage of the fact that zinc-bound carbonic anhydrase binds a class of inhibitors called aryl sulfonamides. These inhibitors are clinically important in treating glaucoma, and thus many homologs have been synthesized (Maren, 1977). We found that the apoprotein has much less affinity for these compounds, and thus the fractional occupancy of the protein with the sulfonamide (at sulfonamide concentrations significantly above its K_d) reflects the occupancy of the active site with zinc, and therefore zinc's concentration. We have found or synthesized several fluorescent aryl sulfonamides. Some of these (DNSA, ABD-N, dapoxyl sulfonamide) exhibit substantial increases in fluorescence lifetime upon binding to the holoprotein, and thus the fraction of zinc-bound protein is equal to the fraction of the long-lived component that is present. For instance, dansylamide has a lifetime of approximately 22 ns when bound, but about 3 ns when free (Chen and Kernohan, 1967).

The final approach also uses an aryl sulfonamide, but a colored one called azosulfamide. In this case, the protein is labeled with a fluor with a suitable overlap of its emission with the azosulfamide's absorbance, such that the azosulfamide can serve as an energy transfer acceptor; an illustration of the approach is shown in Figure 6.10. Thus, if the colored azosulfamide is induced to bind by zinc's presence in the active site, it will quench the donor label's emission with a concomitant reduction in the lifetime; the zinc concentration is thus a simple function of the fractions of short and long lifetimes in the emission. Because most of

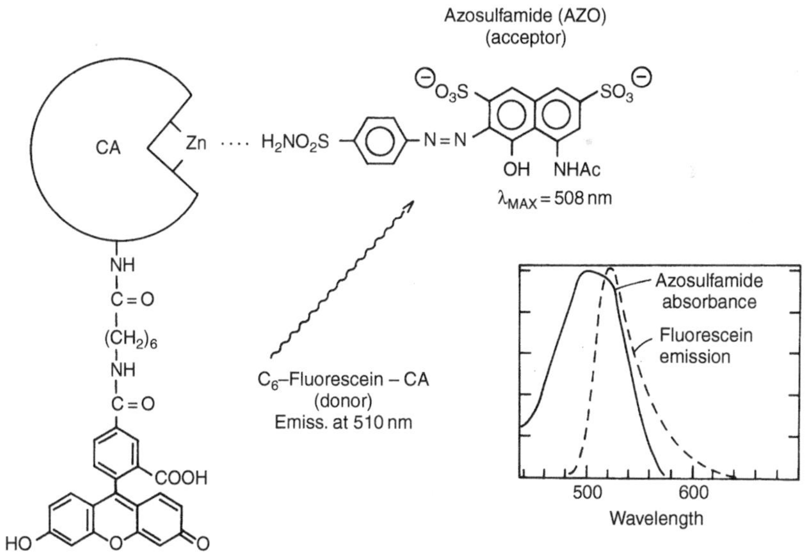

Figure 6.10 Scheme for fluorescence lifetime-based determination of zinc. Binding of zinc to fluorescence-labeled carbonic anhydrase promotes binding of azosulfamide to the zinc enzyme; the azosulfamide is thus brought into close proximity to the fluorescein label on the protein, which is partially quenched by energy transfer to the azosulfamide. The *inset* shows the overlap between fluorescein emission and azosulfamide absorption.

the parameters governing the efficiency of energy transfer by the Förster mechanism can be measured or estimated, the response of the system can be optimized by judicious positioning of the label on the protein and choice of its spectral overlap. A crucial advantage of this approach is that it is more general: the protein may be fluorescently labeled with a donor chosen and positioned to give an ideal level of overlap of the two (Thompson *et al.*, 1996a). For instance, it is convenient to have the azosulfamide binding accompanied by a 75% drop in intensity and lifetime of the donor; thus for a given donor with a calculable spectral overlap integral with the azosulfamide, its position on the protein (fixed by insertion of a cysteinyl residue in the sequence at the right distance by site-directed mutagenesis) should be chosen to be $0.95R_0$ from the azosulfamide bound in the active site. If other colored sulfonamides can be found or made with longer wavelengths, the approach might be extended into the infrared, where fluorescent background is less troubling (Thompson, 1994).

A recent study illustrates the value of FLIM for performing quantitative FRET measurements in cells. Classical (intensity-based) FRET measurements in cells typically suffer from the drawback that to measure the energy transfer efficiency to obtain the fraction bound, one must know the donor intensity in the absence of acceptor and saturated by acceptor. Usually this measurement must be made in a separate cell, and thus the intensity (or intensity ratio) may vary due to differing amounts of one or the other. To avoid this problem, it is customary to measure the donor fluorescence in the absence of acceptor by selectively photobleaching the acceptor, but the excitation may not be uniform. By comparison, to measure FRET by FLIM, one need only know the lifetime of the donor, because the fraction bound can be obtained directly from the fitted data. For example, Shen *et al.* (2007) studied the association between a barley nuclear receptor MLA10, which recognizes a virulence factor of powdery mildew fungus called $AVR_{A10,}$ and a transcription factor called WRKY, which is inactivated when bound and permits a host defense mechanism to operate. In particular, they wished to demonstrate that the association in the cell of MLA10 with WRKY was AVR_{A10} dependent. Fusion proteins of MLA10 and WRKY were constructed and expressed with yellow- (YFP) and cyan-fluorescing (CFP) variants of the green fluorescent protein (GFP), respectively. Because of the large overlap between CFP's emission and YFP's excitation, when genes modified with codes for these proteins were transfected into barley epidermal cells, lifetimes (2.33 ns versus 2.12 ns) were nearly identical for the donor alone (Figure 6.11, *row A*) and donor in the presence of effector (*row B*), respectively. However, in the presence of the effector, the average lifetime of the donor declined to 1.26 ns (*row C*). Substitution of an effector not recognized by MLA10 showed no energy transfer

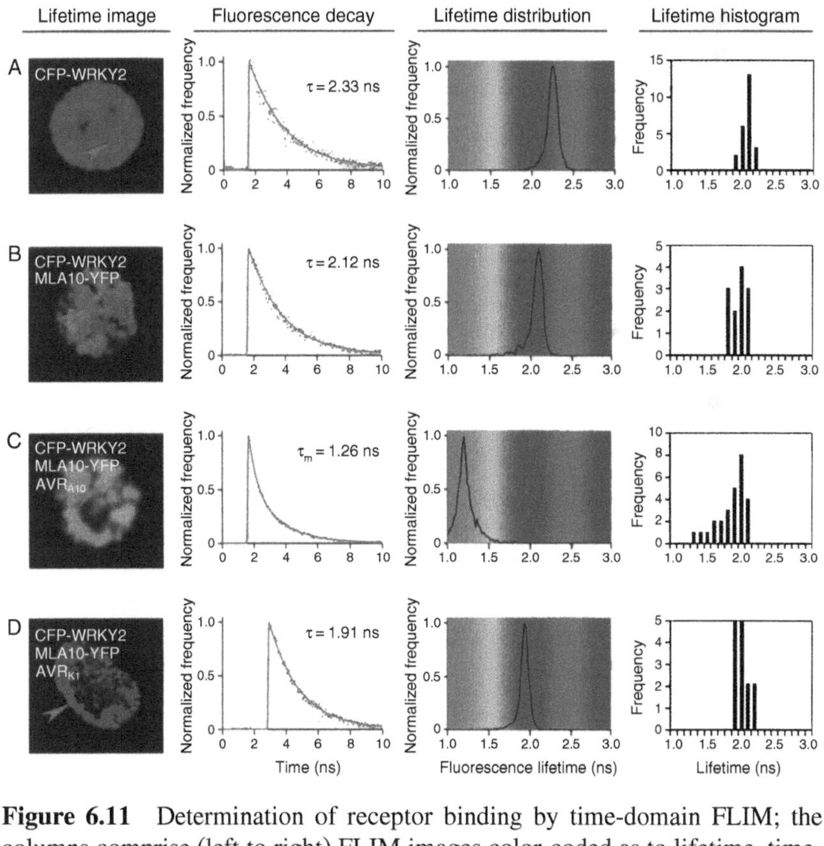

Figure 6.11 Determination of receptor binding by time-domain FLIM; the columns comprise (left to right) FLIM images color-coded as to lifetime, time-resolved intensity at the pixels selected by the red arrows, best fit lifetime distributions, and lifetime histograms for the regions of interest. Reproduced from Shen *et al.* (2007) with permission.

(*row D*). These data indicate the power and accuracy of the lifetime-based approach in comparison to simple intensity measurements.

6.4. Advantages and limitations for use in optical biosensing

There are several key advantages to lifetime-based biosensing, and until recently, one major drawback. Like the widely used ratiometric indicators (Fura-2 and Indo-1 are well-known examples), the measurement of lifetime is essentially independent of intensity, or fluorophore amount (Lakowicz *et al.*, 1993). Thus variations in specimen thickness, fluorophore photobleaching, washout, or fluctuations in excitation do not affect the measurement of lifetime and therefore analyte determinations. It should be understood that these advantages and most others pertain to both time- and frequency-domain measurements, although the latter enjoys some unique advantages in addition (see below). A key advantage over typical sensors or assays of the equilibrium type mentioned in Section 6.3.2 is the lack of a requirement to measure the signal level from both free and bound forms. For instance, an assay based on fluorescence anisotropy (polarization) requires knowing the polarization of the system in bound and free forms to calculate the fraction bound. By comparison, with lifetime-based sensing, one can, in a single determination, directly measure the fraction bound (and thus analyte concentration) by obtaining the α's and τ's. Moreover, the lifetime of the fluorophore is a property of the molecule, and is not a relative measurement, like a fluorescence intensity or absorption. The major disadvantage is the additional complexity of time-resolved instrumentation, in either the time or frequency domains. However, this equipment continues to grow simpler and less expensive.

In addition, there are unique advantages offered by the frequency-domain approach. The most important of these is illustrated for a fluorescent-labeled protein that exhibits a 1.5 ns lifetime in the bound form and a 4 ns lifetime in the free form. The frequency-dependent phases and modulations for different fractional saturations are depicted in Figure 6.12. Note that the curves are not parallel, because the decays are not the arithmetic average of the free and bound values. If we plot the values of phase and modulation at a suitable frequency as a function of fractional saturation, we obtain Figure 6.13.

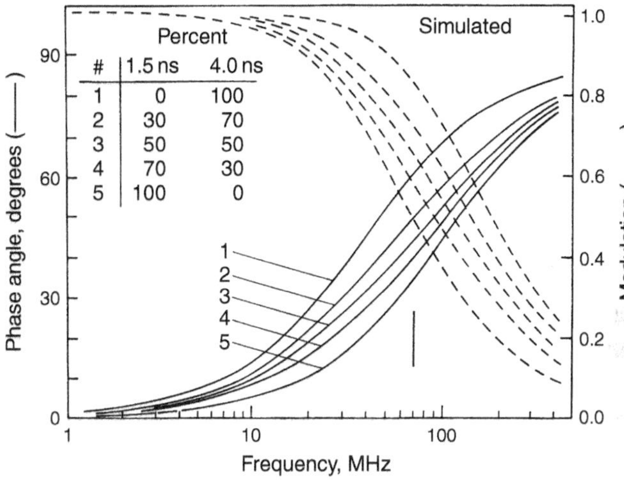

Figure 6.12 Simulated frequency-dependent phase and modulations for mixtures of 1.5 and 4.0 ns emitters in different proportions. Note that unlike Figure 6.4, the shapes of the curves are not all the same, and do not represent the weighted averages of the two lifetimes.

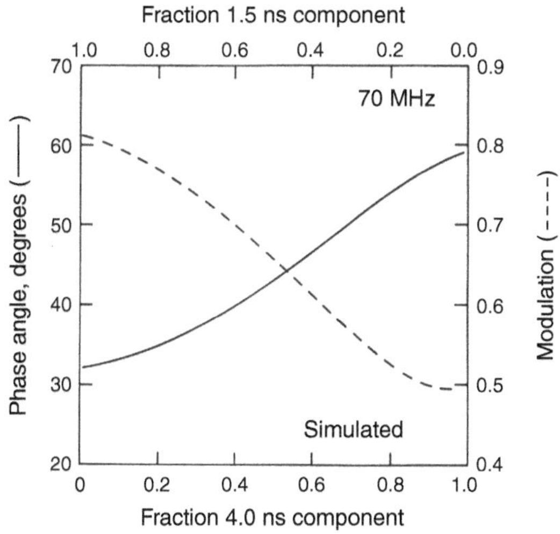

Figure 6.13 Simulated phases (—) and modulations (- - - -) at 70 MHz as a function of the proportion of short and long component, taken from Figure 6.12.

Figure 6.13 is essentially a calibration curve, since the lifetimes of the free and bound forms are properties of the system and (apart from the perturbations induced by temperature and the other influences mentioned above) do not change from one experiment to the next. Thus one could simply measure the phase angle and modulation at 70 MHz and relate it to the fraction bound, and therefore to the analyte concentration using the known binding constant. This procedure is simple and very rapid; with standard instruments precise measurements can be made in a few seconds, and if necessary within a few microseconds (Pinsky *et al.*, 1993).

From our standpoint, a really important advantage is that such phase and modulation measurements can also be done through a length of optical fiber up to hundreds of meters in length (Bright, 1988; Thompson and Lakowicz, 1993; Zeng *et al.*, 2003). By using a single optical fiber to carry the excitation to a fluorescent biosensing transducer at the distal end, one is essentially bringing the analysis to the sample, instead of collecting a sample and bringing it to be analyzed in the laboratory. As was recognized decades ago by Hirschfeld, Polanyi, Walt, Peterson, and others, this approach has significant advantages if the sample is remote, inaccessible, or dangerous to collect or be in close proximity to, such as if it is radioactive (or exposed to ionizing radiation). Perhaps most importantly, fiber optics offer the capability to monitor analyte levels inside the human body, which is otherwise difficult or impossible to do. Being dielectric, fiber optics are inherently safe in the presence of strong electrical fields or flammable atmospheres, and offer no hazard of electric shock when used *in vivo*. It is important to note that while we and others have substantial experience with fiber optic lifetime sensors working in the frequency domain with lifetimes in the nanosecond range, there is no the corresponding level of experience with time-domain methods and fiber optics (especially long fibers). Consequently, it is not known at this time what difficulties might be encountered in the time versus frequency domains.

6.5. Potential for improving performance or expanding current capabilities

While lifetime-based sensing using current technology is already a powerful (and underexploited) approach, there remains scope for significant improvement in the technology underlying the method; this will have the effect of opening up new applications.

6.5.1. Lifetime-based fiber optic sensors

As discussed Section 6.4 an important application of fluorescence lifetime biosensing is environmental monitoring, and particularly that using fiber optics. Classically, determination of water constituents such as metal ions has been done by collecting a sample and performing some analytical chemistry on it, which is time-consuming, expensive, labor intensive, and prone to sample contamination. If the water to be sampled is deep within the water column, merely collecting the sample may consume hours. If the sample collection requires a research vessel, the ship costs ~$10 000 per day to operate without the scientific party and their equipment, and the scarcity of these ships means that sample collection may not be routine or even repeatable for some years. If the analysis cannot be performed on site or on shipboard, the sample must be transported back to the laboratory, with the concomitant risks to the samples. Moreover, if the analytical technique requires preconcentration of the analyte from the sample, the samples may comprise tens of liters of water and be cumbersome and expensive to ship back to the laboratory from some remote site. Recognizing these issues, we (and others, notably David Walt) proposed the use of indwelling sensors that would provide a real-time readout of the analyte levels *in situ*. Recently, we demonstrated that free Cu(II) could be determined in seawater, *in situ*, in real time at picomolar levels using a fluorescence lifetime-based fiber optic sensor (Zeng *et al.*, 2003). Our vision is that ultimately a research vessel could tow a string of fiber optic sensors with their distal end transducers at various depths in the water column, enabling a depth profile to be constructed throughout a transect (Figure 6.14), as has been demonstrated using a "yo-yo-ing" immersed nitrate sensor by Johnson and Coletti (2002). While this vision is attractive, the data in Zeng *et al.*

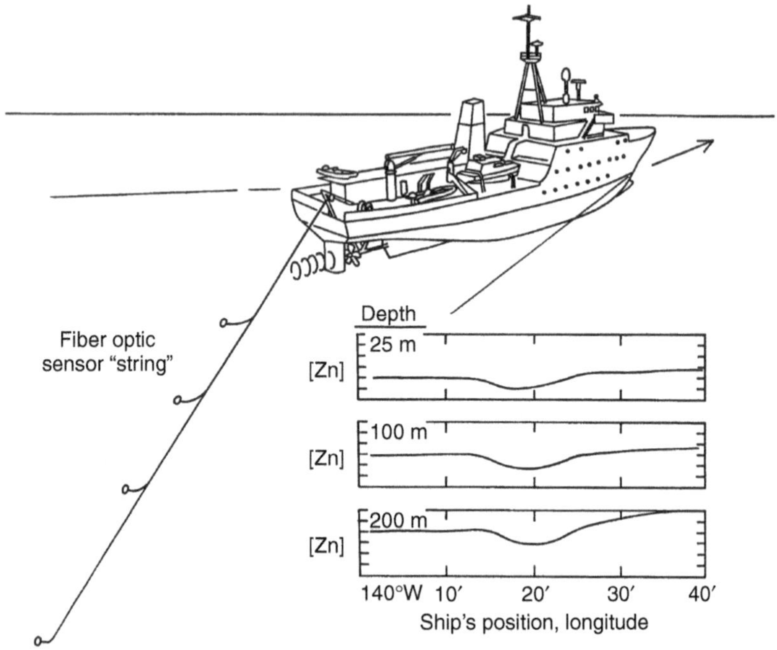

Figure 6.14 Fiber optic sensor string being towed by a research vessel, acquiring metal ion concentrations at various depths throughout a transect. Reproduced with the permission from the Society for Photooptical Instrumentation Engineers.

(2003) were obtained virtually at the surface through only a few meters of fiber. For wider use it will be necessary to use longer lengths of fiber, ideally kilometers long. For such sensors it will probably be necessary to use longer excitation and emission wavelengths due to the strong wavelength dependence of fiber attenuation, and also use single-mode fibers that do not have the modal dispersion that limits the bandwidth of multimode fibers. Ideally, one would like to work at wavelengths in the 1.5 μm telecommunications window, at which attenuation is very low (a few decibels per kilometer) and components (including optical fiber) are readily available and relatively inexpensive. Unfortunately, fluorescent compounds in this wavelength range are not readily available, and reactive analogs suitable for coupling to proteins are unknown.

6.5.2. Lifetime-based imaging

For lifetime-based imaging in the time domain, there are some limitations that will be more difficult to overcome. It should be apparent in the time-resolved (TCSPC) data that the precision and accuracy are crucially dependent on how many events are collected per pixel. Current mode-locked and high repetition rate diode lasers are almost ideal light sources for this purpose, supplying pulses at rates that barely let the fluorescence finish decaying before the next pulse arrives (80 MHz, or 1 pulse/12 ns); thus, data are collected as fast as possible. To acquire a decay with sufficient fidelity to recover a single exponential requires a few hundred events collected, which in turn requires 10 or 20 times as many excitation pulses due to the low inherent duty cycle of TCSPC. Thus a scanning time-resolved microscope might take a few milliseconds per pixel to collect each decay, or minutes for each image in the megapixel regime. To acquire the several thousand events per pixel necessary to resolve multiple decays would evidently require a great deal more time, which is incompatible with the study of living cells under intense laser illumination. Since almost all natural lifetimes are in the nanosecond regime or longer, it seems hard to beat this limitation without going to some parallel acquisition mode. State-of-the-art TCSPC instruments have eight channels, but presumably it would be difficult to add a great many more channels, so this limitation would seem to be difficult to avoid, at least at reasonable cost. On the other hand, the scanning approach is well suited to confocal microscopy for improved resolution, or multiphoton microscopy that perforce avoids prolonged excitation (and consequent photobleaching) of the majority of fluorophores whose emission is not being collected.

By comparison, there seems to be no intrinsic limitation to the time or spatial resolution of the frequency-domain imaging approach. In particular, the major issue is the computational reconstruction of the lifetimes from the phase-sensitive detector data. Thus, collecting six to eight high-resolution images at evenly spaced phase angles might take 100 ms each depending upon exposure time, and then another minute using a modern computer to calculate the lifetime images. While this approach does not permit confocal resolution and would be challenging to perform at video rates, it nevertheless permits multiple images to be made without prolonged illumination of the specimen.

References

Birch, D.J.S. and Imhof, R.E. (1991) Time-domain fluorescence spectroscopy using time-correlated single photon counting. In *Topics in Fluorescence Spectroscopy Vol. 1: Techniques* (J.R. Lakowicz, ed.) New York: Plenum Press, p. 1.

Birks, J.B. and Dyson, D.J. (1961) *J. Sci. Instr.*, **38**, 282.

Bright, F.V. (1988) *Time-Resolved Laser Spectroscopy in Biochemistry* (J.R. Lakowicz, ed.) Vol. 909, Los Angeles, CA: Society of Photooptical Instrumentation Engineers, p. 23.

Chen, R.F. and Kernohan, J. (1967) *J. Biol. Chem.*, **242**, 5813.

Demas, J.N. and DeGraff, B.A. (2001) *Coord. Chem. Rev.*, **211**, 317.

Fierke, C.A. and Thompson, R.B. (2001) *BioMetals*, **14**, 205.

Forster, T. (1948) *Annalen der Physik*, **2**, 55.

Gaviola, Z. (1926) A fluorometer for measuring fluorescence lifetime. *Z. Physik*, **42**, 853.

Grynkiewicz, G., Poenie, M., and Tsien, R.Y. (1985) *J. Biol. Chem.*, **260**, 3440.

Johnson, K.S. and Coletti, L. J. (2002) *Deep Sea Res.*, **I49**, 1291.

Keating, S.M. and Wensel, T.G. (1991) *Biophys. J.*, **59**, 186.

Lakowicz, J.R., Szmacinski, H., Nowaczyk, K. *et al.* (1992) *Anal. Biochem.*, **202**, 316.

Lakowicz, J.R., Szmacinski, H., and Thompson, R.B. (1993) *Proc. SPIE*, **2388**, 2.

Levy, R., Guignon, E.F., Cobane, S. *et al.* (1997) *Proc. SPIE*, **2980**, 81.

Lippitsch, M.E., Pusterhofer, J., Leiner, M.J.P., and Wolfbeis, O.S. (1988) *Anal. Chim. Acta.*, **205**, 1.

Lovgren, T., Hemmila, I., Pettersson, K., and Halonen, P. (1985) Time resolved fluorometry in immunoassay. In *Alternative Immunoassays* (W.P. Collins, ed.) New York: John Wiley and Sons, p. 203.

Maren, T.H. (1977) *Am. J. Physiol.*, **232**, F291.

McCall, K.A. and Fierke, C.A. (2004) *Biochemistry*, **43**, 3979.

Nolan, E.M., Jaworski, J., Okamoto, K.-I. *et al.* (2005) *J. Am. Chem. Soc.*, **127**, 16812.

Nordlund, T.M. (1990) Streak cameras for time-resolved fluorescence. In *Topics in Fluorescence Spectroscopy Vol. 1: Techniques* (J.R. Lakowicz, ed.) Vol. 1, New York: Plenum Press, p. 183.

Ozinskas, A., Malak, H., Joshi, J. *et al.* (1993) *Anal. Biochem.*, **213**, 264.
Pinsky, B.G., Ladasky, J.J., Lakowicz, J.R. *et al.* (1993) *Cytometry*, **14**, 123.
Shen, Q.-H., Saijo, Y., Mauch, S. *et al.* (2007) *Science*, **315**, 1098.
Spencer, R.D. and Weber, G. (1969) Measurements of subnanosecond fluorescence lifetimes with a cross-correlation phase fluorometer. *Ann. N.Y. Acad. Sci.*, **158**, 361.
Szmacinski, H. and Lakowicz, J.R. (1994) Lifetime-based sensing. In *Topics in Fluorescence Spectroscopy Vol. 4: Probe Design and Chemical Sensing* (J.R. Lakowicz, ed.) New York: Plenum Press, p. 295.
Szmacinski, H., Lakowicz, J.R., and Johnson, M.L. (1994) Fluorescence lifetime imaging microscopy: homodyne technique using high-speed gated image intensifier. In *Numerical Computer Methods* (M.L. Johnson and L. Brand, eds) Vol. 240, New York: Academic Press, p. 723.
Thompson, R.B. (1994) Red and near-infrared fluorometry. In *Topics in Fluorescence Spectroscopy Vol. 4: Probe Design and Chemical Sensing* (J.R. Lakowicz, ed.) New York: Plenum Press, p. 151.
Thompson, R.B. (2002) Fluorescence lifetime biosensors. In *Optical Biosensors Present and Future* (F.S. Ligler and C.A.R. Taitt, eds) New York: Elsevier, p. 143.
Thompson, R.B., Ge, Z., Patchan, M.W., and Fierke, C.A. (1996a) *J. Biomed. Optics*, **1**, 131.
Thompson, R.B., Ge, Z., Patchan, M.W. *et al.* (1996b) *Biosens. Bioelectron.*, **11**, 557.
Thompson, R.B. and Lakowicz, J.R. (1993) *Anal. Chem.*, **65**, 853.
Ullman, E.F. and Schwarzberg, M. (1981) *Fluorescence Quenching with Immunological Pairs in Immunoassays*. USA: Syva Company.
vanderOrd, C.J.R., deGrauw, C.J., and Gerritsen, H.C. (2001) *Proc. SPIE*, **4252**, 119.
Verkman, A.S. (1990) *Am. J. Physiol.*, **253**, C375.
Weber, G. (1953) *Adv. Prot. Chem.*, **8**, 415.
Xu, J., SHen, X., and Knutson, J.R. (2003) *J. Phys. Chem.*, **A107**, 8383.
Zeng, H.H., Thompson, R.B., Maliwal, B.P. *et al.* (2003) *Anal. Chem.*, **75**, 6807.
Zhang, Z. and Seitz, W.R. (1984) *Anal. Chim. Acta*, **160**, 47.

Optical Biosensors: Today and Tomorrow (2nd Edition)
Frances S. Ligler and Chris Rowe Taitt (Editors)

Chapter 7

ELECTROCHEMILUMINESCENCE

Mark M. Richter, Ph.D.
Department of Chemistry, Missouri State University,
Springfield, MO 65897, USA.

Electrochemiluminescence (ECL) is the process where species generated at electrodes undergo electron-transfer reactions to form excited states that emit light. Application of a voltage to an electrode in the presence of an ECL luminophore such as $Ru(bpy)_3^{2+}$ (where bpy = 2,2′-bipyridine) results in light emission and detection of the emitter at very low concentrations ($\leq 10^{-11}$ M). By employing ECL-active species as labels on biological molecules, ECL has found application in commercial instruments for immunoassays and DNA analyses. Commercial systems have been developed that use ECL to detect many clinically relevant analytes with high sensitivity and selectivity. The principles, history, applications, advantages, limitations, and possibilities for improving the performance of this technology are outlined in this chapter

7.1. Principles of operation

A wide variety of methods exist for the detection of chemical and biological analytes of interest. One of the most versatile, and one that has been commercially developed for the clinical diagnostic market, is electrochemiluminescence (also called electrogenerated chemiluminescence and abbreviated ECL). Electrochemiluminescence is a means of converting electrical energy into light (radiative energy). It involves the production of reactive intermediates from stable precursors at the surface of an electrode. These intermediates then react under a variety of conditions to form excited states that emit light. It is important to distinguish ECL from chemiluminescence (CL). Both ECL and CL involve the production of light by species that can undergo highly energetic electron-transfer reactions. However, luminescence in CL is initiated by the mixing of reagents and controlled by careful manipulation of fluid flows. In ECL, luminescence is initiated and controlled by switching an electrode voltage.

The first ECL reactions were investigated in the early 1960s on systems composed of polyaromatic organic compounds in highly purified and deaerated nonaqueous solvents such as acetonitrile (CH_3CN) and dimethylformamide (DMF). This early work was followed by studies on luminescent inorganic species, mainly $Ru(bpy)_3^{2+}$ (Tokel and Bard, 1972). Much of this early work has been reviewed (Bard and Faulkner, 1977; Glass and Faulkner, 1982). Adding certain species, called *coreactants*, made it possible to observe ECL in aqueous solution (Rubinstein and Bard, 1981). These developments have resulted in a wide range of analytical applications for ECL (Knight and Greenway, 1994), including commercial applications in clinical diagnostic assays (Blackburn *et al.*, 1991; Bard and Whitesides, 1993a, 1993b; Bard *et al.*, 2000; Bard, 2004; Richter, 2004). Several reviews covering various aspects of ECL have appeared (Faulkner and Bard, 1977; Bard and Faulkner, 1980; Faulkner and Glass, 1982; Knight and Greenway, 1994; Gerardi *et al.*, 1999; Knight, 1999; Aboul-Enein *et al.*, 2000; Bard *et al.*, 2000; Kukoba *et al.*, 2000; Mitschke and Bauerle, 2000; Andersson and Schmehl, 2001; Bard and Faulkner, 2001; Fahnrich *et al.*, 2001; Knight, 2001; Richter, 2002; Kulmala and Suomi, 2003; Richter, 2004; Gorman *et al.*, 2006) including a comprehensive monograph (Bard, 2004), and the interested reader is encouraged to consult both of them and the primary literature for more detailed information than is provided in this chapter.

7.1.1. Annihilation ECL

Traditionally, ECL was generated via annihilation, where the electron-transfer reaction is between an oxidized and a reduced species, both of which are generated at an electrode by alternate pulsing of the electrode potential.

$$A + e^- \rightarrow A^{\bullet -} \text{ (reduction at electrode)} \tag{7.1}$$

$$A - e^- \rightarrow A^{\bullet +} \text{ (oxidation at electrode)} \tag{7.2}$$

$$A^{\bullet -} + A^{\bullet +} \rightarrow A^* + A \text{ (excited-state formation)} \tag{7.3}$$

$$A^* \rightarrow A + h\nu \text{ (light emission)} \tag{7.4}$$

where $h\nu$ is a photon of light. For example, the potential of the working electrode is quickly changed between two different values in order to generate the reduced, $A^{\bullet -}$, and oxidized, $A^{\bullet +}$, species (Eqns (7.1) and (7.2), respectively) that will react near the electrode surface to form the emissive state, A^* (Eqn 7.3)).

A typical reaction involves 9,10-diphenylanthracene (DPA, Figure 7.1). Electrochemiluminescence is generated when a double potential step is applied to an electrode (such as platinum), producing the radical cation ($DPA^{\bullet +}$) upon anodic oxidation and the radical anion ($DPA^{\bullet -}$) upon cathodic reduction. The resulting electrogenerated products can then react and undergo annihilation (i.e., Eqn (7.3)) to produce an excited state (DPA^*) that is able to emit light.

$$DPA - e^- \rightarrow DPA^{\bullet +} \tag{7.5}$$

$$DPA - e^- \rightarrow DPA^{\bullet -} \tag{7.6}$$

$$DPA^{\bullet +} + DPA^{\bullet -} \rightarrow DPA + DPA^* \tag{7.7}$$

$$DPA^* \rightarrow DPA + h\nu \tag{7.8}$$

For DPA, the emission maximum (λ_{max}) occurs at about 512 nm and the ECL spectrum (i.e., a plot of ECL emission versus wavelength)

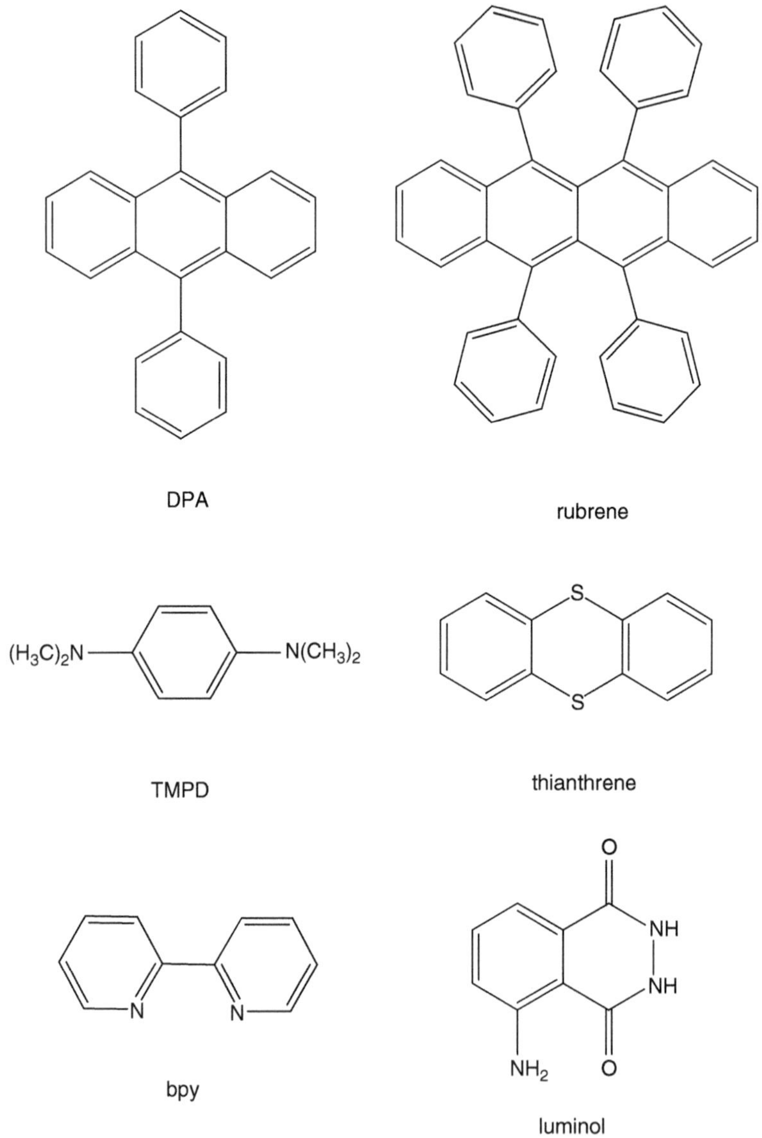

Figure 7.1 Structures of compounds.

is identical to DPA photoluminescence (PL). This indicates that the ultimate product of charge transfer, and hence luminescence, is the lowest singlet DPA species, $^1DPA^*$. There is considerable evidence that the reactions outlined in Eqns (7.7) and (7.8) are oversimplifications and that several mechanistic steps intervene between the electron-transfer and photon emission steps. The identification of these mechanisms has been a focus of past work in the field and is thoroughly reviewed elsewhere (Faulkner and Bard, 1977; Glass and Faulkner, 1982).

DPA and several other polyaromatic hydrocarbons were among the first complexes studied using ECL (Chandross and Visco, 1964; Hercules, 1964; Bader and Kuwana, 1965; Santhanam and Bard, 1965) since they were known to undergo chemically and electrochemically reversible one-electron oxidation and reduction at easily attainable potentials (e.g., $\leq 2\,V$ versus SHE where SHE = standard hydrogen electrode) and display PL efficiencies (ϕ_{em} = number of photons emitted per photon absorbed by the compound) of near unity. These qualities, as well as the stability of the oxidized and reduced forms of the complexes (i.e., the radical anions and cations) and their ability to undergo electron-transfer reactions, are still used as criteria to determine whether a compound shows promise as an ECL luminophore.

DPA is an example of an "energy-sufficient" system. That is, the enthalpy, ΔH^o, of the electron-transfer reaction in Eqn (7.3) is larger than the energy required to produce the excited singlet state from the ground state (Eqn (7.4)). In DPA, one of the products of the reaction is therefore produced with excess energy that can be emitted as light (e.g., $^1DPA^*$). When the luminescence is emitted by a species in an excited singlet state, this process is known as the singlet or S-route. Since the free energy (ΔG^o) available for exciting the product is the energy available from the redox reaction producing the ground-state products (i.e., Eqns (7.1) and (7.2)), ΔH^o can be calculated from the reversible standard potentials of the redox couples:

$$\Delta H^o = \Delta G^o + T\Delta S^o = (E^o_{A+/A} - E^o_{A/A-}) - 0.1\,eV$$

where $E^o_{A+/A}$ and $E^o_{A/A-}$ represent the cyclic voltammetric peak potentials for the oxidative and reductive redox couples, respectively. For example, in DPA:

$$DPA + e^- \rightarrow DPA^{\bullet -} \quad E^o_{DPA/DPA-} = -1.89 \text{ V versus SCE} \tag{7.9}$$

$$DPA - e^- \rightarrow DPA^{\bullet +} \quad E^o_{DPA+/DPA} = +1.35 \text{ V versus SCE} \tag{7.10}$$

$E^o_{DPA+/DPA}$ and $E^o_{DPA/DPA-}$ are the cyclic voltammetric peak potentials for the oxidation and reduction of diphenylanthracene, respectively, and SCE is the saturated calomel reference electrode (Faulkner and Bard, 1977). This results in an enthalpy of 3.14 eV for the electron-transfer reaction. When we compare this to the energy of the emitted light in DPA obtained spectroscopically ($\sim\lambda_{max} = 512$ nm or 3 eV), we see that the emitting state is accessible and $^1DPA^*$ may be populated directly in the reaction.

Another example of an energy-sufficient system is the inorganic species $Ru(bpy)_3^{2+}$ (where bpy = 2,2′-bipyridine, Figure 7.1), whose ECL was first reported in 1972 (Tokel and Bard, 1972).

$$Ru(bpy)_3^{2+} - e^- \rightarrow Ru(bpy)_3^{3+} \quad E^o_{Ru2+/3+} = +1.2 \text{ V versus SCE} \tag{7.11}$$

$$Ru(bpy)_3^{2+} - e^- \rightarrow Ru(bpy)_3^{+} \quad E^o_{Ru2+/1+} = -1.4 \text{ V versus SCE} \tag{7.12}$$

$$Ru(bpy)_3^{3+} - Ru(bpy)_3^{+} \rightarrow Ru(bpy)_3^{2+*} + Ru(bpy)_3^{2+} \tag{7.13}$$

$$Ru(bpy)_3^{2+*} \rightarrow Ru(bpy)_3^{2+} + h\nu \ (\sim 2.1\,\text{eV}, 610\,\text{nm}) \tag{7.14}$$

The excited state formed in the ECL reaction is the same as that formed during photoexcitation (Figure 7.2). In photoexcitation (i.e., PL), an electron is excited from metal-based dπ orbitals to ligand-based π^* orbitals (a metal-to-ligand charge-transfer (MLCT) transition). The excited electron then undergoes intersystem crossing to the lowest triplet state of $Ru(bpy)_3^{2+*}$ from where emission occurs (Demas and Crosby, 1968;

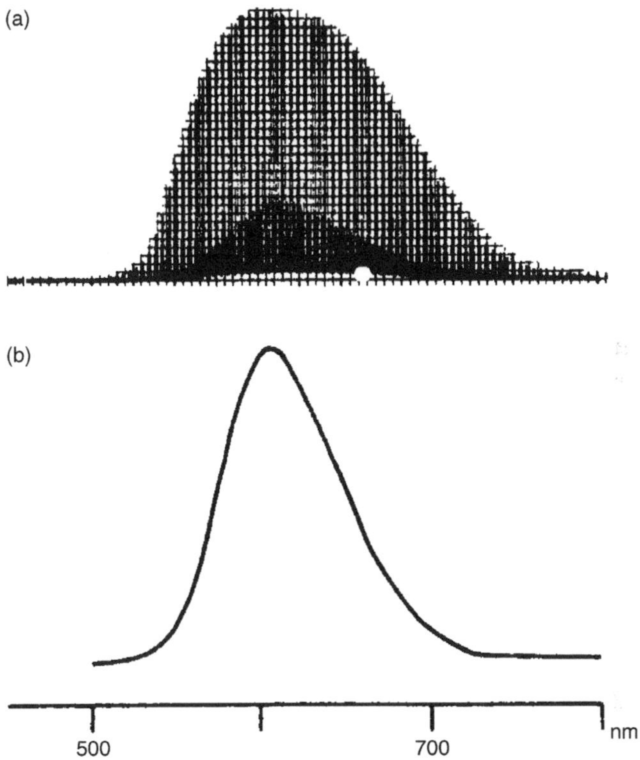

Figure 7.2 (a) Electrochemiluminescence emission spectrum of $Ru(bpy)_3C_{12}$ from 450 to 800 nm using a cyclic square wave at 0.2 Hz between +1.75 and −1.60 V versus Ag reference electrode and (b) fluorescence emission spectrum of $Ru(bpy)_3C_{12}$ in MeCN with excitation at 500 nm. (Reprinted with permission from Tokel and Bard (1972). Copyright 1972, American Chemical Society.)

Van Houton and Watts, 1975). Since the photoluminescent and ECL spectra are nearly identical, the emission process in ECL involves the MLCT state of $Ru(bpy)_3^{2+}$. This state may be formed if an electron is transferred to the π^* orbital of one of the bipyridine ligands. $Ru(bpy)_3^{2+*}$ can then decay to the ground state, producing the same luminescence as obtained from PL spectroscopy (Tokel and Bard, 1972). The reaction given in Eqn (7.14) has about 2.6 V of free energy to place into an excited product. Since the charge-transfer triplet lies at 2.1 eV, it is readily accessible to the homogenous redox process (Glass and Faulkner, 1981; Luttmer and Bard, 1981). No other excited states of appreciable lifetime can be populated, so the opportunities for unusual kinetics and alternate mechanistic pathways in the $Ru(bpy)_3^{2+}$ system are reduced.

$Ru(bpy)_3^{2+}$ is perhaps the most thoroughly studied ECL-active molecule (Tokel-Takvoryan *et al.*, 1973; Itoh and Honda, 1979; Wallace and Bard, 1979; Glass and Faulkner, 1981; Luttmer and Bard, 1981). This is for a number of reasons, including its strong luminescence (Figure 7.2) and solubility in both aqueous and nonaqueous media at room temperature and its ability to undergo reversible one-electron-transfer reactions at easily attainable potentials (Figure 7.3), leading to stable reduced and oxidized species. However, the ECL of several other inorganic compounds has been reported (Hemingway *et al.*, 1975; Faulkner and Glass, 1982; Knight and Greenway, 1994; Richter and Bard, 1996; Richter *et al.*, 1998; Bard, 2004; Richter, 2004). This is not surprising, since many inorganic compounds display the electrochemical and spectroscopic qualities required of ECL luminophores. Furthermore, the overall ECL efficiency (photons produced per redox event) is a product of the PL quantum yield and the efficiency of production of the excited state. $Ru(bpy)_3^{2+}$ has a photoluminescent quantum efficiency (ϕ_{em}) of 0.0682 (Van Houten and Watts, 1976; Caspar and Meyer, 1983) and an ECL efficiency (ϕ_{ecl}) of 0.0500 (Rubinstein and Bard, 1981; White and Bard, 1982). Under certain conditions, the reaction of Eqn (7.13) produces the emitting charge-transfer triplet with an efficiency approaching 100% (Itoh and Honda, 1979; Wallace and Bard, 1979) and is comparable with PL data (Demas and Crosby, 1971; Meyer, 1978), showing that about 5% of the excited states produce luminescence. This demonstrates that PL and electrochemical data can be used to predict compounds that may show promise as ECL light emitters.

It is also possible to use two different precursors to generate an emissive state.

$$A + e^- \rightarrow A^{\bullet -} \tag{7.15}$$

$$D - e^- \rightarrow D^{\bullet +} \tag{7.16}$$

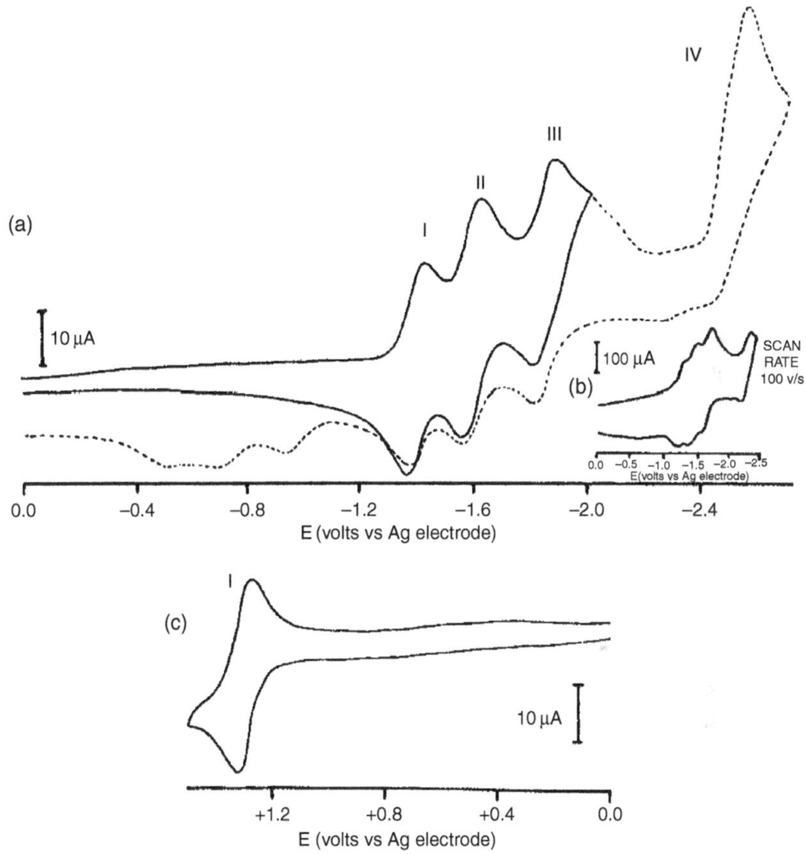

Figure 7.3 Cyclic voltammogram of 1 mM $Ru(bpy)_3(ClO_4)^2$/MeCN/0.1 M TBABF4 at a Pt disk microelectrode: (a) scan from 0.0 to −2.6 V, scan rate = 200 mV/s; (b) scan, scan rate = 100 V/s; (c) scan from 0.0 to +1.4 V, scan rate = 200 mV/s. (Reprinted with permission from Tokel-Takvoryan *et al.* (1973). Copyright 1973, American Chemical Society.)

$$A^{\bullet -} + D^{\bullet +} \rightarrow A^* + D \text{ (or } A + D^*) \quad (7.17)$$

$$A^*(\text{or } D^*) \rightarrow A \text{ (or D)} + h\nu \quad (7.18)$$

A classic example is the electron-transfer reaction and subsequent CL between the anion radical of DPA and the cation radical of

N,*N*,*N*′, *N*′-tetramethyl-*p*-phenylenediamine (Wurster's Blue (TMPD), Figure 7.1) in DMF.

$$\mathrm{DPA} + \mathrm{e}^- \rightarrow \mathrm{DPA}^{\bullet -} \quad E^{\mathrm{o}}_{\mathrm{DPA-/DPA}} = -1.89\,\mathrm{V}\ \text{versus SCE} \tag{7.19}$$

$$\mathrm{TMPD} + \mathrm{e}^- \rightarrow \mathrm{TMPD}^{\bullet +} \quad E^{\mathrm{o}}_{\mathrm{TMPD/TMPD+}} = +0.24\,\mathrm{V}\ \text{versus SCE} \tag{7.20}$$

$$\mathrm{DPA}^- + \mathrm{TMPD}^+ \rightarrow {}^1\mathrm{DPA}^* + \mathrm{TMPD} \tag{7.21}$$

$${}^1\mathrm{DPA}^* \rightarrow \mathrm{DPA} + h\nu \tag{7.22}$$

where ^{1}DPA* is the excited singlet state whose excess energy is emitted as light. In theory, light may be emitted from either ^{1}DPA* or TMPD (e.g., either A* or D* in Eqn (7.17)). However, the emitted light generated via ECL is identical to DPA PL, indicating that ^{1}DPA* is the ultimate product of charge transfer. This is not surprising, since the energy states of TMPD are inaccessible and hence TMPD cannot luminesce (Faulkner and Bard, 1977). The fact that the DPA/TMPD system undergoes ECL at all is surprising since the enthalpy for the electron-transfer reaction is 2.03 eV, much less than that required to reach the emitting singlet excited state for DPA of 3.00 eV. Therefore, the ultimate emitter is not directly populated. Such systems are called "energy deficient," and a more complicated scheme than that depicted in Eqns (7.21) and (7.22) is required to generate the emitter. The most commonly accepted explanation involves triplet intermediates, the so-called triplet or T-route.

$$\mathrm{A}^{\bullet -} + \mathrm{D}^{\bullet +} \rightarrow {}^3\mathrm{A}^* + \mathrm{D} \tag{7.23}$$

$$\mathrm{A}^* + {}^3\mathrm{A}^* \rightarrow {}^1\mathrm{A}^* + \mathrm{A}\ \text{(triplet–triplet annihilation)} \tag{7.24}$$

Equation (7.24) is generally called "triplet–triplet annihilation" (Weller and Zachariasse, 1969; Weller and Zachariasse, 1971; Zachariasse, 1975; Faulkner and Bard, 1977) where the energy from two-electron-transfer reactions is pooled to provide sufficient energy to form the singlet excited state (Parker, 1968; Birks, 1970). For the DPA/TMPD system:

$$\mathrm{TMPD}^{\bullet+} + \mathrm{DPA}^{\bullet-} \rightarrow \mathrm{TMPD} + {}^{3}\mathrm{DPA}^{*} \quad (7.25)$$

$${}^{3}\mathrm{DPA}^{*} + {}^{3}\mathrm{DPA}^{*} \rightarrow \mathrm{DPA} + {}^{1}\mathrm{DPA}^{*} \quad (7.26)$$

Many ECL reactions with different precursors follow this route and several examples are given in the literature (Faulkner and Bard, 1977; Knight and Greenway, 1994; Bard, 2004; Richter, 2004). The T-route may also operate in energy-sufficient systems.

Although analytical uses for ECL are possible with annihilation systems (e.g., display devices), most annihilation systems require the use of rigorously purified and deoxygenated nonaqueous solvents since the available potential range in water is too narrow to generate the required energetic precursors. In essence, the stability range for the electrochemical oxidation and reduction of water is too small to conveniently generate both species (i.e., the radical anion and cation) needed for annihilation ECL. For example, $\mathrm{Ru(bpy)}_3^{2+}$ is oxidized at a Pt electrode to form $\mathrm{Ru(bpy)}_3^{3+}$ at about 1.2 V versus SCE. $\mathrm{Ru(bpy)}_3^{2+}$ is reduced at a Pt electrode in aqueous solution to form $\mathrm{Ru(bpy)}_3^{+}$ at about −1.4 V, a potential not easily attainable at Pt electrodes in aqueous solution without the evolution of large amounts of hydrogen gas. The products of water oxidation and/or reduction interfere with the annihilation reaction (Eqn (7.13)) such that little to no light is observed. Background ECL observed in such systems has been attributed to a light-emitting reaction between $\mathrm{Ru(bpy)}_3^{3+}$ and OH^- ions, rather than the annihilation of the +1 and +3 ruthenium complex (Rubinstein and Bard, 1981).

The ECL systems shown above generate oxidized and reduced species at a single electrode. It is also possible to obtain emission at two different electrodes that are sufficiently close to allow the electrogenerated reactants to interdiffuse and undergo annihilation (e.g., Eqn (7.3)) (Maloy *et al.*, 1971; Brilmyer and Bard, 1980). For example, a rotating ring disk electrode (RRDE) can be employed. One reactant, such as $\mathrm{A}^{\bullet-}$, can be generated at the central disk and the other, $\mathrm{A}^{\bullet+}$, generated at the ring. These are then swept together by diffusion and convection (Bard and Faulkner, 1980), resulting in a ring of light on the inner edge of the ring electrode (Faulkner and Bard, 1977; Maloy and Bard, 1971). Other experiments have employed dual-working-electrode systems (Brilmeyer and Bard, 1980; Bartelt *et al.*, 1992; Fioccabrino *et al.*, 1998) with interdigitated electrodes, thin-layer geometry, or flowing streams to move the reactants together.

7.1.2. Coreactant ECL

Electrochemiluminescence can also be generated in a single potential step utilizing a coreactant. A coreactant is a species that, upon oxidation or reduction, produces an intermediate that can react with an ECL luminophore to produce excited states. Usually, this occurs upon bond cleavage of the coreactant to form strong oxidants or reducants. For example, the oxalate ion ($C_2O_4^{2-}$) is believed to produce the strong reductant $CO_2^{\bullet-}$ upon oxidation in aqueous solution (Rubinstein and Bard, 1981):

$$C_2O_4^{2-} - e^- \rightarrow [C_2O_4^{\bullet-}] \rightarrow CO_2^{\bullet-} + CO_2 \tag{7.27}$$

The oxidizing potential that leads to $CO_2^{\bullet-}$ also oxidizes an ECL luminophore (e.g., D where D is, for example, $Ru(bpy)_3^{2+}$).

$$D - e^- \rightarrow D^{\bullet+} \tag{7.28}$$

$D^{\bullet+}$ and $CO_2^{\bullet-}$ then react to produce an excited state capable of emitting light.

$$CO_2^{\bullet-} + D^{\bullet+} \rightarrow D^* + CO_2 \tag{7.29}$$

$$D^* \rightarrow D + h\nu \tag{7.30}$$

Oxalate is often referred to as an "oxidative–reductive" coreactant due to its ability to form a strong reducing agent upon electrochemical oxidation. Unlike annihilation schemes where a double potential step (e.g., oxidation followed by reduction) is required to generate the highly energetic precursors, in coreactant ECL, the electrode typically only oxidizes *or* reduces the reagents in a *single* potential step. For example, in the oxalate system the electrode oxidizes both the oxalate and the ECL reactant D; the reductant, $CO_2^{\bullet-}$ is then generated upon bond cleavage of oxalate via Eqn (7.27). This strategy is used in most analytical and biotechnology applications, with the reactant D being $Ru(bpy)_3^{2+}$. This methodology has allowed the generation of ECL in aqueous solution, a great advantage in terms of analytical applications. Without this ability, it is doubtful whether ECL would have moved beyond the laboratory phase.

Another example of an "oxidative–reductive" system is the commercially important $Ru(bpy)_3^{2+}$/TPrA system (TPrA = tri-*n*-propylamine). As with the oxalate system, this involves the production of a strong reductant (presumably $TPrA^{\bullet}$) by an initial oxidation sequence (Leland and Powell, 1991; McCord and Bard, 1991).

$$Ru(bpy)_3^{2+} - e^- \rightarrow Ru(bpy)_3^{3+} \tag{7.31}$$

$$TPrA - e^- \rightarrow [TPrA^{\bullet}]^+ \rightarrow TPrA^{\bullet} + H^+ \tag{7.32}$$

$$Ru(bpy)_3^{3+} + TPrA^{\bullet} \rightarrow Ru(bpy)_3^{2+} + \text{products} \tag{7.33}$$

$$Ru(bpy)_3^{2+*} \rightarrow Ru(bpy)_3^{2+} + h\nu \tag{7.34}$$

Electrochemiluminescence is produced upon concomitant oxidation of $Ru(bpy)_3^{2+}$ and TPrA (Figure 7.4). Electrochemical studies of various aliphatic amines have indicated a possible reaction pathway for the oxidation of TPrA (Smith and Mann, 1969). Upon oxidation, the short-lived TPrA radical cation ($TPrA^{\bullet+}$) is believed to lose a proton from an α-carbon to form the strongly reducing intermediate $TPrA^{\bullet}$. This radical can then reduce $Ru(bpy)_3^{3+}$ to $Ru(bpy)_3^{2+*}$. Other reaction mechanisms for production of the excited state have also been proposed (Zu and Bard, 2000; Kanoufi *et al.*, 2001; Miao and Choi, 2002; Miao *et al.*, 2002); for example, reduction of $Ru(bpy)_3^{2+}$ to $Ru(bpy)_3^{1+}$ by $TPrA^{\bullet}$, followed by annihilation:

$$Ru(bpy)_3^{1+} + Ru(bpy)_3^{3+} \rightarrow Ru(bpy)_3^{2+} + Ru(bpy)_3^{2+*} \tag{7.35}$$

Although the details of the coreactant ECL mechanism (Eqns (7.31)–(7.34)) to generate light emission are still under study, the

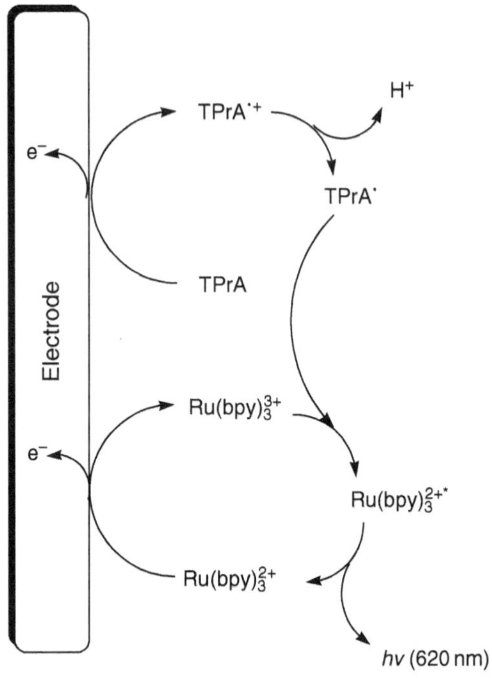

Figure 7.4 Proposed mechanism for $Ru(bpy)_3^{2+}$/TPrA ECL system.

origin of the light emission from $Ru(bpy)_3^{2+}$ has been well documented (Glass and Faulkner, 1981; Faulkner and Glass, 1982; Leland and Powell, 1991; McCord and Bard, 1991). Since the photoluminescent and ECL spectra are nearly identical, the emission process in ECL involves the MLCT state of $Ru(bpy)_3^{2+}$. This state may be formed if the reducing agent (i.e., $TPrA^{\bullet}$) transfers an electron to the π^* orbital of one of the bipyridine ligands. $Ru(bpy)_3^{2+*}$ can then decay to the ground state, producing the same luminescence as obtained from PL spectroscopy. Solution phase coreactant ECL using TPrA and $Ru(bpy)_3^{2+}$ is quite sensitive, with sub-picomolar detection limits achieved (Blackburn *et al.*, 1991; Leland and Powell, 1991; Bard *et al.*, 2000; Bard, 2004).

Other systems use coreactants that are reduced to generate reactive species (i.e., "reductive–oxidative" coreactants). For example, in the case of peroxydisulfate ($S_2O_8^{2-}$), reduction produces the strong oxidant $SO_4^{\bullet-}$, which then undergoes an electron-transfer reaction with an ECL luminophore like $Ru(bpy)_3^{2+}$ to generate light (Bolletta *et al.*, 1981; White and Bard, 1981) as shown below.

$$Ru(bpy)_3^{2+} + e^- \rightarrow Ru(bpy)_3^+ \quad (7.36)$$

$$S_2O_8^{2-} + e^- \rightarrow SO_4^{\bullet-} + SO_4^{2-} \quad (7.37)$$

$$Ru(bpy)_3^+ + SO_4^{\bullet-} \rightarrow Ru(bpy)_3^{2+*} + SO_4^{2-} \quad (7.38)$$

Besides being of practical interest, ECL reactions of this type also demonstrate the intermediacy of species such as $TPrA^\bullet$, $SO_4^{\bullet-}$, and $CO_2^{\bullet-}$.

7.2. History

The first detailed studies on ECL were begun in the mid-1960s but interest in light emitted during electrolysis was generated much earlier. Dufford *et al.* (1927) observed emission at an anode by applying between 500 and 1500 V to a cathode in a solution of Grignard compounds in anhydrous ether. The reaction conditions in these experiments were not very well defined, and it is doubtful whether this process was actually ECL. At such high potentials, electroluminescence (the direct injection and removal of charge with the formation of an electron–hole pair) or electrode processes were probably responsible for the observed light emission. However, 2 years later this initial report was followed by published experiments on luminol (2,3-aminophthalhydrazide) (Figure 7.1) in aqueous/alkaline solution (Harvey, 1929). The potentials used to generate anodic light emission were much lower than those used by Dufford and coworkers (2.8 V versus 500–1500 V, respectively). Several groups followed up on this (Bernanose *et al.*, 1947; Vojir, 1954; Kuwana *et al.*, 1963), and in fact the luminol system continues to generate interest (Haapakka and Kankare, 1980; Haapakka, 1982; Spurlin and Cooper, 1986; Sakura and Imai, 1988; van Dyke and Cheng, 1989; Nakashima *et al.*, 1991; Vitt *et al.*, 1991; Ishida *et al.*, 1992a, 1992b;

Sakura, 1992; Jirka *et al.*, 1993; Steijger *et al.*, 1993; Bowie *et al.*, 1996; Kearney *et al.*, 1996; Ouyang and Wang, 1998; Wilson and Schiffrin, 1998; Zhu and Kok, 1998; Chen and Wang, 1999; Lin *et al.*, 1999; Marquette and Blum, 1999; Dodeigne *et al.*, 2000; Leca and Blum, 2000; Sun *et al.*, 2000; Taylor and Creager, 2000; Guo *et al.*, 2002; Zheng *et al.*, 2002; Su *et al.*, 2005; Dong *et al.*, 2006; Luo and Zhang, 2006; Shi and Cui, 2006;). Practical applications of the luminol system are challenged by a number of factors, most notably the highly non-specific ECL (i.e., background) – possibly due to the formation of oxygen at the anode in aqueous solution, followed by chemiluminescent reactions involving oxygen – and the extremely basic conditions ($pH > 11$) that are needed to generate sufficient light emission.

In the mid-1960s, several research groups (Chandross and Visco, 1964; Hercules, 1964; Bader and Kuwana, 1965; Santhanam and Bard, 1965) studied the luminescence generated during electrolysis of polyaromatic hydrocarbons (e.g., anthracene, diphenylanthracene, thianthrene, rubrene) in aprotic media, Figure 7.1. In essence, they wished to see if excited states could be generated electrochemically as well as photochemically. It was observed, both visually and spectroscopically, that the radiation emitted by sweeping to both negative and positive potentials (annihilation pathway, Eqns (7.1)–(7.4)) was often identical to that generated during PL, indicating formation of the excited singlet state.

Throughout 1960s and 1970s, work continued on the polyaromatic hydrocarbons and was eventually extended to other systems, most notable among them the ruthenium chelates. Since the discovery that $Ru(bpy)_3^{2+}$ is photoluminescent (Paris and Brandt, 1959), a large body of literature aimed at understanding both the ground- and excited-state properties of $Ru(bpy)_3^{2+}$, $Os(bpy)_3^{2+}$, and their polyazine derivatives has appeared (Demas and Crosby, 1971; Meyer, 1978; Sutin and Creutz, 1978; Barigelletti *et al.*, 1991; Roundhill, 1994). Therefore, it is not surprising that these compounds have also played an important role in the development of ECL. The first report of ECL in a metal chelate was in 1972 (Tokel and Bard, 1972), in which the excited state of $Ru(bpy)_3^{2+}$ was generated in aprotic media by annihilation of the reduced, $Ru(bpy)_3^{1+}$, and oxidized, $Ru(bpy)_3^{3+}$, species (Eqns (7.11)–(7.14)), Figures 7.2 and 7.3.

The original coreactant, and thus the first report of ECL in aqueous solution, was oxalate ion ($C_2O_4^{2-}$; Eqns (7.27)–(7.29)) (Rubinstein and Bard, 1981). Subsequently, other species were shown to act as coreactants, among them peroxydisulfate ($S_2O_8^{2-}$; Eqns (7.36)–(7.38)) and tri-*n*-propylamine (TPrA; Eqns (7.31)–(7.34)). The discovery of TPrA (Leland and Powell, 1991) allowed efficient ECL not only in aqueous media, but also at physiological pH. Following the first report on TPrA, other species containing amine groups were proposed, among them many biologically important analytes (e.g., alkylamines, nicotinamide adenine dinucleotide (NADH), antibiotics, L,D-tryptophan, glucose, erythromycin, valine, HIV-gag gene). The list is quite extensive, and compilations up to 2003 have been published (Knight and Greenway, 1994; Knight, 1999; Bard, 2004; Richter, 2004).

To date, ECL has found use in studying the properties of both organic and inorganic systems (Faulkner and Bard, 1977; Knight and Greenway, 1994; Bard, 2004; Richter, 2004). These include polyaromatic hydrocarbons (Faulkner and Bard, 1977; Faulkner and Glass, 1982; Richards and Bard, 1995), exciplexes (Hemingway *et al.*, 1975; Prieto *et al.*, 2001), polymer assemblies (Rubinstein *et al.*, 1983; Downey and Niemann, 1992; Richter *et al.*, 1994), transition metal complexes incorporating metals such as Ru, Os, and Pt (Tokel and Bard, 1972; Faulkner and Glass, 1982; Vogler and Kunkeley, 1984; Kim *et al.*, 1985; Knight and Greenway, 1994; Richter *et al.*, 1998) as well as rare earth chelates (Hemingway *et al.*, 1975; Richter and Bard, 1996), to name a few. $Ru(bpy)_3^{2+}$ is perhaps the most thoroughly studied ECL-active molecule (Tokel-Takvoryan *et al.*, 1973; Itoh and Honda, 1979; Wallace and Bard, 1979; Glass and Faulkner, 1981; Luttmer and Bard, 1981) and, as with other ECL systems, there was particular emphasis on characterizing the nature of the excited state, discerning the mechanisms by which these states were formed, and determining the efficiency of excited-state formation. Various techniques were used and are still being used, including detailed electrochemical studies, spectroscopic and spin-resonance measurements, and magnetic field effects (Faulkner and Bard, 1977, 2001; Faulkner and Glass, 1982; Bard, 2004).

In the early 1980s, Bard and Whitesides (1993a, 1993b, 1994) developed a method for the binding of $Ru(bpy)_3^{2+}$ to biological molecules of interest (e.g., antibodies, proteins, nucleic acids). The interest in using $Ru(bpy)_3^{2+}$ stems from its rather unique properties. Namely, it emits and is soluble at room temperature in aqueous, fluid, solution and undergoes reversible one-electron-transfer reactions at easily attainable potentials. Also, the ligands provide synthetic versatility. For example, by attaching *N*-hydroxysuccinimide (NHS) ester to one of the bipyridine ligands (Figure 7.5), the ECL label can bind to substances containing free amino groups. The amino acid will attack the carboxylate ester, leading to displacement of NHS. BioVeris Corporation (formerly IGEN International, Inc.) began developing ECL for use in biosensor analyses in the early 1980s.

In the early 1990s, the prototype ORIGEN® Analyzer was introduced by BioVeris IGEN International, Inc. The ORIGEN is an ECL-based immunoassay system, incorporating $Ru(bpy)_3^{2+}$-tagged biomolecules and is engineered to be a biomedical research tool for immunoassays and DNA probes. The ORIGEN instrument is semi-automated, and incorporates a flow injection system to allow rapid and reproducible determinations of single samples. The detector is a photomultiplier tube positioned above the working electrode, and light from the electrode is recorded and integrated for each measurement.

Roche Diagnostics licensed the ECL–$Ru(bpy)_3^{2+}$ technology from IGEN International, Inc., in 1992, and subsequently produced the first fully automated instruments (Elecsys® 1010 and 2010), built upon the same flow cell design as its predecessor. The Elecsys systems were launched in Europe in 1996 for use in clinical and reference laboratories (those handling large volumes of samples) and in the USA starting in January of 1997 following FDA approval. Assays that have been developed for these systems will be discussed in more detail below but include α-fetoprotein, digoxin, thyrotropin, protein and steroidal hormones, cytokines, and various antibodies, to name a few (Tables 7.1–7.3).

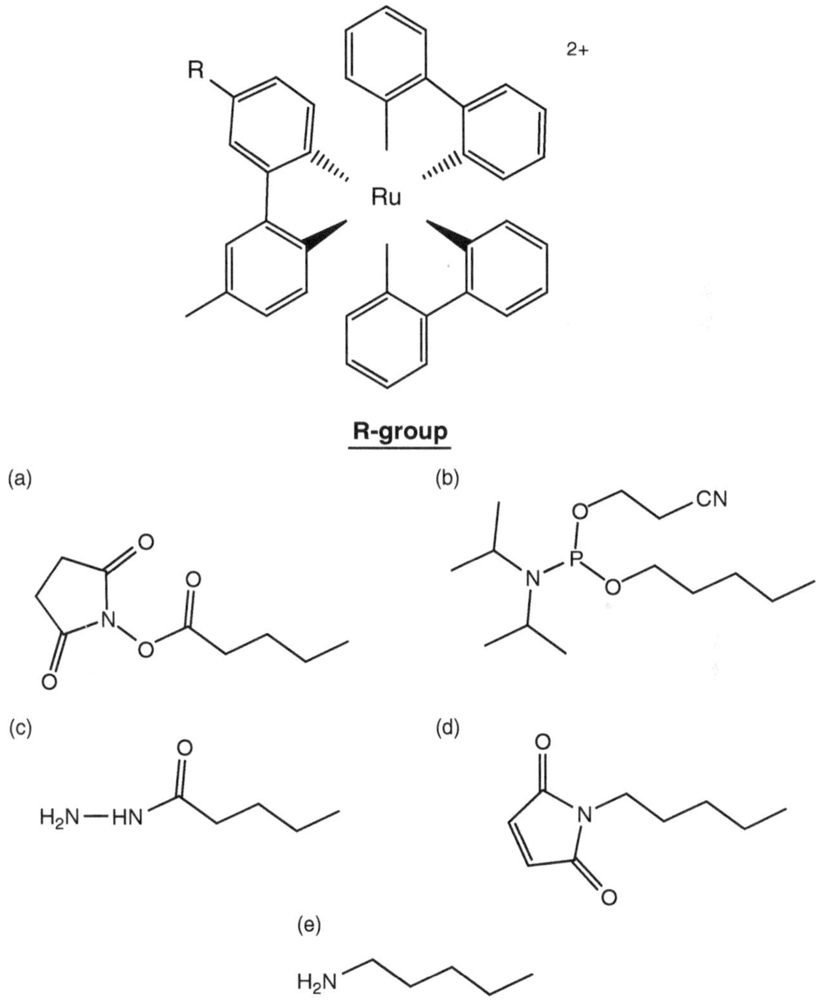

Figure 7.5 $Ru(bpy)_3^{2+}$-functionalized derivatives for labeling biological molecules where R is: (A) $Ru(bpy)_3^{2+}$–NHS ester for ECL labeling of proteins and nucleic acids; (B) $Ru(bpy)_3^{2+}$–phosphoramidite for ECL labeling of oligonucleotides; (C) $Ru(bpy)_3^{2+}$–hydrazide for ECL labeling of carbohydrates; (D) $Ru(bpy)_3^{2+}$–maleimide for ECL labeling of thiols; (E) $Ru(bpy)_3^{2+}$–amine for ECL labeling of caroboxylic acids.

Table 7.1 Assays that relate coreactants to analyte concentration

Analyte	Luminophore/coreactant	References
BSA	$Ru(bpy)_3^{2+}$/DMBD	Yin *et al.*, 2005.
Carbon dioxide	$Ru(bpy)_3^{2+}$/NADH	Jameison *et al*, 1996.
Cholesterol	$Ru(bpy)_3^{2+}/C_2O_4^{2-}$	Jameison *et al*, 1996.
Choline, glucose, glutamate, lysine, urate	Luminol/H_2O_2	Marquette *et al.*, 2003.
Ethanol, glucose	$Ru(bpy)_3^{2+}$/NADH	Jameison *et al*, 1996; Martin and Nieman, 1997.
IgG, immunoglobulin G	$Ru(bpy)_3^{2+}$/DMBD	Yin *et al.*, 2005.
Lactate	$Ru(bpy)_3^{2+}$/NADH	Martin and Nieman, 1997.
β-Lactamase	$Ru(bpy)_3^{2+}$/hydrolyzed β-lactamase	Martin and Nieman, 1997.
Lactate	Luminol/H_2O_2	Sassa *et al.*, 1999.

Table 7.2 Assays where analyte is the coreactant

Analyte	References
$Ru(bpy)_3^{2*}$	
AFP	Blackburn *et al.*, 1991.
Alcohols	Chen *et al*, 1997b, 1998a,b.
Alicyclic tertiary amines	Uchikura and Kirisawa, 1991.
Amines (primary) – derivatized with DVS	Uchikura *et al.*, 1993.
Amino acids	He *et al.*, 1990.
Amitrypyline, chlorpromazine, doxepin, nortriptyline, promazine	Greenway and Dolman, 1999.
Amino acids (dansylated), antihistamines, arginine, ascorbic acid, atropine, bupivicaine, clindamycin-2-phosphate, codeine, dextromethorphan, heroine, leucine, oxalate, oxaprenolol, procyclidine, pyruvate, serine, thiazide compounds, valine	Downey and Nieman, 1992.

Table 7.2 Continued

Analyte	References
Astrovirus	Schutzbank and Smith, 1995.
Carcinoembryonic antigen, thrytrophin	Blackburn *et al.*, 1991.
Citrate, EDTA, lactate, malonate, tartrate,	Knight and Greenway, 1995.
Cystic fibrosis ΔF-508 deletion mutation	Kenten *et al.*, 1992b.
Digoxin	Blackburn *et al.*, 1991.
Dioxopromethazine	Shi *et al.*, 2006.
Emeline dithiocarbamate copper(II), erythromycin, hydrazine, and derivatives,	Knight, 1999.
Erythromycin	Danielson *et al.*, 1989
Glycophosphate and related systems	Ridlen *et al.*, 1997.
Fatty amine ethoxylate surfactants	Alexander and Richter, 1999.
HIV-1 Gag gene	Kenten *et al.*, 1992b.
lignocaine, procaine,	Knight and Greenway, 1996.
Morpholine fungicides (dodemorph and tridemorph)	Gonzalez *et al.*, 2000.
Oxalate	Rubinstein and Bard, 1981
Persulfate	White and Bard, 1981, Yamazaki-Nishida *et al.*, 1990; Yamashita *et al.*, 1991.
Trialkylamines	Noffsinger and Danielson, 1987a,b.
Tri-*n*-propylamine (TPrA)	Leland and Powell, 1991.
L-tryptophan	Uchikura and Kirisawa, 1991b.
Valine	Brune and Bobbitt, 1991.
2-Whiouracil	Chi *et al.*, 2006b.
$[Ru(bpy)_2dcb]Cl_2$	
Amino acids	Dennany *et al.*, 2006b.
Luminol	
Pryogallol	Uchikura and Kirisawa, 1991b.
Digoxin	Qi and Zhang, 2004.

*Grouped according to luminophore used

Table 7.3 Assays that relate emitter to analyte concentration

Analyte	**References**
$Ru(bpy)_3^{2+}$/TPrA (luminophore/coreactant)	
AFP	Namba *et al.*, 1999; Yilmaz *et al.*, 2001
β-Amyloid peptide	De Baer *et al.*, 1999
Amyloid-β(40)	Best *et al.*, 2005
Apo 8-100 gene mutation	Klingler *et al.*, 2000
Anti-Borna Disease antibodies	Yamaguchi *et al.*, 2001; Horii *et al.*, 2001; Fukuda *et al.*, 2001
Cancer antigen 15-3	Yilmaz *et al.*, 2001; Stiebler *et al.*, 2001
Cancer antigen 19-9	Yilmaz *et al.*, 2001
Cancer antigen 72-4	Bruno and Kiel, 2002
Cancer antigen 125	Hubl *et al.*, 1999; Ingen, 1998
Carcinoembryonic antigen	Yilmaz *et al.*, 2001; Stockmann *et al.*, 1998
β-Crosslaps	Seck *et al.*, 2002; Okabe *et al.*, 2001
C-Telopeptides	Scheunert *et al.*, 2001
CKMB, creatine base	Stockmann *et al.*, 1998; Klein *et al.*, 1998
CMV DNA	Boom *et al.*, 1999
CMV mRNA	Debad *et al.*, 2004
Cytokeratin 19	Sanchez-Carbayo *et al.*, 1999
Creatine kinease muscle brain	Stockmann *et al.*, 1998; Ehrhardt *et al.*, 1998
Cytokine	Freebern *et al.*, 2005
Cytomegalovirus, anti-CMV antibodies	Ohlin, 1997
Des-γ-carboxy	Shimizu *et al.*, 2002; Sassa *et al.*, 1999.
Dengue virus RNA	Wu *et al.*, 2001
Enterovirus	Fox *et al.*, 2002
Epstein-Barr virus DNA	Stevens *et al.*, 1999
Ferritin	Yilmaz *et al.*, 2001
Foot and mouth disease	Collins *et al.*, 2002a,b
Glucose	Wang and Hsuan-Jung, 2003
HCG	Stockmann *et al.*, 1998; Ehrhardt *et al.*, 1998
HBsAg, hepatitis B virus surface antigen	Kobayashi *et al.*, 1999; Kashiwagi *et al.*, 1998; Takahashi *et al.*, 1998; Weber *et al.*, 1999
HAsAg, hepatitis A virus surface antigen	Debad *et al.*, 2004

Table 7.3 Continued

Analyte	References
HIV-1 p7 antigen	De Baer *et al.*, 1999
HIV-1 RNA	Van Gemen *et al.*, 1994; Schutzbank and Smith, 1995
HIV DNA	Kenten *et al.*, 1992; Yu *et al.*, 1995; Schutzbank and Smith, 1995; Oprandy *et al.*, 1995
HPIV 1/2/3	Kenten *et al.*, 1992b; Hibbits *et al.*, 2003
IgE, immunoglobulin E	Kobrynski *et al.*, 1996
Influenza virus RNA	Collins *et al.*, 2002a, 2003
Insulin	Sapin *et al.*, 2001; Liebert *et al.*, 2001
Interferon-γ, interleukin (IL)-2, IL-4, Coxsackievirus B3 RNA	Reetoo *et al.*, 1999.
IL (interleukin)-18 binding protein	Novick *et al.*, 2001
IL-6, IL-8	Yilmaz *et al.*, 2001
IL (interleukin)-10	Novick *et al.*, 2001; Swanson *et al.*, 1999; Reetoo *et al.*, 1999
CTC_4, LTB_4, leukotriene, PGE2, PBD2, prostaglandin	Tew *et al.*, 2005
LH, lutropin, prolactin, estradiol, FSH, follitropin	Gassler *et al.*, 2000
Osteocalcin	Scheunert *et al.*, 2001
mRNA	Miyashiro *et al.*, 2001; O'Connell *et al.*, 1999; Taback *et al.*, 2001
Parathyroid hormone	Hermsen *et al.*, 2002
Plasma vascular endothelial growth factor	Willett *et al.*, 2005
Prostrate-specific antigen	Stockmann *et al.*, 1998; Yilmaz *et al.*, 2001
Prothrombin	Shimizu *et al.*, 2002; Sassa *et al.*, 1999
Prothrombin gene mutation	Gellings *et al.*, 2001
Serum interferon-α	Novick *et al.*, 2001
St. Louis encephalitis	Lanciotti and Kerst, 2001
T4, thyroxine	Sapin *et al.*, 2000

(Continued)

Table 7.3 Continued

Analyte	References
T3, T4, thyroxine, triiodothyronine, TSH	Stockmann *et al.*, 1998; Sanchez-Carbayo *et al.*, 1999b; Luppa *et al.*, 1998
TXB2, thromboxane	Tew *et al.*, 2005
Testosterone	Gassler *et al.*, 2000; Sanchez-Carbayo *et al.*, 1998
Troponin	Stockmann *et al.*, 1998; Klein *et al.*, 1998; Hetland and Dickstein, 1998; Ishii *et al.*, 1998; Collinson *et al.*, 2002
Tumor necrosis factor-α	Moreau *et al.*, 1996
E2 and E3 ubiquitin ligases	Davydov *et al.*, 2004
Varicella-zoster virus DNA	Wu *et al.*, 2001
West Nile virus RNA	Lanciotti and Kerst, 2001
ΔF508 deletion	Stern *et al.*, 1995
YFP-GL-GPI	Davydov *et al.*, 2005
$Ru(bpy)_2(dcbpy)^{2+}$/TprA (Luminophore/Coreactant)	
DNA hybridization	Zhang and Dong, 2006
$Ru(bpy)_3^{2+}$/triethylamine (Luminophore/Coreactant)	
Legionella antigen	Yoon *et al.*, 2003
	Terbium chelate
Pancreatic phospholipase A2	Kankare *et al.*, 1992
TSH	Kulmala *et al.*, 2002
Terbium Chelate (Cathodic Luminescence)	
Human C-reactive protein (hCRP)	Ala-Kleme *et al.*, 2006
TAMRA (Cathodic Luminescence)	
Oligonucleotides	Hermsen *et al.*, 2002

7.3. State of the art

7.3.1. Analytical applications of ECL

Coreactant ECL has been used in a wide range of analytical applications (Knight and Greenway, 1994; Bard *et al.*, 2000; Bard, 2004; Richter, 2004) including clinical diagnostics, chromatography, food and water testing, and biowarfare agent detection. Since ECL emission intensity is usually proportional to the concentration of the emitter (Cruser and Bard, 1967) or coreactant (Leland and Powell, 1991), ECL can be used in the analysis of various species (Tables 7.1–7.3). For example, the system of interest is introduced into an electrochemical cell, a voltage is applied to an electrode, and the light intensity and/or ECL spectrum is measured. Electrochemiluminescence in such systems is very sensitive since photon-counting methods can be used to measure very low light levels. For example, ECL from $Ru(bpy)_3^{2+}$ has been used to measure the concentrations of coreactants such as oxalate and peroxydisulfate to levels as low as 10^{-13} M (Ege *et al.*, 1984). In fact, the ability of $Ru(bpy)_3^{2+}$ to undergo "oxidative–reductive" ECL in the presence of coreactants has led to the selective determination of oxalate in synthetic urine samples (Rubinstein *et al.*, 1983), and $Ru(bpz)_3^{2+}$ (where bpz = bipyrazine) has been used for the determination of peroxydisulfate with nanomolar detection limits (Yamashita *et al.*, 1991).

Since the intensity of ECL is a function of both the coreactant and the emitter, ECL can be used to analyze for both. In these examples, ECL was measured in the presence of high, predetermined concentrations of ECL emitters. These types of experiments can then be used as a means to assay for compounds that act as coreactants, including a variety of amines (Noffsinger and Danielson, 1987b; Knight and Greenway, 1994; Bard *et al.*, 2000; Bard, 2004; Richter, 2004; Table 7.2). Electrochemiluminescence assays for amines find many applications since amine groups are prevalent in numerous biologically and pharmacologically important compounds including alkylamines, antibiotics, antihistamines, opiates, nicotinamide, and the reduced form of NADH (i.e., adenine dinucleotide) (Danielson *et al.*, 1989; Knight and Greenway, 1994; Knight and Greenway, 1996; Bard, 2004; Richter, 2004). In general, these compounds contain no chromophore and therefore cannot undergo luminescence unless an ECL-active compound such as $Ru(bpy)_3^{2+}$ is present. As a general rule, the ECL signal from alkylamine coreactants follows the order: 3° > 2° > 1° (Leland and Powell, 1991). Primary amines have been detected using $Ru(bpy)_3^{2+}$ coreactant ECL after prior derivatization with divinylsulfone ($CH_2{=}CH{-}SO_2{-}CH{=}CH_2$). The primary amines undergo a cycloaddition reaction resulting in the formation of acyclic tertiary amines (Uchikura *et al.*, 1993) that then act as efficient coreactants. Other examples of cyclic amines that undergo ECL include nicotine, atropine, and sparteine (Uchikura and Kirisawa, 1991). It is also possible to quantitatively measure amino acids, peptides, and proteins such as proline and valine. In fact, detection limits of 20 pM for proline (He *et al.*, 1990) and 30 pM for valine (Brune and Bobbitt, 1991) using flow injection techniques have been achieved.

Although the ability of numerous amines to act as coreactants makes ECL a versatile technique for their detection, it also makes selectivity for the presence of a specific amine problematic. More recently, Xu and Dong have obtained high selectivity for the measurement of chlorpromazine, a commonly prescribed dopamine inhibitor. Using $Ru(bpy)_3^{2+}$ as the ECL luminophore and chlorpromazine as an oxidative–reductive coreactant, selectivity was achieved by preconcentration of the chlorpromazine at a lauric acid-modified carbon paste electrode with a detection limit of 3.1×10^{-9} M (Xu and Dong, 2000).

Electrochemiluminescence has also been used to monitor enzymatic reactions. In such systems, the reaction is often coupled to the generation or consumption of an ECL coreactant. A good example is the coenzyme NADH. NADH contains an amine moiety that acts as a coreactant for $Ru(bpy)_3^{2+}$. However, the oxidized form (NAD+) is not a coreactant (Downey and Nieman, 1992). Since numerous NADH-dependent enzymes are known, this allows for the detection of a variety of analytes including glucose (Jameison *et al.*, 1996). Another application is the detection of β-lactamase activity (Liang *et al.*, 1996b). Penicillin and its derivatives do not act as coreactants with $Ru(bpy)_3^{2+}$ to produce ECL. However, β-lactamase-catalyzed hydrolysis of penicillin forms a molecule with a secondary amine that can act as a coreactant. The efficiency of the ECL process has been increased by covalent attachment of a β-lactamase substrate to a $Ru(bpy)_3^{2+}$ derivative (Liang *et al.*, 1996a). The ECL of aminopeptidase and esterase cleavage products have also been reported by covalently attaching such species as ligands to bis(bipyridine)ruthenium (II). $(bpy)_2Ru^{2+}$ has little-to-no intrinsic ECL, but attachment of a third ligand leads to enhanced ECL (Dong and Martin, 1996). This method was also used to study dipicolinic acid, a molecule important to biological defense applications since dipicolinic acid can be linked to the presence of anthrax (Tabor *et al.*, 1976). The ECL of the species formed when dipicolinic acid and $(bpy)_2Ru^{2+}$ are allowed to react has been reported using TPrA as a coreactant (Byrd *et al.*, 2006).

Similar methods have been used in detector cells to couple ECL with separation techniques such as high performance liquid chromatography (HPLC) and capillary electrophoresis (CE). In these methods, electrochemical cells combined with some means of light detection (e.g., PMT = photomultiplier tube or CCD = charge-coupled device) are used to detect analytes of interest. This area has recently been reviewed (Lee, 1997; Gorman *et al.*, 2006) and, in almost all cases, $Ru(bpy)_3^{2+}$ is used for the detection of species that act as coreactants such as aliphatic amines (Noffsinger and Danielson, 1987a), antihistamines (Holeman and Danielson, 1994), amino acids (Brune and Bobbitt, 1992; Jackson and Bobbitt, 1994; Lee and Nieman, 1994; Skotty *et al.*, 1996), clindamicine antibiotics (Targrove and Danielson, 1990), oxalate (Skotty and Nieman, 1995), erythromycin (Ridlen *et al.*, 1997), tricyclic antidepressants (Yoshida *et al.*, 2000), and β-blockers (Park *et al.*, 2002), among others (Gorman *et al.*, 2006). For example, one technique that has achieved picomole detection limits uses post-column ECL detection. A solution of $Ru(bpy)_3^{2+}$ is steadily injected into the solution stream containing separated species coming from the HPLC column. The mixed stream flows into an electrochemical flow cell where the ECL reaction occurs and emission can be measured (Holeman and Danielson, 1994). $Ru(bpy)_3^{2+}$ can also be immobilized in a thin film of polymer (e.g., Nafion) deposited on the working electrode (Rubinstein and Bard, 1980). This eliminates the need for a constant stream of $Ru(bpy)_3^{2+}$. In this technique, ECL results when a species that can act as a coreactant is in the solution coming from the HPLC column and reacts with the immobilized $Ru(bpy)_3^{2+}$ in the detector cell (Downey and Nieman, 1992). Electrochemiluminescence in flowing streams has also provided information about the hydrodynamics in the detector cell (Schultz *et al.*, 1996).

Initial studies focused on HPLC, due to low detection and wide linear ranges. However, in the past decade, CE has also been explored (Gilman *et al.*, 1994; Bobbitt and Jackson, 1997; Dickson *et al.*, 1997;

Forbes *et al.*, 1997; Tsukagoshi *et al.*, 1997; Liu *et al.*, 2003). While CE does present certain advantages over HPLC, such as high efficiency of separation (e.g., several hundred thousand theoretical plates), high peak resolution, low instrument cost, and the use of small amounts of reagent (Yin and Wang, 2005), it can suffer from poor sensitivity due to the small dimensions of the cell and sample volume. Capillary electrophoresis using $Ru(bpy)_3^{2+}$ has been used to detect amino acids (Bobbitt *et al.*, 1998; Wang and Bobbitt, 1999), catecholamine (Kang *et al.*, 2005), drug–protein binding (Zhao *et al.*, 2004), and pharmaceuticals (Cao *et al.*, 2002; Liu *et al.*, 2002; Ding *et al.*, 2006a). Electrochemiluminescence coupled with CE has been reviewed (Lee, 1997; Hai *et al.*, 2005; Yan *et al.*, 2005), with special attention devoted to instrument design and applications.

In the methods described above, ECL was measured in the presence of high, predetermined concentrations of ECL emitters. Electrochemiluminescence can also be used to analyze an emitting species (e.g., $Ru(bpy)_3^{2+}$) that often serves as a label on a molecule of interest. The coreactant, typically TPrA, is present in high concentrations, so the amount of luminescence depends on the concentration of the ECL emitter present in the assay. Since the emitters are bound to the analyte of interest, the amount of luminescence can be correlated with the concentration of the analyte (Table 7.3). The most frequently used ECL-active label is $Ru(bpy)_3^{2+}$ for reasons discussed in preceding sections. Also, the emission is intense and fairly stable, and the emission intensity is proportional to concentration over several orders of magnitude (e.g., 10^{-7}–10^{-13} M) (Ege *et al.*, 1984; Leland and Powell, 1991). By attachment of a suitable group to the bipyridine moieties (Figure 7.5), $Ru(bpy)_3^{2+}$ can be linked to biologically interesting molecules, such as antibodies or DNA, where it serves as a label for analysis in a manner analogous to radioactive or fluorescent labels (Blackburn *et al.*, 1991; Bard and Whitesides, 1993a, 1993b). Another advantage of using the $Ru(bpy)_3^{2+}$/TPrA reaction sequence is that, upon formation of the luminescent excited state, $Ru(bpy)_3^{2+*}$, emission of a photon regenerates $Ru(bpy)_3^{2+}$ in its ground state near the electrode surface. This allows a single $Ru(bpy)_3^{2+}$ molecule to participate in multiple ECL reactions and increases the sensitivity of these systems while lowering detection limits.

The most common and, arguably, the most important commercial application to date for ECL is its use in diagnostic assays. These applications typically use ECL emitters as labels in affinity binding assays that attach the ECL emitter to the analyte of interest (Blackburn *et al.*, 1991). The label is physically linked to one of the binding partners in the assay and provides the means for detecting the coupling of the binding partner to the analyte. Several classes of binding partners are used including antibody/antigen, enzyme/inhibitor, carbohydrate/lectin, and nucleic acid/complementary nucleic acid (Wild, 1994). Commercial instruments such as the BioVeris ORIGEN® and M-SERIES®, the Roche Elecsys®, MODULAR, and cobas, the bioMérieux NucliSENS®, the Eisai PicoLumi, and the Meso Scale Discovery SECTOR™ Imager are available are available for ECL assays (Blackburn *et al.*, 1991; Hoyle, 1994; Yang *et al.*, 1994; Bard, 2004; Richter, 2004). The vast majority are based on the use of magnetic bead technology.

The use of magnetic beads for immunomagnetic separations are well known and have been thoroughly reviewed (Olsvik *et al.*, 1994; Uhlen *et al.*, 1994; Safarikova and Forsythe, 1995; Bruno, 1998a, 1998b). In the context of ECL, the use of magnetic beads allows for the separation of the analyte and ECL label onto a solid support (i.e., the bead) followed by collection of the labeled beads on an electrode surface. Most magnetic beads used in ECL systems are paramagnetic (i.e., magnetic only in the presence of an external magnetic field) and consist of a core of magnetite (Fe_3O_4) surrounded by a polystyrene shell. These micron-sized particles may be purchased (e.g., Dynal Corp., Lake Success, NY) with pre-conjugated streptavidin or a variety of surface immobilization chemistries including amines, hydrazides, and long-chain alkyl linkers, to name a few.

Since the "sandwich assay" format is often used for ECL affinity binding assays, our discussion will center on it. The principles of a typical sandwich assay for an antigen are outlined in Figure 7.6. Magnetic beads decorated with an antibody for a particular antigen of interest (e.g., prostrate-specific antigen, PSA), the sample of interest, and $Ru(bpy)_3^{2+}$-labeled antibodies are mixed. If the antigen of interest

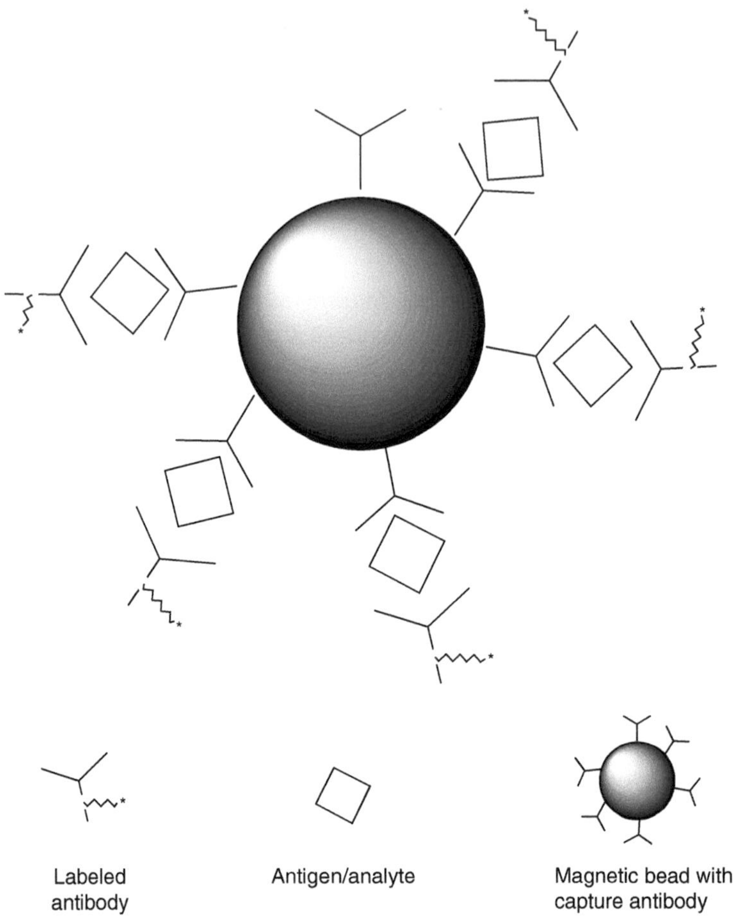

Figure 7.6 Representation of ECL "sandwich" (antibody–antigen/analyte–antibody) assay.

is present, it acts as a bridge to form the "sandwich" structure, and the antibody labeled with ECL luminophore becomes attached to the magnetic bead. If no antigen is present, the labeled antibody does not attach to the bead. These labeled and unlabeled beads are then flushed into an ECL flow cell, where they are captured on the working electrode by positioning a magnet beneath the working electrode. The beads are washed to remove any unattached $Ru(bpy)_3^{2+}$-labeled antibodies as well as other reaction components, and a solution of the appropriate composition containing a coreactant (usually TPrA) is pumped into the cell. The concentration of the coreactant is kept constant and high (e.g., $\geq$50 mM) to maximize the sensitivity of the detection and to prevent fluctuations in concentration of the coreactant from changing the ECL. The electrode is then swept to positive potentials to initiate ECL and the intensity of the emitted light is measured with a photomultiplier tube. In these measurements, the number of ECL labels on the solid phase is directly proportional to the concentration of analyte. The magnetic beads are then washed from the cell, which is cleaned and made ready for the next sample. Currently, more than 75 assays are commercially available (Table 7.1), including those for tumor, cancer and cardiac markers, cell signaling pathways, nucleic acids, immunogenicity assays, analytes relevant to infectious diseases, fertility therapies, and thyroid diseases. A pre-2004 listing of commercial assays has appeared (Debad *et al.*, 2004), and updated lists can be found on company web sites (e.g., BioVeris Corporation, Roche Diagnostics).

Electrochemiluminescence has also been used to develop assays for a variety of biotoxins that are important for both food industry and military applications. Authors have reported extremely sensitive ECL assays for parasites and bacterial species in a variety of matrices (Higgins *et al.*, 1999; Henchal *et al.*, 2001; Bruno and Kiel, 2002; Debad *et al.*, 2004). Species such as bovine luteinizing hormone (Deaver, 1995), anthrax (*Bacillus anthracis*) (Gatto-Menking *et al.*, 1995; Bruno and Yu, 1996; Bruno and Kiel, 1999; Miao and Bard, 2003), *Salmonella* (Yu, 1996; Yu and Bruno, 1996; Debad *et al.*, 2004;), *Staphylococcus aureus* enterotoxins (Kijek, 2000; Yu *et al.*, 2000; Debad *et al.*, 2004), *Cryptosporidium parvum* oocysts (Baeummer *et al.*, 2001; Call *et al.*, 2001; Lee *et al.*, 2001; Kuczynska *et al.*, 2003), *Campylobacter* (Debad *et al.*, 2004) *Listeria monocytogenes* (Debad *et al.*, 2004), and *Escherichia coli* (Yu and Bruno, 1995, 1996; Yu, 1996, 1997; Crawford *et al.*, 2000; Shelton and Karns, 2001; Min and Baeumner, 2002b) have been reported with limits of detection and assay sensitivity equal to or better than conventional assays using flow cytometry, enzyme-linked immunosorbant assays (ELISA), and radioallergosorbent test (RAST).

These assays have been developed in a variety of matrixes using both immunoassay and nucleic acid amplification techniques (Bard, 2004). In all cases, the ECL system incorporated $Ru(bpy)_3^{2+}$ as the label and TPrA as the coreactant. With the increased emphasis on detecting biological threat agents, this area will undoubtedly expand in coming years.

Recently, solid supports such as patterned carbon disposable electrodes have been commercially developed to complement the use of magnetic beads (Bard, 2004). In these systems, analytes can be passively adsorbed on the electrode surface or can be attached using a variety of surface immobilization chemistries. The label can then attach to the analyte or the electrode surface during the assay. A major advantage to patterned disposable electrodes is the formation of multiple binding sites on the electrode surface in an array or microarray format that allows multiple measurements to be performed in parallel on the same sample. The Meso Scale Discovery Sector instruments currently use this technology, where ECL is performed on the surfaces of screen-printed carbon ink electrodes within the wells of multiwall plates. These plates can contain 96-, 384-, and 1536-well formats. These instruments are used for high-throughput screenings (i.e., to test large numbers of potential drug candidates for their effects on biological systems), for proteomics/genomics studies, and for multianalyte analyses (www.meso-scale.com).

Using these formats, tests for screening and diagnosing disease states, for monitoring the effectiveness of drug treatments and surgery, and for diagnosing disease recurrence are actively being developed (see Tables 7.2 and 7.3 for many examples). Specific analytes include cancer and tumor markers (Hoon *et al.*, 2001; Xu *et al.*, 2001; Butch *et al.*, 2002; Haese *et al.*, 2002; Kenten *et al.*, 2005; Vieira *et al.*, 2005;), fertility and related hormones, thyroid function, cardiac markers, heptatitis, bone markers, Alzheimer's disease markers (Khorkova *et al.*, 1998), anemia, diabetes markers, infectious diseases, α-fetoprotein, specific antibodies, cytokines (Sennikov *et al.*, 2003), and des-α-carboxy prothrombinase. Electrochemiluminescence is also being used in life science research laboratories to study detect, discover the causes of, and find cures for diseases (Blohm *et al.*, 1996; Motmans *et al.*, 1996; Horiuchi *et al.*,

1997; Obenauer-Kutner *et al.*, 1997; Hughes *et al.*, 1998; Khorkova *et al.*, 1998; Puren *et al.*, 1998; Shapiro *et al.*, 1998; De Baer *et al.*, 1999; Gopalakrishnan *et al.*, 2000; Mathew *et al.*, 2001; Zhang *et al.*, 2001a, 2001b; Weinreb *et al.*, 2002).

In the past few years, the use of molecular assays involving the hybridization of specific nucleic acid sequences to test for disease states, predisposition for a disease, and infectious diseases has rapidly expanded. Assays for a variety of analytes have been reported in the areas of infectious diseases, tumor markers, metabolism, venous thromboembolism, and cystic fibrosis (de Jong *et al.*, 2000; Klingler *et al.*, 2000; Hoon *et al.*, 2001; Tai *et al.*, 2003; see also Table 7.3). Several studies have coupled ECL with polymerase chain reaction (PCR) amplification to lower detection limits and increase sensitivity (Gudibande *et al.*, 1992; Van Gemen *et al.*, 1994; Heroux and Szczepanik, 1995; Schutzbank and Smith, 1995; Stern *et al.*, 1995; Wilkinson *et al.*, 1995; De Jong *et al.*, 2000; Baeummer *et al.*, 2001). For example, the ECL of HIV-1 gag gene has been reported with detection limits of <10–30 gene copies (Blackburn *et al.*, 1991; Kenten *et al.*, 1992). Coupling ECL with PCR amplification has lowered the detection limit of HIV-1 gag DNA to less than five copies (Schutzbank and Smith, 1995). Other assays and applications incorporating both PCR and ECL for nucleic acid-based analyses have been reported, including the quantitation of varicella zoster DNA in whole blood, plasma, and serum (De Jong *et al.*, 2000) and the detection of viable oocysts of *C. parvum* (Baeumner *et al.*, 2001). This will undoubtedly continue to be an area of intense research activity.

More recently, a sensitive method for the detection of DNA hybridization was reported, where gold nanoparticles carrying multiple probes were prepared (Wang *et al.*, 2006; Figure 7.7). Ruthenium bis(2,2′-bipyridine)(2,2′-bipyridine-4,4′-dicarboxylic acid)-*N*-hydroxysuccinimide ester ($Ru(bpy)_2(dcbpy)NHS$) was used to 5′-label single-strand probe DNAs (ss-DNA) previously self-assembled onto the gold nanoparticle via a 3′-terminal thiol group. The resulting conjugate, ($Ru(bpy)_2(dcbpy)NHS$)-ss-DNA-Au, was used to detect the hybridization event. When target ss-DNA was immobilized on a gold electrode

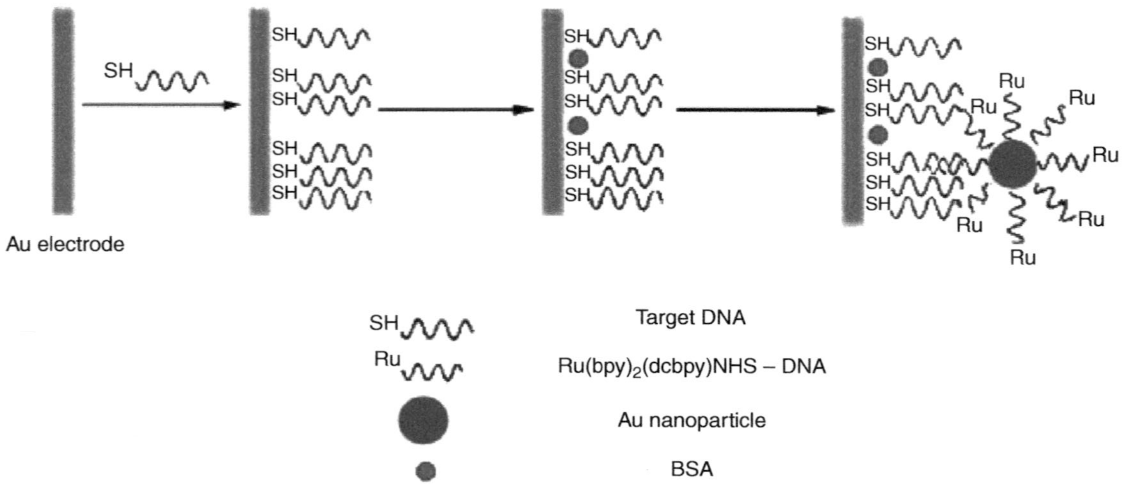

Figure 7.7 Schematic diagram of the ECL detection for DNA hybridization using Ru(bpy)$_2$(dcby)NHS. (Reprinted from *Anal. Chim. Acta*, **575**, Wang, H., Zhang, C., Li, Y., and Qi, H., In *Electrogenerated Chemiluminescence Detection for Deoxyribonucleic Acid Hybridization Based On Gold Nanoparticles Carrying Multiple Probes*, pp. 205–211. Copyright 2006, with permission from Elsevier.)

and hybridized with the ($Ru(bpy)_2$(dcbpy)NHS)-ss-DNA-Au conjugate to form double-stranded DNA, an ECL signal proportional to the concentration of the complementary ss-DNA sequence was generated. The linear range was from 1.0×10^{-11} to 1.0×10^{-8} M, with a detection limit of 5.0×10^{-12} M.

Although $Ru(bpy)_3^{2+}$ and its derivatives are the most widely used ECL luminophores (Tables 7.1–7.3), assays have also been developed for a range of antioxidants that quench anthracene-sensitized ECL upon electrolysis of sodium citrate, methanol, and dissolved oxygen (Chmura and Slawinski, 1994). Indole and tryptophan have also been shown to generate ECL upon electrolysis in the presence of hydrogen peroxide, with detection limits of 0.1 μM for both indole and tryptophan (Chen *et al.*, 1997a) and sensors have been reported for a range of alcohols and saccharides since hydroxyl compounds generate ECL directly (Egashira *et al.*, 1996). Several assays have been developed using the sensitization of luminol in the presence of hydrogen peroxide (Marquette *et al.*, 2003), and a terbium chelate has been used to detect human C-reactive protein (hCRP) when excited using a cathodic luminescence mechanism (Ala-Kleme *et al.*, 2006).

7.3.2. Instrumentation

Experiments focused on annihilation ECL of radical ions are carried out in fairly conventional electrochemical apparatus. However, cells, electrodes, and experimental procedures must be modified to allow electrogeneration of two reactants, rather than one, while taking into account the constraints imposed by optical measurement equipment and the exclusion of stray light (i.e., "light-tight" experiments). In addition, one must pay scrupulous attention to the purity of the solvent/supporting electrolyte system, especially with organic systems (e.g., polyaromatic hydrocarbons). Water and oxygen are particularly harmful to these experiments since they can quench ECL. Thus, cells and electrodes are constructed to allow transfer of solvent and degassing on high-vacuum lines or in inert-atmosphere ("glove") boxes.

Electrochemical apparatus for coreactant ECL are, in many instances, identical to those used in annihilation ECL (Knight and Greenway, 1994; Bard, 2004). However, the constraints of working with nonaqueous systems (e.g., vacuum lines) are alleviated. The earliest experiments were carried out in electrochemical batch cells designed to fit into optical spectrophotometer chambers (Rubinstein and Bard, 1981). As the development of coreactant ECL for use in diagnostics and for flow injection and liquid chromatographic applications increased, many ECL flow cell configurations were developed (Hill *et al.*, 1986; Sakura and Imai, 1988; Downey and Nieman, 1992). Electrode configurations, cells incorporating them and experimental details for both annihilation and coreactant systems have been thoroughly reviewed (Faulkner and Bard, 1977; Faulkner and Bard, 1980; Knight and Greenway, 1994; Bard, 2004).

The ORIGEN™ analyzer was the first commercial instrument that used ECL (Carter and Bard, 1990; Blackburn *et al.*, 1991; Kenten *et al.*, 1991, 1992). This analyzer provides highly sensitive and precise assays in an automated format. It employs a flow injection system that allows rapid and reproducible determinations of sequential samples. The detector is a photomultiplier tube positioned directly above the working electrode so that light from the electrode can be recorded and integrated during each measurement. Typically, the assays use magnetic beads as a solid support and $Ru(bpy)_3^{2+}$/TPrA as the label and coreactant, respectively. However, the instrument may also be operated without the use of beads or other solid supports. A personal computer controls the instrument and aids in the processing and storage of data. The sample and assay reagents are combined in plastic tubes and agitated in a carousel to mix the sample and reagents and allow assay binding reactions to go to completion. The sample solution containing magnetic beads is then flushed automatically into an ECL cell, where they are captured on the working electrode by applying a magnetic field. The beads are washed, a solution containing coreactant is pumped into the cell, and ECL is induced. Following light measurement, the beads are washed from the cell and the cell cleaned in preparation for the next measurement. A typical read and clean cycle requires approximately 1 min.

Roche Diagnostics, a licensee of BioVeris Corporation's technology, has developed the ELECSYS™ series of instruments for conducting immunoassays in centralized hospital and reference laboratories. The fundamental technology and operation of this highly automated instrument is similar to the ORIGEN™ analyzer. The instrument can operate in random access mode and has the capability to produce short turnaround time (STAT) samples. Several other instrument configurations have been developed (Rozhitskii, 1992) including one that incorporates an RRDE for both electrochemical and ECL analyses.

Electrochemiluminescence instrumentation has also followed the trend that has developed in the past decade toward smaller cells and electrodes, and numerous microfabricated electrochemical cells for use in ECL have been reported in the literature (Egashira *et al.*, 1990; Kuhn *et al.*, 1990; Hsueh *et al.*, 1995, 1996, 1998; Kremeskotter *et al.*, 1995; Arora *et al.*, 1997; Haswell, 1997; Fiaccabrino *et al.*, 1998; Egashira *et al.*, 2000; Yoon *et al.*, 2003). For example, *Legionella* antigen was detected using a membrane strip immunosensor utilizing a $Ru(bpy)_3^{2+}$ label (Yoon *et al.*, 2003). With the development of the multiplexing arrays in the Meso Scale Discovery SECTOR instrument and the BioVeris Corporation TRICORDER® detection system (~1/20th the size of the original ORIGEN™ instrument), the microfabrication of ECL cells has hit commercial reality.

Fundamental studies aimed at the development of sensors and probes have also been reported. For example, fiber-optic probes have been developed for ECL analyses (van Dyke and Cheng, 1989; Kuhn *et al.*, 1990; Egashira *et al.*, 1992). In one experiment, a gold-coated fiber-optic probe was used for measurement of $Ru(bpy)_3^{2+}$ in the presence of peroxydisulfate (Kuhn *et al.*, 1990). Also, a miniaturized fiber optic sensor has been developed and applied to the determination of oxalate using $Ru(bpy)_3^{2+}$ (Egashira *et al.*, 1990) in real urine samples with a limit of detection of 3×10^{-5} M. The ultimate goal of many of these studies is the development of portable devices for use in point-of-care clinical analyses (Bard *et al.*, 2000; Bard, 2004) and environmental applications (Bruno, 1998a, 1998b).

7.4. Advantages and limitations

As with other measurements based on the emission of light, ECL labels have distinct advantages over detection methods such as radioactivity. For example, they are sensitive, nonhazardous, inexpensive, diagnostic of the presence of a particular label, linear over a wide range, and incorporate simple and relatively inexpensive equipment. When compared to such light emission techniques as PL and CL, ECL also displays certain desirable qualities. In PL, excited-state formation occurs upon absorption of electromagnetic radiation:

$$\mathrm{R} + h\nu \rightarrow {}^{*}R \tag{7.39}$$

The versatility of this technique lies in the number of species able to luminesce, the quantum efficiency of emission, and the ability to incorporate these molecules into a wide variety of formats. Unfortunately, this versatility also leads to limitations. For example, in clinical situations, typical biological fluids containing analyte may also contain a large number of potential luminophores. In ECL, for a complex to emit, it must meet several stringent criteria, including stable redox chemistry and the ability to undergo energetic electron or energy transfer. Of course, this advantage of ECL is also a potential limitation, in that the number of efficient ECL labels is diminished. Chemiluminescence involves the generation of excited states due to an energetic chemical reaction. In a typical CL reaction, reagents are pumped separately to the reaction site. In ECL, on the other hand, production of reagents occurs electrochemically *in situ* from passive precursors, allowing spatial and temporal localization of the emission near the electrode. This results in enhanced sensitivity since the optics used for light detection can be focused on a relatively small area. Furthermore, amplification is possible in ECL due to the turnover of reactants at or near the electrode surface, resulting in sub-picomole detection limits and a linear dynamic range of greater than six orders of magnitude. $Ru(bpy)_3^{2+}$-labels are also extremely stable and can be stored for over 1 year at room temperature in the dark. Often, the small size of the ECL luminophore allows multiple labels to be attached to the same molecule without affecting the stability, immunoreactivity, or hybridization of the probes. However, as with any electrochemical process, stringent cleaning of the electrodes is required prior to and after each run, or the electrode should be disposed of after use (such as with microfabricated electrodes), to ensure reproducibility.

7.5. Potential for expanding current capabilities

Solution phase coreactant ECL is quite sensitive, with sub-picomolar detection limits achieved (Leland and Powell, 1991). When the ECL luminophore is bound to a magnetic particle (see above), detection limits as low as 10^{-18} M are attainable (Blackburn *et al.*, 1991; Kenten *et al.*, 1991; Bard, 2004). However, there are many systems where even greater sensitivity is needed, such as in environmental (where preconcentration of samples is often necessary) and molecular diagnostics applications, where the detection of as few as 10 molecules would eliminate the need for sample amplification (e.g., via PCR). One approach has been to vary the properties of the ECL luminophore. For example, $Ru(bpy)_3^{2+}$ has an ECL efficiency of 0.050 (Rubinstein and Bard, 1981; Glass and Faulkner, 1982), or ~5% of the $Ru(bpy)_3^{2+}$ molecules that undergo electron transfer generate emission. With the goal of increasing the magnitude of ECL emission, and therefore increasing ECL sensitivity and lowering detection limits, ECL of the bimetallic ruthenium system $[(bpy)_2Ru]_2(bphb)^{4+}$, where bphb = 1,4-bis(4′-methyl-2,2′-bipyridin-4-yl)benzene] has been studied (Richter *et al.*, 1998). The ligand bphb is capable of binding two independent metal centers through a "bridging ligand" framework. This bimetallic species produced more intense emission (two- to threefold) than $Ru(bpy)_3^{2+}$ in aqueous and nonaqueous solution using annihilation and coreactant methods. A key point to this study was that for enhanced ECL to be possible in multimetallic assemblies, there must be small electronic coupling between metal centers via the bridging ligand so that the metal centers are electronically isolated or "valence trapped" (Robin and Day class I Systems; Robin and Day, 1967). This was extended to dendrimeric systems containing eight $Ru(bpy)_3^{2+}$ units at the periphery of a carbosilane dendrimer platform (Zhou and Roovers, 2001). The ECL of the $Ru(bpy)_3^{2+}$ dendrimer is five times that of the reference monometallic species. As with the bimetallic

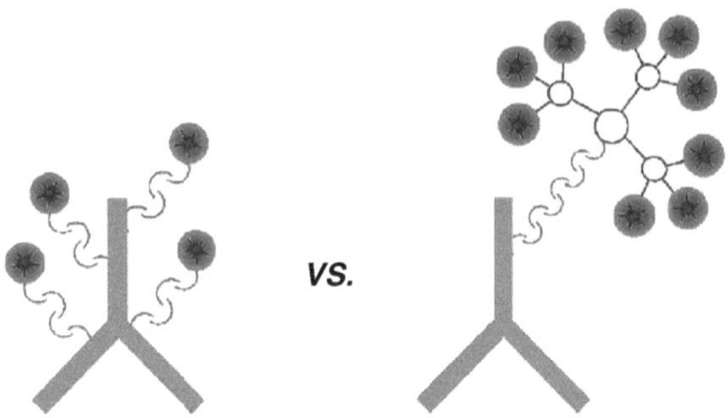

Figure 7.8 Multilabeling a biomolecule at a single site with dendritic label bearing multiple signal-generating units. (Reprinted with permission from Zhou *et al.* (2003). Copyright 2003, American Chemical Society.)

study, spectroscopic and electrochemical studies show that the $Ru(bpy)_3^{2+}$ units do not interact in either the ground or the excited state, showing that ECL (and PL) emission can be amplified by using multimetallic species. A dendritic molecule containing three $Ru(bpy)_3^{2+}$-units linked by a succinimidyl group was synthesized and used to multilabel the protein bovine serum albumin (BSA) at a single site (Zhou *et al.*, 2003, Figure 7.8). Multiple labels are used to increase the signal of the target analyte and the binding of these labels normally occurs at numerous sites. However, such multilabeling can result in the loss of biological activity of the molecules, and in precipitation of target analytes (such as DNA and proteins). It was clear that there was little electronic coupling among the $Ru(bpy)_3^{2+}$-units in the dendrimer, and that enhanced ECL and PL were observed compared to a single $Ru(bpy)_3^{2+}$ molecule. Labeling of the BSA at one NH_2 position was demonstrated with no subsequent loss of biological activity or precipitation of the BSA–dendrimer complex. The ECL of multimetallic systems has also been extended to homometallic complexes containing two and three [$Ru(bpy)_3^{2+}$] units linked by the amino acid lysine ([Ru2-Lys]$^{4+}$) and the dipeptide LysLys, ([Ru3-LysLys]$^{6+}$), (Staffilani *et al.*, 2003); to dendritic complexes containing 2-, 4- and 8-ruthenium units in both homogeneous and heterogeneous assay formats (Staffilani *et al.*, 2003); and to a family of mono- and dinuclear ruthenium polypyridyl complexes containing 5-aryltetrazolate ligands such as the deprotonated form of 4-(1*H*-tetrazol-5-yl)benzonitrile and bis(1*H*-tetrazol-5-yl)benzene (Stagni *et al.*, 2006). Multimetallic compounds such as these show much promise for use in analytical applications.

The ECL of the metallopolymer $[Ru(bpy)_2(PVP)_{10}]^{2+}$, where PVP is poly(4-vinylpyridine), has also been studied (Dennany *et al.*, 2006a). It is well known that immobilizing luminescent materials on electrode surfaces can result in substantial changes in photophysical and electrochemical properties. For $[Ru(bpy)_2(PVP)_{10}]^{2+}$, the overall efficiency of the ECL reaction for the metallopolymer film is almost four times higher than when dissolved in solution.

Electrochemiluminescence in osmium systems has also been reported using coreactant (Bolletta *et al.*, 1981; Bruce *et al.*, 2002; Walworth *et al.*, 2004) and annihilation (Abruna, 1985) methods. Although osmium systems present some advantages over their ruthenium analogues, including greater photostability and oxidation at lower potentials, the ECL of these complexes has been somewhat limited due to the larger spin–orbit coupling in osmium (Creutz *et al.*, 1980; Kober and Meyer, 1982) that typically results in shorter excited-state lifetimes and weaker emission efficiencies. Recently, however, $Os(phen)_2(dppene)^{2+}$ (phen = 1,10-phenanthroline and dppene = bis(diphenylphosphino)ethene) was shown to exhibit twofold greater ECL than $Ru(bpy)_3^{2+}$ ECL in aqueous and mixed aqueous:MeCN solutions using TPrA as coreactant (Bruce *et al.*, 2002). The emission efficiency could be increased up to threefold in the presence of the nonionic surfactant Triton X-100 (Walworth *et al.*, 2004).

Another approach to improving ECL sensitivity that has met with limited success has been to vary the nature of the coreactant. Numerous amine-based coreactants have been studied, including primary, secondary, and tertiary systems (Leland and Powell, 1991), attempts made to understand the electron donating and withdrawing properties that might lead to optimum coreactant efficiency (Knight and Greenway, 1996). To date, TPrA still provides the optimum ECL in the $Ru(bpy)_3^{2+}$ system (Leland and Powell, 1991; McCord and Bard, 1991). Yet another approach is to add a species to the solution that will facilitate excited-state formation and/or lead to increased quantum yields for emission. For example, the ECL intensity of $Ru(bpy)_3^{2+}$ increased slightly ($\leq$5%) in the presence of benzene (Dixon *et al.*, 1993). The reason for the increase is unclear, but the excited states of ruthenium and osmium polypyridyl systems are sensitive to the nature of the environment and are able to detect subtle changes in solution composition (Meyer, 1978). The presence of benzene may lead to decreased interactions between the hydrophobic luminophore and the solvent media, resulting in increased ECL efficiency. The addition of halide species to solution also improves ECL (Zu and Bard, 2000). The halide appears to inhibit the growth of surface oxide films on Pt and Au electrodes, with increases in TPrA oxidation current also observed. The most significant enhancement was at Au electrodes in the presence of bromide. Enhancement was also observed in the commercially available Origen Analyzer, and this was proposed as a simple way to improve the detection sensitivity in $Ru(bpy)_3^{2+}$/TPrA systems (Zu and Bard, 2004).

Although $Ru(bpy)_3^{2+}$ has many properties that make it an ideal ECL luminophore for sensitive and selective analytical methods, it would be useful to have other ECL labels that span a wide range of wavelengths so that simultaneous determination of several analytes in a single sample is possible. For example, $Ru(bpy)_3^{2+}$ has a broad emission spectrum stretching from about 500 to 700 nm ($\lambda_{max} \sim 620$ nm), and this can be a disadvantage in applications where an ECL internal standard or multianalyte determinations are desired. The ECL of a series of europium chelates, cryptates, and mixed-ligand chelate/cryptate complexes has been studied (Richter and Bard, 1996) since many trivalent lanthanides display high PL efficiencies, large Stokes' shifts (~300 nm), and narrow emission spectra (Crosby *et al.*, 1961; Sinha, 1971). Electrochemiluminescence appears to occur by a different mechanism via a "ligand-sensitization" route, where ECL occurs in the organic ligands with subsequent transfer to the f-orbitals of the metal centers. Although it was clear from this work (Richter and Bard, 1996) that the ligands play an integral role in rare earth ECL, very low ECL efficiencies were observed in nonaqueous solvents, with little to no ECL observed in aqueous media.

Electrochemiluminescence in aqueous solution has been observed for the polyaromatic hydrocarbons 9,10-diphenylanthracene-2-sulfonate (DPAS) and 1- and 2-thianthrenecarboxylic acid using TPrA as a coreactant and for DPAS using peroxydisulfate as a reductive–oxidative coreactant (Eqn (7.37)) (Richards and Bard, 1995). These complexes emit in the blue and green regions of the spectrum (e.g., λ_{max} (DPAS) = 430 nm), making them attractive as complementary labels to $Ru(bpy)_3^{2+}$. Electrochemiluminescence is not only limited to the visible region of the spectrum. Near-infrared ECL was described (Lee *et al.*, 1997) for a heptamethine cyanine in acetonitrile using coreactants.

Numerous studies on the ECL of ortho-metalated complexes of Ir(III) have appeared in recent years (Vogler and Kunkely, 1987; Nishimura *et al.*, 2001; Bruce and Richter, 2002; Gross *et al.*, 2002; Kapturkiewicz and Angulo, 2003; Kapturkiewicz *et al.*, 2004, 2006; Muegge and Richter, 2004, 2005; Kim *et al.*, 2005). These complexes display strong visible absorptions and ground- and excited-state redox potentials (King *et al.*, 1985) that, like their Ru(II) counterparts, make them of interest in fundamental and applied ECL. Studies to date have centered on $Ir(ppy)_3$ and its derivatives using both annihilation (Vogler and Kunkely, 1987; Nishimura *et al.*, 2001; Gross *et al.*, 2002; Kapturkiewicz and Angulo, 2003; Kapturkiewicz *et al.*, 2004, 2006; Kim *et al.*, 2005) and coreactant (Bruce and Richter, 2002; Kapturkiewicz and Angulo, 2003; Kapturkiewicz *et al.*, 2004; Muegge and Richter, 2004, 2005; Kim *et al.*, 2005; Kapturkiewicz *et al.*, 2006;) methodologies, and in a variety of matrixes. Although these ECL efficiencies are typically lower than for $Ru(bpy)_3^{2+}$ under identical conditions, the emission maxima of several complexes such as the green emission of $Ir(ppy)_3$ (Bruce and Richter, 2002), and the blue/green emission of F(Ir)pic (Muegge and Richter, 2004; where F(Ir)pic = bis(3,5-difluoro-2-(2-pyridyl)phenyl-(2-carboxypyridyl)iridium III), are far removed from the red-orange emission of $Ru(bpy)_3^{2+}$ that both complexes can be detected in the same sample solution (Figure 7.9). Recently, very efficient ECL was demonstrated by tuning the excited states in ortho-metalated iridium(III) systems by varying the ligands attached to the metal center (Kim *et al.*, 2005). Such variation allowed some control over the relative positions

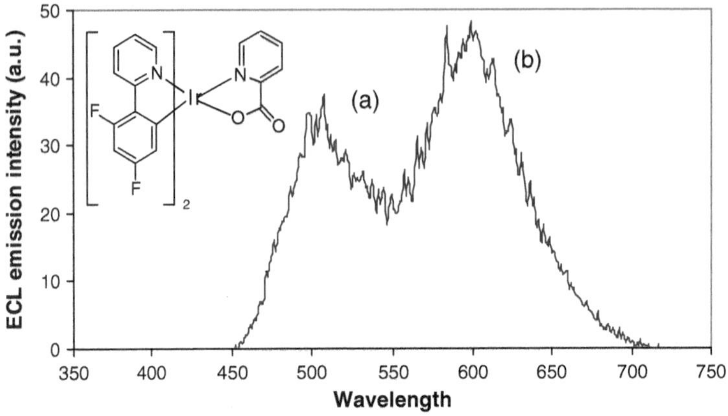

Figure 7.9 Electrochemiluminescence emission spectra of 100 μM (a) F(Ir)pic (structure shown) and (b) $Ru(bpy)_3^{2+}$ in the same MeCN solution with 0.05 M TPrA. (Reprinted with permission from Muegge and Richter (2004). Copyright 2004, American Chemical Society.)

of the oxidation and reduction potentials in the complexes, and hence the separation between the highest occupied molecular orbital (HOMO) and the lowest unoccupied molecular orbital (LUMO) levels and the emission energy. This control resulted in up to 77 times higher ECL from iridium(III) complexes in the presence of TPrA than that of the $Ru(bpy)_3^{2+}$/TPrA system. It was proposed that these systems might find applications as sensors and luminescent devices.

Ionic liquids (ILs) have been proposed as alternative solvents for a variety of applications, including separations and electrochemical devices. ILs are, in essence, molten salts that are liquids below 100°C. Electrochemiluminescence has been reported in 2,5-diphenyloxazole and 2,5-diphenyloxadiazole (Keszthelyi and Bard, 1973). More recently, the electrochemical, ECL, and solvation properties of the room-temperature ILs, tetraalkylammonium (methyltributylammonium bis(trifluoromethylsulfonyl)imide and imidazolium (1-butyl-3-methylimidazolium hexafluorophosphate), were studied (Quinn *et al.*, 2002). Electrochemiluminescence of $Ru(bpy)_3^{2+}$ in both systems was observed, and differences in the response were interpreted in terms of the solvent reactivity and polarity. Considering the increased attention both ECL and ILs are receiving, it will be interesting to see the studies that will emerge in the coming years.

Solubilization of $Ru(bpy)_3^{2+}$ in aqueous nonionic surfactant solutions leads to significant, and potentially useful, changes in the ECL properties (McCord and Bard, 1991; Workman and Richter, 2000). For example, increases in both ECL efficiency ($\geq$ eightfold) and duration of the ECL signal were observed in surfactant media upon oxidation of $Ru(bpy)_3^{2+}$ and TPrA (Workman and Richter, 2000). However, the mechanism of the surfactant effect is still unclear. The effect of surfactants on $Ru(bpy)_3^{2+}$/TPrA (Workman and Richter, 2000; Zu and Bard, 2001) and $Ru(dp\text{-}bpy)_3^{2+}$ and $Ru(dp\text{-}phen)_3^{2+}$ (dp-bpy = 4,4′-biphenyl-2,2′-bipyridyl and dp-phen = 4,7-diphenyl-1,10-phenanthroline, McCord and Bard, 1991), $Os(bpy)_3^{2+}$ (Ouyang and Bard, 1988), and the heptamethine cyanine dye IR-144 ($C_{56}H_{73}N_5O_8S_2$)) (Lee *et al.*, 1998) were attributed to strong hydrophobic interactions between the ECL luminophore and the micellized surfactant. However, recent work (Zu and Bard, 2001) on the effects of electrode hydrophobicity on ECL indicate that adsorption of Triton X-100 on Pt and Au electrodes renders the surface more hydrophobic, facilitating coreactant oxidation and leading to increased ECL intensities in the $Ru(bpy)_3^{2+}$/TPrA system. A study of the effects of nonionic chain lengths on $Ru(bpy)_3^{2+}$/TPrA ECL (Factor *et al.*, 2001) and the enhanced ECL of copper (McCall *et al.*, 2001), iridium (Cole *et al.*, 2003), and osmium (Walworth *et al.*, 2004) systems in the presence of Triton X-100 confirmed these results. Also, the fluorosurfactant Zonyl FSN (Figure 7.10) resulted in 50-fold higher ECL intensity and a 400 mV negative shift in the oxidation potential for the $Ru(bpy)_3^{2+}$/TPrA system compared to the same system in Triton X-100 (Li and Zu, 2004). The fluorosurfactant appears to result in a more hydrophobic electrode compared to Triton X-100 (via adsorption of the hydrophilic polyethylene oxide group to the electrode with the hydrophobic end oriented toward solution; Figure 7.10) and subsequently greater absorption of $Ru(bpy)_3^{2+}$ and TPrA near the surface of the electrode. A significant retardation in the growth of electrode oxide layers was also observed. Although the effects of micelles and discrete complexation of the surfactants with $Ru(bpy)_3^{2+}$ and TPrA cannot be ruled out, these studies

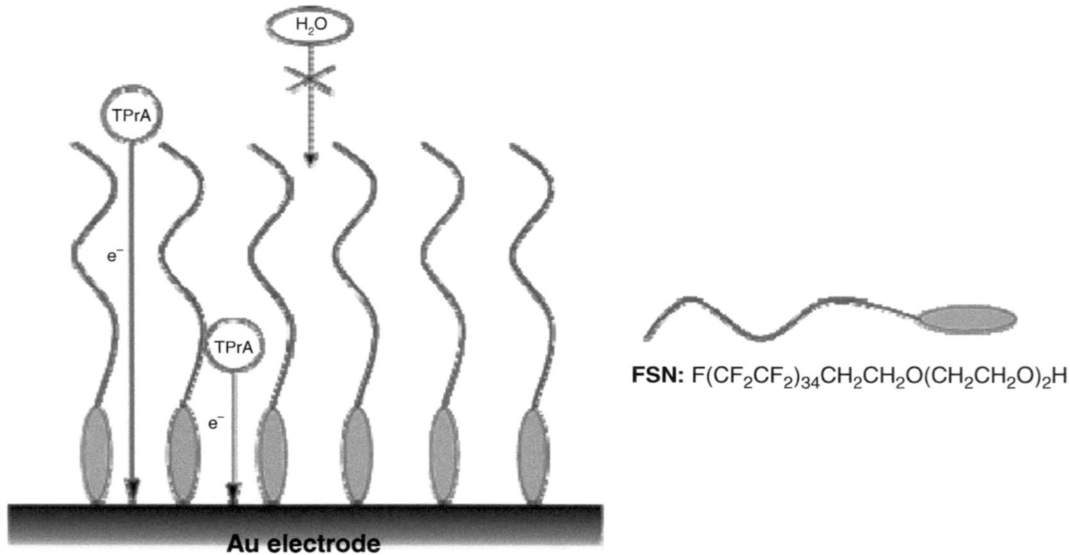

Figure 7.10 Adsorption of FSN molecules at a gold electrode. (Reprinted with permission from Li and Zu (2004). Copyright 2004, American Chemical Society.)

indicate that increases in ECL intensity are probably due to changes in electrode hydrophobicity upon formation of a surfactant adsorption layer and less likely due to micelle interactions (Factor *et al.*, 2001; Zu and Bard, 2001).

Enhanced $Ru(bpy)_3^{2+}$/TPrA ECL is also possible using the anionic surfactant sodium dodecyl sulfate (SDS) and cationic surfactants cetyltrimethylammonium bromide (CTAB), cetyltrimethylammonium chloride, and cetyltrimethylammonium hydrogen sulfate (Xu *et al.*, 2005). Both anionic and cationic surfactants enhanced the ECL, and voltammetric studies confirmed that adsorption of surfactant molecules on the surface of the electrode was responsible. This was surprising, since $Os(bpy)_3^{2+}$ in the presence of oxalate at a gold electrode showed a suppression in ECL by SDS and an attenuation by Triton X-100 and CTAB (Ouyang and Bard, 1988), as did tris(4,4′-biphenyl-2,2′-bipyridine)ruthenium(II) and tris(4,7-diphenyl-1,10-phenanthroline)ruthenium(II) using Triton X-100 or Tween 20 (McCord and Bard, 1991). Such dramatic increases in ECL intensity, coupled with work on more efficient ECL luminophores, coreactants, and solvent conditions, may have a profound impact on the sensitivity of ECL for a variety of applications.

Ultrasonic enhancement of ECL has also been investigated. Ultrasonic irradiation on coreactant and annihilation ECL of $Ru(bpy)_3^{2+}$ in aqueous oxalate solutions and in acetonitrile solutions, respectively, increases the ECL yield over 100%, results in highly stable and reproducible ECL signals, and leads to less electrode fouling (Walton *et al.*, 1992; Malins *et al.*, 1997). The dramatic increases appear to be due to agitation of the system, leading to greater mass transport across the electrode double layer and less dependence on diffusion to get material to the electrode surface. Also, it is speculated that the degassing effects of sonication reduce the aggregation of gas bubbles at the electrode surface and prevent the formation of passivating films. The mechanisms of ECL under sonification are the same as conventional conditions, but the high reproducibility of the signal has allowed the measurement of ECL quenching via oxygen to be measured with greater precision than previously possible (Malins *et al.*, 1997).

The quenching of $Ru(bpy)_3^{2+}$/TPrA ECL in the presence of phenols, hydroquinones, catechols, and benzoquinones has been known for some time (McCall *et al.*, 1999; McCall and Richter, 2000) and has opened up the possibility of using ECL to detect these environmentally, biologically, and pharmacologically important classes of compounds. Recent work involves interacting a range of compounds, including phenols and carboxylic acids, with $Ru(bpy)_3^{2+}$ or $Ru(phen)_3^{2+}$ and TPrA or oxalate as coreactants (Cui *et al.*, 2005). Tetracyclines, gallic acid, adrenaline, noradrenaline, and dopamine have also been detected using ECL quenching (Li *et al.*, 2002, 2003, 2004; Pang *et al.*, 2005;), and glucose can be detected by the quenching of $Ru(bpy)_3^{2+}$/TPrA in the presence of hydrogen peroxide (Wang and Huang, 2003). Another study has found that the most efficient quenching of $Ru(bpy)_3^{2+}$/TPrA ECL occurred in the low oxidation potential region (i.e., at potentials more positive than $\sim$+1.2 V, where oxidation of $Ru(bpy)_3^{2+}$ to $Ru(bpy)_3^{3+}$ occurs), and at low concentrations of TPrA (<5 mM) (Zheng and Zu, 2005).

Environmental applications for ECL are also being explored. For example, the increased ECL emission of $Ru(bpy)_3^{2+}$ in the presence of benzene has led to the proposal of using it to detect aromatic hydrocarbon pollutants (Dixon *et al.*, 1993). Other environmental applications include the detection of toxic metal ions (Bruno, 1998a, 1998b; Taverna *et al.*, 1998), metal ions bonded to aminoaromatics (Bruno and Cornette, 1997), and environmentally important ethoxylate surfactants containing amine groups (Alexander and Richter, 1999). Quenching of $Ru(bpy)_3^{2+}$/TPrA ECL in the presence of phenols, hydroquinones, catechols, and benzoquinones has also been documented (McCall *et al.*, 1999; McCall and Richter, 2000).

The generation of ECL at micro- and ultramicroelectrodes has been known for some time (Collinson and Wightman, 1993; Maness and Wightman, 1995). In fact, an ultramicroelectrode was used to observe individual reaction events of DPA annihilation ECL in nonaqueous solution (Collinson and Wightman, 1995), showing the extreme sensitivity possible with ECL. Electrochemiluminescence with microelectrodes has recently been coupled to scanning probe techniques such as scanning electrochemical microscopy (SECM) to image surfaces (Fan *et al.*, 1998; Maus *et al.*, 1999). High-frequency ECL has also been used to image the surfaces of microelectrodes (Wightman *et al.*, 1998).

More recently, ECL has been used as a light source for near-field scanning optical microscopy (NSOM) using ultramicroelectrodes with effective diameters from 1 μm to less than 100 nm (Zu *et al.*, 2001). This technique was used to image an interdigitated array with resolution comparable to that observed via NSOM. Using ECL for near-field imaging appears to have several advantages over NSOM. In NSOM, a metal-coated fiber-optic probe is used and this leads to fundamental resolution limits due to the finite skin depth of the metal coating. In ECL, tip preparation is easier since standard techniques can be used to generate nanometer-sized electrodes (Zu *et al.*, 2001). In addition, ECL does not require a laser, so there is no heating of samples and tip from absorption of light on metal coatings. Therefore, this approach looks promising for near-field optics in solution. A single faradaic electrode has also been used in ECL (Liu and Bard, 2005). In this experiment, a cell was assembled that contained a Pt microelectrode (the single faradaic electrode) and a polarized electrode consisting of silicon with an insulating silicon dioxide film. The polarized or blocked electrode served as a capacitive counter electrode. By applying a bias between the two electrodes, a current was obtained from a faradaic process at Pt and a capacitive current at the counter electrode without generating any products at the counter electrode. Under pulsed excitation, ECL was clearly observed from a $Ru(bpy)_3^{2+}$/TPrA solution. The use of a single faradaic electrode for the coulometric addition of desired species in nanosystems was proposed (Liu and Bard, 2005).

Ultramicroelectrodes have also been used to determine standard electrochemical potentials for amine redox couples (Badocco *et al.*, 2006). $Ru(bpy)_3^{+}$/$Ru(bpy)_3^{3+}$ annihilation and $Ru(bpy)_3^{2+}$/oxalate systems were used to evaluate an electrically heating controlled cylindrical microelectrode (HME) (Lin *et al.*, 2006). The detection limit for oxalate in $Ru(bpy)_3^{2+}$/oxalate was 3.0×10^{-4} M, when the temperature of the HME was 22°C, and found to be 3.0×10^{-6} M at 80°C, indicating it may be possible to establish sensitive ECL methods using a heated microelectrode.

Annihilation ECL of $Ru(bpy)_3^{2+}$ has been reported in aqueous solution containing no electrolyte (Fioccabrinno *et al.*, 1998). This is possible using a microfabricated interdigitated carbon dual-electrode system. Each electrode is biased to form the reduced, $Ru(bpy)_3^{+}$, or oxidized, $Ru(bpy)_3^{3+}$, species. The electrodes are in close enough proximity (2 μm width and spacing) that the simultaneously produced reactants can diffuse together and undergo annihilation (Eqn (7.13)). Also, carbon is used as the electrode material to prevent formation of water oxidation and reduction products (e.g., oxygen) that tend to quench ECL emission. Annihilation ECL has been used to study aluminum quinolate/triarylamine and related organic complexes used as light-emitting diodes (Anderson *et al.*, 1998; Gross *et al.*, 2000), sol-gel composites containing $Ru(bpy)_3^{2+}$ (Sykora and Meyer, 1999), and diode-like CL in frozen concentration gradients of the ruthenium polymer poly-$[Ru(vbpy)_3](PF_6)_2$ (Maness *et al.*, 1996). ECL in sol-gel-derived glasses (Collinson *et al.*, 1999; Collinson *et al.*, 2000; Zhang *et al.*, 2006), Nafion–silica composite films (Khramov and Collinson, 2000) and gel-entrapped $Ru(bpy)_3^{2+}$ (Collinson *et al.*, 1999) using coreactants has also been observed with the potential for using both coreactant and annihilation ECL in display device technology. Also, the interaction of functionalized $Ru(bpy)_3^{2+}$-derivatives with indium-doped tin oxide electrodes has been investigated (Andersson *et al.*, 2000; Lee *et al.*, 2003; Greenway *et al.*, 2006). For example, a tris(2,2′-bipyridyl)ruthenium(II) derivative containing one bpy ligand functionalized with triethylsilane moieties has been synthesized and covalently attached to a silica surface in a sol-gel, and its CL and ECL have been studied (Greenway *et al.*, 2006). Although commercial applications for these systems have yet to materialize, they have opened up fascinating areas for both fundamental and applied studies.

The development of sensors by changing electrode surfaces (e.g., surface oxidation, adsorption of binding agents or polymer films) has been an area of active study. For example, electrode hydrophobicity was controlled by modifying gold and platinum electrodes with thiol monolayers that contained different terminal groups. Significant increases in TPrA oxidation rate and ECL intensity were observed (Zu and Bard, 2000). In a separate study, $Ru(bpy)_3^{2+}$-derivatized antibodies or antigens interacted with biomolecules that were immobilized on screen-printed gold electrodes (Fahnrich *et al.*, 1999). The biomolecules were incorporated into self-assembled monolayers of thiol or F_c-specific binding protein G. Direct detection of DNA has been demonstrated using a ruthenium polymer adsorbed on an electrode surface as a light emitter and guanine as a coreactant (Dennany *et al.*, 2003). The proposed route for light emission involves the interaction of guanine radicals with Ru^{3+} to generate the Ru^{2+*} excited state and the reduction of Ru^{2+} to Ru^{+} by the guanine radicals, followed by annihilation of Ru^{3+} and Ru^{+}. To determine C-reactive protein (CRP) concentrations in human plasma and serum, biotinylated anti-CRP species were immobilized onto a Au(111) substrate precoated with a layer of avidin covalently linked to a thiol monolayer (Miao and Bard, 2003). CRP and anti-CRP tagged with $Ru(bpy)_3^{2+}$ labels were then conjugated to the surface layer, followed by immersion of the modified electrode in a TPrA solution for ECL generation. CRP is an "acute phase protein" found in human serum and may play a role in predicting the onset of coronary events (e.g., angina). Methods to detect DNA hybridization using a hybridization chip with electric field mismatch discrimination (Spehar-Deleze *et al.*, 2006) and using $Ru(bpy)_3^{2+}$-doped silicon nanoparticles (Chang *et al.*, 2006) have also been published. These techniques open up new approaches for both generating ECL and detecting DNA.

Several papers on ECL with nanocrystals (NCs) and nanoparticles have appeared. The first involved annihilation and coreactant (oxalate and persulfate) ECL in sterically stabilized silicon NCs (Ding *et al.*, 2002). Light emission in Si NCs peaked at 640 nm, significantly lower in energy than the PL maximum of 420 nm, and a mechanism was proposed that involved electron/hole annihilation through electron transfer between NCs, or NCs with redox-active coreactants. The potential use of Si NCs in sensor technologies was proposed.

This work was extended to TOPO-capped CdSe NCs (TOPO = trioctylphosphineoxide) dissolved in methylene chloride containing 0.1 M tetra-*n*-butylammonium perchlorate (Myung *et al.*, 2002). Light emission was observed via annihilation, and the ECL spectrum was substantially red-shifted (~200 nm) compared to the PL spectrum, suggesting that surface states play an important role in the emission process. In these systems, light emission was directly from the NCs. $Ru(bpy)_3^{2+}$ has also been immobilized on gold nanoparticles (via electrostatic interactions between $Ru(bpy)_3^{2+}$ and sulfhydryl groups), and fairly stable and intense ECL was generated when TPrA was present in solution (Sun *et al.*, 2005). Another report of $Ru(bpy)_3^{2+}$ doped in silica nanoparticles (Zhang and Dong, 2006b) had a detection limit of 2.8 nM for TPrA, three orders of magnitude lower than that observed for Nafion-based systems. Platinum nanoparticles have also been treated with $Ru(bpy)_3^{2+}$, and the aggregates that form in aqueous solution result in stable films on solid electrodes (Xun *et al.*, 2006). Electrochemiluminescence was generated upon potential sweep from 0.0 to +1.4 V in aqueous buffered solution containing TPrA. In the absence of TPrA, much weaker ECL was also observed, presumably due to the reaction of $Ru(bpy)_3^{2+}$ with hydroxyl radical. Platinum nanoparticles (PtNPs) were also incorporated into a Eastman AQ55D/ruthenium(II) tris(bipyridine) (PtNPs/AQ/$Ru(bpy)_3^{2+}$) to form a colloidal material that produces ECL with a limit of detection of 1×10^{-15} M in the presence of TPrA (Du *et al.*, 2006). A $Ru(bpy)_3^{2+}$/TPrA ECL sensor has also been fabricated using multilayer films of Nafion-stabilized magnetic nanoparticles (NSMNP; Nafion/Fe_3O_4). As in the commercial systems described above, these nanoparticles are attracted to a platinum electrode surface by using an external magnet (Figure 7.11) (Kim *et al.*, 2005b). The ECL sensor based on these nanoparticles was more sensitive than that based on pure Nafion films, presumably due to the faster mass transport in the NSMNP films. The ability to modify nanoparticle surfaces with specific functional groups will undoubtedly lead to more work in this area and may lead to practical applications in bioanalysis.

The coupling of ECL and carbon nanotubes has become an active area of research. For example, $Ru(bpy)_3^{2+}$-doped silica nanoparticles have been coimmobilized with carbon nanotubes and their ECL generated (Zhang and Dong, 2006). In another report, CdS NCs composited with carbon nanotubes enhanced $Ru(bpy)_3^{+}$/$Ru(bpy)_3^{3+}$ ECL and also lowered

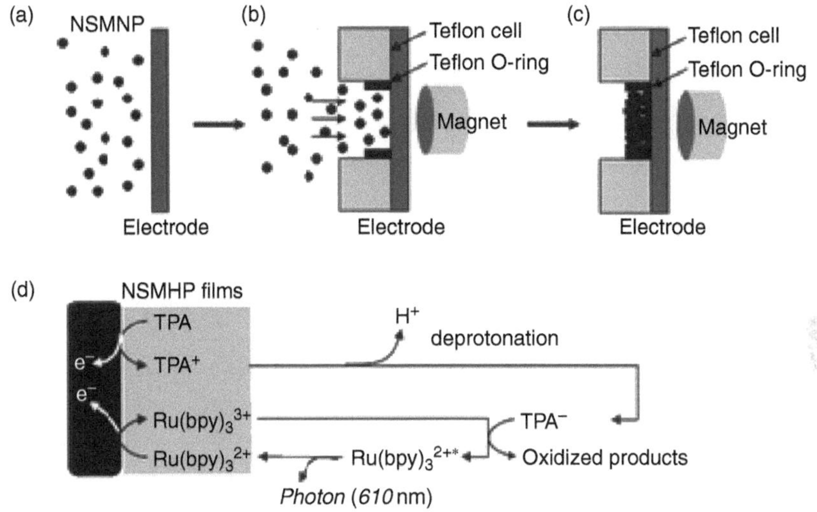

Figure 7.11 Formation of NSMNP (Nafion/Fe_3O_4)-modified electrode (a–c) and the ECL from $Ru(bpy)_3^{2+}$ immobilized in the multilayer films on a Pt electrode surface in the presence of coreactant tripropylamine (TPA) (d). (From Kim *et al.* (2005b). Reproduced with permission from The Royal Society of Chemistry.)

the potential necessary for reactant generation (Ding *et al.*, 2006b). Single-wall and multiwall carbon nanotubes were dispersed in titania–Nafion sol-gel films, with $Ru(bpy)_3^{2+}$ adsorbed on the nanotube surfaces (Choi *et al.*, 2006). A detection limit of 10 nM was obtained for TPrA and the device was stable for greater than 4 months. In another study, a multiwalled carbon nanotube/Nafion composite film-modified electrode was developed and used to study the ECL behavior of $Ru(bpy)_3^{2+}$ in the presence of carbamates (Lin and Chen, 2006). In yet another, a detection limit of 3×10^{-11} M was obtained for $Ru(bpy)_3^{2+}$/TPrA ion-exchanged in Eastman AQ-carbon nanotube composite films adsorbed on a glassy carbon working electrode (Zhang *et al.*, 2006). The $Ru(bpy)_3^{2+}$ ECL of several alkaloids, including berberine, trigonelline, allantoin, and betaine, has been studied in an aqueous alkaline buffer solution (pH 9.5), with the $Ru(bpy)_3^{2+}$ immobilized on an organically modified silicate film on a glassy carbon electrode (Zhao *et al.*, 2006). It will be interesting to see what new systems and sensors for biochemical and chemical analyses based on nanoparticles and nanotubes will be developed in the future.

This chapter has centered on the background and history of ECL, and its development into a biomedical research and clinical diagnostic tool. However, ECL is a versatile detection methodology and is being developed as a sensor and probe for other applications. Since the first detailed studies, over 1000 papers, patents, and book chapters have appeared on ECL, ranging from the very applied to a focus on the underlying science. Since the first edition of this book was published in 2002, dozens of fundamental and applied research articles, several comprehensive reviews, and the first monograph devoted entirely to ECL have appeared. With the interest in using ECL reactions as the basis for highly sensitive and selective analysis, the prediction made by Faulkner and Glass that "continued research in this area will probably stress the development of ECL as a probe rather than as an end in itself" (Faulkner and Glass, 1982) has come to fruition. One wonders what the next 40 years hold for ECL but whatever the improvements and new aspects of ECL that emerge, ECL will continue to show promise for optical biosensing and in other areas of science and technology.

References

Aboul-Enein, H., Stefan, R.I., van Staden, J.F. *et al.* (2000) *Crit. Rev. Anal. Chem.*, **30**, 271–289.

Abruna, H.D. (1985) *J. Electrochem. Soc.*, **132**, 842.

Ala-Kleme, T., Makinen, P., Ylinen, T. *et al.* (2006) *Anal. Chem.*, **78**, 82.

Alexander, C. and Richter, M.M. (1999) *Anal. Chim. Acta*, **402**, 105.

Andersson, A.M., Isovitsch, R., Miranda, D. *et al.* (2000) *Chem. Commun.*, **6**, 505–6.

Andersson, A.M. and Schmehl, R.H. (2001). Sensors based on electrogenerated chemiluminescence. In *Molecular and Supramolecular Photochemistry Vol. 7: Optical Sensors and Switches* (V. Ramamurthy and K.S. Schanze, eds), pp. 153–87.

Anderson, J.D., McDonald, E.M., Lee, P.A. *et al.* (1998) *J. Am. Chem. Soc.*, **120**, 9646.

Arora, A., DeMello, A.J., and Manz, A. (1997) *Anal. Commun.*, **34**, 393.

Bader, J.M. and Kuwana, T. (1965) *J. Electroanal. Chem.*, **10**, 104.
Badocco, D., Zanon, F., and Pastore, P. (2006) *Electrochim. Acta*, **51**, 6442.
Baeummer, A.J., Humiston, M.C., Montagna, R.A., and Durst, R.A. (2001) *Anal. Chem.*, **73**, 1176.
Bard A.J. (ed.) (2004) *Electrogenerated Chemiluminescence*. New York: Marcel Dekker.
Bard, A.J., Debad, J.D., Leland, J.K. *et al.* (2000) *Encyclopedia of Analytical Chemistry* (R.A. Meyers, ed.) Chichester: John Wiley and Sons, pp. 9842–9.
Bard, A.J. and Faulkner, L.R. (1980) *Electrochemical Methods*. New York: John Wiley and Sons, pp. 624–6.
Bard, A.J. and Faulkner, L.R. (2001). *Electrochemical Methods Fundamentals and Applications* (2nd edn) New York: John Wiley and Sons, pp. 736–45.
Bard, A.J. and Whitesides, G.M. (1993a) US Patent 5221605.
Bard, A.J. and Whitesides, G.M. (1993b) US Patent 5238808.
Bard, A.J. and Whitesides, G.M. (1994) US Patent 5310687.
Barigelletti, F., De Cola, L., Balzani, V. *et al.* (1991) *Inorg. Chem.*, **30**, 641.
Bartelt, J.E., Drew, S.M., and Wightman, R.M. (1992) *J. Electrochem. Soc.*, **139**, 70.
Bernanose, A., Bremer, Th., and Goldfinger, P. (1947) *Bull. Soc. Chim. Belg.*, **56**, 269.
Best, J.D., Jay, M.T., Out, F. *et al.* (2005) *J. Pharmacol. Exp. Ther.*, **313**, 902.
Birks, J.B. (1970) *Photophysics of Aromatic Molecules*. New York: John Wiley and Sons.
Blackburn, G.F., Shah, H.P., Kenten, J.H. *et al.* (1991) *Clin. Chem.*, **37**, 1626.
Blohm, S., Kadey, S., McAKeon, K. *et al.* (1996) *Biomed. Prod.*, April.
Bobbit, D.R. and Jackson, W.A. (1997) US Patent 5614073.
Bobbitt, D.R., Jackson, W.A., and Hendrickson, H.P. (1998) *Talanta*, **46**, 565.
Bolletta, F., Rossi, A., and Balzani, V. (1981) *Inorg. Chim. Acta*, **53**, L23.
Boom, R., Sol, C., Weel, J. *et al.* (1999) *J. Clin. Microbiol.*, **37**, 1489.
Bowie, A.R., Sanders, M.G., and Worsfold, P.J. (1996) *J. Biol. Chemilum.*, **11**, 61.
Brilmyer, G.H. and Bard, A.J. (1980) *J. Electrochem. Soc.*, **127**, 104.
Bruce, D. and Richter, M.M. (2002) *Anal. Chem.*, **74**, 1340.
Bruce, D., Richter, M.M., and Brewer, K.J. (2002) *Anal. Chem.*, **74**, 3157.
Brune, S.N. and Bobbitt, D.R. (1991) *Talanta*, **38**, 803.
Brune, S.N. and Bobbitt, D.R. (1992) *Anal. Chem.*, **64**, 166.
Bruno, J.G. (1998a) Broad applications of electrochemiluminescence technology to the detection and quantitation of microbiological, biochemical and chemical analytes. In *Recent Research Developments in Microbiology*, Vol. 1 (S.G. Pandalai, ed.) India: Trivandrum Publishing, pp. 25–46.

Bruno, J.G. (1998b) *J. Biolumin. Chemilumin.*, **13**, 139.
Bruno, J.G., Collard, S.B., and Andrews, A.R.J. (1997) *J. Biolumin. Chemilumin.*, **12**, 155.
Bruno, J.G., Collard, S.B., Kuch, D.J., and Cornette, J.C. (1996) *J. Biolumin. Chemilum.*, **11**, 193.
Bruno, J.G. and Cornette, J.C. (1997) *Microchem. J.*, **56**, 305.
Bruno J.G. and Kiel J.L. (1999) *Biosens. Bioelectron.*, **14**, 457.
Bruno, J.G. and Kiel, J.L. (2002) *Biotechnol. Tech.*, **32**, 178.
Bruno, J.G. and Yu, H. (1996) *Appl. Environ. Microbiol.*, **62**, 3474.
Butch, A.W., Crary, D., and Yee, M. (2002) *Clin. Biochem.*, **35**, 143.
Byrd, J., Bruno, J., and Richter, M.M. (2006) *Luminescence*, **21**, 72.
Call, J.L., Arrowood, M.J., Xie, L.-T. *et al.* (2001), *J. Parasitol.*, **87**, 203.
Cao, W., Liu, J., Qiu, H. *et al.* (2002) *Electroanalysis*, **14**, 1571.
Caspar, J.V. and Meyer, T.J. (1983) *J. Am. Chem. Soc.*, **105**, 5583.
Chandross, E.A. and Visco, R.E. (1964) *J. Am. Chem. Soc.*, **86**, 5350.
Chang, Z., Zhou, J., Zhao, K. *et al.* (2006) *Electrochim. Acta*, **52**, 575.
Chen, C.F. and Wang, C.M. (1999) *J. Electroanal. Chem.*, **466**, 82.
Chen, G.N., Lin, R.E., Zhao, Z.F. *et al.* (1997a) *Anal. Chim. Acta*, **341**, 251.
Chen, X., Sato, M., and Lin, Y.J. (1998a) *Microchem. J.*, **58**, 13.
Chen, X., Jia, L., Wang, X.R. *et al.* (1997b) *Anal. Sci.*, **13**, 71.
Chen, X., Jia, L., and Sato, M. (1998b), *Acta Chim. Sin.*, **56**, 238.
Chi, Y., Duan, J., Lin, S., and Chen, G. (2006) *Anal. Chem.*, **78**, 1568.
Chmura, J. and Slawinski, J. (1994) *J. Biolumin. Chemilumin.*, **9**, 1.
Choi, H.N., Lee, J.-Y., Lyu, Y.-K., and Lee, W.-Y. (2006) *Anal. Chim. Acta*, **565**, 48.
Cole, C., Muegge, B.D., and Richter, M.M. (2003) *Anal. Chem.*, **75**, 601.
Collins, R.A., Ko, L.S., Fung, K.Y. *et al.* (2002a) *Biochem. Biophy. Res. Commun.*, **297**, 267.
Collins, R.A., Ko, L.S., Fung, K.Y. *et al.* (2003) *Biochem. Biophys. Res. Commun.*, **300**, 507.
Collins, R.A., Ko, L.-S., So, K.-L. *et al.* (2002b) *J. Virol. Meth.*, **103**, 213.
Collinson, M.M. and Wightman, R.M. (1993) *Anal. Chem.*, **65**, 2576.
Collinson, M.M. and Wightman, R.M. (1995) *Science*, **268**, 1883.
Collinson, M.M., Novak, B., Martin, S.A., and Taussig, J.S. (2000) *Anal. Chem.*, **72**, 2914.
Collinson, M.M., Taussig, J.S., and Martin, S.A. (1999) *Chem. Mater.*, **11**, 2594.
Collinson, P.O., Jorgensen, B., Sylven, C. *et al.* (2002) *Clin. Chim. Acta*, **307**, 197.

Crawford, C.G., Wijey, C., Fratamico, P. *et al.* (2000) *J. Rapid Methods Autom. Microbiol.*, **8**, 249.
Creutz, C., Chou, M., Netzel, T.L. *et al.* (1980) *J. Am. Chem. Soc.*, **102**, 1309.
Crosby, G.A., Whan, R.E., and Allire, R.M. (1961) *J. Chem. Phys.*, **34**, 743.
Cruser, S.A. and Bard, A.J. (1967) *Anal. Lett.*, **1**, 11.
Cui, H., Li, F., Shi, M.-J. *et al.* (2005) *Electroanalysis*, **17**, 589.
Danielson, N.D., He, L., Noffsinger, J.B., and Trelli, L. (1989) *J. Pharm. Biomed. Anal.*, **7**, 1281.
Davydov, I.V., Kenten, J.H., Safiran, Y.J *et al.* (2005) *Methods. Enzymol.*, **399**, 415.
Davydov, I.V., Woods, D., Safiran, Y.J. *et al.* (2004) *J. Biomol. Screening*, **9**, 695.
Deaver, D.R. (1995) *Nature*, **377**, 758.
Debad, J.D., Glezer, E.M., Leland, J.K. *et al.* (2004). In *Electrogenerated Chemiluminescence* (Bard, A.J., ed.) New York: Marcel Dekker, Chapter 8.
De Baer, M.P., van der Horn, K.H.M., Goudsmit, J. *et al.* (1999) *J. Clin. Microbiol.*, **37**, 63.
De Jong, M.D., Weel, J.F.L., Schuurman, T. *et al.* (2000) *J. Clin. Microbiol.*, **38**, 2568.
Demas, J.N. and Crosby, G.A. (1968) *J. Mol. Spectrosc.*, **26**, 72.
Demas, J.N. and Crosby, G.A. (1971) *J. Am. Chem. Soc.*, **93**, 2841.
Dennany, L., Forster, R.J., and Rusling, J.F. (2003) *J. Am. Chem. Soc.*, **125**, 5213.
Dennany, L., Hogan, C.F., Keyes, T.E., and Forster, R.J. (2006a) *Anal. Chem.*, **78**, 1412.
Dennany, L., O'Reilly, E.J., Keyes, T.E., and Forster, R.J. (2006b) *Electrochem. Commun.*, **8**, 1588.
Dickson, J.A., Ferris, M.M., and Milofsky, R.E. (1997) *J. High Resolut. Chromatogr.*, 20, 643.
Dixon, S.B., Sanford, J., and Swift, B.W. (1993) Electrochemiluminescent sensing of petroleum contamination at trace levels. *Principles and Practices for Petroleum Contaminated Soils* (E.J. Calabrese and P.T. Kosteki, eds) Lewis Publishers, pp. 85–99.
Ding, S.-N., Xu, J.-J., and Chen, H.-Y. (2006a) *Talanta*, **70**, 403.
Ding, S.-N., Xu, J.-J., and Chen, H.-Y. (2006b) *Chem. Commun.*, 3631.
Ding, Z., Quinn, B.M., Haram, S.K. *et al.* (2002) *Science*, **296**, 1293.
Dodeigne, C., Thunus, L., and Lejeune, R. (2000) *Talanta*, **51**, 415.
Dong, L. and Martin, M.T. (1996) *Anal. Biochem.*, **236**, 344.
Dong, Y.-P., Cui, H., and Wang, C-.M. (2006) *J. Phys. Chem. B*, **110**, 18408.
Downey, T.-M. and Nieman, T.A. (1992) *Anal. Chem.*, **64**, 261.

Du, Y., Qi, B., Yang, X., and Wang, E. (2006) *J. Phys. Chem. B*, **110**, 21662.
Dufford, R.T., Nightingale, D., and Gaddum, L.W. (1927) *J. Am. Chem. Soc.*, **49**, 1858.
Egashira, N., Kondoh, N., Kurauchi, Y., and Ohga, K. (1992) *Denki Kagaku*, **60**, 1148.
Egashira, N., Kumasako, H., and Ohga, K. (1990) *Anal. Sci.*, **6**, 903.
Egashira, N., Nabeyama, Y., Kurauchi, Y., and Ohga, K. (1996) *Anal. Sci.*, **12**, 793.
Egashira, N., Piao, J., Hifumi, E., and Uda, T. (2000) *Bunseki Kagaku*, **49**, 1029.
Ege, D., Becker, W.G., and Bard, A.J. (1984) *Anal. Chem.*, **56**, 2413.
Ehrhardt, V., Assman, G., Baetz, O. *et al.* (1998) *Wiener Klinische Wochenschrift*, **110**, 61.
Factor, B., Muegge, G., Workman, S. *et al.* (2001) *Anal. Chem.*, **73(19)**, 4621–4.
Fahnrich, K., O'Sullivan, C.K., and Guilbault, G.G. (1999) SAC99, Dublin, Ireland, July, PC20.
Fahnrich, K.A., Pravda, M., and Guilbault, G.G. (2001) *Talanta*, **54**, 531–59.
Fan, F.-R.F., Cliffel, D., and Bard, A.J. (1998) *Anal. Chem.*, **70**, 2941.
Faulkner, L.R. and Bard, A.J. (1977) Techniques of Electrogenerated Chemiluminescence. *Electroanalytical Chemistry*, Vol. 10 (A.J. Bard, ed.) New York: Marcel Dekker, pp. 1–95.
Faulkner, L.R. and Glass, R.S. (1982) *Chemical and Biological Generation of Excited States* (A. Waldemar and C. Giuseppe, eds) New York: Academic Press, Chapter 6.
Fiaccabrino, G.C., de Rooij, N.F., and Koudelka-Hep, M. (1998) *Anal. Chim. Acta*, **359**, 263.
Fioccabrino, G.C., Koudelka-Hep, M., Hsueh, Y-T. *et al.* (1998) *Anal. Chem.*, **70**, 4157.
Forbes, G.A., Nieman, T.A., and Sweedler, J.V. (1997) *Anal. Chim. Acta.*, **347**, 289.
Fox, D., Han, S., Samuelson, A. *et al.* (2002) *J. Clin. Vir.*, **24**, 117.
Freebern, W.J., Haggerty, C.M., Montano, I. *et al.* (2005) *Pharmacogenomics J.*, **5**, 305.
Fukuda, K., Takahashi, K., Iwata, Y. *et al.* (2001) *J. Clin. Microbiol.*, **39**, 419.
Gatto-Menking, D.L., Yu, H., Bruno, J.G. *et al.* (1995) *Biosens. Bioelectron.*, **10**, 501.
Gassler, N., Peuschel, T., and Pankau, R. (2000) *Clin. Lab.*, **46**, 553.
Gellings, A., Holzem, G., Wielckens, K., and Klingler, K.R. (2001) *Laboratoriumsmedizin*, **25**, 26.
Gerardi, R.D., Barnett, N.W., and Lewis, A.W. (1999) *Anal. Chim. Acta*, **378**, 1.

Gilman, S.D., Silverman, C.E., and Ewing, A.G. (1994) *J. Microcolumn Sep.*, **6**, 97.
Glass, R.S. and Faulkner, L.R. (1981) *J. Phys. Chem.*, **85**, 1160.
Gonzalez, J.M., Greenway, G.M., McCreedy, T., and Qijin, S. (2000) *Analyst*, **125**, 765.
Gopalakrishnan, S.M., Warrior, U., Burns, D., and Groebe, D.R. (2000) *J. Biomol. Screening*, **5**, 369–75.
Gorman, B.A., Francis, P.S., and Barnett, N.W. (2006) *Analyst,* **131**, 616.
Greenway, G.M. and Dolman, S.J.L. (1999) *Analyst*, **124**, 759.
Greenway, G.M., Greenwood, A., Watts, P., and Wiles, C. (2006) *Chem. Commun.*, **1**, 85.
Gross, E.M., Anderson, J.D., Slaterbeck, A.F. *et al.* (2000) *J. Am. Chem. Soc.*, **122**, 4972.
Gross, E.M., Armstrong, N.R., and Wightman, R.M. (2002) *J. Electrochem. Soc.*, **149**, E137.
Gudibande, S., Kenten, J.H., Link, J. *et al.* (1992) *J. Mol. Cell. Probes*, **6**, 495.
Guo, Z., Zheng, X., and Zhang, Z. (2002) *Fenxi Huaxue*, **30**, 461.
Haapakka, K.E. (1982) *Anal. Chim. Acta*, **139**, 229.
Haapakka, K.E. and Kankare, J.J. (1980) *Anal. Chim. Acta*, **118**, 333.
Haese, A., Dworschack, R.T., Piccoli, S.P. *et al.* (2002) *Clin. Chem.*, **48**, 944.
Hai, H., Chen, Z., Cai, P., and Mo, J. (2005) *Fenxi Ceshi Xuebao*, **24**, 111.
Harvey, N. (1929) *J. Phys. Chem.*, **33**, 1456.
Haswell, S.J. (1997) *Analyst*, **122**, R1.
He, L., Cox, K.A., and Danielson, N.D. (1990) *Anal. Lett.*, **23**, 195.
Hemingway, R.E., Park, S.-M., and Bard, A.J. (1975) *J. Am. Chem. Soc.*, **95**, 200.
Henchal, E.A., Teska, J.D., Ludwig, G.V. *et al.* (2001) *Clin. Lab. Med.*, **21**, 661.
Hercules, D.M. (1964) *Science*, **143**, 808.
Hermsen, D., Franzson, L., Hoffman, J.P. *et al.* (2002) *Clin. Lab.*, **48**, 131.
Heroux, J.A. and Szczepanik, A.M. (1995) *PCR Meth. Appl.*, **4**, 327.
Hetland, O. and Dickstein, K. (1998) *Clin. Chem.*, **44**, 1348.
Hibbits, S., Rahman, A., John, R. *et al.* (2003) *J. Virol. Methods*, **108**, 145.
Higgins, J.A., Ibrahim, M.S., Knauert, F.K. *et al.* (1999) *Ann. N. Y. Acad. Sci.*, **894**, 130.
Hill, E., Humphreys, E., and Malcolme-Lawes, D.J. (1986) *J. Chromatogr.*, **370**, 427.
Holeman, J.A. and Danielson, N.D. (1994) *J. Chromatogr. A*, **679**, 277.
Hoon, D.S.B., Kuo, C.T., Wen, S. *et al.* (2001) *Am. J. Pathol.*, **159**, 493.
Horii, Y., Garcia, J.N.P., Novianna, D. *et al.* (2001) *J. Vet. Med. Sci.*, **63**, 921.
Horiuchi, H., Lippé, R., McBride, H.M. *et al.* (1997) *Cell*, **90**, 1149.

Hoyle, N.R. (1994) *J. Biolumin. Chemilumin.*, **9**, 289.
Hsueh, Y.T., Collins, S.D., and Smith, R.L. (1998) *Sens. Actuators B Chem.*, **49**, 1.
Hsueh, Y.T., Smith, R.L., and Northrup, M.A. (1995) *Transducers '95 – Eurosensors IX*. Stockholm, Sweden, June, p. 768.
Hsueh, Y.T., Smith, R.L., and Northrup, M.A. (1996) *Sens. Actuators B Chem.*, **33**, 110.
Ishii, J., Ishikawa, T., Yukitake, J. *et al.* (1998) *Clin. Chim. Acta*, **270**, 183.
Hubl, W., Chan, D.W., Van Ingen, H.E. *et al.* (1999) *Anticancer Res.*, **19**, 2727.
Hughes, S.R., Khorkova, O., Goyal, S. *et al.* (1998) *Proc. Natl. Acad. Sci. USA*, **95**, 3275.
Ingen, H.E., Chan, D.W., Hubl, W. *et al.* (1998) *Clin. Chem.*, **44**, 2530.
Ishida, J., Sonezaki S., and Yamaguchi, M. (1992a) *J. Chromatogr.*, **598**, 203.
Ishida, J., Sonezaki S., Yamaguchi, M., and Yoshitake, T. (1992b) *Analyst*, **117**, 1719.
Itoh, I. and Honda, K. (1979) *Chem. Lett.*, **1**, 99.
Jackson, W.A. and Bobbitt, D.R. (1994) *Anal. Chim. Acta*, **285**, 309.
Jameison, F., Sanchez, R.I., Dory, L. *et al.* (1996) *Anal. Chem.*, **68**, 1298.
Jirka, G.P., Martin, A.F., and Nieman, T.A. (1993) *Anal. Chim. Acta*, **284**, 345.
Kang, J., Yin, X., Yang, X., and Wang, E. (2005) *Electrophoresis*, **26**, 1732.
Kankare, J., Haapakka, K., Kulmala, S. *et al.* (1992) *Anal. Chim. Acta*, **266**, 205.
Kanoufi, F., Zu, Y., and Bard, A.J. (2001) *J. Phys. Chem. B*, **105**, 210.
Kapturkiewicz, A. and Angulo, G. (2003) *Dalton Trans.*, **20**, 3907–13.
Kapturkiewicz, A., Chen, T.-M., Laskar, I.R., and Nowacki, J. (2004) *Electrochem. Commun.*, **6**, 827.
Kapturkiewicz, A., Nowacki, J., and Borowicz, P. (2006) *Z. Phys. Chem.*, **220**, 525.
Kashiwagi, S., Hayashi, J., Asai, T. *et al.* (1998) *Igaku Yakugaku*, **40**, 119.
Kearney, N.J., Hall, C.E., Jewsbury, R.A., and Timmis, S.G. (1996) *Anal. Commun.*, **33**, 269.
Kenten, J.H., Casadei, J., Link, J. *et al.* (1991) *Clin. Chem.*, **37**, 1626.
Kenten, J.H., Davydov, I.V., Safiran, Y.J. *et al.* (2005) *Methods Enzymol.*, **399**, 682.
Kenten, J.H., Gudibande, S., Link, J. *et al.* (1992a) *Clin. Chem.*, **38**, 873.
Kenten, J.H., Gudibande, S.R., Link, J. *et al.* (1992b) *Mol. Cell. Probes*, **6**, 495.
Keszthelyi, C.P. and Bard, A.J. (1973) *J. Electrochem. Soc.*, **120**, 241.
Khorkova, O.E., Pate, K., Heroux, J., and Sahasrabudhe, S. (1998) *J. Neurosci. Methods*, **82**, 159.
Khramov, A.N. and Collinson. M.M. (2000) *Anal. Chem.*, **72**, 2943.

Kim, J., Fan, F.-R.F., Bard, A.J. *et al.* (1985) *Chem. Phys. Lett.*, **121**, 543.
Kim, D.-J., Lyu, Y.-K., Choi, H.N. *et al.* (2005b) *Chem. Commun.*, 2966.
Kim, J.I., Shin, I.-S., Kim, H., and Lee, J.-K. (2005a) *J. Am. Chem. Soc.*, **127**, 1614.
King, K.A., Spellane, P.J., and Watts, R.J. (1985) *J. Am. Chem. Soc.*, **107**, 1431.
Klein, G., Kampmann, M., Baum, H. *et al.* (1998) *Wien. Klin. Wochenschr.*, **110**(Suppl. 3), 40.
Klingler, K.R., Zech, D., and Wielckens, K. (2000) *Clin. Lab.*, **46**, 41.
Knight, A.W. (1999) *Trends Anal. Chem.*, **18**, 47.
Knight, A.W. (2001) Electrogenerated Chemiluminescence. In *Chemiluminescence in Analytical Chemistry* (A.M. Garcia-Campana and W.R.G. Baeyens, eds) New York: Marcel Dekker, pp. 211–47.
Knight, A.W. (1999) *Trends Anal. Chem.* **18**, 47.
Knight, A.W. and Greenway, G.M. (1995) *Analyst*, **120**, 2543.
Knight, A.W. and Greenway, G.M. (1994) *Analyst*, **119**, 879.
Knight, A.W. and Greenway, G.M. (1996) *Anal. Commun.*, **33**, 171.
Kobayashi, Y., Hayakawa, M., and Fukumara, Y. (1999) *Igaku Yakugaku*, **42**, 749.
Kober, E.M. and Meyer, T.J. (1982) *Inorg. Chem.*, **21**, 3967.
Kobrynski, L., Tanimune, L., Pawlowski, A. *et al.* (1996) *Clin. Diagnos. Lab. Immun.*, **3**, 42.
Kremeskotter, J., Wilson, R., Schiffrin, D.J. *et al.* (1995) *Meas. Sci. Technol.*, **6**, 1325.
Kuczynska, E., Boyer, D.G., and Shelton, D.R. (2003) *J. Microbiol. Methods*, **53**, 17.
Kulmala, S. and Suomi, J. (2003) *Anal. Chim. Acta*, **500**, 21–69.
Kulmala, S., Hakansson, M., Spehar, A.-M. *et al.* (2002) *Anal. Chim. Acta*, **458**, 271.
Kuhn, L.S., Weer, A., and Weber, S.G. (1990) *Anal. Chem.*, **62**, 1631.
Kukoba, A.V., Bykh, A.I., and Svir, I.B. (2000) *Frensius' J. Anal. Chem.*, **368**, 439–42.
Kulmala, S., Hakansson, M., Spehar, A.-M. *et al.* (2002) *Anal. Chim. Acta*, **458**, 271.
Kuwana, T., Epstein, B., and Seo, E.T. (1963) *J. Phys. Chem.*, **67**, 2243.
Lanciotti, R.S. and Kerst, A.J. (2001) *J. Clin. Microbiol.*, **39**, 4506.
Leca, B. and Blum, L.J. (2000) *Analyst*, **125**, 789.
Lee, J.-K., Lee, S.-H., Kim, M. *et al.* (2003) *Chem. Commun.*, 1602.
Lee, S.K., Richter, M.M., Strekowski, L., and Bard, A.J. (1997) *Anal. Chem.*, **69**, 4126.

Lee, W.Y. (1997) *Mikrochim. Acta*, **127**, 19.
Lee, W.-Y. and Neiman, T.A. (1994) *J. Chromatogr. A*, **659**, 111.
Lee, Y.M., Johnson, P.W., Call, J.L. *et al.* (2001) *Am. J. Trop. Med. Hyg.*, **65**, 1.
Leland, J.K. and Powell, M.J. (1991) *J. Electroanal. Chem.*, **318**, 91.
Li, F. and Zu, Y. (2004) *Anal. Chem.*, **76**, 1768.
Li, F., Cui, H., and Lin, X.-Q. (2002) *Anal. Chim. Acta*, **471**, 187.
Li, F., Pang, Y.-Q., Lin, X.-Q., and Cui, H. (2003) *Talanta*, **59**, 627.
Liang, P., Dong, L., and Martin, M.T. (1996a) *J. Am. Chem. Soc.*, **118**, 9198.
Liang, P., Sanchez, L., and Martin, M.T. (1996b) *Anal. Chem.*, **68**, 1298.
Liebert, A., Beier, L., Schneider, E., and Kirch, P. (2001) *Chemiluminescence Turn Millennium*, 341.
Lin, Z. and Chen, G. (2006) *Talanta*, **70**, 111.
Lin, X.-Q., Li, F., Pang, Y.-Q., and Cui, H. (2004) **378**, 2028.
Lin, X.Q., Sun, G., and Cui, H. (1999) *Chin. J. Anal. Chem.*, **27**, 497.
Lin, Z., Sun, J., Chen, J. *et al.* (2006) *Anal. Chim. Acta*, **564**, 226.
Liu, C.-Y. and Bard, A.J. (2005) *Anal. Chem.*, **77**, 5339.
Liu, J., Cao, W., Qiu, H. *et al.* (2002) *Clin. Chem.*, **48**, 1049.
Luo, L. and Zhang, Z. (2006) *Anal. Chim. Acta*, **580**, 14.
Luppa, P.B., Reutemann, S., Huber, U. *et al.* (1998) *Clin. Chem. Lab. Med.*, **36**, 789.
Luttmer, J.D. and Bard, A.J. (1981) *J. Phys. Chem.*, **85**, 1155.
Malins, C., Vandeloise, R., Walton, D., and VanderDonckt, E. (1997) *J. Phys. Chem. A*, **101**, 5063.
Maloy, J.T. and Bard, A.J. (1971) *J. Am. Chem. Soc.*, **93**, 5968.
Maloy, J.T., Prater, K.B., and Bard, A.J. (1971) *J. Am. Chem. Soc.*, **93**, 5959.
Maness, K.M. and Wightman, R.M. (1995) *J. Electroanal. Chem.*, **396**, 85.
Maness, K.M., Terrill, R.H., Meyer, T.J. *et al.* (1996) *J. Am. Chem. Soc.*, **118**, 10609.
Marquette, C.A. and Blum, L.J. (1999) *Anal. Chim. Acta*, **381**, 1.
Marquette, C.A., Degiuli, A., and Blum, L.J. (2003) *Biosens. Bioelectron.*, **19**, 433.
Martin, A.F. and Nieman, T.A. (1997) *Biosens. Bioelectron.*, **12**, 479.
Mathew, A., Mathur, S.K., Jolly, C. *et al.* (2001) *Mol. Cell. Biochem.*, **21**, 7163.
Maus, R.G., McDonald, E.M., and Wightman, R.M. (1999) *Anal. Chem.*, **71**, 4944.
McCall, J., Alexander, C., and Richter, M.M. (1999) *Anal. Chem.*, **71**, 2523.
McCall, J., Bruce, D., Workman, S. *et al.* (2001) *Anal. Chem.*, **73**, 4617.
McCall, J. and Richter, M.M. (2000) *Analyst*, **125**, 545.
McCord, P.M. and Bard, A.J. (1991) *J. Electroanal. Chem.*, **318**, 91.

Meyer, T.J. (1978) *Acc. Chem. Res.*, **11**, 94.
Miao, W. and Bard, A.J. (2003) *Anal. Chem.*, **75**, 5825.
Miao, W. and Choi, J.-P. (2004) Coreactants. In *Electrogenerated Chemiluminescence* (A.J. Bard, ed.) Marcel Dekker, pp. 213–71.
Miao, W., Choi, J.-P., and Bard, A.J. (2002) *J. Am. Chem. Soc.*, **124**, 14478–85.
Min, J. and Baeumner, A.J. (2002a) *Anal. Biochem.*, **303**, 186.
Min, J. and Baeumner, A.J. (2002b) *Anal. Biochem.*, **67**, 2908.
Mitschke, U. and Bauerle, P.J. (2000) *Mater. Chem.*, **10**, 1471–507.
Miyashiro, I., Kuo, C., Huynh, K. *et al.* (2001) *Clin. Chem.*, **47**, 505.
Moreau, E., Philippe, J., Couvent, S., and Leroux-Roels, G. (1996) *Clin. Chem.*, **42**, 1450.
Motmans, K., Raus, J., and Vandevyver, C. (1996) *J. Immun. Meth.*, **190**, 107.
Muegge, B.D. and Richter, M.M. (2004) *Anal. Chem.*, **76**, 73.
Muegge, B.D. and Richter, M.M. (2005) *Luminescence*, **20**, 76.
Myung, N., Ding, Z., and Bard, A.J. (2002) *Nanoletters*, **11**, 1315.
Nakashima, K., Suetsuga, K., Yoshida, M. *et al.* (1991) *Anal. Sci.*, **7**, 815.
Namba, Y., Usami, M., and Suzuki, O. (1999) *Anal. Sci.*, **15**, 1087.
Nishimura, K., Hamada, Y., Tsujioka, T. *et al.* (2001) *Jpn. J. Appl. Phys.*, **40**, L945.
Noffsinger, J.B. and Danielson, N.D. (1987a) *J. Chromatogr.*, **387**, 520.
Noffsinger, J.B. and Danielson, N.D. (1987b) *Anal. Chem.*, **59**, 865.
Novick, D., Schwartsburd, B., Pinkus, R. *et al.* (2001) *Cytokine*, **14**, 334.
Obenauer-Kutner, L.J., Jacobs, S.J., Kolz, K. *et al.* (1997) *J. Immun. Meth.*, **206**, 25.
O'Connell, C.D., Juhasz, A., Kuo, C. *et al.* (1999) *Clin Chem.*, **44**, 1161.
Ohlin, M. (1997) *Clin. Diag. Lab.*, **4**, 107.
Ohlin, M., Silvestri, M., Sundqvist, V.-A., and Borrebaeck, C.A.K. (1997) *Clin. Diag. Lab. Immun.*, **4**, 107.
Okabe, R., Nakatsuka, K., Inaba, M. *et al.* (2001) *Clin. Chem.*, **47**, 1410.
Olsvik, O., Popovic, T., Skjerve, E. *et al.* (1994) *Clin. Microbiol. Rev.*, **7**, 43.
Oprandy, J.J., Amemiya, K., Kenten, J.H. *et al.* (1995) Technical Advances in AIDS Research in the Human Nervous System, [Proceedings of an NIH Symposium on Technical Advances in AIDS Research in the Human Nervous System], Washington, DC, October 4–5, 1993, p. 281.
Ouyang, J. and Bard, A.J. (1988) *Bull. Chem. Soc. Jpn.*, 61.
Ouyang, C.S. and Wang, C.M. (1998) *J. Electrochem. Soc.*, **145**, 2654.
Oyama, M., Masuda, T., Mitani, M., and Okazaki, S. (1999) *Anal. Chim. Acta*, **67**, 1211.
Pang, Y.-Q., Cui, H., Zheng, H.-S. *et al.* (2005) *Luminescence*, **20**, 8.
Paris, J.P. and Brandt, W.W. (1959) *J. Am. Chem. Soc.*, **81**, 5001.

Park, Y.-J., Lee, D.W., and Lee, W.-Y. (2002) *Anal. Chim. Acta*, **471**, 51.
Parker, C.A. (1968) *Photoluminescence of Solutions*. Amsterdam: Elsevier.
Prieto, I., Teetsov, J., Fox, M.A. *et al.* (2001) *J. Phys. Chem. A*, **105**, 520.
Puren, A.J., Razeghi, P., Fantuzzi, G., and Dinarello, C.A. (1998) *J. Infectious Diseases*, **178**, 1830.
Qi, H. and Zhang, C. (2004) *Anal. Chim. Acta*, **501**, 31.
Quinn, B.M., Ding, Z., Moulton, R., and Bard, A.J. (2002) *Langmuir*, **18**, 1734.
Reetoo, K.N., Osman, S.A., Illavia, S.J. *et al.* (1999) *J. Virol. Methods*, **82**, 145.
Richards, T.C. and Bard, A.J. (1995) *Anal. Chem.*, **67**, 3140.
Richter, M.M. (2002) Electrochemiluminescence. In *Optical Biosensors: Present and Future* (F.S. Ligler and C.A. Rowe-Taitt, eds) (2nd edn) New York: Elsevier, pp. 173–205.
Richter, M.M. (2004) *Chem. Rev.*, **104**, 3003–36.
Richter, M.M. and Bard, A.J. (1996) *Anal. Chem.*, **68**, 2641.
Richter, M.M., Bard, A.J., Kim, W., and Schmehl, R.H. (1998) *Anal. Chem.*, **70**, 310.
Richter, M.M., Fan, F.-R.F., Klavetter, F. *et al.* (1994) *Chem. Phys. Lett.*, **226**, 115.
Ridlen, J.S., Klopf, G.J., and Nieman, T.A. (1997) *Anal. Chim. Acta*, **341**, 195.
Ridlen, S., Skotty, D.R., Kissinger, P.T., and Nieman, T.A. (1997) *J. Chromatogr. B*, **694**, 393.
Robin, M.B. and Day, P. (1967) *Adv. Inorg. Chem. Radiochem.*, **10**, 247.
Roundhill, D.M. (1994) *Photochemistry and Photophysics of Coordination Complexes*. New York: Plenum, Chapter 5.
Rozhitskii, N.N. (1992) *J. Anal. Chem. USSR*, **47**, 1288.
Rubinstein, I. and Bard, A.J. (1981a) *J. Am.Chem. Soc.*, **103**, 512.
Rubinstein, I. and Bard, A.J. (1981b) *J. Am. Chem. Soc.*, **102**, 6641.
Rubinstein, I., Martin, C.R., and Bard, A.J. (1983) *Anal. Chem.*, **55**, 1580.
Safarikova, S.M. and Forsythe, S.J. (1995) *J. Appl. Bacteriol.*, **78**, 575.
Saji, T. and Bard, A.J. (1977) *J. Am. Chem. Soc.*, **99**, 2235.
Sakura, S. (1992) *Anal. Chim. Acta*, **262**, 49.
Sakura, S. and Imai, H. (1988) *Anal. Sci.*, **4**, 9.
Sanchez-Carbayo, M., Espasa, A., Chinchilla, V. *et al.* (1999a) *Clin. Chem.*, **45**, 1944.
Sanchez-Carbayo, M., Mauri, M., Alfayate, R. *et al.* (1999b) *Clin. Biochem.*, **32**, 395.
Sanchez-Carbayo, M., Mauri, M., Alfayate, R. *et al.* (1998) *Clin. Chem.*, **44**, 1744.
Santhanam, K.S.V. and Bard, A.J. (1965) *J. Am. Chem. Soc.*, **87**, 139.
Sapin, R., Le Galudeck, V., Gasser, F. *et al.* (2001) *Clin. Chem.*, **47**, 602.

Sapin, R., Schlienger, J.-L., Gasser, F. *et al.* (2000) *Clin. Chem.*, **46**, 418.
Sassa, T., Kumada, T., Nakano, S., and Uematsu, T. (1999) *Eur. J. Gastroenterol. Herpatol.*, **11**, 1387.
Scheunert, K., Albrecht, S., Konnegen, V. *et al.* (2001) *Chemiluminescence Turn Millennium*, 347.
Schutzbank, T.E. and Smith, J. (1995) *J. Clin. Microbiol.*, **33**, 2036.
Seck, T., Diel, I., Bismar, H. *et al.* (2002) *Bone*, **30**, 217.
Sennikov, S.V., Kyrsov, S.V., Injelevskaya, T.V. *et al.* (2003) *J. Immunol. Methods*, **275**, 81.
Shapiro, L., Heidenreich, K.A., Meintzer, M.K., and Dinarello, C.A. (1998) *Proc. Natl. Acad. Sci. USA*, **95**, 7422.
Shelton, D.R. and Karns, J.S. (2001) *Appl. Environ. Microbiol.*, **67**, 2908.
Shi, M.-J. and Cui, H. (2006) *Electrochim. Acta*, **3**, 1390.
Shi, L., Liu, X., Li, H., and Xu, G. (2006) *Anal. Chem.*, **78**, 7330.
Shimizu, A., Shiraki, K., Ito, T. *et al.* (2002) *Int. J. Mol. Med.*, **9**, 245.
Sinha, A.P.B. (1971) *Spectrosc. Inorg. Chem.*, **2**, 255.
Skotty, D.R. and Neiman, T.A. (1995) *J. Chromatogr. B*, **665**, 27.
Skotty, D.R., Lee, W.-Y., and Nieman, T.A. (1996) *Anal. Chem.*, **68**, 1530.
Smith, P.J. and Mann, C.K. (1969) *J. Org. Chem.*, **34**, 1821.
Spehar-Deleze, A.M., Schmidt, L., Neier, R. *et al.* (2006) *Biosens. Bioelectron.*, **22**, 722.
Spurlin, S. and Cooper, M.M. (1986) *Anal. Lett.*, **19**, 2277.
Staffilani, M., Hoss, E., Giesen, U. *et al.* (2003) *Inorg. Chem.*, **42**, 7789.
Stagni, S., Palazzi, A., Zacchini, S. *et al.* (2006) *Inorg. Chem.*, **45**, 695.
Steijger, O.M., Ligeman, H., Brinkman, U.A.T. *et al.* (1993) *J. Chromatrogr.*, **615**, 97.
Stern, H.J., Carlos, R.D., and Schutzbank, T.E. (1995) *Clin. Biochem.*, **28**, 470.
Stevens, S.J.C., Vervoort, M.B.H.J., van den Brule, A.J.C. *et al.* (1999) *J. Clin. Microbiol.*, **37**, 2852.
Stiebler, P., Molina, R., Chan, D.W. *et al.* (2001) *Clin. Chem.*, **47**, 2162.
Stockmann, W., Bablok, W., and Luppa, P. (1998) *Wien. Klin. Wochenschr.*, **110**, 10.
Su, Y., Wang, J., and Chen, G. (2005) *Anal. Chim. Acta*, **551**, 79.
Sun, Y.G., Cui, H., and Lin, X.Q. (2000) *Acta Chim. Sin.*, **58**, 567.
Sun, X., Du., Y., Dong, S., and Wang, E. (2005) *Anal. Chem.*, **77**, 8166.
Sutin, N. and Creutz, C. (1978) *Adv. Chem. Ser.*, **168**, 1.
Swanson, S.J., Jacobs, S.J., Mytych, D. *et al.* (1999) *Dev. Biol. Stand.*, **97**, 135.
Sykora, M. and Meyer, T.J. (1999) *Chem. Mater.*, **11**, 1186.
Taback, B., Chan, A.D., Kuo, C.T. *et al.* (2001) *Cancer Res.*, **61**, 8845.

Tabor, M.W., MacGee, J., and Holland, J.W. (1976) *Appl. Environ. Microbiol.*, **31**, 25.
Tai, J.H., Ewert, M.S., Belliot, G. *et al.* (2003) *J. Virol. Methods*, **110**, 119.
Takahashi, M., Hoshino, H., Ohuchi, Y. *et al.* (1998) *Igaku Yakugaku*, **40**, 483.
Targove, M.A. and Danielson, N.D. (1990) *J. Chromatogr. Sci.*, **28**, 505.
Taverna, P.J., Mayfield, H., and Andrews, A.R.J. (1998) *Anal. Chim. Acta*, **373**, 111.
Taylor IV, C.E. and Creager, S.E. (2000) *J. Electroanal. Chem.*, **485**, 114.
Tew, D.J., Johnson, J.K., and Maxey, K.M. (2005) *Genetic Eng. News*, **25**, 34.
Tsukagoshi, K., Miyamoto, K., Saito, E. *et al.* (1997) *Anal. Sci.*, **13**, 639.
Tokel, N. and Bard, A.J. (1972) *J. Am. Chem. Soc.*, **94**, 2862.
Tokel-Takvoryan, N.E., Hemingway, R.E., and Bard, A.J. (1973) *J. Am. Chem. Soc.*, **95**, 6582.
Tsukagoshi, K., Miyamoto, K., Saito, E. *et al.* (1997) *Anal. Sci.*, **13**, 639.
Uchikura, K. and Kirisawa, M. (1991a) *Anal. Sci.*, **7**, 803.
Uchikura, K. and Kirisawa, M. (1991b) *Anal. Sci.*, **7**, 971.
Uchikura, K., Kirisawa, M., and Sugii, A. (1993) *Anal. Sci.*, **9**, 121.
Uhlen, M., Hornes, E., and Olsvik, O. (eds) (1994) *Advances in Biomagnetic Separation*. Eaton Publishing.
van Dyke, D.A. and Cheng, H.Y. (1989) *Anal. Chem.*, **61**, 633.
Van Gemen, B., Van Beuningen, R., Nabbe, A. *et al.* (1994) *J. Virol. Meth.*, **49**, 157.
Van Houten, J. and Watts, R.J. (1975) *J. Am. Chem. Soc.*, **98**, 4853.
Vieira, O.V., Verkade, P., Manninen, A., and Simons, K. (2005) *J. Cell Biol.*, **170**, 521.
Vitt, J.E., Johnson, D.C., and Engstrom, R.C. (1991) *J. Electrochem. Soc.*, **138**, 1637.
Vogler, A. and Kunkeley, H. (1984) *Angew. Chem. Int. Ed. Engl.*, **23**, 316.
Vogler, A. and Kunkely, H. (1987) *ACS Symposium Series No. 333 High Energy Processes in Organometallic Chemistry* (K.S. Suslick, ed.) Washington, DC: American Chemical Society, p. 155.
Vojir, V. (1954) *Collect. Czech. Chem. Commun.*, **19**, 872.
Walworth, J., Brewer, K.J., and Richter, M.M. (2004) *Anal. Chim. Acta*, **503**, 241.
Wang, X. and Bobbitt, D.R. (1999) *Anal. Chim. Acta*, **383**, 213.
Wang, C.-Y. and Huang, H.-J. (2003) *Anal. Chim. Acta*, **498**, 61.
Wallace, W.L. and Bard, A.J. (1979) *J. Phys. Chem.*, **83**, 1350.
Walton, D.J., Phull, S.S., Bates, D.M. *et al.* (1992) *Ultrasonics*, **30**, 186.
Wang, C.-Y. and Hsuan-Jung, H. (2003) *Anal. Chim. Acta*, **498**, 61.
Wang, H., Zhang, C., Li, Y., and Qi, H. (2006) *Anal. Chim. Acta*, **575**, 205.

Watts, D.M., King, C.-C., Murphy, G.S. *et al.* (2001) *J. Clin. Microbiol.*, **39**, 2794.
Weber, B., Bayer, A., Kirch, P. *et al.* (1999) *J. Clin. Microbiol.*, **37**, 2639.
Weinreb, P.H., Yang, W.J., Violette, S.M. *et al.* (2002) *Anal. Biochem.*, **306**, 305.
Weller, A. and Zachariasse, K.A. (1969) Chemiluminescence from radical ion recombination. *Mol. Lumin., Int. Conf.* Meeting Date 1968, 895–905.
Weller, A. and Zachariasse, K.A. (1971) *Chem. Phys. Lett.*, **10**, 153–6.
White, H.S. and Bard, A.J. (1981) *J. Am. Chem. Soc.*, **104**, 6891.
Wightman, R.M., Curtis, C.L., Flowers, P.A. *et al.* (1998) *J. Phys. Chem. B*, **102**, 9991.
Wild, D. (ed.) (1994) *The Immunoassay Handbook*. Macmillan Press Ltd.
Willett, C.G., Boucher, Y., Duda, D.G. *et al.* (2005) *J. Clin. Oncol.*, **23**, 8136.
Wilkinson, E.T., Cheifetz, S., and De Grandis, S.A. (1995) *PCR Meth. Appl.*, **4**, 363.
Wilson, R. and Schiffrin, D.J. (1998) *J. Electroanal. Chem.*, **448**, 125.
Workman, S. and Richter, M.M. (2000) *Anal. Chem.*, **72**, 5556.
Wu, S.-J.L., Lee, E.M., Putvatana, R. *et al.* (2001) *Analyst*, **126**, 1285.
Wu, S.-J.L., Lee, E.M., Putvatana, R. *et al.* (2001) *J. Clin. Microbiol.*, **39**, 2794.
Xu. G. and Dong, S. (2000) *Anal. Chem.*, **72**, 5308.
Xu, G., Pang, X.-L., Xu, B. *et al.* (2005) *Analyst*, **130**, 541.
Xun, X., Du., Y., Zhang, L. *et al.* (2006) *Anal. Chem.*, **78**, 6674.
Yamaguchi, K., Sawada, T., Yamane, S. *et al.* (2001) *Ann. Clin. Biochem.*, **38**, 348.
Yamashita, K., Yamazaki-Nishida, S., Harmia, Y., and Segawa, A. (1991) *Anal. Chem.*, **63**, 872.
Yamazaki-Nishida, S., Harima, Y., and Yamashita, K. (1990) *J. Electroanal. Chem.*, **283**, 455.
Yan, J., Yang, X., and Wang, E. (2005) *Anal. Bioanal. Chem.*, **381**, 48.
Yang, H., Leland, Y.K., Yost, D., and Massey, R.J. (1994) *BioTechnology*, **12**, 193.
Yilmaz, N., Erbagci, A.B., and Aynacioglu, A.S. (2001) *Acta Biochim. Pol.*, **48**, 775.
Yin, X.-B. and Wang, E. (2005) *Anal. Chim. Acta*, **533**, 113.
Yin, X.-B., Qi, B., Sun, X. *et al.* (2005) *Anal. Chem.*, **77**, 3525.
Yoon, C.H., Cho, J.-H., Oh, H.-I. *et al.* (2003) *Biosens. Bioelectron.*, **19**, 289.
Yoshida, H., Hidaka, K., Ishida, J. *et al.* (2000) *Anal. Chim. Acta*, **413**, 137.
Yu, H. (1996) *J. Immunol. Meth.*, **192**, 63.
Yu, H. and Bruno, J.G. (1996) *Appl. Environ. Microbiol.*, **62**, 587.

Yu, H., Bruno, J.G., Cheng, T.-C. *et al.* (1995) *J. Biolumin. Chemilumin.*, **10**, 239.

Yu, H., Raymonda, J.W., McMahon, T.M., and Campagnari, A.A. (2000) *Biosens. Bioelectron.*, **14**, 829.

Zachariasse, K.A. (1975) Exciplexes in chemiluminescent radical ion recombination. *The Exciplex* (M. Gordon and W.R. Ware, eds) New York: Academic Press, pp. 275–303.

Zhang, L. and Dong, S. (2006a) *Electrochem. Commun.*, **8**, 1687.

Zhang, L. and Dong, S. (2006b) *Anal. Chem.*, **78**, 5119.

Zhang, L., Guo, Z., Xu, Z., and Dong, S. (2006) *J. Electroanal. Chem.*, **592**, 63.

Zhang, L., Schwartz, G., O'Donnell, M., and Harrison, R.K. (2001a) *Anal. Biochem.*, **293**, 31.

Zhang, L., Song, L., Terracina, G. *et al.* (2001b) *Biochemistry*, **40**, 5049.

Zhang, L., Xu, Z., and Dong, S. (2006) *Anal. Chim. Acta*, **575**, 52.

Zheng, H. and Zu, Y. (2005) *J. Phys. Chem. B*, **109**, 16047.

Zheng, X., Zhang, Z., Guo, Z., and Wang, Q. (2002) *Analyst*, **127**, 1375.

Zhao, X., You, T., Liu, J. *et al.* (2004) *Electrophoresis*, **25**, 3422.

Zhao, L., Tao, Y., Yang, X. *et al.* (2006) *Talanta*, **70**, 104.

Zhou, M. and Roovers, J. (2001) *Macromolecules*, **34**, 244.

Zhou, M., Roovers, J., Robertson, G.P., and Grover, C.P. (2003) *Anal. Chem.*, **75**, 6708.

Zhu, R.H. and Kok, W.T. (1998) *J. Pharma. Biomed. Anal.*, **17**, 985.

Zu, Y. and Bard, A.J. (2000) *Anal. Chem.*, **72**, 3223.

Zu, Y. and Bard, A.J. (2001) Electrogenerated chemiluminescence. 67. Dependence of light emission of the tris(2,2′-bipyridyl)ruthenium(II)/tripropylamine system on electrode surface hydrophobicity. *Anal. Chem.*, **73(16)**, 3960–4.

Zu, Y., Ding, Z., Zhou, J. *et al.* (2001) *Anal. Chem.*, **73**, 2153.

Optical Biosensors: Today and Tomorrow (2nd Edition)
Frances S. Ligler and Chris Rowe Taitt (Editors)

Chapter 8

PLASMONIC SERS MOLECULAR SENTINELS: A NEW BIOSENSING APPROACH

Tuan Vo-Dinh, Ph.D.
Fitzpatrick Institute for Photonics, Departments of Biomedical Engineering and Chemistry, Duke University, Durham, NC 27708-0281, USA

This chapter describes the detection of specific target DNA sequences using a novel "molecular sentinel" (MS) biosensing approach using surface-enhanced Raman scattering (SERS) detection. The SERS-based MS nanoprobe consists of a metal nanoparticle and a stem–loop DNA molecule tagged with a Raman label. The nanoprobe utilizes the specificity and selectivity of the DNA hairpin probe sequence to detect a specific target DNA sequence of interest. In the normal configuration and in the absence of target DNA, the stem–loop configuration maintains the Raman label in close proximity to the metal nanoparticle, inducing an intense SERS effect that produces a strong Raman signal upon laser excitation. Upon hybridization of a complementary target DNA sequence to the nanoprobe, the stem–loop configuration is disrupted, causing the Raman label to physically separate from the metal nanoparticle, thus quenching the SERS signal. Due to the possibility of performing simple homogeneous bioassays, the SERS-MSs could provide useful diagnostic probes for multiple biological targets. The potential for combining the spectral selectivity and high sensitivity of the SERS process with inherent molecular specificity of MS nanoprobes to diagnose molecular target sequences is discussed.

8.1. Technical concept

8.1.1. Background on SERS and plasmonics

A serious disadvantage with the normal Raman effect for trace detection is it generates a very weak signal. However, Raman spectroscopy has gained increasing interest as an analytical tool with the advent of the surface-enhanced Raman scattering (SERS) technique, which can produce drastic increases in the Raman signal. Experimental evidence suggests that the origin of the enormous Raman enhancement arises from at least two primary mechanisms that contribute to the SERS effect: (a) an electromagnetic effect occurring near metal surface structures associated with large local fields caused by electromagnetic resonances, often referred to as "surface plasmons"; and (b) a chemical effect involving a scattering process associated with chemical interactions between the molecule and the metal surface. According to classical electromagnetic theory, molecules on or near metal nanostructures will experience enhanced fields relative to that of the incident radiation. These fields can be quite large (10^6- to 10^7-fold, even up to 10^{15}-fold enhancement at "hot spots"). When a metallic nanostructured surface is irradiated by an incident electromagnetic field (e.g., a laser beam), conduction electrons are displaced into frequency oscillation equal to the incident light.

These oscillating electrons, called "surface plasmons," produce a secondary electric field, which adds to the incident field. When these oscillating electrons become spatially confined, as is the case for isolated metallic nanospheres or otherwise roughened metallic surfaces (nanostructures), there is a characteristic frequency (the plasmon frequency) at which there is a resonant response of the collective oscillations to the incident field. This condition yields intense localized fields that can interact with molecules in contact with or near the metal surface (Otto, 1978; Gersten *et al.*, 1980; Schatz *et al.*, 1984; Zeeman *et al.*, 1987). In an effect analogous to a "lightning rod" effect, secondary fields can become concentrated at high curvature points on the roughened metal surface. The intensity of the normally weak Raman scattering process is increased by factors as large as 10^{11} for compounds adsorbed onto a SERS substrate, allowing for single-molecule detection (Kneipp *et al.*, 1997; Nie *et al.*, 1997). A commonly investigated class of SERS substrate involves colloidal nanoparticles. Reasons for using colloidal solutions include ease of production and straightforward characterization of colloid solutions by UV–VIS absorption. One common method of producing silver sols is via reduction of silver nitrate. Producing a particle size distribution conducive to the SERS effect can be a challenging task. Nevertheless, several successful procedures have been reported (Lee *et al.*, 1982; Silman *et al.*, 1983; Tarabara *et al.*, 1998; Li *et al.*, 1999).

8.1.2. Molecular sentinels: the new biosensing concept

We describe a novel detection approach that incorporates the "SERS effect modulation" scheme associated with metallic nanoparticles and the DNA hairpin structure (Wabuyele and Vo-Dinh, 2005). The SERS-"molecular sentinel" (SERS-MS) technique uses the stem–loop structure similar to that of molecular beacons (MBs) for DNA recognition. However, the detection scheme is fundamentally different from the MB detection scheme. Molecular beacons consist of oligonucleotide probes having a hairpin structure. The beacon probes consist of a fluorescence molecule attached to the end of one arm and a quencher molecule attached to the end of the other arm. They are designed to report the presence of target DNA sequences (complementary to the hairpin DNA loop) using fluorescence detection. Molecular beacons exploit the principle of fluorescence resonance energy transfer (FRET) between the fluorescent molecule and the quenching molecule by generating a relatively strong fluorescent signal when complementary target sequences are hybridized, thus separating the quencher and the fluorophore. The fluorescence remains low (quenched) in the absence of a complementary sequence.

The change (or modulation) of the SERS effect with the distance between the metallic nanoparticle and the Raman label is utilized in the detection strategy of our SERS-MS probes. In the SERS-MS system (Figure 8.1), MS nanoprobes having a Raman label at one end are immobilized onto a metallic nanoparticle via a thiol group attached on

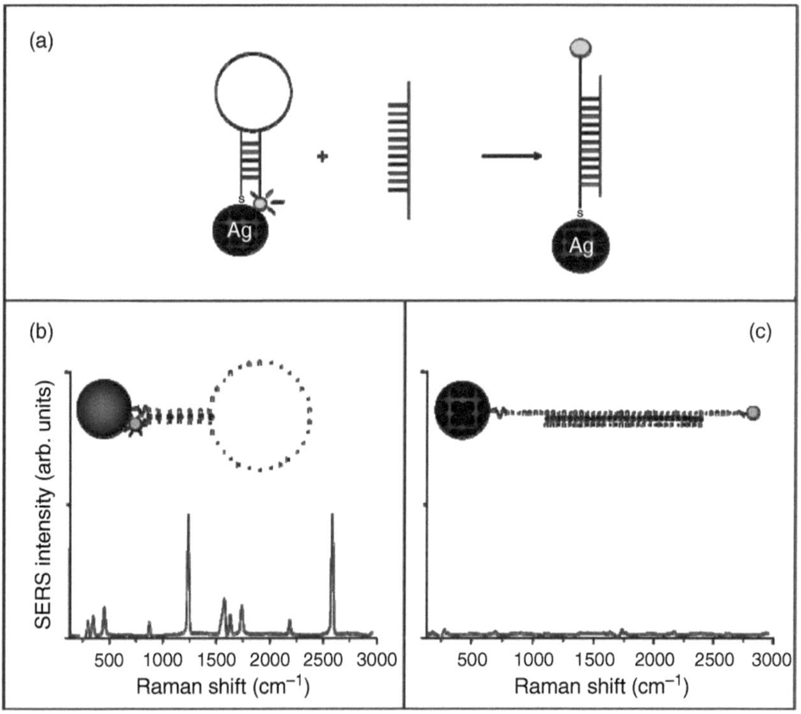

Figure 8.1 Operating Principle of a SERS Molecular Sentinel Nanoprobe. A SERS signal is observed when the MS probe is in the hairpin conformation (closed-state), whereas in the open state, the signal is diminished. [Adapted from Wabuyele and Vo-Dinh, 2005.]

the other end to form a SERS-MS nanoprobe. The metal nanoparticle is used as a signal-enhancing platform for the SERS signal associated with the label. The enhancement is primarily due to electromagnetic effects and chemical effects at the metal surface. Theoretical studies of electromagnetic effects have shown that the SERS enhancement, G factor, falls off as $G = [r/(r+d)]^{12}$ for a single analyte molecule located a distance, d, from the surface of a metal particle of radius, r (Kneipp *et al.*, 1997). Therefore, the electromagnetic SERS enhancement strongly decreases with increasing distance, due to the decay of a dipole over the distance $(1/d)^3$ to the fourth power, thus resulting in a total intensity decay of $(1/d)^{12}$. Because the Raman enhancement field decreases significantly away from the surface, a molecule must be located within a very close range (0–10 nm) of the nanostructure surface in order to experience the enhanced local field. Under normal conditions (i.e., in the absence of target genes), the hairpin configuration has the Raman label in contact or close proximity (<1 nm) to the nanoparticle, thus resulting in a strong SERS effect and indicating that no significant event has occurred (Figure 8.1b). However, when complementary target DNA is recognized and hybridized to the nanoprobes, the SERS signal of these MSs becomes significantly quenched, providing a warning sign of target recognition and capture (Figure 8.1c). Stated differently, the SERS nanoprobes play the role of MSs patrolling the sample solution with their warning light "switched on" when no significant event occurs. Then, when a target species is detected, the MSs extinguish their light, thus providing a warning sign.

8.2. History

Raman spectroscopy is based on vibrational transitions that yield very narrow spectral features characteristic of the investigated samples. Thus, it has long been considered as a valuable tool for the identification of chemical and biological samples as well as the elucidation of molecular structure, surface processes, and interface reactions. Despite these important advantages, Raman scattering applications often have limited application in biosensing due to the extremely poor efficiency of the scattering process. However, discoveries in the late 1970s indicated that the Raman scattering efficiency can be enhanced by factors of up to 10^6 when the sample is adsorbed on or near nanotextured surfaces of special metals such as silver, gold, and transition metals. The technique associated with this phenomenon is known as SERS spectroscopy.

The development of practical and sensitive devices for screening multiple genes related to medical diseases and infectious pathogens is critical for early diagnosis and improved treatments of many illnesses as well as for high-throughput screening (HTS) for drug discovery. To achieve the required level of sensitivity and specificity, it is often necessary to use a detection method that is capable of simultaneously identifying and differentiating a large number of biological constituents in complex samples. One of the most unambiguous and well-known molecular recognition events is the hybridization of a nucleic acid to its complementary target. Thus, the hybridization of a nucleic acid probe to its DNA (or RNA) target can provide a very high degree of accuracy for identifying complementary nucleic acid sequences.

Since we reported the first practical application of the SERS effect in analysis (Vo-Dinh, *et al.*, 1984), the development of SERS-active solid substrates has been an area of in-depth research in our laboratory. We have extensively investigated the SERS technology in the development of nanostructure-based SERS substrates (Vo-Dinh, 1998; Vo-Dinh *et al.*, 2007). These substrates consist of a plate having silver-coated dielectric nanoparticles (Figure 8.2) or isolated dielectric nanospheres (30 nm

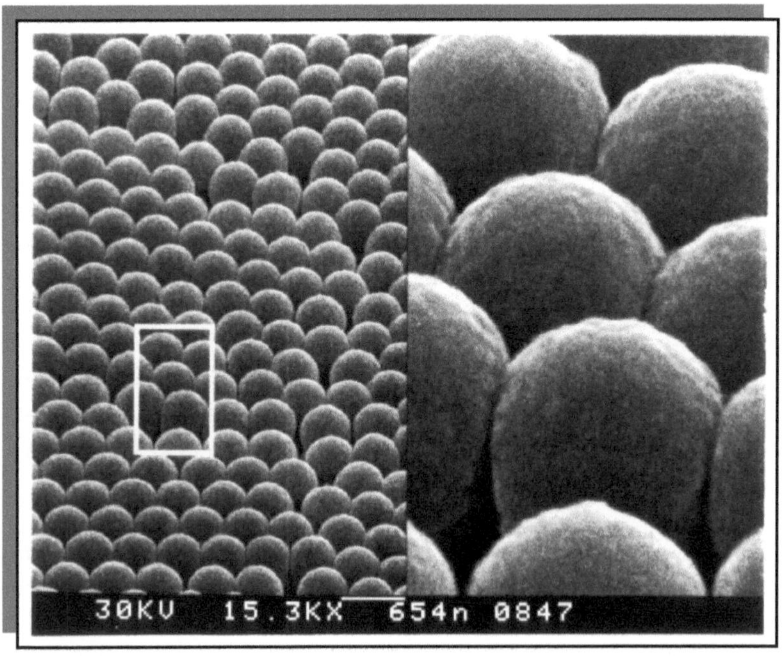

Figure 8.2 SERS-active Medium based on Nanospheres coated with silver. [Adapted from Garrell R.L., Analytical Chemistry, 61, 401A–411A, 1989.]

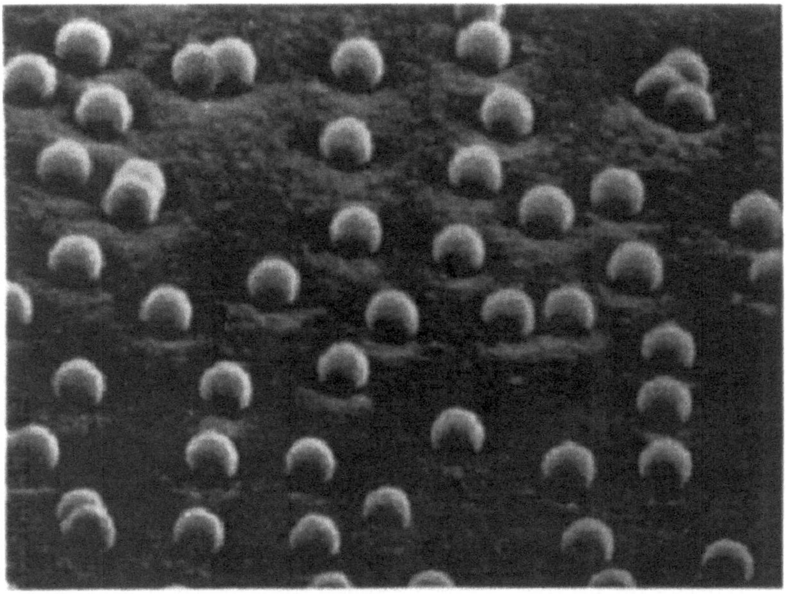

Figure 8.3 Half-Nanoshells consisting of nanospheres coated with silver.

diameter) coated with silver, producing half-nanoshells (Figure 8.3; Vo-Dinh, *et al.*, 1984; Alak and Vo-Dinh, 1987, 1988a, 1988b; Moody *et al.*, 1987; Bello *et al.*, 1989, 1990; Vo-Dinh, 1998). The fabrication process involves depositing nanoparticles (alumina, titanium dioxide, etc.) on a substrate and then coating the nanoparticle base with a 50–150 nm layer of silver via vacuum deposition. The nanoparticle-based SERS technology developed in our laboratory has led to a wide variety of analytical applications including sensitive detection of a variety of chemicals of environmental and toxicological significance, including polycyclic aromatic compounds (Vo-Dinh *et al.*, 1998), organophosphorus compounds (Alak and Vo-Dinh, 1987), chlorinated pesticides (Alak and Vo-Dinh, 1988a, 1988b), fungicides, and other growth inhibitors (Narayanan *et al.*, 1992, 1993). The addition of a polymer coating to a nanoparticle-based SERS substrate was proven to provide the combined advantages of increased detection selectivity and extended substrate shelf life (Pal *et al.*, 1995).

We first reported the use of SERS gene probe technology for DNA detection (Vo-Dinh *et al.*, 1994). Our laboratory has reported the development of SERS-based gene probes and has reported the selective detection of HIV DNA and the breast cancer gene BRCA1 using SERS-active substrates (Isola *et al.*, 1998; Allain *et al.*, 2002; Vo-Dinh *et al.*, 2002; Culha *et al.*, 2003). To demonstrate the SERS gene detection scheme, we used pre-coated SERS-active solid substrates, on which DNA probes were bound and directly used for hybridization. The effectiveness of this scheme for DNA hybridization and SERS gene detection involves several considerations. It is important that the unlabeled DNA fragment does not exhibit any significant SERS signal that might interfere with the label signal. The first step involves selection of a label that is SERS active and compatible with the hybridization platform. First, an ideal label should exhibit a strong SERS signal when used with the SERS-active substrate of interest. Second, the label should retain its strong SERS signal after being attached to a DNA probe. Finally, one should be able to detect the decrease in SERS signal from the labeled probe after hybridization. The use of SERS gene technology was demonstrated for the detection of the HIV gene sequence (Isola *et al.*, 1998) and for the breast cancer BCL2 gene (Culha *et al.*, 2003). The possibility of using Raman and/or SERS labels for extremely sensitive detection of DNA has been demonstrated. Sensitivity for single-molecule DNA detection via surface-enhanced resonance Raman scattering (SERRS) has been reported (Graham *et al.*, 1997).

In recent years, nanoparticles are finding an increasing use as fluorescence and Raman nanoprobes (Graham *et al.*, 1997; Storhoff *et al.*, 1999; Niemeyer, 2001; Pathak *et al.*, 2001; Allain *et al.*, 2002; Chan *et al.*, 2002; Dubertret *et al.*, 2002; Parak *et al.*, 2002; Vo-Dinh *et al.*, 2002; Cao *et al.*, 2003; Culha *et al.*, 2003; Doering *et al.*, 2003; Nam *et al.*, 2003; Docherty *et al.*, 2004). Binding of biological receptor molecules such as single-stranded DNA and antibodies to the surface of nanoparticles has facilitated the development of unique and highly selective nanoprobes for molecular recognition. Gold and silver nanoparticles that are functionalized with proteins, DNA, and antibodies have been used as nanoprobes for biological applications.

8.3. State of the art

We have recently demonstrated the SERS-MS concept for detecting the presence of specific DNA sequences (HIV gene) in a homogeneous assay (Wabuyele *et al.*, 2005). The HIV MS nanoprobes were carefully designed with a stem sequence that allowed the formation of stable hairpin structure at room temperature and incorporated a partial sequence for the HIV-1 isolate Fbr020, the reverse transcriptase (*pol*) gene. The HIV-1 MS nanoprobe (5′-HS-$(CH_2)_6$)-CCTATCACAACAAAGAGCATACATAGG GATAGG-R6G) consisted of a DNA hairpin probe modified with rhodamine 6G on the 3′ end and a thiol substituent at the 5′ end; the 5′ thiol was used for covalent coupling to the surface of silver nanoparticles. The underlined portions of the sequence represent the complementary arms of the MS. The silver colloidal nanoparticles were prepared according to the citrate reduction method, yielding homogenously sized colloids. A 115 bp sequence in the *gag* region of the HIV-1 genome was amplified by PCR, using forward and reverse primers in the gag region of the genome. After gel analysis, the PCR products were hybridized to the SERS-MS nanoprobes. The reaction volume was 40 μl and hybridization was performed at 55°C for 1 min. Surface-enhanced Raman scattering measurements were performed using a Renishaw InVia Raman system equipped with a 50 mW HeNe laser (Coherent, Model 106-1) emitting a 632.8 nm line used as the excitation source.

The presence of a partial sequence of the HIV1 *gag* gene in a homogenous solution was detected using the SER HIV-1 MS nanoprobes as depicted in Figure 8.4. In the absence of the target DNA (Figure 8.4, *top*), the hairpin conformation of the SERS-MS nanoprobes remained stable, resulting in a close proximity of the rhodamine 6G label with the surface of the silver nanoparticles (nanoenhancers), giving rise to a strong SERS signal from rhodamine 6G. However, in the presence of a complementary HIV-1 target sequence (Figure 8.4, *middle curve*), the SERS HIV-1 MS nanoprobes bound to the target DNA, causing the physical separation of the rhodamine 6G label from the surface of the silver nanoparticles and, as a result, the SERS signal was greatly diminished. On the other hand, the presence of non-complementary

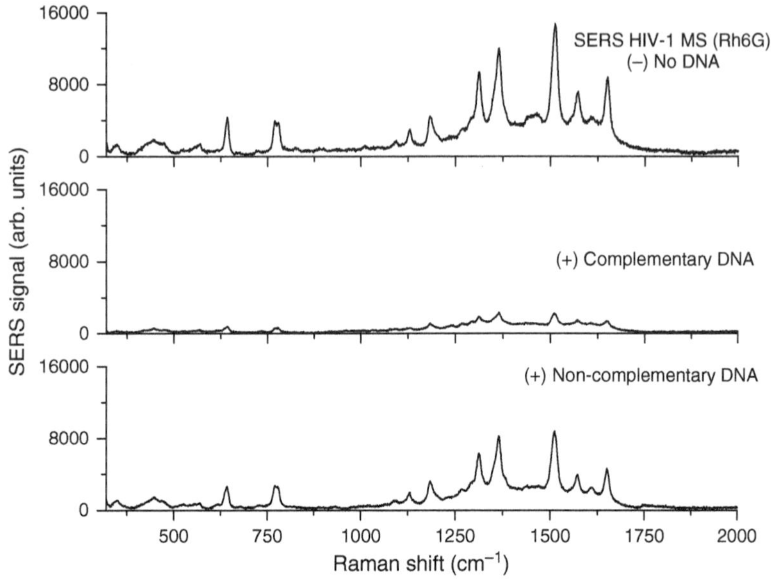

Figure 8.4 SERS spectra of HIV-1 SERS-MS nanoprobe with no target DNA sequence (Top curve) and in presence of a non-complementary DNA target sequence (Negative control: bottom curve) and a complementary HIV-1 DNA target (Positive diagnostic: middle curve). [Adapted from Wabuyele and Vo-Dinh, 2005.]

target DNA template (Figure 8.4, *bottom curve*) did not significantly affect the SERS signal, indicating that the hairpin loop structure of the SERS HIV-1 MS nanoprobes was not disrupted. It is noteworthy that we did observe a slight, but statistically significant, decrease in the SERS signal of the MS nanoprobes when non-complementary DNA template was added. The results of this experiment show that the MS technology can reliably detect specific nucleic acid sequences in a homogenous solution. Using the most intense SERS intensity band at $1521.06\,cm^{-1}$ as the marker band for rhodamine 6G, we estimated the SERS quenching efficiency to be ~75% upon hybridization of the SER HIV-1 MS nanoprobe to the HIV 1 DNA target. It is to be noted, however, that unbound MS nanoprobes do result in a detectable SERS signal thus reducing the sensitivity of the assay. With well-optimized conditions, this background signal can be lowered or reduced, resulting in an increased sensitivity. This result demonstrates the specificity and selectivity of SERS MS nanoprobes and their potential application in selective diagnostics.

8.4. Advantages and limitations for use in optical biosensing

Techniques that use fluorescence probes based on fluorescence energy transfer (e.g., MBs) have been widely applied with great success in biosensing assays. Nevertheless, the need for alternative, rapid, and selective assays has continued to encourage researchers to explore other technologies having comparable sensitivity as fluorescence but having unique and complementary advantages. Raman spectroscopy is an important analytical technique for chemical and biological analysis due to the wealth of information on molecular structures, surface processes, and interface reactions that can be extracted from experimental data. The spectral selectivity associated with the narrow emission lines and the molecular specific vibrational bands of Raman labels make it an ideal tool for molecular genotyping.

The ability to monitor a large number of molecular processes simultaneously is also an important feature in systems biology research. Although detection sensitivities achieved by luminescence techniques are adequate, the overlap of the relatively large bandwidth of fluorescence spectra limits the number of labels that can be used simultaneously. Therefore, alternative techniques with improved multiplex capability must be developed. Due to the narrow bandwidths of Raman bands, the multiplex capability of the SERS-MS probe is excellent in comparison to the other spectroscopic alternatives. For comparison purposes, consider the detection of crystal fast violet (CFV) dye in fluorescence and SERS. The spectral bandwidth of CFV label in the fluorescence spectrum is relatively broad (approximately 50–60 nm halfwidth), whereas the bandwidth of the SERS spectrum is much narrower (<0.5 nm or $3\,cm^{-1}$ halfwidth). In a typical Raman spectrum, a spectral interval of $3000\,cm^{-1}$ can provide 3000/3 or 10^3 available individual spectral "intervals" at any given time. Even allowing a deduction factor of 10 due to possible spectral overlaps, it should be possible to find 10–100 labels that can be used for labeling multiple probes simultaneously. This multiplex advantage is particularly useful in ultra-high-throughput analyses where multiple gene targets can be screened in a highly parallel multiplex modality. For example, MS oligonucleotides with 10–100 labels can be immobilized each element of a 10 000-element microarray to provide a 3D data cube including 10^5–10^6 (1 million) data points. For applications involving multiplexing, Raman spectroscopy has an advantage over fluorescence.

The SERS-MS method is a new concept at its early development stage and further work is needed to develop this technology into a mature bioanalytical tool. Research efforts are needed to improve the reproducibility of the technique by optimizing the experimental conditions. The ruggedness of the MS pin loop structures under various chemical conditions will need to be investigated. The design of MS nanoprobes having multiple labels for various DNA targets is currently under development.

In conclusion, the SERS-MS gene probes could offer a unique combination of performance capabilities and analytical features of merit for use in biosensing. The SERS-MS gene probes are safer than radioactive labels and have excellent specificity due to the inherent specificity of Raman spectroscopy. With Raman scattering, multiple probes can be much more easily selected with minimum spectral overlap. This "label-multiplex" advantage can permit analysis of multiple probes simultaneously, resulting in much more rapid DNA detection, gene mapping, and improved ultra-HTS (uHTS) of small molecules for drug discovery. The SERS-MS probes could be used for a wide variety of applications in areas where nucleic acid identification is involved.

8.5. Potential for improving performance or expanding current capabilities

Preliminary results have demonstrated for the first time the semi-quantitative nature of the SERS response signal from the MS nanoprobes with varying concentration of DNA targets (Wabuyele *et al.*, 2005).

However, there is still need for the technology to be refined for applications that require quantitative analysis. With respect to instrumentation, the possibility of using a single laser source for exciting multiple Raman labels will decrease the complexity of instrumentation in the case where multiple labels are used. The SERS-MS approach involving homogeneous assays greatly simplifies experimental procedures and could potentially be used in assays that require rapid, high-volume identification of genomic materials. Due to these unique properties, the SERS-MS approach could contribute to the development of the next generation of DNA diagnostic tools for molecular screening and molecular imaging.

High-throughput screening is at the core of the drug-discovery process. The rapid increase in the size of compound libraries as well as new targets emerging from the Human Genome Project require progress in uHTS systems. Currently, radiometric assays constitute between 20 and 50 % of all HTS performed worldwide, in spite of safety concerns and other disadvantages such as limited reagent stability, relatively long read-times, etc. Among the optical assay readouts used in biochemical-based assays (including fluorescence, absorbance, luminescence, and scintillation), fluorescence-based techniques are the most commonly used detection approaches; these include fluorescence intensity, fluorescence polarization, and FRET, etc. Raman spectroscopy also offers some distinct features that are important for *in situ* monitoring of complex biological systems. Following laser irradiation of a sample, the observed Raman shifts are equivalent to the energy changes involved in molecular transitions of the scattering species and are therefore characteristic of it. These observed Raman shifts, which correspond to vibrational transitions of the scattering molecule, exhibit very narrow linewidths. For these reasons, Raman spectroscopy has a great potential for multiplexing applications. That is, many organic compounds with distinct Raman spectra may be used as dyes to label biological macromolecules and each labeled molecular species will be able to be distinguished on the basis of its unique Raman spectra. This is not the case with fluorescence, because the broad spectral characteristics of fluorescence excitation and emission spectra result in large spectral overlaps if more than three to four fluorescent dyes are to be detected simultaneously.

Acknowledgments

This work was sponsored by the National Institutes of Health (R01 EB006201). The author acknowledges the contribution of M. Wabuyele.

References

Alak, A.M. and Vo-Dinh, T. (1987) *Anal. Chem.*, **59**, 2149.
Alak, A.M and Vo-Dinh, T. (1988a) *Anal. Chem.*, **61**, 656.
Alak, A.M. and Vo-Dinh, T. (1988b) *Anal. Chim. Acta*, **206**, 333.
Allain, L.R. and Vo-Dinh, T. (2002) *Anal. Chim. Acta*, **469**, 149.
Bello, J.M., Stokes, D.L., and Vo-Dinh, T. (1989) *Appl. Spectrosc.*, **43**, 1325.
Bello, J.M., Stokes, D.L., and Vo-Dinh, T. (1990) *Anal. Chem.*, **62**, 1349.
Cao, Y.C., Jin, R., Nam, J.-M. *et al.* (2003) *J. Am. Chem. Soc.*, **125**, 14676.
Chan, W.C.W., Maxwell, D.J., Gao, X. *et al.* (2002) *Curr. Opin. Biotechnol.*, **13**, 40.
Culha, M., Stokes, D., Allain, L.R., and Vo-Dinh, T. (2003a) *Anal. Chem.*, **75**, 6196.
Culha, M., Stokes, D.L., and Vo-Dinh, T. (2003b) *Expert Rev. Mol. Diagn.*, **3**, 669.
Docherty, T.F., Clark, M., McNay, G. *et al.* (2004) *Faraday Discussions*, **126**, 281.
Doering, W.E. and Nie, S. (2003) *Anal. Chem.*, **75**, 6171.
Dubertret, B., Skourides, P., Norris, D.J. *et al.* (2002) *In vivo* imaging of quantum dots encapsulated in phospholipid micelles. *Science*, **298**, 1759–62.
Gersten, J. and Nitzan, A. (1980) *J. Chem. Phys.*, **73**, 3023.
Graham, D., Smith, W.E., Linacre, A.M.T. *et al.* (1997) *Anal. Chem.*, **69**, 4703
Isola, N.R., Stokes, D.L., and Vo-Dinh, T. (1998) *Anal. Chem.*, **70**, 1352.
Kneipp, K., Wang, Y., Kneipp, H. *et al.* (1997) *Phys. Rev. Lett.*, **78**, 1667–70.
Lee, P.C. and Meisel, D. (1982) *J. Phys. Chem.*, **86**, 3391.
Li, Y.S., Cheng, J.C., and Coons, L.B. (1999) *Spectrochimica Acta Part A*, **55**, 1197.
Moody, R.L., Vo-Dinh, T., and Fletcher, W.H. (1987) *Appl. Spectrosc.*, **41**, 966.
Nam, J., Thaxton, S., and Mirkin, C.A. (2003) *Science*, **301**, 1884.
Narayanan, V.A., Begun, G.M., Bello, J.M. *et al.* (1993) *Analysis*, **21**, 107.
Narayanan, V.A., Begun, G.M., Stokes, D.L. *et al.* (1992) *J. Raman Spectrosco.*, **23**, 281.

Nie, S.M. and Emory, S.R. (1997) *Science*, **275**, 1102.
Niemeyer, C.M. (2001) *Angew. Chem. Int. Ed.*, **40**, 4128.
Otto, A. (1978) *Surf. Sci.*, **75**, L392.
Pal, A., Stokes, D.L., Alarie, J.P., and Vo-Dinh, T. (1995) *Anal. Chem.*, **67**, 3154.
Parak, W.J., Gerion, D., Zanchet, D. *et al.* (2002) *Chem. Mater.*, **14**, 2113.
Pathak, S., Choi, S.K., Arnheim, N., and Thompson, M.E. (2001) *J. Am. Chem. Soc.*, **123**, 4103.
Schatz, G.C. (1984) *Accounts Chem. Res.*, **17**, 370.
Silman, O., Bumm, L.A., Callaghan, R. *et al.* (1983) *J. Phys. Chem.*, **87**, 1014.
Storhoff, J.J. and Mirkin, C.A. (1999) *Chem. Rev.*, **99**, 1849.
Tarabara, V.V., Nabiev, I.R., and Feofanov, A.V. (1998) *Langmuir*, **14**, 1092.
Vo-Dinh, T. (1998) *Tr. Anal. Chem.*, **17**, 557.
Vo-Dinh, T. and Yan, F. (2007). In *Nanotechnology in Biology and Medicine* (T. Vo-Dinh, ed.) New York: Taylor and Francis Publishers.
Vo-Dinh, T., Allain, L.R., and Stokes, D.L. (2002) *J. Raman Spectrosc.*, **33**, 511.
Vo Dinh, T., Hiromoto, M.Y.K., Begun, G.M., and Moody, R.L. (1984) *Anal. Chem.*, **56**, 1667.
Vo-Dinh, T., Houck, K., and Stokes, D.L. (1994) *Anal. Chem.*, **66**, 3379.
Wabuyele, M. and Vo-Dinh, T. (2005) *Anal. Chem.*, **77**, 7810.
Zeeman, E.J. and Schatz, G.C. (1987) *J. Phys. Chem.*, **91**, 634.

Part II

Optical Biosensors: Tomorrow

Optical Biosensors: Today and Tomorrow (2nd Edition)
Frances S. Ligler and Chris Rowe Taitt (Editors)

Chapter 9

CAVITY RING-DOWN BIOSENSING

Peter B. Tarsa, Ph.D.[a] and Kevin K. Lehmann, Ph.D.[b]
[a]Massachusetts Institute of Technology, Cambridge, MA 02139, USA
[b]University of Virginia, Charlottesville, VA 22904, USA

Cavity ring-down spectroscopy (CRDS) is an established technique for gas sensing that is newly emerging in the field of optical biosensing. This approach measures changes in the rate of decay of light injected into an optical resonator and relates the change to optical loss along the length of the resonator. This principle has recently been adapted for use in liquids, providing a highly sensitive method for quantitative biosensing applications. Cavity ring-down spectroscopy, which can incorporate evanescent field sensing elements, has been demonstrated for the detection of near-infrared absorption in liquids and the scattering response of adherent cells. Such results can be extended for measurement of specific recognition and formation of ligand–receptor complexes at the sensor surface. Building on the high sensitivity that CRDS brings to molecular spectroscopy instrumentation, evanescent wave CRDS represents the current state of the art and holds significant potential for a wide range of biosensing applications.

9.1. Technical concept

The interaction of light and matter can be measured in a variety of ways, including molecular fluorescence, absorption, or scattering. Of the wide range of methods that are used to quantify such interactions, each employs a measure of the change in an incident signal that is related to the presence of an analyte. For example, absorption spectroscopy detects the attenuation of light as a function of wavelength-dependent molecular absorption. This attenuation is related to the absorbance of a sample, A, which is described by Beer's law:

$$A = \varepsilon CL \tag{9.1}$$

where ε is the molecular absorption coefficient, C is the analyte concentration, and L is the length of travel of the light through the sample. Variations in concentration linearly affect absorption, resulting in attenuation of an incident beam. This simple relationship between absorbance, concentration, pathlength, and the intrinsic properties of the sample has important implications for the range of molecular spectroscopy approaches. For instance, improved absorption sensitivity can be conveniently realized by increasing the sample cell pathlength. Similarly, very low concentrations can be measured in the mid-infrared or UV spectral regions, where molecular absorption coefficients are typically the largest. Combinations of these approaches have been adapted for atmospheric monitoring of trace quantities of small molecules with a great deal of success (King *et al.*, 2000; Thompson and Myers, 2002; Parkes *et al.*, 2003).

9.1.1. Cavity ring-down spectroscopy

The practical limitations associated with standard absorption spectroscopy techniques can be eased by incorporating a mirrored multipass cell in place of a typical sample cell. When the multipass cell is formed by highly reflecting mirrors, the interaction length can be extended by a factor of several hundred. In traditional designs, such as that of White (1942) and Herriott *et al.* (1964), each pass of the cell was spatially resolved on the input/output mirror, resulting in a volume that was quite large. Instead, when the light radiation is coupled into a stable optical cavity that refocuses the light back on itself on each pass, a very compact cell with an effective number of passes equal to $1/(1\text{-}R)$ (where R is the mirror reflectivity) can be formed. Using a CRDS arrangement such as that shown in Figure 9.1, an effective pathlength of nearly 100 km has

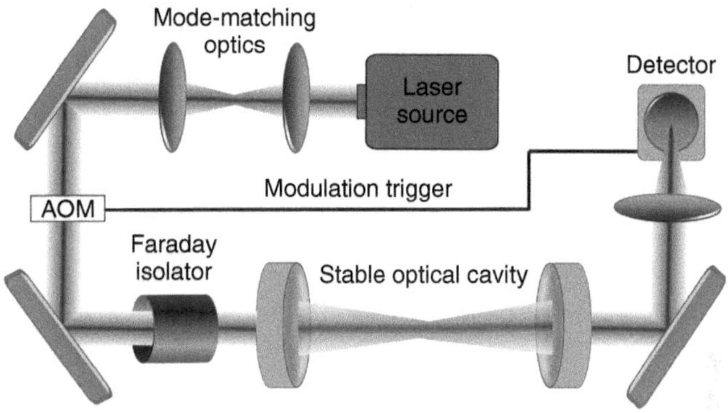

Figure 9.1 Typical CRDS arrangement. This device is based on a semiconductor laser source, which is efficiently coupled into the optical cavity through a mode-matching telescope. Before the cavity, a Faraday isolator prevents feedback from the reflective elements into the laser, ensuring the wavelength stability of the system. Also, before the cavity is an acousto-optic modulator (AOM), which allows the laser to be rapidly switched off when the cavity is fully excited. Upon extinguishing the excitation light, the detector measures the time for the cavity emission to reach $1/e$ of the maximum intensity, also known as the "ring-down time." Other approaches to CRDS use similar arrangements, often with substitution of the cavity mirrors for some other low-loss reflective component.

been realized with a physical cell only 1m in length using a diode laser coupled into a cavity formed by mirrors with a reflectivity of 99.999% (Dudek *et al.*, 2003).

In a cell with an effective pathlength of several kilometers, the resonating light trapped between the mirrors leaks out on a relatively long timescale, $\sim 3\,\mu s$ for each kilometer. By solving the differential rate equation for this resonator, the intensity of the light exiting the cell can be shown to decay as a single exponential (Lehmann and Romanini, 1996):

$$I(t) = I(0)\exp\left[\frac{-tc}{L}(1 - R + \varepsilon CL)\right] = I(0)\exp\left[\frac{-t}{\tau}\right] \qquad (9.2)$$

where $I(t)$ is the time-dependent light intensity leaving the cavity; $I(0)$ is the initial light intensity; c is the speed of light in the medium between the mirrors; C is the concentration of the absorbing species; L is the separation between the mirrors; and τ is the decay time constant of the cavity, or "ring-down" time, defined as the amount of time it takes for the intensity to reach $1/e$ of the initial intensity.

This equation can be rearranged to express the time constant in terms of the characteristic losses of the cavity and the losses due to absorption from a sample analyte:

$$\frac{1}{\tau} = \frac{1}{\tau_0} + \varepsilon C c \qquad \frac{1}{\tau_0} = \frac{c(1-R)}{L} \tag{9.3}$$

in which τ_0 is the intrinsic decay time of the cavity, typically measured in the absence of the sample. Typically on the order of several microseconds long, these cavity ring-down times are entirely independent of laser intensity yet directly related to absorption. Monitoring changes in these cavity decay times is the essence of CRDS, allowing observation of the absorption spectrum or changes in the sample concentration.

The sensitivity of CRDS arises in large part because of the low optical loss of cavities formed from modern optics. In addition, the cavity ring-down time constant is independent of fluctuations in the input light intensity, removing any dependence on laser source intensity, meaning that power fluctuations and external light contributions no longer factor into signal-to-noise considerations. Similarly, cavity ring-down times rely only on losses intrinsic to the cavity, relaxing unwanted contributions from external absorbers, such as in the air path. In practice, the long interaction distances that accompany long decay times produce the very high sensitivity that is characteristic of this technique. For example, if the per pass loss of the sample matches that of the cavity, which can be as low as a few parts per million, the rate of cavity decay is doubled. Because changes in the decay time on the order of 0.01–1% are easily detectable, a loss on the order of 10^{-3} of that of the cavity itself can be routinely measured.

9.1.2. Evanescent wave CRDS

The development of CRDS sensing has traditionally been limited to relatively clean gas samples because of the significant loss caused by interfering absorbers or particulate contaminants. Such factors have also limited the adaptation of CRDS for many biosensing applications, which are typically applied to complex aqueous mixtures. The incorporation of evanescent wave techniques with CRDS technology (Pipino *et al.*, 1997; Tarsa *et al.*, 2004a) provides a versatile platform from which to approach biosensing applications. By taking advantage of the evanescent wave formed at a total internal reflection interface, optical waveguides can be used to sample complex mixtures with minimal background contributions.

Evanescent wave sensors rely on the field that exists beyond a total internal reflection interface, Figure 9.2. This field, which extends

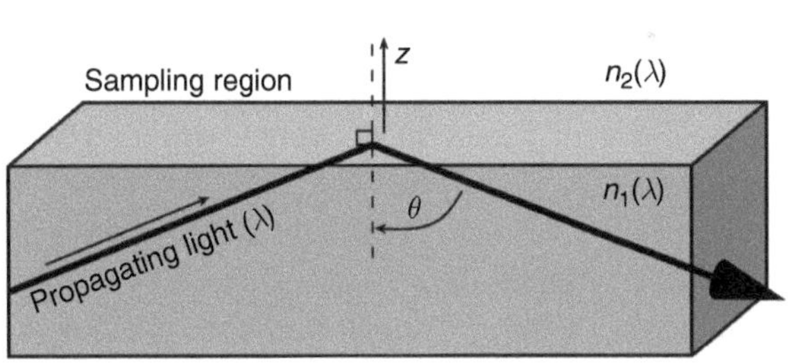

Figure 9.2 Evanescent wave spectroscopy. The key component of CRDS biosensing is the use of evanescent wave sensors, which allow sampling in high-loss matrices that were not previously accessible. This diagram, which is drawn for a slab waveguide, shows the operating principle of evanescent wave sensing. As light hits a total internal reflection interface, a non-propagating field extends normal to the reflection surface. This field decays exponentially over approximately the wavelength of the reflecting light and can interact with samples in close proximity to the surface. This effect is exploited by CRDS biosensing to minimize background losses that are present in complex biological mixtures. Such near-field sensing can be further combined with biochemical surface modifications that bind targeted analytes to the surface to create a highly sensitive and selective biosensor.

perpendicular to the reflection plane, decays exponentially over a distance, z, comparable to the wavelength of the propagating light:

$$E(z) = E_0 \exp\left[\frac{-2\pi n_1(\lambda)\, z}{\lambda}\sqrt{\sin^2\theta - \left(\frac{n_2(\lambda)}{n_1(\lambda)}\right)^2}\right] \tag{9.4}$$

where E_0 is the incident field, λ is the wavelength of the propagating light, θ is the angle of reflection, n_1 is the wavelength-dependent refractive index of the propagation medium, and n_2 is the wavelength-dependent refractive index of the external medium. While the evanescent field propagates no power into the external medium, the external medium can absorb or scatter light to decrease the intensity in the reflected light. Quantification of such power changes can be related to external interactions, including molecular absorption or particulate scattering, using standard spectroscopic instrumentation. Furthermore, sensitivity to these changes can be manipulated through customization of the reflection angle, the wavelength of the propagating light, and the relative refractive indices of the propagation medium and the surrounding environment.

Along with this flexibility in sensor design, adoption of evanescent wave techniques extends the range of sampling conditions accessible to spectroscopic monitoring. In many cases, this extension is accompanied by a loss in sensitivity that results from the shallow penetration depth of evanescent sensors. Decreased sensitivity can be a benefit, however, when sensing is approached in otherwise lossy media. For example, the susceptibility of traditional CRDS to optical loss previously limited its adaptation to liquid environments. With the incorporation of evanescent sensing elements in a CRDS resonator, the effects of background absorption and interfacial scattering are sufficiently minimized.

The first combination of CRDS with evanescent wave techniques was realized through the use of a monolithic prism resonator (Pipino *et al.*, 1997). This device demonstrated the ability to probe thin surface coatings adsorbed onto the surfaces of a total internal reflection interface. In this proof-of-principle device, measurement just beyond the surface of this interface is also accessible, providing sensitivity to biological samples or other immobilized species. Furthermore, variation of prism angle or excitation wavelength allows control over penetration depth, providing an adjustable parameter that can be customized to the sensing environment. A limitation of this original design is that the evanescent field is limited to a few isolated reflections (typically four) on each round trip. Recently, very high Q whispering gallery mode resonators have been developed (Grudinin *et al.*, 2006), where the light circulates around a cylindrical-shaped post with continuous exposure to the surrounding medium, greatly increasing the interaction with the analyte.

The implementation of prismatic elements for evanescent wave CRDS later led to the incorporation of other waveguides in CRDS arrangements. For example, adaptation of optical fiber cavities for evanescent wave CRDS led to the first practical liquid sensing and eventually the first CRDS biosensing of individual cells (Tarsa *et al.*, 2004b). Different approaches for optical fiber CRDS minimize scattering effects that contribute unacceptably high levels of optical loss in traditional mirror-based arrangements. An example, shown schematically in Figure 9.3, contains the entire resonant structure in a length of single-mode fiber, exposing discrete sections of fiber for distributed sensing applications. When employed for single-cell sensing, the cavity ring-down time is measured and related to the number of adhered cells by an expression similar to Eqn (9.5):

$$\frac{1}{\tau} = \frac{1}{\tau_0} + nQc \tag{9.5}$$

where n is the number of scattering cells, Q is a calibration constant related to the sensitivity of the sensing element, and c is the speed of light inside the sensing element. With this apparatus, biological sensing becomes accessible through particle scattering of the evanescent field surrounding the fiber. Because molecular absorption in the surrounding matrix can be minimized through design of the fiber-sensing regions, background effects can be significantly reduced without compromising sensitivity to biological scattering.

The variety of CRDS arrangements, ranging from traditional mirror-based cavities to fiber optic resonators, offers significant latitude in

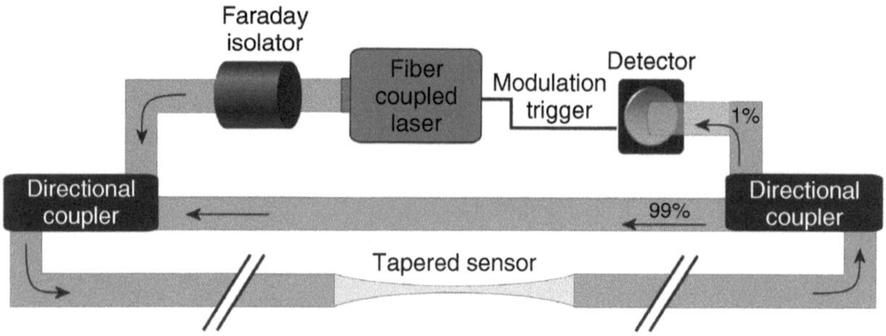

Figure 9.3 Evanescent wave CRDS. This device, which is constructed from several kilometers of single-mode fiber, allows CRDS sensing at the tapered sensing region while eliminating background contributions along the unmodified length of fiber. Similar to a standard CRDS arrangement, this device replaces the traditional mirrors with directional couplers and substitutes single-mode fiber for free space transmission. Because single-mode fiber selects the propagating mode near the laser output, mode-matching optics are not necessary. In evanescent wave CRDS, the relatively high coupling ratios of off-the-shelf tap couplers increase the per-pass loss of this resonator, though the intrinsic loss of the fiber dominates cavity loss over the resonant distance. Despite the relatively high loss of these components, this approach represents the first practical application of CRDS sensing to biological systems.

experimental design and sensing implementation. Experimental arrangements for localized quantification or distributed sensing can be realized, each with different benefits and limitations. By adapting these approaches for biosensing applications, CRDS holds significant potential for rapid detection of a range of biological specimens with both high selectivity and sensitivity. Currently, over 100 papers per year are being published that utilize CRDS or closely related techniques. Rapid development of the method and its applications, including biosensing, can be expected to continue.

9.2. History

The first CRDS devices were designed to measure low levels of optical loss in highly reflective mirror coatings (Anderson, 1984). By coupling a pulsed laser into a cavity formed by these mirrors, it was possible to measure changes in reflectivity on the order of parts per million. It was soon realized that the cavity could also be filled with an absorbing analyte, and absorptive loss could be measured at similar levels (O'Keefe and Deacon, 1988). This initial advance was applied to different systems of weakly absorbing gaseous absorbers, including vibrational absorption of several energy quanta or fundamental absorption of very low concentrations. The high specificity of molecular gas-phase spectroscopy complemented these approaches, yielding a spectral fingerprint in addition to quantitative information about concentration.

Despite the initial success of these first experiments, it was soon discovered that substantial improvements could be realized by replacing the pulsed laser sources with compact continuous wave semiconductor lasers (Romanini *et al.*, 1995 and 1997). The high stability of single-mode diode lasers allowed the selective excitation of the fundamental cavity mode, reducing contributions from higher order modes, which have different decay rates. Combined with beam conditioning optics, this mode-matching approach was shown to improve signal-to-noise by a factor of ~10–100 (Dudek *et al.*, 2003). In addition to substantial improvements in sensitivity, the implementation of continuous wave diode lasers led to the first commercial CRDS instruments (Tiger Optics, 2000). These devices, which can be operated at room temperature with minimal power requirements, were first used for sensitive monitoring of semiconductor process gases. However, an adaptable CRDS platform was soon expanded for health-related applications, such as real-time breath analysis (Dahnke *et al.*, 2001a), and homeland security, including explosive gas detection (Usachev *et al.*, 2001). In addition, CRDS has been used to make precise measurement of isotopic ratios of CO_2 (Crosson *et al.*, 2002) and methane (Dahnke *et al.*, 2001b), which have biological and environmental applications.

While most CRDS experiments have been done in the gas phase, some measurements on liquids have been reported. Xu *et al.* (2002) used an absorption cell inside an optical cavity that was tilted to Brewster's angle to greatly reduce reflection losses. Snyder *et al.* (2003) extended this design to realize Brewster's angle simultaneously for both the glass–air and glass–water interfaces and demonstrated that such a cell can be used for high performance liquid chromatography (HPLC) detection of compounds labeled with strongly absorbing dyes. This work was recently extended by Bechetel *et al.* (2005) with significantly improved sensitivity. Hallock *et al.* (2002) showed that low-loss cavities could be constructed with the dielectric mirrors directly in contact with the solvent and absorption of solvated species detected. Similarly, Bahnev *et al.* (2005) built a miniaturized CRDS cell for HPLC detection, although one important limitation of these measurements is that typical liquid-phase electronic absorption bands are wider than the high reflective band of available low-loss, dielectric mirrors. This does not limit sensitivity, as the sample can be flushed through the cell, but it does limit selectivity as spectral shape of the absorption cannot be used to establish the absorbing species.

While these first instruments were operated in much the same way as traditional spectrometers, new generations of CRDS-based devices were soon invented. The first developments took the form of monolithic prism-based sensors, in which the traditional linear cavity was replaced by a hexagonal or octagonal prism (Pipino *et al.*, 1997). By evanescently coupling light into and out of this resonant structure, CRDS operation was realized both for characterization of the prism material itself and for detection of absorbing species at the prism interfaces. Everest *et al.* (2006) recently used this technique to monitor hemoglobin adsorption on silica. This implementation of evanescent wave CRDS was later extended into the development of a linear prism-based resonator that replaced the traditional coated mirror reflectors with broadband roof-angle prism retroreflectors (Engel *et al.*, 1999). Instead of exploiting the evanescent wave sensing structure, this embodiment took advantage of the wide wavelength range that could be passed through the prism reflectors, allowing laser multiplexing and multiple analyte detection in a single instrument. Other inventions exploited the evanescent wave sensing capabilities of the monolithic prism resonator. The first such approaches incorporated a prismatic element in the middle of a mirror-based resonator (Shaw *et al.*, 2003). By angling the prism entrance interface at the Brewster's angle, where reflection losses are significantly reduced, this approach samples binding events at the external interface of the cavity-integrated element. In addition, the use of whispering gallery mode resonators offers the promise of highly compact and sensitive biosensors based on the CRDS principle (Grudinin *et al.*, 2006). Savchenkov *et al.* (2006) have developed "white light" whispering gallery mode microresonators that have very high Q values (low loss) but have essentially a continuous spectrum, allowing the full sample spectrum to be measured. Furthermore, these prism-based approaches do not have the wavelength limits of dielectric mirrors.

9.3. State of the art

The limited sensitivity of the approach to evanescent wave CRDS that samples at a single discrete point only once per pass was dramatically extended by the invention of optical fiber CRDS. The first demonstration that optical fibers could be used to construct a CRDS resonator explored the effects of cavity stability and optical fiber loss in a linear resonator. By coating the ends of a length of optical fiber, von Lerber and Sigrist (2002) showed that a traditional CRDS resonator could be constructed using a length of fiber as a propagation medium. Similarly, Gupta and O'Keefe (2002) showed that low-loss optical resonators could be constructed using Bragg reflectors built directly into the optical fiber. These first optical fiber CRDS instruments were soon extended with two different approaches using an extended fiber ring as a resonator medium, avoiding the need for reflectors that typically limit bandwidth. In both arrangements, the physical pathlength of the resonator is greatly extended, lengthening cavity ring-down times and resulting in a practical device for measuring a wide variety of new parameters. In the earliest work, the use of a fiber taper greatly increased evanescent field exposure, providing enhanced control over sensitivity to external species. For example, the first implementation of evanescent field sensing with optical fiber CRDS was demonstrated with the spectral resolution of 1-octyne surrounding a modified sensing region of the fiber resonator (Tarsa *et al.*, 2004a). It was also shown that the device could be modified to measure tiny levels of strain on a fiber taper (Tarsa *et al.*, 2004c). In an analogous design, Loock and coworkers incorporated a small microgap into the fiber ring to measure small volumes of liquid (Tong *et al.*,

2003) and have recently reviewed potential applications to microliquid samples (Loock, 2006). These two approaches represented the first practical implementations of optical fiber CRDS that led directly to CRDS biosensing.

Along with these optical fiber CRDS approaches, another area of considerable recent activity with important implications for biosensing is the development of broad bandwidth CRDS. Such advances are in contrast to early CRDS experiments, which used a tunable light source to tune through an absorption spectrum or to monitor the changes in a narrow spectral window as a function of time. The incorporation of broadband sensing capabilities will enable many biosensing applications that require parallel detection across a broad spectral region.

With the ability to perform CRDS absorption measurements in liquid matrices, optical fiber CRDS relaxed previous limitations that prevented the practical measurement of biological samples and liquids. For example, the use of evanescent wave CRDS to measure biological samples was first shown by the resolution of single cells adhered to the surface of a fiber resonator, shown in Figure 9.4. Such single throughput measurement can be further enhanced with broadband scanning technologies, the first demonstration of which was by Engeln and Meijer (1996), who showed that CRDS could be combined with a Fourier transform spectrometer to determine the cavity decay at each mirror spacing. Scherer (1998) further showed that a grating spectrograph could be used to disperse the light as a function of wavelength in one axis of a CCD camera while simultaneously sweeping the signal across along the slit of the spectrograph with a rotating mirror to temporally disperse the cavity decay. In both of these earlier works, dye lasers with modest bandwidth were used. Ball and Jones (2003) expanded this approach with a CCD camera that allows the charge to be moved up and down the chip, allowing temporal dispersion without moving parts and allowing signal averaging on the silicon. Thompson *et al.* (2007) have shown that light emitting diodes can be used as broad bandwidth sources for CRDS measurements, and Fiedler *et al.* (2003) have used incoherent arc lamps. Building on these approaches, P. Johnston and K.K. Lehmann (unpublished data, 2007) are developing a white light spectrometer that

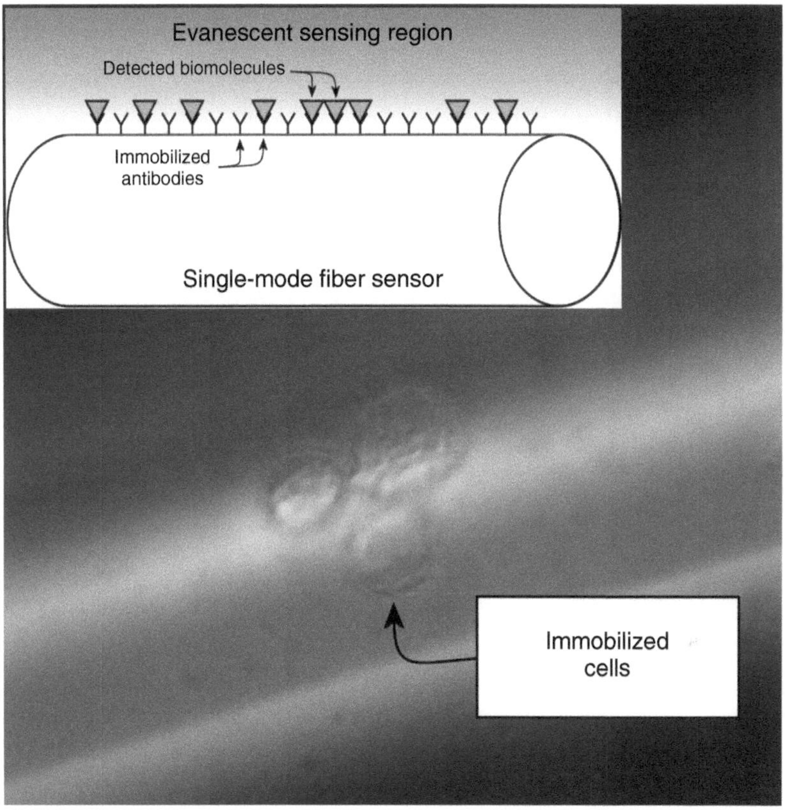

Figure 9.4 The CRDS biosensing. By chemically modifying the optical fiber surface in the sensing region (inset), CRDS biosensing can be realized for samples ranging from small molecule absorbers to individual cells. In this work, adapted from Tarsa *et al.* (2004b), mammalian breast cancer cells were immobilized through a non-specific poly-D-lysine coating. The CRDS signals varied linearly with cell adhesion count, and sensitivity was as low as a single cell. Additional coatings can be easily incorporated into this design, allowing, for example, nucleic acid concentration from cell lysates or specific protein immobilization through antibody coatings.

combines a cavity based upon Brewster angle prisms discussed above and white light super continuum generation in a photonic crystal fiber. The latter allows radiation from ~500 nm to 2 μm to be generated from a low-power (30 μJ per pulse)-diode pumped 1.06 μm laser.

The combination of these broadband techniques and temporal resolution approaches with CRDS technology suggests a strong future for CRDS biosensing. The high sensitivity of CRDS may bring new advances to more common biological approaches, potentially allowing researchers to watch molecular absorption in tiny samples or quantify trace components of complex biological mixtures. In addition, the incorporation of technologies already in use for optical biosensing will help drive CRDS biosensing research forward. These approaches, such as chemical surface modification, can be combined with advances in fiber optic and laser technology to advance CRDS biosensing beyond the current state of the art.

9.4. Potential for improving performance or expanding current capabilities

The relatively recent adaptation of CRDS to biosensing applications leaves significant room for development, both for improved selectivity and for expanded sensitivity. Because CRDS biosensing devices rely on evanescent wave techniques, advanced technology developed for traditional evanescent sensors can be readily adapted for CRDS instrumentation. Such advances in sample concentration and separation can be incorporated with CRDS sensors, providing new approaches for highly sensitive biosensors. Alternatively, intrinsic improvements in the CRDS sensors can enhance performance by exploiting rapidly advancing laser and waveguide technology. The parallel approach of improving sensor construction and adapting surface chemistry will ultimately lead to new generations of CRDS devices for real-time biosensing in a range of detection environments. The fiber loop with a small gap developed by Loock and coworkers as well as the microcavities developed by Bahnev and coworkers can likely be adopted to extend microfluidic technology, which is already highly developed for biological sensing applications.

Chemical modification of optical waveguide surfaces provides a customizable approach to targeting either sample concentration or signal enhancement. Different surface chemistries, which incorporate considerable flexibility for measuring a variety of biological samples, can be employed based on the expected sample matrix composition, providing *in situ* separation of individual species from complex media or concentration of trace species at the sensor surface. Similarly, analyte specificity can be tuned with methods ranging from surface charge manipulation to the incorporation of surface-bound monoclonal antibodies or bacteriophage-displayed recognition species. This wide range of approaches can be used to develop sensors that are specifically tailored to the sensing environment.

Highly specific surface coatings can be combined with advances in waveguide materials and laser emitters to dramatically improve CRDS biosensing. The development of new fiber materials will permit single-mode transmission at shorter visible wavelengths, where optical scattering is enhanced and electronic transitions in biological samples dominate, and at longer infrared wavelengths, where vibrational transitions are strong. In addition, inexpensive laser diodes at newly accessible wavelengths can be interfaced with single-mode fibers to increase the versatility of CRDS biosensing. Such approaches are complemented by broadband white light-scanning CRDS, discussed above, bringing the high sensitivity of CRDS to the diverse range of biological sensing problems.

References

Anderson, D.Z., Frisch, J.C., and Masser, C.S. (1984) *Appl. Optics*, **23**, 1238.
Bahnev, B., van der Sneppen, L., Wiskerke, A.E. *et al.* (2005) *Anal. Chem.*, **77**, 1188.
Ball, S.M. and Jones, R.L. (2003) *Chem. Rev.*, **103**, 5239.
Crosson, E.R., Ricci, K.N., Richman, B.A. *et al.* (2002) *Anal. Chem.*, **74**, 2003.
Dahnke, H., Kleine, D., Hering, P., and Murtz, M. (2001a) *Appl. Phys. B.*, **72**, 971.
Dahnke, H., Kleine, D., Urban, C. *et al.* (2001b) *Appl. Phys. B.*, **72**, 121.
Dudek, J.B., Tarsa, P.B., Valesquez, A. *et al.* (2003) *Anal. Chem.*, **75**, 4599.
Engel, G., Yan, W.B., Dudek, J.B., Lehmann, K.K., and Rabinowitz, P. (1999). In *Laser Spectroscopy XIV Int. Conf.* (R. Blatt, J. Eschner, D. Leibfried, and F. Schmidt-Kaler, eds), p. 314.
Engeln, R. and Meijer, G. (1996) *Rev. Sci. Instr.*, **67**, 2708.

Everest, M.A., Black, V.M., Haehlen, A.S. *et al.* (2006) *J. Phys. Chem. B.*, **110**, 19461.
Fiedler, S.E., Hoheisel, G., Ruth, A.A., and Hese, A. (2003) *Chem. Phys. Lett.*, **382**, 447.
Grudinin, I.S., Ilchenko, V.S., and Maleki, L. (2006) *Phys. Rev. A.*, **74**, 063806.
Gupta, M., Jiao, H., and O'Keefe, A. (2002) *Optics Lett.*, **27**, 1878.
Hallock, A.J., Berman, E.S.F., and Zare, R.N. (2002) *Anal. Chem.*, **74**, 1741.
Herriott, D., Kogelnik, H., and Kompfner, R. (1964) *Appl. Opt.*, **3**, 523.
King, M.D., Dick, E.M., and Simpson, W.R. (2000) *Atmosph. Env.*, **34**, 685.
Lehmann, K.K. and Romanini, D. (1996) *J. Chem. Phys.*, **105**, 10263.
von Lerber, T. and Sigrist, M.W. (2002) *Appl. Optics*, **41**, 3567.
Loock, H.P. (2006) *TrAC-Trends Anal Chem.*, **25**, 655.
O'Keefe, A. and Deacon, D.A.G. (1988) *Rev. Sci. Instr.*, **59**, 2544.
Parkes, A.M., Linsley, A.R., and Orr-Ewing, A.J. (2003) *Chem. Phys. Lett.*, **377**, 439.
Pipino, A.C.R., Hudgens, J.W., and Huie, R.E. (1997) *Chem. Phys. Lett.*, **280**, 104.
Romanini, D., Gambogi, J., and Lehmann, K.K. (1995) Talk RH06, International Symposium on Molecular Spectroscopy (June 15), http://molspect. chemistry. ohiostate.edu/symposium_50/sRH.html
Romanini, D., Kachanov, A.A., and Stoeckel, F. (1997) *Chem. Phys. Lett.*, **270**, 538.
Savchenkov, A.A., Matsko, A.B., and Maleki, L. (2006) *Optics Lett.*, **31**, 92.
Scherer, J.J. (1998) *Chem. Phys. Lett.*, **292**, 143.
Shaw, A.M., Hannon, T.E., Li, F.P., and Zare, R.N. (2003) *J. Phys. Chem. B.*, **107**, 7070.
Snyder, K.L. and Zare, R.N. (2003) *Anal. Chem.*, **75**, 3086.
Tarsa, P.B., Rabinowitz, P., and Lehmann, K.K. (2004a) *Chem. Phys. Lett.*, **383**, 297.
Tarsa, P.B., Brzozowski, D.M., Rabinowitz, P., and Lehmann, K.K. (2004c) *Optics Lett.*, **29**, 1339.
Tarsa, P.B., Wist, A.D., Rabinowitz, P., and Lehmann, K.K. (2004b) *Appl. Phys. Lett.*, **85**, 4523.
Thompson, J.E. and Myers, K. (2007) *Meas. Sci. Technol.*, **18**, 147.
Thompson, J.E., Smith, B.W., and Winefordner, J.D. (2002) *Anal. Chem.*, **74**, 1962.
Tong, Z.G., Jakubinek, M., Wright, A. *et al.* (2003) *Rev. Sci. Instr.*, **74**, 4818.
Usachev, A.D., Miller, T.S., Singh, J.P. *et al.* (2001) *Appl. Spectr.*, **55**, 125.
White, J.U. (1942) *J. Opt. Soc. Am.*, **32**, 285.
Xu, S.C., Sha, G.H., and Xie, J.C. (2002) *Rev. Sci. Instr.*, **73**, 255.

Optical Biosensors: Today and Tomorrow (2nd Edition)
Frances S. Ligler and Chris Rowe Taitt (Editors)

Chapter 10

CANTILEVER BIOSENSORS

Mar Alvarez, Ph.D., Kirill Zinoviev, Ph.D., Miguel Moreno, Ph.D., and Laura M. Lechuga, Ph.D.
Biosensors Group, Centro Nacional de Microelectronica (CNM), Consejo Superior de Investigaciones Científicas (CSIC), Spain

The fabrication of miniaturized and integrated devices as micro- and nano-electro-mechanical systems (MEMS and NEMS) has provided the development of an innovative family of biochemical sensors based on transducers involving mechanical phenomena. Biosensors based on microcantilevers have become a promising tool for directly detecting biomolecular interactions with great accuracy, especially when an optical read-out scheme is applied. The number of applications of these sensors has shown a fast growth in diverse fields, such as genomic, proteomic, environmental, or food quality control, being a promising alternative to the currently exploited biosensor techniques.

10.1. Technical concept

10.1.1. Introduction

This chapter describes the application of nano- and micro-electro-mechanical systems (NEMs and MEMs), and more specifically micro-cantilever structures, as transducers for highly sensitive biosensors. In these devices, named as "nanomechanical biosensors," a biomolecular interaction produces a change in the mechanical behavior of the transducer (a movement at nanometer scale), which can be measured and analyzed in real time. Microcantilevers translate the molecular recognition of biomolecules into a nanomechanical motion that is commonly coupled to an optical read-out system. A rectangular beam, clamped at one end, is the most used transducer as it is the simplest mechanical structure that can be easily batch-fabricated at micrometer scale (Kim *et al.*, 2006). Microcantilevers can be fabricated in arrays of ten to thousands, and for that reason, they are a promising alternative to current biochips as they could permit the parallel, fast, and real-time monitoring of thousands of analytes (proteins, pathogens, DNA strands, etc.) without the need for labeling (Baller and Fritz, 2004).

Nanomechanical sensors are able to detect analytes with picomolar sensitivity and they have the ability for discerning single-base variations in DNA strands. Recently, microcantilever-based biosensors have arisen as a competitive biosensor alternative for measuring, in a direct way, extremely low forces and masses. When fabricated at the nanoscale (*nanocantilevers*), the sensitivity goes up and expected limits of detection are in the femtomole-to-attomole range with the possibility for detection at the single-molecule level in real time.

In this chapter, the main aspects regarding the physics of microcantilever sensors will be described as well the optical read-out techniques. We will review the state-of-the-art, and we will discuss the prospective future directions of this new family of biosensors.

10.1.2. Working principle

Nanomechanical sensors are derived from the microfabricated cantilevers used in atomic force microscopy (AFM) and are based on the bending or resonance change induced in the cantilever when, for example, a biomolecular interaction takes place on one of its surfaces. The cantilever response will depend on its mechanical properties, which are determined mainly by their spring constant and resonance frequency. Both parameters depend on the cantilever material and its geometry.

The spring constant, k, and resonance frequency, f_0, for a rectangular cantilever clamped at one end are given by

$$k = \frac{Ewh^3}{4L^3}$$

$$f_0 = \frac{1}{2\pi}\sqrt{\frac{k}{m}}$$

where E is the Young's modulus, m is the cantilever effective mass, w is the width, h is the thickness, and L is the length, respectively. Interaction or binding of molecules to one of the cantilever surfaces may lead to a change in the angle of cantilever bending or a shift in its resonance frequency. The detection of these responses is usually referred as *static* and *dynamic* modes of operation, respectively. But the mechanical response is also sensitive to different factors, such as temperature, heat, electromagnetic field, stress, and mass. These transducers can be employed for several applications besides biosensing. Figure 10.1

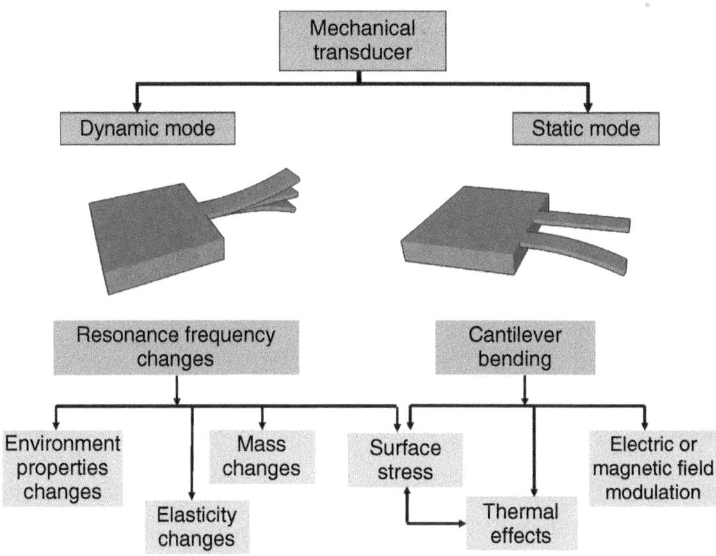

Figure 10.1 Diagram of the two nanomechanical response methods: static and dynamic; and the properties that could be measured with each of them.

shows a diagram with both transducer methods and its relation with the properties that could be measured.

In the sensors working in the *static* mode, the bending arises as a consequence of a surface stress change induced by any molecular reaction, which takes places on only one of the cantilever surfaces. The induced surface stress change could be positive or negative, depending on the surface deformation generated (Shuttleworth, 1950). The cantilever deformation depends mainly on the forces involved in the bioreaction process and is not directly related to the receptor–ligand binding energy. These forces could arise from the bond strength between the receptor and the surface (lineal response) and from the intermolecular interactions between neighboring molecules (non-lineal response). Although the factors and phenomena responsible for this change are still not fully understood, forces coming from electrostatic, steric, and van der Waals interactions, changes in the surface hydrophobicity or conformational changes of the adsorbed molecules could play an important role (Hagan *et al.*, 2002).

The easiest and most extended model to study the surface stress produced on cantilevers is based on the work by Stoney (1909). This model relates the total surface stress change between the top and the bottom sides ($\Delta\sigma_1 - \Delta\sigma_2$) with the cantilever free end displacement, Δz, the Young's modulus, E, the Poisson coefficient, ν, and the cantilever length, L, and thickness, h, represented by

$$\Delta\sigma_1 - \Delta\sigma_2 = \frac{Eh^2}{3L^2(1-\nu)}\Delta z$$

For sensing biomolecular interactions in the *static mode*, only one surface of the microcantilever must be previously biofunctionalized. This can be a complex task when working with cantilever arrays.

In contrast to the static case, the *dynamic* mode does not require the functionalization of only one cantilever surface, as the resonance frequency depends on the total mass adsorbed on both sides. This mass could produce, at the same time, a change in the cantilever spring constant, affecting the final resonance frequency shift (Lu *et al.*, 2005;

Gupta *et al.*, 2006). In this mode, very high sensitivities can be obtained (in the attogram range), superior to those of other similar mass detectors, such as the quartz crystal microbalance (QCM) (Ilic *et al.*, 2004). The microcantilever resonator is also characterized by the quality factor (Q), which quantifies the energy dissipation and is defined as the ratio between the mechanical energy accumulated and dissipated per vibration cycle. The dissipative mechanisms could be both internal and external; the external damping is the dominant factor when working in air or liquids. The quality factor determines the frequency resolution of the system (Lavrik *et al.*, 2004):

$$\Delta f = \sqrt{3}\frac{f_0}{Q}.$$

Operating in liquids, the resonance frequency and the quality factor shift toward much lower values than in air, due to the damping effect of the viscous surround. The quality factor in liquids could be up to 100-fold lower than in air, reducing the frequency resolution. There are different ways to overcome this limitation: by fabricating cantilevers with higher resonance frequency and quality factors, by measuring the resonance frequency in air under controlled humidity before and after the biochemical recognition, by working with higher order vibration modes, or by using external excitations, among others. For the above reasons, this way of operation is more difficult to implement and most of the cantilever biosensors are based on the static mode.

10.1.3. Beam design and sensitivity

Typically, the cantilever dimensions range from tens to hundred of micrometers in length, some tens of micrometers in width, and hundreds of nanometer in thickness. In general, they are made of silicon, silicon nitride, polymers, or piezoelectric materials. The microcantilever properties and sensitivity are determined by the cantilever dimensions and material (Young's modulus and Poisson coefficient). The selection of the appropriate characteristics depends on the working detection method (static or dynamic), the final application, and the available fabrication technology. In general, smaller spring constants give softer cantilevers, which are more sensitive to bending. Making thinner and longer cantilevers could improve the sensitivity, but they are also more fragile and therefore more difficult to handle. In addition, the thermomechanical noise increases.

As an example, the standard silicon microfabrication technology allows fabricating micrometer-sized cantilevers with a high length:thickness ratio in a reproducible and inexpensive way. However, the noise arising from the cantilever thermal motion could limit this ratio value (Alvarez *et al.*, 2006). On the other hand, reducing the cantilever length and thickness results in a sensitivity increase for the dynamic cantilever sensors, reducing the sensitivity for the surface stress sensors. Polymers, such as SU-8, have a lower Young's modulus than silicon and could be more sensitive for static deflection measurements (even if they do not have a high length:thickness ratio); their fabrication process is relatively inexpensive, fast, and reliable.

The final sensitivity could be determined by previously modeling the system to determine the range of dimensions needed for a specific material and working method. Modification of the microcantilevers' shape could also improve the sensor sensitivity, but the final sensitivity will also depend on the mechanical variabilities between microcantilevers in one array, the resolution of the detection set-up, and the total capacity for integration.

10.1.4. Detection method

A read-out system capable of monitoring changes with subnanometer resolution is crucial for detecting the nanomechanical motion induced by the biochemical recognition process. In addition to supplying high sensitivity and accuracy, the signal read-out is critical in the real-time measurements acquisition and, in the final biosensor, integration of microsystems. To avoid the influence of external factors, such as non-specific binding or temperature changes, a reference cantilever is typically used. For that reason, and for detecting several analytes at the same time, the read-out schemes have to enable the use of arrays of microcantilevers.

There are several techniques suitable for the cantilever response read-out: optical beam deflection, piezoresistivity, piezoelectricity, interferometry, and capacitance are among the most important (Lavrik *et al.*, 2004; Ziegler, 2004). However, the majority of the biochemical applications carried out with cantilever-based sensors are based on the *optical beam deflection method*, due to its high sensitivity. A variation of this optical method has been recently developed using an array of a new type of optical waveguide microcantilevers (Zinoviev *et al.*, 2007). These cantilevers act as a waveguide for conducting light, and it is possible to collect the exit light with a second waveguide or with a photodetector. In this case, the cantilever bending is related to the reduction in intensity of the collected light with respect to the input light.

10.1.4.1. Optical beam deflection

The *optical beam deflection* method is simple to implement and shows a linear response with sub-angstrom resolution. Movement of the cantilever's free end is detected by measuring the reflected laser beam displacement into a position-sensitive photodetector (PSD, Figure 10.2). The laser beam displacement over the photodetector, X, is related with the cantilever free end deflection, Δz, by a simple algebraic equation

$$\Delta z = \frac{XL}{2D}$$

where L is the cantilever length and D is the photodetector–microcantilever distance.

The read-out implementation of array platforms is technologically challenging, as it requires an array of laser sources with the same number of elements as the cantilever array. The laser's displacement could be measured using an array of photodetectors, adding alignment complications, or using just one photodetector and sequentially switching on and off each laser source to avoid the overlapping of the reflected beams (Lang *et al.*, 1998). Using a CCD camera and an array of microcantilevers with paddles at its end (Yue *et al.*, 2004) or a scanning laser source to sequentially illuminate each microcantilever (Alvarez *et al.*, 2004) are new ways to overcome the problem of overlap.

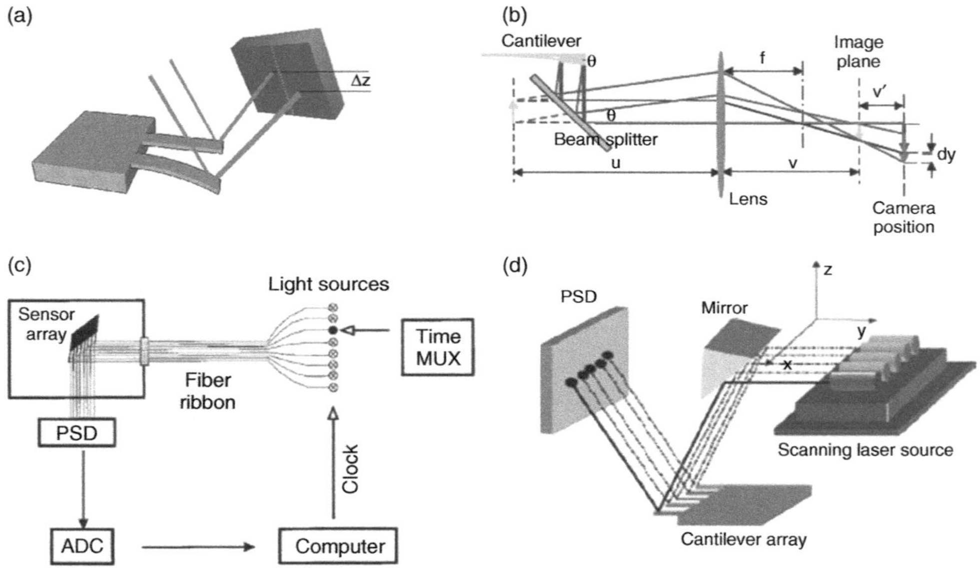

Figure 10.2 (a) Optical beam deflection read-out configuration. (b) Ray optics principles underlying optical detection of cantilever motion by using a CCD camera. (Reproduced with permission from TechScience Press [Yue *et al.*, 2004]). (c) Schematic of the quasi-simultaneous read-out of eight sensors achieved by time-multiplexing (MUX) of eight light sources which are guided by an optical fiber-ribbon onto the sensor array located in the analysis chamber. (Reproduced with permission from H.P. Lang [Lang *et al.*, 1998]. Copyright 1998, American Institute of Physics). (d) Scheme of the sequential illumination by using a scanning laser source (Alvarez *et al.*, 2004).

10.1.4.2. Optical waveguide microcantilever

This technique, based on butt-coupled optical waveguides, was proposed as a new alternative to solve some of the limitations of optical beam detection, the most serious of which is integration in array platforms. The use of a cantilever as a waveguide has been demonstrated by different groups (Wu and Frankena, 1992; Ollier *et al.*, 1999; Budakian and Putterman, 2002; Wang *et al.*, 2002; Xu *et al.*, 2005; Zinoviev *et al.*, 2006b; Nordin *et al.*, 2007). The sub-micron thickness of the cantilevers provides a high sensitivity, while the overall dimensions help to miniaturize the device and make it suitable for further integration in lab-on-chip applications. A schematic view of the optical sensor is shown in Figure 10.3. The device can be operated in both the visible and the infrared ranges.

The "heart" of the sensor is an optically transparent cantilever of sub-micron thickness. Light from the cantilever is injected into the output waveguide, called the "receptor," separated from the free end of the cantilever by a short gap (see Figure 10.3). Both the cantilever and the receptor are total internal reflection waveguides. If the gap is in the order of several microns, the energy transmitted into the receptor changes dramatically with the displacement of the cantilever's free end. The idea is to monitor the changes in the power transmitted into the output waveguide in order to register the deflection of the cantilever caused by any biomolecular interaction occurring on its surface. The device presented by Zinoviev *et al.* (2006a) was fabricated using standard silicon technologies. To obtain a straight cantilever free end aligned with the output waveguide, with precision better than 1 μm, a thermally grown silicon oxide film was used. The input waveguide used for delivering the light into the cantilever should be made out of high temperature low-pressure chemical vapor deposition (LPCVD) silicon nitride because the silica layer cannot work as an optical waveguide on the silicon substrate, due to its lower refractive index. If the input waveguide is fabricated thin enough and utilizes a taper, all the energy can be transmitted into the cantilever, due to the large evanescent field tail of the waveguide mode propagating along the silicon nitride layer.

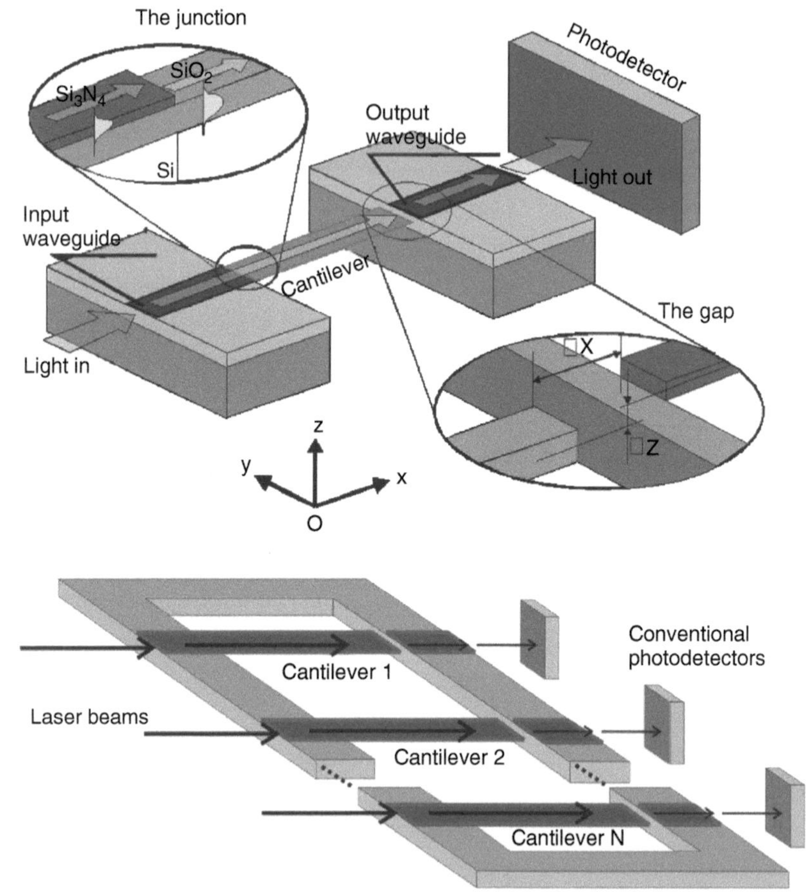

Figure 10.3 (*Top*) Schematic view of a waveguide optical microcantilever; (*bottom*) Schematic view of an array of optical cantilevers.

10.1.5. Microcantilever surface functionalization

Immobilization of bioreceptors on the cantilever surface strongly influences the quality of the bioanalysis to be performed. The efficiency of biomolecule attachment, the accessibility to its targets, and the degree of non-specific binding have to be taken into account. The immobilization process should avoid any change in the biological properties of the receptors, but, at the same time, must keep the chemical and physical properties of the cantilever as uniform as possible in order to generate a large surface stress. The current interest in nanometer-sized cantilever is demanding new immobilization techniques for the manipulation at nanometer scales; newer strategies from synthetic chemistry will become essential for the manipulation of biosensor surfaces and subsequent assembly of receptors. Controlled chemical functionality can be accomplished using self-assembled monolayers (SAMs). The strong affinity of sulphur compounds (thiols, thioethers, and disulfides) for gold and other noble metals make them excellent candidates for nanomechanical biosensing (Nuzzo and Allara, 1983; Ferretti *et al.*, 2000), since the cantilever surface usually has one surface coated with a thin layer of gold (20–100 nm) to increase reflectance of the laser beam. The immobilization of both DNA and proteins on gold surfaces using thiol-SAMs has become the most widely employed method.

The covalent immobilization of proteins on the gold surfaces of the cantilevers can be achieved by a wide variety of chemical procedures, ensuring reproducibility and stability of the protein coating (see Figure 10.4). The main drawback is that some of the functional groups could randomize the orientation of the active sites of the protein. Thus, the appropriate immobilization procedure must be chosen in order to avoid inactivation or a decrease in the activity of the protein receptor. Several

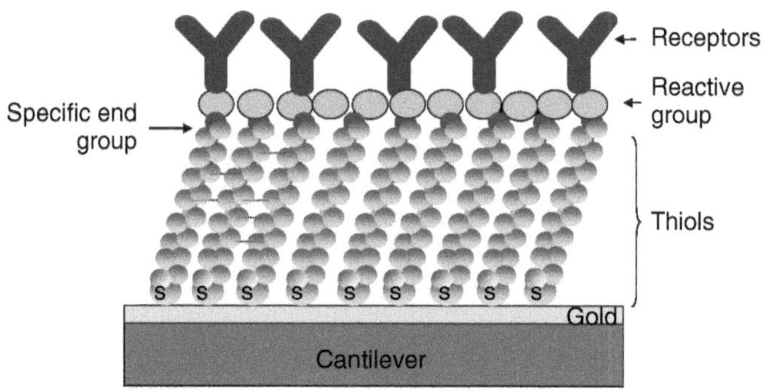

Figure 10.4 Scheme of covalent immobilization by using organic multilayers.

strategies could be employed which have already been useful in many other analytical systems, such as surface plasmon resonance (Boozer *et al.*, 2004), QCM (Mannelli *et al.*, 2005), or electrochemical measurements (Hianik *et al.*, 2001). A very interesting one is to covalently immobilize carboxylate-terminated alkanethiols onto the gold surface, followed by the esterification of the carboxylic groups with 1-ethyl-3-(3-dimethylaminopropyl)-carbodiimide (EDC) and *N*-hydroxysuccinimide (NHS) (Staros *et al.*, 1986; Grabarek and Gergely, 1990; Patel *et al.*, 1997). Another alternative is using a cystamine modified with glutaraldehyde and the subsequent attachment of the protein through an amine group (Alvarez *et al.*, 2003; Pavlov *et al.*, 2004). In addition, we can immobilize biotinylated proteins on avidin monolayers previously attached to the surface using EDC or cystamine chemistry. The immobilization procedure developed by Park and Kim employs the reaction of sulfosuccinimidyl 6-[3-(2-pyridyldithio)pro-pionamido] hexanoate (sulfo-LC-SPDP) with the protein NH_2-groups to give amide linkages. The subsequent addition of dithiothreitol (DTT) reduces the disulfide to give a thiol, which finally self assembles onto the gold surface (Park *et al.*, 2002). Alternatively, a protein, such as antibody, protein A, or avidin, can be thiolated, with the thiol group subsequently used for immobilization on the metal surface; however, there is a high risk of denaturing the protein and therefore losing its biological function (Pyun *et al.*, 2005).

Nucleic acid immobilization is much easier, since chemical modifications can be included in their *in vitro* synthesis, avoiding the risk of losing its functionality. The direct coupling of DNA probes by self assembly of thiol-labeled oligonucleotides is widely used on microcantilever gold surfaces and most of the DNA biosensing applications performed with microcantilever technology are based on this strategy (Biswal *et al.*, 2006; Lechuga *et al.*, 2006). Herne and coworkers demonstrated that a mixed-layer with mercaptohexanol enhanced the hybridization rate and minimized the physical adsorption of DNA probes on gold surfaces by eliminating the weaker adsorptive contacts between the nucleotide chain and gold. In this way, the majority of the immobilized DNA probes are accessible for hybridization with the complementary strand. To prevent non-specific binding to the silicon side of the cantilever, a PEG-silane coating is used (Biswal *et al.*, 2006).

For silicon optical waveguide microcantilevers, immobilization chemistry offers a multitude of well-established and relatively simple attachment strategies. Frequently used silicon modifications, such as aldehyde activation or coating with poly-lysine or nitrocellulose, are typical for the biomolecule attachment (MacBeath and Schreiber, 2000). The highly reactive epoxysilane surface, for example, reacts not only with amino groups, but also with other nucleophilic moieties on protein surfaces like alcohol, thiol, and acid groups, exhibiting a high binding capacity (Zhu *et al.*, 2001). Other methods employ amine-terminal silanes such as aminopropyl-triethoxysilane to provide a reactive amine group at the nitride surface. Thus, alkylamine coupling using glutaraldehyde is simple and fast, giving an aldehyde derivative that can form an imine linkage with primary amines on the protein (Williams and Blanch, 1994). The direct immobilization via thiol-terminal silanes is also a well-known methodology. Using crosslinkers such as EDC and NHS, it is possible to control the directional immobilization with more than 90% of specific binding to the surface (Shriver-Lake *et al.*, 1997).

The immobilization of different bioreceptors on each cantilever of an array is a complicated task. There are several commercial platforms devoted to the specific functionalization of individual cantilevers. Several commercial platforms have been reported using different strategies. For example, the *Autodrop* platform (Microdrop) (Bietsch *et al.*, 2004; www.microdrop.com) uses the principle of the ink jet printing technology and employs up to eight dispensers in an area of $200 \times 200 \times 80$ mm. The core of the dispensing head consists of a glass capillary which is surrounded by a tubular piezo actuator. At one end, the capillary forms a nozzle. By applying a voltage pulse, the piezo actuator creates a pressure wave which propagates through the glass into the liquid. This platform has been used for uniform deposition of thiolated DNA and others alkenothiols (Bietsch *et al.*, 2004). The *Nanojet* technology, also from Microdrop (www.microdrop.com), forms a liquid jet by means of a dispenser system controlled by time and pressure. Provided that the micro valve switches at a very fast rate, volumes down to 200 nl can be dispensed. Moreover, a combination of the Autodrop and Nanojet systems, called "Dropjet," is also available, allowing the generation of a large number of drops with well-defined size (www.microdrop.com).

Cantion has developed the *Canti*™*Spot* platform, which can be used for delivering reagents onto individual cantilevers as well as onto controlled small areas. Its spotter, capable of delivering 100 pl droplets, is connected to a computer-controlled syringe pump for adjusted aspiration and dispensing of reagents (www.cantion.com). Although the *Cantisens*® *FU-401* platform from Concentris is currently used only for cantilever functionalization, it offers a novel way to immobilize biomolecules on cantilever arrays using capillary techniques; up to four cantilevers can be functionalized simultaneously (www.concentris.ch). Finally, a completely different approach under development in several laboratories uses a microfluidics system with an independent flow path for each cantilever in an array (Figure 10.5). With this type of system, immobilization can be performed on each cantilever independently by flowing the appropriate reagents through the individual flow cells.

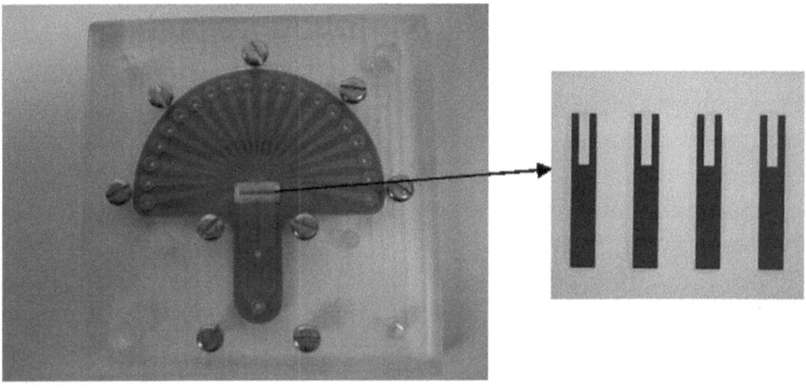

Figure 10.5 Flow cell with 20 independent flow channels for the measurement of an array of 20 microcantilevers with a pitch of 250 microns (flow cell photograph courtesy of ESD University of Southampton; microcantilevers photograph courtesy of CNM-CSIC).

10.2. History

The first chemical sensor based on a macroscopic bimetallic plate was proposed by Norton (1943) for hydrogen detection. In the following years, the concept of a device able to provide a direct conversion between the chemical stimulus and the mechanical energy was developed, focusing attention on miniaturized devices (Kuhn *et al.*, 1950; Steinberg *et al.*, 1966). At the end of 1960s, some integrated miniaturized devices were developed (Newell, 1968; Newell *et al.*, 1968). However, the idea of fabricating floating structures was given up due to the microelectronics fabrication limitations at that time. As early as 1986, with the appearance of AFM (Binnig *et al.*, 1986) and improvements in microelectronic technology, micro-mechanical transducers started to be used routinely and mass production at low cost was possible. Atomic force microscopy arose as a surface characterization technique, in air and subsequently in liquid, measuring the cantilever deflection induced by the interaction between a tip placed at its free end and the surface of the sample. Expansion of AFM techniques showed the microcantilever sensitivity for measuring intermolecular forces between two complementary biological molecules (Force Spectroscopy), such as the interaction between proteins (avidin/biotin) (Lee *et al.*, 1994), hybridization between complementary single-stranded DNAs (Lee *et al.*, 1994), or antigen–antibody interactions (Dammer *et al.*, 1996).

The sensitivity of microcantilevers for measuring intermolecular forces, the commercial availability of cantilevers, and their fabrication using standard microelectronic technology resulted, around 1994, in a new type of sensor where the transducer system is based on a silicon microcantilever with a tipless free end (Figure 10.6) (Gimzewski *et al.*, 1994; Chen *et al.*, 1995). Biochemical applications for this type of sensor have been specifically developed for bending-based modes of measurement, with an optical read-out, due to the complexity required for working with the dynamic mode in liquids. The first experiments in solution were focused on ion detection and measurement of changes in surface stress induced by changes in pH or ion concentrations. It was found that the cantilever response depended upon both pH and the ionic strength of the aqueous medium (Butt, 1996).

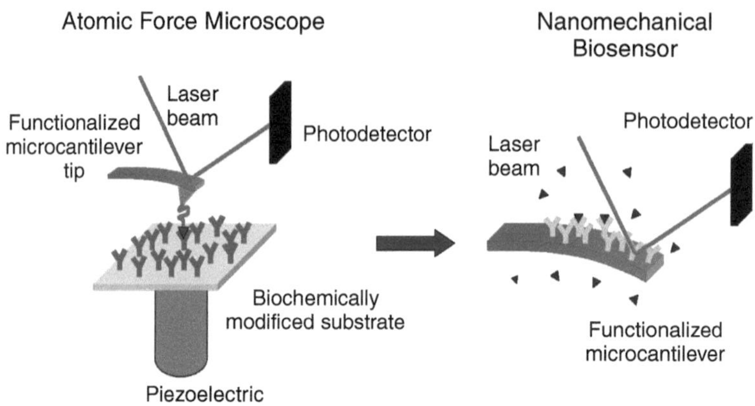

Figure 10.6 Schematic view of (*left*) force spectroscopy and (*right*) nano-mechancial biosensor working principles.

Among other biochemical applications, induction of surface stress by formation of SAMs (Berger *et al.*, 1998; Datskos and Sauers, 1999; Fritz *et al.*, 2000b) or by non-specific adsorption of proteins such as BSA or lipoproteins has been studied (Moulin *et al.*, 2000). However, the largest growth in cantilever-based biosensor research has occurred after the landmark paper of Fritz *et al.* (2000a), in which sensitive discrimination of single-base variations in DNA strands was demonstrated without using fluorescent labels. This paper had a wide impact and marked the beginning of a major research effort in this field. Shortly afterwards, microcantilever sensors were also employed to detect diverse biomolecular interactions. Several commercial platforms (Carrascosa *et al.*, 2006) based on microcantilever array sensors are now available.

10.3. State of the art

The extensive development that cantilever-based biosensing has experienced during the last 10 years has been mainly focused on demonstrating its suitability and sensitivity in a wide range of different fields: genomic, proteomic, environmental control, clinical diagnosis, etc. This section reviews some of the biosensing applications of the microcantilever-optical read-out configurations, as well as the different approaches proposed to increase cantilever biosensor sensitivity.

10.3.1. Biosensing applications: optical read-out configuration

The *bending* detection method is very sensitive to changes in the surrounding environment, and has been applied to detection of variations in pH or salt concentration using silicon microcantilevers modified and non-modified with SAMs (Butt *et al.*, 1995; Fritz *et al.*, 2000b; Ji *et al.*, 2001). The effect of functionalized cantilevers with SAMs, such as alkylthiols with different chemical termini or lengths, on surface stress has also been studied (Berger *et al.*, 1998; Datskos and Sauers, 1999; Godin *et al.*, 2004; Kohale *et al.*, 2007). Likewise, different functionalization schemes have shown that cantilevers were able to detect different ions, such as Ca^{2+} (Ji and Thundat, 2002) or CrO_4^{2-} in concentrations as low as 10^{-11} M (Zhang *et al.*, 2003), among many others.

Genomics is one of the fields where the cantilever-based biosensors have generated both interest and controversy. The first publication in this field demonstrated the detection of a single-base mismatch of single-stranded DNA chains comprised of 12 nucleotides, with a detection limit of 10 nM. An array of two cantilevers was used, employing one as a reference; the final signal was determined as the difference in bending of the two cantilevers (Fritz *et al.*, 2000a). A subsequent study reported the discrimination of single nucleotide polymorphisms with a single cantilever (Hansen *et al.*, 2001; Wu *et al.*, 2001b). This opened a discussion about the possibility of detecting the DNA hybridization and single nucleotide polymorphisms without a reference cantilever (Arntz *et al.*, 2003; Alvarez *et al.*, 2004).

McKendry *et al.* (2002) detected 75 nM of target oligonucleotide using an array of eight microcantilevers previously functionalized using microcapillaries containing a 40 μM solution of thiolated DNA probes. Thermal denaturation has also been checked (Biswal *et al.*, 2006), reporting a decrease in the melting temperature with chain length and salt concentration; lower melting temperatures were also reported for surface-grafted DNA than for DNA in solution. More recently, this technology has been applied for the validation of mRNA biomarkers in total cellular RNA, with sensitivity in the picomolar range without target amplification (Zhang *et al.*, 2006). Another novel application is the design of synthetic DNA motors (Shu *et al.*, 2005), which generate a micro-cantilever nanoscale motion via controlled conformational changes (see Figure 10.7). The forces exerted by the conformational change of precise duplex to non-classical i-motif were shown to induce a compressive surface stress of 32 ± 3 mN/m (which means a single motor force of approximately 11 pN/m). The cantilever direction and amplitude were controlled by buffer pH and the ionic strength, signifying the importance of electrostatic forces in surface stress generation.

In the proteomics field, numerous studies have used antibodies as receptors for detecting their complementary proteins. Wu *et al.* (2001a) measured two isoforms of prostate-specific antigen (cPSA and fPSA) over a range of concentrations from 0.2 ng/ml to 60 μg/ml, in a background of human serum albumin (HSA) and human plasminogen (HP) at 1 mg/ml (Figure 10.8); these isoforms have proven to be a useful biomarker for early detection of prostate cancer. The limit of detection obtained was at the current limit of ELISA (0.2 ng/ml), with the advantage of being label free. Many other applications have been described, such as the detection of 100 nM of streptavidin with biotin functionalized cantilevers (Raiteri *et al.*, 2001) or the real-time recognition of multiple different cardiac biomarker proteins (creatin kinase and myoglobin) (Arntz *et al.*, 2003). A cantilever-based sensor sensitivity of $\approx$1 nM was achieved using single-chain Fv (scFv) antibody fragments as receptor molecules, covalently attached to the gold-coated sensor interfaces in directed orientation (Backmann *et al.*, 2005).

The potential of the cantilever-based biosensor for detection conformational changes in proteins has also been explored. Conformational change-induced surface stress was studied for the protein calmodulin. The cantilever bent when the calmodulin bound to Ca^{2+}, in contrast to the cantilevers modified with other proteins, such as hemoglobin and myoglobin, which do not exhibit change conformations upon binding with analytes (Yan *et al.*, 2006). The ability of cantilever-based sensors for detection of ligand–protein interactions and conformational

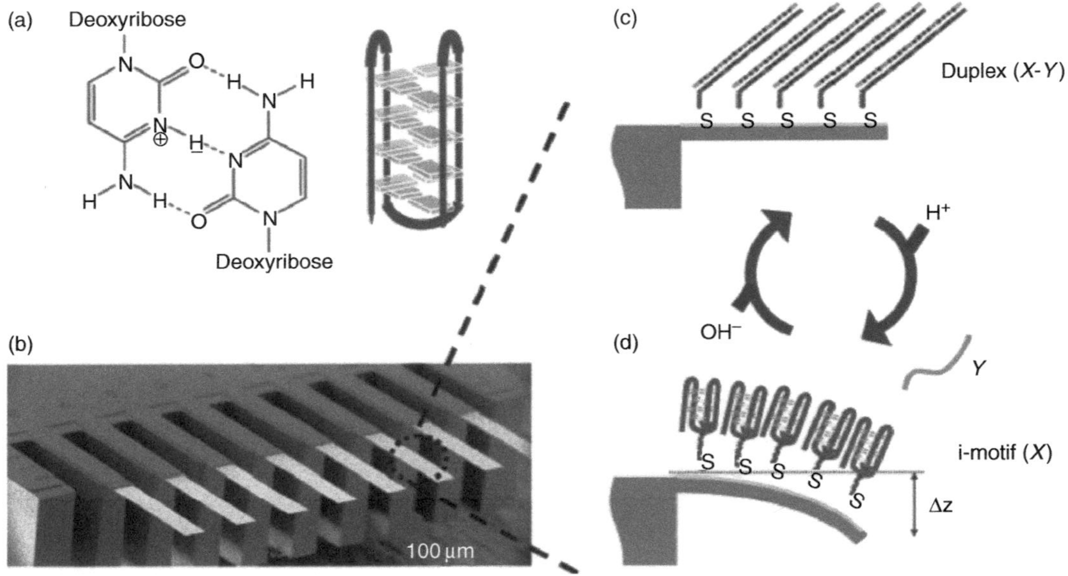

Figure 10.7 Conformational changes from duplex to i-motif. (a) Schematic diagram to show the intramolecular interdigitation of strand **X** to form the i-motif (at pH 5.0). (b) Scanning electron microscope image of an array of eight rectangular silicon cantilevers. (c) Schematic diagram to show the hybridization between the surface-tethered **X** and the strand **Y** in solution (at pH > 6.7). (d) Formation of the **X** strand self-folded i-motif, inducing repulsive in-plane surface forces (at pH 5.0). (Reproduced with permission from Shu *et al.* (2005). Copyright 2005, American Chemical Society.)

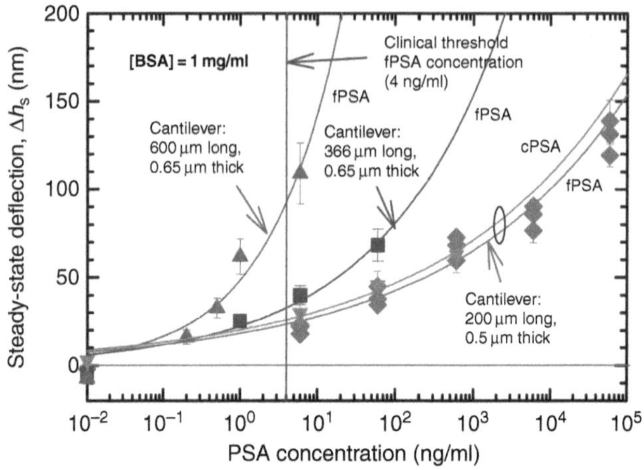

Figure 10.8 Steady-state cantilever deflections as a function of fPSA and cPSA concentrations for three different cantilever geometries (*longer cantilevers produce larger deflections for the same PSA concentration*). (Reproduced with permission from Macmillan Publishers Ltd: Nature Biotechnology [Wu *et al.*, 2001a], Copyright 2001.)

changes of membrane proteins (such as bacteriorhodopsin) has been demonstrated as well (Braun *et al.*, 2006).

Applications have expanded to new fields, such as detection of pathogens such as *Escherichia coli* for food quality control (Zhang and Ji, 2004) or environmental control. In this latter field, the interactions between the herbicide 2,4-dichlorophenoxyacetic (Raiteri *et al.*, 1999) and the organochlorine insecticide compound dichlorodiphenyltrichloroethane (DDT) (Alvarez *et al.*, 2003) with their corresponding monoclonal antibodies have been measured. Direct detection of DDT was achieved with a competitive assay, measuring DTT concentrations as low as 10 nM. In addition, the novel development for early osteosarcoma discovery, sensing the interactions between vimentin antibodies and antigens with a single cantilever-based biosensor, proved that cantilever biosensors can provide a suitable platform for life sciences research (Milburn *et al.*, 2005). Also of interest is application of microcantilevers for detection of the feline coronavirus type I (Velanki and Ji, 2006), which demonstrated the feasibility of detecting severe acute respiratory syndrome associated coronavirus (SARS-CoV).

As previously mentioned, the application of *resonant* mechanical sensors for biochemical recognition in a liquid medium is a great challenge, as a consequence of the cantilever's resonant frequency and quality factor reduction. In spite of this limitation, there are some works that demonstrate detection of biomolecules by measuring the cantilever resonance before and after the binding event. Using this methodology, DNA hybridization has been sensed using DNA strands linked to the cantilever by gold–thiol covalent bonding and gold nanoparticle-modified probes. After hybridization, the gold nanoparticles act as a nucleating agent for the growth of silver particles, which leads to a detectable frequency shift by increasing the cantilever effective mass. This method can detect at least 0.05 nM and is able to discriminate a single-base mismatch (Su *et al.*, 2003).

The number of applications performed using the dynamic mode grows when working with pathogens, cells, viruses, etc. that have a large mass. Ilic *et al.* (2001) demonstrated the ability to detect both an *E. coli* antibody monolayer and a single *E. coli* cell, operating in air and measuring the cantilever thermal noise with the optical bending method. The mass sensitivities for a 15 and 25 μm long silicon beam were 1.1 and 7.1 Hz/fg, respectively. Measurement of growth of pathogens such as *Aspergillus niger* spores (Nugaeva *et al.*, 2005) and *E. coli* cells (Gfeller *et al.*, 2005), in a humidity-controlled environment, has also been reported. In these experiments, the resonance frequency shifts as a function of the increasing mass on the cantilever.

Reducing the cantilever size is a widespread alternative for increasing the frequency resolution, in both air and liquid. With a silicon cantilever 4–5 μm long, 1–2 μm wide, and 20–30 nm thick, Gupta *et al.* (2004) established detection of a single vaccinia virus particle with an average mass of 9.5 fg (under ambient conditions, and by measuring the cantilever thermal noise). Later, they found that the protein attachment increases with cantilever size (Figure 10.9), resulting in an increase in the effective spring constant and a corresponding anomalous increase in frequency for a certain class of cantilevers (Gupta *et al.*, 2006).

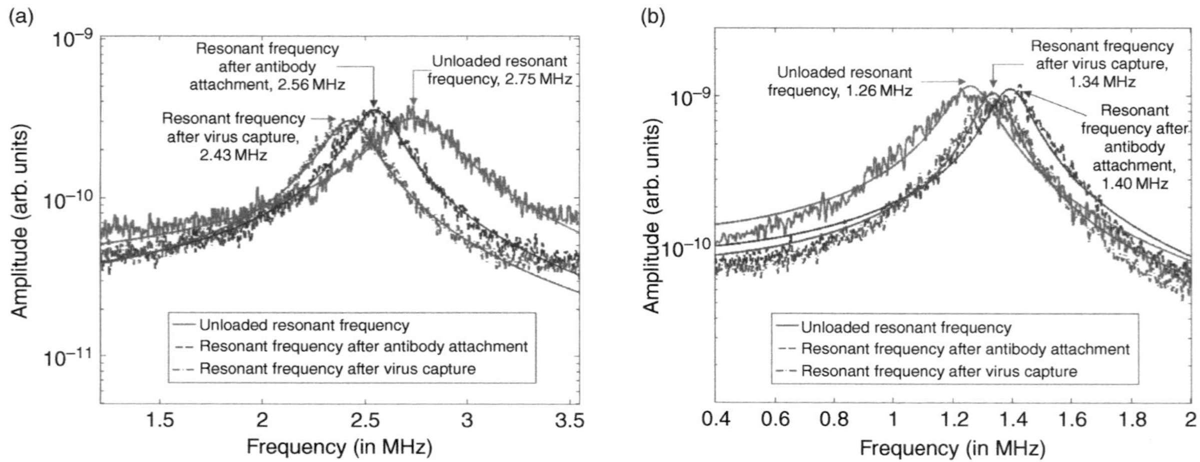

Figure 10.9 Thermal noise-induced frequency spectra of cantilever beams, at various stages of the biosensor analysis. For the same process, the resonance frequency decreases (a) or increases (b) depending on the cantilever dimensions. (Reproduced with permission from Gupta *et al.* [Gupta *et al.*, 2005], Copyright 2005, IEEE.)

The interaction of biotin with streptavidin has been measured in liquids by working with higher cantilever resonance harmonics (Ghatkesar *et al.*, 2004). As an alternative, the integrated piezoelectric microcantilevers are becoming increasingly popular, showing quality factors one order of magnitude greater than thermal-driven cantilevers, allowing for mass virus detection in the femtogram range (5–10 fg) in air (Johnson *et al.*, 2006). A nanomechanical lead zirconium titanate or PZT thin film cantilever, composed of SiO_2/Ta/Pt/PZT/Pt/SiO_2 on a SiN_x supporting layer, was used for detection of different concentrations (1–100 ng/ml) of prostate-specific antigen (PSA) in liquids (Hwang *et al.*, 2004). For these experiments, the antibody was previously immobilized with calixcrown SAMs on a gold surface deposited on the cantilever; the subsequent antigen detection was executed in a fabricated poly(dimethylsiloxane) (PDMS) flow cell with a 200 μm width channel and a 20 μl volume reaction chamber, by measuring the resonant frequency with a laser Doppler vibrometer (Figure 10.10).

10.3.2. Other cantilever-based biosensor approaches

Currently, there are many different and alternative ways to increase the sensitivity of cantilever-based biosensors, depending on the sensor working mode. For example, production of cantilever surface nanostructures has been demonstrated as a good method for amplifying the bending signal, due to the increased effective surface area (Lavrik *et al.*, 2001; Headrick *et al.*, 2003).

Exploitation of polymers for cantilever fabrication results in a cheaper alternative to increase the cantilever static deflection, as the Young's moduli of polymers are lower than that of silicon. At present, one of the most widely employed alternative materials for new sensor designs is the SU-8 polymer, due to its favorable properties. A non-vacuum fabrication process to produce arrays of SU8 cantilevers has been reported, demonstrating their application as chemical sensors (Ransley *et al.*, 2006). Zhang and Xu (2004) demonstrated the suitability of 6 μm thick polyethylene terephthalate films for microcantilever fabrication using laser micromachining techniques. In this work, DNA hybridization experiments showed the capability of detecting 12 base oligonucleotides

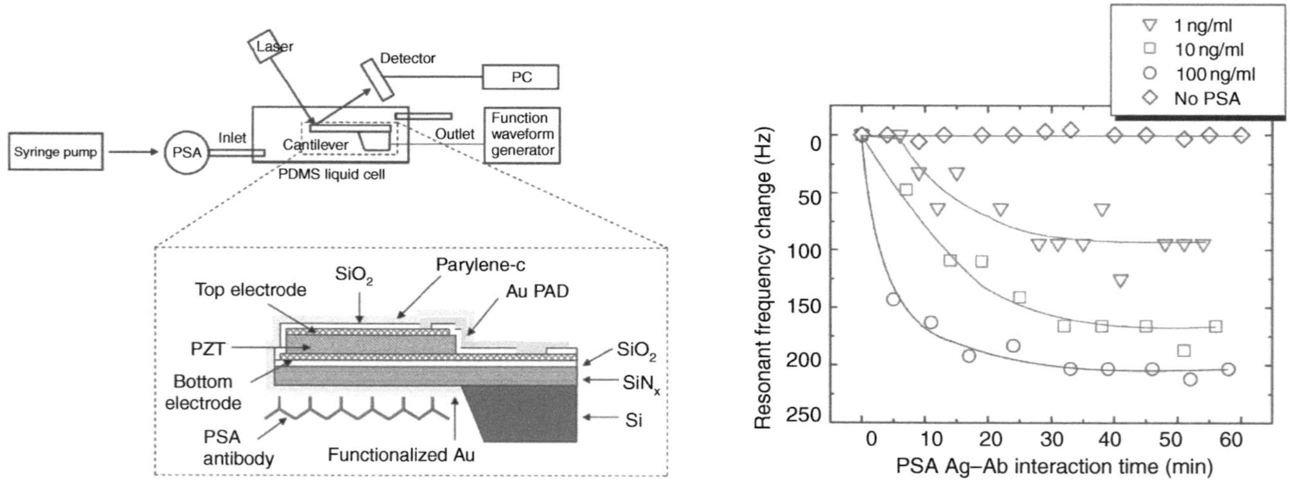

Figure 10.10 Schematics of the experimental measurement system and in-situ resonant frequency change as a function of PSA antigen concentrations (1, 10, and 100 ng/ml[1], respectively). ([Hwang *et al.*, 2004] – Reproduced by permission of the Royal Society of Chemistry.)

at concentrations as low as 0.01 μm. Both experiments were carried out with the optical beam deflection method. Another widely employed polymer is PDMS. With this polymer, three-dimensional thin structures with different geometries on each side have been fabricated (Park *et al.*, 2006). In addition to polymer cantilever fabrication, this work demonstrated the mechanical shear stress enhancement of self-organized cardiomyocytes on three-dimensional grooved surfaces compared to that on two-dimensional surfaces (Figure 10.11), revealing quantitative mechanical changes in cells in real time between the relaxated state and the contracted state of cardiomyocytes.

Other works propose new designs and cantilever configurations fabricated using standard silicon microtechnology to improve sensitivity or solve existing drawbacks. A configuration based on two adjacent cantilevers forming a sensor/reference pair was used to demonstrate the capability of a DNA aptamer–protein binding event to generate changes in surface stress (Savran *et al.*, 2004). This configuration allowed the researchers to directly detect the differential tip deflection between the two cantilevers.

For sensors based on an array of microcantilevers and standard optical detection, either with a single or an array of PSD, a very low variability

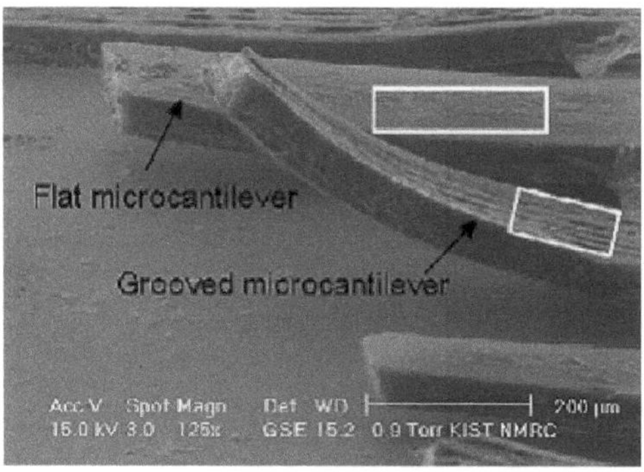

Figure 10.11 ESEM (environmental scanning electron microscope) image of cardiomyocytes cultured on flat and grooved microcantilever structures. [Reproduced with permission from IOP Publishing Limited (Park *et al.*, 2006).]

in the initial position of cantilever array is desired. A reduction in the initial angular offset and angle deviation between the cantilevers of an array is achieved by fabricating T-shaped cantilevers (Plaza *et al.*, 2006). This geometry consists of a sensing cantilever joined to a doubly clamped beam (Figure 10.12), which allows that all the cantilevers remain flat and parallel to each other in spite of possible non-uniform initial stresses at the anchor region.

New designs, such as an array of cantilevers with a common or discrete window (Lechuga *et al.*, 2006) or the aforementioned optical waveguide microcantilevers, offer a new and interesting approach for further integration in "lab-on-a-chip" microsystems. Figure 10.13 shows the device produced by Zinoviev *et al.*, which contains cantilever transducers with lengths of several hundreds of microns fabricated in an array of 20 (Zinoviev *et al.*, 2006a, 2007). In this work, characterization

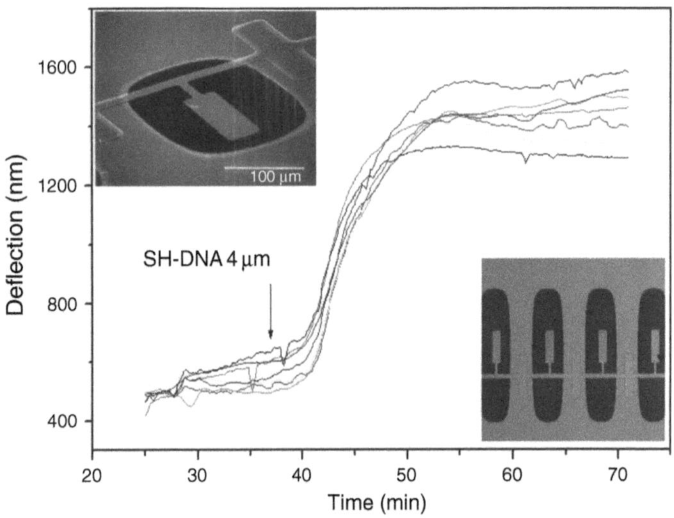

Figure 10.12 Real-time deflection of seven T-shaped microcantilevers due to the immobilization of thiol-modified DNA chains (SH ssDNA). Scanning electron microscopy images of T-shaped cantilevers. (Reproduced with permission from J.A. Plaza [Plaza *et al.*, 2006]. Copyright 2006, American Institute of Physics.)

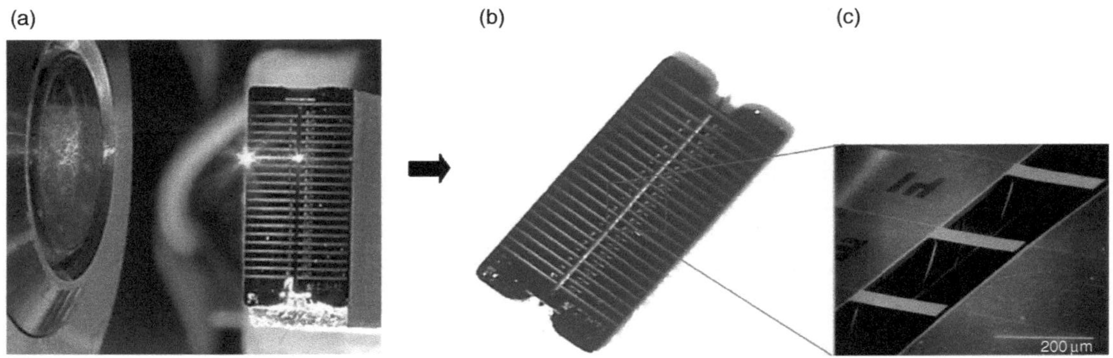

Figure 10.13 (a) Picture showing a chip with optical waveguide cantilevers when light is coupled in one of its channels. Light was coupled using direct focusing with an objective lens. (b) Fabricated chip with dimensions of 3×7 mm (c) SEM photograph of the fabricated optical cantilevers.

of the cantilevers' resonance frequency and bending was performed using the optical beam technique for monitoring the waveguide output signal as a function of the cantilever absolute displacement. The sensitivity of both methods, resonant frequency and bending, allows resolution of cantilever displacement to less than 0.1 nm. The effect of covering the cantilevers with gold for biosensing applications has also been studied, showing a strong adsorption depending on the cantilever length and thickness as well as the gold layer thickness. In contrast, the higher order modes are getting filtered, which has the advantage of making the cantilevers single mode in the transverse direction. An additional property of the metallized cantilevers is their ability to deform with light absorption, which helps to adjust the initial position of the cantilever free end with respect to the output waveguide.

10.4. Advantages and limitations for use in optical biosensing

The increasing number of applications of microcantilevers as biosensors has established these systems as a versatile platform for real-time and *in situ* measurements of physical, chemical, and biochemical interactions. As previously mentioned, one of the main advantages is their ability to detect molecular interactions without any kind of label, reducing the time and cost of sample preparation. Another advantage is the low cost of fabrication and mass production of extremely sensitive devices, as well as the flexibility in fabrication techniques to change the size and shape depending on the required application. Hence, it is possible to fabricate cantilevers with very low spring constants that are very sensitive for bending detection, or reduce its size (stiffer cantilevers) for mass changes measurements. A disadvantage related with the cantilever dimensions is the mechanical–thermal noise, which could limit the fundamental biosensor detection limit.

The microcantilever can be operated in vacuum, gasses, and liquids, although the dynamic resolution is reduced when working in liquids, due to the damping effect over the resonance response. This limitation has rendered the static mode as the most attractive method for biosensing platforms. The main disadvantage related to bending detection is the complex relation between the measured signal and the factors producing it, due to the number of forces acting in the biorecognition processes. Although the working principle of mechanical response could be more difficult to understand than in other types of sensors, a new type of information is reported and represents a new alternative for the discrimination of interactions, such as DNA single-base mismatch polymorphisms or single-molecule detection, otherwise not possible with other established biosensing techniques. Among other advantages, microcantilevers present a reduced sensor area, is compatible with the complementary metal oxide semiconductor (CMOS) technology, and can be easily scaled up, being currently fabricated in arrays of tens to thousands of microcantilevers. This could allow the parallel, fast, and real-time monitoring of thousands of analytes (DNA strands, proteins, pathogens, cells, etc.) without the need for labels. Moreover, the bending and resonance response (frequency, amplitude, Q-factor or phase), when simultaneously detected, can supply valuable information.

Currently, excellent results have been obtained with the optical bending read-out configuration; however, this set-up is limited by environmental optical properties which confound its applications with real samples such as blood.

In order to obtain an industrial sensor system, low power consumption and compact portable devices are required. These requirements have led to an increase in the development of integrated detection read-outs with a CMOS electric circuitry incorporated (Lechuga *et al.*, 2006). In addition to an integrated read-out, the biotechnology field requires that biosensors be capable of working with small amounts of reagents, with highly reproducible surface functionalization and subsequent recognition processes. For this purpose, an integrated microfluidics system that allows liquid handling in an easy and reproducible approach is needed (Bietsch *et al.*, 2004; Yue *et al.*, 2004; http://www.protiveris.com/products/pages/sensorcartridge.html).

10.5. Potential for improving performance or expanding current capabilities

Some of the most desired improvements on the cantilever-based biosensors are the increase in the sensitivity and reproducibility, a more in-depth knowledge of the working principles, and a means of integration of these sensors into a portable and miniaturized system.

There are different ways to increase the biosensor sensitivity. Some are related to optimization of the cantilever material and dimensions, noise reduction, or improvements in the detection methods. The development of a highly sensitive and integrated detection method is one of the most important unresolved matters in this field where new approaches, such as the optical waveguide cantilevers, arise as a great alternative to standard optical beam methods. Other techniques based on piezoresistive and piezoelectric microcantilevers allow an easy and more integrated readout and are becoming important during the past several years, although their sensitivities are presently inferior to those obtained with the optical method.

Other ways to improve sensitivity are focused on the biochemistry, looking for new methods to functionalize the cantilever reproducibly, reducing both the non-specific interactions and the volume of reagents. SAMs are especially useful for binding assays, allowing surface regeneration for subsequent assays. Development of an integrated microfluidic system would be especially valuable for improving the signal reproducibility, reducing the volume required and the total system size. The possibility of working with several cantilevers at the same time, in completely independent microfluidics channels, would allow the detection, in real time, of many different reagents simultaneously, providing this sensor with a greater capability of measurement and higher versatility than other current techniques (such as ELISA or DNA microarrays).

A more fundamental understanding about the cantilever working principles and the factors involved in the cantilever response, both static and dynamic, are needed. This knowledge will provide new useful information for increasing the biosensor sensitivity and for controlling the binding process itself (Lu *et al.*, 2005; Reed *et al.*, 2006; Craighead, 2007; Watari *et al.*, 2007).

Acknowledgements

The authors would like to acknowledge the European Union for funding and their partners working within the European project Optonanogen (IST-2001-37239).

References

Alvarez, M. and Tamayo, J. (2004) *Sens. Actuators B*, **106**, 687.
Alvarez, M., Calle, A., Tamayo, J. *et al.* (2003) *Biosens. Bioelectron.*, **18**, 649.
Álvarez, M., Carrascosa, L.G., Moreno, M. *et al.* (2004) *Langmuir*, **20**, 9663.
Álvarez, M., Tamayo, J., Plaza, J. *et al.* (2006) *J. Appl. Phys.*, **99**, 024910.
Arntz, Y., Seelig, J., Lang, H. *et al.* (2003) *Nanotechnology*, **14**, 86.
Backmann, N., Zahnd, C., Huber, F. *et al.* (2005) *Proc. Natl. Acad. Sci. USA*, **102**, 14587.
Baller, M.K. and Fritz, J. (2004) *Protein Microarray Technology: Nanomechanical Cantilever Sensors For Microarrays* (D. Kambhampatiand, ed.) Wiley-VCH Verlag GmbH & Co. KGaA, p. 195.
Berger, R., Delamarche, E., Lang, H.P. *et al.* (1998) *Appl. Phys. A Mat. Sci. Proc.*, **66**, S55.
Bietsch, A., Hegner, M., Lang, H.P., and Gerber, C. (2004) *Langmuir*, **20**, 5119.
Bietsch, A., Zhang, J.Y., Hegner, M. *et al.* (2004) *Nanotechnology*, **15**, 873.
Binnig, G., Quate, C.F., and Gerber, C. (1986) *Phys. Rev. Lett.*, **56**, 930.
Biswal, S.L., Raorane, D., Chaiken, A. *et al.* (2006) *Anal. Chem.*, **78**, 7104.
Boozer, C., Ladd, J., Chen, S.F. *et al.* (2004) *Anal. Chem.* **76**, 6967.
Braun, T., Backmann, N., Vogtli, M. *et al.* (2006) *Biophys. J.*, **90**, 2970.
Budakian, R. and Putterman, S.J. (2002) *Appl. Phys. Lett.*, **81**, 2100.
Butt, H.J. (1996) *J. Colloid Interface Sci.*, **180**, 251.
Butt, H.J., Jaschke, M., and Ducker, W. (1995) *Bioelectrochem. Bioenerg.*, **38**, 191.
Carrascosa, L.G., Moreno, M., Alvarez, M., and Lechuga, L.M. (2006) *Trends Anal. Chem.*, **25**, 196.

Chen, G., Thundat, T., Wachter, E., and Warmack, R. (1995) *J. Appl. Phys.*, **77**, 1.

Craighead, H. (2007) *Nat. Nanotechnol.*, **2**, 18.

Dammer, U., Hegner, M., Anselmetti, D. *et al.* (1996) *Biophys. J.*, **70**, 2437.

Datskos, P.G. and Sauers, I., (1999) *Sens. Actuators B*, **61**, 75.

Ferretti, S., Paynter, S., Russell, D.A. *et al.* (2000) *Trends Anal. Chem.*, **19**, 530.

Fritz, J., Baller, M.K., Lang, H.P. *et al.* (2000a) *Science*, **88**, 316.

Fritz, J., Baller, M.K., Lang, H.P. *et al.* (2000b) *Langmuir*, **16**, 9694.

Gfeller, K.Y., Nugaeva, N., and Hegner, M. (2005) *Biosens. Bioelectron.*, **21**, 528.

Ghatkesar, M., Barwich, V., Braun, T. *et al.* (2004) *Proc. IEEE Sens*, p. 1060.

Gimzewski, J.K., Gerber, C., Meyer, E., and Schlittler, R.R. (1994) *Chem. Phys. Lett.*, **217**, 589.

Godin, M., Williams, P.J., Tabard-Cossa, V. *et al.* (2004) *Langmuir*, **20**, 7090.

Grabarek, Z. and Gergely, J. (1990) *Anal. Biochem.*, **185**, 131.

Gupta, A., Akin, D., and Bashir, R. (2004) *Appl. Phys. Lett.*, **84**, 1976.

Gupta, A., Akin, D., and Bashir, R. (2005) MEMS 2005, 18th IEEE International Conference, p. 746.

Gupta, A.K., Nair, P.R., Akin, D. *et al.* (2006) *Proc. Natl Acad. Sci. USA*, **103**, 13362.

Hagan, M.F., Majumdar, A., and Chakraborty, A.K. (2002) *J. Phys. Chem. B*, **106**, 10163.

Hansen, K.M., Ji, H., Wu, G. *et al.* (2001) *Anal. Chem.*, **73**, 1567.

Headrick, J., Sepaniak, M., Lavrik, N., and Datskos, P. (2003) *Ultramicroscopy*, **97**, 417.

Hianik, T., Gajdos, V., Krivanek, R. *et al.* (2001) *Bioelectrochemistry*, **53**, 199.

Hwang, K.S., Lee, J.H., Park, J. *et al.* (2004) *Lab Chip*, **4**, 457.

Ilic, B., Czaplewski, D., Zalalutdinov, M. *et al.* (2001) *J. Vac. Sci. Technol. B*, **19**, 2825.

Ilic, B., Craighead, H.G., Krylov, S. *et al.* (2004) *J. Appl. Phys.*, **95**, 3694.

Ji, H., Hansen, K., Hu, Z., and Thundat, T. (2001) *Sens. Actuators B*, **72**, 233.

Ji, H.F. and Thundat, T. (2002) *Biosens. Bioelectron.*, **17**, 337.

Johnson, L., Gupta, A.K., Ghafoor, A. *et al.* (2006) *Sens. Actuators B*, **115**, 189.

Kim, T.S., Lee, J.H., and Yoon, D.S. (2006) *Micromanufacturing and Nanotechnology: Nanomechanical Cantilever for Biological Sensors* (N.P. Mahalikand, ed.) Berlin, Heidelberg:Springer, p. 299.

Kohale, S., Molina, S., Weeks, B. *et al.* (2007) *Langmuir*, **23**, 1258.

Kuhn, W., Hargitay, B., Katchalsky, A., and Eisenberg, H. (1950) *Nature*, **65**, 514.
Lang, H.P., Berger, R., Andreoli, C. *et al.* (1998) *Appl. Phys. Lett.*, **72**, 383.
Lavrik, N., Sepaniak, M., and Datskos, P. (2004) *Rev. Sci. Instrum.*, **75**, 2229.
Lavrik, N., Tipple, C., Sepaniak, M., and Datskos, P. (2001) *Biomed. Microdevices*, **3**, 35.
Lechuga, L.M., Tamayo, J., Álvarez, M. *et al.* (2006) *Sens. Actuators B*, **118**, 2.
Lee, G.U., Chrisey, L.A., and Colton, R.J. (1994) *Science*, **266**, 771.
Lee, G.U., Kidwell, D.A., and Colton, R.J. (1994) *Langmuir*, **10**, 354.
Lu, P., Lee, H.P., Lu, C., and O'Shea, S.J. (2005) *Phys. Rev. B*, **72**, 085405.
MacBeath, G. and Schreiber, S.L. (2000) *Science*, **289**, 1760.
Mannelli, F., Minunni, A., Tombelli, S. *et al.* (2005) *Bioelectrochemistry*, **66**, 129.
McKendry, R., Zhang, J., Arntz, Y. *et al.* (2002) *Proc. Natl Acad. Sci. USA*, **99**, 9783.
Milburn, C., Zhou, J., Bravo, O. *et al.* (2005) *J. Biomed. Nanotech.*, **1**, 30.
Moulin, A., O'Shea, S., and Welland, M. (2000) *Ultramicroscopy*, **82**, 23.
Newell, W.E. (1968) *Science*, **161**, 1320.
Newell, W.E., Wickstrom, R.A., and Page, D.J. (1968) *IEEE Trans. Electron Dev.*, **Ed15**, 411.
Nordin, G.P., Noh, J.W., and Kim, S. (2007) *SPIE Proc. Photonics West*, 6447.
Norton, F.J. (1943) *Gas Analyzer*, General Electric Co. US Patent 2,307,800.
Nugaeva, N., Gfeller, K., Backmann, N. *et al.* (2005) *Biosens. Bioelectron.*, **21**, 849.
Nuzzo, R.G. and Allara, D.L. (1983) *J. Am. Chem. Soc.*, **105**, 4481.
Ollier, E., Philippe, P., Chabrol, C., and Mottie, P. (1999) *J. Lightwave Technol.*, **17**, 26.
Park, J., Kim, J., Roh, D. *et al.* (2006) *J. Micromech. Microeng.*, **16**, 1614.
Park, S.J., Taton, T.A., and Mirkin, C.A. (2002) *Science*, **295**, 1503.
Patel, N., Davies, M.C. Hartshorne, M. *et al.* (1997) *Langmuir*, **13**, 6485.
Pavlov, V., Xiao, Y., Shlyahovsky, B., and Willner, I. (2004) *J. Am. Chem. Soc.*, **126**, 11768.
Plaza, J.A., Zinoviev, K., Villanueva, G. *et al.* (2006) *Appl. Phys. Lett.*, **89**, 094109.
Pyun, J.C., Kim, S.D., and Chung, J.W. (2005) *Anal. Biochem.*, **347**, 227.
Raiteri, R., Grattarola, M., Butt, H., and Skládal, P. (2001) *Sens. Actuators B*, **79**, 115.
Raiteri, R., Nelles, G., Butt, H. *et al.* (1999) *Sens. Actuators B Chem.* **61**, 213–7.

Ransley, J.H.T., Watari, M., Sukumaran, D. *et al.* (2006) *Microelectron. Eng.*, **83**, 1621.

Reed, J., Wilkinson, P., Schmit, J. *et al.* (2006) *Nanotechnology*, **17**, 3873.

Savran, C.A., Knudsen, S.M., Ellington, A.D., and Manalis, S.R. (2004) *Anal. Chem.*, **76**, 3194.

Shriver-Lake, L.C., Donner, B., Edelstein, R. *et al.* (1997) *Biosens. Bioelectron.*, **2**, 1101.

Shu, W., Liu, D., Watari, M. *et al.* (2005) *J. Am. Chem. Soc.*, **127**, 17054.

Shuttleworth, R. (1950) *Proc. Phys. Soc. London B*, **63**, 374.

Staros, J.V., Wright, R.W., and Swingle, D.M. (1986) *Anal. Biochem.*, **156**, 220.

Steinberg, I., Oplatka, A., and Katchals, A. (1966) *Nature*, **210**, 568.

Stoney, G. (1909) *Proc. R. Soc. London*, **82**, 172.

Su, M., Li, S.U., and Dravid, V.P. (2003) *Appl. Phys. Lett.*, **82**, 3562.

Velanki, S. and Ji, H. (2006) *Meas. Sci. Technol.*, **17**, 2964.

Wang, W., Fauver, M., Ho, J.N. *et al.* (2002) *Sens. Actuators A*, **102**, 165–75.

Watari, M., Galbraith, J., Lang, H. *et al.* (2007) *J. Am. Chem. Soc.*, **129**, 601.

Williams, R.A. and Blanch, H.W. (1994) *Biosens. Bioelectron.*, **9**, 159.

Wu, G., Datar, R., Hansen, K.M. *et al.* (2001a) *Nat. Biotechnol.*, **19**, 856.

Wu, G., Ji, H., Hansen, K. *et al.* (2001b) *Proc. Natl Acad. Sci. USA*, **98**, 1560.

Wu, S. and Frankena, H.J. (1992) *SPIE – Integr. Opt. Microstruct.*, **1793**, 83.

Xu, T., Chang, R., Bachman, M., and Li, G. (2005) *IEEE Sensors*, **4**, 963.

Yan, X., Hill, K., Gao, H., and Ji, H. (2006) *Langmuir*, **22**, 11241.

Yue, M., Stachowiak, J., and Majumdar, A. (2004) *Mol. Cell. Biomech.*, **1**, 211.

Zhang, J. and Ji, H.F. (2004) *Anal. Sci.*, **20**, 585.

Zhang, J., Lang, H.P., Huber, F. *et al.* (2006) *Nat. Nanotechnol.*, **1**, 214.

Zhang, X.R. and Xu, X. (2004) *Appl. Phys. Lett.*, **85**, 2423.

Zhang, Y.F., Ji, H.F., Brown, G.M., and Thundat, T. (2003) *Anal. Chem.*, **75**, 4773.

Zhu, H., Bilgin, M., Bangham, R. *et al.* (2001) *Science*, **293**, 2101.

Ziegler, C. (2004) *Anal. Bioanal. Chem.*, **379**, 946.

Zinoviev, K., Dominguez, C., Plaza, J.A. *et al.* (2006a) *J. Lightwave Technol.*, **24**, 2132.

Zinoviev, K., Domínguez, C., Plaza, J., and Lechuga, L.M. (2006b) *Appl. Opt.*, **45**, 229.

Zinoviev, K., Plaza, J.A., Cadalso, V. *et al.* (2007) *Proc. SPIE*, **6477**, 64771A-1.

Optical Biosensors: Today and Tomorrow (2nd Edition)
Frances S. Ligler and Chris Rowe Taitt (Editors)

Chapter 11

PROTEIN MICROARRAY TECHNOLOGIES: AN ARRAY OF APPLICATIONS

Thomas O. Joos, Ph.D.[a], Jutta Bachmann, Ph.D.[b], and James W. Jacobson, Ph.D.[c]
[a]NMI, Natural and Medical Sciences Institute at the University of Tübingen, Markwiesenstr. 55, 72770 Reutlingen, Germany
[b]Jutta Bachmann, Bachmann Consulting, Nøkkefaret 12, 1450 Nesoddtangen, Norway
[c]Luminex Corporation, 12212 Technology Blvd, Austin, TX 78727, USA

Within the last years, protein microarray-based research has moved from being technology-based to application-oriented. Protein microarrays have been successfully used in a variety of applications such as the identification, quantification, and functional analysis of proteins in basic and applied proteome research. Array-based assay systems analyze hundreds of molecular ligand-binding events in a single experiment. Such miniaturized and parallelized assay systems have the potential to replace singleplex analysis systems. Numerous analytical platforms have been developed that are likely to evolve into key technologies for the characterization of complex samples. Many of the systems developed have demonstrated robustness, automation, and the appropriate level of sensitivity. In this overview, we summarize the current stage of protein microarray technology with a special focus on applications enabling the simultaneous analysis of a broad range of parameters from minute samples.

11.1. Technical concept

Microarrays are solid phase–based assay systems consisting of an array of miniaturized test sites arranged in rows and columns. These microspots are usually less than 250 μm in diameter (Chipping Forcast II, 2002). This arrangement allows the user to perform many tests simultaneously or in parallel. The most common format is a two-dimensional format in which the miniaturized test sites are placed on a microscopic glass slide. However, multiplexed bead-based assay platforms, i.e. three-dimensional formats, bead-based or liquid arrays are becoming increasingly popular. Protein microarray technologies generate an enormous amount of quantitative information with considerable savings in labor and sample volumes (Stoll *et al.*, 2005; Kricka *et al.*, 2006; Master *et al.*, 2006).

11.1.1. Surface chemistry

In protein arrays, capture molecules need to be immobilized in a functional state on a solid support. In principle, the format of the assay system does not limit the choice of appropriate surface chemistry. The same immobilization procedure can be used for both planar and bead-based systems. Proteins can be immobilized on various surfaces (e.g., nitrocellulose, polystyrene, poly-lysine, or aminosilane-coated surfaces) with standard physical adsorption methods (Stoll *et al.*, 2002). Covalent attachment is achieved using reactive groups such as aldehyde, epoxy, thiol, or active ester groups (Hermanson, 1996; Nakanishi *et al.*, 1996; Shriver-Lake *et al.*, 1997; Blawas and Reichert, 1998; Schlottmann *et al.*, 2006). In addition, affinity-binding reagents such as proteins A, G, and L have proved suitable for the immobilization of antibodies (Seong, 2002), streptavidin for biotinylated proteins (Pritchard *et al.*, 1995; Dontha *et al.*, 1997; Finckh *et al.*, 1998; Gaber *et al.*, 1999), Ni^{2+} chelate for His-tagged proteins (Zhu *et al.*, 2001; Lauer and Nolan, 2002), anti-GST antibodies for GST fusion proteins (Waterboer *et al.*, 2005), and oligonucleotides for cDNA or mRNA-protein hybrids (Wacker *et al.*, 2004).

11.1.2. Arraying technology

Capture molecules can be printed onto chip surfaces with contact printing arrayers equipped with tiny needles to place sub-nanoliter sample volumes directly onto the surface. Alternatively, non-contact deposition technologies which employ capillaries or ink jet technology to deposit nanoliter to picoliter droplets onto the surface can be used. The TopSpot™ developed by IMTEK (University of Freiburg, Germany) is a very efficient depositing technology, using a microstructured device where arrays of tiny capillaries (up to 96) are directly linked to distinct macroscopic fluid reservoirs (Gutmann *et al.*, 2005). A piezo-actuator hits the back of the microcapillary device, thus generating a steep air pressure ramp to the open upper side of the liquid reservoirs. This air pressure change results in the simultaneous dispensing of microdroplets from each capillary tip, with volumes down to 1 nl. The TopSpot™ device is a very useful and quick way to produce large quantities of identical DNA or protein arrays inexpensively. Approximately 20 000 droplets can be generated from 20 μl of sample solution with no loss of quality.

Micropatterned protein arrays have also been produced with photolithographic methods (Vail *et al.*, 2006) scanning probe-based lithography (Schwarz *et al.*, 1998) or soft-lithography technologies like microcontact printing (Zhao *et al.*, 1997; Bernard *et al.*, 1998; McDonald *et al.*, 2000; Unger *et al.*, 2000). This last technique uses hydrogel stamps for the parallel deposition of probes (James *et al.*, 1998; Kane *et al.*, 1999). In addition, several high-end technologies have been applied to generate sophisticated protein arrays. Protein solutions can be deposited on slightly conductive surfaces using electrospray deposition (ESD) (Morozov and Morozova, 1999). Atomic clusters prepared by physical methods represent a nanotechnology-based approach to immobilize and orientate proteins onto distinct surfaces (Palmer and Leung, 2007). Such approaches may be relevant for future protein microarrays in which no more than one specific protein molecule will be immobilized at distinct locations to study single-molecule interactions.

11.1.3. Microarray detection

A number of different detection technologies have been discussed and employed for microarray experiments. Captured targets are mainly detected by fluorescence using CCD cameras or laser scanners with confocal detection optics. A highly sensitive alternative to confocal optics is the application of planar waveguide excitation devices combined with CCD cameras or photomultipliers as detectors (Pawlak *et al.*, 1998, 2002; Rowe *et al.*, 1999; Lundgren *et al.*, 2000; Alexandre *et al.*, 2001; Graves *et al.*, 2002; Shi *et al.*, 2005; Timlin, 2006). This technology shows greater sensitivity with regard to signal intensity, linearity, signal-to-noise ratio, and background. Zeptosens AG, now a division of Bayer Technologies (www.zeptosens.com, Witterswil, Switzerland), has developed the Zepto™ READER which is based on the planar waveguide technology. Capture molecules are immobilized in a microarray format on a thin (100–200 nm) film (planar waveguide) which consists of material with a high refractive index (e.g., Ta_2O_5) deposited on a transparent support. A laser beam is optically coupled into the planar waveguide via a diffractive grating and is propagated in the thin layer, thereby creating a strong, surface-confined evanescent electromagnetic field. The penetration depth of this evanescent field into the adjacent medium is limited to about 200 nm. Thus, only surface-confined fluorophores are excited and emit fluorescent light. Fluorophores in the bulk medium are not excited and therefore not detectable. A CCD camera is used to detect fluorescent light with high spatial resolution. Parallel excitation and parallel detection of binding events on different spots is both highly selective and highly sensitive, even in solution.

A different detection technology was developed by Meso Scale Discovery (MSD) (www.mesoscale.com, Gaithersburg, Maryland, USA). Meso Scale Discovery uses a proprietary combination of electrochemiluminescence (ECL) detection and patterned arrays. Electrochemiluminescence detection offers a unique combination of sensitivity, dynamic range, and convenience. Electrochemiluminescence is a special form of chemiluminescence in which the light-emitting chemiluminescent reaction is preceded by an electrochemical reaction (Richter, 2004; see chapter by Richter). Electrochemiluminescent labels such as ruthenium chelates (tris(2,2′-bipyridyl)ruthenium (II) (Ru(II)(bpy)) are non-radioactive, stable, and emit electrochemically triggered light. The electrochemical reaction controls the time and position of the light-emitting reaction. By controlling the electrical stimulation of the emission, the generation of light can be delayed for as long as required for the antibody or enzyme-catalyzed reactions to take place. Multiplexing is possible with electrode arrays combined with appropriate optics. MSD's instruments use ultrasensitive photodetectors to collect and quantitatively measure light emitted from microelectrode arrays on the bottom of microtiter plates. Only labels near the electrode are excited and detected; washing of the assays is therefore not necessary (Debad, 2004).

11.1.4. Bead-based microarrays

Planar microarray-based systems are perfectly suited to screen for a large number of target proteins. However, bead-based systems have emerged as very promising alternatives, especially in cases when the number of parameters to be analyzed simultaneously is comparably low (Templin *et al.*, 2004). There is a growing list of commercially available, ready-to-use, multiplexed bead-based sandwich immunoassays which can quantify cytokines and cell-signaling molecules and can also analyze kinase activity (Biochipnet, www.biochipnet.de). Figure 11.1 shows a schematic of the commercial system Luminex, which performs multiplexed assays on up to 100 different bead sets. Each bead set is encoded with two dyes in a specific ratio for identification; assays are performed using a third, distinct fluorescent dye, which is used to quantify target bound to the relevant bead. Liquid array-based assays have the clear advantage over planar assays in that they are more flexible, more robust, and in a more advanced state of automation. Bead-based technology enables scientists to screen thousands of samples within a very short time. Hence, high-throughput applications using minimal amounts of samples are well within the capabilities of bead-based systems

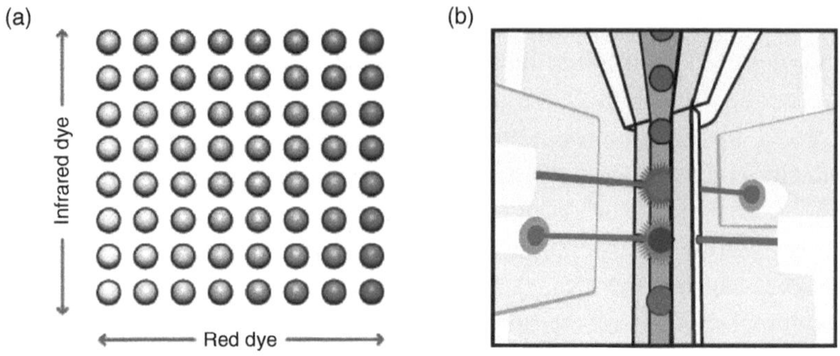

Figure 11.1 (a) Image represents bead regions dyed with 10 concentrations of two different classification fluorophores – red and near infrared – to produce 100 distinct bead "codes." Color coding enables each microsphere set to be classified individually and to be multiplexed with other microsphere sets. (b) Once inside the instrument, fluidics cause the microspheres to line up in single file as they pass by two lasers – a red laser to classify each color-coded microsphere for determining which assay is being carried out on that particle, and the green laser to measure the assay result on its surface. The presence and abundance of the reporter tag quantifies the occurring reactions precisely. (see Plate 3)

(Heuer *et al.*, 2004, 2005; Fath *et al.*, 2005; Perper *et al.*, 2006; Singh and Johnson, 2006; Toy *et al.*, 2006).

11.1.5. Array formats

11.1.5.1. Forward-phase arrays

Currently, forward-phase protein microarrays (Figure 11.2, *upper panel*) are the most frequently used microarray assay formats. They can be used for the simultaneous analysis of a large number of parameters from distinct samples, which are incubated on arrays of numerous microspots, each containing a single, well-defined, immobilized capture molecule. Examples of forward-phase protein microarrays include antibody microarrays and protein affinity arrays. Antibody microarrays are used to identify and quantify target proteins of interest. Protein affinity assays are used to study the interactions between proteins and immobilized binding molecules such as proteins, peptides, low molecular weight compounds, oligosaccharides, or DNA (MacBeath and Schreiber, 2000; Templin *et al.*, 2002).

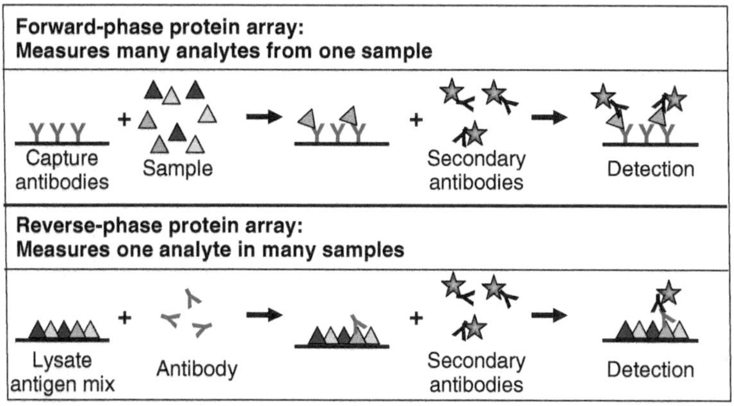

Figure 11.2 Model of microarray-based forward and reverse protein microarrays. Forward assays involve specific protein-capture molecules immobilized on distinct microspots. The entire array is incubated with one protein mixture and allows the measurement of many analytes from one single sample. Reverse arrays rely on the sample itself. Different samples are immobilized in a microarray format and probed with a specific antibody that binds to the immobilized antigen target present in the sample spot. This allows the determination of one specific analyte in many samples simultaneously.

11.1.5.2. Reverse-phase arrays

Reverse-phase arrays (Figure 11.2, *lower panel*) are a recent development in the field of protein microarrays. They can be used to determine a distinct set of parameters in a large collection of tissue or cell samples, or sample fractions that are immobilized in an array format on a solid support. Highly specific antibodies are incubated onto a reverse array and used to screen these tiny spots simultaneously for the presence or absence of distinct target proteins (Paweletz *et al.*, 2001a; Petricoin *et al.*, 2002; Yan *et al.*, 2003).

11.2. Historical background

The basic principles of miniaturized ligand-binding assays were first described in the early 1980s by Roger Ekins' Ambient Analyte Theory (Ekins, 1989). Miniaturized ligand-binding assays can achieve excellent levels of sensitivity. A small number of immobilized capture molecules capture a small proportion of analytes or target proteins onto a microspot. Therefore, the actual concentration of the analyte molecules in the sample does not change significantly, not even in the case of very low analyte concentrations and high affinity binding reactions. Microspot binding assays effectively determine the actual concentration of the analyte because the amount of analyte captured from the solution directly reflects its concentration in the solution. In the investigation of analytes such as thyroid-stimulating hormone or hepatitis B surface antigen, femtomolar concentrations could be detected (Finckh *et al.*, 1998). However, what pushed the development of microarrays forward was the need to process large amounts of information in the field of genomics. This was only possible by testing for all possible analytes simultaneously ("massive parallel testing") in which thousands of parameters could be determined in one single experiment. DNA microarrays, built from different capture oligonucleotides on individual "features" at a density of tens of thousands per square centimeter, are nowadays well-established systems that are able to analyze the whole transcriptome in a single experiment (Chipping Forcast II, 2002).

11.3. Applications for protein microarrays

11.3.1. Protein expression analysis – antibody arrays

Microarray-based protein expression analysis can be performed using the same two-color labeling approach – an approach which has been successfully used in differential mRNA expression analysis (Chipping Forcast II, 2002). Two different protein lysate samples derived from treated and non-treated cell lines or diseased and healthy tissues are labeled with two different fluorophores and are incubated simultaneously on a microarray containing a large number of different capture antibodies. Bound molecules are visualized using dual wavelength fluorescence. The difference in target protein concentration is immediately revealed. Figure 11.3 summarizes the parameters of antibody microarray experiments used for this type of protein expression analysis.

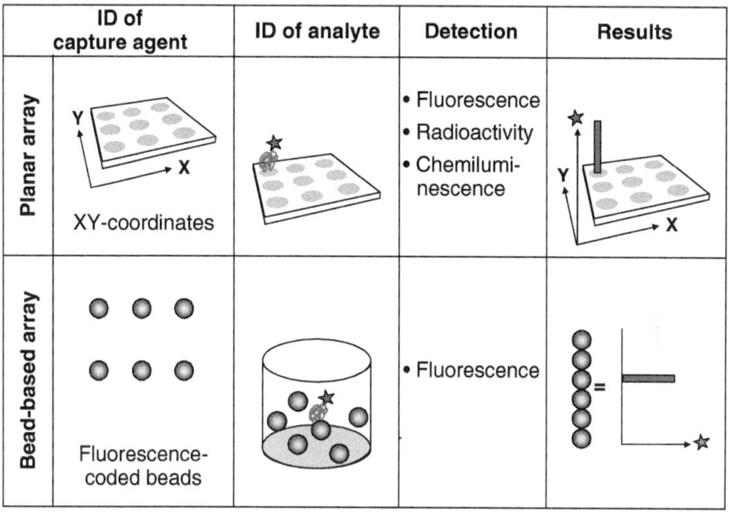

Figure 11.3 Comparison of planar and bead-based microarray formats. On planar microarrays, each specific capture agent is identified by its xy-coordinates. In bead-based systems, each capture agent is identified by the color or size-code of the bead to which the capture agent is attached. Different detection principles have been established for the identification of analytes on planar microarrays, including fluorescence-, chemiluminescence-, or radioactivity-based readout systems. Bead-based microarrays usually use fluorescence-based readout technology.

Using a direct labeling approach, theoretically, array-based proteome analysis can be carried out by using thousands of antibodies. However, as previously mentioned, the lack of appropriate highly specific capture molecules, combined with limited sensitivity for low abundance proteins, is the major drawback of this approach. Low abundance proteins may not be detectable due to the huge amount of label present in the system, resulting in a lower signal-to-noise ratio. Furthermore, the current high costs per experiment prevent the broad use of these types of antibody arrays in screening procedures.

In contrast, miniaturized, multiplexed sandwich immunoassays are much more advanced and provide an interesting and feasible solution for the growing demand in research and clinical applications for the simultaneous analysis of an increasing number of target proteins. Different types of miniaturized and parallelized immunoassays offer appropriate sensitivity and throughput (Figure 11.2, *upper panel*). The sandwich immunoassay format, which employs a labeled second, not necessarily highly specific, antibody, provides appropriate sensitivity. But one has to be aware that the higher the concentration of detection antibodies within a specific assay system, the higher the non-specific interaction, which might result in an increase in background. In practice, this means that unlimited multiplexed sandwich immunoassays are impractical. However, in a multiplexed assay system used for the detection of antibodies directed against hundreds of specific allergens or autoantibodies, the detection antibody only has to bind to one species of analyte, namely anti-human-antibody (Figure 11.2, *lower panel*). Therefore, protein arrays allow the screening of large numbers of different antibodies present in patient samples including autoantibodies (Robinson *et al.*, 2002; Gilburd *et al.*, 2004; Shovman *et al.*, 2005a, 2005b) or pathogens (Mezzasoma *et al.*, 2002). Some of the assay systems used for antibody analysis have already been cleared by the FDA for patient treatment.

11.3.2. Protein expression analysis – reverse-phase microarrays

Reverse-phase microarrays (Figure 11.2, *lower panel*) are an alternative assay format for a protein profiling approach. A multitude of different samples such as tissue or cell lysates are immobilized in a microarray format. Each microspot contains the whole proteome repertoire, equivalent to a single cell of a tissue or cell line. Single highly specific antibodies are used to simultaneously screen these microspots for the presence or absence of distinct target proteins. Sets of target proteins in large collections of tissue or cell samples, or sample fractions, can be determined using replicates of reverse-phase microarrays. The advantage of this screening method is the low amount of sample required and the possibility of screening a large number of tissue or cell lysates simultaneously (Paweletz *et al.*, 2001b; Petricoin and Liotta, 2002; Bouwman *et al.*, 2003; Chan *et al.*, 2004; Aguilar-Mahecha *et al.*, 2006). Proteins representing a millionth of the microspot can be detected in a complex protein mixture. The efficient analysis of molecules at very low concentrations remains problematic and it is hoped that a PCR-like amplification system will be developed for proteins. The high complexity of the lysate samples is another major problem experienced with low concentration molecules. Sensitivity could be increased by integrating a fractionation or enrichment step into the sample preparation procedure (Nam *et al.*, 2003; Yan *et al.*, 2003).

It is vital to optimize surfaces for protein-binding and suitable array substrates have been developed (Nijdam *et al.*, 2007). One advantage of reverse-phase microarrays is that there is no need to label the sample proteins and that hundreds of samples can be screened simultaneously – an approach that is able to support the early drug profiling process. Zeptosens' "CelLyA" cell lysate arrays have been shown to be suitable for screening defined sets of proteins using multiplexed, direct affinity assays in a much higher throughput than is possible with traditional Western blot technology. Currently, the most advanced reverse screening method, tissue microarrays enable the simultaneous screening of hundreds of paraffin-embedded tissues, whereas the traditional histological analysis of tissue specimens is rather slow and labor-intensive (Kononen *et al.*, 1998; Aguilar-Mahecha *et al.*, 2006; Persson *et al.*, 2006).

11.3.3. Protein interaction studies

Protein microarrays have become complementary tools to the yeast two-hybrid system and tandem affinity purification to study protein interactions *in vivo*. Arrays generated from fusion proteins can be used for the analysis of protein–protein, enzyme–substrate, protein–DNA, protein–oligosaccharide, and protein–drug interactions. Low and high density protein, peptide, and small molecule arrays have been used to investigate the binding of small chemical ligands, proteins, DNA, and RNA to their immobilized binding partners (Zhu *et al.*, 2001; Lizcano *et al.*, 2002; Michaud *et al.*, 2003, 2006; Satoh *et al.*, 2006). Whereas protein interaction studies involving full-length proteins allow the identification of interaction partners under experimental conditions, they do not provide information about the interaction sites. This kind of information can be provided by protein domain arrays or peptide arrays (Espejo *et al.*, 2002; Houseman *et al.*, 2002; Liu *et al.*, 2002; Burns-Hamuro *et al.*, 2003; Newman and Keating, 2003; Chamnongpol and Li, 2004; Mah *et al.*, 2005).

Miniature multiplexed assay systems are also suited for measuring kinase activity and specificity in a single experiment (Rychlewski *et al.*, 2004; Mah *et al.*, 2005; Ptacek *et al.*, 2005; Ptacek and Snyder, 2006). Another promising application is carbohydrate microarrays. The interactions of proteins and carbohydrates, the key components for glycoproteins, glycolipids, and proteoglycans, are essential for many biological processes (tissue growth and repair, cell–cell adhesion and inflammation, fertilization, viral replication, tumor cell motility, progression, etc.). They are, therefore, likely to be useful tools in the determination of different kinds of infections (Fukui *et al.*, 2002; Feizi *et al.*, 2003; Feizi and Chai, 2004; Wang *et al.*, 2005; Anjum *et al.*, 2006; Lee *et al.*, 2006). Protein microarrays have also been shown to be valuable tools in drug screening processes, in which the identification of drug candidates depends on the immobilization of small organic compounds and the subsequent screening for receptor–ligand interactions (MacBeath *et al.*, 1999; Jones *et al.*, 2006; Uttamchandani *et al.*, 2006).

Label-free detection systems are of general interest for analyzing the binding of a single purified biomolecule to immobilized capture candidates. Apart from the fact that no label has to be introduced into the biomolecule, label-free detection systems provide kinetic information as well as enable *in situ* identification of unexpected binding events. The lack of sensitivity, which is generally associated with any of the label-free detection systems, is not an important issue because well-defined amounts of purified binding molecules can be used. A recent article by Yu *et al.* (2006) summarizes the most important label-free detection methods such as surface plasmon resonance imaging, atomic force microscope applications, electrochemical impedance, and mass spectrometry.

11.4. Advantages and limitations of protein microarrays

Nowadays, after many years of development, microarrays are robust, reliable research tools with which a multitude of parameters can be screened from minimal amounts of sample. The acceptance of protein microarrays, due to the efficiency of sandwich immunoassays, is constantly growing, and they have become useful screening tools in biomarker screening programs, where panels of disease-specific biomarkers are generated. Furthermore, protein microarrays have been introduced into clinical trials to screen for potential adverse effects of potential drug candidates. Depending on the number of validated disease-specific biomarkers, as well as on their therapeutic relevance, such assays are performed either on a protein microarray or a bead-based platform. If only a few parameters have to be analyzed from a sample, which is often the case in clinical settings, then microarrays are not the first method of choice because the current highly automated diagnostic systems readily allow an increase in throughput.

Immune or allergy diagnostics, involving the screening of patient sera for the presence or absence of a relatively small number of different types of autoantibodies or antibodies directed against parasitic and viral antigens or allergens, are beginning to enter the diagnostic market. However, high content multiplexed analyses, containing hundreds of different capture agents, are still far from becoming a routine process. We will see whether the large efforts in biomarker identification, combined with the growing understanding of systems biology and cellular networks, will result in an increase of the number of parameters which have appropriate therapeutic relevance.

Medical need and clinical utility, combined with an overall cost reduction, must become the driving forces for protein arrays to gain a substantial share of the *in vitro* diagnostic market. Only then will protein arrays be able to live up to the high expectations they have raised.

References

Aguilar-Mahecha, A., Hassan, S., Ferrario, C., and Basik, M. (2006) *Omics*, **10**, 311.

Alexandre, I., Hamels, S., Dufour, S. *et al.* (2001) *Anal. Biochem.*, **295**, 1.

Anjum, M.F., Tucker, J.D., Sprigings, K.A. *et al.* (2006) *Clin. Vaccine Immunol.*, **13**, 561.

Bernard, A., Delamarche, E., Schmid, H. *et al.* (1998) *Langmuir*, **14**, 2225.

Blawas, A.S. and Reichert, W.M. (1998) *Biomaterials*, **19**, 595.
Bouwman, K., Qiu, J., Zhou, H. *et al.* (2003) *Proteomics*, **3**, 2200.
Burns-Hamuro, L.L., Ma, Y., Kammerer, S. *et al.* (2003) *Proc. Natl. Acad. Sci. USA*, **100**, 4072.
Chamnongpol, S. and Li, X. (2004) *Methods Mol. Biol.*, **264**, 183.
Chan, S.M., Ermann, J., Su, L. *et al.* (2004) *Nat. Med.*, **10**, 1390.
Chipping Forecast II (2002) *Nat. Genet.*, **32**, 461.
Debad, J.D., Glezer, E.N., and Wohlstadter, J.N. (2004) *Electrogenerated Chemiluminescence*. New York: Marcel Dekker, pp. 43–78.
Dontha, N., Nowall, W.B., and Kuhr, W.G. (1997) *Anal. Chem.*, **69**, 2619.
Ekins, R.P. (1989) *J. Pharm. Biomed. Anal.*, **7**, 155.
Espejo, A., Cote, J., Bednarek, A. *et al.* (2002) *Biochem. J.*, **367**, 697.
Fath, M.A., Mullins, R.F., Searby, C. *et al.* (2005) *Hum. Mol. Genet.*, **14**, 1109.
Feizi, T. and Chai, W. (2004) *Nat. Rev. Mol. Cell Biol.*, **5**, 582.
Feizi, T., Fazio, F., Chai, W., and Wong, C.H. (2003) *Curr. Opin. Struct. Biol.*, **13**, 637.
Finckh, P., Berger, H., Karl, J. *et al.* (1998) *Proc. UK NEQAS Meeting*, **3**, 155.
Fukui, S., Feizi, T., Galustian, C. *et al.* (2002) *Nat. Biotechnol.*, **20**, 1011.
Gaber, B.P., Martin, B.D., and Turner, D.C. (1999) *Chemtech*, **29**, 20.
Gilburd, B., Abu-Shakra, M., Shoenfeld, Y. *et al.* (2004) *Clin. Dev. Immunol.*, **11**, 53.
Graves, D.J., Su, H.J., Addya, S. *et al.* (2002) *Biotechniques*, **32**, 346.
Gutmann, O., Kuehlewein, R., Reinbold, S. *et al.* (2005) *Lab Chip*, **5**, 675.
Hermanson, G.T. (1996) *Bioconjugate Techniques.* San Diego, CA: Academic Press.
Heuer, J.G., Cummins, D.J., and Edmonds, B.T. (2005) *Expert Rev. Proteomics*, **2**, 669.
Heuer, J.G., Sharma, G.R., Gerlitz, B. *et al.* (2004) *Crit. Care Med.*, **32**, 1570.
Houseman, B.T., Huh, J.H., Kron, S.J., and Mrksich, M. (2002) *Nat. Biotechnol.*, **20**, 270.
James, C.D., Davis, R.C., Kam, L. *et al.* (1998) *Langmuir*, **14**, 741.
Jones, R.B., Gordus, A., Krall, J.A., and MacBeath, G. (2006) *Nature*, **439**, 168.
Kane, R.S., Takayama, S., Ostuni, E. *et al.* (1999) *Biomaterials*, **20**, 2363.
Kononen, J., Bubendorf, L., Kallioniemi, A. *et al.* (1998) *Nat. Med.*, **4**, 844.
Kricka, L.J., Master, S.R., Joos, T.O., and Fortina, P. (2006) *Ann. Clin. Biochem.*, **43**, 457.
Lauer, S.A. and Nolan, J.P. (2002) *Cytometry*, **48**, 136.
Lee, M.R., Park, S., and Shin, I. (2006) *Bioorg. Med. Chem. Lett.*, **16**, 5132.
Liu, M.Y., Cai, S., Espejo, A. *et al.* (2002) *Cancer Res.*, **62**, 6475.

Lizcano, J.M., Deak, M., Morrice, N. *et al.* (2002) *J. Biol. Chem.*, **277**, 27839.
Lundgren, J.S., Watkins, A.N., Racz, D., and Ligler, F.S. (2000) *Biosens. Bioelectron.*, **15**, 417.
MacBeath, G., Koehler, A., and Schreiber, S. (1999) *J. Am. Chem. Soc.*, **121**, 7967.
MacBeath, G. and Schreiber, S.L. (2000) *Science*, **289**, 1760.
Mah, A.S., Elia, A.E., Devgan, G. *et al.* (2005) *BMC Biochem.*, **6**, 22.
Master, S.R., Bierl, C., and Kricka, L.J. (2006) *Drug Discov. Today*, **11**, 1007.
McDonald, J.C., Duffy, D.C., Anderson, J.R. *et al.* (2000) *Electrophoresis*, **21**, 27.
Mezzasoma, L., Bacarese-Hamilton, T., Di Cristina, M. *et al.* (2002) *Clin Chem.*, **48**, 121.
Michaud, G.A., Salcius, M., Zhou, F. *et al.* (2003) *Nat. Biotechnol.*, **21**, 1509.
Michaud, G.A., Samuels, M.L., and Schweitzer, B. (2006) *IDrugs*, **9**, 266.
Morozov, V.N. and Morozova, T.Y. (1999) *Anal. Chem.*, **71**, 3110.
Nakanishi, K., Muguruma, H., and Karube, I. (1996) *Anal. Chem.*, **68**, 1695.
Nam, M.J., Madoz-Gurpide, J., Wang, H. *et al.* (2003) *Proteomics*, **3**, 2108.
Newman, J.R. and Keating, A.E. (2003) *Science*, **300**, 2097.
Nijdam, A.J., Ming-Cheng Cheng, M., Geho, D.H. *et al.* (2007) *Biomaterials*, **28**, 550.
Palmer, R.E. and Leung, C. (2007) *Trends Biotechnol.*, **25**, 48.
Paweletz, C.P., Charboneau, L., Bichsel, V.E. *et al.* (2001a) *Oncogene*, **20**, 1981.
Paweletz, C.P., Liotta, L.A., and Petricoin, E.F. 3rd (2001b) New technologies for biomarker analysis of prostate cancer progression: Laser capture microdissection and tissue proteomics. *Urology. Apr.*, **57** (4 Suppl 1), 160–3.
Pawlak, M., Grell, E., Schick, E. *et al.* (1998) *Faraday Discuss*, **111**, 273; discussion 331.
Pawlak, M., Schick, E., Bopp, M.A. *et al.* (2002) *Proteomics*, **2**, 383.
Perper, S.J., Browning, B., Burkly, L.C. *et al.* (2006) *J. Immunol.*, **177**, 2610.
Persson, A., Hober, S., and Uhlen, M. (2006) *Curr. Opin. Mol. Ther.*, **8**, 185.
Petricoin, E.E., Paweletz, C.P., and Liotta, L.A. (2002) *J. Mammary Gland Biol. Neoplasia*, **7**, 433.
Petricoin, E.F. and Liotta, L.A. (2002) *Trends Biotechnol.*, **20**, S30.
Pritchard, D.J., Morgan, H., and Cooper, J.M. (1995) *Anal. Chem.*, **67**, 3605.
Ptacek, J. and Snyder, M. (2006) *Trends Genet.*, **22**, 545.
Ptacek, J., Devgan, G., Michaud, G. *et al.* (2005) *Nature*, **438**, 679.
Richter, M.M. (2004) *Chem. Rev.*, **104**, 3003.
Robinson, W.H., DiGennaro, C., Hueber, W. *et al.* (2002) *Nat. Med.*, **8**, 295.
Rowe, C.A., Scruggs, S.B., Feldstein, M.J. *et al.* (1999) *Anal. Chem.*, **71**, 433.

Rychlewski, L., Kschischo, M., Dong, L. *et al.* (2004) *J. Mol. Biol.*, **336**, 307.
Satoh, J., Nanri, Y., and Yamamura, T. (2006) *J. Neurosci. Methods*, **152**, 278.
Schlottmann, S.A., Jain, N., Chirmule, N., and Esser, M.T. (2006) *J. Immunol. Methods*, **309**, 75.
Schwarz, A., Rossier, J.S., Roulet, E. *et al.* (1998) *Langmuir*, **14**, 5526.
Seong, S.Y. (2002) *Clin. Diagn. Lab Immunol.*, **9**, 927.
Shi, L., Tong, W., Su, Z. *et al.* (2005) *BMC Bioinformatics*, **6**(Suppl 2), S11.
Shovman, O., Gilburd, B., Barzilai, O. *et al.* (2005a) *Ann. NY Acad. Sci.*, **1050**, 380.
Shovman, O., Gilburd, B., Zandman-Goddard, G. *et al.* (2005b) *Autoimmunity*, **38**, 105.
Shriver-Lake, L.C., Donner, B., Edelstein, R. *et al.* (1997) *Biosens. Bioelectron.*, **12**, 1101.
Singh, M. and Johnson, L. (2006) *Clin. Cancer Res.*, **12**, 5312–28.
Stoll, D., Templin, M.F., Bachmann, J., and Joos, T.O. (2005) *Curr. Opin. Drug Discov. Develop.*, **8**, 239.
Stoll, D., Templin, M.F., Schrenk, M. *et al.* (2002) *Front Biosci.*, **7**, c13.
Templin, M.F., Stoll, D., Bachmann, J., and Joos, T.O. (2004) *Comb. Chem. High Throughput Screen*, **7**, 223.
Templin, M.F., Stoll, D., Schrenk, M. *et al.* (2002) *Trends Biotechnol.*, **20**, 160.
Timlin, J.A. (2006) *Methods Enzymol.*, **411**, 79.
Toy, D., Kugler, D., Wolfson, M. *et al.* (2006) *J. Immunol.*, **177**, 36.
Unger, M.A., Chou, H.P., Thorsen, T. *et al.* (2000) *Science*, **288**, 113.
Uttamchandani, M., Wang, J., and Yao, S.Q. (2006) *Mol. Biosyst.*, **2**, 58.
Vail, T.L., Cushing, K.W., Ingram, J.C., and St Omer, I. (2006) *Biotechnol. Appl. Biochem.*, **43**, 85.
Wacker, R., Schroder, H., and Niemeyer, C.M. (2004) *Anal. Biochem.*, **330**, 281.
Wang, R., Liu, S., Shah, D., and Wang, D. (2005) *Methods Mol. Biol.*, **310**, 241.
Waterboer, T., Sehr, P., Michael, K.M. *et al.* (2005) *Clin. Chem.*, **51**, 1845.
Yan, F., Sreekumar, A., Laxman, B. *et al.* (2003) *Protemics*, **3**, 1228.
Yu, X., Xu, D., and Cheng, Q. (2006) *Proteomics*, **6**, 5493.
Zhao, X.M., Xia, Y.N., and Whitesides, G.M. (1997) *J. Mat. Chem.*, **7**, 1069.
Zhu, H., Bilgin, M., Bangham, R. *et al.* (2001) *Science*, **2001**, 26.

Optical Biosensors: Today and Tomorrow (2nd Edition)
Frances S. Ligler and Chris Rowe Taitt (Editors)
Published by Elsevier B.V.

Chapter 12

SINGLE-DOMAIN ANTIBODIES: RUGGED RECOGNITION ELEMENTS FOR TOMORROW'S BIOSENSORS

Jinny L. Liu, Ph.D.[a], George P. Anderson, Ph.D.[a], Andrew Hayhurst, Ph.D.[b], and Ellen R. Goldman, Ph.D.[a]
[a]Center for Bio/Molecular Science and Engineering, Naval Research Laboratory, Washington, DC 20375-5348, USA
[b]Department of Virology and Immunology, Southwest Foundation for Biomedical Research, San Antonio, TX 78227, USA

Camelids and sharks possess a class of unconventional immunoglobulins consisting of heavy-chain homodimers; antigen binding is mediated through a single variable domain. When expressed recombinantly, these variable domains, referred to as "single-domain antibodies" (sdAb), comprise the smallest known antigen-binding fragments. Having good solubility, thermal stability, and complex sequence variation, these sdAb fragments make promising alternatives to traditional antibodies. Displayed libraries of sdAb derived from immunized and naïve camels, llamas, and sharks have been constructed and mined for binders to a variety of protein and small molecule targets. These antibody fragments have several properties that give them an advantage over conventional antibody fragments. They often possess long CDR3s allowing them to bind to epitopes not accessible to conventional antibodies. In addition, shark and camelid sdAb have been reported to function after exposure to temperatures over 90°C, and several groups have reported the binding of llama sdAb to their targets at elevated temperatures. SdAb are able to refold after exposure to chemical denaturants implying that sensor surfaces constructed using sdAb capture reagents could be regenerated. These sdAb represent a new generation of durable detection reagents and could potentially be integrated into any antibody-based biosensor. SdAb may one day permit biosensors to function at elevated temperatures and under environmental extremes for long periods of time.

12.1. Technical concept

12.1.1. Conventional antibodies

Many immunoassays rely on monoclonal or polyclonal antibodies (IgG) as recognition elements. Functional IgGs are comprised of four polypeptide chains, two identical heavy (H) chains, and two identical light (L) chains, linked by disulfide bonds. Each antibody has two antigen-binding domains formed by the interaction of adjacent variable (V) domains from the H and L chains. The antigen recognition moiety is maximally composed of six complementarity-determining regions (CDRs), three residing in each of the V_H and V_L protein domains. The interaction of a selection of these CDR loops of varying sizes and sequences allows the formation of diversified antigen-binding surfaces with the topologies to recognize a wide range of antigenic targets. Although sensitive and specific, conventional antibodies are generally time-consuming and expensive to develop and have limited stability (Vermeer and Norde, 2000). Figure 12.1 shows a schematic representation of an IgG as well as the cloned binding derivatives. Cloned derivatives of conventional IgG, comprising just the V_H and V_L to form a minimal antigen-binding construct, the Fv, were first expressed in *Escherichia coli* by Skerra and Plückthun (1988). Owing to the difficulties expressing stoichiometrically matched amounts of the separate domains and the relative instability of the resulting Fv, the two domains were expressed as one gene, linked by a flexible polypeptide linker (Bird *et al.*, 1988; Huston *et al.*, 1988). These more stable single-chain antibodies (scFv) are easily expressed in bacteria and can be modified by protein engineering to tailor their functionality and physicochemical properties. ScFv, however, are frequently less stable than the parental

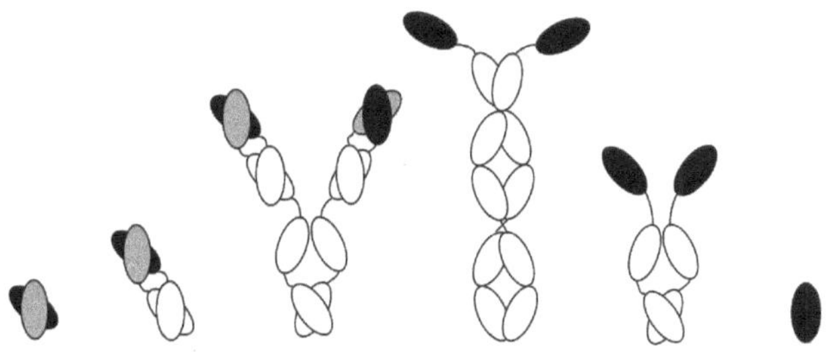

Figure 12.1 Schematic of conventional antibodies, Shark and camelid heavy-chain antibodies, and their cloned binding fragments.

full-length antibodies and often aggregate irreversibly at elevated temperatures due to their two-domain structure. Ideally, utilization of a single-domain structure capable of antigen binding would avert aggregation upon heating, and thereby facilitate the application of biosensors at elevated environmental temperatures or for continuous use over long periods of time.

12.1.2. Camelid and shark antibodies

In the mid-1990s, it was discovered that in addition to standard antibody isotypes containing both H and L chains, certain animals such as camelids (i.e., camels, llamas) and sharks possess a class of immunoglobulins consisting of heavy-chain homodimers where antigen binding is mediated through a single V domain (Figure 12.1, Hamers-Casterman *et al.*, 1993; Greenberg *et al.*, 1995). These V domains, when cloned separately as single-domain antibodies (sdAb), comprise the smallest known antigen-binding fragments (12–15 kDa). Despite their small size, sdAb display a high level of specificity and affinity for their antigens (Lauwereys *et al.*, 1998) and have been shown to have nanomolar affinities (K_D) for haptens (Spinelli *et al.*, 2000) and proteins (Nuttall *et al.*, 2003; Huang *et al.*, 2005; Zarebski *et al.*, 2005). SdAb can refold to bind antigen after chemical or heat denaturation, enabling them to retain the ability to bind antigen after exposure to elevated temperatures (Perez *et al.*, 2001; Dumoulin *et al.*, 2002). Several studies have found sdAbs to be inherently thermostable, demonstrating antigen binding at elevated temperatures (Van der Linden *et al.*, 1999; Ladenson *et al.*, 2006), which suggests they will be well suited for long-term field applications where refrigeration is often not possible. Recognition elements based on sdAb should offer the specificity of conventional antibodies with the advantage of use and storage at elevated temperatures and the capability of allowing regeneration of sensor surfaces.

12.2. History

12.2.1. Generation of single V_H domains

Prior to the discovery of sdAb, researchers experimented with antigen-binding fragments composed of the V_H domains of conventional antibodies (Ward *et al.*, 1989). These V_H fragments often exhibit specificity for antigen, as their CDR3 makes the major contribution to antigen binding. The affinity of these V_H domains, however, drops in comparison to the intact scFv (Borrebaeck *et al.*, 1992). In addition, it is difficult to produce soluble protein from isolated V_H domains; the large hydrophobic surface previously at the interface with the V_L domain is now exposed and greatly enhances aggregation. These factors make conventional V_H fragments a poor choice as an alternate recognition element. One approach to overcoming this limitation has been the "camelization" of conventional V_H fragments. By substituting more hydrophilic residues found in camelid heavy-chain V_Hs for the more hydrophobic residues normally involved with V_L interaction, this process has resulted in the production of functional and soluble binding elements derived from conventional antibodies (Davies and Riechmann, 1994).

12.2.2. Alternative binding elements and scaffolds

Artificial binding elements composed of random peptides expressed individually or incorporated into protein scaffolds, derived from non-immunoglobulin sources, have been used as recognition elements in biosensors. Alternative protein scaffolds have been investigated for potential advantages they offer over conventional immunoglobulin-based scaffolds. Desirable scaffold characteristics include small size, ability to access epitopes not otherwise accessible to antibodies, stability, and efficient protein production. Engineering sufficient molecular diversity into these scaffolds to produce binders to a wide range of antigens while conserving parental scaffold characteristics is not trivial. Alternatives to protein-based binding elements, such as aptamers and molecularly imprinted polymers, are detailed in chapters by Ellington and colleagues and by Turner and Piletsky, respectively.

12.2.2.1. Peptides

Artificial binding elements can be synthetically prepared by creating a pool of peptides with random sequences. These sequences can vary in length and be either constrained by fixed loops or disulfide bonds, or unconstrained. The latter unconstrained sequences have the advantage of being free to explore a large variety of conformational space, but are perhaps less stable. Peptide-binding elements are usually genetically encoded to confer selectability and have been used as recognition elements in biosensors (Goldman *et al.*, 2000).

12.2.2.2. Protein scaffolds

To address the issue of peptide conformational stability, random peptides can be incorporated into folded protein scaffolds. A variety of proteins have been utilized as alternative scaffolds including α-amylase inhibitor, thioredoxin, green fluorescent protein, staphylococcal nuclease, and fibronectin III. All are at least 50 amino acids in length and many have multiple disulfide bridges and one to several loops, which are the sites where the random peptide sequences are inserted. Protein A is also a good scaffold example; it contains 58 residues and has had variable amino acid sequences inserted within two α-helical segments (Nord *et al.*, 1995; Tolmachev *et al.*, 2007). A zinc finger scaffold of 26 residues with a bound Zn^{2+} has also been used and has a variable α-helix (Nygren and Skerra, 2004). Libraries consisting of strings of A-domains, termed "avimers," have been used to isolate binding elements that proved to have good affinity and specific while retaining thermostability, and protease resistance (Silverman *et al.*, 2005). The use of such a wide variety of scaffolds gives a huge assortment of stable tertiary structures on which to build the binding interaction necessary for high affinity, highly selective interactions. Examples of use of the these binding elements in biosensors include the integration of binders based on protein A scaffolds into microarrays (Renberg *et al.*, 2005) and green fluorescent protein derivatives that undergo a change in fluorescence on target binding (Doi and Yanagawa, 1999). In a sense, these approaches have been artificial attempts to achieve what nature has accomplished with sdAb.

12.2.3. Single-domain antibodies in camels and sharks

The heavy-chain-only antibodies of camelids and sharks bind their antigen through a single V domain. Interestingly, while binding regions of both species consist of a single domain, the shark and camel sdAb differ in both size and structure.

Camelid sdAb have up to 75% amino acid identity with other mammalian V regions (Roux *et al.*, 1998). Since sdAbs bind antigen in the absence of the extra structural information borne by a light chain V domain, they compensate in several important ways. The CDR3 loops of sdAbs are on average longer than those of conventional antibodies. These long CDR3 loops can protrude from the sdAb exposing the antigen to a three-dimensional "finger" of structural information akin to a long constrained peptide. Charged residues have replaced hydrophobic residues at the interface where a light V domain would have interacted with this region. These mutations enhance the solubility of the sdAb and prevent aggregation after denaturation to allow quantitative refolding (Perez *et al.*, 2001; Ewert *et al.*, 2002). SdAb genes have been shown to undergo high rates of somatic mutations, which can result in improved contact residues within CDRs and improved overall CDR geometry through framework mutations. Structural analysis has shown that although the scaffold architecture of sdAb and V_H are identical, the sdAb CDRs exhibit a larger structural repertoire than conventional V_H, possessing many deviations from the canonical structures (Decanniere *et al.*, 2000).

Shark heavy-chain-only antibody represents a new member of the immunoglobulin superfamily designated new antigen receptor (NAR) due to conserved amino acid residues involved in forming core of immunoglobulin and T-cell receptor V regions (Greenberg *et al.*, 1995). They are H chain homodimers with each H chain containing one single variable (V) and five constant (C) domains. Shark-derived sdAb (also designated NAR V) have only about 25% identity to conventional V_H (Roux *et al.*, 1998) showing higher sequence homology with immunoglobulin V_L chains (Stanfield *et al.*, 2004). V domain diversity is due to both rearrangement and somatic diversification mechanisms.

NAR V domains were first isolated and identified from the nurse shark, *Ginglymostoma cirratum*. By taking advantage of conserved framework residues (FR) at the N- and C-termini, NAR V from other shark species, such as spotted wobbegong shark (*Orectolobus maculates*), spiny dogfish shark (*Squalus acanthias*), and smooth dogfish shark (*Mustelus canis*), were subsequently cloned (Greenberg *et al.*, 1995; Nutall *et al.*, 2001; Liu *et al.*, 2007). NAR V lacks CDR2, making it the smallest antigen-binding domain among all recombinant antibodies (~12 kDa). Its complexity resides both in the CDR1 region with sequence variation within nine residues and in the extended CDR3 region, which varies in length (5–23 residues) and in amino acid composition.

Structural analysis by electron microscopy, 3D modeling, and recent crystal structure revealed that there are three NAR isotypes (Roux *et al.*, 1998; Diaz *et al.*, 2002; Stanfield *et al.*, 2004; Strelstov *et al.*, 2004). Type 1 NAR is characterized by a unique disulfide linkage between FR 2 and 4 and usually one or two disulfide linkages within extended CDR3 containing two or four Cys residues. Type 1 NARs have to date been observed only in nurse shark. Type 2 NARs are usually identified by an interloop disulfide linkage forming between CDR1 and CDR3. They appear as a dominant isotype in shark NARs. Type 3 NARs are recognized by a conserved Trp (W) 31 within CDR1 and display limited diversity in both the size and composition of CDR regions (Diaz *et al.*, 2002). They were found primarily in embryonic nurse sharks, but recently they have also been identified in adult spiny dogfish shark (Liu *et al.*, 2007). Shark NARs uniquely have additional hypervariable (HV) regions, HV2 and HV4; structural analysis demonstrated that HV also participate in antigen binding along with CDRs (Dooley *et al.*, 2006).

12.2.4. Isolating recombinant sdAb

Libraries of sdAb for the selection of specific recognition elements have been generated by cloning them into display systems. A variety of methods have been developed for the selection of recombinant binders. These methods include systems such as phage display (Scott and Smith, 1990; McCafferty *et al.*, 1990; Petrenko and Vodyanoy, 2003; Bradbury and Marks, 2004), yeast display (Boder and Wittrup, 2000; Feldhaus and Siegel, 2004), *E. coli* display (Francisco *et al.*, 1993), baculovirus/insect cell display (Grabherr and Ernst, 2001), and ribosome display (Schaffitzel *et al.*, 1999). Phage display has been the most commonly reported method of developing sdAb. First, total RNA is isolated from lymphocytes of immunized or non-immunized animals. The total RNA can then be reverse transcribed using oligo-dT primers. The resulting reverse transcripts are subsequently used as templates to amplify sdAb fragments flanked with desired restriction enzyme sites at both ends using polymerase chain reaction (PCR) (Ghahroudi *et al.*, 1997; Nuttall *et al.*, 2001). The amplified sdAb genes are then subjected to restriction enzyme digestion and cloned into appropriate M13 phage vectors. With proper *E. coli* hosts, such as XL 1 blue or TG1, sdAb fragments will be expressed and displayed on the surface of M13 phage. Each M13 phage can take up only one sdAb gene sequence and express a few copies of sdAb protein on its tail. Pools of M13 phage, each displaying a unique sdAb, are assembled. The complexity of the constructed library depends on the number of M13 phage with unique sequences inserted. Panning techniques allow the isolation of antigen-binding clones from the background of non-binders. SdAb genes isolated from animals previously immunized with the antigens of interest generate what is known as a "biased or immune library." Camels (Ghahroudi *et al.*, 1997; Lauwereys *et al.*, 1998), llamas (Dekker *et al.*, 2003), and nurse sharks (Dooley *et al.*, 2003) have all been immunized to yield high affinity binders. The process can take many months for the animals to raise a sufficiently high immune response. This route is often unavailable when targeting potentially lethal and transmissible agents.

Non-immunized animals can be used to generate unbiased, naïve, or non-immune libraries from which antigen binders can also be selected.

Since no *in vivo* affinity maturation has occurred, naïve libraries need to have a highly diverse antibody population to enhance the probability of having representatives capable of binding to the desired target antigen (Tanha *et al.*, 2002; Yau *et al.*, 2003; Zhang *et al.*, 2004; Liu *et al.*, 2007). However, germline diversity of sharks, llamas, and camels is considerably more limited than, for example, humans, thus the resulting binders are typically of low affinity. To improve the likelihood of isolating specific binding clones from a naïve sdAb library, CDRs may be randomized after the initial cloning of the sdAb genes to confer more diversity than occurs naturally (Nuttall *et al.*, 2001, 2003; Jobling *et al.*, 2003; Goldman *et al.*, 2006; Shao *et al.*, 2007). This process is described in Section 12.3.2. Once made, such "single-pot" non-immune libraries can be stored indefinitely and used to generate binding clones against an unlimited number of antigens since the library can be reamplified and the "pot" replenished with relatively simple techniques.

The ability to isolate and manipulate the genes of these recognition elements is a powerful asset. It enables the evolution of improved variants that have higher target affinities, improved specificities, or greater stability. Isolated proteins can also be engineered for improved function in biosensors, for example through the creation of linker variants.

12.2.4.1. First selections of antigen-specific camelid sdAb

The first camelid sdAb were selected from a phage display library prepared from a camel immunized with two model antigens, tetanus toxoid and lysozyme (Ghahroudi *et al.*, 1997). Binders toward both targets were isolated and found to be specific toward their target antigen, as well as to have increased stability when compared to conventional antibody fragments.

A non-immune llama-based library was used to select an anti-idiotypic sdAb to a monoclonal antibody that recognizes *Brucella O*-polysaccharide (Tanha *et al.*, 2002). Although conventional V_H fragments, not sdAb, were selected from the library, they showed good solubility and stability, and were able to bind target after incubations at temperatures up to 90°C.

12.2.4.2. First selections of antigen-specific shark sdAb
In the first report of the selection of shark-derived sdAb, the NAR V from wobbegong shark was used as a scaffold for the construction of a small library (3×10^7 clones) in which the CDR3 loop was randomized (Nuttall *et al.*, 2001). Specific binders were isolated toward the protease gingipain K, but not against lysozyme and the malarial apical membrane antigen-1 (AMA1). This work showed the potential of using a naïve shark-based sdAb library for the isolation of specific recognition elements.

Libraries generated from immunized nurse shark were employed for the isolation of lysozyme-binding shark-derived sdAb (Dooley *et al.*, 2003). The isolated sdAb were shown to be highly specific for their selected target, binding to hen egg lysozyme but not to irrelevant antigens such as turkey egg lysozyme. These V fragments were also shown to retain greater than 20% functionality after a 3-h incubation at 97°C.

12.2.5. Properties of sdAb

SdAb have been shown to have unique properties not commonly found in conventional antibody fragments. In addition to their small size, these features may offer advantage for use as recognition elements.

12.2.5.1. Expanded epitopes
SdAb panned against several enzyme targets were found to bind in the enzyme's active site and were shown to be excellent enzyme inhibitors (Lauwereys *et al.*, 1998). The sdAb's extended CDR3 "fingers" are able to penetrate the active sites, in contrast to the binding sites of conventional antibodies whose CDRs create a more planar, two-dimensional binding surface. This was documented through the crystal structure of an anti-lysozyme sdAb that also acted as an enzyme inhibitor (Desmyter *et al.*, 1996). The 24 amino acid CDR3 was observed extending into the active site of the lysozyme. It is envisaged that sdAb will be able to bind epitopes in clefts or cavities on an antigen's surface that are inaccessible to conventional antibodies (Stijlemans *et al.*, 2004). This may expand their role beyond that of simple recognition elements to that of active therapeutics.

12.2.5.2. Stability and regeneration of sdAb

The stability and ability to refold after denaturation are traits that make sdAb attractive as recognition elements. Several reports claim functional binding activity of isolated sdAb at temperatures up to 90°C (Van der Linden *et al.*, 1999; Ladenson *et al.*, 2006). One study showed that protein folding of sdAb was induced at high temperature (80°C) on addition of antigen (Dolk *et al.*, 2005a). Perez *et al.* (2001) showed that llama-derived sdAb can refold into their native conformation after thermal denaturation and many researchers have reported the function of sdAb after previous exposure to elevated temperatures. These results suggest that sdAb reagents should possess a good shelf life and may not require refrigerated storage. In addition to refolding after thermal denaturation, camelid-derived sdAb have also been shown to be active after chemical denaturation (Dumoulin *et al.*, 2002). This ability to refold after heat or chemical denaturation can lead to reusable sensor surfaces, an important trait for continuous monitoring applications. The potential for repetitive use of sensor surfaces was demonstrated by the use of a sdAb selected against an ice structuring protein as a capture reagent in immunoaffinity chromatography; the column was regenerated over 2000 times without loss of capacity (Verheeson *et al.*, 2003).

The ability of some sdAb to function in non-physiological complex matrices offers the opportunity to select recognition elements in the environment in which they are needed to perform for real-world applications. For example, sdAb selected in shampoo containing nonionic and anionic surfactants were more stable to these harsh conditions than those selected under physiological selection conditions (Dolk *et al.*, 2005b). The shampoo-tolerant sdAb were found to also be more stable in the presence of urea and guanidine HCl.

12.3. State of the art

12.3.1. Multivalent constructs

Both camelid- and shark-derived sdAb have been incorporated into multivalent and/or bivalent constructs. These reagents have the advantage that through avidity, namely the binding of multiple elements, they can have a higher apparent affinity than a monomeric sdAb.

The first multivalent sdAb constructs were based on camel sdAb and consisted both of bivalent and bispecific constructs in which two sdAb were tethered using the structural upper hinge of a natural llama antibody (Conrath *et al.*, 2001). The bivalent constructs recognized two separate epitopes on the same target, while the bispecific constructs recognized epitopes on separate antigens. As with the parental sdAb, all of the bispecific and bivalent constructs were found to have good solubility. The bivalent sdAb constructs reported in this work showed a decrease in the off-rate, leading to a fivefold avidity enhancement. Constructs retained stability, being able to function for extended periods of time after 37°C incubations.

Pentamers of sdAbs have been constructed through linking to an oligomerization domain, the B-subunit of *E. coli* verotoxin (Zhang *et al.*, 2004). The verotoxin B-subunit spontaneously self-associates to form pentamers. A peptide-specific sdAb from a naïve llama library was evaluated as a pentamer construct. The multimer construct was well expressed in *E. coli*, was soluble, and an avidity gain of three-to-four orders of magnitude was realized. The pentamer was found to be thermostable, and showed good resistance to digestion with several proteases. This pentamerization strategy was used in conjunction with sdAb selected against α-cobrotoxin, resulting in much improved functional affinity, allowing toxin detection in an ELISA format (Stewart *et al.*, 2006). A bispecific decavalent assembly with good solubility has recently been constructed through a construct linking sdAb on both sides of the verotoxin B (Stone *et al.*, 2007).

A number of strategies have been employed toward the construction of bivalent sdAb derived from shark (Simmons *et al.*, 2006). Generation of constructs with increased functional affinity was accomplished through the introduction of a C-terminal cysteine in the natural hinge from the shark NAR. Although these constructs showed increased avidity over the monomeric sdAbs, the level of protein expression was reduced 20–50-fold.

12.3.2. Protein engineering to improve stability and affinity

Strategies including DNA shuffling, random PCR, and mutational hotspot randomization have been used to generate and select sdAb variants with improved protein expression, stability, and affinity (Van der Linden *et. al.*, 2000; Nuttall *et al.*, 2004; Yau *et al.*, 2005; Harmsen *et al.*, 2006). Alternatively, researchers have searched for a stable camelid sdAb framework onto which CDRs derived from less stable sdAb could be grafted (Saerens *et al.*, 2005b). The ability to design or tailor sdAb to meet an application's particular needs is a strong advantage that favors their future adoption in detection schemes.

12.3.2.1. DNA shuffling

The technique of DNA shuffling has been used to improve the functional characteristics of proteins (Stemer, 1994). The technique works by randomizing sequences from different homologous genes and reassembling them to create a library that contains novel genes with combined characteristics of the parental sequences. DNA shuffling has been successfully employed to produce sdAb with improved production levels in yeast, increased affinity toward targets, improved thermal stability, and improved proteolytic stability.

Llama-derived sdAb specific for the azo-dye reactive red-6 (RR-6) were the starting point for DNA shuffling for the purpose of improving protein production in yeast (Van der Linden *et al.*, 2000). Three sdAb sequences with significant differences in expression and functional characteristics were chosen as the starting sequences. Sixteen shuffled RR-6-specific clones were characterized and all found to be composed of two or three parental sequences; three had additional point mutations not found in the parental sequences. All clones maintained specificity toward target and two of the clones showed improved protein production over the best expressing parental sdAb. One of these clones also showed a fourfold higher affinity toward target than the original sequences while the other showed improved thermal stability.

DNA shuffling was also utilized to isolate llama sdAb specific for *E. coli* f4 fimbriae with improved proteolytic stability (Harmsen *et al.*,

2006). Three parental DNA sequences were shuffled to create a secondary phage display library. This secondary library was subjected to a single round of selection after phage incubation in jejunal or gastric fluid and four clones with increased stability were isolated. Although the ultimate purpose of these experiments was the development of an immunotherapeutic for the prevention of piglet diarrhea, the development of proteolytically stable reagents could also be of importance for biosensor applications.

12.3.2.2. Error-prone PCR

The technique of error-prone PCR introduces random mutations to a parental sequence (Fromant *et al.*, 1995), leading to a library of variants that can be screened for improved qualities. This technique can be used when there are limited binding sequences and it makes no assumptions about the importance of any one region of the protein sequence.

A shark-derived sdAb specific for the AMA1 from *Plasmodium falciparum* served as the parental sequence for a secondary library created by error-prone PCR (Nuttall *et al.*, 2004). Interestingly, the mutations in the clones isolated from the secondary library mapped to the CDR1 and CDR3 loops of the sdAb. The best variant showed a 10-fold enhanced affinity. The enhanced affinities were shown to be due to slower off-rates.

Error-prone PCR has also been used to diversify a naïve library in order to increase the probability of being able to select binders to any target (Goldman *et al.*, 2006). A combination of error-prone PCR and splice overlap extension PCR was used in the construction of a semi-synthetic llama library. This library has generated binders to all toxin and viral targets against which it has been panned.

12.3.2.3. Mutational hotspot randomization

DNA shuffling and error-prone PCR offer non-targeted approaches for affinity maturation, while hotspot randomization is a targeted method. This method randomizes short sequences within CDRs that are the preferred site of mutations during *in vivo* somatic hypermutation events

(Betz *et al.*, 1993). The amino acids encoded by these codons are often involved in antigen binding.

Using this technique nine "hotspot codons," AGY/RGYW (R = A or G; Y = C or T; W = A or T) within CDR2 and CDR3 of an anti-parathyroid hormone (PTH) sdAb were randomized (Yau *et al.*, 2005). One sdAb was identified having a ~30-fold improved affinity over that of the parental sequence. From their data, the authors hypothesize that some of the mutational hotspot codons impact directly sdAb solubility and structure, as well as deletion/insertion events.

12.3.2.4. Development of a stable universal sdAb framework

SdAb vary in their stability as well as in their ability to be expressed and purified as soluble proteins. It would be advantageous to have an sdAb framework with both good thermal stability and high protein production on which to graft CDR loops from any sdAb. An anti-β-lactamase sdAb with good expression and stability was chosen as a candidate for use as a universal scaffold (Saerens *et al.*, 2005b). CDR loops from camel subfamily 2-derived sdAb, with different expression levels and low (nM) dissociation constants for their targets were grafted onto the anti-β-lactamase sdAb framework. The expression of all the chimeras was comparable or higher than the level of the loop donor; the expression of a chimera, which included the loops from a low expressing sdAb, was drastically improved. The equilibrium dissociation constants of the chimeras and original sdAb were similar, and most of the chimeras were more stable than the parental sdAb, although not as stable as the native scaffold. Grafts of CDRs from sdAb from different CDR subfamilies failed to produce high affinity binding chimeras, showing that the concept of a single universal V_H scaffold may not be feasible.

12.3.3. Biosensing using sdAb recognition elements

The biophysical properties of sdAb make them attractive recognition elements in biosensors. Interestingly, although there has been a fair amount of characterization of sdAb, examples of their use in biosensors for real-world applications are limited.

The specificity and stability of sdAbs was exploited in their use in sensing human prostate-specific antigen (hPSA, Huang *et al.*, 2005; Saerens *et al.*, 2005a). This work focused on using the BIAcore surface plasmon resonance sensor. Different sdAb constructs were immobilized onto sensor surfaces through metal chelation, biotin–streptavidin, or covalent coupling, and examined for their ability to detect hPSA (Saerens *et al.*, 2005a). Engineering a specific biotinylation sequence at the C-terminus of the sdAb allowed directional immobilization on streptavidin surfaces. Using this configuration, the researchers saw a fivefold improvement in antigen detection; regeneration cycles removed only 0.1% of the immobilized sdAb. Direct chemical coupling was also accomplished with sdAb to which extra lysine residues had been added at the C-terminus in an effort to improve the chance of directional immobilization. Unfortunately, the additional lysines raised the pI of the sdAb and led to suboptimal protein coupling to the sensor chip. Covalent coupling of the standard (unmodified) sdAb led to the best limits of detection and regeneration removed less than 0.01% of the immobilized sdAb. Using the covalently attached sdAb, 0.19ng/ml hPSA could be detected. By employing a combination of multiple sdAb, clinically significant concentration ranges of free hPSA were determined; the capture reagent recognized both free and complexed antigen, and the tracer reagent was an sdAb reconstituted in an intact heavy-chain antibody form that specifically bound to free hPSA. Use of sdAb as capture reagents led to a large increase in the amount of functional sdAb after each regeneration compared to the amount of functional conventional antibody after regeneration cycles.

An anti-caffeine sdAb has been characterized and used to measure the levels of caffeine in assorted hot and cold beverages, laying the groundwork toward the development of a dipstick format immunoassay (Ladenson *et al.*, 2006). Conventional anti-caffeine antibodies lost virtually all activity after heating for 20 min at 70°C, while the sdAb retained over 90% of its binding activity after 20 min incubations at higher temperatures (up to 90°C). A thermally stable biotinylated nanopeptide was conjugated to caffeine for testing of the sdAb binding at 70°C. Samples containing sdAb and the caffeine–peptide conjugate were incubated at 70°C, after which the sdAb bound to the caffeine conjugate was removed using streptavidin-coated magnetic beads, and the remaining sdAb measured. Results showed that the sdAb bound target at the elevated temperatures. A competition assay was developed to measure caffeine in beverages. Results agreed well with HPLC measurements.

Our group has used both shark- and llama-derived sdAb to detect toxins using the Luminex 100 platform, a specialized flow cytometer for performing multiplexed sandwich fluoroimmunoassays on the surface of coded microspheres (Goldman *et al.*, 2006; Liu *et al.*, 2007). We selected binders toward ricin, cholera toxin (CTX), and staphylococcal enterotoxin B (SEB) from a semi-synthetic non-immune llama sdAb library (Goldman *et al.*, 2006). For each of the toxin targets, we examined direct binding, employing the Luminex 100 to ensure the specificity of the isolated sdAb (Figure 12.2). We compared direct binding of sdAb to both their cognate antigen and irrelevant antigens. In addition to showing the toxin specificity, we evaluated binding to immobilized A and B chains of ricin and showed that the sdAb binds to the ricin B chain. Toxin-binding sdAb were shown to recognize antigen in solution as well as immobilized on a surface.

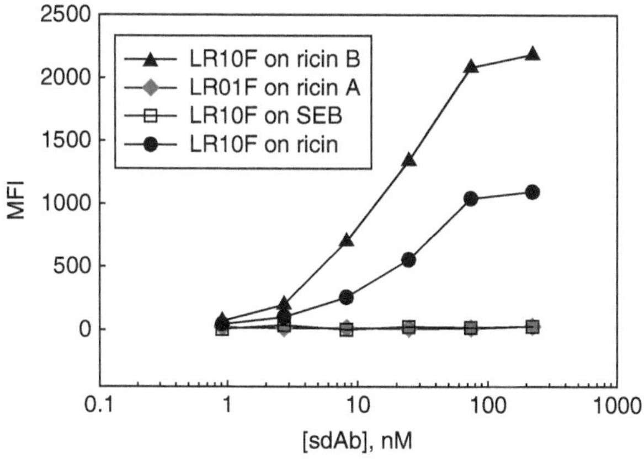

Figure 12.2 Specificity of anti-ricin sdAb. The sdAb shows binding to ricin and the ricin B chain but not the ricin A chain or the irrelevant toxin SEB.

Thermal stability was initially assayed by heating to 95°C for 5 min, allowing them to cool, and then performing direct binding assays at room temperature. The majority of sdAb retained greater than 90% of their binding activity after this treatment while most of the conventional antibodies retained only 10% binding ability. As a further test of thermal stability and an indicator of improved shelf life of the sdAbs, sdAb and conventional antibodies were heated to 95°C for various lengths of time up to 90 min (Figure 12.3). All examined sdAb proved to be more stable than conventional antibodies when subjected to prolonged heating. Some retained 100% of their binding ability even after heating to 95°C for 45 min. Other sdAb retained close to 100% binding activity for the first ~10 min and then slowly lost activity over the course of 60–90 min. The majority of conventional antibodies lost ~90% of their binding activity after the first 5 min of heating, while the best conventional antibody lasted no more than 20 min at 95°C prior to losing ~90% of its binding activity. Our characterized sdAb were tested as both capture and reporters in sandwich immunoassays. Although we were able to find sdAb specific for each toxin that functioned well as immobilized capture reagents, some of the sdAbs tested did not function in this role, potentially because they were immobilized in a way that precluded binding of the target toxin. The sdAb functioned better as

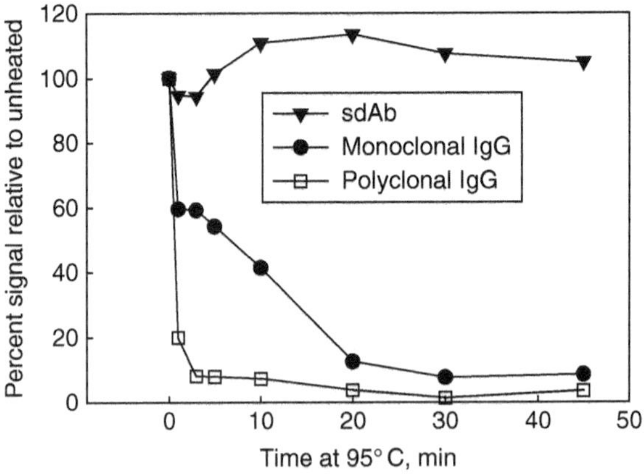

Figure 12.3 Binding ability of an anti-ricin sdAb, monoclonal anti-ricin, and polyclonal anti-ricin after heating to 95°C for various time points.

tracer reagents. These first generation binders were found to have fast off-rates; thus their performance as reagents should be improved with affinity maturation or multimeric constructs that effectively decrease the off-rates.

Binders toward CTX were selected from a naïve shark library (Liu *et al.*, 2007). We isolated binders that were specific for target and able to be competed by soluble toxin. Thermal stability and ability to properly refold were assessed for both soluble sdAbs and conventional polyclonal anti-CTX by incubation at 42, 65, and 95°C for times ranging from 10 to 60 min, followed by cooling and subsequent direct binding assay. At 42°C, there was no decrease in sdAb-binding activity in 60 min and only a slight decrease of binding signal for polyclonal anti-CTX. Interestingly, at 65°C, we noticed that there was a substantial increase in sdAb binding, but a substantial decrease in polyclonal anti-CTX binding after 60 min of heating. The heat denaturation at 65°C might unfold misfolded sdAb, which then, when cooled may refold to a more active configuration. Binding to irrelevant proteins (ricin and SEB) remained negligible. After 10 min at 95°C, we saw a total loss of polyclonal anti-CT-binding activity and 85% residual binding activity for sdAb. There was about 15% residual sdAb-binding activity retained after 60 min of heating at this temperature. Our results suggested that sdAb can endure heat denaturing at temperatures up to 95°C for short periods, with complete recovery upon cooling; however, long-term incubation of the unfolded sdAb eventually leads to permanent denaturation, likely due to an aggregation process. Isolated anti-CTX binders were used as reporter elements in sandwich assays; however, none of these first anti-CTX shark-derived sdAb functioned well when immobilized on the sensing surface.

12.4. Advantages and limitations for use in optical biosensing

Single-domain antibodies have many advantages over conventional antibodies and other alternate recognition elements. Highly specific sdAb reagents against almost any target can be isolated from naïve libraries in a matter of days. Their small size can be a big advantage as capture reagents, allowing a higher density of capture element per unit area. The ability to refold after denaturation can allow for multiple uses of a sensor surface, saving reagents. Thermal stability may translate into increased shelf life and the ability to operate sensors long term in hot climates, and to store reagents without refrigeration. Increased stability could also lead to the ability to measure in environmental matrices without dilution or the addition of buffer.

Currently, there are several limitations to the use of sdAb in biosensors. Although sdAb can be rapidly developed from naïve libraries, the highest affinity reagents have been derived from immunized animals. Using animals adds both time and expense to reagent development. Attempts at directional immobilization of sdAb through the His tag, C-terminal biotinylation sites, or C-terminal lysines have, to date, shown only limited success (Saerens *et al.*, 2005a). Although covalent immobilization has led to functionalized sensor surfaces (Saerens *et al.*, 2005a; Goldman *et al.*, 2006), many of the sdAb may be immobilized in orientations unsuitable for antigen binding. We have found that several sdAb that work well as tracer reagents fail when covalently immobilized on sensor surfaces.

12.5. Potential for improving performance or expanding current capabilities

Most of the perceived limitations of sdAb recognition elements can be overcome. The construction of improved naïve libraries combined with efficient affinity maturation protocols can provide high affinity reagents without the need for animal immunizations. Affinities of scFv and Fab have been improved by several orders of magnitude through affinity maturation; the same should be possible with sdAb. On the other hand, in many situations the creation of immunized libraries may be a viable route for obtaining the desired high affinity binders.

In either case, the continued development of constructs for efficient directional immobilization should provide the key to the preparation of highly stable, reusable sensor surfaces with a high coverage of capture reagents. Protein engineering offers the potential to create fusions of sdAb with protein domains to impart specific functionalities, such as the construction of monomolecular protein lattices that can be used to pattern surfaces (Pleschberger *et al.*, 2003, 2004); this cannot currently be accomplished with conventional IgG. Directed evolution enables the selection of sdAb variants with desirable traits such as improved thermodynamic stability and proteolytic stability (Jung *et al.*, 1999; Van der Linden *et. al.*, 2000; Udea *et al.*, 2004; Harmsen *et al.*, 2006). These features combined make sdAb ideal reagents for use in the development of tomorrow's rugged biosensors.

Acknowledgements

Our work was supported by a grant from the USAA Foundation (a charitable trust), SFBR donor funds, NIH Research Facilities Improvement Program Grant #CO6 RR12087, the JSTO CBD, and the Office of Naval Research. The views, opinions, and/or findings described in this report are those of the authors and should not be construed as official Department of the Navy positions, policies, or decisions.

References

Betz, A.G., Rada, C., Pannell, R. *et al.* (1993) *Proc. Natl Acad. USA*, **90**, 2385.

Bird, R.E., Hardman, K.D., Jacobson, J.W. *et al.* (1988) *Science*, **242**, 423.

Boder, E.T. and Wittrup, K.D. (2000) *Methods Enzymol.*, **328**, 430.

Borrebaeck, C.A.K., Malmborg, A.C., Furebring, C. *et al* (1992), *Biotechnology*, **10**, 697.

Bradbury, A.R.M. and Marks, J.D. (2004) *J. Immunol. Methods*, **290**, 29.

Conrath, K.E., Lauwereys, M., Wyns, L., and Muyldermans, S. (2001) *J. Biol. Chem.*, **276**, 7346.

Davies J. and Riechmann, L., (1994) *FEBS Lett.*, **339**, 285.

Decanniere, K., Muyldermans, S., and Wyns, L. (2000) *J. Mol. Biol.*, **300**, 83.

Dekker, S., Toussaint, W., Panayotou, G. *et al.* (2003) *J. Virol.*, **77**, 12132.

Desmyter, A., Transue, T.R., Ghahroudi, M.A. *et al.* (1996) *Nat. Struct. Biol.*, **3**, 803.

Diaz, M., Stanfield, R.L., Greenberg, A.S., and Flajnik, M.F. (2002) *Immunogenetics*, **54**, 501.

Doi, N. and Yanagawa, H. (1999) *FEBS Lett.*, **453**, 305.

Dolk, E., van Vliet, C., Perex, J.M.J. *et al.* (2005a), *Proteins Struct. Funct. Bioinformatics*, **59,** 555.

Dolk, E., van der Vaart, M., Hulsik, D.L. *et al.* (2005b) *Appl. Environ. Microbiol.*, **71**, 442.

Dooley, H., Flajnik, M.F., and Porter, A.J. (2003) *Mol. Immunol.*, **40**, 25.

Dooley H., Stanfield, R.L., Brady, R.A., and Flajnik, M.F. (2006) *PNAS*, **103**, 1846.

Dumoulin, M, Conrath, K., Van Meirhaeghe, A. *et al.* (2002) *Protein Sci.*, **11**, 500.

Ewert, S., Cambillau, C., Conrath, K., and Pluckthun, A. (2002) *Biochemistry*, **41**, 3628.

Feldhaus, M.J. and Siegel, R.W. (2004) *J. Immunol. Methods.*, **290**, 69.

Francisco, J.A., Campbell, R., Iverson, B.L., and Georgiou, G. (1993) *Proc. Natl Acad. Sci. USA*, **90**, 10444.

Fromant, M., Blanquet, S., and Plateau, P. (1995) *Anal. Biochem.*, **224**, 347.

Ghahroudi, M.A., Desmyter, A., Wyns, L. *et al.* (1997) *FEBS Lett.*, **414**, 521.

Goldman, E.R., Anderson, G.P. Liu, J.L. *et al.* (2006) *Anal. Chem.*, **78**, 8245.

Goldman, E.R., Pazirandeh, M.P., Mauro, J.M. *et al.* (2000) *J. Mol. Recognit.*, **13**, 382.

Grabherr, R. and Ernst, W. (2001) *Comb Chem. High Throughput Screen.*, **4**, 185.

Greenberg, A.S., Avila, D., Hughes, M. *et al.* (1995) *Nature*, **374**, 168.

Hamers-Casterman, C., Atarhouch, T., Muyldermans, S. *et al.* (1993) *Nature*, **363**, 446.

Harmsen, M.M., van Solt, C.B., van Zijderveld-van Bemmel, A.M. *et al.* (2006) *Appl. Microbiol. Biotechnol.*, **72**, 544.

Huang, L., Reekmans, G., Saerens, D. *et al.* (2005) *Biosens. Bioelectron.*, **21**, 483.

Huston, J.S., Levinson, D., Mudgetthunter, M. *et al.* (1988) *Proc. Natl. Acad. Sci. USA*, **85**, 5879.

Jobling, S.A. Jarman, C., Teh, M.M. *et al.* (2003) *Nat. Biotechnol.*, **21**, 77.

Jung, S, Honegger, A., and Pluckthun, A. (1999) *J. Mol. Biol.*, **294**, 163.

Ladenson, R.C., Crimmins, D.L., Landt, Y., and Ladenson J.H. (2006) *Anal. Chem.*, **78**, 4501.

Lauwereys, M., Ghahroudi, M.A., Desmyter, A. *et al.* (1998) *EMBO J.*, **17**, 3512.

Liu, J.L., Anderson, G.P., Delehanty, J.B. *et al.* (2007) *Mol. Immunol.*, **44**, 1775.
McCafferty, J., Griffiths, A.D., Winter, G., and Chiswell, D.J. (1990) *Nature*, **348**, 552.
Nord, K., Nilsson, J., Nilsson, B. *et al.* (1995) *Protein Eng.*, **8**, 601.
Nuttall, S.D., Humberstone, K.S., Krishnan, U.V. *et al.* (2004) *Proteins Struct., Funct. Bioinformatics*, **55**, 187.
Nuttall, S.D., Krishnan, U.V., Doughty, L. *et al.* (2003) *Eur. J. Biochem.*, **270**, 3543.
Nuttall, S.D., Krishnan, U.V., Hattarki, M. *et al.* (2001) *Mol. Immunol.*, **38**, 313.
Nygren, P.A and Skerra, A. (2004) *J. Immunol. Methods*, **290**, 3.
Perez, J.M.J., Renisio, J.G., Prompers, J.J. *et al.* (2001) *Biochemistry*, **40**, 74.
Petrenko, V.A. and Vodyanoy, V.J. (2003) *J. Microbiol. Meth.*, **53**, 253.
Pleschberger, M., Neubauer, A., Egelseer, E.M. *et al.* (2003) *Bioconjugate Chem.*, **14**, 440.
Pleschberger, M., Saerens, D., Weigert, S. *et al.* (2004) *Bioconjugate Chem.*, **15**, 664.
Renberg, B., Shiroyama, I., Engfeldt, T. *et al.* (2005) *Anal. Biochem.*, **341**, 334.
Roux, K.H., Greenberg, A.S., Greene, L. *et al.* (1998) *Proc. Natl. Acad. Sci. USA*, **95**, 11804.
Saerens, D., Frederix, F., Reekmans, K. *et al.* (2005a) *Anal. Chem.*, **77**, 7547.
Saerens, D., Pellis, M., Loris, R. *et al.* (2005b) *J. Mol. Biol.*, **352**, 597.
Schaffitzel, C., Hanes, J., Jermutus, L., and Pluckthun, A. (1999) *J. Immunol. Methods*, **231**, 119.
Scott, J.K. and Smith, G.P. (1990) *Science*, **249**, 386.
Shao, C.Y., Secombes, C.J., and Porter, A.J. (2007) *Mol. Immunol.*, **44**, 656.
Simmons, D.P., Abregu, F.A., Krishnan, U.V. *et al.* (2006) *JIM*, **315**, 171.
Silverman, J., Lu, Q., Bakker, A. *et al.* (2005) *Nat. Biotechnol.*, **23**, 1556.
Skerra, A. and Pluckthun, A. (1988) *Science*, **240**, 1038.
Spinelli, S., Frenken, L.G.J., Hermans, P. *et al.* (2000) *Biochemistry*, **39**, 1217.
Stanfield R.L., Dooley, H., Flajnik, M.F., and Wilson, I.A. (2004) *Science*, **305**, 1770.
Stijlemans, B., Conrath, K., Cortez-Retamozo, V. *et al.* (2004) *J. Biol. Chem.*, **279**, 1256.
Stemmer, W.P.C. (1994) *Nature*, **370**, 389.
Stewart, C.S., MacKenzie, C.R., and Hall, C.J. (2007) *Toxicon.*, **49**, 699.
Stone, E., Hiram. T., Tanha, J. *et al.* (2007) *JIM*, **318**, 88.
Streltsov, V.A., Varghese, J.N., Carmichael, J.A. *et al.* (2004) *Proc. Natl. Acad. Sci. USA*, **101**, 12444.

Tanha, J., Dubuc, G., Hirama, T. *et al.* (2002) *JIM*, **263**, 97.
Tolmachev, V., Orlova, A., Nilsson, F.Y. *et al.* (2007) *Expert Opin. Biol. Th.*, **7**, 555.
Ueda, H., Kristensen, P., and Winter, G. (2004) *J. Mol. Catal. B Enzym.*, **28**, 173.
Van der Linden, R.H.J., de Geus, B., Frenken, L.G.J. *et al.* (2000) *J. Biotechnol*, **80**, 261.
Van der Linden, R.H.J., Frenken, L.G.J., de Geus, B. *et al.* (1999) *Biochim. Biophys. Acta*, **1431**, 37.
Verheesen, P., ten Haaft, M.R., Lindner, N. *et al.* (2003) *BBA*, **1624**, 21.
Vermeer, A.W.P. and Norde, W. (2000) *Biophys. J.*, **78**, 394.
Ward, E.S., Gussow, D., Griffiths, A.D. *et al.* (1989) *Nature*, **341**, 544.
Yau, K.Y.F., Groves, M.A.T., Li, S.H. *et al.* (2003) *J. Immunol. Methods*, **281**, 161.
Yau, K.Y.F., Dubuc, G., Li, S.H. *et al.* (2005) *JIM*, **297**, 213.
Zarebski, L.M., Urrutia, M., Goldbaum, F.A. (2005) *J. Mol. Biol.*, **349**, 814.
Zhang, J.B., Tanha, J., Hirama, T. *et al.* (2004) *J. Mol. Biol.*, **335**, 49.

Optical Biosensors: Today and Tomorrow (2nd Edition)
Frances S. Ligler and Chris Rowe Taitt (Editors)

Chapter 13

NUCLEIC ACIDS FOR REAGENTLESS BIOSENSORS

Eun Jeong Cho, Ph.D.[a], Joo-Woon Lee, Ph.D.[b], Manjula Rajendran, Ph.D.[c], and Andrew D. Ellington, Ph.D.[a]

[a]Department of Chemistry and Biochemistry, University of Texas at Austin, Austin, TX 78712, USA

[b]Division of Liberal Arts & Sciences, Chungju National University, Chungju, Chungbuk 380-702, Korea

[c]Althea Technologies, Inc., 11040 Roselle Street, San Diego, CA 92121, USA

In vitro selection has yielded a range of nucleic acid-binding species (aptamers) and catalysts (ribozymes) whose ligand-binding affinities and activation parameters rival those of proteins. Precisely because functional nucleic acids can be engineered based largely on an appreciation of their secondary structures and chemically synthesized in bulk, it has proven remarkably easy to incorporate aptamers and ribozymes into reagentless biosensors that directly transduce ligand recognition to optical signals. Aptamers undergo conformational changes upon interaction with their cognate ligands, and, by appending fluorophores to aptamers, it has proven possible to generate "signaling aptamers." Ribozymes can act on fluorogenic substrates to generate fluorescent signals; any molecules or processes that affect ribozyme catalysis can therefore be reported. Aptamers can also be appended to ribozymes to generate aptazymes, or effector-activated ribozymes, that transduce molecular recognition to ribozyme catalysis. Aptazymes have proven remarkably plastic and can be activated by metals, small organic molecules, peptides, and proteins. By simply cleaving apart or adjoining fluorophores and quenchers on oligonucleotide substrates, one can potentially construct reagentless ribozyme or aptazyme chips that could simultaneously report the concentration of multiple different analytes. Prototypes of such chips have now been made. While reagentless nucleic acid biosensors may ultimately prove less sensitive or robust than reagentless protein biosensors, it is nonetheless likely that nucleic acid biosensors will prove much more amenable to generation by high throughput selection methods, and thus may be the best vehicle for developing chips that can acquire organismal proteomes and metabolomes.

13.1. The concept of a reagentless biosensor

Reagentless biosensors are sensors that can detect a target analyte in a homogenous format; that is, without the addition of reagents other than the sample. A reagentless biosensor has several advantages over conventional diagnostic assays. The first and most obvious advantage of a reagentless biosensor is its simplicity and ease of use. Reagentless biosensors should prove to be intrinsically practical in a variety of settings, from the detection of hazardous biological agents in the field to monitoring blood glucose levels in a hospital setting to providing real-time readouts of cellular states in a research laboratory. Second, since the detection process does not require the addition of reagents, the system remains unperturbed throughout the course of the assay. This is of particular importance for the *in vivo* detection and quantitation of analytes. Third, because reagentless biosensors directly transduce the sensitivity and specificity of biomolecular detection to the production of a signal, they can potentially be integrated with any number of detection platforms. Finally, as a consequence of their ease of use, utility, and adaptability, it is expected that reagentless biosensors may prove to be significantly more cost-effective than conventional diagnostic assays.

Reagentless biosensors can be constructed by integrating a signaling or reporter component with a biological macromolecule, allowing the biomolecule to directly transduce the molecular recognition event into a detectable signal (e.g., an optical or electrochemical signal). The simplest reagentless optical biosensors can be envisioned as fluorescently labeled macromolecules. For example, many proteins undergo ligand-induced conformational changes. Fluorophores incorporated at specific sites on such molecules can be used to transduce molecular recognition events with high sensitivity; as an example, the maltose-binding protein has been derivatized with fluorescent dyes, and the hinge-bending motion that occurs upon interaction with maltose can be used to quantify maltose levels in solution (Marvin *et al.*, 1997; Marvin and Hellinga, 1998). The reason such sensors work is that fluorophores are frequently sensitive to minute changes in their chemical environments, and such changes are reported as changes in fluorescence intensity, wavelength, or anisotropy. Alternately, changes in fluorescence-resonance energy transfer between pairs of fluorophores, or between fluorophores and quenchers, can be used to detect binding.

Similarly, reagentless electrochemical biosensors have been made by electrostatic self-deposition of redox polyelectrolyte mediators and enzymes (Leech and Daigle, 1998; Mulchandani and Pan, 1999; Narvaez *et al.*, 2000). The enzymes turnover analytes (substrates) and produce electrons that are in turn coupled to electrodes via the adjacent mediators. Such integrated molecular sensors are not only inherently more practical than non-integrated system in terms of size and expense, but also have enhanced sensitivity and speed of response.

13.2. History – *in vitro* selection

Aptamers are single-stranded nucleic acids (RNA, ssDNA, modified RNA, or ssDNA), capable of binding tightly and specifically to their targets. They are isolated from combinatorial oligonucleotide libraries by a process known as "*in vitro* selection." *In vitro* selection mimics the process of natural evolution in that a pool of nucleic acids is sieved for a desired functional property, such as the ability to bind to a target or catalyze a reaction. Once functional species have been isolated, they are preferentially amplified via conventional molecular biology techniques, such as reverse transcription, polymerase chain reaction (PCR), and *in vitro* transcription. Over multiple rounds of selection and amplification, quite large populations (>10^{13} different sequences) can be sieved and those few "fittest" nucleic acid species isolated.

The *in vitro* evolution of nucleic acids was first reported by Sol Spiegelman in the 1960s. Spiegelman and his co-workers established a cell-free system in which the genomic RNA of bacteriophage Qβ could be evolved. For example, in a serial dilution experiment using purified Qβ replicase, the genomic RNA was evolved to replicate more quickly (Mills *et al.*, 1967; Spiegelman, 1971). While other phenotypes, such as resistance to ethidium bromide (EB) (Spiegelman, 1971) were also evolved, these *in vitro* selection methods may have ultimately been limited in scope because the diversity of the RNA population was dictated solely by the error rate of the polymerase.

More practical approaches to *in vitro* selection became possible in the 1990s with the advent of solid-phase DNA synthesis, and the invention of the PCR. Using solid-phase DNA synthesis, synthetic oligonucleotides with randomized regions could be generated and large nucleic acid libraries (up to 10^{15} species) that contained extremely diverse sequences could be obtained. In 1989, Kevin Struhl and co-workers reported the isolation and identification of double-stranded DNA-binding species from a random library (Oliphant *et al.*, 1989). In 1990, two separate groups reported the *in vitro* selection of RNA-binding species. Tuerk and Gold (1990) started with a library derived from the natural RNA substrate of T4 DNA polymerase and returned both wild-type and non-wild-type winners. Ellington and Szostak (1990) evolved nucleic acid ligands for targets with no previously known affinity for nucleic acids, starting from a library with 100 randomized positions. The first selection of a ribozyme from a randomized population was reported by the Joyce group in the same year (Robertson and Joyce, 1990); they evolved a natural ribozyme, the *Tetrahymena* self-splicing intron, to carry out a novel DNA cleavage reaction.

Contemporary *in vitro* selection experiments begin with the chemical synthesis of single-stranded DNA libraries (Figure 13.1). The inclusion

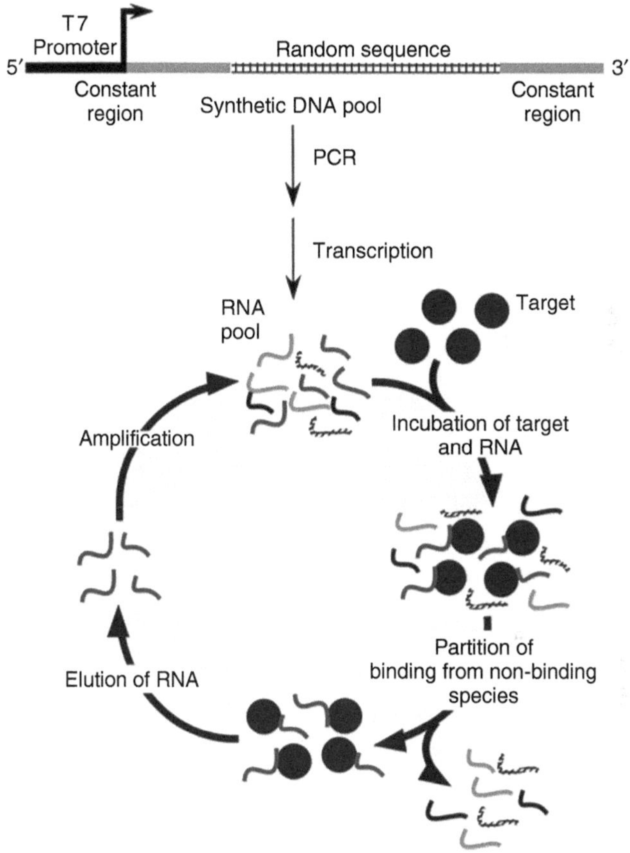

Figure 13.1 Scheme for *in vitro* selection. A chemically synthesized, single-stranded DNA pool is PCR amplified and transcribed *in vitro* to generate an RNA pool. Target molecules are incubated with the pool. Those sequences and shapes that can interact with the target are sieved from the population by one of a variety of methods (affinity chromatography, filter capture). The captured sequences are eluted, amplified, and subjected to additional rounds of selection and amplification.

of constant sequence regions flanking the random sequence core allows the amplification of such libraries via PCR, and the conversion of libraries to different chemical forms, for example the conversion of a double-stranded DNA template to single-stranded RNA via *in vitro* transcription. DNA or RNA libraries can be sieved for a variety of functions, the simplest of which is binding. By passing an RNA library over an affinity column or by collecting protein:RNA complexes via filtration on modified cellulose filters, binding species can be sieved from non-binding species. Eluted binding species can then be amplified by a combination of reverse transcription, PCR, and *in vitro* transcription. Multiple cycles of selection and amplification generally result in the purification of those binding species (also known as "aptamers") that have the highest affinity and specificity for a given target. A plurality of the selected population is often found to bind a target after 6–18 rounds of selection. Individual aptamers can be cloned and sequenced, and sequence comparisons typically reveal one or more families related by common ancestry (mutational variants of a single, original sequence) or by similarity (similar sequence or structural motifs that were selected in parallel, from different original sequences). While the *in vitro* selection procedure often requires from several weeks to months to perform, automation of the methods has been accomplished and can potentially deliver aptamers within only a few days (Cox *et al.*, 1998, 2002).

Aptamers possess many properties, which make them potential candidates for biosensor applications (Brody *et al.*, 1999; Jayasena, 1999; Wilson and Szostak, 1999; Brody and Gold, 2000; Famulok *et al.*, 2000; Hesselberth *et al.*, 2000a). They are the only reagents that rival antibodies in their universal molecular recognition properties and have been selected against an amazingly wide range of targets, ranging from metal ions (Ciesiolka and Yarus, 1996; Kawakami *et al.*, 2000) to complex cellular structures such as the ribosome (Ringquist *et al.*, 1995) and even to whole cells (Morris *et al.*, 1998), viral particles (Pan *et al.*, 1995), and live pathogenic protozoa (Homann and Goringer, 1999). Also, since aptamers are selected *in vitro*, they can potentially be raised against pathogens, toxins (e.g., ricin; Hesselberth *et al.*, 2000b), and purportedly to biological warfare agents, targets which frequently prove problematic for *in vivo* immunization procedures.

In addition to their comprehensive binding properties, aptamers have high binding affinities and also remarkable specificities. The binding affinities of aptamers are typically in the nanomolar to picomolar range for protein targets and in the micromolar to nanomolar range for small organic targets. The affinities of selected aptamers can be modulated by varying the concentration of the target, the buffer conditions employed during the binding reaction, or the number and type of wash steps used during each round of selection. Aptamers can discriminate between targets on the basis of subtle differences such as single amino acid changes in protein targets (Conrad *et al.*, 1994; Hirao *et al.*, 1998), or the presence or absence of a methyl (Jenison *et al.*, 1994; Haller and Sarnow, 1997) or a hydroxyl group (Sassanfar and Szostak, 1993; Mannironi *et al.*, 1997) in small organic targets. The specificities of aptamers can also be controlled during selection; for example, negative selections against related analytes or the matrices used for target immobilization can remove cross-reactive aptamers from a population (Jenison *et al.*, 1994). Finally, since aptamers can frequently be minimized to relatively small (30–50 nucleotides) oligonucleotides, they can be chemically synthesized in bulk and modified during chemical synthesis for conjugation or sensor function. The primary limitation on the use of aptamers as recognition and/or transduction elements in biosensors has been the perception that they are unstable and highly susceptible to degradation in biological media. However, the incorporation of modified nucleotides either pre- or post-selection can protect aptamers from nuclease degradation (Lin *et al.*, 1994; Eaton and Pieken, 1995; Green *et al.*, 1995; Pagratis *et al.*, 1997), and the conjugation of aptamers to supermolecular carriers such as PEG or liposomes (Willis *et al.*, 1998; Tucker *et al.*, 1999) can greatly increase their stability and retention in biological fluids.

Because aptamers have been shown to discriminate between even closely related isomers or different conformational states of the same protein (Conrad, *et al.*, 1994; Seiwert *et al.*, 2000), they are becoming an increasingly popular tool for molecular recognition that may eventually rival antibodies. Their utility has now been demonstrated in a number of analytical applications, such as flow cytometry (Davis *et al.*, 1996; Davis *et al.*, 1998), affinity probe capillary electrophoresis (German *et al.*, 1998; Pavski and Le, 2001; Haes *et al.*, 2006), sandwich assays (Drolet *et al.*, 1996; Centi *et al.*, 2007; Li *et al.*, 2007), capillary electrochromatography (Kotia *et al.*, 2000; Rehder and McGown, 2001; Joyce and McGown, 2004), affinity chromatography (Romig *et al.*, 1999; Deng *et al.*, 2001; Connor *et al.*, 2006; Connor and McGown, 2006), and more generally as biosensors (Kleinjung *et al.*, 1998; Hesselberth, *et al.*, 2000a; Rajendran and Ellington, 2002; Kirby *et al.*, 2004; McCauley *et al.*, 2003; Knudsen *et al.*, 2006; Lin *et al.*, 2006, 2007; Lai *et al.*, 2007; Zuo *et al.*, 2007).

Nucleic acid pools can also be sieved for catalytic function (for a review, see Wilson and Szostak, 1999). For example, nucleic acid pools can be immobilized on columns, and catalytically active species will cleave themselves from the column and can be collected in the eluate (Breaker and Joyce, 1994). Conversely, ribozyme ligases can be selected from random sequence pools based on their ability to append particular sequences to themselves that can be captured on affinity columns and used as primer-binding sites for PCR amplification (Bartel and Szostak, 1993). A variety of other ribozymes have been selected based on variations on these two themes: cleave away or add to. Alkyl transferase ribozymes have been selected that can add an activated biotin to themselves (Wilson and Szostak, 1995), tRNA synthetase-like ribozymes have been selected that can aminoacylate themselves (Illangasekare *et al.*, 1995), and amide synthetases have been selected by selectively modifying the 5′ end of a pool with an amine and then identifying those ribozymes that can conjugate an activated carboxylate to themselves (Lohse and Szostak, 1996).

It has also proven possible to meld aptamer and catalytic selections. Just as catalytic antibodies can be selected by identifying antibody variants that bind transition-state analogues of a given reaction, aptamers that bind transition-state analogues have proven to have catalytic activity. Peter Schultz and co-workers initially isolated ribozymes that could catalyze the isomerization of a biphenyl compound to its diastereomer by using a transition-state analogue as a target (Prudent *et al.*, 1994). Similarly, using *N*-alkylated porphyrin transition-state analogues, both catalytic RNA (Conn *et al.*, 1996) and catalytic DNA (Li and Sen, 1996) have been selected that catalyze porphyrin metalation. More recently, an aptamer having cholesterol esterase activity was isolated by *in vitro* selection of RNA using a phosphate ester transition-state analogue of cholesterol ester hydrolysis as a target (Chun *et al.*, 1999).

13.3. State of the art

13.3.1. Signaling aptamers

13.3.1.1. Designed signaling aptamers

Nucleic acid biosensors that can directly transduce the molecular recognition of other nucleic acids into optical signals have previously been described. Kramer and his co-workers originally designed "molecular beacons" (Figure 13.2) that juxtaposed fluorophores and quenchers in a stem structure (Tyagi and Kramer, 1996; Tyagi *et al.*, 1998). A loop capping the stem was complementary to some RNA or DNA target; upon interaction with the target, the formation of a new, stable helical structure resulted in the original stem being pried apart, which in turn freed the fluorophore from the adjacent quencher, resulting in a strong optical signal. Molecular beacons have now become mainstays in the diagnostics industry and have been adapted to a variety of applications: for allele discrimination in real-time PCR assays of genomic DNA (Piatek *et al.*, 1998), for detection of target genes (for example, detection of drug resistance in *Mycobacterium tuberculosis*; Piatek *et al.*, 1998), as sensitive DNA biosensors (Liu and Tan, 1999; Liu *et al.*, 2000), for studying protein–DNA interactions (Fang *et al.*, 2000a; Tan *et al.*, 2000; Bar-Ziv and Libchaber, 2001), and for real-time detection

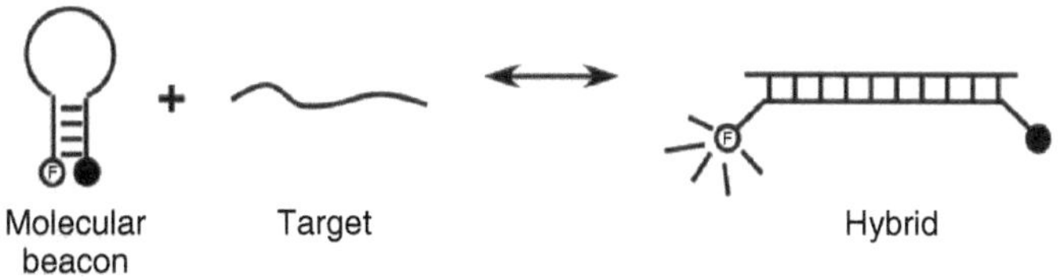

Figure 13.2 Molecular Beacon. A stem-loop structure contains a fluorophore (open circle with "F") and a quencher (closed circle). Sequence-specific interactions between the loop and a target nucleic acid results in the formation of an extended helix that pries apart the fluorophore and quencher, resulting in a target-specific increase in fluorescent signal.

of DNA–RNA hybridization in living cells in antisense research (Sokol *et al.*, 1998).

While sequence recognition alone can potentiate a wide range of diagnostic and other applications, the utility of nucleic acids as reagentless biosensors would be greatly expanded if they could also signal the presence of non-nucleic acid analytes, such as proteins or small organics. To this end, there are several different schemes that can be imagined for converting aptamers to biosensors or "signaling aptamers" (Figure 13.3). Each of these models has been realized in practice and will be described in turn.

Just like the reagentless protein biosensors described above, aptamers can be adapted to signal the presence of non-nucleic acid analytes.

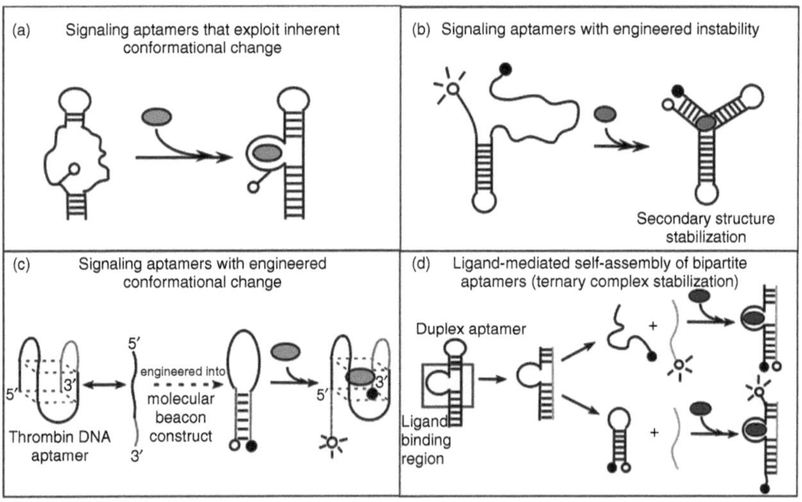

Figure 13.3 Strategies for signaling aptamers. (a) Exploitation of small conformational changes. (b) Exploitation of larger secondary structural conformational changes. (c) Exploitation of tertiary structural conformational changes. (d) Exploitation of quarternary structural conformational changes. In the case of (a), the conformational change may be inherent to the aptamer. In (b)–(d), the conformational changes have been engineered into the aptamer by altering its secondary, tertiary, and/or quarternary structure.

Structural studies have shown that aptamers frequently undergo small but significant conformational changes or reorganizations upon binding their cognate ligands (Hermann and Patel, 2000; Patel and Suri, 2000). By incorporating a fluorophore into a conformationally labile region of an aptamer, the binding event can lead to a change in the chemical environment of the fluorophore and hence to a change in fluorescence intensity (Figure 13.3a). Based on the known three-dimensional structures of anti-adenosine RNA and DNA aptamers (Dieckmann *et al.*, 1996; Jiang *et al.*, 1996; Lin and Patel, 1997), fluorescent dyes were introduced in the proximity of the adenosine-binding site. The resultant signaling aptamers not only showed an ATP-dependent increase in fluorescence, but were also selective for ATP relative to other nucleotides and could track ATP concentrations in solution (Jhaveri *et al.*, 2000a). Interestingly, while the sites of fluorophore insertion were chosen to interfere with ATP binding as little as possible, the apparent K_ds of the designed signaling aptamers (~30 μM for the DNA signaling aptamer and ~300 μM for the RNA signaling aptamer) were much higher than those of the parental aptamers (Sassanfar and Szostak, 1993; Huizenga and Szostak, 1995) (~6 μM for the anti-adenosine DNA aptamer and 6–8 μM for the anti-adenosine RNA aptamer). This may indicate that the design process must be greatly refined, or it may be that there is an inherent loss of binding affinity during conformational transduction.

Effective signaling (two- to threefold increase) was also obtained when the anti-ATP-binding aptamer was internally modified with a bis-pyrene-modified propanediol phosphoramidite (Yamana *et al.*, 2003). Since it has been established that the excimer (480 nm) and monomer (380 nm) fluorescence emissions of the bis-pyrene fluorophore are highly sensitive to the local structural change caused by base-pairing and/or nucleotide sequence variations in oligonucleotide duplexes (Yamana *et al.*, 2002), it was anticipated that bis-pyrene-labeled aptamers would also exhibit a large fluorescence change upon binding to a cognate ligand.

Fluorescent nucleotide analogues can also be used to generate signaling aptamers. The quantum yields (0.39–0.88) of nucleotide analogues such as 2-aminopurine (2AP, adenosine analogue), 4-amino-6-methylpteridone (6MAP, adenosine analogue), or 3-methylisoxanthopterin (3MI, guanosine analogue) are highly dependent on their local chemical environment, in particular upon the extent of base-stacking between the nucleotide analogue and its neighbors (Hawkins, 2001). These modified nucleotides have thus been used to study nucleic acid structure and dynamics (Menger *et al.*, 1996; Stivers, 1998). Several DNA aptamers (anti-thrombin, anti-immunoglobulin E (IgE), and anti-platelet-derived growth factor (PDGF)) have been modified with fluorescent nucleotide analogues (Katilius *et al.*, 2006), and the resulting signaling aptamers yielded a specific, binding-induced increase in fluorescence signal of up to 30-fold. The positions of the modifications were chosen by modeling the known aptamer tertiary structure (for the anti-thrombin aptamer) or by simply screening a set of modified aptamers (for the anti-IgE and anti-PDGF aptamers) when no tertiary structural information was available. This approach has decided advantages over other methods for the production of signaling aptamers, including larger protein-dependent changes in fluorescence, little apparent decrease in binding affinity, and high resistance to photobleaching.

The conformational changes that aptamers undergo upon ligand binding should significantly change the electrostatic environment near 2′-ribose position in nucleic acids (Merino *et al.*, 2005). Therefore, a 2′-O-pyrene-modified nucleoside (Yamana *et al.*, 1991) was also used to label an anti-ATP DNA aptamer at specific residues adjacent to the ligand-binding site (Kamekawa *et al.*, 2006). The aptamer could readily signal the presence of ATP and retained its binding affinity (K_{d1}, 0.12 mM and K_{d2}, 0.16 mM), in contrast with the bis-pyrene described above (Yamana, *et al.*, 2003), which lost almost 10-fold in affinity (K_{d1}, 1.7 mM and K_{d2}, 1.68 mM). Similarly, a BODIPY fluorophore has also been incorporated at the 2′ position in three different aptamers (anti-AMP, anti-tyrosinamide, and anti-argininamide), converting them into signaling aptamers (Merino and Weeks, 2005). Modified 2′ aminocytidine nucleotides were serially incorporated at each of the internal cytidine positions in each aptamer, and subsequently derivatized with the environmentally sensitive BODIPY dye. The resultant fluorescent aptamers were screened for signaling. Those aptamers with 2′-derivatization at the junctions between base-paired and flexible regions yielded a change in fluorescence quantum yield of up to a 3.7-fold (in the case of the argininamide-signaling aptamer). Using this strategy, many aptamers can potentially be converted into signaling aptamers with little or no pre-existing structural information.

Changes in the fluorescence anisotropy of single fluorophores can also be used to monitor binding events. Fluorescence anisotropy has been used to study numerous biomolecular phenomena, including conformational changes in proteins and the self-association of peptides and proteins (Morrison, 1999). In fact, fluorescence anisotropy may be ideally suited for studying the associations between proteins and appropriately labeled aptamers. Many target proteins are relatively large molecules when compared with aptamers, especially minimized aptamers. Therefore, protein binding to an aptamer should bring about a significant change in the overall molecular weight and size of the aptamer and in turn greatly alter the rotational diffusion rate of any fluorophore appended to the aptamer, resulting in a detectable variation in fluorescence anisotropy.

A short (15-mer) DNA quadruplex aptamer that bound to the blood-clotting factor thrombin has been adapted to fluorescence anisotropy assays on surfaces (Potyrailo *et al.*, 1998). The 5′ end was labeled with fluorescein isothiocyanate (FITC) and the 3′ end contained an alkyl amine that could be selectively immobilized. The labeled aptamer could detect thrombin at concentrations as low as 0.7 amol in a 140 pl interrogated volume and was selective for thrombin compared with elastase, another serine protease with an isoelectric point and molecular weight similar to that of thrombin.

Similar fluorescence anisotropy measurements have been made using fluorescently labeled anti-PDGF aptamers (Fang *et al.*, 2001). The anisotropy change (twofold) upon PDGF-binding was complete within only a few seconds. As little as 0.22 nM PDGF could be detected in homogeneous solution in real time. The sensitivity demonstrated by this method should be suitable for the detection of PGDF in serum samples and in biological fluids surrounding tumors (0.4–0.7 nM or higher).

While signaling aptamers that rely on changes in fluorescence intensity are simple in conception and simple to use, the advantage of fluorescence anisotropy measurements is that fluorescence anisotropy is a relative (or ratiometric) detection technique. As a result, common problems associated with fluorescence intensity assays, such as bleaching and non-uniform emission of the fluorophore during imaging, are not major concerns. Finally, since the reporter rather than the analyte is the signal transducing molecule, fluorescence anisotropy should be capable of monitoring real-time changes in analyte concentrations, and may eventually be useful for *in vivo* measurements.

While the conformational changes that the signaling aptamers described above underwent are relatively small, much larger conformational changes are possible. For example, the ligand-dependent organization of aptamer secondary structure can be envisaged. A secondary structure could be poised so that it would be largely unstructured in the absence of analyte, but substantially stabilized upon analyte binding (Figure 13.3b). An anti-cocaine DNA aptamer has been converted into a signaling aptamer using this strategy (Stojanovic *et al.*, 2001). One of the stems of a three-way junction constituting the cocaine-binding region was destabilized by truncation. At the same time, the stem was labeled with a 5′ fluorophore and a 3′ quencher. Ligand binding stabilized the engineered stem and the aptamer went from a fluorescent, unliganded form to a quenched, ligand-bound form. The signaling aptamer was not only able to measure cocaine concentration in the concentration range from 10 μM to 2.5 mM, but was also robust enough to report cocaine concentrations in serum. The aptamer appeared selective for cocaine relative to cocaine derivatives, such as benzoyl ecgonine, but it was possible that the observed selectivity was for hydrophobicity rather than for a defined chemical structure. In another example, an anti-PDGF aptamer was converted into an aptamer beacon by removing several base pairs from a paired stem and labeling one end of the aptamer with FAM (5′ end) and the other with DABCYL (3′ end; Fang *et al.*, 2003). In the absence of PDGF, the aptamer is largely denatured, and FAM can emit. Upon interaction with PDGF, the aptamer forms a secondary structure in which the fluorophore and quencher are in apposition and fluorescence is quenched. The PDGF aptamer beacon could detect PDGF concentrations as low as 110 pM in biological samples. One potential problem with such quenching aptamer beacons is that there are a variety of ligands or solvents that might interfere with quenching, resulting in a false-positive signal.

Signaling aptamers that rely on tertiary structural transitions can also be engineered. In this instance, an extant secondary structural element is not merely destabilized, but instead an entirely new conformation is pre-engineered into the aptamer (Figure 13.3c). A short, anti-thrombin aptamer that was known to form a quadruplex structure (Bock *et al.*, 1992; Macaya *et al.*, 1993; Schultze *et al.*, 1994) served as the starting point for the design of a simple "aptamer beacon." Sequences were added to the 5′ terminus of the anti-thrombin aptamer that were complementary to critical residues within the thrombin-binding structure (Hamaguchi *et al.*, 2001). In consequence, the dominant structure in solution was not a quadruplex, but rather was a hairpin stem. A fluor was added to the 5′ end of the hairpin, while a quencher was added to the 3′ end. The addition of thrombin shifts the equilibrium from the quenched stem-loop form to the thrombin-bound form, resulting in dequenching of the fluorophore and the creation of an optical signal. The thrombin "beacon" could detect thrombin concentrations as low as 5 nM and was able to discriminate against other serine proteases, such as factors IX and Xa.

Tan's group (Li *et al.*, 2002) has also taken advantage of conformational changes in the thrombin aptamer. However, instead of engineering conformational changes, these researchers noted that, even in the absence of the protein target, equilibrium exists between the random coil and quadraplex states of the anti-thrombin aptamer. Target binding shifts the equilibrium in favor of the quadraplex state. The anti-thrombin aptamer was modified to act as an aptamer beacon by incorporating either a fluorophore (fluorescein) or a quencher (DABCYL), or two fluorophores (coumarin (donor) and fluorescein (acceptor)) at the two termini of the aptamer. Thrombin binding was monitored via either quenching of a single fluorophore (fluorescein–DABCYL aptamer, maximal fluorescence decrease ~60%) or via increasing fluorescence of an acceptor fluorophore (fluorescein–coumarin aptamer, signal enhancement factor = ratio of acceptor to donor intensity before and after binding = ~14-fold). The K_d and limits of detection for the quenching aptamer beacon were 5.20 ± 0.49 nM and 373 ± 30 pM, while for the two-fluorophore aptamer beacon, these values were 4.87 ± 0.55 nM and 429 ± 63 pM. Fluorescence resonance energy transfer (FRET)-type aptamer beacons incorporating two fluorophores may be particularly useful for real-time analysis of proteins and could possibly be used in living specimens in conjunction with ratiometric imaging. An advantage of this tertiary structural rearrangement strategy, as opposed to the secondary structural rearrangement strategy, is that the signal is a dequenched increase in fluorescence, rather than a quenched decrease in fluorescence. While there are numerous analytes in a complex mixture or biological sample that might lead inadvertently to fluorescence quenching, there should be relatively few compounds other than the target analyte that should lead to an increase in fluorescence intensity.

Finally, aptamer quaternary structure can also be engineered to yield analyte-dependent changes in optical signals (Figure 13.3d). In this case, aptamers are split into separate pieces that can self-assemble in the presence of a cognate ligand. Again, fluorophores are used to label each of the aptamer pieces. In the absence of the target ligand, the two oligomers exist as individual units in solution, but target binding brings the oligomers together and leads to ternary complex stabilization, ultimately resulting in a quenched optical signal (or in a FRET signal). Anti-cocaine and anti-rATP aptamers have been converted into signaling aptamers using this strategy (Stojanovic *et al.*, 2000). The signaling aptamers were again not only reproducibly sensitive, but also selective for their cognate ligands. Within the concentration ranges of 10 μM–1 mM ATP and 10–150 μM cocaine, it proved possible to simultaneously report the concentrations of the two analytes using the two signaling aptamers via fluorescence changes. In a similar set of experiments, an aptamer that binds the Tat protein of HIV has been converted into a signaling aptamer by this method. In this case, one of the two oligomers was designed to be a molecular beacon construct, which opens up on ligand binding to generate a signal. The obvious advantage of this method is again that it allows a "positive" optical signal to be generated, rather than looking for a "negative" signal against a highly fluorescent background. The anti-Tat aptamer had an extremely low K_d for Tat (~120 pM), and the adapted biosensor could quantitate Tat samples as low as 100 nM (Yamamoto *et al.*, 2000).

Aptamer beacons that undergo quarternary structural changes can also be designed by using an "antisense strategy" (Nutiu and Li, 2003) in which an additional complementary DNA sequence is used to denature the aptamer (Figure 13.4). Upon addition of ligand, the native aptamer structure is stabilized, and the equilibria are concomitantly shifted away from the denatured, antisense duplex and toward the native structure. The aptamer can be labeled with a fluorophore, and the antisense oligonucleotide with a quencher (denoted QDNA), leading to

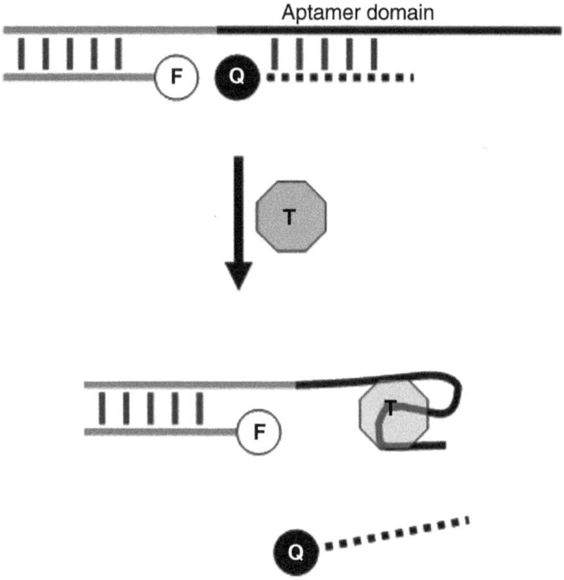

Figure 13.4 Aptamer beacons that use an antisense strategy. Nutiu and Li (2003) showed that two antisense oligonucleotides containing a fluor (**F**) and quencher (**Q**) could be aligned adjacent to one another on a denatured aptamer. When an analyte (**T**) that binds to the aptamer is introduced, the equilibrium is altered in favor of the native state of the aptamer, and a fluorescent signal is produced.

a quenched fluorescent signal in the absence of target, similar to the unimolecular aptamer beacons described above. Li and his co-workers (Nutiu and Li, 2003) have demonstrated the potential of this approach by making antisense aptamer beacons that can detect ATP ($K_d = 600\,\mu M$) and thrombin ($K_d = 400\,nM$). Like the unimolecular aptamer beacons, this strategy requires no foreknowledge of aptamer structure.

Compared to organic dyes that generally have low photostability, narrow excitation spectra, and broad emission bands, quantum dots (QDs) and other nanoparticles offer a number of advantages over standard fluorescent dyes for monitoring biological systems in real time, including greater photostability, larger effective Stokes shifts, longer fluorescence lifetimes, and sharper emission bands than traditional organic fluorophores. In addition, QDs all respond to the same excitation wavelength, but emit at different wavelengths. These properties potentially make QDs useful reagents for multi-spectral *in vivo* diagnostic imaging of cells, tissues, and living animals (Lim *et al.*, 2003; Michalet *et al.*, 2005). Recent work has demonstrated that QDs can be used in the construction of biosensors that signal by FRET. For example, a QD-based molecular beacon has been described in which the non-fluorescent dye DABCYL was used to quench the fluorescence of the QD. In the presence of a target DNA, the opening of the molecular beacon resulted in an approximately fivefold increase in fluorescence of the QD (Kim *et al.*, 2004).

While numerous aptamer beacons that utilize organic fluorophores have previously been developed as described above, the adaptation of these strategies to the development of a QD aptamer beacon (QDB) was not straightforward, primarily because the inorganic QD is much larger and brighter than its organic partner. The development of QD beacons therefore relied upon two important design features ensuring that: a substantive conformational change would occur upon analyte binding, and the QD was efficiently quenched. Therefore, multiple aptamer beacons in which a quencher was displaced upon analyte binding were immobilized on the QD surface (Levy *et al.*, 2005). The QDB construct showed a 19-fold increase in fluorescence in the presence of the cognate protein target. This observed increase in fluorescence was similar to that observed for a similar construct designed on traditional organic fluorophores (19-fold versus 12-fold; Nutiu and Li, 2003).

13.3.1.2. Dye displacement signaling aptamers
In a number of instances, the conjugation of fluorophores to aptamers has been shown to compromise aptamer affinities. As an alternative, it has proven possible to develop signal transduction methods based on the displacement of non-covalently bound dyes from aptamers. There are a number of commercially available nucleic acid stains, such as cyanine dyes or EB, that are typically non-fluorescent in solution but form fluorescent intercalation complexes with nucleic acids (Glazer and Rye, 1992; Bunkenborg *et al.*, 2000; Laib and Seeger, 2004). Upon ligand binding, some of the bound dyes are displaced, leading to a diminution in fluorescence. For example, the cyanine dye, TOTO (1,1-(4,4,8,8-tetramethyl-4,8-diazaundecamethylene)-bis-4-(3-methyl-2,3-dihydro(benzo-1,3-thiazole)-2-methylidene) quinolinium tetraiodide), has been used to generate aptamer–TOTO signaling complexes. When TOTO was applied to an anti-PDGF aptamer, PDGF-BB led to dye displacement and could be detected in physiological buffer with high selectivity and a detection limit as low as 0.1 nmol l^{-1}. Despite the simplicity of the method, this was in fact better than other reported aptamer-based methods for PDGF detection (2.0 nM using fluorescence anisotropy; (Fang *et al.*, 2001) and 1.0 nM using a light-switching Ru complex (Jiang *et al.*, 2004)). Stojanovic and Landry (2002) have similarly reported that cyanine can be used to detect cocaine in a complex with an anti-cocaine aptamer in the concentration range of 2–600 μM.

A water-soluble cationic polythiophene dye with conformation-dependent optical properties has also been used as part of a signaling scheme (Ho *et al.*, 2002). The polymeric dye was complexed with the anti-thrombin DNA aptamer and addition of thrombin resulted in a significant spectral shift, even when the aptamer was immobilized on a surface (Ho and Leclerc, 2004). This biophotonic tool has been reported to detect as little as two femtomoles (2×10^{-15} mol) of human thrombin in a few minutes. However, these experiments were carried out in water, while the aptamer was originally selected in a buffered solution, so it is unclear whether interactions with the quadruplex aptamer structure were being detected.

Metal complexes as well as organic dyes have been used in displacement assays (Jiang, *et al.*, 2004). $[Ru(phen)_2\text{–}(dppz)]^{2+}$ (phen = 1, 10-phenanthroline, dppz = dipyrido[3, 2-*a*:2′, 3′-*c*]phenazine) has no luminescence in solution but exhibits strong luminescence upon intercalation into a nucleic acid duplex. A 37-nucleotide DNA aptamer against IgE was used as a model system, and upon binding to IgE, the release of the ruthenium complex led to luminescence in solution. IgE could be detected at concentrations as low as 100 pM. The generality of the method was demonstrated with other anti-protein aptamers, including a DNA aptamer against PDGF-BB and RNA aptamers against α-thrombin (Kubik *et al.*, 1994) as well as anti-ATP DNA aptamer (Wang *et al.*, 2005).

These approaches based on dye displacement can eliminate relatively complex synthesis or modification procedures, and thus may allow the simple and economical conversion of many different aptamers into signaling aptamers. The generality of the method should further potentiate the use of aptamers in high throughput screening in drug and environmental monitoring (Liu and Danielsson, 2005).

13.3.2. Nucleic acid catalysts as biosensors

Aptamers function as reagentless biosensors because their ability to signal is embedded within the receptor itself. This concept can also be carried over to catalytic nucleic acids (ribozymes) by simply embedding the ability to signal within the catalyst, its substrate, or some aspect of the catalytic mechanism. For example, John Burke and his co-workers have generated variants of the hairpin ribozyme that cleave an RNA substrate containing both a fluorophore and a quencher (Vitiello *et al.*, 2000). In the presence of the ribozyme, a fluorescent signal is produced and the kinetics of the ribozyme can be readily followed. Famulok and his co-workers have developed a similar signaling system for the hammerhead ribozyme (Jenne *et al.*, 2001), and Krupp and his co-workers have followed the kinetics of the Group I self-splicing ribozyme by monitoring the ligation-mediated release of a fluorescent dye and concomitant changes in fluorescence polarization (Singh *et al.*, 2000). In each instance, the system is homogenous and tracks ribozyme activity.

However, since ribozyme activity is itself dependent upon a number of cofactors, notably metals, these reactions can also be viewed as biosensors for any of the reaction components that lead to catalysis. In this respect, it is interesting to note that the metal dependence of ribozymes and deoxyribozymes can be altered or *de novo* engineered, seemingly at will. For example, it has long been known that yeast tRNA (Phe) undergoes site-specific cleavage in the presence of lead ions. Pan and Uhlenbeck (1992) exploited this property to select for variants that were even better "leadzymes." One variant looked completely unlike the original tRNA, yet still used lead hydroxide to cleave the phosphodiester bond, generating a 2′,3′-cyclic phosphate, which was in turn hydrolyzed (Pan *et al.*, 1994). The structure of this leadzyme has now been solved (Wedekind and McKay, 1999), and it appears as though lead coordinates to the 2′-hydroxyl of the scissile residue, a mechanism that has previously been seen for protein enzymes. Breaker and Joyce (1994) were able to select a lead-dependent deoxyribozyme from a random sequence pool that could cleave a substrate with a single ribotide. The same technique was further generalized to the selection of deoxyribozymes that were dependent on other ions, such as magnesium, manganese, and zinc (Breaker and Joyce, 1995). Li *et al.* (2000) repeated Breaker and Joyce's experiments in order to better describe zinc-binding and catalytic motifs, but obtained sequences similar to those that had already been found by Breaker and Joyce. Li and Lu (2000) have taken advantage of the ability to track ribozyme kinetics using fluorescent reporters by developing a lead-sensing ribozyme biosensor. Substrate cleavage could be monitored in real time using kinetic fluorescence spectroscopy, and there was a modest specificity for lead (80-fold relative to other divalent ions).

Amazingly, the chemistry of nucleic acid catalysts can also be extended beyond simple metal dependence. Since most catalysis is pH dependent, it is not surprising that ribozymes can be evolved to possess different pH optima (Jayasena and Gold, 1997). Other variables that affect catalysis can also be probed by selection. Breaker and his co-workers have selected a deoxyribozyme that relies upon copper and peroxide to cleave a DNA substrate (Carmi *et al.*, 1996). The radical cleavage induced by the "DNAzyme" is mechanistically identical to nucleic acid scission mediated by other radical-generating reagents, such as iron:EDTA complexes. However, the deoxyribozyme locally produces radicals and directs their attack at a restricted set of sites on the substrate. Similarly, Roth and Breaker (1998) have shown that a ribose moiety embedded within a DNA strand can be cleaved by a deoxyribozyme selected in the presence of histidine. The imidazole ring apparently functions as a general base. As a biosensor, the deoxyribozyme is surprisingly specific, eschewing a variety of histidine analogues, including such closely related compounds as D-histidine and 3-methyl-L-histidine. Ribozymes can also report on the presence of inhibitors. Most recently, Famulok and co-workers (Jenne *et al.*, 2001) have used the fluorescence-based assay they originally developed to screen hammerhead ribozymes for novel inhibitors. Such technologies could potentially even be used to identify inhibitors of other RNA sequences that were appended to or activated by ribozymes.

13.3.3. Aptazymes

Another variable that clearly impinges on catalytic mechanism is ribozyme structure. While ribozyme structures can be subtly altered by metals or pH or other conditions, more global alterations can be induced by changes in base pairing. Lizardi engineered an allosteric hammerhead ribozyme that initially folded into an inactive conformation which was in turn relieved by the addition of an oligonucleotide effector (Porta and Lizardi, 1995). This strategy is most similar to that shown in Figure 13.3c for signaling aptamers. However, allosteric activation was only about 10-fold.

More recently, Taira and co-workers have generated novel allosteric ribozymes, called "maxizymes," which form active quaternary structures following oligonucleotide recognition (Kuwabara *et al.*, 2000a, 2000b; Warashina *et al.*, 2000). This strategy is most similar to that shown in Figure 13.3d. A hammerhead ribozyme dimer was broken into two pieces such that both ends present "arms" that can hybridize to a specific nucleic acid sequence. Once the ribozyme is brought together by hybridization of the substrate-binding "arms" at one end to an mRNA sequence, a second mRNA or another portion of the same mRNA can hybridize to the substrate-binding "arms" at the other end, leading to cleavage at both sites. Alternatively, a hammerhead heterodimer can be dissected such that one end of stem-loop II serves as a binding (but not cleavage) site for a particular sequence, while the other end still cleaves a desired target. In this way, the maxizyme can serve as a biosensor.

It has also proven possible to construct ribozymes that are structurally responsive to effectors other than oligonucleotides. As we have seen in earlier sections, functional nucleic acids undergo conformational changes upon interactions with their cognate ligands. It therefore seemed reasonable to suppose that by appending nucleic acid aptamers to nucleic acid catalysts (ribozymes), it might prove possible to alter the conformation, and hence the catalytic activity, of a nucleic acid catalyst in a ligand-dependent fashion. Ron Breaker and his co-workers were the first to attempt this feat by swapping an anti-adenosine aptamer with a stem of the hammerhead ribozyme whose sequence was known to be relatively unimportant for catalysis, yet was juxtaposed with the catalytic core (Tang and Breaker, 1997) (Figure 13.5a). The activity of the resultant chimeric "aptazyme" was in fact modulated by ATP.

While there were various mechanisms by which it could be imagined that the conformational change of the aptamer regulated the activity of the ribozyme, one likely hypothesis was that the joining region between the aptamer and the ribozyme strongly affected the structure of the catalytic core. In order to further probe this hypothesis, Soukup and Breaker (1999) randomized the joining region and selected for ribozymes that were either activated or inhibited in the presence of FMN (Figure 13.5b). The generality of the selection procedure (Figure 13.3) was such that both types of aptazymes could be derived from the same random sequence pool. In one instance, ribozymes that reacted prior to the addition of FMN were removed from the population, then the FMN-dependent ribozymes were harvested; in the other instance, FMN was initially added, those ribozymes that reacted were removed, and the remaining catalytically active population was then harvested. The best

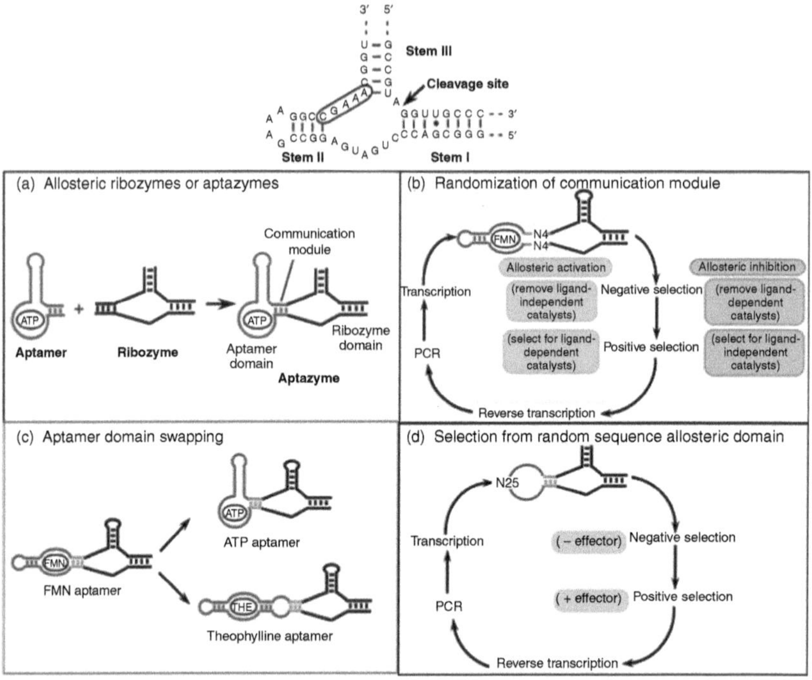

Figure 13.5 Strategies for adapting the hammerhead ribozyme to be an allosteric enzyme. *Top*: Sequence and structure of the hammerhead ribozyme. (a) Appending an aptamer to the hammerhead ribozyme in place of a non-essential stem. (b) Randomization of the joining region between the catalytic core of the hammerhead ribozyme and the appended aptamer. Selection is for inhibition or activation by the aptamer's cognate ligand, and typically involves coupled negative and positive selection steps. (c) Changing the effector-binding domain on an allosteric ribozyme. The pre-identification of a "communication module" by the method described in (b) sometimes assists in this process. (d) Complete randomization of the effector-binding domain and selection for inhibition or activation of the ribozyme.

aptazymes that were selected by these procedures had activities that were increased several 100-fold in the presence of FMN. Moreover, it was found that the "communication modules" that emerged from the randomized joining regions could actually mediate between different ligand-binding domains and the hammerhead ribozyme (Figure 13.5c).

For example, when an anti-theophylline aptamer was joined to the hammerhead ribozyme via a communication module originally selected for its ability to transduce FMN binding to catalysis, the cleavage activity of the hammerhead was again activated, only this time by theophylline.

Koizumi *et al.* (1999a) went beyond the modular joining of aptamers and catalysts by randomizing the entire allosteric domain (Figure 13.5d). A random sequence pool of 25 residues in length was appended to the catalytic core of the ribozyme via the same stem that had previously proven useful for the addition of aptamers, and a selection for ribozymes whose activities were modulated by cyclic mononucleotides was initiated. All four cyclic mononucleotides were included in the selection, and after 16 rounds of selection and amplification the population was observed to be somewhat dependent upon cGMP for activity. This effector was then removed from the population and the selection was continued. Ultimately, ribozymes were found that were dependent upon three of the four cyclic mononucleotides; no cUMP-specific ribozymes were ever discovered. Individual aptazymes could be activated up to 5000-fold by their cognate effectors.

Interestingly, the selected allosteric domains have many of the same characteristics as selected binding domains. A designed, theophylline-dependent hammerhead ribozyme was partially randomized, and variants that could be activated by other effectors were selected (Soukup *et al.*, 2000). A single mutation altered the specificity of the aptamer from theophylline to 3-methylxanthine, which differs from theophylline by the absence of a single methyl group. In fact, selected allosteric domains can be detached from ribozymes and function as aptamers in their own right (Soukup *et al.*, 2001). The cNMP-binding domains originally selected by Koizumi *et al.* (1999b) were probed by random mutagenesis and re-selection (Koizumi *et al.*, 1999a). Based on the relative degree of sequence conservation and variation at different positions, it proved possible to predict which residues were involved in effector binding and which were more likely involved in catalysis. The predicted allosteric domains were separately synthesized and largely retained the ligand-binding properties that they exhibited within the aptazymes.

While the hammerhead ribozyme has obviously proven to be an excellent platform for the design and selection of aptazymes, it was originally unclear whether this ribozyme and the design principles that were built around it were unique. Our laboratory therefore undertook similar experiments with a ribozyme ligase (L1) that had been selected from a random sequence pool (Robertson and Ellington, 1999). The L1 ligase initially (and fortuitously) proved to be highly dependent (10 000-fold activation) on an oligonucleotide effector that was present during the selection, and thus it seemed likely that it might be adapted to other types of effector dependence as well. To this end, anti-adenosine, anti-theophylline, and anti-flavin aptamers were adjoined to the ligase in place of a stem structure that was relatively unimportant for catalysis, yet was adjacent to the catalytic core (Robertson and Ellington, 2000). As with the hammerhead ribozyme, these L1 ligase chimeras proved to be ATP- and theophylline-dependent. However, flavin-dependence was initially minimal, but was readily optimized by randomization and selection of the "communication module" connecting the aptamer and ribozyme.

We were most interested in generating aptazymes that were protein-dependent. However, the design principles that had previously proven effective in identifying aptazymes that were modulated by small organic effectors did not readily yield protein-dependent aptazymes. Anti-protein aptamers directly conjoined to the L1 ligase did not impart protein-dependence, nor did randomization and selection of the communication module, nor did randomization of the entire allosteric domain followed by a coupled negative and positive selection for protein-dependence (Figure 13.6). However, when both the allosteric domain and a portion of the catalytic domain were randomized and screened for protein-dependence, it proved possible to identify aptazymes that were highly dependent on their cognate proteins (Robertson and Ellington, 2001). For example, one aptazyme was isolated that was 75 000-fold dependent on tyrosyl tRNA synthetase from *Neurospora* mitochondria (Cyt18), and another was found to be 3500-fold dependent on hen egg white lysozyme. The protein-dependent aptazymes had many of the characteristics previously observed for protein-dependent aptamers, in that they could readily distinguish between cognate and non-cognate proteins, including between the native and denatured states of the same protein.

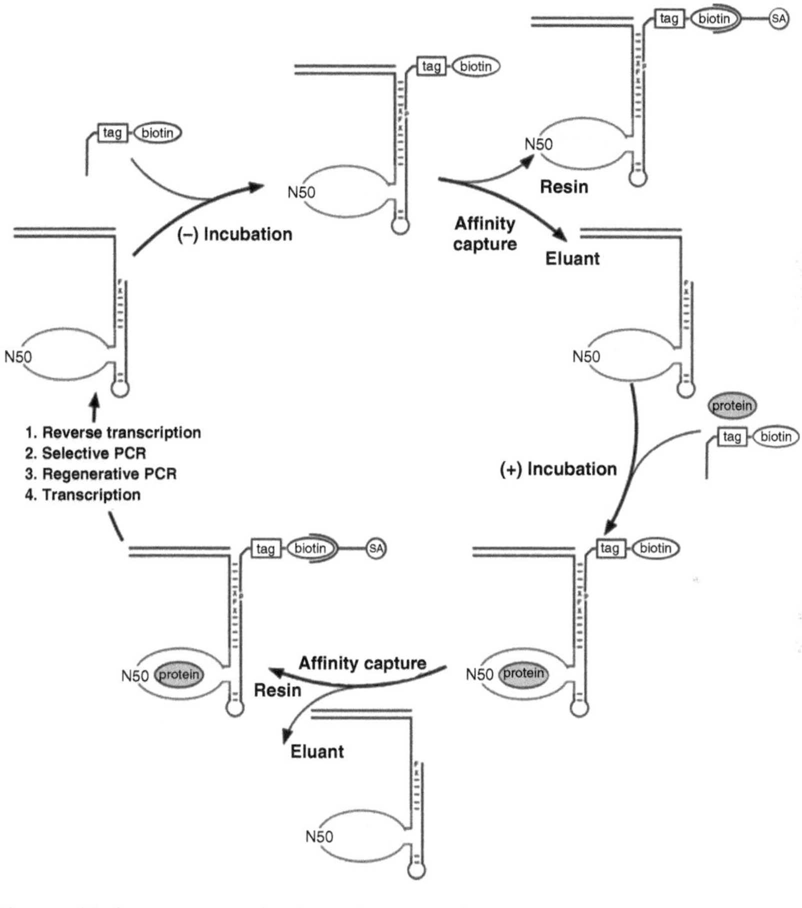

Figure 13.6 *In vitro* selection of protein-dependent aptazyme ligases. This scheme is similar to that shown for the isolation of effector-dependent hammerhead ribozymes (Figure 13.3d). However, in this instance, ligation of a substrate oligonucleotide to a random sequence population allows the capture of active catalysts via two mechanisms: binding to an oligonucleotide affinity column, and preferential amplification via PCR. It is interesting to note that to achieve the selection of protein-dependent aptazymes both the effector-binding domain and the catalytic core of the ribozyme had to be randomized.

Given that the design principles originally elaborated by Breaker and his co-workers seemed to be generalizable, it is easy to imagine the development of a wide variety of aptazymes with a wide variety of catalytic functionalities. For example, by mounting two effector domains on the same ribozyme, it should be possible to make aptazymes that are doubly dependent on their effectors and that function as molecular "and" gates (Figure 13.7). This concept was originally put into practice by Breaker and his co-workers (Jose *et al.*, 2001), who mounted two aptamers on the hammerhead ribozyme in series. The resultant aptazyme was in fact dependent upon both ligands for full activity, and appeared to exhibit cooperative interactions between the effector-binding domains. We have similarly mounted two aptamers on the L1 ligase, only in this instance the aptamers were appended in parallel, rather than in series. Nonetheless, the resultant aptazyme was again dependent upon both ligands for full activity (Michael Robertson, personal communication).

13.3.4. Nucleic acid biosensor chips

In the initial discussion of reagentless biosensors, it was clear that the utility of such sensors was in their ability to be adapted to multiple types of detection platforms. For reagentless biosensors based on nucleic acids, the ease of *in vitro* selection procedures offers the possibility that multiple different aptamers, ribozymes, or aptazymes could be selected and modularly adopted to a single type of detector. Ultimately, it may be possible to develop chip arrays of nucleic acid biosensors that would be suitable for "large" biological problems such as the acquisition of information about whole proteomes, metabolomes, or environmental dispositions of organisms and compounds (Brody *et al.*, 1999).

Initial work on adapting aptamers to optical biosensors was modeled after antibody diagnostics. Aptamers have been used in a sandwich ELISA-like format (called ELONA, enzyme-linked oligonucleotide assay) to quantitate human vascular endothelial growth factor (VEGF) in serum samples (Drolet *et al.*, 1996). The ELONA assay was found to be reproducible, could be used to measure VEGF concentrations as low as 25 pg/ml, and had a dynamic range of over three orders of magnitude. Capillary electrophoresis with laser-induced fluorescence detection

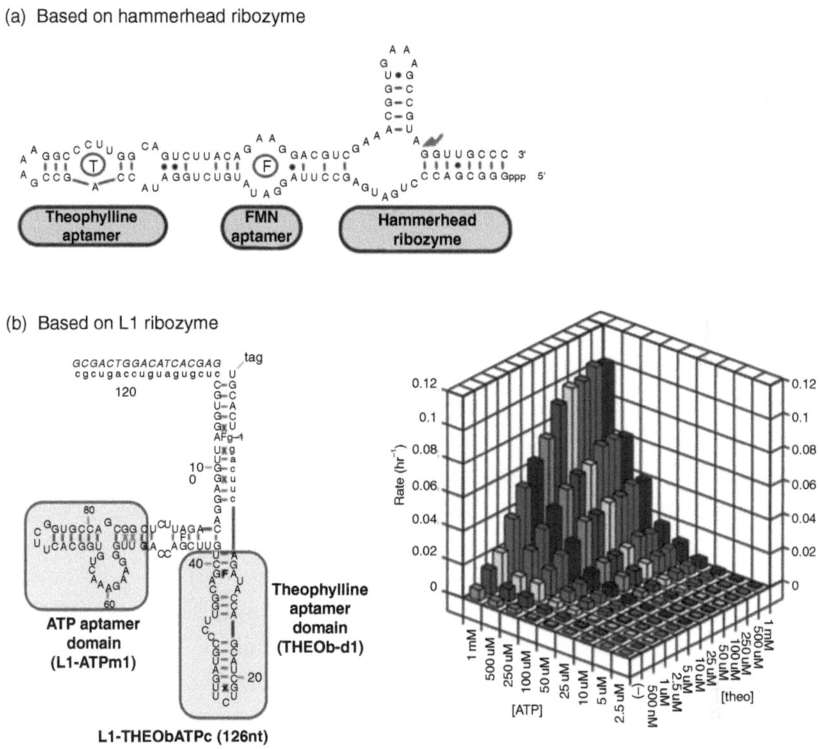

Figure 13.7 Different "dual effector" aptazymes. (a) A dual effector hammerhead aptazyme designed by Breaker and co-workers. The two aptamers are joined in series. Stabilization of one ligand-binding domain results in stabilization of the adjacent domain, and overall stabilization of this structure results in stabilization of the catalytic core. (b) A dual effector ligase aptazyme designed by Michael Robertson and co-workers. The two aptamers are mounted in parallel on different stems of the L1 ligase. The responsivity profile as a function of effector concentration is shown to the right. Maximal activity of the nucleic acid "and gate" is only seen when both effectors are present at high concentrations. In the presence of single effectors, activation is minimal. (see Plate 4)

(CE-LIF) has been used to sensitively detect IgE in solution using fluorescently labeled anti-IgE aptamer as a selective fluorescent probe (German *et al.*, 1998). The method was highly sensitive, with a mass detection limit of 37 zmol of IgE, and a dynamic range of 10^5. Aptamers have also been used as probes in flow cytometry: a fluoresceinated anti-human neutrophil elastase (HNE) aptamer has been compared with an anti-HNE antibody in detecting HNE coated on beads (Davis *et al.*, 1996). Similarly, aptamers against human CD4 have been conjugated to different fluorophores and used to stain human CD4 expressed on cells by flow cytometry (Davis *et al.*, 1998), while aptamers against prostate-specific membrane antigen (PSMA) have been labeled with QDs and used for the *in vitro* imaging of prostate tumor cell lines (Chu *et al.*, 2006).

These somewhat simplistic antibody substitutions have been followed up by more sophisticated attempts to adapt compact, easily labeled aptamers to more system-specific analytic detection techniques and instrumentation. A fiber-optic microarray biosensor has been developed for thrombin by immobilizing anti-thrombin DNA aptamer at the distal tip of an imaging fiber coupled to a modified epifluorescence microscope system (Lee and Walt, 2000). The system has a detection limit of 1 nM and was used to measure thrombin concentrations in the range from nanomolar to low micromolar. Other examples include an aptamer biosensor for L-adenosine based on total internal reflection fluorescence detection which could detect L-adenosine in the submicromolar range (Kleinjung *et al.*, 1998) and one for thrombin based on evanescent wave-induced fluorescence detection (Potyrailo *et al.*, 1998). The thrombin sensor had a dynamic range of three orders of magnitude, and was highly sensitive with a detection limit of 0.7 amol in a 140 pl volume. Fluorescence anisotropy chips have also been used for the parallel detection of multiple proteins by the company Archemix (McCauley *et al.*, 2003). The sensor array could be used to specifically detect thrombin, basic fibroblast growth factor, inosine monophosphate dehydrogenase, and VEGF against a background of complex biological media. The sensitivity of detection for thrombin ($K_d = 15.5\,\text{nM}$) was similar to that previously demonstrated for other aptamer biosensors.

While these systems and adaptations have variable potential for the eventual development of chip arrays, the very fact that aptamers are readily adapted across such diverse platforms bodes well for the eventual use of aptamers in virtually any material or detection modality that comes to the fore. For example, an ATP signaling aptamer has been adapted to function in paper, which is extensively used as a barrier material in medical face masks and protective clothing (Su *et al.*, 2007). The same ATP signaling aptamer has also been immobilized into sol-gel-derived silica (Rupcich *et al.*, 2006). Although the response was slow due to the diffusional barriers to mass transport of the analyte through the silica matrix, encapsulation of aptamer in the silica matrix resulted in 5.5-fold slower nuclease degradation compared to the aptamer in solution.

Alternatively, nano- as well as macrostructures based on DNA tile self-assembly can be modified with signaling aptamers. Since aptamers are nucleic acids, they can be much more readily incorporated into nanostructures than other types of biomolecules, such as proteins. Additionally, self-assembled two-dimensional lattices can be from micro- to millimeter sized and contain from millions to trillions of individual building blocks (He *et al.*, 2005a, 2005b). Because of their regular and information-rich surfaces, DNA lattices can be uniquely used to assemble molecular probes with precisely controlled distances and relatively fixed spatial orientations (Winfree *et al.*, 1998; LaBean *et al.*, 2000; Yan *et al.*, 2003). For instance, thrombin-signaling aptamers containing 3MI, a fluorescent analogue, were linked to DNA tiles and self-assembled into a thrombin-signaling aptamer array (Lin *et al.*, 2006). By using different aptamers with different fluors on different tiles, a multiplex array could be assembled that simultaneously measured ATP and thrombin (Lin *et al.*, 2007).

The fact that aptazymes directly transduce molecular recognition into catalysis may allow them to function as reagentless biosensors in simple but robust chip arrays. As an example, Seetharaman *et al.* (2001) immobilized various radioactive hammerhead aptazymes on a gold surface via a 5′ phosphorothioate moiety. Following the addition of effectors or effector mixtures, appropriately activated hammerheads cleaved themselves away from the surface and could be detected by simply transferring the supernatant to a new microtiter plate. Most remarkably, a cAMP-sensing aptazyme could be used to accurately quantitate the amount of cAMP in the culture medium of various *E. coli* strains that had alterations to adenosine metabolism. We have developed similar

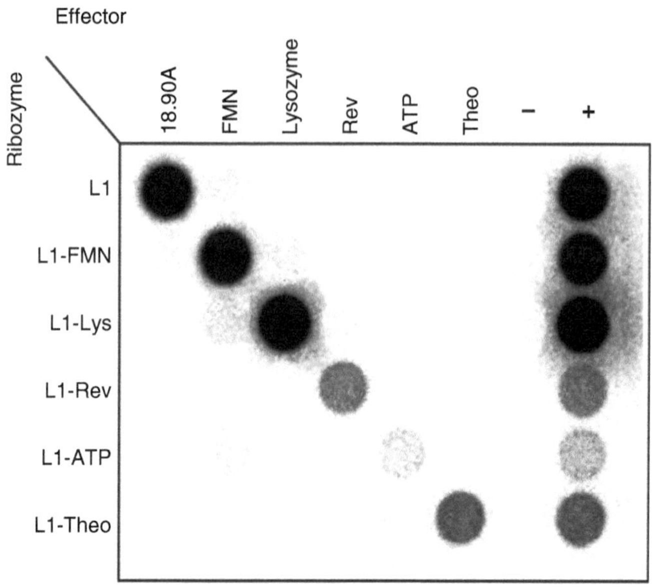

Figure 13.8 Aptazyme ligase chip. Radiolabeled aptazymes derived from the L1 ligase (see ligand specificities to the left) were incubated with biotinylated substrates. Following reaction, the mixtures were incubated with a streptavidin microtitre plate and unbound material was washed away. While this is not a reagentless application, the figure above does show the ability of aptazymes to specifically detect small molecules (FMN, ATP, theophylline), oligonucleotides (18.90, the effector for the original L1 ligase, see Figure 13.7, sequence in italics), peptides (the arginine-rich motif of Rev), and proteins (Lys = lysozyme). "–" is the no effector control, and "+" is all effectors mixed together.

chips based on the L1 ligase. In this manifestation, however, activated aptazyme ligases conjugate themselves to substrates and coimmobilize radiolabels onto a surface (Figure 13.8). One of the advantages of aptazymes for chip-based applications is that they are already known to recognize a chemically diverse set of substrates. The ligase chip simultaneously detects the presence of oligonucleotide, small organic molecules, peptides, and proteins, a feat that would not be possible in a conventional ELISA assay.

13.4. Advantages and limitations

While the use of signaling aptamers and aptazymes in biosensors holds great potential, it is nonetheless true that nucleic acids are generally not as effective as proteins both as binding reagents and as catalysts. While aptamers typically bind their ligands in the nanomolar to micromolar range, antibodies can generally bind ligands well into the femtomolar range. Similarly, the few protein-sensing aptazymes developed so far can sense proteins into the nanomolar range, whereas ELISA assays can sense as few as 10^8 molecules in a sample (Rogers, 2000). Thus, an important question is whether the seemingly superior recognition elements, proteins, can function in a reagentless system in much the same way that aptamers and aptazymes can.

Reagentless optical biosensors based on fluorescently labeled proteins are known (Hellinga and Marvin, 1998; Brennan, 1999). Many bacterial periplasmic binding proteins such as the phosphate-binding protein (Brune *et al.*, 1994), maltose-binding protein (Gilardi *et al.*, 1994; Marvin, *et al.*, 1997), glucose/galactose-binding protein (Hellinga and Marvin, 1998), and glutamine-binding protein (Dattelbaum and Lakowicz, 2001) have been converted into receptors for reagentless biosensors by the incorporation of single fluorophores that report ligand-dependent conformational changes. Such protein-based biosensors are usually generated by incorporating fluorophores (via covalent coupling to site-specific single-point cysteine mutations) either in close proximity to the ligand-binding site, so that they can directly report ligand binding, or in distal regions of the protein such that the fluor can give an indirect readout based on domain movements in proteins which involve an allosteric coupling mechanism. This strategy is much the same as we have described above for signaling aptamers.

As an example, when reporter fluors were incorporated into different positions in *E. coli* glucose-binding protein, two of the variants were found to signal well (Hellinga and Marvin, 1998). One of the variants had the fluorescent label in an allosterically linked site and showed a twofold decrease in fluorescence upon ligand binding with minimal effect on the sugar binding constant [K_d (glc) increased by a factor of 2, and K_d (gal) by a factor of ~1.5]. A variant with the reporter fluor in the ligand-binding pocket showed a fourfold increase in fluorescence, but with a much larger associated loss of binding; the K_d (glc) increased ~100-fold, and K_d (gal) increased ~500-fold. While these glucose-binding proteins can potentially be used to measure glucose concentrations in the micromolar range, a composite maltose biosensor obtained by mixing four similarly engineered maltose-binding proteins measured maltose concentrations over a range of 0.1 μM–20 mM with an accuracy of 5% (Marvin *et al.*, 1997). The most responsive maltose-binding protein had a fluorophore incorporated in an allosteric site and showed a greater than fourfold increase in fluorescence. The remaining three were made by mutating residues in the binding pocket known to interact with maltose; this decreased the affinity for maltose without effecting signaling ability.

These latter results suggest that molecular recognition and allosteric signal transduction can be independently manipulated in protein-based biosensors. Therefore, it should be possible to change the binding specificity of a protein without affecting signaling. Hellinga's group recently converted a maltose sensor into a zinc biosensor by changing the specificity of the maltose-binding protein using a rational design strategy (Marvin and Hellinga, 2001). By employing an iterative progressive design strategy, they were able to increase the zinc affinity of the maltose-binding protein (one of their final constructs had a $K_d = 350$ nM for zinc); the mutant also showed a greater than 17-fold increase in fluorescence upon zinc binding, indicating an alteration of the initial structure. Protein biosensors based on other metalloproteins such as zinc finger peptides (Godwin and Berg, 1996; Walkup and Imperiali, 1996, 1997) and carbonic anhydrase (Elbaum *et al.*, 1996) have also been made in a similar manner by incorporating environmentally sensitive fluors.

Alternately, the intrinsic fluorescence of reporter proteins, such as the green fluorescence protein (GFP), has been exploited through the introduction of engineered analyte-binding sites. Calcium biosensors have been made by fusing two variants of GFP with calmodulin, and calmodulin-binding peptide. The introduction of Ca^{2+} induced the binding of calmodulin around the calmodulin-binding peptide, ultimately resulting in a fluorescence energy transfer between the flanking GFPs (Miyawaki *et al.*, 1997). This construct showed a 70% increase in the ratio of ultraviolet-excited emissions at 510 and 445 nm (1.8-fold change in the maximal ratio), and with the introduction of mutations into calmodulin, was able to report calcium concentrations over a very wide range from 10^{-8} to 10^{-2} M. In another study, a GFP-based calcium sensor was made by again joining together two GFP variants using the calmodulin-binding peptide (Romoser *et al.*, 1997). This sensor showed a sixfold decrease in the ratio of emissions at 505:440 nm and was able to report calcium concentrations from 50 nM to 1 μM.

Peter Schultz and co-workers have developed a totally different strategy for modulating protein–protein and protein–nucleic acid interactions using small molecules (Guo *et al.*, 2000). The method is based on first creating a cavity at a protein–protein interface such that it results in a reduction in binding affinity, and then screening libraries of small molecules to identify ligands that can bind the cavity and restore the interaction. This method is similar to one of the strategies for aptazyme construction (Figure 13.5d). However, rather than adapting the randomized allosteric domain to a particular effector, a random set of effectors is adapted to a particular allosteric domain. Two amino acids at the interface between human growth hormone (hGH) and the hGH receptor were mutated to glycines, creating a cavity at the interface and decreasing the binding affinity between the two by a factor of 10^6. By screening a library of indole derivatives, a ligand was identified which increased the affinity of mutated hGH for its mutated receptor more than 1000-fold.

13.5. Potential for improving performance

Although proteins with their wider array of functional groups may be inherently better biopolymer-binding species and catalysts than nucleic acids, to date, they have not proven to be much better for reagentless sensing. For example, initially designed signaling aptamers showed approximately 25–45% increase in fluorescence upon interaction with ATP, while the first selected signaling aptamers showed an 80% increase

(Jhaveri *et al.*, 2000b); these results are similar to those that have been exhibited by the designed protein-based biosensors described above. Further improvements in signal transduction are being achieved as improved strategies are introduced. For example, Yamana *et al.* (2003) have now shown two- to threefold ATP-dependent activation of fluorescence with site-specifically incorporated bis-pyrene, while Nutui and Li (2005) showed a four- to fivefold increase by selecting for signaling; see also Section 13.5.1. In addition, the sensitivities of these nucleic acid biosensors are the same or perhaps even better than those exhibited by the reagentless protein biosensors we have examined. For example, the anti-thrombin aptamer formulated as a signaling aptamer showed a sensitivity of 5 nM in solution (Hamaguchi *et al.*, 2001), while the same aptamer mounted on a glass platform also showed a sensitivity of 5 nM (the biosensor could detect 0.7 amol of thrombin in a 140 pl volume) (Potyrailo *et al.*, 1998). Finally, we have seen that aptazymes can be activated over a 1000-fold by their cognate effectors, results that are as good as the engineered allostery that Schultz and co-workers achieved with hGH and its receptor (Guo *et al.*, 2000). Thus, at the current time, nucleic acid and protein biosensors seem to be equivalant with regards to function, and the relative newness of aptamers, ribozymes, and aptazymes in the sensor arena means that virtually any new experiment can potentially result in greatly improved performance.

Overall, nucleic acids for use in reagentless biosensors are essentially only as good as three inherent parameters: binding affinity, binding specificity, and signaling relative to background. Binding affinity can potentially be improved by augmenting the chemically simple complement of canonical nucleotides with modified nucleotides. For example, would the inclusion of a uridine residue that contained a branched, hydrophobic (isoleucine-like) group at the 5′ position enable the selection of structures that could better recognize more hydrophobic epitopes and small molecules? The binding specificities exhibited by aptamers and aptazymes have so far been quite good, and can likely be greatly improved by a continued focus on negative selection experiments with highly related targets. One of the great advantages of nucleic acid selection relative to *in vivo* immunization is the ability to discretely and exactly control selection stringencies and parameters in order to deliver up molecules with just the right functional properties. As indicated above, it is the activation parameters of nucleic acid receptors in reagentless biosensors that truly shine relative to that of protein counterparts, and it is likely that sly manipulations of nucleic acid conformational changes by design or selection will continue to tweak these numbers upward for sometime to come. Paradoxically, though, increases in activation seem to come at a cost in sensitivity: signaling aptamers show higher apparent K_ds for their ligands than do their parental aptamers; aptazymes show higher K_ds for their effectors than do corresponding aptamers. In both instances, it is not unreasonable to suspect that ligand-binding energy is transduced into conformational changes, and that the greater the conformational change, the more ligand-binding energy must be diverted to that conformational change. In turn, the more ligand-binding energy that is diverted to a conformational change, the lower the intrinsic affinity of a given signaling aptamer or aptazyme will be for its cognate ligand. The question thus becomes whether there is an experimental resolution of this seeming thermodynamic paradox; is it possible to design or (more likely) select reagentless nucleic acid biosensors that simultaneously have both higher affinities and higher activation parameters? More detailed empirical answers to this question are becoming available as more and more nucleic acid biosensors have been generated and assayed. For example, Ron Breaker has found that it is indeed possible to select for increasingly lower apparent K_ds for cyclic nucleotide activated hammerhead aptazymes without loss of activation (Koizumi *et al.*, 1999a).

13.5.1. **In vitro** *selection of signaling aptamers*

In many of the above examples, the generation of signaling aptamers required prior knowledge of the secondary or even tertiary structure of the aptamer; they were all "designed" as signaling aptamers. Obviously, the need to understand the detailed structure of an aptamer may limit the applicability of these methods, especially in the development of large-scale sensor arrays (see also Section 13.3.4.). Therefore, it is important to determine whether other methods might be developed that would directly couple aptamer selection to signal transduction. In other words, can methods be devised for the direct selection of signaling aptamers?

As a first attempt, we have incorporated modified, fluorescent nucleotides directly into selection experiments (Jhaveri *et al.*, 2000b). A single-stranded DNA pool was synthesized that contained largely A, G, and C, and only a small fraction of T. *In vitro* transcription was used to generate a RNA pool that completely incorporated a uridine analogue, fluorescein UTP, at all positions that would have normally contained uridine. The random region in the pool was skewed and the uridine ratio was kept low to avoid an intrinsic background which could mask signaling by selected aptamers. Aptamers were isolated that could bind to the target analyte ATP. Several different families emerged from the selection, and all contained a relatively small number of (or even numbers) uridine residues. Individual families were then screened for their ability to signal the presence of ATP by a ligand-dependent change in fluorescence intensity. One family showed excellent signaling abilities; the best signaling aptamer contained only one uridine, could sense ATP concentrations as low as 25 μM, was selective for ATP relative to other nucleotides, and was stable enough in complex mixtures to quantitate ATP. The presence of signaling aptamers in the selected population suggested that fluors may be present during the selection of aptamers, rather than added later.

A general method for the direct selection of aptamer beacons also has been developed (Rajendran and Ellington, 2003). This method relies upon ligand-dependent elution of immobilized nucleic acids from an affinity column. As illustrated in Figure 13.9a, a fluoresceinated, single-stranded DNA pool with 20 randomized positions was annealed with a twofold molar excess of a biotinylated capture oligonucleotide which was in turn immobilized on streptavidin agarose beads. The oligonucleotide affinity column was then developed with an equimolar mixture of two 16-mer oligonucleotide targets (OT1 and OT2). The eluant from the column was amplified by PCR, transcribed to generate RNA, and reverse transcribed. The enriched single-strand cDNA labeled with fluorphore was purified by RNase digestion and gel purification, and was used for the next round of selection. After nine rounds of selection

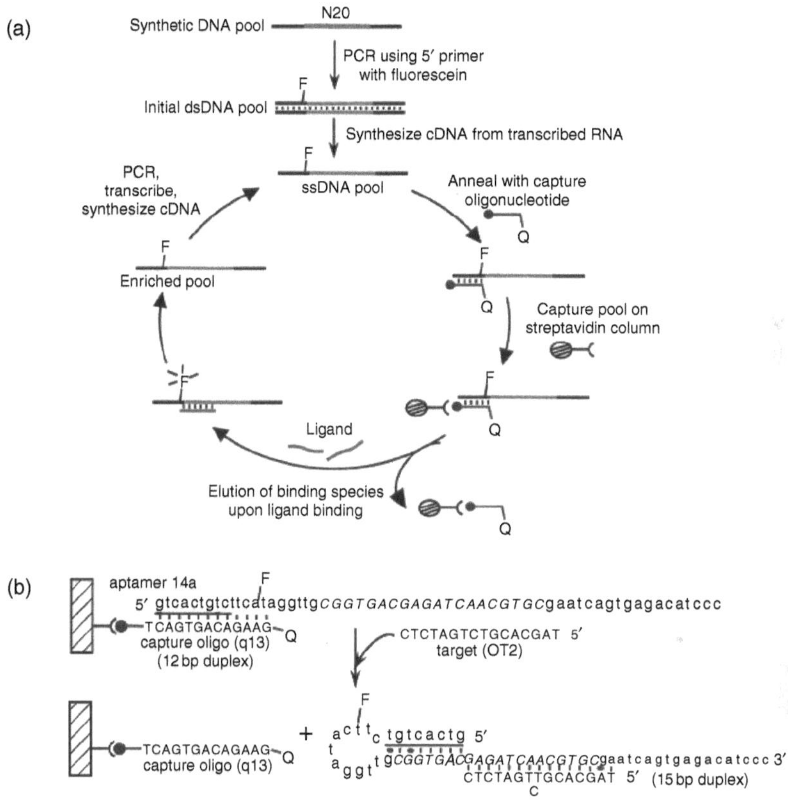

Figure 13.9 *In vitro* selection of molecular beacons. (a) Selection cycle. F indicates fluorescein, and Q, DABCYL. The closed circle at the termini of the capture oligonucleotide represents biotin. (b) Correlation of the mechanism of elution and fluorescence responsivities in selected molecular beacons. Hybridization of the oligonucleotide target OT2 stabilizes the formation of an internal hairpin stem and disrupts interactions with the capture oligonucleotide.

and amplification, the population was enriched in variants that could be specifically eluted by target OT2 (but not target OT1). The selected molecular beacons bore an FAM moiety on residue T11 within their 5′ constant regions and when hybridized with a capture oligonucleotide containing a DABCYL moiety at its 5′ end, their fluorescence was correspondingly quenched (Figure 13.9b). Addition of the oligonucleotide target (OT2) resulted in the same conformational change that was originally selected for the displacement of the capture oligonucleotide bearing the quencher, and thus also resulted in an increase in fluorescence intensity. The predominant selected beacons showed 9.5- and 16.5-fold increases in fluorescence in the presence of a 2-fold of molar excess of OT2, and the fluorescence intensity increased as a function of OT2 concentration. The gain in fluorescence increase was similar to that observed in many designed molecular beacons (Tyagi and Kramer, 1996; Sokol, *et al.*, 1998; Poddar, 1999; Tan, *et al.*, 2000; Yamamoto, *et al.*, 2000). The apparent K_ds and LODs for the selected molecular beacons were 37 ± 11 nM and 14 nM for beacon 14a and 34 ± 8 nM and 3.6 nM for beacon 16c, values which were again similar to those previously demonstrated for designed molecular beacons (Liu and Tan, 1999; Fang *et al.*, 2000b; Steemers *et al.*, 2000). The predicted mechanism for the selected molecular beacons was quite different from that of designed molecular beacons. In selected molecular beacons, a hairpin stem is formed, rather than disrupted, in the presence of the target oligonucleotide, resulting in a loss of the capture oligonucleotide. Although the selected molecular beacons have performance characteristics comparable with those of designed molecular beacons, the ability to select beacons may prove useful for identifying available sites on complex targets, such as mRNAs. More importantly, the general method for selection can now be further extended to virtually any target class, potentially yielding selected aptamer beacons. Indeed, this has now been done by Nutiu and Li (2005), who have used a very similar method to generate ATP- and GTP-signaling aptamers. The original library contained constant regions that could hybridize to oligonculeotides carrying fluors and quenchers, and selected aptamers could therefore be immediately converted into a signaling molecule in which addition of analyte resulted in release of the quencher. The signaling aptamers that were eventually selected showed moderate fluorescence intensity enhancements of 4.5-fold with the best ATP-binding aptamer and 2-fold with the best GTP-recognizing aptamer (comparable to rationally designed signaling aptamers (1.5-fold (Jhaveri *et al.*, 2000a); 3-fold (Yamamoto *et al.*, 2000; Yamana *et al.*, 2003); 5-fold (Stojanovic and Kolpashchikov, 2004)).

13.5.2. Computation design of signaling aptamers

While empirical answers suffice for the moment, computational methods are becoming increasingly tractable and at some point may provide highly quantitative design methods for nucleic acid biosensors. Predictive algorithms have advanced to the point where likely nucleic acid sequences can be rapidly paired with nucleic acid secondary structures (Zuker, 2000). Likewise, probabilistic methods have been developed to sample populations of possible secondary structures and to predict structural stability (McCaskill, 1990; Ding and Lawrence, 2003). There have also been theoretical treatments of what computational design criteria will maximize sensitivity and specificity in nucleic acid sensors (Dirks *et al.*, 2004). As these methods come to fruition, the computational design of biosensors should be inherently faster than experimental selection methods, and it may be possible to control the design process with high precision–generating biosensors with optimal sensitivities, signal-to-noise ratios, and dynamic ranges.

As a harbinger of things to come, the Breaker laboratory has developed a computational method for developing allosteric ribozymes activated by nucleic acid sequences (Penchovsky and Breaker, 2005). In this work, a virtual pool of RNA molecules was created that contained a randomized region inserted into the middle of the hammerhead ribozyme. An algorithm was then used to predict the dominant secondary structures in the presence and absence of an oligonucleotide effector that was complementary to a given sequence in the randomized region. For example, using this algorithm, it proved possible to predict the sequence of ribozymes that would first be inactive in the absence of an oligonucleotide effector (due to internal pairings that disrupted the ribozymes), and then would undergo structural rearrangement in the presence of the oligonucleotide effector, returning the ribozymes to their native structures.

Computational methods have also been applied to the more difficult problem of controlling nucleic acid conformational change by non-nucleic acid effectors. Hall *et al.* (2007) have explored an algorithm for assessing a “slip structure” (Figure 13.10) model in which alternative

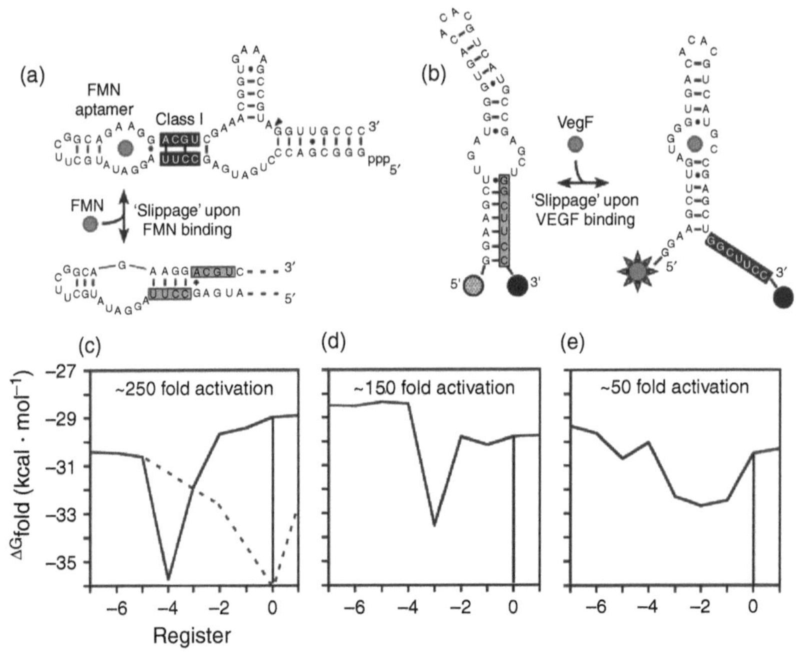

Figure 13.10 Energetic profile model. (a and b) Schematic of "slip structure" activation. In the absence of ligand (FMN or VEGF), the joining region is trapped in an inactive conformation (gray highlighted). A structural rearrangement occurs upon ligand binding, shifting the joining region to the "active register" (inverted letters highlighted in black) and potentiating either catalysis (cleavage site denoted by filled arrow in (a)) or binding. (c–e) Energetic profiles of three previously selected, FMN-responsive hammerhead aptazymes (Class I, Class III, and Class V, respectively). The free energies of folding (ΔG-fold) in the absence of ligand are plotted versus the register of the joining region. Position "0" (vertical line) denotes the "active" register of the ribozyme. The dashed line in (c) is a hypothetical energetic profile following ligand binding.

sets of base pairs are formed in the absence and presence of ligand, and they have applied this algorithm to both designed aptazymes and designed aptamer beacons (Hall *et al.*, 2007). Computational automation of the algorithm allows the structures and minimum free energies of millions of different sequences to be computed in hours on a typical desktop computer. In this work, the slip structure model proved to be generalizable, and could be applied with equal facility to computationally generate aptazymes that proved to be experimentally activated by other ligands (theophylline) or that contained other catalytic cores (hairpin ribozyme). Moreover, the slip structure model could be applied to the prediction of a ligand-dependent aptamer beacon biosensor in which the addition of the protein VEGF led to a 10-fold increase in fluorescent signal.

Overall, perhaps the greatest advantage and potential that aptamers and aptazymes currently have relative to protein reagents is the prospect for the high throughput generation (by either selection or design) of multiple different receptors against multiple different targets. The ability to generate large numbers of nucleic acid recognition elements at will should make it possible to plumb the constitution and functionality of organismal proteomes and metabolomes.

References

Bartel, D.P. and Szostak, J.W. (1993) *Science*, **261**, 1411.
Bar-Ziv, R. and Libchaber, A. (2001) *Proc. Natl. Acad. Sci. USA*, **98**, 9068.
Bock, L.C., Griffin, L.C., Latham, J.A. *et al.* (1992) *Nature*, **355**, 564.
Breaker, R.R. and Joyce, G.F. (1994) *Chem. Biol.*, **1**, 223.
Breaker, R.R. and Joyce, G.F. (1995) *Chem. Biol.*, **2**, 655.
Brennan, J.D. (1999) *J. Fluoresc.*, **9**, 295.
Brody, E.N. and Gold, L. (2000) *J. Biotechnol.*, **74**, 5.
Brody, E.N., Willis, M.C., Smith, J.D. *et al.* (1999) *Mol. Diagn.*, **4**, 381.
Brune, M., Hunter, J.L., Corrie, J.E., and Webb, M.R. (1994) *Biochemistry*, **33**, 8262.
Bunkenborg, J., Gadjev, N.I., Deligeorgiev, T., and Jacobsen, J.P. (2000) *Bioconjug. Chem.*, **11**, 861.
Carmi, N., Shultz, L.A., and Breaker, R.R. (1996) *Chem. Biol.*, **3**, 1039.
Centi, S., Tombelli, S., Minunni, M., and Mascini, M. (2007) *Anal. Chem.*, **79**, 1466.
Chu, T.C., Shieh, F., Lavery, L.A. *et al.* (2006) *Biosens. Bioelectron.*, **21**, 1859.
Chun, S.M., Jeong, S., Kim, J.M. *et al.* (1999) *J. Am. Chem. Soc.*, **121**, 10844.
Ciesiolka, J. and Yarus, M. (1996) *RNA*, **2**, 785.
Conn, M.M., Prudent, J.R., and Schultz, P.G. (1996) *J. Am. Chem. Soc.*, **118**, 7012.

Connor, A.C., Frederick, K.A., Morgan, E.J., and McGown, L.B. (2006) *J. Am. Chem. Soc.*, **128**, 4986.
Connor, A.C. and McGown, L.B. (2006) *J. Chromatogr. A*, **1111**, 115.
Conrad, R., Keranen, L.M., Ellington, A.D., and Newton, A.C. (1994) *J. Biol. Chem.*, **269**, 32051.
Cox, J.C., Hayhurst, A., Hesselberth, J. *et al.* (2002) *Nucleic Acids Res.*, **30**, e108.
Cox, J.C., Rudolph, P., and Ellington, A.D. (1998) *Biotechnol. Prog.*, **14**, 845.
Dattelbaum, J.D. and Lakowicz, J.R. (2001) *Anal. Biochem.*, **291**, 89.
Davis, K.A., Abrams, B., Lin, Y., and Jayasena, S.D. (1996) *Nucleic Acids Res.*, **24**, 702.
Davis, K.A., Lin, Y., Abrams, B., and Jayasena, S.D. (1998) *Nucleic Acids Res.*, **26**, 3915.
Deng, Q., German, I., Buchanan, D., and Kennedy, R.T. (2001) *Anal. Chem.*, **73**, 5415.
Dieckmann, T., Suzuki, E., Nakamura, G.K., and Feigon, J. (1996) *RNA*, **2**, 628.
Ding, Y. and Lawrence, C.E. (2003) *Nucleic Acids Res.*, **31**, 7280.
Dirks, R.M., Lin, M., Winfree, E., and Pierce, N.A. (2004) *Nucleic Acids Res.*, **32**, 1392.
Drolet, D.W., Moon-McDermott, L., and Romig, T.S. (1996) *Nat. Biotechnol.*, **14**, 1021.
Eaton, B.E. and Pieken, W.A. (1995) *Annu. Rev. Biochem.*, **64**, 837.
Elbaum, D., Nair, S.K., Patchan, M.W. *et al.* (1996) *J. Am. Chem. Soc.*, **118**, 8381.
Ellington, A.D. and Szostak, J.W. (1990) *Nature*, **346**, 818.
Famulok, M., Mayer, G., and Blind, M. (2000) *Acc. Chem. Res.*, **33**, 591.
Fang, X., Cao, Z., Beck, T., and Tan, W. (2001) *Anal. Chem.*, **73**, 5752.
Fang, X., Li, J.J., Perlette, J. *et al.* (2000a) *Anal. Chem.*, **72**, 747A.
Fang, X., Li, J.J., and Tan, W. (2000b) *Anal. Chem.,* **72**, 3280.
Fang, X., Sen, A., Vicens, M., and Tan, W. (2003) *ChemBioChem.*, **4**, 829.
German, I., Buchanan, D.D., and Kennedy, R.T. (1998) *Anal. Chem.*, **70**, 4540.
Gilardi, G., Zhou, L.Q., Hibbert, L., and Cass, A.E. (1994) *Anal. Chem.*, **66**, 3840.
Glazer, A.N. and Rye, H.S. (1992) *Nature*, **359**, 859.
Godwin, H.A. and Berg, J.M. (1996) *J. Am. Chem. Soc.*, **118**, 6514.
Green, L.S., Jellinek, D., Bell, C. *et al.* (1995) *Chem. Biol.*, **2**, 683.
Guo, Z., Zhou, D., and Schultz, P.G. (2000) *Science*, **288**, 2042.
Haes, A.J., Giordano, B.C., and Collins, G.E. (2006) *Anal. Chem.*, **78**, 3758.

Hall, B., Hesselberth, J.R., and Ellington, A.D. (2007) *Biosens. Bioelectron.*, **22**, 1939.
Haller, A.A. and Sarnow, P. (1997) *Proc. Natl. Acad. Sci. USA*, **94**, 8521.
Hamaguchi, N., Ellington, A., and Stanton, M. (2001) *Anal. Biochem.*, **294**, 126.
Hawkins, M.E. (2001) *Cell Biochem. Biophys.*, **34**, 257.
He, Y., Chen, Y., Liu, H. *et al.* (2005a) *J. Am. Chem. Soc.*, **127**, 12202.
He, Y., Tian, Y., Chen, Y. *et al.* (2005b) *Angew. Chem. Int. Ed.*, **44**, 6694.
Hellinga, H.W. and Marvin, J.S. (1998) *Trends Biotechnol.*, **16**, 183.
Hermann, T. and Patel, D.J. (2000) *Science*, **287**, 820.
Hesselberth, J., Robertson, M.P., Jhaveri, S., and Ellington, A.D. (2000a) *J. Biotechnol.*, **74**, 15.
Hesselberth, J.R., Miller, D., Robertus, J., and Ellington, A.D. (2000b) *J. Biol. Chem.*, **275**, 4937.
Hirao, I., Spingola, M., Peabody, D., and Ellington, A.D. (1998) *Mol. Divers.*, **4**, 75.
Ho, H.A. and Leclerc, M. (2004) *J. Am. Chem. Soc.*, **126**, 1384.
Ho, H.A., Boissinot, H., Bergeron, M.G. *et al.* (2002) *Angew. Chem. Int. Ed.*, **41**, 1548.
Homann, M. and Goringer, H.U. (1999) *Nucleic Acids Res.*, **27**, 2006.
Huizenga, D.E. and Szostak, J.W. (1995) *Biochemistry*, **34**, 656.
Illangasekare, M., Sanchez, G., Nickles, T., and Yarus, M. (1995) *Science*, **267**, 643.
Jayasena, S.D. (1999) *Clin. Chem.*, **45**, 1628.
Jayasena, V.K. and Gold, L. (1997) *Proc. Natl. Acad. Sci. USA*, **94**, 10612.
Jenison, R.D., Gill, S.C., Pardi, A., and Polisky, B. (1994) *Science*, **263**, 1425.
Jenne, A., Hartig, J.S., Piganeau, N. *et al.* (2001) *Nat. Biotechnol.*, **19**, 56.
Jhaveri, D., Kirby, R., Conrad, R. *et al.* (2000a) *J. Am. Chem. Soc.*, **122**, 2469.
Jhaveri, S., Rajendran, M., and Ellington, A.D. (2000b) *Nat. Biotechnol.*, **18**, 1293.
Jiang, F., Kumar, R.A., Jones, R.A., and Patel, D.J. (1996) *Nature*, **382**, 183.
Jiang, Y., Fang, X., and Bai, C. (2004) *Anal. Chem.*, **76**, 5230.
Jose, A.M., Soukup, G.A., and Breaker, R.R. (2001) *Nucleic Acids Res.*, **29**, 1631.
Joyce, M.V. and McGown, L.B. (2004) *Appl. Spectrosc.*, **58**, 831.
Kamekawa, N., Shimomura, Y., Nakamura, M., and Yamana, K. (2006) *Chem. Lett.*, **36**, 660.
Katilius, E., Katiliene, Z., and Woodbury, N.W. (2006) *Anal. Chem.*, **78**, 6484.
Kawakami, J., Imanaka, H., Yokota, Y., and Sugimoto, N. (2000) *J. Inorg. Biochem.*, **82**, 197.

Kim, J.H., Morikis, D., and Ozkan, M. (2004) *Sens. Actuators, B Chemical*, **102**, 315.
Kirby, R., Cho, E.J., Bayer, T. *et al.* (2004) *Anal. Chem.*, **76**, 4066.
Kleinjung, F., Klussmann, S., Erdmann, V.A. *et al.* (1998) *Anal. Chem.*, **70**, 328.
Knudsen, S.M., Lee, J., Ellington, A.D., and Savran, C.A. (2006) *J. Am. Chem. Soc.*, **128**, 15 936.
Koizumi, M., Kerr, J.N., Soukup, G.A., and Breaker, R.R. (1999a) *Nucleic Acids Symp. Ser.*, **42**, 275.
Koizumi, M., Soukup, G.A., Kerr, J.N., and Breaker, R.R. (1999b) *Nat. Struct. Biol.*, **6**, 1062.
Kotia, R.B., Li, L., and McGown, L.B. (2000) *Anal. Chem.*, **72**, 827.
Kubik, M.F., Stephens, A.W., Schneider, D. *et al.* (1994) *Nucleic Acids Res.*, **22**, 2619.
Kuwabara, T., Warashina, M., and Taira, K. (2000a) *Trends Biotechnol.*, **18**, 462.
Kuwabara, T., Warashina, M., and Taira, K. (2000b) *Curr. Opin. Chem. Biol.*, **4**, 669.
LaBean, T.H., Yan, H., Kopatsch, J. *et al.* (2000) *J. Am. Chem. Soc.*, **407**, 1848.
Lai, R.Y., Plaxco, K.W., and Heeger, A.J. (2007) *Anal. Chem.*, **79**, 229.
Laib, S. and Seeger, S. (2004) *J. Fluoresc.*, **14**, 187.
Lee, M. and Walt, D.R. (2000) *Anal. Biochem.*, **282**, 142.
Leech, D. and Daigle, F. (1998) *Analyst*, **123**, 1971.
Levy, M., Cater, S.F. and Ellington, A.D. (2005) *ChemBioChem*, **6**, 2163.
Li, J.J., Fang, X. and Tan, W. (2002) *Biochem. Biophys. Res. Commun.*, **292**, 31.
Li, Y., Lee, H.J., and Corn, R.M. (2007) *Anal. Chem.*, **79**, 1082.
Li, Y. and Sen, D. (1996) *Nature Struct. Biol.*, **3**, 743.
Lim, Y.T., Kim, S., Nakayama, A. *et al.* (2003) *Mol. Imaging*, **2**, 50.
Lin, C., Katilius, E., Liu, Y. *et al.* (2006) *Angew. Chem. Int. Ed.*, **45**, 5296.
Lin, C., Liu, Y., and Yan, H. (2007) *Nano Lett.*, **7**, 507.
Lin, C.H. and Patel, D.J. (1997) *Chem. Biol.*, **4**, 817.
Liu, X., Farmerie, W., Schuster, S., and Tan, W. (2000) *Anal. Biochem.*, **283**, 56.
Liu, Y. and Danielsson, B. (2005) *Anal. Chem.*, **77**, 2450.
Lin, Y., Qiu, Q., Gill, S.C., and Jayasena, S.D. (1994) *Nucleic Acids Res.*, **22**, 5229.
Lohse, P.A. and Szostak, J.W. (1996) *Nature*, **381**, 442.

Macaya, R.F., Schultze, P., Smith, F.W. *et al.* (1993) *Proc. Natl. Acad. Sci. USA*, **90**, 3745.
Mannironi, C., Di Nardo, A., Fruscoloni, P., and Tocchini-Valentini, G.P. (1997) *Biochemistry*, **36**, 9726.
Marvin, J.S. and Hellinga, H.W. (1998) *J. Am. Chem. Soc.*, **120**, 7.
Marvin, J.S. and Hellinga, H.W. (2001) *Proc. Natl. Acad. Sci. USA*, **98**, 4955.
Marvin, J.S., Corcoran, E.E., Hattangadi, N.A. *et al.* (1997) *Proc. Natl. Acad. Sci. USA*, **94**, 4366.
McCaskill, J.S. (1990) *Biopolymers*, **29**, 1105.
McCauley, T.G., Hamaguchi, N., and Stanton, M. (2003) *Anal. Biochem.*, **319**, 244.
Medintz, I.L., Clapp, A.R., Mattoussi, H. *et al.* (2003) *Nat. Mater.*, **2**, 630.
Menger, M., Tuschl, T., Eckstein, F., and Porschke, D. (1996) *Biochemistry*, **35**, 14710.
Merino, E.J. and Weeks, K.M. (2005) *J. Am. Chem. Soc.*, **127**, 12766.
Merino, E.J., Wilkinson, K.A., Coughlan, J.L., and Weeks, K.M. (2005) *J. Am. Chem. Soc.*, **127**, 4223.
Michalet, X., Pinaud, F.F., Bentolila, L.A. *et al.* (2005) *Science*, **307**, 538.
Mills, D.R., Peterson, R.L., and Spiegelman, S. (1967) *Proc. Natl. Acad. Sci. USA*, **58**, 217.
Miyawaki, A., Llopis, J., Heim, R. *et al.* (1997) *Nature*, **388**, 882.
Morris, K.N., Jensen, K.B., Julin, C.M. *et al.* (1998) *Proc. Natl. Acad. Sci. USA*, **95**, 2902.
Morrison, L.E. (1999) *J. Fluorescence*, **9**, 187.
Mulchandani, A. and Pan, S. (1999) *Anal. Biochem.*, **267**, 141.
Narvaez, A., Suarez, G., Popescu, I.C. *et al.* (2000) *Biosens. Bioelectron.*, **15**, 43.
Nutiu, R. and Li, Y. (2003) *J. Am. Chem. Soc.*, **125**, 4771.
Nutiu, R. and Li, Y. (2005) *Angew. Chem. Int. Ed.*, **44**, 1061.
Oliphant, A.R., Brandl, C.J., and Struhl, K. (1989) *Mol. Cell Biol.*, **9**, 2944.
Pagratis, N.C., Bell, C., Chang, Y.F. *et al.* (1997) *Nat. Biotechnol.*, **15**, 68.
Pan, T., Dichtl, B., and Uhlenbeck, O.C. (1994) *Biochemistry*, **33**, 9561.
Pan, W., Craven, R.C., Qiu, Q. *et al.* (1995) *Proc. Natl. Acad. Sci. USA*, **92**, 11509.
Patel, D.J. and Suri, A.K. (2000) *Rev. Mol. Biotechnol.*, **74**, 39.
Pavski, V. and Le, X.C. (2001) *Anal. Chem.*, **73**, 6070.
Penchovsky, R. and Breaker, R.R. (2005) *Nat. Biotechnol.*, **23**, 1424.
Piatek, A.S., Tyagi, S., Pol, A.C. *et al.* (1998) *Nat. Biotechnol.*, **16**, 359.
Poddar, S.K. (1999) *J. Virol. Methods*, **82**, 19.
Porta, H. and Lizardi, P.M. (1995) *Biotechnology*, **13**, 161.

Potyrailo, R.A., Conrad, R.C., Ellington, A.D., and Hieftje, G.M. (1998) *Anal. Chem.*, **70**, 3419.
Prudent, J.R., Uno, T. and Schultz, P.G. (1994) *Science*, **264**, 1924.
Rajendran, M. and Ellington, A.D. (2002) *Comb. Chem. High Throughput Screen*, **5**, 263.
Rajendran, M. and Ellington, A.D. (2003) *Nucleic Acids Res.*, **31**, 5700.
Rehder, M.A. and McGown, L.B. (2001) *Electrophoresis*, **22**, 3759.
Ringquist, S., Jones, T., Snyder, E.E. *et al.* (1995) *Biochemistry*, **34**, 3640.
Robertson, D.L. and Joyce, G.F. (1990) *Nature*, **344**, 467.
Robertson, M.P. and. Ellington, A.D. (1999) *Nat. Biotechnol.*, **17**, 62.
Robertson, M.P. and Ellington, A.D. (2000) *Nucleic Acids Res.*, **28**, 1751.
Robertson, M.P. and Ellington, A.D. (2001) *Nat. Biotechnol.*, **19**, 650.
Rogers, K.R. (2000) *Mol. Biotechnol.*, **14**, 109.
Romig, T.S., Bell, C., and Drolet, D.W. (1999) *J. Chromatogr. B Biomed. Sci. Appl.*, **731**, 275.
Romoser, V.A., Hinkle, P.M., and Persechini, A. (1997) *J. Biol. Chem.*, **272**, 13270.
Rupcich, N., Nutiu, R., Li, Y., and Brennan, J.D. (2006) *Angew. Chem. Int. Ed.*, **45**, 3295.
Sassanfar, M. and Szostak, J.W. (1993) *Nature*, **364**, 550.
Schultze, P., Macaya, R.F., and Feigon, J. (1994) *J. Mol. Biol.*, **235**, 1532.
Seetharaman, S., Zivarts, M., Sudarsan, N., and Breaker, R.R. (2001) *Nat. Biotechnol.*, **19**, 336.
Seiwert, S.D., Stines Nahreini, T., Aigner, S. *et al.* (2000) *Chem. Biol.*, **7**, 833.
Singh, K.K., Rucker, T., Hanne, A. *et al.* (2000) *Biotechniques*, **29**, 344.
Sokol, D.L., Zhang, X., Lu, P., and Gewirtz, A.M. (1998) *Proc. Natl. Acad. Sci. USA*, **95**, 11538.
Soukup, G.A., DeRose, E.C., Koizumi, M., and Breaker, R.R. (2001) *RNA*, **7**, 524.
Soukup, G.A., Emilsson, G.A., and Breaker, R.R. (2000) *J. Mol. Biol.*, **298**, 623.
Spiegelman, S. (1971) *Q Rev. Biophys.*, **4**, 213.
Steemers, F.J., Ferguson, J.A., and Walt, D.R. (2000) *Nat. Biotechnol.*, **18**, 91.
Stivers, J.T. (1998) *Nucleic Acids Res.*, **26**, 3837.
Stojanovic, M.N. and Kolpashchikov, D.M. (2004) *J. Am. Chem. Soc.*, **126**, 9266.
Stojanovic, M.N. and Landry, D.W. (2002) *J. Am. Chem. Soc.*, **124**, 9678.
Stojanovic, M.N., de Prada, P., and Landry, D.W. (2000) *J. Am. Chem. Soc.*, **122**, 11547.

Stojanovic, M.N., de Prada, P., and Landry, D.W. (2001) *J. Am. Chem. Soc.*, **123**, 4928.

Su, S., Nutiu, R., Filipe, C.D. *et al.* (2007) *Langmuir*, **23**, 1300.

Tan, W., Fang, X., Li, J., and Liu, X. (2000) *Chemistry*, **6**, 1107.

Tang, J. and Breaker, R.R. (1997) *Chem. Biol.*, **4**, 453.

Tucker, C.E., Chen, L.S., Judkins, M.B. *et al.* (1999) *J. Chromatogr. B Biomed. Sci. Appl.*, **732**, 203.

Tuerk, C. and Gold, L. (1990) *Science*, **249**, 505.

Tyagi, S., Bratu, D.P., and Kramer, F.R. (1998) *Nat. Biotechnol.*, **16**, 49.

Tyagi, S. and Kramer, F.R. (1996) *Nat. Biotechnol.*, **14**, 303.

Vitiello, D., Pecchia, D.B., and Burke, J.M. (2000) *RNA*, **6**, 628.

Walkup, G.K. and Imperiali, B. (1996) *J. Am. Chem. Soc.*, **118**, 3053.

Walkup, G.K. and Imperiali, B. (1997) *J. Am. Chem. Soc.*, **119**, 3443.

Wang, J., Jiang, Y., Zhou, C., and Fang, X. (2005) *Anal. Chem.*, **77**, 3542.

Warashina, M., Kuwabara, T., and Taira, K. (2000) *Structure*, **8**, R207.

Wedekind, J.E. and McKay, D.B. (1999) *Nat. Struct. Biol.*, **6**, 261.

Willis, M.C., Collins, B.D., Zhang, T. *et al.* (1998) *Bioconjug. Chem.*, **9**, 573.

Wilson, C. and Szostak, J.W. (1995) *Nature*, **374**, 777.

Wilson, D.S. and Szostak, J.W. (1999) *Annu. Rev. Biochem.*, **68**, 611.

Winfree, E., Liu, F., Wenzler, L.A., and Seeman, N.C. (1998) *Nature*, **394**, 539.

Yamamoto, R., Baba, T., and Kumar, P.K. (2000) *Genes Cells*, **5**, 389.

Yamana, K., Iwai, T., Ohtani, Y. *et al.* (2002) *Bioconjug. Chem.*, **13**, 1266.

Yamana, K., Ohashi, Y., Nunota, K. *et al.* (1991) *Tetrahedron Lett.*, **32**, 6347.

Yamana, K., Ohtani, Y., Nakano, H., and Saito, I. (2003) *Bioorg. Med. Chem. Lett.*, **13**, 3429.

Yan, H., Park, S.H., Finkelstein, G. *et al.* (2003) *Science*, **301**, 1882.

Zuker, M. (2000) *Curr. Opin. Struct. Biol.*, **10**, 303.

Zuo, X., Song, S., Zhang, J., Pan, D., Wang, L., and Fan, C. (2007) *J. Am. Chem. Soc.*, **129**, 1042.

Optical Biosensors: Today and Tomorrow (2nd Edition)
Frances S. Ligler and Chris Rowe Taitt (Editors)

Chapter 14

IMPRINTED POLYMERS AND THEIR APPLICATION IN OPTICAL SENSORS

Sergey A. Piletsky, Ph.D. and Anthony P.F. Turner, Ph.D., D.Sc.
Cranfield Health, Cranfield University, Silsoe, Bedfordshire
MK45 4DT, UK

Molecular imprinting is the process of template-induced formation of specific recognition sites (binding or catalytic) in a material where the template directs the positioning and orientation of the material's structural components by a self-assembly mechanism. Synthetic receptors prepared using molecular imprinting possess a unique combination of properties, such as high affinity, specificity, low price, and robustness, which make them an attractive alternative to natural receptors, enzymes, and antibodies used in biosensors. This chapter gives a brief overview of the technology, with specific emphasis on the mechanisms underlying the ability of imprinted polymers to perform highly selective functions such as recognition and transformation of a binding event into a detectable optical signal. The problems associated with the application of molecularly imprinted polymers (MIPs) in sensors are highlighted. Possible solutions to these problems are discussed and recommendations made about where commercial application of imprinted sensors seems most feasible in the near future.

14.1. Molecular imprinting: technical concepts

The molecular imprinting approach exploits the formation of a complex between a template molecule and functional monomers, which is fixed by copolymerization with cross-linker into a growing polymer network (Figure 14.1). Following removal of the template, binding sites are left in the polymers, which have the shape and orientation of functional groups complementary to those of the template molecule (Wulff, 1995; Mayes and Mosbach, 1997).

14.1.1. Different formats used for design of imprinted materials

The typical recipe for molecularly imprinted polymer (MIP) preparation includes mixing the target compound–template with a functional monomer (most frequently – methacrylic acid) and cross-linker (e.g., ethylene glycol dimethacrylate) in an appropriate solvent (e.g., chloroform, acetonitrile) and polymerizing this mixture using UV or chemical initiation (O'Shannessy *et al.*, 1989). The template can be extracted from the polymer by washing or by electrophoresis (Piletsky

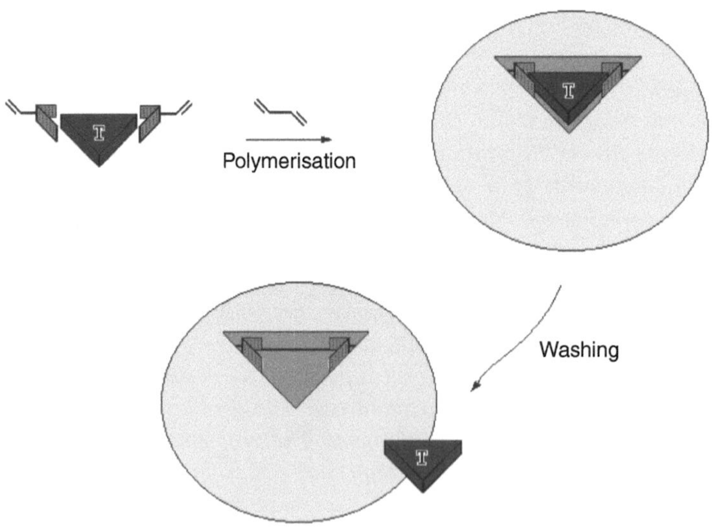

Figure 14.1 Scheme of molecularly imprinted polymerization.

et al., 1992). Subsequent polymer grinding and washing yield polymer particles with receptor sites on the accessible surface.

Other formats of molecular imprinting include:

1. polycondensation of silica acid in the presence of a template (Katz and Davis, 2000);
2. electropolymerization (Malitesta *et al.*, 1999);
3. formation of two-dimensional templated monolayers onto a SiO_2, metal oxide, or gold surface (Starodub *et al.*, 1992; Mirsky *et al.*, 1999);
4. grafting of imprinted polymers to the inert solid surface (Dhal *et al.*, 1995; Piletsky *et al.*, 2000a) (Figure 14.2);
5. templating of a pre-formed polymer structure by precipitation or cross-linking in the presence of a template (Braco *et al.*, 1990; Peissker and Fischer, 1999);
6. formation of imprinted poly/oligomers (e.g., peptides) in the presence of a template (Giraudi *et al.*, 2000; Piletska *et al.*, 2000); and
7. preparation of soluble and colloidal imprinted polymers by living polymerization (Piletsky *et al.*, 2005).

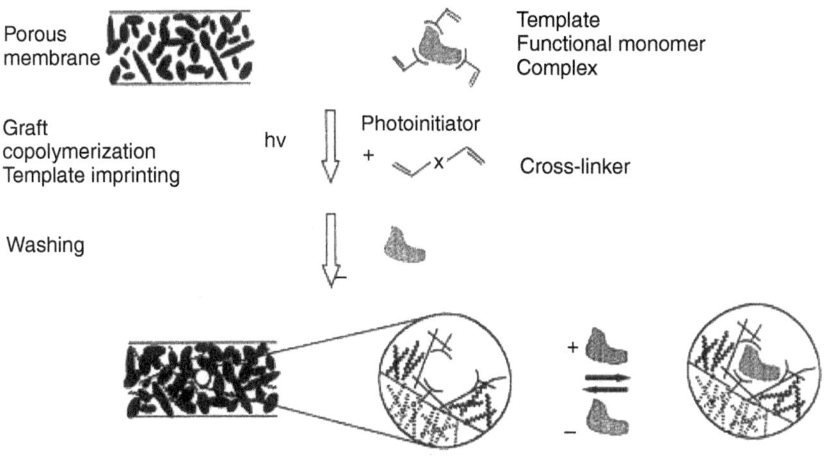

Figure 14.2 Scheme of MIP synthesis via surface photografting onto porous polymeric substrate (Piletsky *et al.*, 2000a).

Historically, formation of imprinted silica gels was the first example of molecular imprinting (Polyakov, 1931). Despite the fact that this specific technique reached its peak in the 1960s and is now in the process of gradual decline due to limited flexibility of the method, it still remains the most popular choice for the preparation of specific zeolites (Dong *et al.*, 2000).

Electropolymerization faces the same type of problem as silica imprinting due to the limited number of polymerizable functional monomers available, which are selected mainly from the group of aniline, phenol, pyrrole, and thiophene (Panasyuk *et al.*, 1999; Blanchard *et al.*, 2000). Electropolymerization retains its attraction, however, because it provides a means for precise deposition of a sensitive layer on an electrode surface, which is extremely important for microsensor and multianalyte sensor production. Electropolymerized MIPs have been used almost exclusively in potentiometric (Boyle *et al.*, 1989; Vinokurov *et al.*, 1990) and amperometric sensors (Piletsky *et al.*, 1994a).

Two-dimensional MIPs or imprinted monolayers were developed and used in optical sensors by Andersson *et al.* (1988). They used Tabushi's method (Tabushi *et al.*, 1987) to produce a surface sensitive to vitamin K_1. Andersson *et al.* (1988) chemosorbed octadecylchlorosilane in the presence of inert template hosts (*n*-hexadecane) to a silicon dioxide surface. After extraction of the hosts, the analyte, vitamin K_1 was detected by ellipsometry. Although imprinted monolayers are not real polymers, we regard them as two-dimensional MIPs because of the similarity in their synthesis and behavior.

Expanding this method for the preparation of monolayers imprinted with water-soluble templates, we developed materials selective for amino and nucleic acids (Piletsky and Starodub, 1992). This approach involves two steps: first, adsorption of the template on the surface of SiO_2 or metal oxide and, second, treatment of the surface with the adsorbed template by trimethyl chlorosilane from the gas phase. In another similar approach, a gold surface was imprinted with a cholesterol-specific monolayer using coadsorption of the template with hexadecylmercaptane (Piletsky *et al.*, 1999a).

Despite some advantages, such as fast sensor response and easy preparation, these systems, however, suffer from lack of stability. The lateral mobility of the components of imprinted monolayer is responsible for steady decrease in the specificity of imprinted cavities. An important improvement in sensor stability was achieved by coimmobilization of the template in the imprinted layer. Using a new approach called "spread bar architecture design," it was possible to develop stable monolayers, consisting of a template, thiobarbituric acid, and a functional monomer, hexadecylmercaptane. A depression in the hexadecylmercaptane layer formed by the template was able to accommodate barbituric acid, changing the electrode capacitance in the binding process (Mirsky *et al.*, 1999). This two-dimensional format for MIP design is particularly attractive for sensing based on evanescent wave measurements (e.g., surface plasmon resonance, SPR).

Several reports on the preparation of MIPs by surface grafting have appeared where a thin imprinted layer, most frequently a monolayer, is formed on a solid support (Dhal *et al.*, 1995; Lele *et al.*, 1999; Piletsky *et al.*, 2000a). Grafting can be performed using chemical, UV, or plasma initiation (Shi *et al.*, 1999; Piletsky *et al.*, 2000b). The advantage of this approach lies in the possibility of modifying a very inert surface (polystyrene, polypropylene, etc.) with specific polymers. The additional attraction of the electropolymerization and grafting methods is their convenient format, which does not require an additional processing step. Molecularly imprinted polymer synthesis and immobilization is performed as a one-step procedure, directed by applied potential or by exposing the monomer mixture-coated detector to UV light (Figure 14.2).

A further approach, frequently called "bioimprinting," involves precipitation or cross-linking of biological molecules (proteins) in the presence of a template (Braco *et al.*, 1990; Peissker and Fischer, 1999). The conformation adopted by interacting biopolymer around the template remains fixed after template extraction with an appropriate solvent. Although the authors are unaware of any examples where bioimprinting has been used for sensor design, this technique could potentially be useful for introducing either additional recognition sites into enzymes or catalytic sites into antibodies. These chimeric molecules might possess the combined characteristics of antibodies and enzymes and, in this way, be useful for the development of new, label-free types of assays and sensors.

A further format of molecular imprinting is template-directed synthesis. This process includes the formation of a new substance by a chemical modification of the substrate or by the coupling of two or more molecules in the presence of a template to serve as a pattern for the formation of a new structure. The most well-known example of this process is gene replication. An important issue is that the synthesized molecule always has a structure complementary to that of the template, which can be exploited for the synthesis of biospecific ligands or to obtain information about the structure and properties of the template molecule. This approach is actively pursued in molecular biology (gene sequencing) and in DNA sensors where complementary DNA or RNA chains are synthesized using transcription facilitated by enzymes such as DNA polymerase or reverse transcriptase.

Unfortunately, a similar technique does not exist for the analysis of molecules other than DNA, such as proteins and polysaccharides. It is possible, however, to produce a complementary ligand for a target molecule using a synthetic approach (Giraudi *et al.*, 2000; Piletska *et al.*, 2000). The method involves the formation of oligomers (e.g., peptides) in the presence of a template. Prior to the initiation of polymerization and during polymerization, the monomers, which could be amino acids or nucleotides, spatially distribute themselves around the template molecules in accordance with the size, polarity, and functionality of the template. The monomers are polymerized into linear, water-soluble oligomers specific for the template. The advantage of this approach is the possibility of obtaining water-soluble ligands, which can be treated in the same manner as antibodies and other natural receptors.

One of the most exciting new approaches to molecular imprinting is the preparation of soluble and colloidal imprinted polymers by living polymerization (Piletsky *et al.*, 2006). The conditions of living polymerization ensure a relatively small size of synthesized molecules, which have

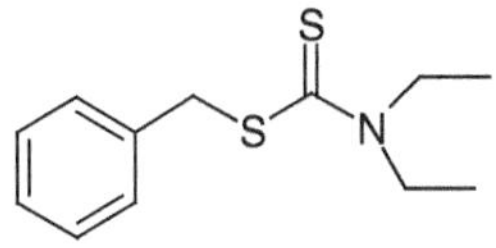

Figure 14.3 Structure of the iniferter diethyldithiocarbamic acid benzyl ester.

a higher affinity to the template than the original monomers. The living free-radical polymerization techniques, such as iniferter (Figure 14.3) polymerization, nitroxide-mediated radical polymerization, atom-transfer radical polymerization (ATRP), and reversible addition–fragmentation chain transfer (RAFT) polymerization, open new routes for the synthesis of polymers with controlled relatively low molecular weights. Controlled/living polymerization techniques are based on a delicate balance between dormant and active species that effectively reduces the concentration of free radicals in the system and minimizes the extent of termination. Living polymerization could be free of side reactions such as termination and chain transfer and thus can generate polymers with well-defined molecular weight distribution and structure.

Living polymerization has been used previously to produce bulk-grafted MIPs (Ruckert *et al.*, 2002; Hattori *et al.*, 2004). Soluble polymers were also produced by living polymerization and used later in MIP production (Li *et al.*, 2005). However, no one so far has developed soluble MIPs by living polymerization. In the Cranfield work, the polymerization reaction is terminated at an early stage (prior to the precipitation stage shown in Figure 14.4) when the size of synthesized molecules is relatively small (30–100 kDa). The product of such a process can exist in a soluble or colloidal form, which is stable in solution (Figure 14.4). This is, of course, distinctive from a gelled form that would normally arise from allowing the polymerization to run its full course. Affinity separation is used to isolate a complementary polymer fraction with specific affinity to the template. This is akin to affinity purification of polyclonal antibodies. The "nanoMIPs" created by this process are in the form of discrete particles that are less than 1 μm in diameter and typically around 200 nm. A nanoMIP for atrazine was synthesized and a fraction of nanoMIPs with a molecular weight of 90 kDa purified by affinity

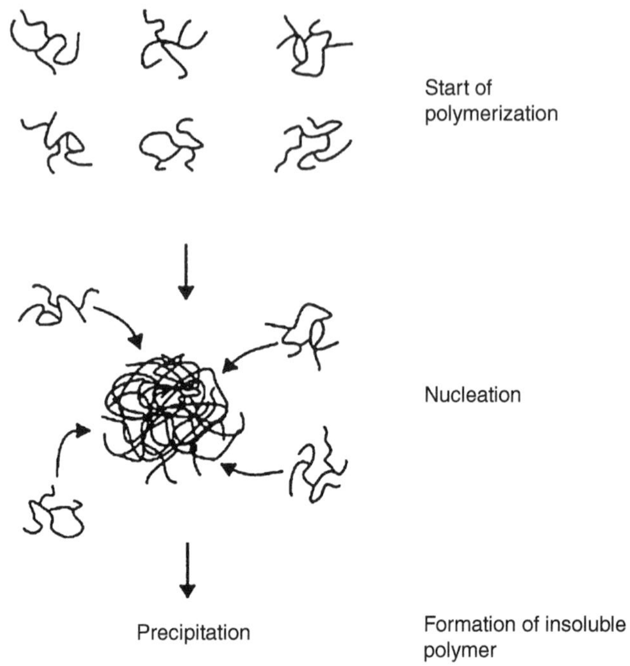

Figure 14.4 Schematic representation of polymerization process leading to formation of cross-linked insoluble polymer.

chromatography. The resulting nanoMIP had a K_d of 7.7×10^{-8} M compared to a blank with a K_d of 1.9×10^{-5} M. This was equivalent to approximately one binding site per molecule. These characteristics make nanoMIPs an exciting stable alternative to antibodies and other affinity elements in optical sensors.

Notwithstanding the new methods detailed above, traditional bulk polymerization remains the most popular choice for the preparation of MIPs for theoretical study and practical application in separation and sensing.

14.1.2. Mechanism of template recognition by imprinted polymer

Three major factors determine the recognition process: the quantity of the functional groups participating in the interaction, their correct arrangement within the cavity, and the shape of the cavity itself. The types of interactions explored in molecular imprinting include reversible covalent bonds (Wulff and Haarer, 1991), electrostatic interactions (ionic and hydrogen bonds) (Piletsky *et al.*, 1990a; Nicholls *et al.*, 1995), van der Waals (Dickert *et al.*, 1998), hydrophobic interactions (Yu *et al.*, 1997), and metal chelation (Matsui *et al.*, 1996) (Figure 14.5). The shape of the cavity alone can provide specificity (Yoshizako *et al.*, 1998) although the specificity is substantially better when the template interacts with one or more properly oriented functional monomers (Ramstrom *et al.*, 1993). The required strength of the monomer–template interaction varies depending on the size and the structure of the template. For a small template molecule, the presence of strong interactions, preferably ionic and/or hydrogen bonds, is critically important. For a large molecule such as a protein or nucleic acid, successful results can be achieved with a combination of multiple weak interactions (Hjerten *et al.*, 1997). The choice of solvent depends on the type of interaction. Thus, if template recognition depends on hydrogen bond formation, better results can be achieved if both polymer synthesis and re-binding

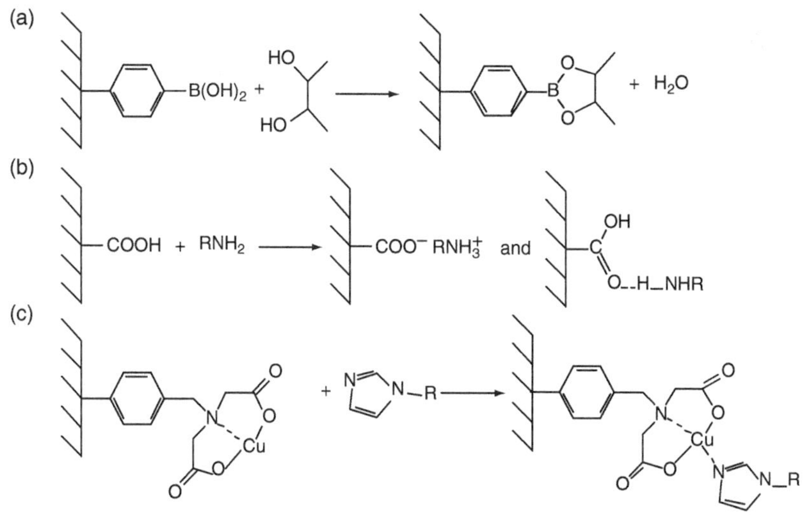

Figure 14.5 Different types of interactions explored in molecular imprinting: (a) reversible covalent bond formation; (b) electrostatic interactions; and (c) metal chelation.

take place in a hydrophobic solvent, where hydrogen bonds are stronger (Andersson, 1996; Yu and Mosbach, 2000).

The equilibrium dissociation constants (K_d) for the binding of ligands to their corresponding polymers have been estimated by Scatchard plot analysis of binding data. Mostly, non-linear plots were obtained because of multiple K_d values, varying in range from micromolar to nanomolar in the majority of cases. In a similar way to polyclonal antibodies, imprinted polymers contain a heterogeneous population of binding sites (Wulff, 1995).

One of the important components of the recognition mechanism observed in MIP systems is the conformational change in the polymer induced by template interaction (Piletsky *et al.*, 1992; Watanabe *et al.*, 1998; Wolfbeis *et al.*, 1999). Depending on the experimental conditions (solvent, temperature, and types of the monomer–template and monomer–monomer interactions), the polymer matrix can shrink or swell in the presence of a template. The mechanism is similar to the "induced fit" observed for natural enzymes and receptors (Koshland, 1995; Agmon, 2000). The importance of this effect for sensor technology lies in the possibility of using the effect for measuring template concentration (Piletsky *et al.*, 1998).

An additional factor contributing to MIP recognition properties is the presence of nanopores in the polymer structure with specificity for the template molecules (Piletsky *et al.*, 1990b; MathewKrotz and Shea, 1996). Membranes prepared by molecular imprinting possess selective permeability for the imprinted species and can be used for purification of desirable analytes or removal of potential interfering compounds.

Molecularly imprinted polymers are capable of recognizing small variations in the structure of the template and the specificity of imprinted polymers under optimized conditions is often equal to or even superior to that of natural enzymes and receptors (Andersson *et al.*, 1995). However, quite often MIPs demonstrate a high level of non-specific binding.

Hence, imprinted polymers could be considered as "plastic" antibodies or receptors with specific advantages and disadvantages (Table 14.1).

Table 14.1 Comparison of natural antibodies and receptors with MIPs

Property	Natural biomolecules	MIPs
Stability	Low	Stable at low/high pHs, pressure, and temperature
Cost	High	Inexpensive and easy preparation
Integration into multisensor unit	Integration of natural biomolecules in multisensor unit is difficult due to different operational requirements of these molecules (pH, ionic strength, temperature, substrate)	Flexible MIP design allows preparation of MIPs against many combinations of analytes
Compatibility with micromaching technology/ miniaturization	Poor	Fully compatible
Spectrum of analytes	Limited	Practically unlimited

14.1.3. "Pluses" and "minuses" in MIP technology and their comparison with natural enzymes and receptors

Being purely synthetic materials, it is natural that the imprinted polymers have a much higher stability than enzymes and receptors. The reason for this lies first of all in the high level of cross-linking, which provides adequate protection for binding sites created in the polymer by imprinting. Imprinted polymers can withstand harsh treatments with acidic and basic solutions or with organic solvent. They are stable under both high and low pressure, and, as well as at extreme temperatures (Kriz and Mosbach, 1995; Svenson and Nicholls, 2001).

Imprinting polymerization is a very inexpensive procedure for the development of artificial receptors. In the majority of cases, the price of an

MIP depends almost entirely on the price of the template used. Furthermore, if the templates themselves are expensive, it may be possible to recover the template and use it again. Alternatively, inexpensive template analogues can be used for the preparation of MIPs. Generally speaking, MIP preparation is three to four orders less expensive than production of the equivalent natural receptor, and this makes the technology very competitive.

The possibility of using MIPs in organic solvents opens new areas of application such as biomimetic sensing and catalysis in chemical and pharmaceutical manufacturing. Quality control and online monitoring of manufacturing processes are particularly attractive.

One of the most challenging problems associated with the development of multisensors is related to the significant differences in the performance of natural enzymes and receptors. All these biological materials have different stabilities, activities, and sensitivities and, in many cases, they require different substrates and buffers with different ionic strengths and pHs. Due to such factors, the integration of naturally occurring biomolecules in one single unit may be problematic. Since MIP design is flexible and a variety of monomers are available for their preparation, it is possible to develop a set of polymers specific for a range of templates that will have almost identical operational requirements (solvent, temperature, pH, etc.).

An additional benefit comes from the possibility of processing MIPs in the same way as traditional photoresist materials. Molecularly imprinted polymers can be immobilized at precise spots on the detector surface using masks and photopolymerization. The compatibility of MIPs with micromachining technology makes MIP-based multisensors feasible.

Last, but not the least, is the ability to develop MIPs for practically any type of compound. Examples of templates for which MIPs have been successfully produced include inorganic ions, drugs, nucleic acids, proteins, and even cells (Table 14.2). Although antibodies can also be prepared for a broad range of analytes, they have two disadvantages

Table 14.2 Examples of templates used in molecular imprinting

Template	Application	Reference
Amino acids and derivatives	Separation, sensors	Kempe and Mosbach, 1995; Vidiasankar *et al.*, 1997; Piletsky *et al.*, 1998
Aniline, phenol, derivatives	Sensing	Vinokurov and Grigoreva, 1990; Morita *et al.*, 1997
Drugs	Separation, sensing	Levi *et al.*, 1997; Wang *et al.*, 1997; Mirsky *et al.*, 1999; Andersson, 2000
Flavanoids	Sensing	Suárez-Rodríguez and Díaz-García, 2000
Herbicides	Separation, sensing	Kroger *et al.*, 1999; Sergeyeva *et al.*, 1999, 2001
Inorganic ions	Separation, sensing	Hutchins and Bachas, 1995; Yoshida *et al.*, 2000; Kimaro *et al.*, 2001.
Micro-organisms	Recognition	Alexander and Vulfson, 1997; Dickert *et al.*, 2001
Nucleic acids and derivatives	Separation, sensing	Piletsky *et al.*, 1990a, 1990b; MathewKrotz and Shea, 1996
Polynuclear aromatic hydrocarbons	Sensing	Dickert *et al.*, 1998
Proteins	Separation, recognition	Hjerten *et al.*, 1997; Shi *et al.*, 1999
Steroids	Separation, detection	Hishiya *et al.*, 1999; Rachkov *et al.*, 2000
Sugars, sugar derivatives	Separation, sensing	Wulff and Haarer, 1991; Piletsky *et al.*, 1998
Toxins and narcotics	Separation, sensing	Kriz and Mosbach, 1995; Matsui *et al.*, 1996; Takeuchi *et al.*, 2001.
Volatile compounds	Sensing	Ji *et al.*, 200; Dickert *et al.*, 2001.

when compared to MIPs. Firstly, small compounds often have to be derivatized in order to generate antibodies. This necessitates an additional synthetic step, which can sometimes drastically change the recognition characteristics. Secondly, flexibility in antibody preparation is limited to 20 naturally occurring amino acids. In the case of MIPs, the large number of synthetic monomers available makes it possible to engineer binding sites with a variety and flexibility unmatched by nature.

As with any other technology, molecular imprinting has shortcomings. Among them are (i) the absence of a general technology for MIP design; (ii) poor performance of MIPs in aqueous environments; (iii) high levels of non-specific binding, which produce too low a signal-to-noise ratio in sensors; (iv) poor processability of MIPs; and (v) difficulty in transforming binding events into electrical signals.

Several attempts have been made in the past to develop a general procedure for the rational design of imprinted polymers with predictable properties (Nicholls, 1995; Whitcombe *et al.*, 1998; Lanza and Sellergren, 1999; Takeuchi *et al.*, 1999). In the best examples, workers have produced rules or hints, indicating how MIPs should be made in order to possess a certain level of specificity. The most important conclusion is that the stability of the monomer–template complex formed during polymerization determines the affinity of the resulting polymer. Thus, it is known that polymerization should be performed in a hydrophobic solvent in order to produce a material able to interact with the template through electrostatic interactions. At the same time, the choice of the monomer, solvent, and polymerization conditions generally depends on common knowledge, one's personal experience, or available information describing the behavior of the similar systems (Piletsky *et al.*, 2002, 2004, 2006).

Recently, we developed a method believed to be a general solution for MIP design (Piletsky *et al.*, 2000c, 2001; Chianella *et al.*, 2002). The method involves computational screening of a virtual library of functional monomers against a target molecule. The monomers giving the best score in virtual binding experiments are then brought into contact with the template and left to equilibrate. The composition of the monomer shell surrounding the template after equilibration provides the information on the type and quantity of monomers, and should be used for polymer design. Commercially available software permits calculations to be performed using different dielectric constants, reflecting the polarity of the environment (solvent) where the polymers are prepared and used. Polymers designed using this computational approach have proved to have excellent affinity and specificity for the target compound, surpassing those of polyclonal antibodies (Table 14.3).

The possibility of tailoring MIPs for specific target analytes and specific operational conditions is very attractive since it permits polymers to be developed with optimized characteristics and shortens the time needed for design, preparation, and testing of the polymers. These computer simulation and molecular modeling approaches could also help to solve a second major problem associated with MIPs – their poor performance in aqueous environments. The majority of monomers used so far in polymer design form hydrogen and ionic bonds in the process of template recognition. These interactions are less effective in polar solvents and, as a result, the use of such MIPs is restricted mainly to hydrophobic solvents such as chloroform, toluene, and acetonitrile. Although MIPs capable of forming hydrophobic and van der Waals interactions with the template under aqueous conditions have been developed (Dickert *et al.*, 1999), the design of such polymers is much more difficult than the design of MIPs that exploit electrostatic interactions. The reason lies in the complex nature of factors contributing to hydrophobic and van der Waals interactions. Computer simulation and molecular modeling

Table 14.3 Affinity and sensitivity range of computationally designed molecularly imprinted polymer in comparison with antibodies for the template – microcystine-LR (Chianella *et al.*, 2003)

Receptor	K_d, (nM)	Sensitivity range ($\mu g\ l^{-1}$)
Computational MIP	0.3 ± 0.08	0.1–100
Monoclonal antibody	0.03 ± 0.004	0.025–5
Polyclonal antibody	0.5 ± 0.07	0.05–10

can, in principle, solve this problem and help to select the monomers ideally suited for the recognition of the template in water.

Typically, the ratio of functional monomer:template used in molecular imprinting is 4:1–10:1. Therefore, the resulting polymer contains large amounts of monomers outside of the specific binding sites; these are capable of non-specific interaction with molecules other than the template. Additionally, the cross-linker itself can interact with a variety of analytes in aqueous media. A combination of these factors, together with the large surface area (80–200 m^2/g) of the polymer, is responsible for a high level of non-specific binding, which hinders the development of MIP-based affinity materials and sensors. It is possible to overcome this problem, however, by further optimization of the polymerization procedure and by rational selection of monomers capable of forming stoichiometric complexes with templates (Lubke *et al.*, 2000).

Detection of binding can be achieved with the help of optical devices if the template has, for example, fluorescent properties. At least three general characteristics of MIPs can be used for the design of MIP-based sensors (Figure 14.6): first, by substituting MIPs for the antibodies in immunosensors (affinity sensors); second, by exploiting the receptor properties of imprinted polymers (receptor sensors); and, third, by combining MIPs possessing catalytic properties with traditional electrochemical or optical transducers (catalytic sensors).

The majority of biosensors produced to date use enzymes as a biorecognition element (Turner, 1999). The reason for this lies in the amplification effect achieved as result of multiple turnovers of catalytic processes. Many of the unique characteristics of enzymes are connected with their polymeric nature and this fact attracts attention to the methods of development of MIP catalysts (Srikovsky *et al.*, 2000). The application of catalytically active MIPs for sensor development seems to be promising and attractive as the most direct way of achieving the replacement of current biosensors by more stable devices. Nevertheless, no practical examples exist for the integration of MIPs that mimic natural enzymes into sensors. Although essential progress has been made in MIP catalysis, imprinted polymers still have properties inferior to natural enzymes,

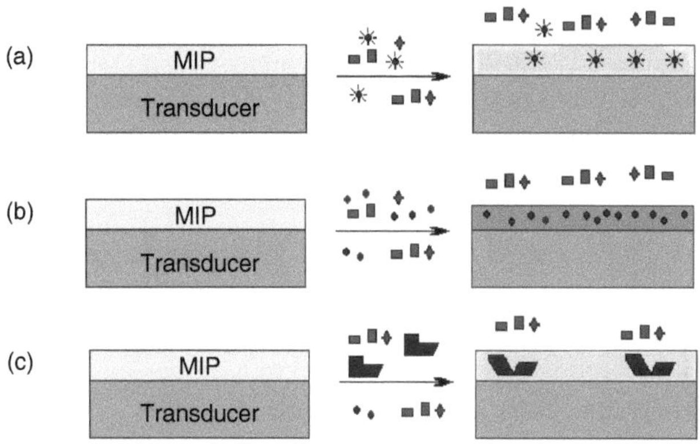

Figure 14.6 Three principal types of MIP sensors: (a) affinity sensor, where response is produced by accumulation of template on MIP surface; (b) receptor sensor where response is generated by changes in polymer characteristics, induced by its interaction with template; and (c) enzyme-mimicking sensor responding to the change in the chemical environment induced by MIP-mediated catalytic reaction.

i.e., much lower activity and turnover. Success in the preparation of more effective MIP-based catalysts, with high turnover and reacting with water-soluble and practically important analytes, will change this situation. The recent successes with development of high-efficiency MIP catalysts indicate that this is indeed possible (Svenson *et al.*, 2004; Liu and Wulff, 2004).

Due to the very limited information available on the development of MIP-based catalytic sensors, this chapter will concentrate on the development of the two remaining types of sensors: affinity and receptor-mimicking devices. Integration of imprinted polymers with detectors remains a difficult issue. Despite significant improvement in MIP technology, the processability of these materials remains challenging. The high level of cross-linking necessary for maintaining the polymer's specificity makes them extremely hard, solid, and fragile materials.

One solution to this problem is the use of plasticizers, such as oligourethane acrylates (Sergeyeva *et al.*, 1999). As a result, polymers can be made in the form of thin and stable membranes that can be used directly in sensors. Alternatively, imprinted polymers can be grafted (Mirsky *et al.*, 1999; Lotierzo *et al.*, 2004) or electropolymerized (Boyle *et al.*, 1989) onto the detector surface.

14.2. Development of MIP-based optical sensors: history and state of the art

The majority of published papers related to MIP sensors deal with electrochemical or piezoelectric devices. However, optical and, in particular, fluorescent sensors are attracting increasing attention in the literature (Henry *et al.*, 2005). This is largely due to the great flexibility that fluorescent detection offers to MIP technology. The account below highlights current achievements and prospects for the development and commercialization of MIP sensors that use optical detection for template recognition.

14.2.1. Affinity sensors

The most common type of MIP sensor is the affinity, immunosensor-type device. The detection principle is based on the measurement of the concentration of the template adsorbed by MIP immobilized on the detector surface. The first example of this type of device was the development of a two-dimensional MIP sensor for vitamin K_1 (Andersson *et al.*, 1988). Ellipsometry was used for the measurement of template concentration. Although this work was very preliminary and suffered from lack of appropriate controls, it demonstrated the possibility for direct detection of a template adsorbed by an imprinted monolayer.

Another direct method of monitoring binding of analyte by a MIP is the use of SPR. Lotierzo *et al.* (2004), for example, described an SPR sensor for the toxin that causes amnesic shellfish poisoning, domoic acid, based on a grafted imprinted polymer. This approach not only provides a route to sensor design, but also provides a valuable tool for the characterization of the binding characteristics of sensing layers. Surface plasmon resonance can also, of course, be used to characterize structural and conformational changes in polymers. Björk *et al.* (2005), for example, used SPR to study the dynamics of complex formation between biological and luminescent conjugated polyelectrolytes and noted considerable changes in the underlying polymer matrix with different buffers. Surface plasmon resonance offers the major advantage of being label free, but lack of sensitivity is a major drawback. However, sub-picomolar sensing of delta-opioid receptor ligands by MIPs has been achieved using SPR spectroscopy (Devanathan *et al.*, 2005). The authors suggested that the sensitivity and robustness of a MIP sensor made it suitable for applications ranging from biowarfare agent detection to pharmaceutical screening. Nopper *et al.* (2003) described the use of reflectometric interference spectroscopy to study the interaction of polymers imprinted with either (R,R)- or (S,S)-2,3-di-O-benzoyltartaric acid. Chiral separation with a separation factor of 1.2 could be achieved with these sensors whereas a reference polymer resulted in no separation. Hence, a rudimentary measurement is possible.

Steinke *et al.* (1996) proposed an interesting variant of an optical sensor device based on MIPs. The completely transparent imprinted polymer prepared for their experiments had anisotropic properties and provided a particular orientation of bound template molecules. The polymers, therefore, showed a pronounced dichroism in UV light, which enabled specific binding to be recognized. This work, in particular, could be applicable for the detection of optical isomers.

An optical sensor specific for the fluorescent substance dansyl-L-phenylalanine was developed using a dansyl-L-phenylalanine-imprinted polymer and a fiber-optic sensing device (Kriz *et al.*, 1995). Accumulation of fluorescent template in the polymer matrix resulted in an increase in fluorescence that could be used to detect 10 mg/l of substrate within 4 h. In another example, fluorescent polycyclic aromatic hydrocarbons were selectively enriched and detected using optical sensors based on imprinted polyurethanes (Figure 14.7) (Dickert and Tortschanoff, 1999).

A fundamental problem associated with broadening the scope of these methods is the limited quantity of fluorescent substances, which are practically important and can be used as templates in the preparation of MIPs for sensor technology. To overcome this problem, sensors that

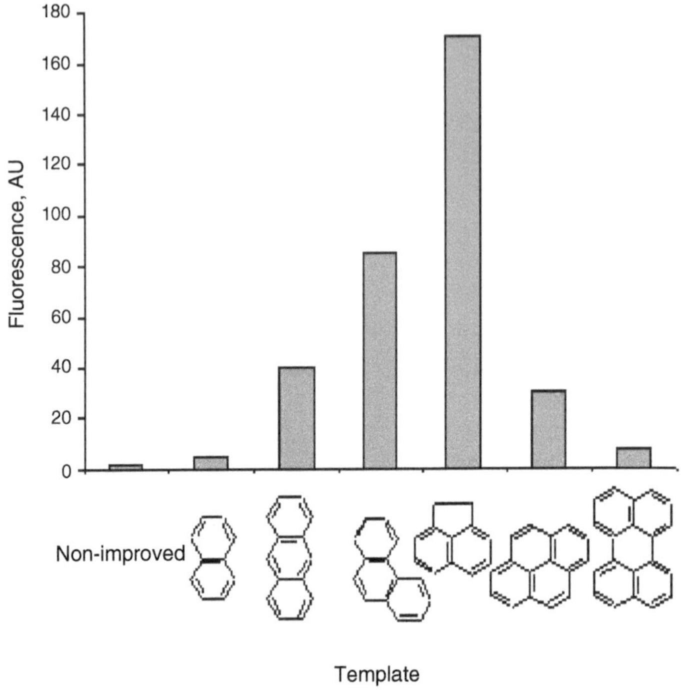

Figure 14.7 Selectivity pattern of pyrene detection by fluorescence, using polyurethanes imprinted with polyaromatic hydrocarbons of different sizes (Dickert and Tortschanoff, 1999).

operate in a competitive mode can be developed. The important question was whether the binding sites in imprinted polymers are capable of recognizing template molecules that are labeled with a fluorescent dye or enzyme. This possibility was successfully demonstrated for polymer imprinted with triazine (Piletsky *et al.*, 1997). The competition between fluorescein-labeled and unlabeled template was used to measure 10^{-8}–10^{-5} M concentrations of free template dissolved in ethanol (Figure 14.8). The polymer was able to discriminate between the template and other triazines (e.g., atrazine) and triazinone (simazine). Later, competitive assays with enzyme-labeled templates were developed for epinephrine (Piletsky *et al.*, 2000b) and 2,4-dichlorophenoxyacetic acid (Surugiu *et al.*, 2001).

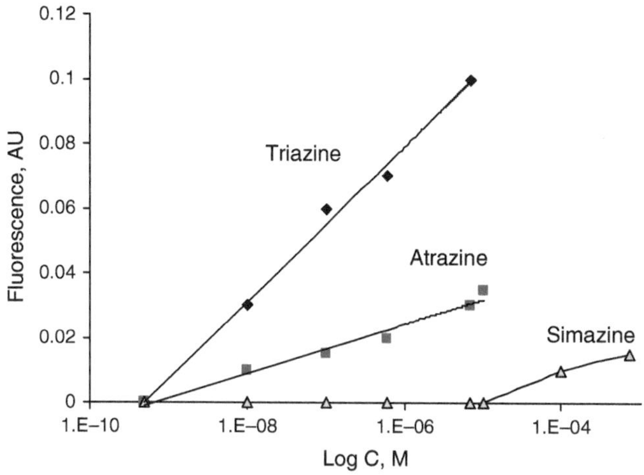

Figure 14.8 Displacement of fluorescein–triazine from triazine-imprinted polymer by analytes (Piletsky *et al.*, 1997).

The displacement format has been used for the development of an MIP sensor for chloramphenicol (CA) (Levi *et al.*, 1997). The sensor included a high performance liquid chromatography (HPLC) column with CA-specific MIPs. A constant flow of dye-labeled CA (CA–Methyl Red) at a concentration of 0.5 μg/ml was run through the column under equilibrium conditions. When analyte containing free CA was injected, it displaced the adsorbed conjugate, giving a peak with an area proportional to the CA concentration (Figure 14.9). Successful analysis of CA was achieved in model and real samples (blood serum).

Unfortunately, not all analytes can be easily modified with dyes and the modification itself can change the affinity of the analyte. Recently, the displacement of non-specific dyes from an MIP has been used for the detection and quantification of ligand–polymer binding events (Piletsky *et al.*, 1999b). Rhodamine B solution was passed through an HPLC column with L-Phe-amide-specific MIP. When the template was injected, part of the dye was competitively replaced by the analyte from the MIP. This displacement peak was three times higher for the template than for the opposite enantiomer. This approach can be considered as general and suitable for different kinds of templates, dyes, and polymers. A similar

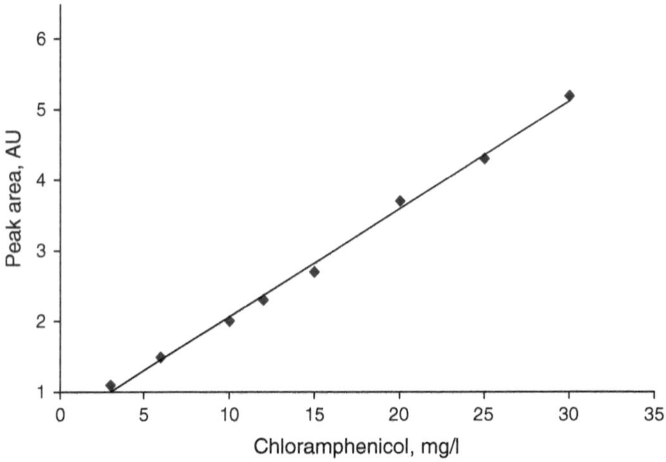

Figure 14.9 Displacement of CA–Methyl Red from a CA-imprinted polymer upon injection of template (Levi *et al.*, 1997).

displacement principle was used also in combination with electrochemical measurements for template detection (Kroger *et al.*, 1999). It is proposed that the displacement of non-specific indicator molecules from a set or array of MIPs could be used to develop multianalyte sensors.

The analyte may also be characterized after absorption on a MIP. Zhou *et al.* (2005) describe a chemiluminescence sensor for detection of salbutamol in urine. The sample was selectively bound by the MIP and a stream of luminol and potassium ferricyanide flowed through the column to react with adsorbed salbutamol and produced strong chemiluminescence. The assay was linear with salbutamol concentration from 5.0×10^{-8} to 1.0×10^{-5} g/ml and with a detection limit of 1.6×10^{-8} g/ml (3σ).

Ng and Narayanaswamy (2006) reported an optical method using a MIP for Al^{3+} ion detection using 8-hydroxyquinoline sulfonic acid ligand as a fluorescence tag. The polymer was synthesized using acrylamide as monomer, 2-hydroxyethyl methacrylate as comonomer, and ethylene glycol dimethacrylate as cross-linker. The imprinted polymer was fluorometrically characterized using a fiber-optic attachment in a flow cell.

The dynamic range of the system was linear up to 1.0×10^{-4} M with a limit of detection of 3.62 μM.

In an alternative strategy designed for penicillin G analysis in pharmaceutical formulations, the interaction of fluorescently labeled beta-lactams with a library of MIPs against penicillin were evaluated using fluorescence competitive assays (Benito-Pena *et al.*, 2006). The highly fluorescent competitors were engineered to contain pyrene or dansyl labels while keeping intact the 6-aminopenicillanic acid moiety for efficient recognition by the cross-linked polymers. Pyrenemethylacetamidopenicillanic acid was selected for the development of a fluorescence competitive assay for penicillin G analysis and delivered a dynamic range of 3–890 μM in 99:1 acetonitrile–water solution.

Grafe *et al.* (2006) developed an optical polymer-based sensor for amines in organic solvents. Thin polymer membranes were synthesized incorporating a chromogenic functional dye, 4-trifluoroacetyl-4′-[*N*-(methacryloxyethyl)-*N*-(ethyl)amino]-azobenzene with a polymerizable methacrylate group, that showed a significant color change during a reversible chemical reaction with the analyte. While the authors did not find any enhancement in selectivity through imprinting in the presence of different analytes, this suggests an interesting line of development.

Johnson-White *et al.* (2007) demonstrated the potential for porphyrin-embedded periodic mesoporous materials to be used for the detection of volatile organic compounds using simple optical interrogation. A greater degree of selectivity was obtained when porphyrin was combined with periodic mesoporous organosilica material than when combined with gel-immobilized porphyrins. Moreover, they showed that molecular imprinting could be used to obtain a higher degree of selectivity.

A Monte Carlo model was developed by Chen *et al.* (2004) to analyze the sensitivity and the performance of a fluorescence-based MIP sensor. The MIP sensor consisted of highly cross-linked polyurethane containing anthracene-binding sites coated on a transparent substrate. They concluded that to make a fluorescence-based MIP sensor that is capable of detecting one part per billion analyte concentration with a

200-μm-thick MIP film, the imprinted polymer would need to have an absorption coefficient less than 0.001 cm^{-1} or have a quantum yield 10^5 times lower than that of the analyte at the detection wavelength.

The affinity sensors described above are able to detect templates that possess a specific property such as optical absorbance, fluorescence, or electrochemical activity. However, direct detection of "inert" templates can be realized in receptor sensors.

14.2.2. Receptor sensors based on MIPs

Two approaches exist for the development of receptor-like MIP sensors. One is connected with an MIP's ability to change conformation upon binding with a template, leading to change in a measurable property such as conductivity, permeability, or surface potential (Piletsky *et al.*, 1998). A second principle is based on the ability of a functional monomer to change its property upon interaction with a template, most frequently fluorescence (Rathbone *et al.*, 2000).

Receptor properties of imprinted materials were first reported in 1992 (Piletsky *et al.*, 1992). It was shown that templates such as amino acids, nucleic acids, and cholesterol increase the transport of ions passing through the imprinted membranes (Piletsky *et al.*, 1994b, 1998). This so-called gate effect has been used for quantification of the concentration of templates. Most often, MIP-based receptor sensors measure the change in membrane electroconductivity, induced by specific interaction of MIPs with template molecules. Sensors specific for L-phenylalanine, cholesterol, sialic acid, and atrazine have shown high selectivity and sensitivity at the micromolar and even nanomolar range (Sergeyeva *et al.*, 1999). The "gate effect" can also be probed using optical detection (Piletsky *et al.*, 1996). In this work, imprinted polymer based on allylamine was imprinted with sialic acid. When polymer suspension was brought into the contact with OPA reagent, a mixture of *o*-phthaleic dialdehyde and mercaptoethanol, a fluorescent complex was formed. The kinetics of complex formation depended on the presence of the template, sialic acid, which modulated the diffusion of soluble components

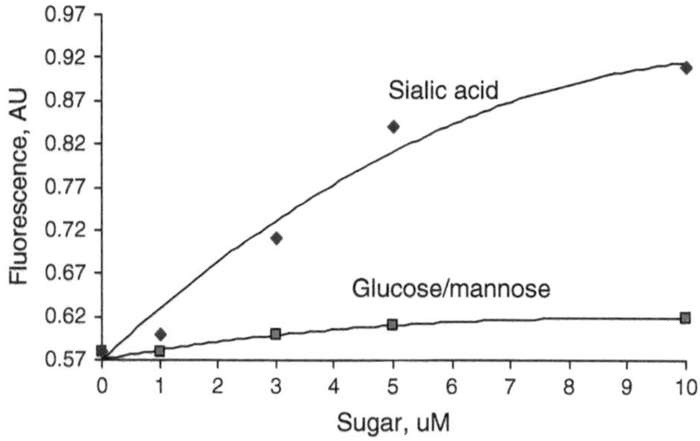

Figure 14.10 Influence of sugars on the formation of fluorescent isoindole complex between o-phthalaldehyde (OPA) reagent and amino functionality in the polymer imprinted with sialic acid (Piletsky *et al.*, 1996).

to the reactive sites (Figure 14.10). The polymer was able to discriminate sialic acid from other sugars such as glucose and mannose.

Potentially, it should be possible to design a sensor where the "gate effect" would be used for direct monitoring of conformational changes in imprinted polymers.

Imprinted polymer can be labeled with two different chromophores: one being the donor, the other the acceptor. The light energy adsorbed by the chromophore may be dissipated non-radiatively via a mechanism known as fluorescence resonance energy (Förster) transfer (FRET), which is sensitive to intra- and intermolecular interactions (Lakowicz *et al.*, 1993). Fluorescence resonance energy transfer occurs through induction of a dipole oscillation in the unexcited acceptor by the excited-state donor chromophore. The rate of energy transfer between the chromophores is a function of the sixth power of the distance between the donor and the acceptor. Because of the strong distance dependence of energy transfer, monitoring the fluorescent intensity in a system of labeled polymers can be used to quantify the concentration of a template. In our experiments, imprinted polymers specific for L-phenylalanineamide were labeled with fluorescein (donor), eosin (acceptor), and their mixture (Wolfbeis *et al.*, 1997). The emission spectra of the resulting polymers are shown in Figure 14.11, *top*. It can be seen that there is strong FRET occurring in the polymer labeled with both chromophores. The added template decreased the distance between the donor and the acceptor as a result of the polymer shrinking, which led to an increase in fluorescent emission (Figure 14.11, *bottom*).

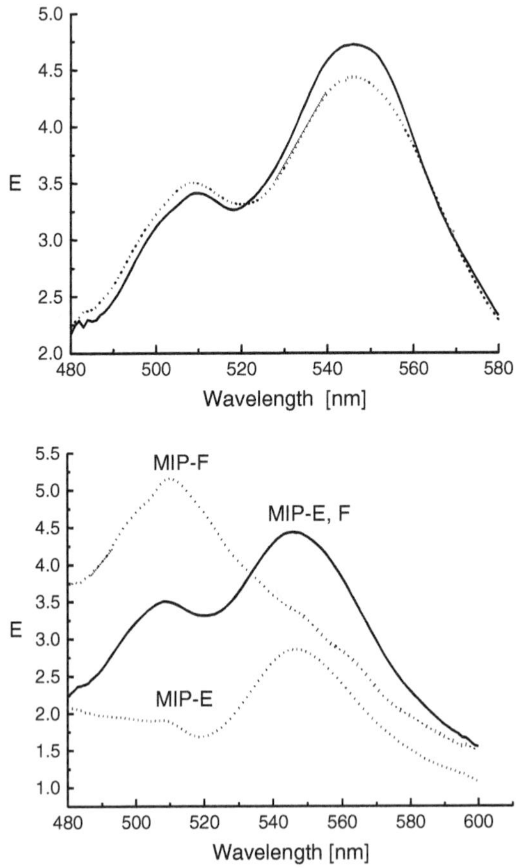

Figure 14.11 Relative emission spectra of polymers containing fluorescein (MIP-F), eosin (MIP-E), and their mixture (MIP-E,F) (*top*); and spectral changes of donor and acceptor emission in MIP-E,F induced by 10 μM concentration of L-phenylalanineamide (*bottom*) (Wolfbeis *et al.*, 1997).

The sensitivity of the system can be further improved using donor–acceptor pairs with different Förster distances.

An interesting approach for the design of signaling polymers and their use in sensors was proposed by Cooper *et al.* (1997). They used an environment-sensitive functional monomer (**I**) integrated into a cross-linked matrix, which was able to change its fluorescent properties in the presence of compounds with proton donor properties.

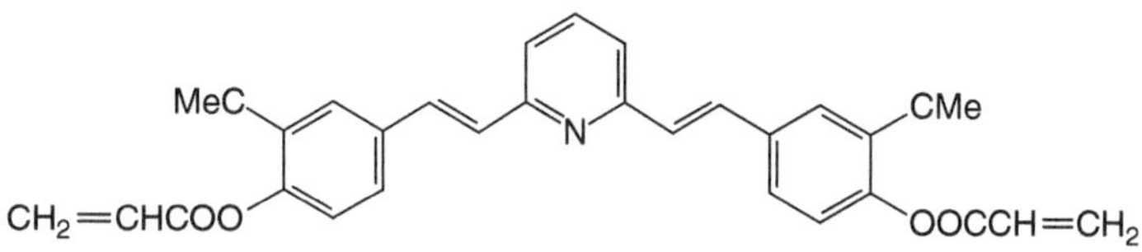

I

Strong quenching of fluorescent emission induced by hydrogen bonding might be used for recognition of polar templates. Similarly, betaine dyes with strong hypsochromic effect for protic solvents were used for gas-phase analysis (Figure 14.12) (Dickert *et al.*, 2000).

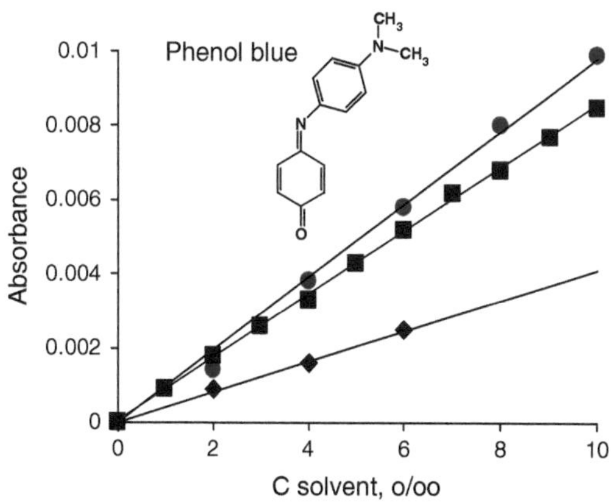

Figure 14.12 Absorbance versus solvent concentration in air for a polymer with polymer-immobilized phenol blue exposed to tetrahydrofuran (•), chloroform (■), and ethanol (♦) (Dickert *et al.*, 2000).

Environmentally sensitive dye has been used in the design of a fluorescent sensor for cAMP detection (Turkewitsch *et al.*, 1998). In this case, a fluorescent dye, trans-4-[*p*-(*N*,*N*-dimethylamino)styryl]-*N*-vinylbenzylpyridinium chloride, was copolymerized with cross-linker and template. The resulting polymer displayed two functions simultaneously: template recognition and sensing (Figure 14.13). In a later publication (Wandelt *et al.*, 2004), the same group developed this concept further and used fluorescence techniques to study the cAMP-imprinted polymer in a thin-layer film. They compared the performance of the thin layers with that of bulk-imprinted polymers and concluded that the quenching of fluorescence was much more effective for the thin-layer MIPs than for bulk-imprinted polymers. However, the selectivity of adsorption of the template on the thin-layer film of MIPs was not entirely satisfactory.

A new type of proximity scintillation assay has been developed for (*S*)-propranolol (Ye and Mosbach, 2001). A scintillation monomer,

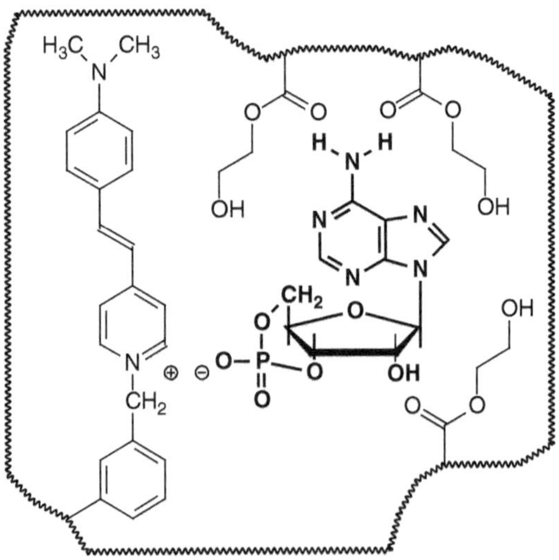

Figure 14.13 Schematic representation of polymer binding site for cAMP with signaling trans-4-[*p*-(*N*,*N*-dimethylamino)styryl]-*N*-vinylbenzylpyridinium monomer (Turkewitsch *et al.*, 1998).

4-hydroxymethyl-2,5-diphenyloxazole acrylate (**II**), has been covalently incorporated into MIP microparticles during the imprinting reaction. This monomer is capable of transforming β-radiation from the bound tritium-labeled template into a fluorescent signal. The small size of the particles (0.6–1 μm) guarantees that the reporter group, randomly distributed throughout the polymer matrix, is located in close proximity to the MIP binding site for signal generation.

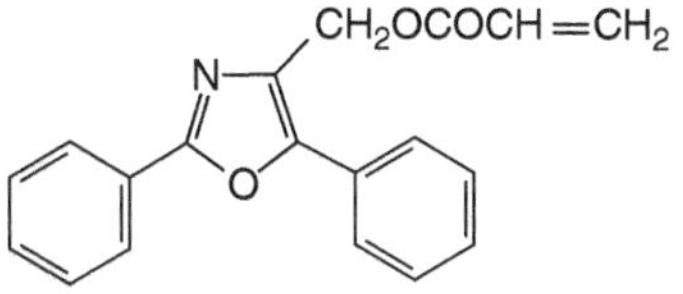

II

A reverse scenario can be used, in principle, for the quantification of the concentration of environment-sensitive templates (Matsui *et al.*, 2000). Fluorescent spectra of the cinchona alkaloids exhibit a characteristic shift through binding to these polymer particles, containing an acidic monomer – 2-(trifluoromethyl)acrylic acid (TFMAA). The authors demonstrated the possibility of using TFMAA-based imprinted polymers as polymer reagents for the analysis of the cinchona alkaloid bound to the polymers without bound/free separation (Figure 14.14).

Fluorescent polymers containing monomer (**III**) were imprinted with various N^1-benzylidene pyridine-2-carboxamidrazones (Rathbone and Ge, 2001). Dramatic quenching of fluorescence approaching background levels was observed for most cases where the "empty" MIP was re-exposed to its template. The authors claimed that this approach is suitable for high-throughput screening.

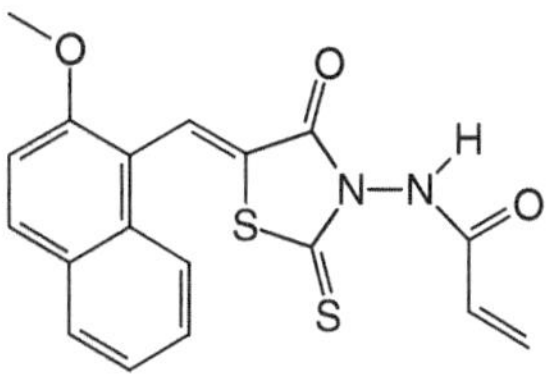

III

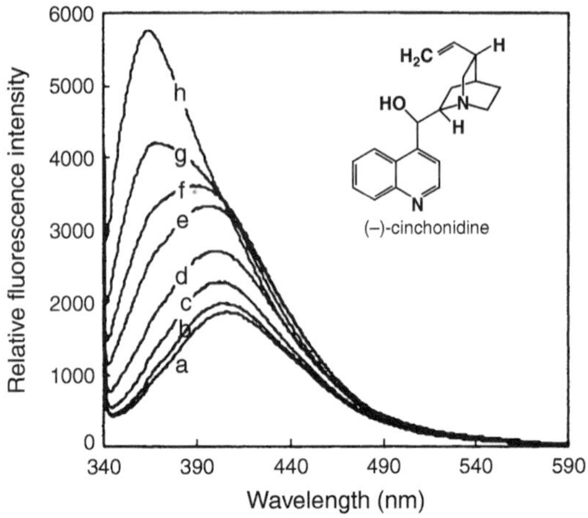

Figure 14.14 Fluorescence spectra of MIP suspension incubated with cinchonidine in chloroform/acetonitrile at concentrations of (a) 0; (b) 0.005; (c) 0.02; (d) 0.05; (e) 0.10; (f) 0.15; (g) 0.25; (h) 0.50 mM. ($\lambda_{ex} = 330$ nm) (Matsui *et al.*, 2000).

Jenkins *et al.* (1998) developed a very sensitive lanthanide-based luminescent sensor for sarin and soman with a detection limit of 7 ppt. The sensor functions by selectively and reversibly binding the phosphonate hydrolysis product of this agent to an MIP containing a coordinatively bound Eu^{3+} ion. This binding leads to the appearance of a narrow luminescence band in the 610 nm region of the Eu^{3+} spectrum, which can be monitored using a miniature spectrometer. A high degree of selectivity is obtained by combining both chemical and spectroscopic selectivities.

Very promising combinations of group-specific fluorescent reporters with template-specific MIPs were reported for sugars (Wang *et al.*, 1999), carboxylic acids (Zhang *et al.*, 2001), and primary amines (Subrahmanyam *et al.*, 2000). In the first two cases, the anthracene reporting group was modified in order to introduce polymerizable and recognition functionalities.

The interactions of boronic acid (**IV**) with *cis*-dioles and guanidine (**V**) with carboxylic acids are non-specific by their very natures.

Nevertheless, they could be made specific by incorporating these monomers into specific binding sites created by imprinting.

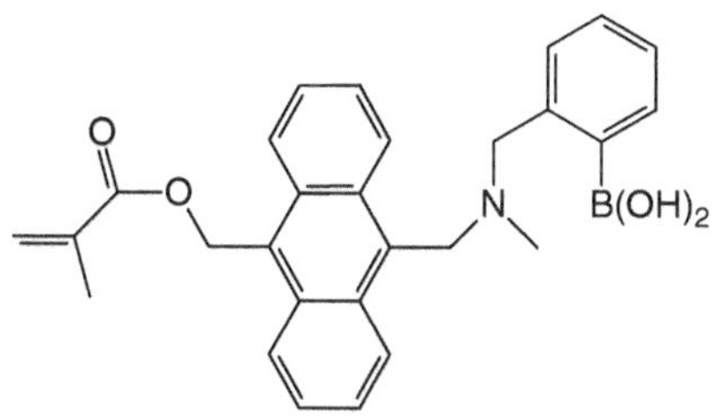

IV

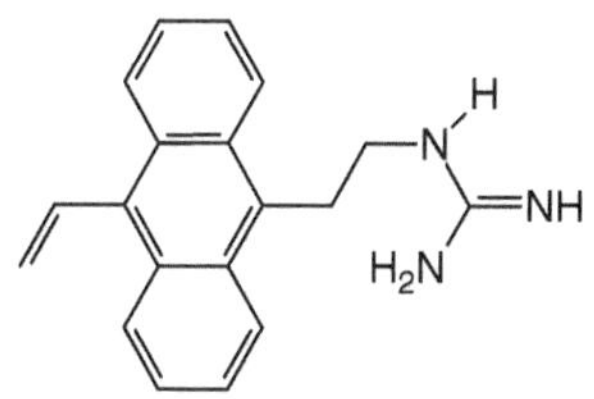

V

Similarly, non-specific interaction between thioacetal and primary amines, which leads to formation of fluorescent isoindole complex, was made specific for creatine by imprinting in the presence of methylated analogue of the template (Subrahmanyam *et al.*, 2000) (Figure 14.15).

Diode lasers are instruments of choice for the development of fluorescent sensors: their light flux is coherent and allows better integration with waveguides, and they are inexpensive and small. However, diode lasers necessitate polymerizable fluorescent markers with long-wavelength adsorption and emission, which at present are scarce. A further problem associated with the development of MIP-based optical sensors is light scatter due to heterogeneity in polymer structure.

One way to overcome this problem is optimization of the polymerization conditions (first of all by choice of solvent and polymerization temperature) that will lead to synthesis of optically transparent and homogeneous materials. Another way is to measure the decay time of

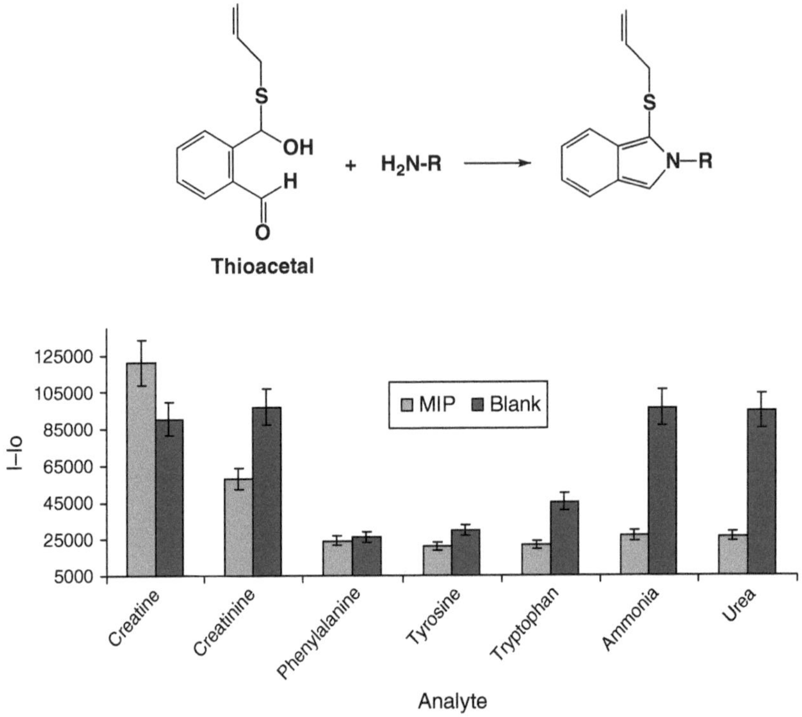

Figure 14.15 Complex formation between polymerizable thioacetal and primary amine, and recognition of analytes by blank polymer and polymer imprinted with methylated creatine analogue (Subrahmanyam *et al.*, 2000).

luminescence rather than its intensity. This approach is highly advantageous because measurements of decay time are less affected by light scattering, analyte concentration, and detector sensitivity.

14.3. Market potential of MIP sensors and future prospects

Three particular properties make commercial application of MIP sensors attractive: (i) polymers are highly stable and can be autoclaved; (ii) they are fully compatible with microfabrication technology; and (iii) the low cost of the materials and easy processes of polymer preparation in comparison with natural and other artificial receptor systems.

The most promising areas of MIP sensor applications are:

1. chemical and pharmaceutical manufacturing: using MIP sensors in extreme conditions (high and low pHs, toxic solvents, and high temperature, pressure, and radiation);
2. medicine and pharmaceuticals: application of MIPs mimicking natural receptors for drug screening and for *in vivo* monitoring;
3. environment: remote sensing, continuous emissions sensors, and point-source monitors;
4. defense and security: rapid detection of chemical and biological agents and toxins, and law enforcement applications such as detection of drugs of abuse; and
5. deep ocean and space exploration: sensors for analysis of extreme environments.

Several key problems associated with MIP development need to be addressed, however, before successful commercialization can commence. The issues include:

1. development and validation of a general protocol for MIP design;
2. development of MIPs capable of effective functioning in water;
3. the need for a substantial increase in polymer affinity and improvement of the ratio between specific and non-specific bindings; and
4. development of effective immobilization protocols.

With further progress in polymer science and engineering, we can expect to see the appearance of a new generation of MIP sensors that will gradually replace traditional biosensors and chemical sensors in many areas of biotechnology and pharmacology, environmental, clinical, and food analysis.

14.4. New developments

During the last 3 years, more than 50 original papers and patent applications were published in the area of MIP sensors, with a significant proportion of these related to the development of optical sensors. Worthy of particular mention is the development of a chemiluminescence chip for terbutaline by He *et al.* (2006). The sensor allowed accurate detection of terbutaline in model solutions over a concentration range of 8–100 ng/ml. This chapter is an expansion of previously published work by the same group on the development of a microchip for the detection of salbutamol (Zhou *et al.*, 2005).

A further new important development was the creation of fiber-optic luminescent sensors for chemical agents EA2192, VX, sarin, and soman (Jenkins and Bae, 2005). The instrument has been designed for monitoring of water supplies and allowed detection of 11–33 ppt of toxins in water.

An interesting new direction in design of optical sensors has been proposed by Sanchez-Barragan *et al.* (2005). They have used room-temperature phosphorescence induced by heavy atoms, such as iodide, in target polycyclic aromatic hydrocarbons to monitor these compounds in spiked water samples. The detection limit for fluoranthene was 35 ng/l.

A novel variant of plasmon absorption was reported by Matsui *et al.* (2004). A sensing layer comprising gold nanoparticles embedded in an MIP was constructed. The variable proximity of the gold nanoparticles caused by swelling of the polymer after binding of the analyte caused a blue shift in the plasmon absorption band of the immobilized nanoparticles. Adrenaline was detected at 5 μM using this system, which was proposed as a general strategy for constructing MIP sensors.

A new promising protocol for integration of binding and signaling functionalities in MIP nanoparticles has been proposed by Southard *et al.* (2007). The luminescence of europium (III) integrated into the polymer core was used for measuring polymer interaction with the template, dicrotophos.

These recent developments indicate that research into MIP-based sensors will continue to expand. More exciting achievements in optical sensors can be expected to follow as a result of this activity.

References

Agmon, N. (2000) *J. Phys. Chem. B*, **104**, 7830.

Alexander, C. and Vulfson, E.N. (1997) *Adv. Mater.*, **9**, 751.

Andersson, L.I. (1996) *Anal. Chem.*, **68**, 111.

Andersson, L.I. (2000) *J. Chromatogr. B*, **739**, 163.

Andersson, L.I., Mandenius, C.F., and Mosbach, K. (1988) *Tetrahydron Lett.*, **29**, 5437.

Andersson, L.I., Muller, R., Vlatakis, G., and K. Mosbach (1995) *Proc. Natl. Acad. Sci USA*, **92**, 4788.

Benito-Pena, E., Moreno-Bondi, M.C., Aparicio, S. *et al.* (2006) *Anal. Chem.*, **78**, 2019.

Björk, P., Persson, N., Peter, K. *et al.* (2005) *Biosens. Bioelectron.*, **20**, 1764.

Blanchard, P., Huchet, L., Levillain, E., and Roncali, J. (2000) *Electrochem. Commun.*, **2**, 1.

Boyle, A., Genies, E.M., and Lapkowski, M. (1989) *Synth. Metals*, **28**, C769.

Braco, L., Dabulis, K., and Klibanov, A.M. (1990) *Proc. Natl. Acad. Sci. USA*, **87**, 274.

Chen, Y.C., Brazier, J.J., Yan, M.D. *et al.* (2004) *Sens. Actuat. B-Chem.*, **102**, 107.

Cooper, M.E., Hoag, B.P., and Gin, D.L. (1997) *Polym. Prepr.*, **38**, 209.

Chianella, I., Lotierzo, M., Piletsky, S.A. *et al.* (2002) *Anal. Chem.*, **74**, 1288.

Chianella, I., Piletsky, S.A., Tothill, I.E. *et al.* (2003) *Biosens. Bioelectron.*, **18**, 119.

Devanathan, S., Salamon, Z., Nagar, A. *et al.* (2005) *Anal. Chem.*, **77**, 2569.

Dhal, P.K., Vidyasankar, S., and Arnold, F.H. (1995) *Chem. Mater.*, **7**, 154.

Dickert, F. and Tortschanoff, M. (1999) *Anal. Chem.*, **71**, 4559.

Dickert, F., Geiger, U., Lieberzeit, P., and Reutner, U. (2000) *Sens. Actuators B*, **70**, 263.

Dickert, F.L., Besenbock, H., and Tortschanoff, M. (1998) *Adv. Mater.*, **10**, 149.

Dickert, F.L., Hayden, O., and Halikias, K.P. (2001) *Analyst*, **126**, 766.

Dong, J., Lin, Y.S., Hu, M.Z.-C. *et al.* (2000) *Micropor. Mesopor. Mat.*, **34**, 241.

Giraudi, G., Giovannoli, C., Tozzi, C. *et al.* (2000) *Chem. Commun.*, **13**, 1135.

Hattori, K., Hiwatari, M., Iiyama, C. *et al.* (2004) *J. Membr. Sci.*, **233**, 169.

He, D., Zhujun, Zh., Houjiang, Zh., and Huang, Y. (2006) *Talanta*, **69**, 1215.

Henry, O.Y.F., Cullen, D.C., and Piletsky, S.A. (2005) *Anal. Bioanal. Chem.*, **382**, 947.

Hishiya, T., Shibata, M., Kakazu, M. *et al.* (1999) *Macromolecules*, **32**, 2265.

Hjerten, S., Liao, J.L., Nakazato, K. *et al.* (1997) *Chromatographia*, **44**, 227.

Hutchins, R.S. and Bachas, G. (1995) *Anal. Chem.* **67**, 1654.

Jenkins, A.L. and Bae, S.Y. (2005)*Anal. Chim. Acta*, **542**, 32.
Jenkins, A.L., Uy, O.M., and Murray, G.M. (1998) *Anal. Chem.,* **71**, 373.
Ji, H.S., McNiven, S., Lee, K.H. *et al.* (2000) *Biosens. Bioelectron.*, **15**, 403.
Johnson-White, B., Zeinali, M., Shaffer, K.M. *et al.* (2007) *Biosens. Bioelectron.*, **22**, 1154.
Katz, A. and Davis, M.E. (2000) *Nature*, **403**, 286.
Kempe, M. and Mosbach, K. (1995) *J. Chromatogr. A*, **691**, 317.
Kimaro, A., Kelly, L.A., and Murray, G.M. (2001) *Chem. Commun.*, 1282.
Koshland, D.E. (1995) *Angew. Chem. Int. Ed.*, **33**, 2375.
Kriz, D. and Mosbach, K. (1995) *Anal. Chim. Acta*, **300**, 71.
Kriz, D., Ramstrom, O., Svensson, A., and Mosbach, K. (1995) *Anal. Chem.*, **67**, 2142.
Kroger, S., Turner, A.P.F., Mosbach, K., and Haupt, K. (1999) *Anal. Chem.*, **71**, 3698.
Lakowicz, J.R., Wiczk, W., Gryczynsky, I. *et al.* (1993) *Macromolecules*, **26**, 349.
Lanza, F. and Sellergren, B., (1999) *Anal. Chem.*, **71**, 2092.
Lele, B.S., Kulkarni, M.G., and Mashelkar, R.A. (1999) *React. Functional Polym.*, **39**, 37.
Levi, R., McNiven, S., Piletsky, S.A. *et al.* (1997) *Anal. Chem.*, **69**, 2017.
Li, Z., Day, M., Ding, J.F., and Faid, K. (2005)*Macromolecules*, **38**, 2620.
Liu, J.Q. and Wulff, G. (2004) *Angew. Chem. Int. Ed.,* **43**, 1287.
Lotierzo, M., Henry, O.Y.F., Piletsky, S. *et al.* (2004) *Biosens. Bioelectron.*, **20**, 145.
Lubke, C., Lubke, M., Whitcombe, M.J., and Vulfson, E.N. (2000) *Macromolecules*, **33**, 5098.
Malitesta, C., Losito, I., and Zambonin, P.G. (1999) *Anal. Chem.*, **71**, 1366.
MathewKrotz, J. and Shea, K.J. (1996) *J. Am. Chem. Soc.*, **118**, 8154.
Matsui, J., Akamatsu, K., Nishiguchi, S. *et al.* (2004) *Anal. Chem.*, **76**, 1310.
Matsui, J., Kubo, H., and Takeuchi, T. (2000) *Anal. Chem.*, **72**, 3286.
Matsui, J., Nicholls, I.A., Takeuchi, T. *et al.* (1996) *Anal. Chim Acta*, **335**, 71.
Mayes, A.G. and Mosbach, K. (1997) *TrAC*, **16**, 321.
Mirsky, V.M., Hirsch, T., Piletsky, S.A., and Wolfbeis, O.S. (1999) *Angew. Chemie. Int. Ed.*, **38/8**, 1108.
Morita, M., Niwa, O., and Horiuchi, T. (1997) *Electrochim. Acta*, **42**, 3177.
Ng, S.M. and Narayanaswamy, R. (2006) *Anal. Bioanal.Chem.*, **386**, 1235.
Nicholls, I.A. (1995) *Chem. Lett.*, 1035.
Nicholls, I.A., Ramstrom, O., and Mosbach, K. (1995) *J. Chromatogr. A*, **691**, 349.

Nopper, D., Lammershop, O., Wulff, G., and Gauglitz, G. (2003) *Analyt. Bioanalyt. Chem.*, **377**, 608.

O'Shannessy, D.J., Ekberg, B., and Mosbach, K. (1989) *Anal. Biochem.*, **177**, 144.

Panasyuk, T.L., Mirsky, V.M., Piletsky, S.A., and Wolfbeis, O.S. (1999) *Anal. Chem.*, **71**, 4609.

Peissker, F. and Fischer, L. (1999) *Bioorg. Med. Chem.*, **7**, 2231.

Piletska, E.V., Piletsky, S.A., Subrahmanyam, S. *et al.* (2000) *Proc. 1st Int. Workshop on Molecular Imprinting*, Cardiff, UK, 2000, p. 87.

Piletsky, S.A., Butovich, I.A., and Kukhar, V.P. (1992) *Zh. Anal. Khim.* **47**, 1681 (in Russian).

Piletsky, S.A., Day, R.M., Chen, B. *et al.* (2000c) UK Patent Application PCT/GB01/00324.

Piletsky, S.A., Dubey, I.Y., Fedoryak, D.M., and Kukhar, V.P. (1990b) *Biopolym. Cell* **6**, 55 (in Russian).

Piletsky, S.A., Fedoryak, D.M., and Kukhar, V.P. (1990a) *Dokl. Acad. Sci. Ukraine B*, **4**, 53 (in Russian).

Piletsky, S.A., Guerreiro, A., Piletska, E.V. *et al.* (2004) *Macromolecules*, **37**, 5018–5022.

Piletsky, S.A., Guerreiro, A., Piletska, E.V. *et al.* (2005) *Preparation of Soluble and Collodial Imprinted Polymers by Living Polymerization.* UK Patent Application 0511116.6. Filing date: 01/06/05. PCT/GB06/001986, Publication Date: 07/12/06, Publication No: WO2006/129088.

Piletsky, S.A., Karim, K., Piletska, E.V. *et al.* (2001) *Analyst*, **126**, 1826.

Piletsky, S.A., Kurys, Y.I., Rachkov, A.E., and El'skaya, A.V. (1994a) *Russ. J. Electrochem.*, **30**, 1090 (in Russian).

Piletsky, S.A., Matuschewski, H., Schedler, U. *et al.* (2000a) *Macromolecules*, **33**, 3092.

Piletsky, S.A., Mijangos, I., Guerreiro, A. *et al.* (2005) *Macromolecules*, **38**, 1410–4.

Piletsky, S.A., Parhometz, Y.P., Panasyuk, T.L., and El'skaya, A.V. (1994b) *Sens. Actuators B*, **18/19**, 629.

Piletsky, S.A., Piletska, E.V., Chen, B. *et al.* (2000b) *Anal. Chem.*, **72**, 4381.

Piletsky, S.A., Piletska, E.V., El'skaya, A.V. *et al.* (1997) *Anal. Lett.*, **30**, 445.

Piletsky, S.A., Piletska, E.V., Karim, K. *et al.* (2002) *Macromolecules*, **35**, 7499–7504.

Piletsky, S.A., Piletska, E.V., Yano, K. *et al.* (1996) *Anal. Lett.*, **29**, 157.

Piletsky, S.A., Piletskaya, E.V., Panasyuk, T.L. *et al.* (1998) *Macromolecules*, **31**, 2137.

Piletsky, S.A., Piletskaya, E.V., Sergeeva, T.A. *et al.* (1999a) *Sens. Actuators B*, **60**, 216.
Piletsky, S.A. and Starodub, N.F. (1992) *Zh. Anal. Khim.* **47**, 623 (in Russian).
Piletsky, S.A., Terpetschnig, E., Andersson, H.S. *et al.* (1999b) *Fresenius J. Anal. Chem.*, **364**, 512.
Polyakov, M.V. (1931) *Zhur. Fiz. Khim.*, **2**, 799 (in Russian).
Rachkov, A., McNiven, S., El'skaya, A.V. *et al.* (2000) *Anal. Chim. Acta*, **405**, 23.
Ramstrom, O., Andersson, L.I., and Mosbach, K. (1993) *J. Org. Chem.*, **58**, 7562.
Rathbone, D.L. and Ge, Y. (2001) *Anal. Chim. Acta*, **435**, 129.
Rathbone, D.L., Su, D., Wang, Y., and Billington, D.C. (2000) *Tetrahedron Lett.*, **41**, 123.
Ruckert, B., Hall, A.J., and Sellergren, B. (2002) *J. Mater. Sci.*, **12**, 2275.
Sanchez-Barragan, I., Costa-Fernandez, J.M., Pereiro, R. *et al.* (2005) *Anal. Chem.*, **77**, 7005.
Sergeyeva, T.A., Matuschewski, H., Piletsky, S.A. *et al.*. (2001) *J. Chromatogr. A*, **907**, 89.
Sergeyeva, T.A., Piletsky, S.A., Brovko, A.A. *et al.* (1999) *Anal. Chim. Acta*, **392**, 105.
Shi, H.Q., Tsai, W.B., Garrison, M.D. *et al.*. (1999) *Nature*, **398**, 593.
Southard, G.E., Van Houten, K.A., and Murray, G.M. (2007) *Macromolecules*, **40**, 1395.
Starodub, N.F., Piletsky, S.A., Lavryk, N.V., and El'skaya, E.V. (1992) *Sens. Actuators B*, **13–14**, 708.
Steinke, J.H.G., Dunkin, I.R., and Sherrington, D.C. (1996) *Macromolecules*, **29**, 407.
Strikovsky, A.G., Kasper, D., Grün, M. *et al.* (2000) *J. Am. Chem. Soc.*, **122**, 6295.
Suárez-Rodríguez, J.L. and Díaz-García, M.E. (2000) *Anal. Chim. Acta*, **405**, 67.
Subrahmanyam, S., Piletsky, S.A., Piletska, E.V. *et al.*. (2000) *Adv. Mater.*, **12**, 722.
Surugiu, I., Danielsson, B., Ye, L. *et al.*. (2001) *Anal. Chem.*, **73**, 487.
Svenson, J. and Nicholls, I.A. (2001) *Anal. Chim. Acta*, **435**, 19.
Svenson, J., Zheng, N., and Nicholls, I.A. (2004) *J. Am. Chem. Soc.*, **126**, 8554.
Tabushi, I., Kurihara, K., Naka, K. *et al.* (1987) *Tetrahedron Lett.*, **28**, 4299.
Takeuchi, T., Fukuma, D., and Matsui, J. (1999) *Anal. Chem.*, **71**, 285.
Takeuchi, T., Seko, A., Matsui, J., and Mukawa, T. (2001) *Instrum. Sci. Technol.*, **29**, 1.

Turner, A.P.F. (1999) *Biosensor: McGraw-Hill Yearbook of Science and Technology*. New York: McGraw-Hill.

Vidiasankar, S., Ru, M., and Arnold, F.H. (1997) *J. Chromatogr. A*, **775**, 51.

Vinokurov, I.A. and Grigoreva, M.A. (1990) *Zh. Anal. Khim.*, **45**, 1009 (in Russian).

Wandelt, B., Mielniczak, A., and Cywinski, P. (2004) *Biosens. Bioelectron.*, **20**, 1031.

Wang, H.Y., Kobayashi, T., Fukaya, T., and Fujii, N. (1997) *Langmuir*, **13**, 5396.

Wang, W., Gao, S., and Wang, B. (1999) *Org. Lett.*, **1**, 1209.

Watanabe, M., Akahoshi, T., Tabata, Y., and Nakayama, D. (1998) *J. Am. Chem. Soc.*, **120**, 5577.

Whitcombe, M.J., Martin, L., and Vulfson, E.N. (1998) *Chromatographia*, **47**, 457.

Wolfbeis, O.S., Terpetschnig, E., Piletsky, S.A., and Pringsheim, E. (1999) Fluorescence techniques for probing molecular interactions. In imprinted polymers. In *Imprinted Polymers: Applied Fluorescence in Chemistry, Biology and Medicine* (W. Rettig, B. Strehmel, S. Schrader, and H. Seifert, eds) Berlin-Heidelberg: Springer.

Wulff, G. (1995) *Angew. Chem. Int. Ed. Engl.*, **34**, 1812.

Wulff, G. and Haarer, J. (1991) *Makromol. Chem.*, **192**, 1329.

Ye, L. and Mosbach, K. (2001) *J. Am. Chem. Soc.*, **123**, 2901.

Yoshida, M., Hatate, Y., Uezu, K. *et al.* (2000) *J. Polym. Sci. Pol. Chem.*, **38**, 689.

Yoshizako, K., Hosoya, K., Iwakoshi, Y. *et al.* (1998) *Anal. Chem.*, **70**, 386.

Yu, C. and Mosbach, K. (2000) *J. Chromatogr. A*, **888**, 63.

Yu, C., Ramstrom, O., and Mosbach, K. (1997) *Anal. Lett.*, **30**, 2123.

Zhang, H., Verboom, W., and Reinhoudt, D.N. (2001) *Tetrahedron Lett.*, **42**, 4413.

Zhou, H., Zhang, Z., He, D., and Xiong, Y. (2005) *Sens. Actuat. B Chemical* **107**, 798.

Optical Biosensors: Today and Tomorrow (2nd Edition)
Frances S. Ligler and Chris Rowe Taitt (Editors)

Chapter 15

NANOPARTICLES FOR BIOSENSORS

Huaizhi Kang, Lin Wang, Ph.D.,
Meghan O'Donoghue, Y. Charles Cao, Ph.D.,
and Weihong Tan, Ph.D.
Department of Chemistry, University of Florida, Gainesville, FL 32611, USA

The demand for highly sensitive and selective optical sensors without the need for complex instrumentation and processing has driven the development of novel nanomaterials for optical sensing applications. Nanoparticles have the potential to be used for biosensing in a diversity of fields and could be further developed into multifunctional sensors able to offer sensitive, specific, rapid, and cost-effective solutions for modern biological research and clinical practice. By utilizing the unique properties of a variety of nanoparticles for biosensing functions, effective biosensors have been developed and applied. As an attractive alternative to conventional dyes, fluorescent nanoparticles have greatly increased the sensitivity in a variety of biosensor formats. This chapter gives a brief overview of nanoparticles and nanotechnology with a specific emphasis on the underlying mechanisms, functionality, modifications, and applications relevant to optical sensing. Nanoparticle-based sensing is an emerging field, and the further development of nanomaterials for sensor applications will provide numerous advanced tools with both increased sensitivity and improved capability for unique applications in molecular biology, drug discovery, and clinical diagnosis.

15.1. Technical concepts

15.1.1. Nanoparticles and nanomaterials

Nanoscience has led to a revolution in both traditional and still-developing areas of biotechnology, material sciences, chemistry, medicine, electronics, computation, and information technology. The development of new and better nanomaterials is at the leading edge of the rapidly developing field. The unique size-dependent physical and chemical properties of nanomaterials make them superior and indispensable in many areas of science and technology research and development.

"Nano" is derived from the Greek word "nanos," which means a dwarf or extremely small creature. Nanoparticles are discrete clusters of atoms generally in the range of 1–100 nm, although the term is often extended to include nanoparticles of up to 1000 nm. This size range is larger than for individual atoms but smaller than for bulk solids. In some nanomaterials, neither absolute quantum chemistry nor the laws of classical physics can be applied. Thus nanoparticles display properties that differ markedly from the bulk solid. It is interesting to note that typical biomolecular components are in the range of 5–200 nm, and therefore have a large overlap with nanoparticles. Probes with dimensional similarities to their biological targets interfere less with the target of study. Nanomaterials have unique features to be useful as sensors for a variety of applications.

15.1.2. Optical sensors

Sensors can be broadly grouped in different categories depending on (a) how the sensing is accomplished; (b) the physical extent of the sensing; and (c) the role of the sensor in the sensing process. Optical sensing has been one of the most thoroughly investigated sensing platforms. Optical sensors have significant advantages compared to other sensor types such as thermal, electromagnetic, and mechanical in sensitivity, specificity, and dynamic range. They also offer both point and distributed configurations, are free from electromagnetic interference, and have great multiplexing capabilities. When designing an optical sensing measurement system, there are some criteria that need to be optimized:

1. Signal and background noise
2. Selectivity
3. Quantitation and linearity
4. Speed of response for the desired bandwidth
5. Stability
6. Compatibility

A simplified description of molecular recognition and signal generation is shown in Scheme 15.1. Basically, optical sensors perform three functions – sensing molecular recognition, signal generation and amplification, and reading. For example, taking advantage of the capacity of optical fibers to send and receive optical signals over long distances, a current trend is to create networks of sensors, or sensor arrays. This avoids having to convert between electronics and photonics separately at each sensing site, thereby reducing costs and increasing flexibility (Gambling, 2000). Overall, there have been many applications of optical sensors in a variety of settings. These optical techniques are utilized in a wide range of fields, including biomedicine, environmental sensing, mechanical measurement, and even art preservation.

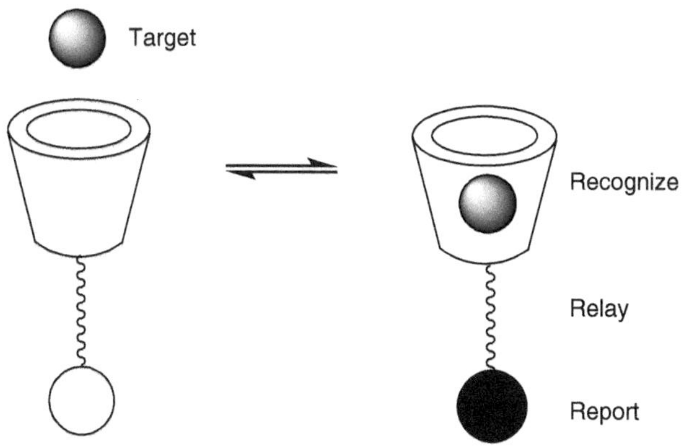

Scheme 15.1 The scheme of molecular recognition and signal generation.

However, as with other types of sensors, optical sensors suffer from limitations such as ultrasensitivity, reproducibility, and ease in sensing recognition.

Conventional materials are composed of particles of several hundreds of microns to millimeters in size, and display their intrinsic chemical and physical properties of the bulk materials. In nanomaterials, some of these properties may be significantly different. Possessing fascinating properties, nanomaterials have been exploited for a variety of technological applications according to their unique structural and non-structural attributes. The main reasons for these phenomena are surface effects and quantum-size effects that only occur when the size of the particles is small (Emil *et al.*, 2006). The different behaviors of nanomaterials (Figure 15.1) compared to bulk materials make them unique for many optical sensor applications.

Nanomaterials have been developed at a tremendous speed over the past decade. Starting from original sphere-shaped nanoparticles, more materials broadly defined as nanomaterials such as nanotubes, nanowires, and quantum dots (QDs) have been prepared. These nanoparticles have diverse shapes and structures. Several applications have used the unique properties of these nanoparticles, such as next-generation computer chips, high-sensitivity sensors, and long-lasting medical implants.

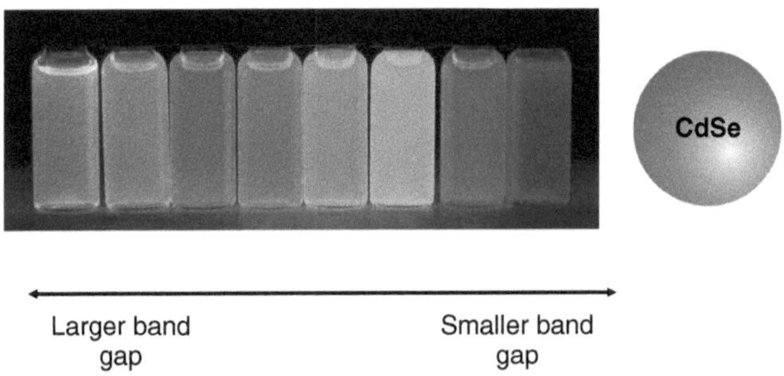

Figure 15.1 Optical properties of semiconductor nanocrystals. As the nanoparticles' size changes from 2 to 6 nm, there is a significant red shift in the luminescence of the CdSe QDs (see Plate 5).

Among these applications, nanomaterials-based biosensors are one of the most significant.

As multiplex bioanalysis emerges as a more powerful technique for analyzing several biological properties simultaneously, limitations in optical sensors based on organic/inorganic dye molecules become more significant. These fluorophores often exhibit weak optical signals and photobleaching due to their intrinsic photophysical properties. Furthermore, the overlapping emissions of conventional fluorophores, which have a spectral width of 50–200 nm in the visible region, limit the use of multiple fluorescent dyes. Nanomaterials and nanotechnologies provide new opportunities for biosensor development to have superior diagnostic, targeting, and therapeutic functions (Marcel *et al.*, 1998; Nicewarner-Pena *et al.*, 2001; Niemeyer *et al.*, 2001; Brigger *et al.*, 2002; Emerich *et al.*, 2003; Larson *et al.*, 2003; Zhao *et al.*, 2004). The adoption of nanoparticles for optical biosensing and imaging in research and clinic applications has been a major focus of interest over the past 5 years and has yielded promising schemes using a highly diverse collection of nanomaterials, including silica nanoparticles (Santra *et al.*, 2001a; Tan *et al.*, 2004; Wang *et al.*, 2006), QDs (Bruchez *et al.*, 1998; Han *et al.*, 2001), polystyrene nanoparticles (Härmä *et al.*, 2000; Soukka *et al.*, 2001), metallic nanoparticles (Rosi *et al.*, 2006), magnetic nanoparticles (Ivo Šafařík *et al.*, 2002), and core–shell nanoparticles (Marie-Christine *et al.*, 2004).

Dye-doped nanoparticles are one of the most interesting types of nanoparticles with their ability for diverse surface modifications. When large quantities of dye molecules are trapped inside a polymer or silica matrix, they offer intense optical intensity and excellent luminescent properties over conventional organic dyes (Meallet-Renault *et al.*, 1999, 2000; Santra *et al.*, 2001a; Gao *et al.*, 2002).

Quantum dots can also be used for ultrasensitive, multiplexed analysis. By varying their size and composition, the emission wavelength can be tuned from blue to near-infrared (NIR) (Michalet *et al.*, 2005). The narrow emission of QDs provides the capability for multiplexed optical coding. This combined with QD's high quantum yield and photostability make them useful for high-throughput screening and biomedical imaging.

Polystyrene latex nanoparticles containing high concentrations of fluorescent dye have been prepared and used for highly sensitive time-resolved fluoroimmunoassays (TR-FIA) (Gedanken *et al.*, 2000). Using these materials, non-specific light scattering can be eliminated due to the longer lifetime fluorescence of the nanoparticle system, thus providing ultrasensitive signals.

Gold nanoparticles consist of small granules of gold transition metal in a stable and uniform state. The shift of the gold nanoparticles from blue to red upon aggregation is attributed to surface plasma resonance (Dusemund *et al.*, 1991). Gold nanoparticles can be readily prepared with little deviation in size and exhibit a sharp plasmon absorption band that has been utilized by other non-oligonucleotide-based strategies for organizing nanoparticles into aggregate structures (Brust *et al.*, 1995).

Magnetic nanoparticles are paramagnetic; due to their size, they only exhibit magnetic properties when in a magnetic field. This paramagnetic property has introduced some unique applications in biomedical research (review: Pankhurst *et al.*, 2003). An additional external control can be fulfilled by the interaction of the particles with the magnetic field, and thus can enhance sensitive detection functions.

Recently, core–shell nanoparticles have been constructed and have elicited significant interest. This type of nanostructure consists of a dielectric spherical core surrounded by a thin, uniform metallic shell (Oldenburg *et al.*, 1998). The nanoshells can absorb light, which is self-enhanced by the core–shell structure to trigger a photothermal conversion process. This conversion can increase the surrounding temperature enough to kill cells that are attached to the nanoshells. Photothermal treatments based on plasmonically heated gold nanoshells are under development (Dakrong *et al.*, 2006) to be used in combination with other therapies.

15.1.3. Chemistry of nanoparticles

In 1968, Stöber *et al.* (1968) reported a base-catalyzed sol-gel method for the preparation of spherical silica particles with sizes covering almost the whole colloidal range. Thereafter, the Stöber method has been widely used to synthesize silica nanoparticles and several other types of nanoparticles. Although this method is relatively simple and particles can be prepared in only a few hours, in some cases, non-uniform nanoparticles are obtained. An alternative method, reverse-micelle microemulsion is performed by adding a surfactant and a water-immiscible solvent to the base-catalyzed sol-gel (Yamauchi *et al.*, 1989; Osseo-Asare *et al.*, 1990; Lindberg *et al.*, 1995). The main advantage of the microemulsion method is size tunability of nanoparticles by varying the microemulsion parameters such as water-to-surfactant molar ratio and the ability to trap small molecules in a stable silica matrix. Although there are some other branches of synthesis such as the gas-phase method, the Stöber and microemulsion methods are the two most common ones to obtain nanoparticles in solution.

In actuality, different types of nanoparticles not only require specific methods of synthesis, characterization, and modification, but each type also needs optimized experimental conditions. Here, we list the synthesis methods of three types of nanoparticles: dye-doped silica nanoparticles, QDs, and magnetic nanoparticles.

15.1.3.1. Dye-doped silica nanoparticles

Silica nanoparticles have been synthesized and doped with either inorganic dyes such as Rubpy and Osbpy or organic dyes such as rhodamine 6G (R6G) and tetramethylrhodamine (TAMRA). To make inorganic dye-doped silica nanoparticles, Rubpy dye molecules are dispersed in aqueous cores of a reverse microemulsion system (Figure 15.2). The silica matrix then encapsulates the dye molecules as it polymerizes (Santra *et al.*, 2001a). Through this method, luminescent nanoparticles can be obtained with a variety of uniform sizes. To synthesize common

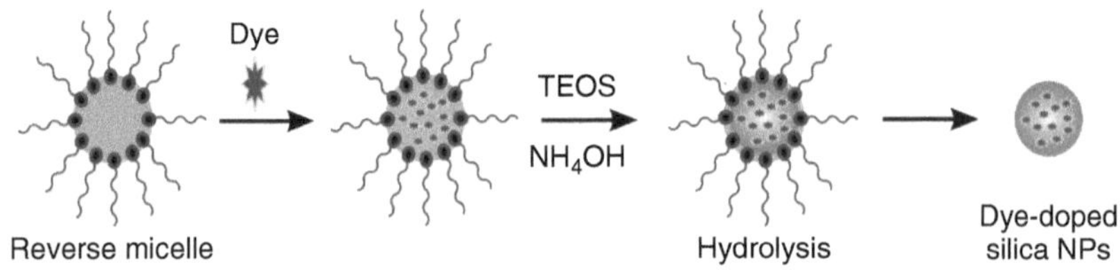

Figure 15.2 Microemulsion method to make dye-doped silica nanoparticle. Published with permission from CRC Press LLC (Wang *et al.*, 2007).

dye-doped nanoparticles to have higher quantum yield, two issues have to be considered:

1. Common dyes cannot be easily doped inside the hydrophilic silica shell due to their hydrophobic nature.
2. The neutral charge of organic dye molecules decreases the electrostatic attraction between the silica shells and the dye molecules, aiding in the retention of the dye.

The Stöber method has been generally applied for this purpose (Tapec *et al.*, 2002). To circumvent the broad size distribution that sometimes results with the Stöber method, improved reverse micelle medium such as TAMRA-doped silica nanoparticles was developed for the synthesis (Zhao *et al.*, 2004).

15.1.3.2. Quantum dots

Quantum dots were first discovered in doped silicate glasses by Ekimov and Onuschenko (Ekimov, 1996; Ekimov *et al.*, 1980, 1985a, 1985b, 1993; Ekimov and Onuschenko, 1981, 1982, 1984). The breakthrough in synthesizing high-quality colloidal semiconductor QDs can be traced to the work of Murray *et al.* (1993). In their work, dispersed (8–11%) highly crystalline CdSe QDs could be synthesized at high temperatures using a mixture of organometallic precursors and trioctyl phosphine/trioctyl phosphine oxide (TOP/TOPO) growth solvent/ligands.

After the first reaction to form the CdSe core, further coating with wider bandgap semiconductors such as ZnS and CdS can be performed, which increases the photoluminescence yield of the QDs (Dabbousi *et al.*, 1997). In recent years, Peng *et al.* (2001) made further refinements to this scheme by using less pyrophoric precursors such as CdO and Cd acetate, which can significantly optimize the QDs for biological applications. Additionally, other QDs including ZnS, CdS, CdTe, and PbSe with emissions ranging from the UV to the IR have also been synthesized (Murphy, 2002; Ozkan, 2004; Michalet *et al.*, 2005).

15.1.3.3. Magnetic nanoparticles

The basic principle of chemical synthesis of magnetic nanoparticles is to initiate chemical reactions, such as coprecipitation reactions, and control the nucleation and growth of the reaction products. Among various chemical synthesis methods, two approaches are successful in creating stable colloids with good size dispersion. Microemulsion-based syntheses are typically carried out at low temperatures and are widely used in industry. However, microemulsion-based synthesis has the disadvantage that the particles are sometimes less crystalline and more polydisperse because of the slow nucleation rate. Organometallic-based syntheses have been adopted recently to synthesize particles with high crystallinity and monodispersity. This method requires high temperatures and results in high-quality magnetic nanoparticles. Nevertheless, the toxicity of the solvents, ligands, and precursors used in these types of syntheses limits their applications to a certain extent (Peng *et al.*, 2001).

Silica-coated magnetic nanoparticles are one important type of magnetic nanoparticles and have been used in bioassays and biosensors. These magnetic nanoparticles are generally prepared using the reverse microemulsion method (Santra *et al.*, 2001b). Iron oxide nanoparticles ($Fe_3O_4/Fe_2O_3/SiO_2$) are first formed by coprecipitation of ferrous and ferric salts with inorganic base, followed by silica coating for further biomodification. To evaluate these nanoparticles, samples can be analyzed by transmission electron microscopy (TEM) and dynamic light scattering. Under these methods, they are found to be very uniform and diameter can be well controlled from 2 to 30 nm. Evaluation of iron oxide–silica nanoparticles using superconducting quantum interference device (SQUID) shows that the nanoparticles exhibit a behavior similar to paramagnetic materials even when at temperatures below the

Curie or the Néel temperature, close to those for superparamagnetic materials. The results are used to optimize the synthesis conditions to make magnetic nanoparticles useful for bioanalysis.

15.2. State of the art

In this chapter we will outline a few major classes of optical biosensors. Their applications for detection and diagnostic screening for biomolecules will be discussed (Figure 15.3). Moreover, we will highlight some of the key fundamental properties of certain nanostructures that make them ideal for these types of specific biosensing application. It is important to understand that the development of nanotechnology and nanomaterials can significantly enhance biosensing function by introducing novel functional structures on the nanometer scale. Although there are several other sensing devices under development, such as cantilevers, nanotubes, and nanowires (McKendry *et al.*, 2002; Cui *et al.*, 2003; Wang *et al.*, 2004), they are beyond the scope of this chapter.

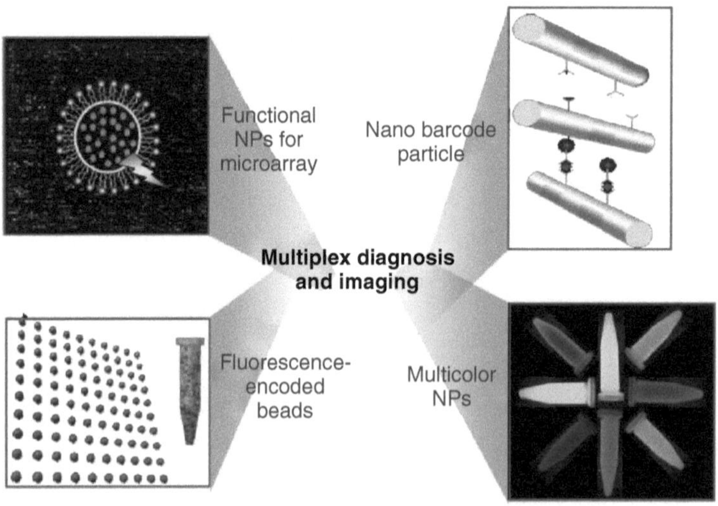

Figure 15.3 Representative schematic of nanoparticle-based multiplex diagnostics and imaging applications.

15.2.1. Silica nanoparticles for imaging and bioanalysis

Silica nanoparticles have been developed with new modification techniques and many doping materials. Compared with polymer nanoparticles, silica nanoparticles have less aggregation and are easy to separate via centrifugation during and after particle preparation, surface modification, and structural modification. Silica nanoparticles are generally highly stable under a diverse set of conditions. More importantly, silica nanoparticles are hydrophilic and biocompatible in biological environments. Combining them with luminescent materials such as organic fluorophores, dye-doped silica nanoparticles have broadly extended the application in imaging and bioanalysis. Single-dye-doped silica nanoparticles have been used for ultrasensitive DNA analysis down to sub-fM concentrations (Zhao *et al.*, 2003), and for single bacterium monitoring within 20 min (Zhao *et al.*, 2004).

15.2.1.1. Luminescent multiple dye-doped silica nanoparticles

Luminescent dye-doped silica nanoparticles have been developed for multiple analyses. These nanoparticles are non-blinking, multicolored, highly luminescent, photostable, biocompatible, hydrophilic, chemically inert, and resistant to microbial degradation. Therefore, dye-doped nanoparticles have been used as novel substrates for multiplexed optical signaling. These nanoparticles display advantages over currently available luminescence probes (Wang *et al.*, 2005, 2006).

By encapsulating multiple fluorophores simultaneously into the silica nanoparticles at controlled ratios, multiplexed bioanalysis is possible under proper conditions. For example, two-dye nanoparticles (Wang *et al.*, 2005) and three-dye nanoparticles (Wang *et al.*, 2006) have been prepared. By varying concentrations of the dyes within the nanoparticles, excitation with a single wavelength leads to a series of differentiated emission spectra, permitting simultaneous detection of multiple targets in a single procedure. Wang *et al.* (2005) studied dual-dye-doped silica nanoparticles using ruthenium bipyridine (Rubpy) and osmium bipyridine (Osbpy) dyes as encoding elements. The two dyes have similar octahedral structures and share a broad overlapping excitation spectrum but have distinct maximum emission spectra. By encapsulating the two dyes at precisely controlled ratios into nanoparticles, unique spectra were obtained using a single excitation wavelength. A schematic representation of the dual-dye-encoding system and its application for multiplex bacterial/cell detection within a flow system is shown in Figure 15.4. Antibody-conjugated nanoparticles with varying dye ratios were able to

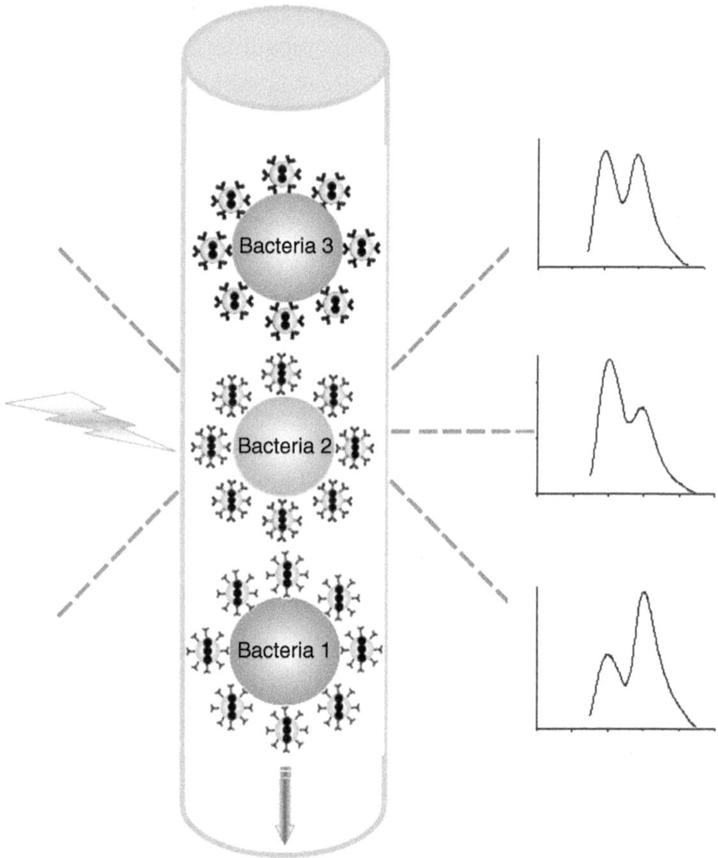

Figure 15.4 Schematic representation of the dual-dye-encoding system and multiplexing potential. As the bacteria pass through the channel, the fluorescence signals correspond to the different encoding nanoparticles specifically attached to the target bacteria. Published with permission from Elsevier (Wang *et al.*, 2005).

specifically recognize and bind to the corresponding antigen-expressing bacteria. When the bacteria–nanoparticle complex passed through a fluorescence detector, each bacteria–nanoparticle complex exhibited a unique spectrum pattern. Therefore, a rapid, sensitive, and selective detection method could be achieved.

15.2.1.2. Multicolor fluorescence resonance energy transfer

Florescent nanoparticles that exhibit multiple emissions by single-wavelength excitation can be used for molecular imaging and multiplex bioanalysis (Wang *et al.*, 2006). By using a modified Stöber method, three different organic dyes were doped inside silica nanoparticles. By varying the doping ratios of the dyes, fluorescence resonance energy transfer (FRET) emissions were tuned to display multiple colors with single-wavelength excitation with a large Stokes shift for FRET (Wang *et al.*, 2006). Molecular recognition moieties can be added to these nanoparticles' surface, allowing specific recognition of a variety of targets. The significance of this study was the method of applying fluorescent dye-doped nanoparticles for simultaneous, highly efficient, and sensitive detection of multiple analytes. More importantly, it provided a way to design a multifunctional biosensor based on silica nanoparticles, which have been investigated for future applications (Figures 15.5 and 15.6).

15.2.1.3. Microarray detection

Platforms such as DNA microarrays or DNA biochips for multiplex biomolecule detection have utilized silica nanoparticles as the signaling molecule. In a microarray for genetic analysis, the location of the probe DNA in a microarray can identify the targeted analytes, while the probe intensity is roughly proportional to the target concentration. This powerful technique has been applied for whole-genome expression analysis (Gasch *et al.*, 2000), cancer research (Alizadeh *et al.*, 2000), drug discovery (Wick *et al.*, 2005), vaccine development (Grifantini *et al.*, 2002), and single nucleotide polymorphism (SNP) analyses (Fan *et al.*, 2000). It has been found that metal-nanoparticle-based DNA detection exhibits a significant sensitivity enhancement compared with individual fluorophores (Taton *et al.*, 2000). Fluorescent nanoparticles have also been

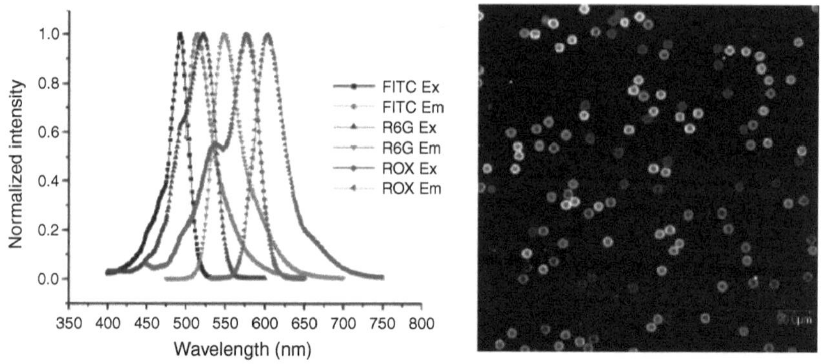

Figure 15.5 (a) Normalized excitation and emission spectra of FITC, R6G, and ROX dyes in pH 7.4 phosphate buffer (*left*). (b) Confocal fluorescence images of a mixture of five types of microsphere–nanoparticle complexes (*right*). Published with permission from the American Chemical Society (Wang *et al.*, 2006) (see Plate 6).

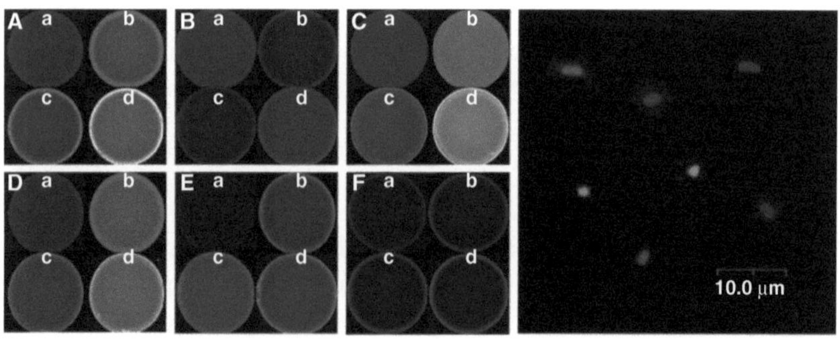

Figure 15.6 (*Left panel*, A–F) Fluorescence images of FRET nanoparticles taken under a confocal microscope: (a) FITC emission channel; (b) R6G emission channel; (c) ROX emission channel; (d) combinatorial color of the three detection channels. The six types of nanoparticles (dye doping ratios: 1:1:1, 1:0:1, 4:4:1, 0:1:1, 0:1:3, 0.5:0.5:3) exhibit six different resultant colors. (*Right panel*) Confocal image of three bacteria species (*Escherichia coli*, *Salmonella typhimurium*, and *Staphylococcus aureus*) specifically covered with three types of monoclonal antibody-labeled nanoparticles. Published with permission from the American Chemical Society (Wang *et al.*, 2006) (see Plate 7).

linked to probe DNA for hybridization with target molecules. In a typical 60 nm nanoparticle, more than 10^4 Rubpy molecules can be trapped and homogeneously dispersed, reducing the detection limit to around 50 fM and increasing signal stability (Wang *et al.*, 2007). Similar strategies are shown in Figure 15.7. Basically, streptavidin-coated nanoparticles will bind to biotinylated target DNA (gene chip) and a biotinylated detection antibody (protein chip). Therefore, highly fluorescent nanoparticles will produce amplified signals from trace amounts of analytes in samples, solving the major sensitivity limitation of microarray technology.

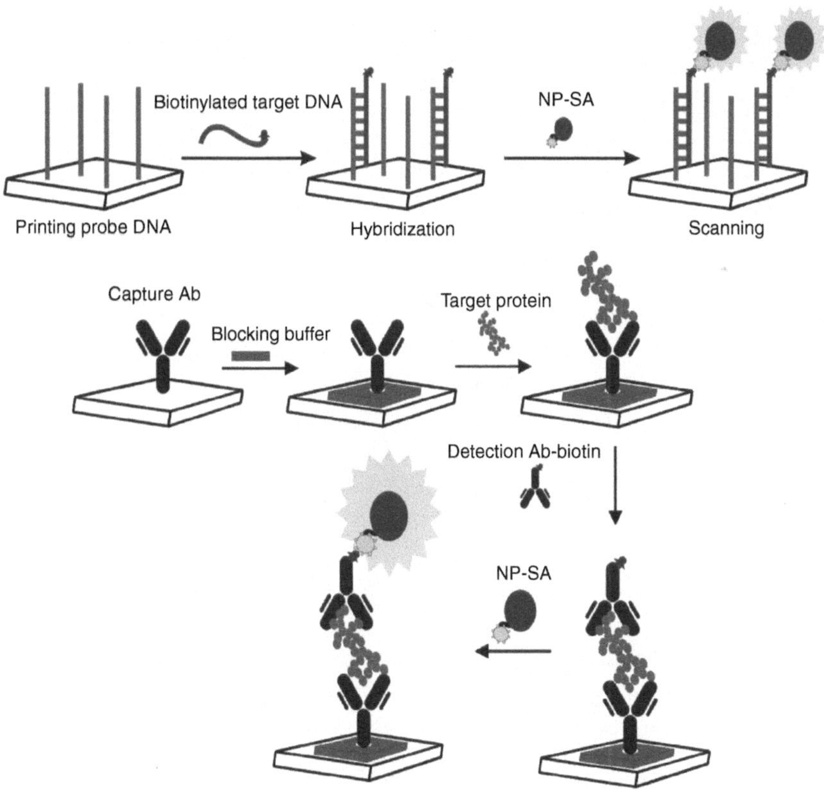

Figure 15.7 Strategies of nanoparticle-based labeling for (a) DNA microarray and (b) protein microarray technology. Published with permission from CRC Press LLC (Wang *et al.*, 2007).

Although the nanoparticle-based microarray technology shows promising results and potential for high throughput, high sensitivity, and stability in biosensing, the complex chemistry of the nanoparticle surface may lead to difficulties in applications with biological samples. Additionally, the extensive profiling and genotyping capabilities are limited using microarrays for the assessment of gene transcripts at low abundance levels due to the low signal relative to background. Nanoparticles with high optical intensities may be highly useful to further microarray applications in genotyping and gene profiling.

15.2.1.4. Surface-enhanced Raman scattering probes for multiplexing

In addition to fluorescent nanoparticles, multiplex bioassays can also be visualized by surface-enhanced Raman scattering (SERS) (Grubisha *et al.*, 2002, 2003). Since the Raman spectrum is composed of spectroscopic lines that are not as broad as fluorescence spectra, it is one of the most investigated techniques in chemical analysis (Petry *et al.*, 2003). Surface-enhanced Raman scattering could provide a specific labeling probe when dyes such as Texas Red are doped on nanoparticles, with Raman and fluorescent detection (Santra *et al.*, 2001a; Lim *et al.*, 2006). As a representative example of Raman tags for multiplexing, six dyes with distinct Raman spectra were incorporated into oligonucleotides, allowing for the detection of the six dissimilar DNA targets using a sandwich assay (Cao *et al.*, 2002). The presence of the target was confirmed by silver staining in advance. The SERS signal of the Raman dyes near the nanoparticle surface further identified the targets (Figure 15.8).

There are several groups who investigated Raman tags based on SERS for multiplex bioassays (Isola *et al.*, 1998; Graham *et al.*, 2002). Raman tags are generally coated with silica for stability and modified to be functional for biosensing purposes. Recently, a novel nanomaterial named COIN (composite organic–inorganic nanoparticle) was synthesized and investigated. The nanomaterial has encoded SERS-like signatures by chemical reactions (Su *et al.*, 2005). In this case, a large quantity of Raman compounds was conjugated to COINs to display a series of signatures. With three Raman labels, 37 signatures can be obtained simultaneously. It is plausible that hundreds of thousands of unique

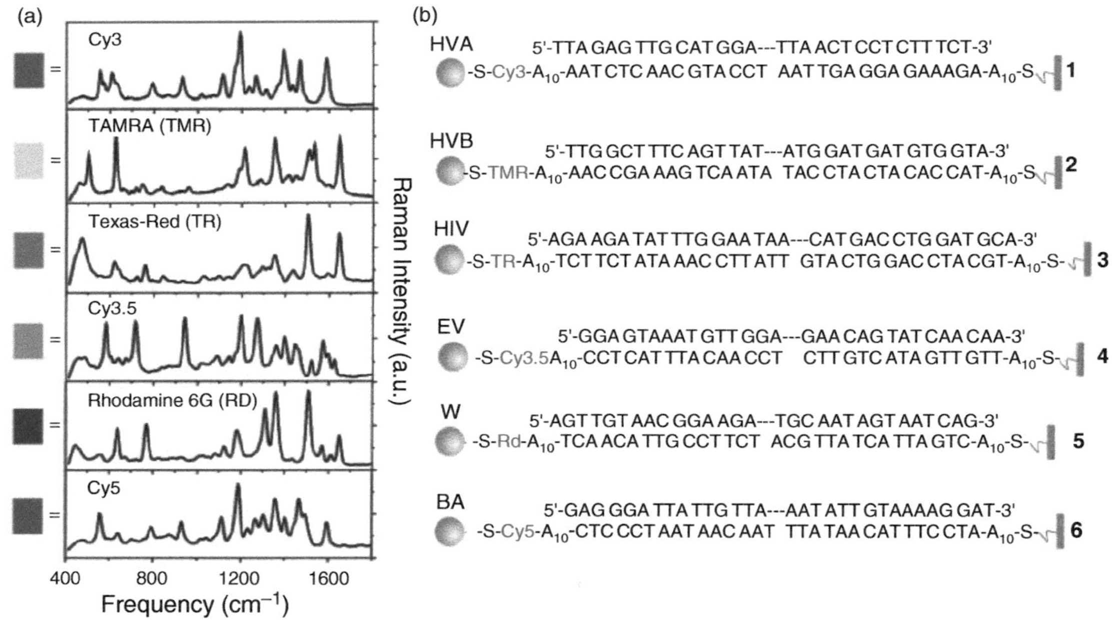

Figure 15.8 Surface-enhanced Raman scattering probes for multiplex DNA sandwich assays. (a) The Raman spectra of six dye-labeled nanoparticle probes after silver enhancing (after background subtraction). Each dye correlates with a different color. (b) Six DNA sandwich assays with corresponding target analysis systems. Published with permission from *Science* (Cao *et al.*, 2002).

signatures can be made with just a dozen or so Raman labels that differ by 15 wavenumbers between peaks.

Gold-nanoparticle probes with different Raman dyes have been used to detect multiple oligonucleotide targets, and to label different oligonucleotides to distinguish the sequence (Graham *et al.*, 2000). By coating individual gold particles with an extrinsic layer of Raman reporter molecules and then a layer of monoclonal antibodies to form a triple-layer structure, Raman tags were generated. Picomolar concentrations of complex forms of prostate-specific antigen (PSA) were detected in human serum (Grubisha *et al.*, 2003).

Other systems applied the same concepts involving conventional probes as labels in sandwich immunoassays (Cai *et al.*, 2000) and tissue labeling (Gao *et al.*, 2002). Although the variety of applications is wide and positive results have been documented, SERS requires comparably complicated skills as well as laboratory equipment. The more signatures we expect to obtain, the more critical the exact experimental conditions become. This is one of the major barriers in commercializing this technique for use in simple tests.

15.2.2. Quantum dots for optical sensing

Based on a semiconductor nanostructure that confines the motion of conduction band electrons, QDs have been studied for their chemical and physical aspects and further developed with modification and doping processes. The complex photophysical properties have been a major focus of many studies due to the semiconductive materials in their composition. Quantum dots possess several interesting properties and have been illustrated to be superior to traditional dyes because of their broad excitation spectra, size-tunable luminescent properties, long fluorescence lifetime, and photostability.

Quantum dots are comparably small nanoparticles usually in the range of 2–10 nm (Chan *et al.*, 1998). The remarkable brightness ($\sim$20 $\times$ brighter than general organic dyes), long lifetime with high stability under diverse circumstances, and tunable florescence spectra have proved QDs to be useful tools as a luminescence source for optical sensing events (Dahan *et al.*, 2001). By changing the size and composition of QDs, the emission wavelength can be precisely tuned so that the emission spectra have the large overlap with absorption spectra of acceptor (Hines *et al.*, 1998). Table 15.1. presents an overview of QD properties compared with organic and protein-based fluorophores.

Table 15.1 Comparison of the properties of organic/protein fluorophores and QDs

Property	**Fluorophores**	**Quantum dots (QDs)**	**Reference**
Absorption spectra	Narrow, mirror as emission spectra	Broad, steady increase toward UV	Haugland (2005) Dabbousi *et al.* (1997) Striolo *et al.* (2002)
Molar extinction coefficient	Small, mostly less than 200,000 $M^{-1} cm^{-1}$	Large, mostly more than 2,000,000 $M^{-1} cm^{-1}$	Dabbousi *et al.* (1997) Striolo *et al.* (2002) Leatherdale *et al.* (2002)
Emission spectra	Broad, asymmetric tail-like spectra	Narrow	Dabbousi *et al.* (1997)
Stocks shift (effective)	Mostly less than 100 nm	Can be more than 200 nm	Michalet *et al.* (2005) Ozkan (2004)
Quantum yield	Wide range	Mostly high	Haugland (2005) Michalet *et al.* (2005) Ozkan (2004)
Florescence lifetime	Less than 5 ns	More than 10 ns	Michalet *et al.* (2005) Ozkan (2004)
Photostability	Variable, some are poor	Very stable, no or little photobleaching for various conditions and timescales	Wu *et al.* (2003) Jaiswal and Simon (2004)
Multiphoton cross section	Variable, some are poor	2–3 orders of magnitude of dyes	Larson *et al.* (2003)

(Continued)

Table 15.1 Continued

Property	**Fluorophores**	**Quantum dots (QDs)**	**Reference**
Multiplex capability	Rare	Excellent	Michalet *et al.* (2005) Ozkan (2004)
FRET	Variable, low efficient donor–acceptor energy transfer	Tunable emission permits efficient FRET, under investigation	Clapp *et al.* (2004, 2005)
Chemical resistance	Variable	Very stable	
Reactivity	Variable, multiple reactivity	Variable, less reactivity comparably	Ozkan (2004)
Size	Less than 0.5 nm	Mostly 2–10 nm	Michalet *et al.* (2005)
Electro-chromicity	Rare	Often	Guyot-Sionnest (2001)

One typical QD used for biological purposes (molecular recognition and signal generation) is CdSe/ZnS QD (shown in Figure 15.1). These QDs are highly biocompatible and are ready for surface conjugation (Figure 15.9). CdSe/ZnS-based materials range from 2 nm (~480 nm

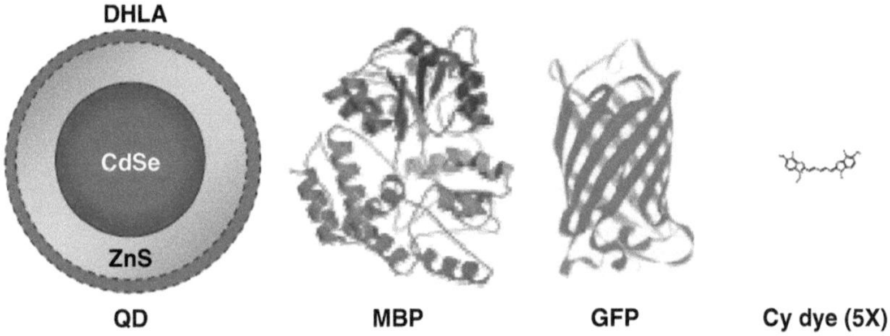

Figure 15.9 Comparison of the size of a representative DHLA-capped CdSe/ZnS QD, diameter ~6 nm, to a MBP molecule (mw ~ 44 000), green fluorescent protein (GFP, mw ~ 30 000), and a cyanine dye (Cy, mw ~ 700). Published with permission from Molecular Diversity Preservation International (Sapsford, K.E. *et al.*, 2006).

emission) to 8 nm (~660 nm emission) in diameter, while its derivative,the CdTe–CdSe QD, can range from 4 nm (~650 nm emission) to more than 9 nm (~850 nm emission) in diameter. These QDs have their finely tuned emission spectra dictated by the change of particle size and dimension – larger-sized QDs induce red shifting of emission spectrum. The high extinction coefficients, broad excitation spectra, and tunable fluorescence emission spectra spanning from UV to NIR range make CdSe/ZnS QDs ideal for optical sensing functions (Evident Technologies, 2005) (Figure 15.10).

Quantum dots can also be used as energy donors in combination with other dye acceptors for quantitative detection of proteins (Clapp *et al.*, 2005; see chapter by Medintz and Delehanty.). This type of sensing assembly is based on FRET and can detect multiple targets simultaneously. Quantum dots offer several advantages when used as FRET donors. Their narrow emission can prevent the overlap of the donor emission spectrum with that of the acceptor. For example, multicolor oligonucleotide-functionalized QDs were used to analyze multiple DNA sequences by measuring individual nanoprobes after assembly (Ho *et al.*, 2005). It is also reported that multicolor QDs can be conjugated to second or even a third type of antibodies to increase the sensitivity and efficiency (Goldman *et al.*, 2004). The FRET-based maltose-binding assays were developed by coating CdSe/ZnS QDs capped with dihydrolipoic acid (DHLA) with maltose-binding protein (MBP). The FRET assay for maltose was based on the interaction between the QDs and the acceptor rhodamine red (Medintz *et al.*, 2004). A hybrid QD-antibody FRET-based sensor for TNT was developed as well (Goldman *et al.*, 2005). It has been reported recently that gold nanoparticle-coated QDs have been prepared as collagenase activity probes (Chang *et al.*, 2005).

Recently, FRET-based sensors were reported for the measurement of extracellular matrix metalloproteinases (MMPs) activity in normal and cancer cells (Shi *et al.*, 2006). A novel QD-based probe was prepared by exchanging the TOPO-capping ligands of CdSe/ZnS with tetrapeptide RGDC molecules and labeling the peptide with a rhodamine acceptor.

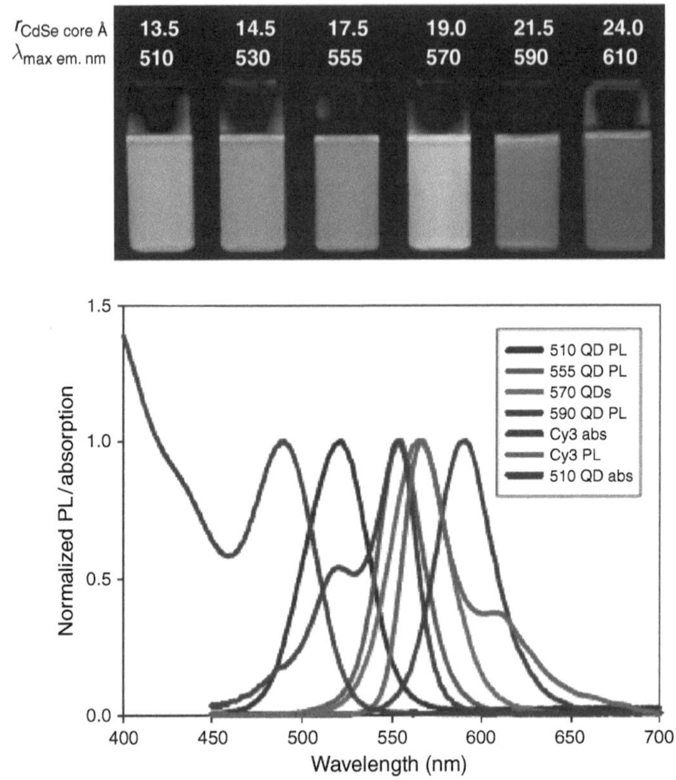

Figure 15.10 Properties of QDs (Ozkan *et al.*, 2004). (Above) Color photo demonstrating the size-tunable fluorescence properties and spectral range of six CdSe/ZnS core/shell QD dispersions. All samples were excited at 365 nm. For the 610 nm emitting QDs, this translates into an effective Stokes shift of ~250 nm. Figure reprinted with permission of the Nature Publishing Group. (Bottom) Absorption of 510 nm QDs and emission of 510, 555, 570, and 590 nm QDs superimposed over the absorbance and emission spectra of Cy3 dye. Published with permission from Molecular Diversity Preservation International (Sapsford, K.E. *et al.*, 2006) (see Plate 8).

Upon enzymatic cleavage of the peptide molecules, the rhodamine (acceptor) molecules no longer provided an efficient energy transfer channel to the QDs. As a result, the emission color of the QDs changed back to green (see Scheme 15.2).

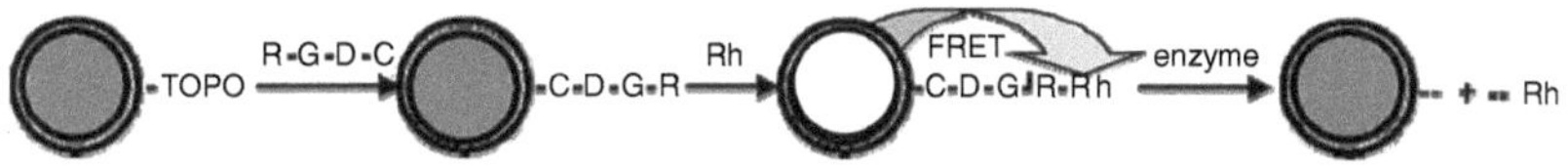

Scheme 15.2 The QD-based FRET for sensing applications. Published with permission from the American Chemical Society (Shi *et al.*, 2006).

The specific FRET probes were used to monitor the proteolytic activity of MMPs in normal and cancer cells. The fast response time (less than 15 min) was the main advantage. Additionally, this method can be extended to other applications involving overexpression of proteolytic activity, measuring the activity of specific proteolytic enzymes, or enabling high-throughput screening of protease inhibitors and activators in an array format.

In addition to their application for *in vitro* multiplex bioassays, QDs have also been applied to study cellular processes after chemical modification to increase their biocompatibility (Michalet *et al.*, 2005). The less toxic QDs were delivered into live cells to conduct multicolor optical coding (Mattheakis *et al.*, 2004). This encoding system provides a unique method for a mixture of living cells to be differentiated by barcoding.

There have been other QD-based biosensing applications in both *in vitro* and *in vivo* use. Quantum dot bioconjugates formed with the enzyme organophosphorus hydrolase (OPH) have been demonstrated to detect paraoxon in solution (Ji *et al.*, 2005). The luminescence of QD–OPH bioconjugates can be switched on and off by the presence of paraoxon. The change of luminescence signals from the QDs correlates with the conformational change in the attached enzyme structure, as monitored by circular dichroism. A nanomolar concentration of paraoxon has been detected using this system.

Although there are numerous applications of QDs, QDs have several disadvantages such as solubility issues, difficult bioconjugations, aggregation, blinking, and sometimes low quantum yield. Efforts have been made to solve these problems by coating QDs with hydrophilic layers or ligands (Gerion *et al.*, 2001) or conjugating them with biocompatible molecules (Chan *et al.*, 2002). Several associated areas need to be explored in order to understand the mechanism of surface functionalization and bioconjugation more completely. Furthermore, toxicity of QDs is a serious issue for *in vivo* biosensing and multicolor detection; a thorough study on the safe range of QD concentrations for practical use is required.

15.2.3. Metallic nanoparticles as sensors

Many reports on the biological applications of metallic nanoparticles as sensors have recently surfaced. Unlike QDs and other inorganic nanostructures, metallic nanoparticles were incorporated or used in many biological systems – especially biomedical applications – in the 1970s and 1980s (Horisberger *et al.*, 1977, 1992). For example, gold nanoparticles have been used as contrast agents in electron microscopy or as detection probes for dipsticks. In recent years, the manipulation of the size, shape, and aggregation-dependent optical properties of metallic nanoparticles and potential applications have made this material one of the most exciting areas of bionanotechnology research.

The Mirkin group studied the absorbance and scattering of light by gold and silver nanoparticles displaying size and aggregation dependence (Jin *et al.*, 2001). Within the size range of 1–100 nm, the luminescence properties of particles can be tuned by changing the material's dimensions (Figure 15.11).

A solution of gold nanoparticles appears ruby red, but the color of the solution changes to blue when the nanoparticles are close together or aggregated. In this colorimetric diagnostic system, the biorecognition molecule oligonucleotide is coated onto the surface of the gold nanoparticles. Therefore, when a media containing a gene sequence matches the oligonucleotide sequence on the gold, hybridization occurs. The nanoparticles are brought close together by this hybridization, and the solution change can be detected.

Metallic nanostructures can also be a platform for SERS. Raman spectroscopy is an analytical technique that measures the vibrational frequencies of chemical bonds upon optical excitation. The major advantage of Raman spectroscopy is the simplicity of detection since tag

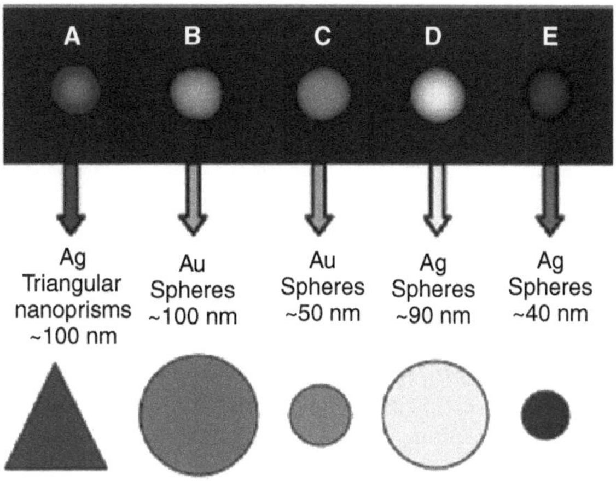

Figure 15.11 Rayleigh light scattering of metallic nanostructures of different sizes and shapes. Published with permission from *Science* (Jin *et al.*, 2001).

and label are not required although the sensitivity is lower than for fluorescence spectroscopy. Since the adsorption of targeted molecules conjugated onto metallic nanoparticles enhances the detection capability of Raman spectroscopy, the SERS has been widely used in molecular sensing and for single-molecule detection with sensitivity comparable to that of fluorescence (Emory *et al.*, 1998).

By studying the properties of nanoparticles functionalized with DNA, many novel structures can be identified, providing useful components for new biodetection schemes. Mirkin's group studied nanoparticle-based DNA detection methods with modified metallic materials. By studying the properties of nanoparticle functionalized with DNA, many novel structures can be identified, providing useful components in new biodetection schemes. A highly selective, colorimetric polynucleotide detection method based on modified metallic nanoparticles probes was developed. There was a pinkish/purple color change associated with DNA detection visually on a reverse-phase silica plate (Elghanian *et al.*, 1997). In another project, the sharp melting profiles of DNA–nanoparticle aggregates provides extremely selective detection of oligonucleotides as discrimination of single-base-pair mismatches

(Taton *et al.*, 2000). Other metallic nanoparticle-based biosensors for real-time DNA detection are also under investigation (Bailey *et al.*, 2003). The high sensitivity and selectivity afforded by nanoparticle probes and the experimental simplicity and wavelength-dependent resonant enhancement features are combined and applied for sensing technology.

Therapeutic applications of metallic nanoparticles are also possible. The photothermal properties of metal nanoparticles and the optical triggering processes can be engineered into the nanoparticles for laser ablation therapy (Pissuwan, 2006). Surface plasmon resonance (SPR) can also be used for optical sensing. The surface phenomenon is particularly useful for highly sensitive detection of proteins and other biological molecules. While we have highlighted some of the conventional and recent development applications of metallic nanoparticles, there are, however, many other applications (e.g., molecular rulers and drug delivery) for metallic nanoparticles.

15.2.4. Magnetic nanoparticle-based biosensors

15.2.4.1. Magetoresitive sensors

Based on the physical effects related to electrical spin and magnetic fields, diversified magnetic nanomaterials have been developed into magnetic chips and functional magnetic nanoparticles (Guenter *et al.*, 2005). Magnetoresistive biosensors use a new detection method on magnetic markers and giant magnetoresistive (GMR) or tunneling magnetoresistive (TMR) sensors. Magnetoresistive biochips mainly include GMR biochips (Baselt *et al.*, 1998) and TMR biosensors (Brzeska *et al.*, 2004). They have been developed in microengineering fields as lab-stage biosensors. Magnetic nanoparticles were first developed for separating a target from an arbitrary solution by specific binding of the target to the particle surface and subsequent separation by magnetic fields. In recent years, nanosized magnetic particles have drawn much attention for their highly specific binding properties. Accordingly, therapeutic *in vivo* applications such as drug targeting, cancer therapy, and bioimaging have been successfully demonstrated (Häfeli *et al.*, 1997).

(a)

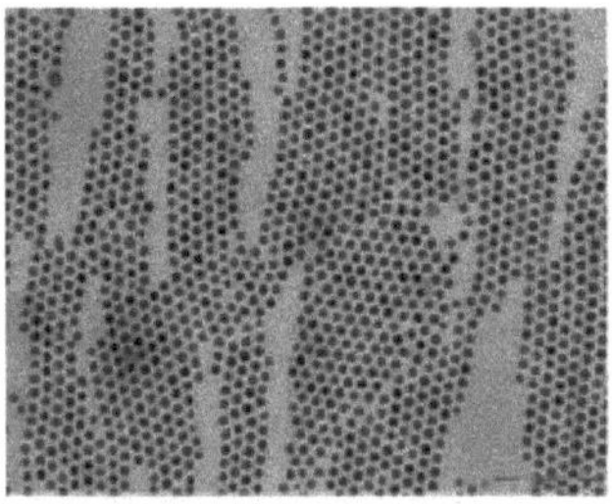

(b)

Composition	Ms [kA/m]	D_{SP} [nm]
bcc-Fe	1714	16.0
fcc-Co	1420	15.8
$Fe_{50}Co_{50}$	1910	23.6
Fe_3Co	1993	17.0
Fe_3O_4	415	28.0
Fe_2O_3	380	34.9

Figure 15.12 (a) Transmission electron microscopy (TEM) image of 10-nm-diameter Co particles and (b) values for the saturation magnetization for different compositions in comparison to the usually used iron oxides. Published with permission from Materials Research Society (Guenter Reiss *et al.*, 2005).

Compared with conventional iron oxide particles, nanoparticles composed of replacement materials such as ferromagnetic cobalt (Puntes *et al.*, 2001) and iron–platinum (Sun *et al.*, 2000) have been developed as carriers for *in vitro* and *in vivo* study (Figure 15.12). Cobalt-based magnetic nanoparticles have very narrow size distribution. Additionally, the ferromagnetic and magnetic properties are superior to those of iron oxide particles in terms of long-term stability. These properties provide the conditions for magnetoresistive biosensors to detect biomolecules (Vries *et al.*, 2004).

15.2.4.2. Magnetically modulated optical nanoprobes

Modulated optical nanoprobes (MOONs) are microscopic photonic probes that look like moons in which the northern hemisphere is colored with a fluorescent indicator dye. They rotate in response to thermal fluctuations in a fluid (Brownian MOONs) or to rotating magnetic fields (magnetically modulated optical nanoprobes, MagMOONs) (Figure 15.13) – blinking like the changing phases of the moon (Figure 15.14).

Kopelman's group studied the MagMOONs and related systems (Anker *et al.*, 2005). Magnetically modulated optical nanoprobes emit a fluorescent signal in response to varying the external magnetic field,

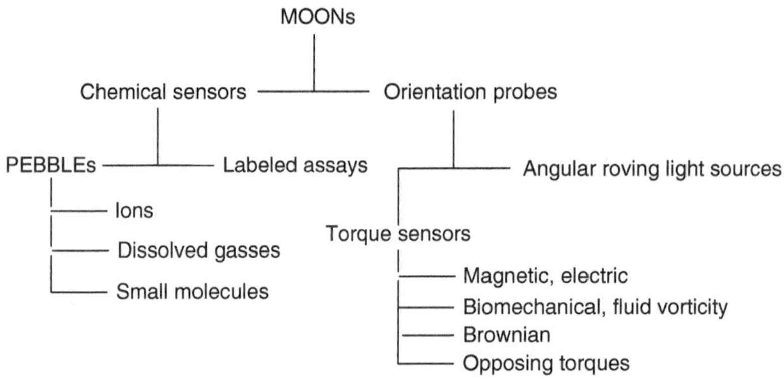

Figure 15.13 Application branches for MagMOONs and Brownians.

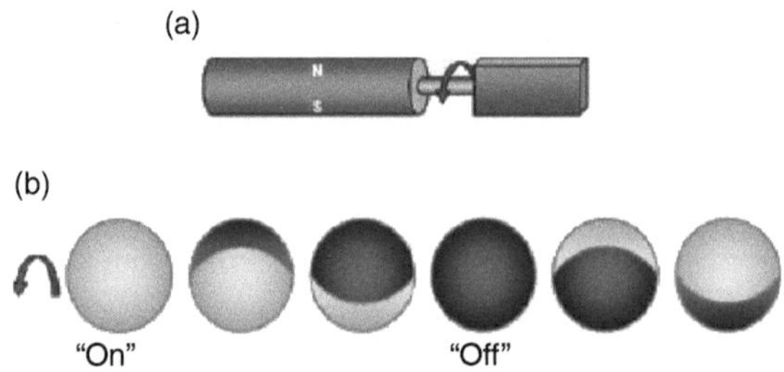

Figure 15.14 (a) Graphic of rotating magnetic used for rotating MagMOONs in aqueous solution. (b) The MagMOONs appearing to blink as they rotate. Published with permission from Elsevier (Teresa *et al.*, 2005).

allowing rapid *in situ* background subtraction and detection of low concentrations of analyte (Roberts *et al.*, 2005). Additionally, MagMOONs extend the detection of standard fluorophores into the UV region. These modifications allow the local optical and chemical properties to be more sensitive and specific. Subtracting unmodulated background fluorescence from the modulated MagMOON signal greatly increases the signal-to-background ratio, thus lowering detection limit.

Overall, MagMOONs provide novel tools for ultrasensitive biosensors for intracellular imaging and highly sensitive immunoassays. One exciting application for these probes is the simultaneous imaging of physical and chemical changes within biological systems such as living cells and tissues. Future work using MagMOONs will be focused on improving applications for biomedical fields and monitoring of various targets within intracellular environments to study the step-by-step activities in biological systems.

15.2.5. Optically encoded microbeads

Microbeads usually function as carriers hosting small molecules such as dye molecules or nanoparticles with identified properties for multiplex encoding. Since the size of the microbeads is in the range of several micrometers, they can accommodate large number of small molecules for biosensing and provide multiple binding sites (Han *et al.*, 2001). Therefore, optical-encoded microbeads have been regarded as one alternative to microarrays. Even compared with planar DNA chips, it has been demonstrated that microbeads have higher efficiency binding kinetics and shorter hybridization times (Wilkins Stevens *et al.*, 1999). From early 2000, encoded microbead techniques have been commercialized and optimized as a fast method for multiple purposes such as drug discovery, drug screening, disease diagnostics, and biosensing (Finkel *et al.*, 2004). Some applications include kinase and allergy testing, measurement of cytokine and thyroid levels, screening for cystic fibrosis, and antigen typing in human lymphocytes (Martins *et al.*, 2002; Prabhakar *et al.*, 2002; Riegger *et al.*, 2006).

The biggest limitation of encoded microbeads is controlling the loading. The different ratios of dyes are one crucial parameter for providing multiplex capability; the accuracy and reproducibility of the loading are necessary for consistent results (Kettman *et al.*, 1998). Measures to prevent leakage of loaded components, such as further coating with another layer (Gao *et al.*, 2003), are also crucial. Additionally, modifications to loading techniques as well as loading variable materials are other aspects of current investigation (Cao *et al.*, 2006). Only limited types of materials can be effectively loaded; these materials must be less than 50 nm in size.

15.2.6. Striped metallic barcodes

Another approach utilizing comparably large nanoparticles (a few micrometers) for biosensing employs stripped metallic barcodes. Generally, metallic barcodes are fabricated using metal ions such as gold, platinum, and silver by electrochemical deposition into a porous alumina template (Braeckmans *et al.*, 2002). As opposed to microbeads with a spherical shape, metallic barcodes usually resemble a rod several micrometers long and several hundreds of nanometers wide. A detailed description of optical biosensing function using this type of nanostructure has been discussed by Finkel *et al.* (2004), and simultaneous sandwich immunoassays have been demonstrated on two different barcoding particles (Nicewarner-Peña *et al.*, 2001).

Instead of relying on the spectrum width of the particles by varying the electrochemical conditions, it is possible to program the codes to obtain numerous variations. For example, with 3 metals and 13 strips, a total of 8×10^{15} code patterns could be theoretically identified (Han *et al.*, 2001). Simultaneous sandwich immunoassays have been demonstrated on two different barcoding particles (Nicewarner-Peña *et al.*, 2001). Recently, the approach was utilized for simultaneous analysis of SNPs in human samples (Sha *et al.*, 2006). The advantage of this method is that it is relatively fast, broadly applicable for biosensing and biodetection, and low cost.

While metallic barcodes can provide many advantages, the leakage of the contents at the interface between the stripes is the main problem in application. Furthermore, aggregation of barcodes can prevent accurate identification. Thus, it is necessary to fine-tune each parameter to obtain stable, uniform barcodes.

15.3. Advantages and disadvantages

The advantages and disadvantages of using each type of particle has been summarized in the relevant sections above. A summary comparing the physical properties and applications of multifluorophore-doped silica nanoparticles, QDs, optically encoded microbeads, SERS probes, and striped metallic barcodes is listed in Table 15.2. It is clear that each type

Table 15.2 Brief summary of different types of nanomaterials in multiplex bioanalyisis (Yao *et al.*, 2006)

Name and size	Particle description	Applications
QDs (2–10 nm without outside coating)	• Semiconductor nanocrystals with unique spectral properties due to quantum confinement effects • Core/shell QDs with improved biocompatibility and conjugation ability	• *In vitro* multiplex detection of proteins, DNA, and toxins • *In vivo* simultaneous detection of multiple mRNA
Multifluorophore-doped silica nanoparticles (20–70 nm)	• A large number of multiple fluorophores doped in silica matrix • Tunable emission signatures by varying dye doping ratios • Different spectral chemical properties by FRET	• *In vitro* detection of multiple analytes such as antibodies and bacteria • Multiple imaging modality such as optical and magnetic imaging
Optically encoded microbeads (1–10 μm)	• Coding elements with identifiable features hosted inside or onto microbeads	• *In vitro* biomolecule detection • Drug discovery and screening • Disease diagnostics
SERS probes (30–100 nm)	• Raman specific labels adsorbed on metallic nanoparticles	• *In vitro* multiplex detection of DNA and protein
Striped metallic barcodes (several micrometers in diameter: 0.3 μm)	• Striped, cylindrically shaped metal nanoparticles	• Simultaneous *in vitro* sandwich immunoassays and SNP analysis

of nanomaterial has its own strengths and weaknesses. The selection of specific nanomaterials for molecular sensing depends on the specific applications.

15.4. Future perspectives

Although nanoparticle-based optical sensors have made advances in high-throughput molecular screening and detecting, they have not yet been practically useful in complex biological systems and clinical fields. Most of the high-throughput multiplex bioanalyses using nanomaterials are still in the stage of demonstrating the principles. There are quite a few technical and environmental hurdles to overcome before these nanomaterials and nanotechnologies can be used effectively in real life.

For example, in spite of the efforts in nanoparticle surface modification to render them water-soluble, chemically stable, and biocompatible in physiological media, strategies are needed to improve the properties of the nanoparticle support matrices and surfaces. Surface modification can help reduce non-specific binding and facilitate the subsequent attachment of biological moieties, which will improve the binding kinetics and affinities of the nanoparticles for their target molecules. Non-specific binding and nanoparticle aggregation are still major issues blocking or slowing our progress in realizing the power of nanomaterials for ultrasensitive, multiplexed bioanalysis. Total elimination of non-specific binding is a difficult or probably impossible task, especially when the nanoparticles are used in a biological milieu. This problem demands that the scientific community design new strategies to reduce nanoparticle background signal due to non-specific binding. Additionally, the capacity to attach multiple functional groups to nanomaterials and to manipulate the matrix itself is necessary for diagnostics and therapeutics (Figure 15.15).

Although researchers have demonstrated interesting applications of biocompatible nanoparticles in clinical use, there is a huge gap between the results obtained in research labs and the requirements of real life and clinical applications. It is necessary to push the limits in low-cost,

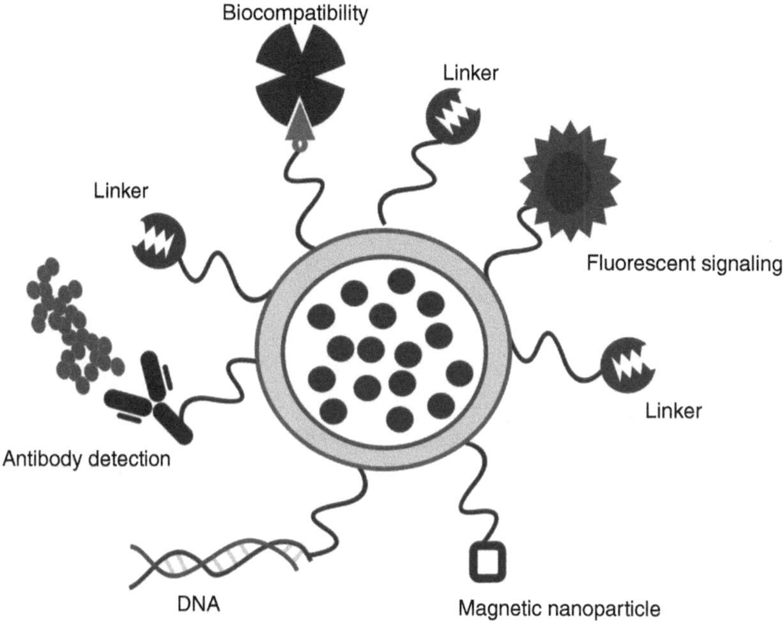

Figure 15.15 Multifunctional nanoparticles applied to biosensing and bioanalysis. Published with permission from CRC Press LLC (Lin Wang *et al.*, 2007).

large-scale nanoparticle production and portable detection systems to rapidly and automatically decode the optical signals from nanoparticles. These requirements, in addition to the cost of the technologies, will probably determine which candidate will be broadly useful and feasible for commercial applications. Today's signaling instrumentation and imaging facilities may not be capable of simultaneously detecting and decoding the multiplex information. Instrumentation to acquire and process a large amount of information is necessary. Large data storage and imaging analysis are a must for simultaneous imaging of a few analytes. Both call for novel technological developments from molecular imaging, signal processing, electrical and computer engineering, informatics, and other instrumentation and data analysis-related fields.

Finally, future applications of optical biosensors will emphasize the development of *in vivo* sensors for multiplex bioassays. While microbeads are generally useful for *in vitro* multiplex bioassays, they are not suitable for the staining or labeling of subcellular components or intracellular measurements as they are relatively large in size. Nano-sized devices are envisaged to be ingested or injected into the body, where they could act as reporters for diseased tissues and organs in the human body. Before this happens, detailed studies need to be performed on the *in vivo* behavior of such probes including their biodistribution, non-specific uptake, cellular toxicity, and pharmacokinetics. To use nanoparticles for encoding cellular assays, the nanoparticles tested must not interfere with normal cell processes such as signal transduction, receptor trafficking, and membrane function. When these nanoparticles are better integrated into the complex biological system, we will witness an explosion in the use of nanoparticle-based multiplex assays for faster, more sensitive, and accurate disease detection – first in laboratory tests and then in the patients. While we should be optimistic about the future of nanotechnology-based optical biosensing and imaging, a significant uncertainty and many technical challenges are still there.

References

Adler, J., Jayan, A., and Melia, C.D. (1999) *J. Pharm. Sci.*, **88**, 371.

Alivisatos, A.P., Johnsson, K.P., Peng, X. *et al.* (1996) *Nature*, **382**, 609.

Alizadeh, A.A., Eisen M.B., Davis R.E. *et al.* (2000) *Nature*, **403**, 503.

Anker, J.N., Behrend, C.J., and Kopelman, R. (2005) *J. Magn. Magn. Mater.*, **293**, 655.

Anker J.N. and Kopelman, R. (2003) *Appl. Phys. Lett.*, **82**, 1102.

Bailey, R.C., Nam, J.-M., Mirkin, C.A., and Hupp, J.T. (2003) *J. Am. Chem. Soc.*, **125**, 13541.

Baselt, D.R., Lee, G.U., Natesan, M. *et al.* (1998) *Bioelectronics*, **13**, 731.

Bergey, E.J., De, T., Rodman, D.J. *et al.* (2002) *Mol. Cryst. Liq. Cryst.*, **374**, 589.

Bourel, D., Rolland, A., Leverge, R., and Genetet, B. (1988) *J. Immunol. Methods*, **106**, 161.

Braeckmans, K., De Smedt, S.C., Leblans, M. *et al.* (2002) *Nat. Rev. Drug Discovery*, **1**(6), 447.

Brigger, I., Dubernet, C., and Couvreur, P. (2002) *Adv. Drug Delivery Rev.*, **54**, 631.

Bruchez, Jr., M., Moronne, M., Gin, P. *et al.* (1998) *Science*, **281**, 2013.
Brust, M. *et al.* (1995) *Adv. Mater.*, **7**, 795.
Brzeska, M., Panhorst, M., Kamp, P.B. *et al.* (2004) *J. Biotechnol.*, **112**, 25.
Cai, H., White P.S., and Torney, D. *et al.* (2000) *Genomics*, **66**, 135.
Cao, Y.C., Huang, Z.L., Liu, T.C. *et al.* (2006) *Anal. Biochem.*, **351**, 193.
Cao, Y.C., Jin, R., and Mirkin, C.A. (2002) *Science*, **297**, 1536.
Chan, W.C.W., Maxerll, D.J., Gao, X. *et al.* (2002) *Curr. Opin. Biotechnol.*, **13**, 40.
Chan, W.C.W. and Nie, S. (1998) *Science*, **281**, 2016.
Chang, E., Miller, J.S., Sun, J. *et al.* (2005) *Biochem. Biophys. Res. Commun.*, **334**(4), 1317.
Clapp, A.R., Medintz, I.L., Mauro, J.M. *et al.* (2004) *J. Am. Chem. Soc.*, **126,** 301.
Clapp, A.R., Medintz, I.L., Uyeda, H.T. *et al.* (2005) *J. Am. Chem. Soc.*, **127**, 18212.
Cui, Y., Wei, Q., Park, H., and Lieber, CM. (2003) *Science,* **293**, 1289.
Dahan, M., Laurence, T., Pinaud, F. *et al.* (2001) *Opt. Lett.*, **26**, 825.
Dabbousi, B.O., Rodriguez-Viejo, J., Mikulec, F.V. *et al.* (1997) *J. Phys. Chem. B. 1997*, **101**, 9463.
Daniel, M.-C. and Astruc, D. (2004) *Chem. Rev.*, **104**, 293.
Day, Y.S.N., Baird, C.L., Rich, R.L., and Myszka, D.G. (2002) *Protein Sci.*, **11**, 1017.
Doering, W.E. and Nie, S. (2003) *Anal. Chem.*, **75**, 6171.
Dusemund, B. *et al.* (1991) *Z. Phys. D*, **20**, 305.
Ekimov, A.I. (1996) *J. Luminesc.*, **70**, 1.
Ekimov, A.I., Efros, A.L., and Onuschenko, A.A. (1985b) *Sol. Stat. Comm.*, **56**, 921.
Ekimov, A.I., Hache, F., Schanneklein, M.C. *et al.* (1993) *J. Opt. Soc. Am. B Opt. Phys.*, **10**, 100.
Ekimov, A.I. and Onuschenko, A.A. (1981) *JETP Lett.*, **34**, 345.
Ekimov, A.I. and Onuschenko, A.A. (1982) *Sov. Phys. Semicond.*, **16**, 775.
Ekimov, A.I. and Onuschenko, A.A. (1984) *JETP Lett.*, **40**, 337.
Ekimov, A.I., Onuschenko, A.A., Pluhkin, A.G., and Efros, A.L. (1985a) *JETP Lett.*, **88**, 1490.
Ekimov, A.I., Onuschenko, A.A., and Tsekhomskii, V.A. (1980) *Fiz. Khim. Stekla*, **6**, 511.
Elghanian, R., Storhoff, J.J., Mucic, R.C. *et al.* (1997) *Science*, **277**, 1078.
Emerich, D.F. and Thanos, C.G. (2003) *Expert Opin. Biol. Ther.*, **3**, 655.
Emory, S.R. and Nie, S. (1998) *J Phys Chem B.*, **102**, 493.

Evident Technologies (2005) Quantum dot core concepts, http://www.evidenttech.com (12 December).

Fan, J.B., Chen, X., Halushka, M.K. *et al.* (2000) *Genome Res.*, **10**, 853.

Finkel, N.H., Lou, X.H., Wang, C.Y., and He, L. (2004) *Anal. Chem.*, **76**(19), 352A.

Gambling, W.A. (2000) *IEEE J. Selected Top. Quantum Electron.*, **6**(6), 1084.

Gao, H., Zhao, Y., Fu, S. *et al.* (2002) *Colloid Polym. Sci.*, **280**, 653.

Gao, X. and Nie, S. (2003) *J. Phys. Chem. B*, **107**(42), 11575.

Gao, X., Chan, W.C., and Nie, S. (2002) *Biomed. Opt.*, **7**, 532.

Gasch, A.P., Spellman, P.T., Kao, C.M. *et al.* (2000) *Mol. Biol. Cell*, **11**, 4241.

Gedanken, A., Reisfeld, R., Sominski, L. *et al.* (2000) *Appl. Phys. Lett.*, **77**, 945.

Gerion, D., Pinaud, F., Williams, S.C. *et al.* (2001) *J. Phys. Chem. B*, **105**, 8861.

Goldman, E.R., Clapp, A.R., Anderson, G.P. *et al.* (2004) *Anal. Chem.*, **76**, 684.

Goldman, E.R., Medintz, I.L., Whitley, J.L. *et al.* (2005) *J. Am. Chem. Soc.*, **127**, 6744.

Graham, D. and Mallinder, B.J. (2000) *Angew. Chem. Int. Ed.*, **39**, 1061.

Graham, D., Mallinder, B.J., and Whitcombe, D. *et al.* (2002) *Anal. Chem.*, **74**, 106.

Grifantini, R., Bartolini, E., and Muzzi, A. *et al.* (2002) *Nat. Biotechnol.*, **20**, 914.

Grubisha, D.S., Lipert, R.J., and Park, H.Y. *et al.* (2003) *Anal. Chem.*, **75**(21), 5936.

Guyot-Sionnest, P. (2001) *Science*, **201**, 2390.

Häfeli, U., Schütt, W., and Teller, J., (1997) *Scientific and Clinical Applications of Magnetic Carriers*. New York: Plenum.

Han, M., Gao, X., Su, J.Z., and Nie, S. (2001) *Nat. Biotechnol.*, **19**, 631.

Hansen, J.A., Wang, J., Kawde, A.-N. *et al.* (2006) *J. Am. Chem. Soc.*, **128**, 2228.

Härmä, H., Soukka, T., Lönnberg, S. *et al.* (2000) *Luminescence*, **15**, 351.

Haugland, R.P. (2005) *The Handbook. A Guide to Fluorescent Probes and Labeling Technologies* (10th edn) San Diego: Invitrogen Corporation.

Hayat, M.A. (1989) *Colloidal Gold: Principles, Methods and Applications*. New York: Academic Press.

Hines, M.A. and Guyot-Sionnest, P. (1998) *J. Phys. Chem. B*, **102**, 3655.

Ho, Y.P., Kung, M.C., Yang, S., and Wang, T.H. (2005) *Nano Lett.*, **5**, 1693.

Holm, B.A., Bergey, E.J., De, T. *et al.* (2002) *Mol. Cryst. Liq. Cryst.*, **374**, 589.

Horisberger, M. (1992) *Int. Rev. Cytol.*, **136**, 227–87.

Horisberger, M., and Rosset, J. (1977) *J. Histochem. Cytochem.*, **25**, 295.

Ipe, B.I., Lehnig, M., and Niemeyer, C.M. (2005) *Small*, **1**, 706.
Isola, N.R., Stokes, D.L., and Vo-Dinh, T. (1998) *Anal. Chem.*, **70**, 1352.
Jaiswal, J.K., and Simon, S.M. (2004) *Trends Cell Biol.*, **14**, 497.
Ji, X.J., Zheng, J.Y., Xu, J.M. *et al.* (2005) *J. Phys. Chem. B.*, **109**, 3793.
Jin, R., Cao, Y., Mirkin, C.A. *et al.* (2001) *Science*, **294,** 1901.
Kettman, J.R., Davies, T., Chandler, D. *et al.* (1998) *Cytometry*, **33**, 234.
Larson, D.R., Zipfel, W.R., Williams, R.M. *et al.* (2003) *Science*, **300**, 1434.
Leatherdale, C.A., Woo, W.K., Mikulec, F.V., and Bawendi, M.G. (2002) *J. Phys. Chem. B.*, **106**, 7619.
Levy, L., Sahoo, Y., Kim, K.S. *et al.* (2002) *Chem. Mater.*, **14**, 3715.
Lidke, D.S. and Arndt-Jovin, D.J. (2004) *Physiology*, **19**, 322.
Lim, Y.T., Kim, J.K., Shin, Y.B., and Chung, B.H. (2006) *Adv. Funct. Mat.*, **16**, 1015.
Lindberg, R., Sjöblom, J., and Sundholm, G. (1995) *Colloids Surf. A*, **99**, 79.
Loessner, M.J., Kramer, K., Ebel, F., and Scherer, S. (2002) *Mol. Microbiol.*, **44**, 335.
Marcel, B., Bruchez, M., Moronne, M. *et al.* (1998) *Science*, **281**, 2013.
Mattheakis, L.C., Dias, J.M., Choi, Y.J. *et al.* (2004) *Anal. Biochem.*, **327**(2), 200–8.
Martins, T.B. (2002) *Clin. Diagn. Lab. Immun.*, **9**, 41.
McKendry, R. (2002) *Proc. Natl. Acad. Sci. USA*, **99**, 9983.
Meallet-Renault, R., Denjean, P., and Pansu, R.B. (1999) *Sens. Actuators B Chem.*, **59**, 108.
Meallet-Renault, R., Yoshikawa, H., Tamaki, Y. *et al.* (2000) *Polym. Adv. Technol.*, **11**, 772.
Medintz, I.L., Konnert, J.H., Clapp, A.R. *et al.* (2004) *Proc. Natl. Acad. Sci. USA*, **101**, 9612.
Michalet, X., Pinaud, F.F., and Bentolila, L.A. *et al.* (2005) *Science*, **307**, 538.
Michalet, X., Pinaud, F.F., Bentolila, L.A. *et al.* (2001) *Science*, **294**, 137.
Michalet, X., Pinaud, F.F., Bentolila, L.A. *et al.* (2002) *Anal. Chem.*, **74**, 520A.
Michalet, X., Pinaud, F.F., Bentolila, L.A. *et al.* (2005) *Science*, **307**, 538.
Milliron, D. and Alivisatos, A.P. (2002) *J. Phys. Chem. B.*, **106**, 5500.
Mirkin, C.A., Letsinger, R.L., Mucic, R.C., and Storhoff, J.J. (1996) *Nature*, **382**, 607.
Mulvaney, S.P., Musick, M.D., Keating, C.D. *et al.* (2003) *Langmuir*, **19**, 4784.
Murray, C.B., Norris, D.J., and Bawendi, M.G., (1993) *J. Am. Chem. Soc.*, **115**, 8706.
Nam, J.M., Thaxton, C.S., and Thaxton, C.S. (2003) *Science*, **301**, 1884.
Nathaniel, L.R. and Mirkin, C.A. (2005) *Chem. Rev.*, **105**, 1547.

Nicewarner-Peña, S.R., Freeman, R.G., Reiss, B.D. *et al.* (2001) *Science*, **294**, 137.
Niemeyer, CM. (2001) *Angew Chem. Int. Ed. Engl.*, **40**, 4128.
Oldenburg, S.J., Averitt, R.D., Westcott, S.L., and Halas, N. (1998) *J. Chem. Phys. Lett.*, **288**, 243.
Osseo-Asare, K. and Arriagada, F.J. (1990) *Colloids Surf.*, **50**, 321.
Ozkan, M. (2004) *Drug Discov. Today*, **9**, 1065.
Pankhurst, Q.A., Connolly, J., Jones, S.K., and Dobson, J. (2003) *J. Phys. D Appl. Phys.*, **36**, R167.
Patolsky, F., Gill, R., Weizmann, Y. *et al.* (2003) *J. Am. Chem. Soc.*, **125**, 13918.
Pelkkikangas, A.-M., Jaakohuhta, S., Lovgren, T., and Harma, H. (2004) *Anal. Chim. Acta*, **517**, 169.
Peng, Z.A. and Peng, X.(2001) *J. Am. Chem. Soc.*, **123**, 183.
Petry, R., Schmitt, M., and Popp, J. (2003) *Chemphyschem.*, **4**, 14.
Phillipse, A.P., Bruggen, M.P.B.V., and Pathmamanoharan, C. (1994) *Langmuir*, **10**, 92.
Pissuwan, D., Valenzuela, S.M., and Cortie, M.B. (2006) *Trends Biotechnol.*, **24**, 2, 62.
Prabhakar, U., Eirikis, E., and Davis, H.M. (2002) *J. Immunol. Methods*, **260**, 207.
Puntes, V.F., Krishnan, K.M., and Alivisatos, A.P. (2001) *Science*, **291**, 2115.
Qhobosheane, M., Santra, S., Zhang, P., and Tan, W. (2001) *Analyst.*, **126**, 1274.
Reiss, G., Brueckl, H., Huetten, A. *et al.* (2005) *J. Mater. Res.*, **20**, 12.
Riegger, L., Grumann, M., Nann, T. *et al.* (2000) *Sensors and Actuators A*, **126**, 455.
Roberts, T.G., Anker, J.N., and Kopelman R. (2005) *J. Magn. Magn. Mater.*, **293**, 715.
Rosi, N.L., Giljohann, D.A., Thaxton, C.S. *et al.* (2006) *Science*, **312**, 1027.
Šafařík, I. and Šafaříková, M. (2002) *Monatsh. Chem.*, **133**, 737.
Sapsford, K.E., Pons, T., Medintz, I.L., and Mattoussi, M. (2006) *Sensors*, **6**, 925.
Santra, S., Tapec, R., Theodoropoulou, N. *et al.* (2001) *Langmuir*, **17**, 2900.
Santra, S., Wang, K., Tapec, R., and Tan, W. (2001) *J. Biomed. Opt.*, **6**, 160.
Santra, S., Zhang, P., Wang, K. et al. (2002) *Anal. Chem.*, **73**, 4988.
Sha, M.Y., Walton, I.D., and Norton, S.M. *et al.* (2006) *Anal. Bioanal. Chem.*, **384**, 658.
Shi, L., De Paoli, V., Rosenzweig, N., and Rosenzweig, Z. (2006) *J. Am. Chem. Soc.*, **128**, 10378.
Soukka, T., Paukkunen, J., Härmä, H. *et al.* (2001), *Clin. Chem.*, **47**, 1269.

Stöber, W., Fink, A., and Bohn, E. (1968) *J. Colloid Interface Sci.*, **26**, 62.
Striolo, A., Ward, J., Prausnitz, J.M. *et al.* (2002) *Int. J. High Speed Electron. Syst.*, **12**, 1039.
Su, X., Zhang, J.W., and Sun, L. *et al.* (2005). Composite Organic-Inorganic Nanoparticles (COINs) with chemically encoded optical signatures. *Nano Lett.*, **5(1)**, 49–54.
Sun, S., Murray, C.B., Weller, D. *et al.* (2000) *Science*, **287**, 1989.
Vries, A. de *et al.* (2004) University of Twente, Enschede, Ph.D. thesis.
Wang, L. and Tan, W. (2006) *Nano Lett.*, **6**, 84.
Wang, L., Wang, K., and Swadeshmukul, S. *et al.* (2006) *Anal. Chem.*, **78**(3), 646A.
Wang, L., Yang, C., and Tan, W. (2005) *Nano Lett.*, **5**, 37.
Wang, C.J., Shim, M., and Guyot-Sionnest, P. (2001) *Science*, **291**, 2390.
Wang, J., Liu, G., and Jan, M.R. (2004) *J. Am. Chem. Soc.*, **126**, 3010.
Wang, J., Liu, G., and Merkoci, A. (2003) *J. Am. Chem. Soc.*, **125**, 3214.
Wang, L., Loften, C., and Tan, W. (2007) Bioconjugated nanoparticles for biotechnology and bioanalysis. *Nanotechnol. Biol. Medi.* 19/1–19/16.
Wick, I., and Hardiman, G. (2005) *Curr. Opin. Drug Discov. Dev.*, **8**(3), 347.
Wilkins Stevens, P., Henry, M.R., Kelso, D.M. (1999) *Nucleic Acids Res.*, **27**, 1719.
Wu, A.M., Gambhir, S.S., and Weiss, S. (2005) *Science*, **307**, 538.
Wu, X., Liu, H., Liu, J. *et al.* (2003) *Nat. Biotech.*, **21**, 41.
Yamauchi, H., Ishikawa, T., and Kondo, S (1989) *Colloids Surf.*, **37**, 71.
Yang, L., Tran, D.K., and Wang, X. (2001) *Genome Res.*, **11**, 1888.
Yao, G., Wang, L., Wu, Y. *et al.* (2006) *Anal Bioanal. Chem.* **385**(3), 518.
Zhao, X., Bagwe, R.P., and Tan, W. (2004) *Adv. Mater.*, **16**, 173.
Zhao, X., Dytocio, R.T., and Tan, W. (2003) *J. Am. Chem. Soc.*, **125**(38), 11474.
Zhao, X., Hilliard, L.R., and Mechery, S.J. *et al.* (2004) *Proc. Natl. Acad. Sci. USA*, **101**(42), 15027.

Optical Biosensors: Today and Tomorrow (2nd Edition)
Frances S. Ligler and Chris Rowe Taitt (Editors)
Published by Elsevier B.V.

Chapter 16

FLUORESCENCE-BASED INTRACELLULAR SENSING

Igor L. Medintz, Ph.D. and James B. Delehanty, Ph.D.
Center for Bio/Molecular Science and Engineering, Naval Research Laboratory, Washington, DC 20375-5348, USA

Intracellular or in situ biosensors will have tremendous impact upon research into the understanding of basic cellular structure and function. In this chapter, we examine the state-of-the-art biosensors available for such use and focus on those that utilize fluorescence and fluorescence resonance energy transfer in particular. Intracellular targets of interest for monitoring include: nucleic acids, small molecule nutrients and ligands, post-translational modifications, intracellular localization, and a variety of biochemical reactions involved in signal transduction cascades. The benefits and liabilities of particular sensor designs are discussed along with potential developments which may improve their performance. Beyond their current intracellular utility, successful development of these biosensors will find applications in pharmaceutical research, environmental and defense monitoring as well as clinical diagnostics.

16.1. Technical concepts

Understanding the myriad structure and function of all the cellular components, and, more importantly, how they function in concert, is one of the ultimate goals of biology. The complexity within a single cell is staggering; each cell is able to regulate thousands of simultaneous chemical reactions, respond to its environment, grow and replicate as well as defend itself from stress and pathogens (Alberts *et al.*, 2002; Watson *et al.*, 2004). A basic understanding of the cellular unit will lead directly to a hierarchical understanding of tissue and organ function, development and regulation with all the predicted impact on concomitant disease diagnosis and therapeutics. For example, a fundamental understanding of each cell type that makes up the nervous system will help us to visualize the system as a whole and, it is hypothesized, lead to an eventual understanding of consciousness and thought. Although several approaches to this field have been initiated, including systems biology, it is clear that this endeavor will require a biosensing toolset capable of monitoring each individual component directly within the natural cellular milieu. Ultimately, this is a tremendous biosensing challenge, and an entire field of biosensor development and subsequent applications are devoted almost exclusively to these *in situ analyses.* Such biosensors are our principal focus here.

Although still unmet technically in most regards, the functional requirements of such a biosensor family are easily envisioned. They include: (1) a generally applicable structural/functional design that allows targeting of many disparate analytes; (2) facile delivery to the cell or to a particular subcellular compartment; (3) specificity and sensitivity; (4) facile detection within the heterogeneous cellular environment; (5) long-term stability; and perhaps most importantly (6) amenability to "multiplexing" or the simultaneous monitoring of multiple analytes. The latter can allow spaciotemporal correlation of multiple concurrent events within a cell and open the doorway to a real understanding of the complexities of cellular function.

For our purposes, we limit our definition of *in situ* to the monitoring of an event within the cellular environment in real time, as all other hierarchical biosensing stems from this most basic application. This can also be considered the driving force for this application; namely the desire to monitor myriad intracellular events including biochemical reactions (i.e., transcription, synthesis, catabolism), analyte flux, second messengers and signal transduction, etc. (Alberts *et al.*, 2002; Zhang *et al.*, 2002;

Miyawaki, 2003; Watson *et al.*, 2004; Tsien, 2005; Giepmans *et al.*, 2006). Although other signal transduction modalities are in limited use, electrochemistry for example, we limit our discussion to fluorescence-based biosensing and fluorescence resonance energy transfer (FRET) in particular, as these are not only the most applied but also the most promising techniques at the present time.

The power of fluorescence-based biosensing is derived from the ability to selectively label a particular biomolecule and then discretely excite and observe it in a manner that is clearly distinguishable from the sample background (Lakowicz, 2006). The popularity of fluorescence analysis in biosensing is directly attributable to three inter-related reasons. The first is the wide variety of commercial and research grade fluorophores available for incorporation into such biosensors including organic dyes, inorganic nanocrystals, and rare-earth chelates, all of which are constantly being expanded in number and improved in performance (Haugland, 2005; Wolfbeis, 2005; Sapsford *et al.*, 2006). The second reason is the continuing improvement of fluorescent proteins. As more are cloned and mutationally improved, our understanding of their structure and function is increasing and their applications and availability are concomitantly growing (Tsien, 1998; Baird *et al.*, 2000; Campbell *et al.*, 2002; Zhang *et al.*, 2002; Shaner *et al.*, 2004, 2005; Sapsford *et al.*, 2006). Their principal benefit is that they can be genetically appended to virtually any target protein and co-expressed *in vivo* in target cells ranging from bacteria to higher eukaryotes and plants. The third reason is the continuous updating and availability at a reasonable cost of evermore sensitive and sophisticated instrumentation targeting almost all aspects of fluorescent analysis (Lakowicz, 2006). Further, this instrumentation has directly benefited from the revolution in microanalytical formats (Nagl *et al.*, 2005; Alper, 2006; Mukhopadhyay, 2006; Wolfbeis, 2006).

Within the "super-group" of fluorescent analysis, one prominent biosensing technique is Förster or FRET. Fluorescence resonance energy transfer is defined as the non-radiative process whereby an excited-state donor fluorophore (D) transfers energy to a proximal ground-state acceptor (A) via long-range dipole coupling (Figure 16.1) (Lakowicz,

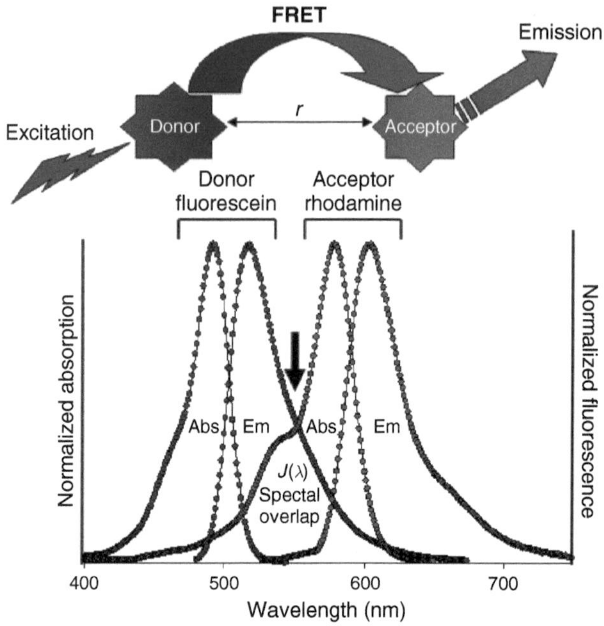

Figure 16.1 Schematic describing the FRET process. Upon excitation, the excited state donor molecule transfers energy non-radiatively to a proximal acceptor molecule located at distance *r* from the donor. The acceptor may either remit the energy fluorescently or dissipate it through non-radiative channels. The spectra shows the absorption (Abs) and emission (Em) profiles of one of the most commonly used FRET organic dye pairs: fluorescein donor–rhodamine acceptor. Fluorescein can be efficiently excited at 480 nm and emit at ~520 nm. The spectral overlap (between fluorescein emission and rhodamine absorption) occurs from 500 to 600 nm. The Förster distance R_0 for this pair is 55 Å. Thus, in an optimal configuration (donor–acceptor separation distance $r<55$ Å), excitation of fluorescein at <500 nm can result in significant rhodamine emission at >600 nm due to efficient FRET. Figure adapted from Sapsford *et al.* (2006). (see Plate 9)

2006; Sapsford *et al.*, 2006). The energy transfer rate between a single D–A pair has an inverse sixth power dependence upon the distance between the two and is usually expressed in terms of the Förster distance, R_0. This is the distance corresponding to 50% energy transfer efficiency (Lakowicz, 2006). R_0 is defined as:

$$R_0 = 9.78 \times 10^3[\kappa^2 n^{-4} Q_D J(\lambda)]^{1/6} \tag{16.1}$$

where n is the refractive index of the medium, Q_D is the photoluminescence quantum yield of the donor, $J(\lambda)$ is the spectral overlap integral, and κ^2 is the dipole orientation factor (Lakowicz, 2006). The FRET efficiency, E, is usually calculated from experimental data using the expression:

$$E = \frac{(F_D - F_{DA})}{F_D} \tag{16.2}$$

where F_D and F_{DA} are the fluorescent intensities (alternatively, the excited state fluorescent lifetimes) of the donor in the absence and presence of acceptors, respectively (Lakowicz, 2006). Donor–acceptor dye separation distances, r, can be estimated by using the FRET efficiency(s), Förster theory, and the number of acceptors n per donor with:

$$E(n) = \frac{n}{n + (r/R_0)^6} \tag{16.3}$$

The interested reader is referred to Lakowicz's (2006) text for an excellent discussion of all aspects pertinent to FRET. For biosensing in particular, the power of this technique is derived from the unparalleled ability to monitor molecular scale interactions including binding events, structural rearrangements, and dissociation events. No other analytical technique available to date compares to FRET for its consistent ability to monitor both inter- and intra-molecular interactions (Lakowicz, 2006; Sapsford *et al.*, 2006). As FRET is intrinsically sensitive to nanoscale changes in the D–A separation, appropriate labeling of a biomolecule(s) allows a sensing event or change in the biomolecule state to be transduced by a change in FRET efficiency and almost all FRET-based biosensing relies upon this. Thus, labeling of two proteins with appropriate D–A dyes can allow the association/dissociation to be monitored (Sapsford *et al.*, 2005, 2006). More important to the current focus, labeling the termini of a single protein or peptide with a D–A pair can allow monitoring of binding-induced conformational changes. This mechanism forms the basis for many of the intracellular sensors discussed below.

Since our focus here is predominantly intracellular sensors, we divide *in situ* biosensors into two primary classes which we term "fluorescent biosensors" and "fluorescent protein sensors." Fluorescent sensors include biomolecules such as proteins, peptides, or DNA that is labeled with one or two fluorophores and binding of target analyte results in a change in fluorescence or FRET (fluorescence may also emanate from intrinsic components, i.e. tryptophan fluorescence in proteins). This classification of sensors can also include completely abiotic elements such as ion-sensitive dyes (Haugland, 2005). Fluorescence resonance energy transfer-based protein sensors include proteins or peptides appended with two fluorescent proteins, usually on a plasmid, and expressed *in vivo* for cellular sensing. For the first class, the sensor is expressed, purified, and labeled *ex vivo* (extrinsic) and must still be introduced intracellularly by transfection or microinjection, while the latter class can be expressed and targeted to a particular cellular organelle for sensing function (intrinsic). A related subset of intracellular protein sensors are appended with single fluorescent proteins to allow monitoring of intracellular changes in localization; for example, monitoring of nucleocytoplasmic transport kinetics has allowed an estimation of nuclear pore size (Chen *et al.*, 2006). A related field where *in situ* sensors are of particular interest is cell-based biosensors. There, a cell's response to an analyte in its environment forms the initial "biosensing" event and this is usually transduced by monitoring the signal-transduction cascade or the induction of a reporter gene. The primary benefit here is that the cell can act as a very powerful biofilter to discriminate between a target and other material in its environment (Stenger *et al.*, 2001; Rider *et al.*, 2003).

One of the continuing challenges in this field, however, remains the labeling of biological molecule(s), both *in vivo* and *in vitro*, in such a manner that useful data can be derived from a biosensing configuration (Sapsford *et al.*, 2006). Labeling proteins or peptides at known distinct locations is the most desirable for both generic fluorescent- and FRET-based sensing. This can ensure that labeling does not interfere with binding, recognition, or subsequent function and also allows optimal placement of D–A dyes for efficient FRET. However, a variety of considerations go into any biomolecule labeling and optimal placement is not always available (Sapsford *et al.*, 2006). For example, recombinant introduction of a single cysteine residue can provide a unique thiol for monovalent labeling of a protein or peptide. However, should other cysteine(s) already be present and/or necessary for structure, the new residue can induce "thiol-scrambling" of the protein structure and result in loss of function (Medintz *et al.*, 2005b). Alternatively, targeting lysine residues with amine-reactive dyes is probably the most common method of labeling proteins, but their ubiquitous presence, including in many binding sites, can also result in loss of function.

16.2. History

Anecdotally, there have been references to fluorescent phenomena dating back to the ancient Chinese (~700 BC) and Aristotle (350 BC) (Jameson, 2006). Although countless observations and contributions were made by many luminaries including Boyle and Galileo, George Gabriel Stokes is credited with coining the term "fluorescence" in 1852. Stokes also observed that fluorescence is of a longer wavelength than the excitation light which led to the optical displacement between the two being termed the "Stokes shift." In 1856, William Henry Perkin synthesized an aniline dye while trying to make quinine, and histologists began using it for cellular staining a decade later marking the introduction of fluorescent staining. More importantly in the 1870s, Adolph von Beyer synthesized fluorescein which still remains to this day one of the most ubiquitous fluorescent dyes available. Shortly before World War I, Helmstaedt and Lehmann developed the first fluorescence microscope which was utilized by Von Provazek in 1914 to study dye binding in living cells. In 1941, Albert Coons labeled antibodies with fluorescein isothiocyante (FITC, fluorescein functionalized with an amine-reactive group that targets primarily lysine groups) and started the field of immunofluorescence (Jameson, 2006). In the 1950s, Gregorio Weber synthesized dansyl chloride for attachment to proteins and was able to study protein hydrodynamics with fluorescence polarization. These seminal experiments started the field of quantitative biological fluorescence, and Weber is still considered by many the most important pioneer in the field (Lakowicz, 2006). Most of the detailed principles and formalism of fluorescence were concomitantly worked out from the 1920s onwards by many scientists including Perrin, Gaviola, Wavilov, Förster, Stern, Jablonski, and others (Jameson, 2006; Lakowicz, 2006).

The development of FRET can be traced to Perrin, who first proposed a mechanism for resonance energy transfer in 1918. In 1928, Kallman and London developed the concomitant quantum theory of dipole coupling, and the R_0 parameter was used for the first time. In 1932, Perrin put forward a quantum mechanical theory of energy transfer, and this was followed up in the late 1940s by a quantitative theory by Theodor Förster (Lakowicz, 2006). It is this treatment and his name that have become synonymous with this fluorescent phenomena and its analysis. Interestingly, the common parlance "*fluorescence* resonance energy transfer" is technically incorrect, as no fluorescence is transferred during this process. Stryer and Haugland (1967) are credited with performing the key experiments that eventually led to FRET being widely applied in biology. Using a series of poly-L-proline oligomers as spacers of defined length, they labeled both the carboxyl and amino termini of the peptides with D–A dyes, one of which was Weber's dansyl chloride. They then observed that the correlation of FRET efficiency with the peptide size or D–A distances was in excellent agreement with Förster's predictions and proposed that it could be used as a biomolecular and "spectroscopic ruler" (Stryer and Haugland, 1967). These results, in conjunction with the development of an ever-increasing commercial fluorescent dye library, with many choices in both absorption/emission properties (different D–A combinations) and a wide range of targeted reactivities, have led to the steady growth of FRET applications in biology from the early 1970s onwards (Hermanson, 1996; Haugland, 2005; Sapsford *et al.*, 2006).

Three areas of protein science have been key to developing both fluorescent and fluorescent protein biosensors: (1) understanding protein and in particular periplasmic-binding protein (PBP) structure and function; (2) recombinant DNA technology coupled to computational modeling/design; and (3) the development of fluorescent proteins. Beyond classical immunofluorescent staining, the growth in understanding protein structure/function over the last ~50 years has led to the ability to exploit their properties in biosensing; this benefit is exemplified by research on the PBP family (Marvin and Hellinga, 2001b; De Lorimier *et al.*, 2002; Dwyer and Hellinga, 2004). This diverse superfamily, found in bacteria primarily but also in species ranging up to higher eukaryotes, targets ligands ranging from sugars, amino acids, and peptides to even small ions. Although very dissimilar at the primary sequence level, PBPs share a common structure consisting of two domains surrounding a central ligand-binding pocket which are joined by a "hinge" region (Figure 16.2) (Quiocho and Ledvina, 1996; Medintz and Deschamps, 2006). Upon recognizing and binding a specific ligand, the two lobes undergo a conformationally coupled structural transition whereby they bend and twist about the hinge region from an open to a closed/bound configuration in a manner analogous to a "venus-fly-trap" closing (Sharff *et al.*, 1992, 1993; Quiocho and Ledvina, 1996; Medintz and Deschamps, 2006).

Functionally, PBPs have two intrinsic characteristics that are available for biosensor exploitation – namely, recognition specificity and conformational changes. These can be used either separately or may function in concert. Recognition specificity has been utilized for fluorescence- and FRET-based sensing while conformational changes have been utilized for FRET-, electrochemical-, fluorescent-, and enzymatic-based sensing (Medintz and Deschamps, 2006). The Quicho lab has been responsible for pioneering the determination of the open and closed crystallographic structures of many of these proteins, including maltose-binding protein (MBP), which is regarded as both the structural and the functional archetype (Sharff *et al.*, 1992, 1993; Quiocho and Ledvina, 1996; Medintz and Deschamps, 2006). This structure/function understanding led to the seminal demonstration by Cass that MBP could be mutated to express a unique cysteine residue near the binding pocket which, when labeled with an environmentally sensitive fluorophore, could transduce maltose binding-induced conformational changes by substantial concentration-dependent fluorescent increases (Gilardi *et al.*, 1994). Using a similar strategy, Kobatake coupled a similarly modified

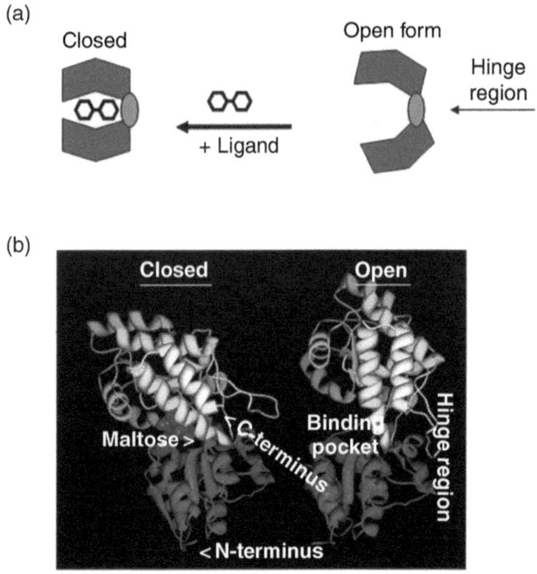

Figure 16.2 MBP structure and function. (a) A prototypic PBP structure is depicted; this has two lobes (green) that create the ligand-binding pocket and are joined by the hinge-binding region (pink). Upon binding the ligand, the PBP undergoes a transition to the closed structure. (b) Ribbon and string rendering of MBP structure in the open form and ligand-bound, closed form (Spurlino *et al.*, 1991; Sharff *et al.*, 1992, 1993). MBP dimensions are ~30 Å × 40 Å × 65 Å (Sharff *et al.*, 1992; Fehr *et al.*, 2002). The two domains (lobes) are highlighted in green and grey. Upon binding maltose (purple), the lobes rotate ~35° and twist laterally ~8° relative to each other. Overall the amino and carboxy termini move ~7 Å closer to each other after binding. Note the change in conformation of the overall structure upon binding maltose. Figure reprinted from Medintz and Deschamps (2006) with permission from Elsevier. (see Plate 10)

glutamine-binding protein to a designer hydrophobic peptide, allowing surface adherence and demonstrated sensing in a quasi-array format (Wada *et al.*, 2003). Other more complex surface-attached MBP sensors with tunable binding affinities have also been demonstrated (Medintz *et al.*, 2004). Using the PBP family as a working platform, in conjunction with computationally intensive modeling and an iterative mutational improvement approach, the development of sensors for more than 10 different analytes has been realized (De Lorimier *et al.*, 2002). These include sensors for further sugars beyond maltose (glucose, galactose, arabinose, and ribose), small ligands such as Fe(III), sulfate, and phosphate, and amino acids or dipeptides (De Lorimier *et al.*, 2002). An entire research field has slowly grown around this sensing strategy whereby specific proteins, most often drawn from the PBP family, are utilized to accomplish sensing (Dwyer and Hellinga, 2004; Deuschle *et al.*, 2005).

In seminal research, Hellinga has shown that proteins drawn from the PBP family can be redesigned/reconfigured to allow targeting to other disparate and unnatural analytes. Using computationally intensive modeling along with an iterative mutation process, the binding/recognition sites of PBPs are redesigned to target a new analyte (Marvin and Hellinga, 2001a; Dwyer *et al.*, 2003, 2004; Allert *et al.*, 2004). Examples of proteins that have undergone this re-engineering process include: (1) MBP to form a zinc sensor (Marvin and Hellinga, 2001a); (2) glucose and ribose-binding proteins (RBP) to recognize the primary hydrolytic product of the nerve agent soman (Allert *et al.*, 2004); (3) RBP to bind the explosive TNT (Looger *et al.*, 2003); and (4) RBP into a catalytically active enzymatic isomerase (Dwyer *et al.*, 2004). This breakthrough technology signals an alternate method to develop biosensors specific for many different analytes in addition to those that PBPs already target or that use antibodies for recognition specificity; the latter are often limited when targeting small molecules. Beyond this, a variety of conformationally dependent sensors utilizing nucleic acid, aptamer, peptide, organic dye, and polymer scaffolds have also been developed and function in a similar manner – namely, target binding alters conformation/fluorescence (Tan *et al.*, 2004).

For the development of fluorescent proteins, the key event not only for biosensing but also for all biological usages was the demonstration that the genetic sequence alone contained all the necessary information for expression and maturation in a variety of prokaryotic and eukaryotic cell lines (Tsien, 1998; Schmid and Neumeier, 2005). The prototypical member of this fluorescent family is green fluorescent protein (GFP) which is derived from the jellyfish *Aequorea victoria.* Although it was first described more than 40 years ago, it was not cloned until the early 1990s (Tsien, 1998). Mutational analysis allowed derivatives of GFP with many different emission spectra to be selected along with a better understanding of chromophore structure and function. There are now a wide variety of fluorescent proteins commercially available on plasmids from many different species including coral, allowing them to be appended either to the amino or carboxyl terminus on almost any cloned protein sequence (Tsien, 1998; Zhang *et al.*, 2002; Shaner *et al.*, 2004, 2005; Giepmans *et al.*, 2006). The availability of multiple fluorescent protein gene sequences, along with an understanding of their basic properties, can be considered the enabling technology that allowed Tsien to design FRET sensors that incorporate D–A fluorescent proteins for intracellular calcium sensing and Frommer to create a variety of intracellular sensors for small molecule nutrients and ligands, as described below (Miyawaki *et al.*, 1997; Fehr *et al.*, 2004a; Deuschle *et al.*, 2005; Lalonde *et al.*, 2005).

16.3. State of the art

16.3.1. Fluorescent sensors

As stated previously, fluorescent sensors as defined herein are sensing moieties that are labeled with one or two fluorophores wherein the binding of target analyte results in a change in fluorescence or FRET. Alternatively, these sensors can take the form of fluorophores that fluoresce or quench in response to specific stimuli.

16.3.1.1. Fluorescent reagents for sensing cellular homeostasis

Numerous fluorescent reagents aimed at monitoring intracellular homeostasis have been developed. In this section, we highlight some of the more commonly used fluorophore-based reagents for monitoring events and characteristics that have roles in the onset of apoptosis: intracellular pH, intracellular calcium levels, and the production of free radicals. The acetoxymethyl (AM) ester derivatives of the pH-sensitive dyes BCECF and carboxy SNARF-1 have been used successfully to investigate the role of cytosolic acidification during apoptosis (Gottlieb *et al.*, 1996). BCECF, for example, is membrane permeant, has a pK_a (7.0) within the normal range of cytoplasmic pH (6.0–8.0), and has a pH-dependent emission profile. Changes in intracellular calcium levels have also been shown to trigger apoptosis. The AM derivatives of the cell-permeant calcium-sensitive dyes Fluo-4 and Fura-2 remain the standard dyes for visualizing changes in intracellular calcium levels. Both fluorophores show very large increases in fluorescence intensity upon binding to calcium and have become the reagents of choice for monitoring calcium flux in living cells in real time. In many cell types, oxidative stress is a critical part of the apoptotic process (Jabs, 1999; Ye *et al.*, 1999). Accordingly, numerous fluorescent probes have been developed for the detection of lipid peroxidation and intracellular reactive oxygen species. The fatty acid *cis*-parinaric acid, for example, is a structural analog of intrinsic membrane fatty acids that possesses a very large fluorescence Stokes shift (~100 nm). When oxidized by lipid hydroperoxides, the fluorescence of *cis*-parinaric acid is quenched.

16.3.1.2. Molecular beacons

The ability to visualize in real time the expression level and localization of specific RNAs in living cells offers tremendous opportunities for biological and disease studies. Until recently, the detection of RNAs has relied heavily on the use of mostly semiquantitative and destructive techniques such as polymerase chain reaction (PCR), northern blotting, differential display, and DNA microarrays. In the mid-1990s, however, Tyagi and Kramer provided the seminal description of oligonucleotide probes that fluoresce upon hybridization to target nucleotide sequences (Tyagi and Kramer, 1996). Termed "molecular beacons," these hairpin nucleic acid probes have the potential to be highly specific and sensitive in living cell RNA detection. These fluorescent probes are designed to adopt a stem-loop conformation in the absence of a complementary target such that fluorescence of the fluorophore is quenched by a proximally located quencher. Hybridization with the target nucleic acid sequence opens the hairpin, physically separating the fluorophore from the quencher, resulting in a fluorescence signal. Though useful, traditional molecular beacons are often subject to false-positives due to the presence of cellular nucleases and non-specific binding to nucleic acid-binding proteins. To overcome this limitation, dual FRET molecular beacons have been developed (Tsourkas *et al.*, 2003; Santangelo *et al.*, 2004). In this scenario, a *pair* of molecular beacons, each labeled with a donor and acceptor fluorophore, is used. The probes are designed to hybridize to adjacent regions on a single RNA target. This modified molecular beacon design has markedly improved the rate of false-positive signals as the FRET-derived signal is obtained only upon the hybridization of *both* probes to proximal locations on the same RNA target. In sum, the ability to transduce target recognition directly into a fluorescence signal with high signal-to-background ratio has allowed molecular beacons to find utility in applications such as multiple analyte detection, cancer cell detection, and mRNA transcript detection (Vet *et al.*, 1999; Santangelo *et al.*, 2004; Peng *et al.*, 2005).

16.3.2. Fluorescent protein sensors

Fluorescent protein sensors typically contain a centrally located protein- or peptide-based substrate moiety appended with two fluorescent proteins that function as an efficient FRET D–A pair. Upon the binding of target analyte or the modification of the substrate sequence, a change in fluorescence or FRET is measured.

16.3.2.1. Phosphorylation

The phosphorylation of proteins by protein kinases is the most important way that individual proteins are post-translationally modified to modulate their function, with major implications in both physiology and the pathological progression of disease. Monitoring kinase function, therefore, is a major thrust of pharmaceutical drug development. To study kinase function, methods are required to image not only their localization but also their activities inside living cells. Tsien's laboratory spearheaded the development of FRET-based monitoring of protein conformations and interactions using mutants of GFP (Miyawaki and Tsien, 2000). A series of FRET-based sensors was developed to monitor the activity of serine kinases (protein kinase A (PKA)) (Zhang *et al.*, 2001) and tyrosine kinases (Src, epidermal growth factor receptor, and Abl) (Ting *et al.*, 2001). Operationally, these sensors consist of four domains that are encoded in a concatenated fashion and expressed within mammalian cells via transient transfection: (1) cyan fluorescent protein (CFP) as FRET donor; (2) an SH2 domain capable of binding a phosphorylated peptide sequence; (3) an appropriate kinase substrate peptide sequence; and (4) yellow fluorescent protein (YFP) as FRET acceptor (Figure 16.3). Upon activation of the kinase to be monitored, the substrate peptide sequence is phosphorylated and the SH2 domain forms an intramolecular complex with the phosphate-bearing peptide sequence. This conformational change alters the distance and/or relative orientation between the CFP and the YFP to generate a FRET change.

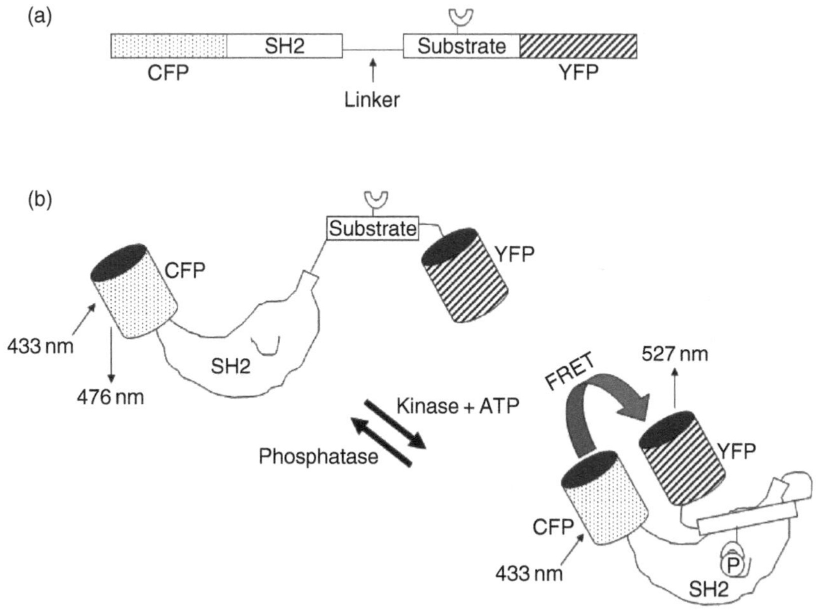

Figure 16.3 FRET-based phosphorylation sensor. (a) A schematic depicting the functional domains of the phosphorylation sensor. The FRET pair consists of CFP (donor) and YFP (acceptor) separated by a concatenated sequence of (1) an SH2 domain, (2) a linker region, and (3) a phosphorylation substrate sequence. (b) In the unphosphorylated state, the CFP and YFP are sufficiently separated so as to have no FRET response. Upon phosphorylation of the substrate sequence by the kinase under study, the SH2 domain intramolecularly complexes with the phosphate-bearing peptide sequence, inducing a conformational and distance change that is sufficient to mediate FRET between CFP and YFP.

The modular design of the recombinant construct allows for the design of phosphorylation sensors to target any kinase for which a suitable substrate sequence can be designed. More recently, Green and Alberola-Ila (2005) reported a variation on this theme in their development of a sensor for extracellular-regulated kinase (ERK). In this arrangement, the presence of an SH2 domain was not required; phosphorylation of the cognate peptide substrate sequence was sufficient to induce an observable FRET change. Collectively, these phosphorylation sensors have been used to demonstrate the non-destructive imaging of dynamic protein tyrosine and serine kinases with high spatial and temporal resolution in single living cells.

16.3.2.2. Proteolysis

In vivo sensors for monitoring proteolytic activity are another example where both FRET and pharmaceutical research have had a strong impact on development. The estimated 561 proteases in the human genome are generally recognized as critical participants in normal and aberrant biochemical processes that manifest from the cellular to the organismal level (Puente and Lopez-Otin, 2004). Proteases have roles in regulating wound healing, necrosis, and apoptosis (Cohen, 1997; Thornberry *et al.*, 1997; Puente *et al.*, 2003; Puente and Lopez-Otin, 2004), as well as the normal development of tissues and organs (Artal-Sanz and Tavernarakis, 2005; Green and Lund, 2005). Proteases are also key participants in many diseases including stroke (Sheehan and Tsirka, 2005), cancer (Ala-Aho and Kahari, 2005; Handsley and Edwards, 2005; Vihinen *et al.*, 2005), and more than 50 genetic disorders (Puente *et al.*, 2003; Richard, 2005). Furthermore, many infectious microorganisms, including viruses, parasites, and bacteria, utilize proteases as essential virulence factors (Shao *et al.*, 2002; Anand *et al.*, 2003; Imamura, 2003; Puente *et al.*, 2003; Wu *et al.*, 2003). This has meant that the superfamily of proteases has become an important pharmaceutical target, thus driving the development of biosensors to monitor their activity both *in vivo* and *in vitro*.

Almost all of the currently available assays or sensors for monitoring protease activity rely on monitoring the cleavage of an appropriate substrate for signal transduction (Richardson, 2002; Nagai and Miyawaki, 2004). Labeled peptide substrates, where donor and acceptor fluorophores flank a minimal substrate sequence are among the more popular protease substrates and rely on changes in FRET between the two fluorophores for signal transduction (Mahajan *et al.*, 1999; Rodems *et al.*, 2002; Nagai and Miyawaki, 2004). The fluorophores can be either organic dyes attached to peptidyl substrates (extrinsically prepared) or fluorescent proteins (intrinsically biosynthesized *in vivo*). Since extrinsically prepared peptide substrates labeled with organic dyes are easily synthesized but still need intracellular delivery, they are more utile for *in vitro* assays (Rodems *et al.*, 2002). This format has been especially useful for screening large libraries of inhibitor molecules and determining enzymatic kinetics; this type of assay has even been extended to coupled formats where the presence or absence of phosphorylation determines sensitivity to proteolysis (Rodems *et al.*, 2002). The latter provides an indirect method for monitoring *in vitro* kinase/phosphatase activity and screening potential inhibitors against particular target enzymes of this class. Recently developed protease substrates also include quantum dots (QDs) as central donor fluorophores, decorated with several acceptor dye-labeled peptide substrates (Medintz *et al.*, 2003a; Shi *et al.*, 2007).

Monitoring intracellular proteolysis has focused heavily on basic research applications, particularly the elucidation of cellular processes. The Miyawaki laboratory has been responsible for developing optimized sensors specific for caspases that consist of ECFP- and YFP-Venus separated by an appropriate caspase-specific peptide substrate (Nagai and Miyawaki, 2004). These sensors have been used to monitor the spaciotemporal activation of caspase-3 within single cells and a similar sensor design has also investigated the role of the caspase signaling cascade in single HeLa cells during TNFα-induced apoptosis (Luo *et al.*, 2003; Takemoto *et al.*, 2003). These type of sensors have proven useful for elucidating the role of proteolysis in generating the amyloid-beta protein which is deposited in senile plaques in Alzheimer's disease (Kinoshita *et al.*, 2001, 2002). In general, many of the protease sensors are used for proprietary drug screening and related research and thus are not as well described in the literature as other types of sensor or assay formats.

16.3.2.3. Protein translocation

The expression of a protein under study as a fusion with a fluorescent protein partner has become a very convenient and powerful way to monitor the intracellular location and activity of a number of proteins ranging from transcription factors to ligand receptors. In this section, we highlight two elegant examples which exemplify the power of fluorescent protein fusions in monitoring cellular protein location and function. Nelson *et al.* (2002) conducted detailed studies of the nuclear factor κB (NF-κB) family of transcription factors. Proteins of this family normally reside in the cytoplasm of cells in a complex with IκB inhibitor proteins. Stimulation with TNFα leads to proteosomal degradation of the IκB proteins and nuclear translocation of the NF-κB proteins. The expression of RelA (an NF-κB subunit) and IκBα fused to fluorescent proteins was used to measure the dynamics of these processes in transfected HeLa cells. The authors were able to simultaneously visualize RelA-dsRed nuclear translocation and IκBα-EGFP degradation in response to TNFα stimulation. In later studies, this group was able to use this same system to demonstrate the NF-κB-directed oscillations in IκBα expression levels (Nelson *et al.*, 2004).

Martinez *et al.* (2005) used a fluorescently tagged chimera of the glucocorticoid (GR) and estrogen (ER) receptors to develop a cell-based assay screen for ligands of the ER. Their GR–ER–GFP chimeric receptor took advantage of the observation that when fused to GFP, GR is found primarily in the cytoplasm. Upon addition of the ER domain, however, the new chimeric protein not only localized to the cytoplasm but also translocated to the nucleus in response to all ERα ligands, while remaining unresponsive to glucocorticoids. This construct performed equally well when used in either transient transfection or when stable cells lines were generated that constitutively expressed the chimeric receptor.

16.3.2.4. Sugars, amino acids, and small ligands

An alternative name for this section could very well be "Frommers Sensors," due to this laboratory's seminal contributions to their development

(Medintz, 2006). By combining fluorescent proteins with PBP specificity, Frommer created a series of FRET-based intracellular sensors based on a common structural design. An enhanced cyan fluorescent (ECFP) donor and enhanced yellow fluorescent (YFP) acceptor protein termed "Venus" are fused at the termini of selected PBPs. Upon binding targeted analyte, the obligatory PBP conformational change alters the donor/acceptor separation distance and thus the FRET efficiency in a quantifiable manner (Figure 16.4) (Fehr *et al.*, 2004a). Some sensors function by bringing the donor/acceptor fluorophores closer together and increasing FRET efficiency while others decrease this distance and efficiency; however, it appears that this functional choice cannot yet be engineered into sensors. This sensor design can be considered the closest thing to a first generation biosensing "toolset" by virtue of the fact that the same ECFP/YFP donor/acceptor pairs are consistently reused and target specificity is determined in the choice of PBP inserted between the two. Further, semi-rational manipulation of the linker size coupled with optimized chromophore placement in the sensors and mutation-directed alteration of binding site affinity have created a set of sensors with wide-ranging specificity and sensitivity (Deuschle *et al.*, 2005).

The first sensor based on this design utilized MBP to target uptake of the nutrient maltose in yeast cells (Fehr *et al.*, 2002). Maltose binding resulted in an increase in FRET efficiency. Selective mutation of the MBP-binding site altered binding kinetics to create a series of sensors that could monitor sugar concentrations from the low micromolar to low millimolar concentrations. Continuing with this same strategy, a variety of other sensors were created, almost all of which also utilize selective mutations to alter target affinity and thus increase the effective sensing concentrations. These include sensors for ribose (Lager *et al.*, 2003), glucose (Fehr *et al.*, 2004b), glutamate (Okumoto *et al.*, 2005), sucrose (Lager *et al.*, 2006), and phosphate (Gu *et al.*, 2006). As all of these sensors are encoded on plasmids, a number of mechanistic functions can be added or deleted as desired. For example, addition of both a leader sequence and a transmembrane domain to the glutamate sensor directed the expressed sensor to the secretory pathway and anchored the final mature protein in the extracellular plasma membrane (Figure 16.4) (Okumoto *et al.*, 2005). This allowed monitoring of glutamate

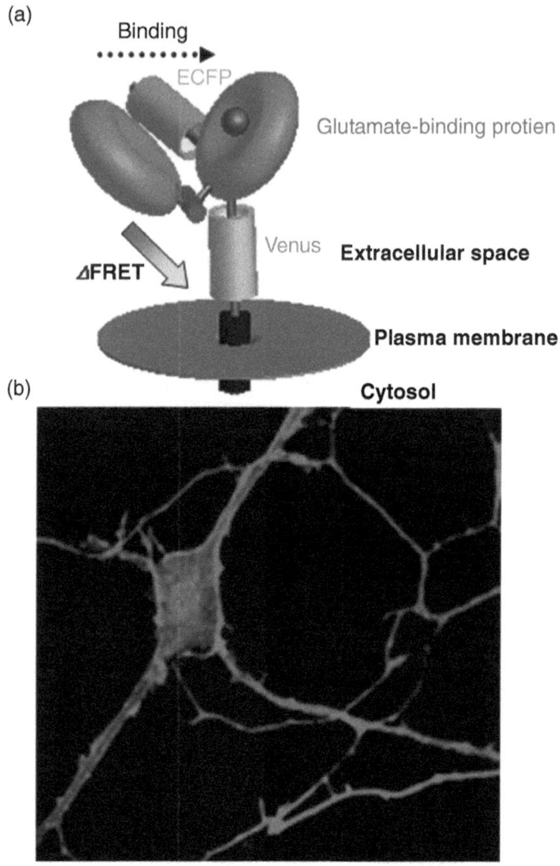

Figure 16.4 Membrane immobilized glutamate sensor (Okumoto *et al.*, 2005). (a) Model of the glutamate sensor, as displayed on the cell surface. The two lobes of the bacterial glutamate-binding protein YBEJ are shown in green, with glutamate (red) in the central binding pocket. The fluorescent proteins ECFP (blue) and Venus (yellow) are fused to one of the lobes. Upon binding glutamate the two lobes close relative to each other, which alters the FRET efficiency between the two fluorescent proteins and signals that binding has occurred. (b) Confocal image of hippocampal neurons expressing a 600 nM affinity surface-localized glutamate nanosensor. The fluorescence is highest at the plasma membrane. Figure courtesy of W. Frommer and reproduced from Okumoto *et al.* (2005) with permission from the National Academy of Sciences, USA. (see Plate 11)

neurotransmitter levels in rat hippocampal cells and showed its accumulation prior to depolarization. Frommer has also been able to extend his sensors from yeast and mammalian cells to plants. Introduction of a glucose nanosensor into a transgene-silencing mutant strain of *Aribidopsis thaliana* allowed monitoring of glucose levels in both roots and leaves and showed the presence of steep sugar gradients. This strongly suggests that within this plant the steady-state concentrations of this carbon source are not under tight homeostatic control (Deuschle *et al.*, 2006). As a general class, these sensors have an important research role to play especially in understanding intracellular nutrient flux. Given the large number of available PBPs to choose from, and the ability to rationally redesign their binding specificity/sensitivity, it is almost assured that we will continue to see significant growth in the targeting capabilities of this sensing "toolbox" coming out of the Frommer Laboratory.

16.3.2.5. Carbohydrate modification of proteins

Carrillo *et al.* (2006) have reported one of the few sensors that allow monitoring of post-translational modification of proteins with a carbohydrate. This sensor consists of four components: (1) an enhanced CFP donor; (2) a *O*-*N*-acetyl-D-glucosamine (*O*-GlcNAc)-binding domain derived from the monomeric *Escherichia coli* bacterial lectin GafD; (3) a peptide substrate specific for *O*-GlcNAc modification from casein kinase II; and (4) a YFP-Venus acceptor. Upon glycosylation of the peptide moiety, the GafD domain binds the sugar substrate, bringing the two fluorophores into close contact and increasing FRET efficiency. Interestingly, this sensor allows reversible monitoring of both glycosylation and its enzymatic removable in marked contrast to many other similar but single-reaction-modification sensor designs. Although the *in vitro* kinetics are slow (up to 20 h, similar to many available commercial chemical reagents for monitoring similar modifications), *in vivo* transfection with this genetic sensor allowed intracellular monitoring of glycosylation dynamics (Carrillo *et al.*, 2006).

16.3.2.6. Lactamase activity

An interesting enzyme–chemical substrate sensing combination that has had tremendous impact on pharmaceutical research deserves some discussion. The enzyme consists of the β-lactamase derived from the *E. coli* ampicillin resistance gene which efficiently cleaves penicillin and cephalosporins. Indeed, it has been shown in some configurations that this efficiency can approach diffusion limitations. Appending this lactamase gene sequence onto an appropriately sensitive promoter sequence allows the quantitative expression and interaction of transcriptional and regulatory factors to be monitored through concomitant downstream enzymatic activity (Zlokarnik *et al.*, 1998). However, the key to this sensor's function was the synthesis of a complex, multifunctional cell-permeable FRET-based chemical substrate (Figure 16.5). This was accomplished by attaching a coumarin donor and fluorescein acceptor to a central cephalosporin and functionalizing the construct with several ester functionalities. The esters allow the substrate to cross the cell membrane and enter the cytoplasm where ubiquitous esterases hydrolyze them and the increased polarity now traps the chemical intracellularly. Further modifications allowed the substrate to be more pH insensitive and more polar, and optimized the spacing for lactamase recognition. In the uncleaved form, almost all of the donor excitation energy is efficiently transferred to the fluorescein acceptor where about a third of this energy is re-emitted fluorescently. Lactamase hydrolysis results in a spontaneous rearrangement of the structure where the donor emits about three-quarters of its absorbed excitation energy and the fluorescein is almost completely quenched by the free thiol (Zlokarnik *et al.*, 1998). Use of this unique enzyme–substrate combination allowed dose–response and kinetic analysis of gene transcription at both the single-cell and population levels and almost all the signaling pathways tested resulted in either expression, repression, or both of specific response genes and elements when stimulated (Zlokarnik *et al.*, 1998). Thus, this type of sensor continues to be an important player in pharmaceutical screening, as it allows monitoring of transcriptional readouts including not only expression, but also quantitative inhibition by drug candidates. Indeed using flow cytometry, this sensor has already been applied in a genome-wide functional assay to identify and isolate cell clones and genetic elements responsive to specific drug or stress stimuli (Whitney *et al.*, 1998), as well as low-expression or down-regulated genes (Knapp *et al.*, 2003). A variety of other enzyme reporter–chemical substrate combinations

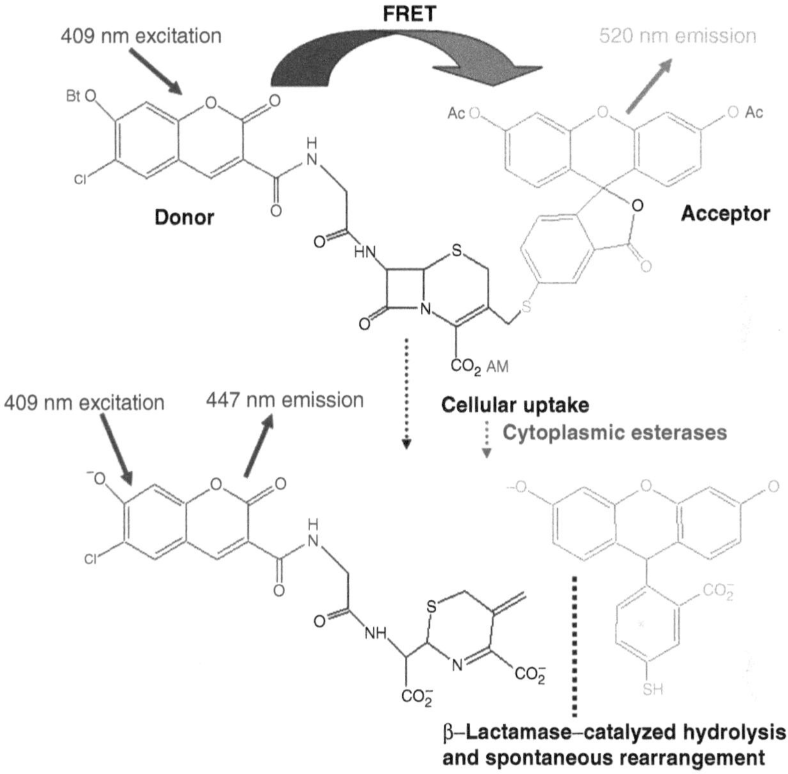

Figure 16.5 Cell permeable FRET-based substrate for monitoring intracellular β-lactamase activity. The donor coumarin (blue) and acceptor fluorescein (green) are joined by a central cephalosporin (black). Red designates ester functionalities which include butyryl (Bt) and acetoxymethyl (AM) groups. Upon exposure to cells, the ester groups allow cellular uptake across and are then cleaved by cytoplasmic esterases, trapping the substrate within the cell. β-lactamase expression and activity catalyzes the hydrolysis and structural rearrangement of the cephalosporin, altering FRET efficiency. Structures adapted from Zlokarnik *et al.* (1998). (see Plate 12)

exist, including alkaline phosphatase, luciferase, and β-galactosidase, although they are, in general, far less catalytically efficient than the above described β-lactamase/substrate combination (Zlokarnik *et al.*, 1998).

16.3.2.7. Zinc

As a first step in creating an intracellular Zn(II) sensor, Van Dongen *et al.* (2006) were able to create a subnanomolar affinity Zn(II) sensor while trying to design a copper sensor. In this case, CFP and YFP were again used as donor and acceptor fluorophores, respectively. CFP was fused to Atox1 and YFP was fused to WD4, each of which is a copper-binding domain from a protein involved in human copper homeostasis. Copper binding induces formation of a dimer between these proteins increasing FRET, but is easily disrupted by any thiol. However, the presence of very low concentrations of Zn(II), even with thiols present, forms a stable dimer structure. The sensor is also able to detect Co(II), Cd(II), and Pb(II) but is not cross-reactive with other physiologically relevant metals (Van Dongen *et al.*, 2006). Improving the choice of fluorescent proteins and their placement along with fusing the entire sensor into a single protein construct should allow intracellular applications with this sensor (Van Dongen *et al.*, 2006). The principal benefit of this design is that it can be completely encoded and expressed from a plasmid and is not sensitive to endogenous cellular thiols. Further, the affinity is similar to Zn(II)-binding peptides and orders of magnitude better than previous genetically encoded versions of similar sensors (Van Dongen *et al.*, 2006).

16.3.2.8. Phospholipase activity

An interesting example of a sensor that builds upon many of the concepts applied to proteases and lactamases has been developed for monitoring cleavage activity of phospholipase A_2 (PLA_2) (Wichmann *et al.*, 2006). The PLA_2 class of enzymes plays an important role in cellular signaling and inflammation, as they release arachidonic acid from phospholipids during synthesis of both prostoglandins and leukotrines. Wichman and colleagues synthesized a modified phosphatidylethanolamine (PE) substrate functionalized with nitrobenzoxazole (NBD) donor and Nile red acceptor dyes. The uncharged nature of both the dyes and PE substrate used allowed the probe to be inserted into lipid bilayers. *In vitro* testing of substrate entrapped in micelles with PLA_2 derived from bee venom showed a 30-fold change in FRET ratio after 30 min and almost 90% cleavage. A variety of cell types was then incubated with this fluorescent probe and after intracellular activity had been confirmed, intracellular dose-dependent inhibition with known inhibitors was also demonstrated. Based upon these initial findings, the probe was microinjected into an early fish embryo and time-dependent lipase activity was monitored during the developmental process, although the particular lipase enzyme isoform responsible was not apparent. The use of this probe and others based upon its design can help elucidate the complex intracellular role of lipids and provide insight into a field which still remains largely uninvestigated. The reader will have noted that the protease, lacatamase, and phospholipase sensors described above share a common substrate design. This not only reflects that they share a common "cleavage" type functionality but it is also a reflection of the power of FRET for monitoring such activity.

16.4. Advantages and limitations for use in optical biosensing

Each of the biosensors or sensing strategies described above comes with its own set of benefits and liabilities. The liabilities can be generally categorized as arising from the fluorophore(s) photophysical properties, fluorophore/sensor stability, or intracellular sensor delivery. The reader will note that for each sensor, one or more of these same liabilities may be applied, albeit in different combinations. For all extrinsically prepared sensors, appropriate intracellular delivery to target organelle(s) remains a major challenge. This can be accomplished by using either unassisted transfection, assisted delivery, or microinjection. Unassisted transfection is exemplified by the uptake of the lactamase chemical substrate described above, which is primarily diffusion-driven and the substrate readily crosses the cellular membrane (Zlokarnik *et al.*, 1998). Assisted delivery includes utilizing cellular delivery peptides to facilitate membrane crossing (Nitin *et al.*, 2004; Delehanty *et al.*, 2006), lipid-mediated transfection reagents or electro/mechanical means such as electroporation and high-velocity gene-gun delivery (El-Andaloussi *et al.*, 2005; Torchilin, 2006; Toub *et al.*, 2006). The use of cellular delivery peptides, however, usually results in endosomal entrapment and, unless this is the actual targeting site for sensing, warrants endosomal escape of the sensor into the cytoplasm. The electro/mechanical means physically stress the cells, often causing damage to some part of the cellular population and often necessitating a long recovery period. Lastly, as microinjection is almost always manual, only a very limited number of cells can be addressed.

As a general class, fluorescent biosensors (excluding FRET protein sensors) have two major drawbacks. The first is that they are usually prepared *in vitro* and thus must be reintroduced *in vivo* for any intracellular monitoring. The second stems from their labeling with environmentally sensitive fluorophores. In general, these dyes do not have long-term aqueous stability and thus sensors utilizing them must be prepared and used relatively quickly. However, in their favor, large amounts of these materials can be prepared and delivered as desired, and the choice of dye(s) can be altered as needed. Using the PBP family as a sensing platform comes with several inherent benefits. Many have been cloned and are available on plasmids targeting a wide variety of disparate analytes, although these are mostly nutrients and biochemical precursors. Almost all PBPs can be easily prepared recombinantly and their affinities semi-rationally modified to allow sensing over a wide concentration range; a rich literature on this subject already exists (De Lorimier *et al.*, 2002). Frommer's fluorescent protein–PBP combination appears to be one of the most promising sensor designs currently available, but development of such combinations has shown that although many such PBP sensors work, there is no single "design and go" recipe on-hand and iterative testing and redesign is key to improving overall capabilities. Hellinga's method for redesigning these PBPs to target unnatural analytes remains promising; however, this computationally intensive process is limited to a few very skilled laboratories and has not yet been widely applied. Beyond protein-based sensors, those consisting of peptides or nucleic acids have similar issues of uptake, dye labeling, and intracellular delivery to address. Further, many cells do not tolerate exogenous nucleic acids or small peptides as part of what is believed to be a putative innate defense against perceived pathogens or pathogenic products.

The primary benefit of using fluorescent proteins, especially in any FRET-based design, is that the complete sensor can be genetically assembled and encoded on a single plasmid and introduced into almost any cell for expression. However, most fluorescent proteins are closely related both in structure and function, often slow to mature, have pH and ion sensitivity, and, more importantly, small Stokes shifts (Tsien, 1998). This means that for any FRET configuration most of the D–A pairs used have overlapping absorptions and the acceptor will thus receive significant direct excitation, necessitating the use of spectral deconvolution and sensitive instrumentation (Sapsford *et al.*, 2006). Recent research has focused on mutationally optimizing a CFP–YFP D–A combination for improved FRET-based sensing, which may alleviate some of these issues (Nguyen and Daugherty, 2005). Another consideration is the physical size of the fluorescent proteins themselves. As these usually fall into the 25–35 kDa range, excluding obligate dimers and tetramers (Baird *et al.*, 2000), they are often equal to or greater in size than the protein they are appended to. This can adversely affect any chimeric sensor's function due to steric hindrance and other allosteric interactions. Lastly, due to inherent photophysical issues such as small Stoke's shifts combined with broad absorption and emission spectra, almost all the sensors described above are not amenable to "multiplexing." This inability to multiplex has significantly hindered studies that try to correlate multiple physiological responses or phenomena, something that is a critical next step in this field (Table 16.1).

16.5. Potential for improving performance or expanding current capabilities

Four areas will contribute the most to improving the capabilities of all classes of *in situ* fluorescent sensors in the near future: new fluorophores and instrumentation, *de novo* sensor protein design, new stable "chemically" based sensors, and improved targeted intracellular delivery. The many different classes of fluorophores currently available are constantly being updated and expanded. Fluorescent proteins continuously undergo improvement to optimize their properties for intracellular probing and biosensing in general (Campbell *et al.*, 2002; Nguyen and Daugherty, 2005). New chemical reactivities are being developed that allow site-specific chemical dye-labeling of proteins *in vivo*. For

Table 16.1 Selected intracellular sensors

Sensor class	Sensing target	Fluorophore/construct	Reference
Fluorescent sensors	*Cellular Homeostasis*		
	–pH	BCECF, SNARF-1	Gottlieb *et al.* (1996)
	–Calcium	Fluo-4, Fura-2	Harkins *et al.* (1993)
	–Lipid peroxidation	*cis*-Parinaric acid	
	Nucleic acids		
	–Single FRET Molecular beacon	Single hairpin FRET probe	Tyagi and Kramer (1996)
	–Dual FRET	Dual hairpin FRET probes	Tsourkas *et al.* (2003); Santangelo *et al.* (2004)
Fluorescent protein sensors	Molecular beacon		
	Phosphorylation	FP–peptide FRET construct	Zhang *et al.* (2001); Ting *et al.* (2001); Green and Alberola-Ila (2005) Nagai and Miyawaki (2004)
	Proteolysis	FP–peptide FRET construct	Nelson *et al.* (2002); Martinez *et al.* (2005)
	Protein translocation	Target protein–FP fusion	Fehr *et al.* (2004a); Fehr *et al.* (2004b); Lager *et al.* (2006); Gu *et al.* (2006)
	Sugars, small ligands	PBP-based FRET construct	Carrillo *et al.* (2006) Zlokarnik *et al.* (1998)
	Glycosylation	FP–peptide FRET construct	Van Dongen *et al.* (2006)
	Lactamase activity	FRET chemical substrate	Wichmann *et al.* (2006)
	Zinc	FP–FRET construct	
	Phospholipase activity	Phosphatidylethanola mine-based FRET substrate	

example, the FlAsH method allows coupling of non-fluorescent, cell-permeable, biarsenical fluorophores to intracellular proteins, expressing an optimized polycysteine sequence, with only the conjugated fluorophore emitting light (Griffin *et al.*, 1998). Luminescent semiconductor quantum dots (QDs) are a promising new class of nanoscale fluorophore material whose unique properties may help address the "multiplexing" deficiency of the current generation of *in situ* fluorescent biosensors. Photophysically, these properties include: (1) the unparalleled ability to size-tune their narrow-symmetrical photoluminescent emissions; (2) their broad absorption spectra which increase dramatically toward the blue from their first band-edge; (3) molar extinction coefficients on the order of several millions; and (4) high quantum yields and high resistance to photobleaching (Murray *et al.*, 2000; Alivisatos *et al.*, 2005; Medintz *et al.*, 2005a; Michalet *et al.*, 2005). Combining these properties means that a mixture of many different QD populations/emissions can be efficiently excited at a wavelength far removed from their selective emissions and the resultant multiplex spectra are easily collected and deconvoluted (Clapp *et al.*, 2005; Chattopadhyay *et al.*, 2006). New types of organic and chemical fluorophores with improved stability are also being continuously added to the library of what is commercially available (Haugland, 2005).

The field of *de novo* protein design and engineering has only just begun and will obviously have much to offer to biosensing (Floudas *et al.*, 2006). Preliminary examples of what can be accomplished include Hellinga's semi-rational design strategies for creating new biorecognition elements for non-natural targets (Marvin and Hellinga, 2001a; Dwyer *et al.*, 2003, 2004; Allert *et al.*, 2004), as well as rigid linker elements for biosensor surface attachment and anchoring of ordered sensor arrays (Medintz *et al.*, 2006). It is clear that as the science of *de novo* protein engineering matures, we can expect protein-based sensors with enhanced stability, sensitivity, and specificity to a wide variety of both natural and synthetic targets. Additionally, the development of chemically based sensing elements such as peptide-nucleic acid-based molecular beacons (Corey, 2003; Zhilina *et al.*, 2005; Shakeel *et al.*, 2006) and amplified fluorescent polymer sensors will undoubtedly add to the biosensor "toolbox" (McQuade *et al.*, 2000; Disney *et al.*, 2004; Thomas *et al.*, 2005). Another field with much to contribute is that of targeted cellular delivery. Relevant issues to be dealt with here include targeted cell or subcellular organelle-specific delivery, control of concentration, long-term stability, and cellular toxicity. Fortunately, a great deal of research is being carried out by the pharmaceutical industry for targeted drug delivery and gene replacement, and the findings will undoubtedly spill over into improving intracellular biosensor capabilities.

Lastly, it is well worth considering that although the biosensing technology discussed here is primarily targeted at intracellular or *in situ* sensing applications, *in situ* biosensing can and will be applied to other important areas: pharmaceutical assays, environmental and defense concerns, and clinical diagnostics, to name but a few. This paradigm can be easily highlighted by using Hellinga's research for example. A fundamental understanding of how to utilize MBP and other PBPs' innate structure and function have led to biosensors specific for nerve agents (Allert *et al.*, 2004; Dwyer and Hellinga, 2004). In a similar vein, we have investigated new biosensing modalities beginning with MBP as our test "platform," and this research led to several different sensor configurations specific for explosives (Medintz *et al.*, 2003a, 2003b, 2004, 2005b; Goldman *et al.*, 2005). Clearly, we can expect that in the future fluorescent biosensors will illuminate far more than the intracellular environment.

Acknowledgements

The authors acknowledge the US Naval Research Laboratory and the Office of Naval Research for continuing support.

References

Ala-Aho, R. and Kahari, V.M. (2005) *Biochimie*, **87**, 273.
Alberts, B., Johnson, A., Lewis, J. *et al.* (2002) *Molecular Biology of the Cell.* New York: Garland Science Publishers.
Alivisatos, A.P., Gu, W., and Larabell, C.A. (2005) *Ann. Rev. Biomed. Eng.*, **7**, 55.

Allert, M., Rizk, S.S., Looger, L.L., and Hellinga, H.W. (2004) *Proc. Natl. Acad. Sci. USA*, **101**, 7907.
Alper, J. (2006) *Anal. Chem.*, **78**, 5249.
Anand, K., Ziebuhr, J., Wadhwani, P. *et al.* (2003) *Science*, **300**, 1763.
Artal-Sanz, M. and Tavernarakis, N. (2005) *FEBS Lett.*, **579**, 3287.
Baird, G.S., Zacharias, D.A., and Tsien, R.Y. (2000) *Proc. Natl. Acad. Sci. USA*, **97**, 11984.
Campbell, R.E., Tour, O., Palmer, A.E. *et al.* (2002) *Proc. Natl. Acad. Sci. USA*, **99**, 7877.
Carrillo, L.D., Krishnamoorthy, L., and Mahal, L.K. (2006) *J. Am. Chem. Soc.*, **128**, 14768.
Chattopadhyay, P.K., Price, D.A., Harper, T.F. *et al.* (2006) *Nat. Med.*, **12**, 972.
Chen, Y., MacDonald, P.J., Skinner, J.P. *et al.* (2006) *Microsc. Res. Tech.*, **69**, 220.
Clapp, A.R., Medintz, I.L., Uyeda, H.T. *et al.* (2005) *J. Am. Chem. Soc.*, **127**, 18212.
Cohen, G.M. (1997) *Biochem. J.*, **326**, 1.
Corey, D.R. (2003) *Lett. Pept. Sci.*, **10**, 347.
De Lorimier, R.M., Smith, J.J., Dwyer, M.A. *et al.* (2002) *Prot. Sci.*, **11**, 2655.
Delehanty, J.B., Medintz, I.L., Pons, T. *et al.* (2006) *Bioconj. Chem.*, **17**, 920.
Deuschle, K., Chaudhuri, B., Okumoto, S. *et al.* (2006) *Plant Cell*, **18**, 2314.
Deuschle, K., Okumoto, S., Fehr, M. *et al.* (2005) *Prot. Sci.*, **14**, 2304.
Disney, M.D., Zheng, J., Swager, T.M., and Seeberger, P.H. (2004) *J. Am. Chem. Soc.*, **126**, 13343.
Dwyer, M.A. and Hellinga, H.W. (2004) *Curr. Opin. Struc. Biol.*, **14**, 495.
Dwyer, M.A., Looger, L.L., and Hellinga, H.W. (2003) *Proc. Natl. Acad. Sci. USA*, **100**, 11255.
Dwyer, M.A., Looger, L.L., and Hellinga, H.W. (2004) *Science*, **304**, 1967.
El-Andaloussi, S., Holm, T., and Langel, U. (2005) *Curr. Pharm. Des.*, **11**, 3597.
Fehr, M., Ehrhardt, D.W., Lalonde, S., and Frommer, W.B. (2004a) *Curr. Opin. Plant Biol.*, **7**, 345.
Fehr, M., Frommer, W.B., and Lalonde, S. (2002) *Proc. Natl. Acad. Sci. USA*, **99**, 9846.
Fehr, M., Lalonde, S., Ehrhardt, D.W., and Frommer, W.B. (2004b) *J. Fluoresc.*, **14**, 603.
Floudas, C.A., Fung, H.K., McAllister, S.R. *et al.* (2006) *Chem. Eng. Sci.*, **61**, 966.
Giepmans, B.N.G., Adams, S.R., Ellisman, M.H., and Tsien, R.Y. (2006) *Science*, **312**, 217.

Gilardi, G., Zhou, L.Q., Hibbert, L., and Cass, A.E.G. (1994) *Anal. Chem.*, **66**, 3840.
Goldman, E., Medintz, I., Whitley, J. *et al.* (2005) *J. Am. Chem. Soc.*, **127**, 6744.
Gottlieb, R.A., Nordberg, J., Skowronski, E., and Babior, B.M. (1996) *Proc. Natl. Acad. Sci. USA*, **93**, 654.
Green, H.M. and Alberola-Ila, J. (2005) *BMC Chem. Biol.*, **5**, 1.
Green, K.A. and Lund, L.R. (2005) *BioEssays*, **27**, 894.
Griffin, B.A., Adams, S.R., and Tsien, R.Y. (1998) *Science*, **281**, 269.
Gu, H., Lalonde, S., Okumoto, S. *et al.* (2006) *FEBS Lett.*, **580**, 5885.
Handsley, M.M. and Edwards, D.R. (2005) *Int. J. Cancer*, **115**, 849.
Haugland, R.P. (2005) *The Handbook: A Guide to Fluorescent Probes and Labeling Technologies*. San Diego: Invitrogen Corp.
Harkins, A.B., Kurebayashi, N., and Baylor, S.M. (1993) *Biophys. J.*, **65**, 365.
Hermanson, G.T. (1996) *Bioconjugate Techniques*. San Diego: Academic Press.
Imamura, T. (2003) *J. Periodontol.*, **74**, 111.
Jabs, T. (1999) *Biochem. Pharm.*, **57**, 231.
Jameson, D. (2006) A Nano-History of Fluorescence, *1st LFD Workshop in Advanced Fluorescence Imaging and Dynamics*. University of California, Irvine.
Kinoshita, A., Whelan, C.M., Smith, C.J. *et al.* (2001) *J. Neurosci.*, **21**, 8354.
Kinoshita, A., Whelan, C.M., Smith, C.J. *et al.* (2002) *J. Neurochem.*, **82**, 839.
Knapp, T., Hare, E., Feng, L.X. *et al.* (2003) *Cytometry Part A*, **51A**, 68.
Lager, I., Fehr, M., Frommer, W.B., and Lalonde, S.W. (2003) *FEBS Lett.*, **553**, 85.
Lager, I., Looger, L.L., Hilpert, M. *et al.* (2006) *J. Biol. Chem.*, **281**, 30875.
Lakowicz, J.R. (2006) *Principles of Fluorescence Spectroscopy*. New York: Springer.
Lalonde, S., Ehrhardt, D.W., and Frommer, W.B. (2005) *Curr. Opin. Plant Biol.*, **8**, 574.
Looger, L.L., Dwyer, M.A., Smith, J.J., and Hellinga, H.W. (2003) *Nature*, **423**, 185.
Luo, K.Q., Yu, V.C., Pu, Y.M., and Chang, D.C. (2003) *Biochem. Biophys. Res. Commun.*, **304**, 217.
Mahajan, N.P., Harrison-Shostak, D.C., Michaux, J., and Herman, B. (1999) *Chem. Biol.*, **6**, 401.
Martinez, E.D., Rayasam, G.V., Dull, A.B. *et al.* (2005) *J. Steroid Biochem. Mol. Biol.*, **97**, 307.
Marvin, J.S. and Hellinga, H.W. (2001a) *Proc. Natl. Acad. Sci. USA*, **98**, 4955.
Marvin, J.S. and Hellinga, H.W. (2001b) *Nat. Struct. Biol.*, **8**, 795.
McQuade, D.T., Pullen, A.E., and Swager, T.M. (2000) *Chem. Rev.*, **100**, 2537.

Medintz, I.L. (2006) *Tr. Biotechnol.*, **24**, 539.
Medintz, I.L., Anderson, G.P., Lassman, M.E. *et al.* (2004) *Anal. Chem.*, **76**, 5620.
Medintz, I.L., Clapp, A.R., Mattoussi, H. *et al.* (2003a) *Nat. Mat.*, **2**, 630.
Medintz, I.L. and Deschamps, J.R. (2006) *Curr. Opin. Biotechnol.*, **17**, 17.
Medintz, I.L., Goldman, E.R., Lassman, M.E., and Mauro, J.M. (2003b) *Bioconj. Chem.*, **14**, 909.
Medintz, I.L., Goldman, E.R., Lassman, M.E. *et al.* (2005b) *Anal. Chem.*, **77**, 365.
Medintz, I.L., Sapsford, K.E., Clapp, A.R. *et al.* (2006) *J. Phys. Chem. B*, **110**, 10683.
Medintz, I.L., Uyeda, H.T., Goldman, E.R., and Mattoussi, H. (2005a) *Nat. Mater.*, **4**, 435.
Michalet, X., Pinaud, F.F., Bentolila, L.A. *et al.* (2005) *Science*, **307**, 538.
Miyawaki, A., Llopis, J., Heim, R. *et al.* (1997) *Nature*, **388**, 882.
Miyawaki, A. and Tsien, R.Y. (2000) *Meth. Enzymol.*, **327**, 472.
Miyawaki, A. (2003) *Dev. Cell*, **4**, 295.
Mukhopadhyay, R. (2006) *Anal. Chem.*, **78**, 7379.
Murray, C.B., Kagan, C.R., and Bawendi, M.G. (2000) *Ann. Rev. Mater. Sci.*, **30**, 545.
Nagai, T. and Miyawaki, A. (2004) *Biochem. Biophys. Res. Comm.*, **319**, 72.
Nagl, S., Schaeferling, M., and Wolfbeis, O.S. (2005) *Microchim. Acta*, **151**, 1.
Nelson, D.E., Ihekwaba, A.E., Elliott, M. *et al.* (2004) *Science*, **306**, 704.
Nelson, G., Paraoan, L., Spiller, D.G. *et al.* (2002) *J. Cell Sci.*, **115**, 1137.
Nguyen, A.W. and Daugherty, P.S. (2005) *Nat. Biotech.*, **23**, 355.
Nitin, N., Santangelo, P.J., Kim, G. *et al.* (2004) *Nucleic Acids Res.*, **32**, e58.
Okumoto, S., Looger, L.L., Micheva, K.D. *et al.* (2005) *Proc. Natl. Acad. Sci. USA*, **102**, 8740.
Peng, X.H., Cao, Z.H., Xia, J.T. *et al.* (2005) *Cancer Res.*, **65**, 1909.
Puente, X.S. and Lopez-Otin, C. (2004) *Genome Res.*, **14**, 609.
Puente, X.S., Sanchez, L.M., Overall, C.M., and Lopez-Otin, C. (2003) *Nat. Rev. Genet.*, **4**, 544.
Quiocho, F.A. and Ledvina, P.S. (1996) *Molec. Microbiol.*, **20**, 17.
Richard I. (2005) *J. Med. Genet.*, **42**, 529.
Richardson, P.L. (2002) *Curr. Pharm. Des.*, **8**, 2559.
Rider, T.H., Petrovick, M.S., Nargi, F.E. *et al.* (2003) *Science*, **301**, 213.
Rodems, S.M., Hamman, B.D., Lin, C. *et al.* (2002) *Assay Drug Dev. Tech.*, **1**, 9.
Santangelo, P.J., Nix, B., Tsourkas, A., and Bao, G. (2004) *Nucleic Acids Res.*, **32**, e57.

Sapsford, K.E., Berti, L., and Medintz, I.L. (2005) *Minerva Biotec*, **16**, 253.
Sapsford, K.E., Berti, L., and Medintz, I.L. (2006) *Angew. Chem. Int. Ed.*, **45**, 4562.
Schmid, J.A. and Neumeier, H. (2005) *Chem. Biochem*, **6**, 1149.
Shakeel, S., Karim, S., and Ali, A. (2006) *J. Chem. Technol. Biotech.*, **81**, 892.
Shaner, N.C., Campbell, R.E., Steinbach, P.A. *et al.* (2004) *Nat. Biotech.*, **22**, 1567.
Shaner, N.C., Steinbach, P.A., and Tsien, R.Y. (2005) *Nat. Methods*, **2**, 905.
Shao, F., Merritt, P.M., Bao, Z.Q. *et al.* (2002) *Cell*, **109**, 575.
Sharff, A.J., Rodseth, L.E., and Quiocho, F.A. (1993) *Biochemistry*, **32**, 10553.
Sharff, A.J., Rodseth, L.E., Spurlino, J.C., and Quiocho, F.A. (1992) *Biochemistry*, **31**, 10657.
Sheehan, J.J. and Tsirka, S.E. (2005) *Glia*, **50**, 340.
Shi, L., Rosenzweig, N., and Rosenzweig, Z. (2007) *Anal. Chem.*, **79**, 208.
Spurlino, J.C., Lu, G.Y., and Quiocho, F.A. (1991) *J. Biol. Chem.*, **266**, 5202.
Stenger, D.A., Gross, G.W., Keefer, E.W. *et al.* (2001) *Tr. Biotech.*, **19**, 304.
Stryer, L. and Haugland, R.P. (1967) *Proc. Natl. Acad. Sci. USA*, **58**, 719.
Takemoto, K., Nagai, T., Miyawaki, A., and Miura, M. (2003) *J. Cell Biol.*, **160**, 235.
Tan, W.H., Wang, K.M., and Drake, T.J. (2004) *Curr. Opin. Chem. Biol.*, **8**, 547.
Thomas, S.W., Amara, J.P., Bjork, R.E., and Swager, T.M. (2005) *Chem. Comm.*, **36**, 4572.
Thornberry, N.A., Ranon, T.A., Pieterson, E.P. *et al.* (1997) *J. Biol. Chem.*, **272**, 17907.
Ting, A.Y., Kain, K.H., Klemke, R.L., and Tsien, R.Y. (2001) *Proc. Natl. Acad. Sci. USA*, **98**, 15003.
Torchilin, V.P. (2006) *Ann. Rev. Biomed. Eng.*, **8**, 343.
Toub, N., Malvy, C., Fattal, E., and Couvreur, P. (2006) *Biomed. Pharmacother.*, **60**, 607.
Tsien, R.Y. (1998) *Ann. Rev. Biochem.*, **67**, 509.
Tsien, R.Y. (2005) *FEBS Lett.*, **579**, 927.
Tsourkas, A., Behlke, M.A., Xu, Y., and Bao, G. (2003) *Anal. Chem.*, **75**, 3697.
Tyagi, S. and Kramer, F.R. (1996) *Nat. Biotechnol.*, **14**, 303.
Van Dongen, E., Dekkers, L.M., Spijker, K. *et al.* (2006) *J. Am. Chem. Soc.*, **128**, 10754.
Vet, J.A., Majithia, A.R., Marras, S.A. *et al.* (1999) *Proc. Natl. Acad. Sci. USA*, **96**, 6394.
Vihinen, P., Ala-Aho, R., and Kahari, V.M. (2005) *Curr. Cancer Drug Targets*, **5**, 203.

Wada, A., Mie, M., Aizawa, M. *et al.* (2003) *J. Am. Chem. Soc.*, **125**, 16228.

Watson, J.D., Baker, T.A., Bell, S.P. *et al.* (2004) *Molecular Biology of the Gene*. Reading, MA: The Benjamin Cummings Publishing Co.

Whitney, M., Rockenstein, E., Cantin, G. *et al.* (1998) *Nat. Biotechnol.*, **16**, 1329.

Wichmann, O., Wittbrodt, J., and Schultz, C. (2006) *Angew. Chem. Int. Ed.*, **45**, 508.

Wolfbeis, O.S. (2005) *J. Mater. Chem.*, **15**, 2657.

Wolfbeis, O.S. (2006) *Anal. Chem.*, **78**, 3859.

Wu, Y.M., Wang, X.Y., Liu, X., and Wang, Y.F. (2003) *Gen. Res.*, **13**, 601.

Ye, J., Wang, S., Leonard, S.S. *et al.* (1999) *J. Biol. Chem.*, **274**, 34974.

Zhang, J., Campbell, R.E., Ting, A.Y., and Tsien, R.Y. (2002) *Nat. Rev. Mol. Cell Biol.*, **3**, 906.

Zhang, J., Ma, Y., Taylor, S.S., and Tsien, R.Y. (2001) *Proc. Natl. Acad. Sci. USA*, **98**, 14997.

Zhilina, Z.V., Ziemba, A.J., and Ebbinghaus, S.W. (2005) *Curr. Topics Med. Chem.*, **5**, 1119.

Zlokarnik, G., Negulescu, P.A., Knapp, T.E. *et al.* (1998) *Science*, **279**, 84.

Optical Biosensors: Today and Tomorrow (2nd Edition)
Frances S. Ligler and Chris Rowe Taitt (Editors)

Chapter 17

MICROFLUIDICS

Abraham D. Stroock, Ph.D.
School of Chemical and Biomolecular Engineering,
Cornell University, Ithaca, NY 1485, USA

Microfluidic technology offers a growing set of tools for manipulating small volumes of fluids to control chemical, biological, and physical processes that are relevant to sensing. The development of these tools based on lithographic techniques enables integration with electronic and optical elements necessary for the construction of operational sensor. This chapter focuses on two themes: (1) The presentation of the physical and technical context of microfluidic technology with an emphasis on the importance of specific flow properties in ensuring efficient chemical reactions in the bulk and at surfaces. (2) A presentation of the use of microfluidics to control living systems for biomedical applications such as tissue engineering and wound healing. Opportunities and challenges for implementation of microfluidics for sensors are discussed.

17.1. Technical concept

Microfluidics refers to fluid-handling technologies in which at least one dimension of the typical elements is less than 1 mm and internal volumes are less than 100 μl. This technology has been developed for printing (e.g., ink-jet printers), analytical chemistry, synthetic chemistry, and biotechnology. In this section, we will discuss the physical and technical aspects of microfluidic systems.

17.1.1. Physical characteristics of the microfluidic regime

The basic governing equations of fluid flow and mass transfer are the same in the microfluidic regime as in macroscopic systems. Nonetheless, there are several important features that distinguish dynamics in microstructures from that in common macroscopic systems (e.g., pipes and flasks). As the characteristic dimension of the flow (~volume/surface area of container or conduit) decreases, the importance of forces that act on the volume of the fluid, such as gravity, diminishes relative to that of forces that act at the surfaces, such as surface tension. The relative importance of inertial forces relative to viscous forces acting on the fluid also diminishes. The ratio of inertial to viscous forces is expressed as the Reynolds number of the flow, *Re*:

$$Re = \frac{\sigma v w}{\eta},$$

where ρ (kg/m^3) is the density of the fluid, v (m/s) is the velocity, w (m) is the characteristic dimension of the flow, and η (kg/ms) is the dynamic viscosity of the fluid. In microchannels, $Re < 100$, so flows are laminar rather than turbulent; flows in channels are typically laminar for $Re < 2000$. Co-flowing streams in a laminar flow intermix only by diffusion because no spontaneous eddies carry mass and momentum between them. This characteristic of laminar flows allows spatial control of solute within the flow, but it hinders rapid mixing and efficient mass transfer to solid reactive boundaries.

17.1.2. Technical characteristics of the microfluidic regime

The most common methods for driving flows in microchannels are with (1) pressure gradients applied via external reservoirs; (2) pressure-driven flow generated by peristaltic pumps on the device; and (3) externally applied electric fields. Capillary, acoustic, and magnetic forces have also been used to move fluids in microchannels. In pressure-driven flow in channels, the flow speed, v,

$$v \sim -\frac{w^2 \nabla P}{\eta} (m/s),$$

where ∇P is the applied pressure gradient; to maintain a given flow speed, the required pressure gradient grows rapidly as the channel shrinks. In pressure-driven flows, the flow speed varies from zero at the wall to its maximum value in the middle of the channel. This variation leads to dispersion (spreading) of solute along the direction of the flow; this dispersion is unfavorable for transporting narrow bands of solute such as in a chemical separation.

Electric fields can be applied by placing electrodes in the inlet and outlet of a microchannel. In electrolyte solutions (e.g., buffered water), flows are generated by the interaction of the applied electric field with ions that accumulate in a thin ($<1\ \mu$m) layer of fluid, the Debye layer (DL), adjacent to channel walls; these flows are called electroosmotic (EO) flows. Fluid in the DL moves due to the electrical body force and entrains the remainder of the fluid. The flow speed in the bulk is:

$$v_{\mathrm{eo}} \sim \mu_{\mathrm{eo}} E,$$

where μ_{eo} is the EO mobility that depends on the surface charge density, the concentration of ions in the liquid, and the viscosity of the liquid ($|\mu_{\mathrm{eo}}| \sim 1 - 5 \times 10^{-8}$ m^2/(Vs)) and E is the magnitude of the electric field. In uniformly charged channels, v_{eo} is constant over the cross section except in the DL. This property allows bands of a single type of solute to be transported along the channel with little dispersion; capillary electrophoresis is a useful tool for chemical analysis that exploits this behavior and is performed in microfluidic devices (www.calipertech.com). The flow speed, v_{eo}, is also independent of the dimension of the channel down to the thickness of the DL; EO pumping is therefore appropriate in channels down to sub-micron dimensions. Disadvantages of EO pumping include the requirement for high voltages (V $\sim$ 1 kV), sensitivity to the chemical characteristics of the walls, and the tendency for species to separate electrophoretically in the applied field (this is an advantage for analysis, but a disadvantage for transporting general chemical mixtures). All microfluidic systems discussed below involve pressure-driven flow.

17.1.3. Technical approaches

Most methods of fabricating microfluidic systems are adaptations of techniques for fabricating microelectronic structures; photolithography is the core of these methods. These techniques lead to a flat, chip-like format in which microchannels are constrained in planar layers on a flat substrate. This format facilitates integration of electronics and optics, but it restricts the geometries that are accessible for the design of fluidic elements. The channel structures are etched into a flat surface of silicon or glass, or molded into a soft or hard plastic sheet; the channels are closed by sealing the structured material to another flat surface.

The most widely used lithographic technique for forming microfluidic structure is referred to as soft lithography (Xia and Whitesides, 1998). In soft lithography, elastomeric stamps, microfluidic channels, and membranes are prepared by casting or spin coating the liquid prepolymer of an appropriate elastomer against a master that has a patterned relief structure (Figure 17.1). Most of the research based on soft lithography has used poly(dimethylsiloxane) (PDMS) as the elastomer, because PDMS is biocompatible, permeable to gases (and can thus be used for cell culture), and inexpensive. Poly(dimethylsiloxane) also has good optical characteristics; the cured polymer is transparent from 235 nm to the near infrared (Wu and Whitesides, 2001) and can make tight, weakly scattering seals around embedded optical elements such as optical fibers (Chabinyc *et al.*, 2001). The interfacial properties of PDMS can be readily modified by plasma oxidation and silanization (Chaudhury and Whitesides, 1991). Poly(dimethylsiloxane) structures can be used many times in transferring patterns (we have used the same PDMS stamp in microcontact printing approximately 100 times over a period of several months without any noticeable degradation in its performance), and each master can be used to make a large number of stamps or membranes. The access to photolithographic equipment required (to fabricate masters) in soft lithography is therefore minimal.

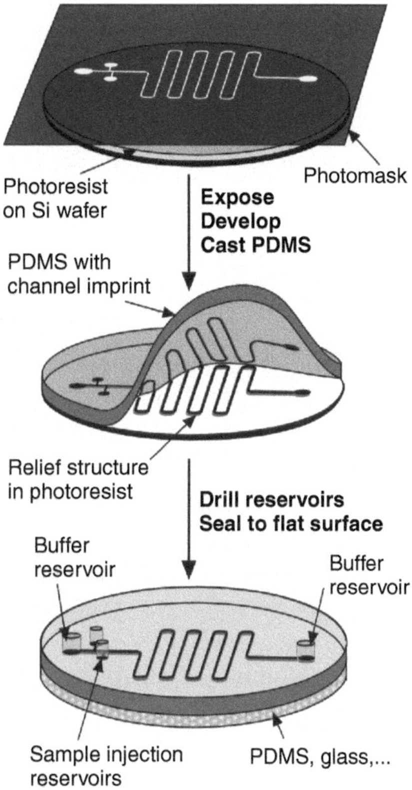

Figure 17.1 Schematic diagrams of the soft lithographic approach to fabricate microfluidic channels. Epoxy photoresist is spun onto a silicon wafer, exposed through a photomask, and developed to create a master structure. A negative of the structure on the master can be molded many (>100) times in PDMS, an elastomeric polymer. To form a closed channel, the PDMS mold is sealed to a flat surface either covalently by oxidizing the surfaces in a low-temperature plasma or non-covalently by applying pressure (Xia and Whitesides, 1998; McDonald *et al.*, 2000; Whitesides and Stroock, 2001). Adapted from Whitesides and Stroock (2001).

Soft lithography also facilitates the fabrication of complex structures such as three-dimensional (3D) networks of channels (Anderson *et al.*, 2000; Love *et al.*, 2001). Chiu *et al.* demonstrated an example of a multilevel network of microchannels made by stacking a stamp and a membrane (Chiu *et al.*, 2000). The elastomeric character of PDMS enables the simple integration of a variety of thin organic materials such as filters and dialysis membranes (Chiu *et al.*, 2001; Ismagilov *et al.*, 2001). Soft lithography is also the basis for fabrication of highly integrated microfluidic structures (Thorsen *et al.*, 2002).

17.2. History

In the initial development stage of microfluidics (Manz *et al.*, 1991) fabrication techniques adapted from the electronics industry were used. With these techniques, channel structures are formed in hard materials such as glass and silicon using photolithography followed by etching; the channels are typically sealed by anodic bonding (Kovacs, 1998). These steps are slow, expensive, and require a cleanroom environment. Hard plastics are also used (see for example, Micronics, www.micronics.net). Most of the early work in microfluidics focused on using EO flow (Harrison *et al.*, 1993; Jacobson *et al.*, 1994). More recently, Whitesides and others have been developing alternative methods based on soft lithography to fabricate microfluidic devices (Figure 17.1) (Delamarche *et al.*, 1997; Duffy *et al.*, 1998; Beebe *et al.*, 2000; Quake and Scherer, 2000). These methods, which use PDMS as the principal material, are simple, and inexpensive, and can be performed in a standard laboratory environment. The mechanical flexibility of PDMS makes it appropriate for the fabrication of the movable components that are often required for the control of pressure-driven flows (Unger *et al.*, 2000). The use of pneumatically controlled valves and peristaltic pumps has allowed for the integration of hundreds of functional elements onto a single microfluidic chip (Thorsen *et al.*, 2002).

Academic activity in microfluidics has been thoroughly reviewed by several authors. See reviews by Manz (Auroux *et al.*, 2002; Reyes *et al.*, 2002; Vilkner *et al.*, 2004; Dittrich *et al.*, 2006) for a very complete survey of the literature with little analysis. See reviews by Stone *et al.* (2004) and Squires and Quake (2005) for a more complete analysis of physical phenomena.

In the past several years, microfluidic elements have been incorporated in a number of commercial instruments designed for bioanalysis. Agilent was an early entrant with its Bioanalyzer 2100 (www.chem.agilent.com) in which electrophoretic or press-driven separations are performed in glass chips and with fluorescence detection. A number of companies, including Abbott (www.abbottpointofcare.com) and Roche (www.roche.com), have introduced small-scale analysis systems for blood and protein analyses; these systems are designed to provide "point-of-care" analysis to the clinician.

17.3. State of the art

17.3.1. Controlled mass transfer for sensors

In microchannels of common dimensions, $w > 10\,\mu m$, the motion of solute by diffusion across the laminar flow is often slow relative to the flow speed along the channel. In this situation, mixing is often the slow step in a chemical process, unless transverse flows that stir the fluid are purposely induced. (Note: in turbulent flow, these transverse flows occur spontaneously.) An efficient stirring flow stretches and folds the fluid such that the interface between unmixed regions grows exponentially in time; this type of flow is called chaotic (Stone *et al.*, 2004). Active mixers create transverse motion in the principal flow, with local, oscillatory forces generated with bubbles, applied electric or magnetic fields, or flows in cross-channels; for an example, see Dodge *et al.* (2005). Active mixers have the potential to be very efficient, but require sophisticated controls. Passive mixers use fixed geometrical features in the channel to induce transverse components in the principal flow; for an example, see Stroock *et al.* (2002). These designs are simple to operate but may not achieve full mixing as quickly as active designs.

In the first example of a manipulation that depends on efficient mass transfer, we consider measurements of the kinetics of fast chemical reactions, such as an enzyme-catalyzed event or the folding of a protein. In this case, the homogenization of the mixture of reagents must be fast relative to the rate of the reaction that one wishes to study. Conventional techniques (e.g., stopped flow) rely on high flow speeds and turbulence in order to achieve this fast mixing event. While these techniques allow for high temporal resolution (sub-millisecond), they require specialized equipment and significant volumes of reagent solutions (>1 ml).

Figure 17.2 presents a microfluidic-based alternative developed by the Ismagilov laboratory that provides millisecond temporal resolution with

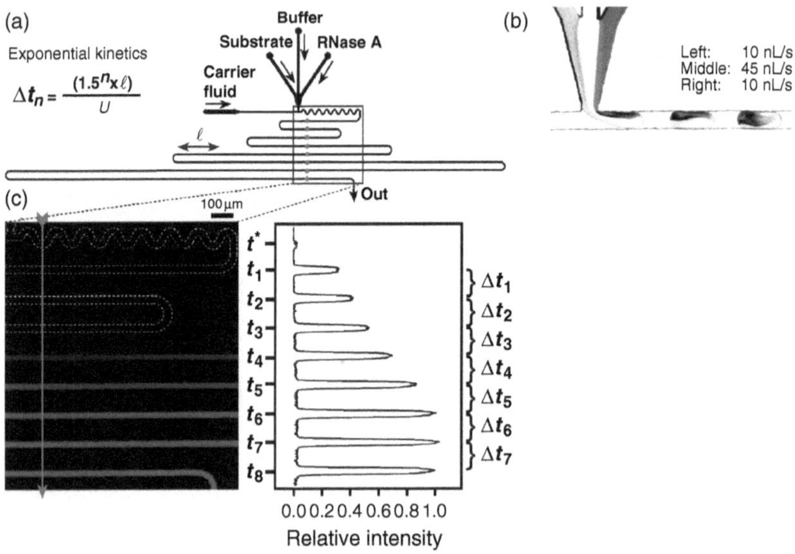

Figure 17.2 Enzyme kinetics measured with millisecond resolution in a microfluidic device. (a) Schematic diagram of a microfluidic device in which a substrate, an enzyme, and buffer are combined into a droplet that travels along the principal microchannel in a stream of inert oil. Chaotic mixing is induced in the drop in the wavy section of the channel. Concentration of product is observed downstream. (b) Optical micrograph of the injection region showing the formation of droplets. Initial mixing within the droplets is evident. (c) Fluorescence micrograph of downstream region (*left* – 4 s exposure time) and line scan of fluorescence intensity (*right*). The spatially continuous fluorescence is the result of the passage of many droplets. Intensity of fluorescence is proportional to the concentration of the product. Total temporal range is 450 ms, with an initial resolution of ~1–2 ms. Adapted from Song and Ismagilov *et al.* (2003).

minute volumes of sample (< μl) by generating efficient, non-turbulent mixing (Song and Ismagilov, 2003). In this approach, the reagents are transferred into droplets flowing in a stream of inert oil (Figure 17.2b). The shear in the oil between the droplets and the walls of the channel stir the contents of the droplets. This stirring motion is particularly efficient as a droplet passes through a zig-zagging section of microchannel (Figure 17.2a). In this region, the flow folds the fluid into itself in an exponential process known as a Baker's transformation (i.e., as in kneading dough); in fluid mechanics, such a flow is called *chaotic*. The reagents are nearly perfectly homogenized as the droplets leave this mixing region. Since the creation, mixing, and flow of droplets is steady, the temporal evolution of the reaction (RNase A cleaving a fluorogenic substrate in this case) can be read off of a single time-lapse fluorescence image of the remainder of the channel (Figure 17.2c). In this steady mode of operation, temporal resolution can be increased by simply raising the flow speed and the strength of the measured signal (fluorescence intensity in this case) can be increased with a longer time-lapse exposure time.

In the second example, we consider the management of mass transfer in assays such as enzyme linked immunosorbent assay (ELISA) in which the crucial binding event must occur between a surface-immobilized species and a solute. In these assays, the generic difficulties associated with mass transfer in small volumes are exacerbated by the fact that the chemical binding event must occur at a solid boundary in the fluid; at such solid–liquid interfaces, there is no motion in the fluid due to the interaction of the fluid with the solid. As binding of a solute occurs, it is depleted from the solution near the boundary, and more solute can only reach the boundary by diffusion across this depleted layer. The growth of this depleted layer is illustrated in Figure 17.3a–c for a simple ELISA-like assay in which a flow of analyte is driven over the binding surface through a simple microchannel (Figure 17.3a). Figure 17.3b, c show cross-sectional views of the computed distribution of the concentration of a solute in a solution at the beginning of the channel (Figure 3a) and 3 cm downstream (Figure 3b). Due to the growth of the depleted layer, only a small faction (14%) of the solute in the incoming solution has a chance to bind to the detection surface before the solution

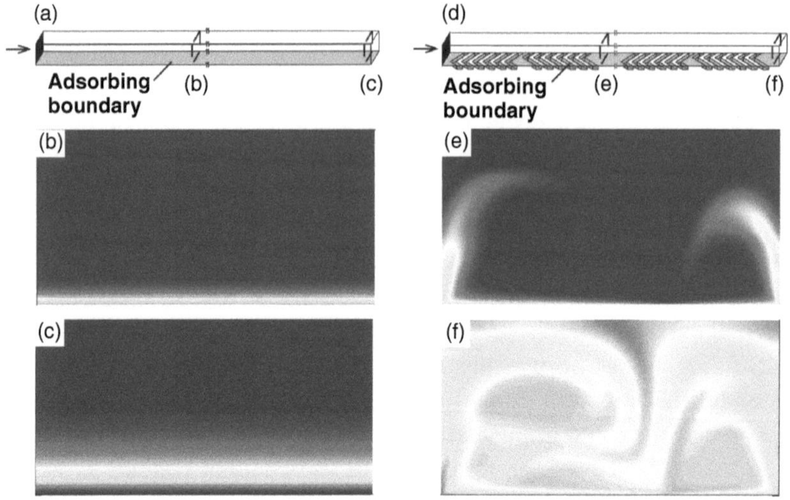

Figure 17.3 Calculations of the effects of micromixing on the delivery of analyte to surface-immobilized ligands. (a–c) Delivery of analyte solution in a simple microchannel (schematic diagram in (a)). A depleted layer forms above the adsorbing boundary as seen in the distribution of the adsorbing species across the cross section of the channel (b) 0.2 cm and (c) 2 cm downstream. (d–f) Delivery of analyte in a microchannel containing the SHM (grooves shown schematically in (d)). The stirring generated by the mixer sweeps out the depleted solution near the boundary, as seen in the cross section (e) 0.2 cm and (f) 2 cm downstream. These calculations correspond to a channel depth of 100 μm, a flow speed of 1 mm/s, a solute diffusivity of 10^{-11} m^2/s, and infinitely fast binding at the surface.

leaves the channel. This inefficient delivery of analyte to the surface means that the reagents may be wasted (although recirculation is an option), and the time required to achieve a detectable signal may be unnecessarily long. In order to improve the efficiency of delivery to a binding surface, one must stir the fluid so as to sweep fresh solution into the depleted region near the boundary. Figure 17.3 illustrates how this stirring of the depleted layer can be achieved with a strategy based on the Staggered Herringbone Mixer (SHM; Stroock *et al.*, 2002). The SHM is a passive (no moving parts) micromixer in which appropriately positioned and oriented grooves in the adsorbing boundary drive a chaotic stirring motion in fluid adjacent to the boundary. The calculations in

Figure 17.3b, c, e, and f show the concentration distribution of a reactive solute; these are the output of a finite difference integration of convection diffusion in an analytical model of the flow in the SHM (Stroock and McGraw, 2004). As seen by comparing Figure 17.3c and f, substantially more solute can be delivered to the adsorbing boundary (40% in this case) with the mixer than without it under the same flow conditions. This increased rate of delivery to the boundary can improve the efficiency with which reagents are used and increase the speed at which the assay is performed. We have made a detailed study of the impact of efficient mixing flows on rates of diffusion-limited surface reactions (Kirtland *et al.*, 2006). A number of recent studies have focused on mixing for DNA and RNA hybridization reactions in the contexts of biochip-based assays (Adey *et al.*, 2002; Raynal, 2004; Liu *et al.*, 2006).

The above examples highlight the enhancement of biochemical processes brought about by mixing in order to eliminate non-uniformity in the distribution of solute. In certain contexts, however, it can be beneficial to establish and control non-uniform distributions of reagents. One such context is a tedious sample preparation procedure employed throughout the course of many assays: serial dilutions. A microfluidic tool developed by Jiang *et al.* (2003) provides researchers with a cost-efficient way to shorten preparation time for dilutions of a sample. A schematic diagram of a microdiluter network (μDN) is shown in Figure 17.4a. The authors illustrate the function of the μDN by building a microfluidic version of the traditional ELISA assay for IgG antibodies in HIV+ human serum. The microfluidic network performs a serial dilution in a continuous fashion by repeatedly splitting the diluted stream of sample (serum) into two and mixing one of the resulting streams with a stream of diluent (buffer with bovine serum albumin (BSA)). The incorporation of SHMs (cf. Figure 17.3) ensures efficient mixing at each stage. In this system, one controls the dilution ratio in each stage by simply tuning the relative flow rates of the incoming streams of sample and diluent. For example, for equal incoming flow rates, the dilution ratio is 1:1, and the μDN produces steady streams with two-, four-, eight-fold, etc., dilutions of the original sample. The authors demonstrate the exponential dilution of BSA-FITC over 10 stages with a 1:1 ratio (Figure 17.4b, c).

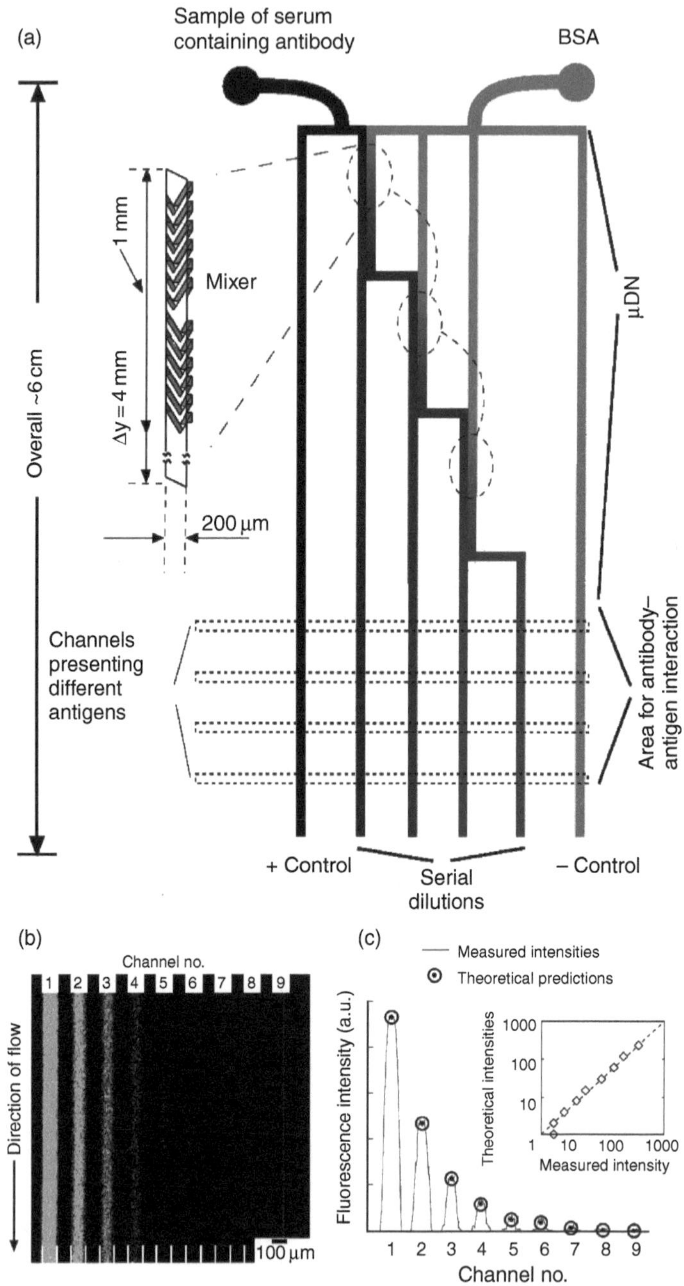
(a)
Sample of serum containing antibody
BSA
Overall ~6 cm
1 mm
Mixer
Δy = 4 mm
200 μm
μDN
Channels presenting different antigens
Area for antibody–antigen interaction
+ Control
Serial dilutions
– Control
(b)
Channel no.
1 2 3 4 5 6 7 8 9
Direction of flow
100 μm
(c)
Measured intensities
Theoretical predictions
Fluorescence intensity (a.u.)
Theoretical intensities
Measured intensity
Channel no.

We note that these and other authors have employed microfluidic strategies related to the μDN in order to generate steady-state gradients for the study of the chemotaxic response of mammalian cells such as neurons (Dertinger *et al.*, 2002) and neutrophils (Jeon *et al.*, 2002).

17.3.2. Integration

Integration of functions is not a focus area of this chapter, but it is worth noting a few important efforts on this front. With regard to integration of chemical function, the laboratory of Stephen Quake has been the clear leader, building on the pneumatic valve and pump mentioned in Section 17.1. Recently, he and collaborators have demonstrated the integration of all of the necessary components – mixers, separation columns, reactors – required for a multistep synthesis of an organic species (Lee *et al.*, 2005). They have also implemented polymerase chain reaction (PCR) (Ottesen *et al.*, 2006) and protein crystallization (Anderson *et al.*, 2005) in a high-throughput manner.

Another notable trend in integration that has particular relevance for this book involves the coupling of microfluidic and optical elements; this work was reviewed recently (Psaltis *et al.*, 2006). This effort aims both to exploit tunable properties and dynamics of moving fluids to control optical effects and to couple optical elements to microfluidics for

---

Figure 17.4 Microdiluter network (μDN) for performing a microfluidic ELISA assay. (a) Schematic representation of the μDN and ELISA assay. The μDN serially dilutes the antibody sample and each dilution crosses over a patch of immobilized antigen. An expanded view of the 4-mm-long SHM present at each junction of buffer and sample streams is also shown. The channels are defined in PDMS and the antigens are immobilized on a polycarbonate membrane. (b, c) Experimental characterization of the μDN operating at a 1:1 dilution ratio in nine channels. The fluorescence measured in (b) is plotted in (c) and indicates an exponentially decreasing concentration of analyte in each sequential channel. The inset shows the correlation between measured and predicted intensities. Adapted from Jiang *et al.* (2003).

detection and analysis. Whitesides and co-workers have used laminarly flowing streams of distinct fluids to form dynamic waveguides (Vezenov *et al.*, 2005b). They have further exploited these systems to form tunable lasers (Vezenov *et al.*, 2005a). Psaltis and colleagues have demonstrated a microscope integrated within a microchannel (Heng *et al.*, 2006) as well as a fluid-tunable photonic crystal (Erickson *et al.*, 2006).

17.3.3. Microfluidic interface with cellular systems

Many researchers have been developing microfluidic platforms for performing *in vitro* culture of mammalian cells. Notable early work was performed by Takayama and Whitesides, in which they demonstrated the ability to control the chemical environment of cultured cells with single-cell resolution (Takayama, 1999, 2001). Shuler has initiated an interesting effort to mimic the interconnected physiology of organs in an organism for studies of pharmacokinetics and toxicology that depend on biochemical connections between different tissues (Viravaidya and Shuler, 2004; Viravaidya *et al.*, 2004). In this so-called microcell culture analogue (μCCA), they culture three or more cell types in monolayers, each in a distinct chamber; all chambers are connected to a single fluidic loop such that the different cells share one volume of growth medium. They have demonstrated application of the μCCA to test naphthelene toxicity due to enzymatic activity of liver cells (Viravaidya *et al.*, 2004) and bioaccumulation in fibroblastic cells (Viravaidya and Shuler, 2004).

17.3.4. Microfluidic applications in biomedicine

Biomedical technologies such as prosthetics, scaffolds for tissue engineering, wound dressings, and vehicles for drug delivery require synthetic materials to interact with complex biological systems. Recent developments in microfluidic technology suggest a route to integrate function into synthetic biomaterials to control their internal and interfacial chemistry with high spatial and temporal resolution. Microfluidic paths permeating the material and connected to external reservoirs could allow a user to control the exchange of solutes with the bulk and surfaces of the material (Figure 17.5). The essential roles of the embedded microfluidic structure would parallel those of the vascular system in

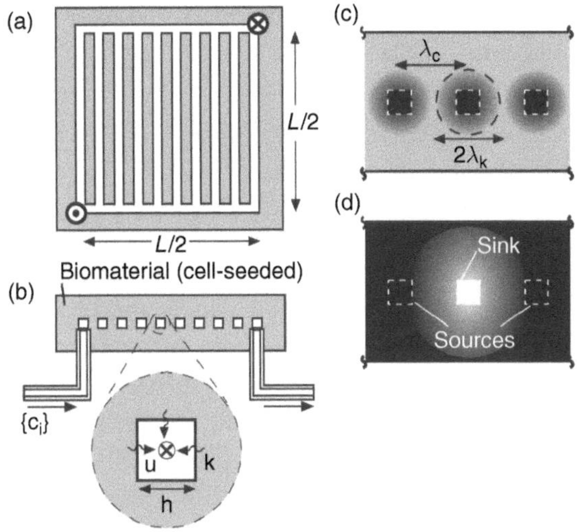

Figure 17.5 Microfluidic biomaterial. (a, b) Schematic diagram of a rudimentary microfluidic material in (a) top and (b) cross-sectional views. (c, d) Characteristic modes of operation for (c) solutes that are consumed in the bulk material, and (d) solutes that are not consumed. Parameters: concentrations of solutes in flow (Ajdari) (mol/m^3), flow speed in microchannel, u (m/s), mass transfer coefficient, k (m/s), channel dimension, h (m), interchannel spacing, λ_c (m), penetration depth of a consumed solute, λ_K (m). Adapted from Stroock and Cabodi (2006).

living systems: (1) efficient convective mass transfer to all volumes within the material; (2) centralized processing to refresh and monitor the working fluid; and (3) spatial heterogeneity. The major challenges in this development of microfluidic infrastructure within biological materials include the choice of materials that are compatible both with the fabrication and with the biology, development of fabrication processes that are biocompatible, and the design of microfluidic networks that allow for efficient operation.

17.3.4.1. Tissue engineering

The ambitious, and largely unrealized, goal of tissue engineering is to grow physiologically appropriate human tissues – muscle, cartilage, bone, and others – for implantation into patients or for use as sophisticated models for pharmacological studies or basic research (Langer and Vacanti, 2000). A dominant strategy in this pursuit is to template the growth process by seeding cells into a biomaterial; this scaffold can help define the geometry of the tissue as well as the chemical environment of the evolving cells during the growth process either *in vitro* or *in vivo* (e.g., within a laboratory animal acting as temporary host, or in the target patient). Important progress has been made in the past two decades on the chemical character of the materials that serve as scaffolds (Ratner and Bryant, 2004) and methods of controlling the growth process. Despite this progress, there still exist significant limitations on the size and complexity of engineered tissues. These limitations are due in part to the lack of infrastructure for solute exchange within existing scaffolds; infrastructure in the form of microfluidic structure could serve to both supply the tissue with adequate fluxes of metabolites and define spatially and temporally varying chemical signals with which to influence the behavior – phenotype, metabolic activity, movement – of the embedded cells.

In collaboration with the laboratory of Lawrence Bonassar, we have recently created a microfluidic network entirely within a biocompatible hydrogel, calcium alginate (Figure 17.6). We have adapted an injection molding technique that Bonassar developed to form macroscopic scaffolds (Chang *et al.*, 2001); we use the molding technique as part of a soft lithographic strategy (Cabodi *et al.*, 2005). Briefly, we flow a solution of sodium alginate onto a microstructured master, and cross-link it *in situ* by adding calcium ions. We spray the resulting calcium alginate gel (with microfeatures) with a sodium citrate solution to dissolve a thin surface layer, and placed it in contact with another alginate gel containing fluidic connections; the tubing connections are supported by a glass substrate. We re-gel the interface to produce a sealed bilayer with an enclosed microfluidic network.

This development is important in two ways. Firstly, microfluidic structure can be formed in calcium alginate gels of low solid fraction (4 wt.% in Figure 17.6). Due to this low solid fraction, the gel is permeable to diffusion of small and large molecules. As illustrated by the experiment shown in frames 6c—e, this property allows the flows through the

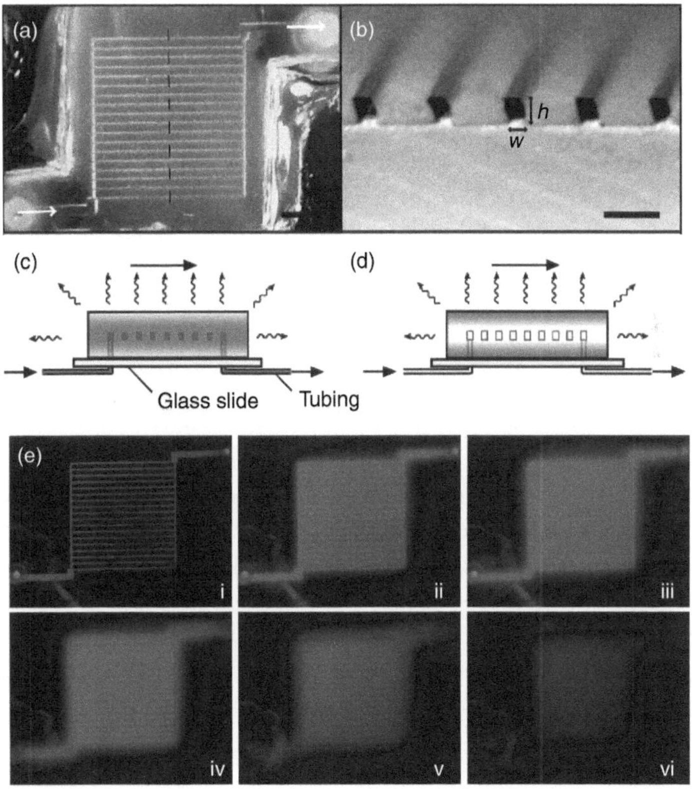

Figure 17.6 Microfluidic alginate gel. (a, b) Optical micrographs from (a) top and (b) cross-sectional view of a sealed double-layer microfluidic network in calcium alginate. (c, d) Schematic depictions of operation of microfluidic gel submerged in a well-stirred bath. The microfluidic network either delivers solute to the gel (c) or extracts solute from the gel (d). (e) Time series of fluorescence micrographs of delivery (i–iii) and extraction (iv–vi) of (rhodamine isothiocyanate) RITC-dextran (MW 70 kDa) from gel via microfluidic network. Adapted from Cabodi *et al.* (2005).

microfluidic structure to exchange solute with the 3D volume defined by the gel; this exchange is crucial to the function of the microfluidic structure as a supply network of the material. Secondly, the entire fabrication process is performed under conditions that are mild enough to allow pre-seeding of the alginate with cells. By pre-seeding the cells, they end up uniformly dispersed in the gel that defines the channels. The viability of chondrocytes (cartilage cells) through the key steps in this fabrication process is well established (Chang *et al.*, 2001; Rowley *et al.*, 1999). We are using this microfluidic hydrogel to direct the development of cells growing in the 3D volume defined by the gel. In this application, the channels can act either as an entirely synthetic vascular system or with endothelial cells plated on their inner walls.

17.3.4.2. Active wound dressing

In recent years, clinicians have been interested in the development of dressings that could be used to treat difficult wounds in a mechanically and chemically active manner. Such active dressings could be used to inhibit infection by clearing pathogens and delivering antibiotics, provide diagnostic information on the wound in real time, and stimulate healthy healing responses. In order to provide control of fluid motion within the wound bed, we have developed a fluidic dressing with which we can control mass transfer with the wound bed (Cabodi *et al.*, 2007). Our approach, illustrated in Figure 17.7a–c, focuses on the creation of a porous material for contact with the wound bed that has two key characteristics: (1) a well-defined geometry and (2) the potential to be patterned on the micrometer scale. We create a sponge of fixed micro- and macro-geometry ($\sim$20 μm pores in a 500 μm thick layer) via phase-separation polymerization of hydroxyethyl methacrylate. This material is compatible with UV photo-patterning. The well-defined mass transfer provided by this microstructured dressing enables applications in diagnostics and drug delivery during treatment, and mechanistic studies of the distinct roles played by mass transfer and mechanical stresses in the efficacy of active wound dressings (AWDs). Figure 17.7d shows photographs of a model wound bed formed from a hydrogel on which an active dressing is applied to perform an exchange of solutes: the gel is originally saturated with a solution of fluorescein (Figure 17.7d(i)); following 150 min of operation of the dressing with clean buffer, all of the dye has been extracted from the gel; following an additional 150 min of operation of the dressing with a solution of phenol red, the gel is impregnated with the new solute.

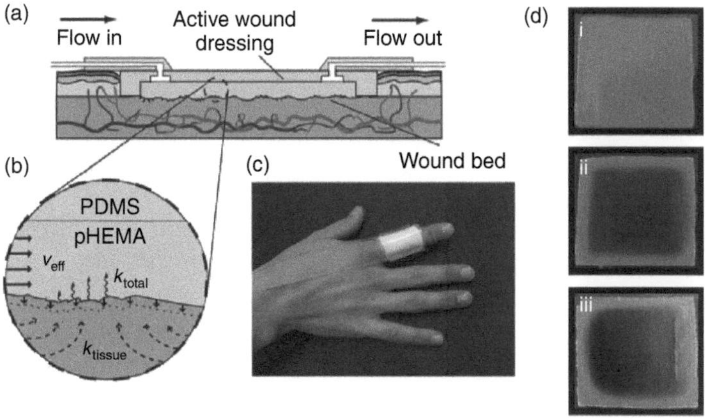

Figure 17.7 Active wound dressing (AWD). (a) Schematic cross-sectional view of an AWD placed in a wound bed, with input and output flow connections. (b) Expanded view of the interface between the AWD and the wound bed illustrating modes of action of AWD: the wiggly arrows represent convective mass transfer with mass transfer coefficient k_{total} into a flow through the porous pHEMA sponge; the dashed arrows represent flow through the tissue (which has hydraulic permeability κ_{tissue}); the bold arrows pointing into the tissue represent mechanical stress. (c) Image of an AWD adhered to a curved substrate (an uninjured index finger). (d) Images showing a top view of a hydrogel model of a wound bed at various stages of a sequential process of extraction and delivery (scale bar = 1 cm): (i) before extraction ($t = 0$ min); (ii) completed extraction ($t \sim 150$ min); (iii) completed delivery ($t \sim 320$ min). Adapted from Cabodi *et al.* (2007). (see Plate 13)

17.4. Advantages and limitations for use in optical biosensing

Microfluidics enables the manipulation of small volumes of fluid in a precise manner. This capability is important for the preparation and delivery of small synthetic or biological molecules for sensing. Microfluidic manipulations can also increase the rate of mass transfer in order to increase the rate of a fast chemical process that may be related to the sensing mechanism; this capability is especially important for sensors that require a binding reaction at a solid surface, as in DNA and protein chips. The implementation of microfluidics in biologically compatible materials allows for the creation of an intimate interface with living systems, such as cell cultures or full organisms. The formation of this biological interface may open new applications for sensors in biomedical and biochemical defense. The microfluidic format is compatible with integration of both electronic and optical elements that might be necessary in a sensor.

However, the implementation of microfluidics within a sensing scheme necessarily adds complexity relative, for example, to a solid-state sensor that directly senses from the ambient atmosphere. The integration of fluids with electronic and optical elements can be complicated by the need to isolate sensitive elements from conductive and corrosive fluids. Microfluidic processes will also require added power and control systems.

17.5. Potential for improving performance or expanding current capabilities

The continued development of microfluidic elements is clearly necessary to generalize the chemical processes for which it can be used. This development is under way in a large number of laboratories. Two specific directions of interest are (1) hybrid synthetic and living microfluidic devices that can act as long-term living sensors and (2) nanofluidic systems in which the characteristic size is less than a micrometer. In the development of hybrid synthetic and living microfluidic systems, the challenge is to tune the interplay between synthetic structure and imposed stimuli and the natural developmental processes of living cells. A central piece of physiology that would be valuable for capture is the wall of the blood vessels that mediate the exchange between tissues and flowing fluids. A piece of artificial tissue with native-like blood vessels could act as a sensor for chronic exposure to provide high chemical and biological specificity. In the nanofluidic regime, pressure may not be a useful driving force, whereas electroosmosis likely will be. Diffusion will often transport solute molecules as rapidly as the flow does; mixing will not be a problem, but careful delivery of solute will. Whereas the design principles for microfluidics were largely borrowed from macroscopic chemical systems (pumps, channels, etc.), the best guides to designing nanofluidic systems may be biological systems such as cells: transport will be controlled on the molecular rather than the hydrodynamic level with molecular motors carrying chemical building blocks along definite paths between distinct chemical environments. There are many opportunities for exciting developments in this young field.

References

Adey, N.B., Lei, M., Howard, M.T., *et al.* (2002) *Anal. Chem.*, **74**, 6413.

Ajdari, A. (2000) *Phys. Rev. E*, **61**, R45.

Anderson, J.R., Chiu, D.T., Jackman, R.J. *et al.* (2000) *Anal. Chem.* **72**, 3158.

Anderson, M.J., Hansen, C.L., and Quake, S.R. (2005) *Biophys. J.*, **88**, 55A.

Auroux, P.-A., Iossifidis, D., Reyes, D.R., and Manz, A. (2002) *Anal. Chem.*, **74**, 2637.

Beebe, D.J., Moore, J.S., Yu, Q. *et al.* (2000) *Proc. Natl. Acad. Sci. USA*, **97**, 13488.

Cabodi, M., Choi, N.W., Gleghorn, J.P. *et al.* (2005) *J. Am. Chem. Soc.*, **127**, 13788.

Cabodi, M., Cross, V.L., Qu, Z. *et al.* (2007) *J. Biomed. Mat. Res. B* (online).

Chabinyc, M.L., Chiu, D.T., McDonald, J.C. *et al.* (2001) *Anal. Chem.*, **73**, 4491.

Chang, S.C.N., Rowley, J.A., Tobias, G. *et al.* (2001), *J. Biomed. Mat. Res.*, **55**, 503.

Chaudhury, M.K. and Whitesides, G.M. (1991) *Langmuir*, **7**, 1013.

Chiu, D.T., Jeon, N.L., Huang, S. *et al.* (2000) *Proc. Natl. Acad. Sci. USA*, **97**, 2408.

Chiu, D.T., Pezzoli, E., Wu, H. *et al.* (2001) *Proc. Natl. Acad. Sci. USA*, **98**, 2961.

Delamarche, E., Bernard, A., Schmid, H. *et al.* (1997) *Science*, **276**, 779.

Dertinger, S.K.W., Jiang, X.Y., Li, Z.Y. *et al.* (2002) *Proc. Natl. Acad. Sci. USA*, **99**, 12542.

Dittrich, P.S., Tachikawa, K., and Manz, A. (2006) *Anal. Chem.*, **78**, 3887.

Dodge, A., Hountondji, A., Jullien, M.C., and Tabeling, P. (2005) *Phys. Rev. E*, **72**, 056312.

Duffy, D.C., McDonald, J.C., Schueller, O.J.A, and Whitesides, G.M. (1998) *Anal. Chem.*, **70**, 4974.

Erickson, D., Rockwood, T., Emery, T. *et al.* (2006) *Opt. Lett.*, **31**, 59.

Harrison, D.J., Fluri, K., Seiler, K. *et al.* (1993) *Science,* **261**, 895.
Heng, X., Erickson, D., Baugh, L.R. *et al.* (2006) *Lab Chip*, **6**, 1274.
Ismagilov, R.F., Ng, J.M.K., Kenis, P.J.A., and Whitesides, G.M. (2001) *Anal. Chem.*, **73**, 5207.
Jacobson, S.C., Hergenroder, R., Koutny, L.B., and Ramsey, J.M. (1994) *Anal. Chem.*, **66**, 2369.
Jeon, N.L., Baskaran, H., Dertinger, S.K.W. *et al.* (2002) *Nat. Biotechnol.*, **20**, 826.
Jiang, X.Y., Ng, J.M.K., Stroock, A.D. *et al.* (2003) *J. Am. Chem. Soc.,* **125**, 5294.
Kirtland, J.D., McGraw, G.J., and Stroock, A.D. (2006) *Phys. Fluids*, **18**,.
Kovacs, G.T.A. (1998). *Micromachined Transducers Sourcebook.* Boston: WCB/McGraw-Hill.
Langer, R. and Vacanti, J. (eds) (2000). *Principles of Tissue Engineering.* San Diego, CA: Academic Press.
Lee, C.C., Sui, G.D., Elizarov, A. *et al.* (2005) *Science*, **310**, 1793.
Liu, J., Williams, B.A., Gwirtz, R.M. *et al.* (2006) *Angew. Chem. Int. Ed. Engl.*, **45**, 3618.
Love, J.C., Anderson, J.R., and Whitesides, G.M. (2001) *MRS Bull.*, **26**, 523.
Manz, A., Harrison, D.J., Verpoorte, E.M.J. *et al.* (1991) *Chimia*, **45**, 103.
McDonald, J.C., Duffy, D.C., Anderson, J.R. *et al.* (2000) *Electrophoresis*, **21**, 27.
Ottesen, E.A., Hong, J.W., Quake, S.R., and Leadbetter, J.R. (2006) *Science*, **314**, 1464.
Psaltis, D., Quake, S.R., and Yang, C.H. (2006) *Nature*, **442**, 381.
Quake, S.R. and Scherer, A. (2000) *Science*, **290**, 1536.
Ratner, B.D. and Bryant, S.J. (2004) *Annu. Rev. Biomed. Eng.*, **6**, 41.
Raynal, F., Plaza, F., Beuf, A. *et al.* (2004) *Phys. Fluids*, **16**, L63.
Reyes, D.R., Iossifidis, D., Auroux, P.-A., and Manz, A. (2002) *Anal. Chem.*, **74**, 2623.
Rowley, J.A., Madlambayan, G., and Mooney, D.J. (1999) *Biomaterials*, **20**, 45.
Song, H. and Ismagilov, R.F. (2003) *J. Am. Chem. Soc.*, **125**, 14613.
Squires, T.M. and Quake, S.R. (2005) *Rev. Mod. Phys.*, **77**, 977.
Stone, H.A., Stroock, A.D., and Ajdari, A. (2004) *Annu. Rev. Fluid Mech.,* **36,** 381.
Stroock, A.D. and Cabodi, M. (2006) *MRS Bull.*, **31**, 114.
Stroock, A.D., Dertinger, S.K.W., Ajdari, A. *et al.* (2002) *Science*, **295**, 647.
Stroock, A.D. and McGraw, G.J. (2004) *Philos. Trans. Roy. Soc. A*, **362**, 971.
Takayama, S., McDonald, J.C., Ostuni, E. *et al.* (1999) *Proc. Natl. Acad. Sci. USA.,* **96**, 5545.

Takayama, S., Ostuni, E., LeDuc, P. *et al.* (2001) *Nature*, **411**, 1016.
Thorsen, T., Maerkl, S.J., and Quake, S.R. (2002) *Science*, **298**, 580.
Unger, M.A., Chou, H., Thorsen, T. *et al.* (2000) *Science*, **288**, 113.
Vezenov, D.V., Mayers, B.T., Conroy, R.S. *et al.* (2005a) *J. Am. Chem. Soc.*, **127**, 8952.
Vezenov, D.V., Mayers, B.T., Wolfe, D.B., and Whitesides, G.M. (2005b) *App. Phys. Lett.*, **86**, 041104.
Vilkner, T., Janasek, D., and Manz, A. (2004) *Anal. Chem.*, **76**, 3373.
Viravaidya, K. and Shuler, M.L. (2004) *Biotech. Prog.*, **20**, 590.
Viravaidya, K., Sin, A., and Shuler, M.L. (2004) *Biotech. Prog.*, **20**, 316.
Whitesides, G.M. and Stroock, A.D. (2001) *Phys. Today*, **54**, 42.
Wu, M.H. and Whitesides, G.M. (2001) *Appl. Phys. Lett.*, **78**, 2273.
Xia, Y. and Whitesides, G.M. (1998) *Angew. Chem. Int. Ed. Engl.*, **37**, 550.

Index

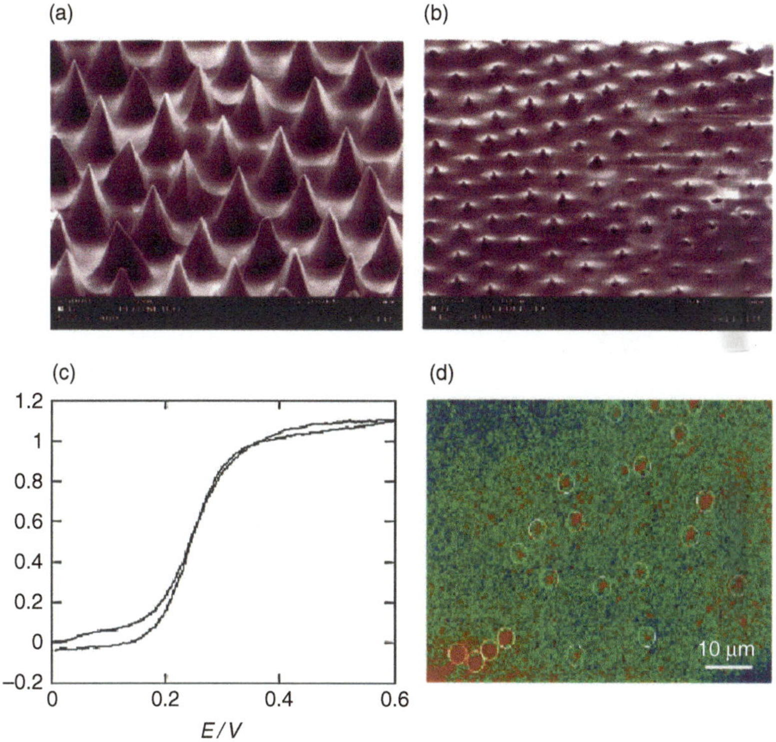

Plate 1 (a) SEM images of an etched and gold-coated fiber array and (b) an etched, gold-coated, and resin-insulated electrode array; (c) Cyclic voltammogram of the resin-insulated electrode array in 10 mM $Fe(CN)_6$ at $0.1\ V\ s^{-1}$; (d) ECL image monitored with an applied potential at selected areas of an etched, gold-deposited, and partially polymer-insulated electrode array. White circles indicate identified ECL spots. 1.2 V versus Ag/AgCl applied for 60 s, acquisition time 2 s, in 2 mM $Ru(bpy)_3^{2+}$/100 mM TPrA/pH 7 phosphate buffer. The red color corresponds to the most intense ECL signal; green and blue corresponds to little or no ECL signal. Reprinted with permission from the American Chemical Society (see Figure 1.6, p. 15).

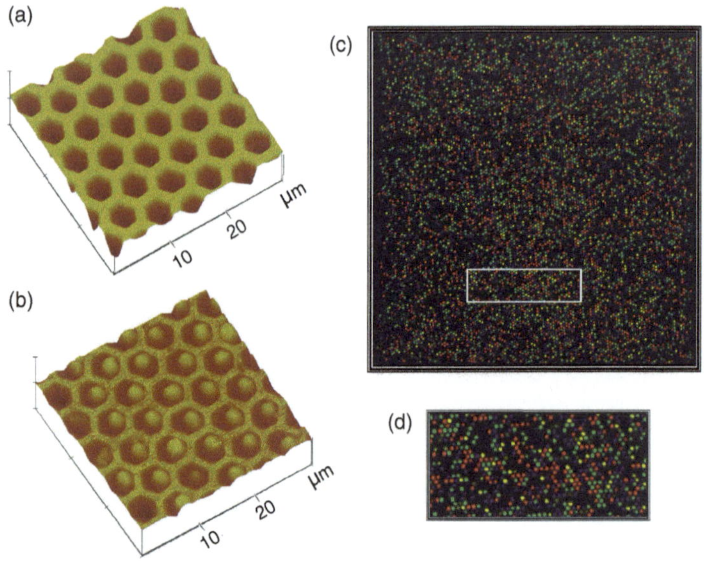

Plate 2 High-density multianalyte bio-optrode composed of microsphere array on an imaging fiber. (a) SFM of microwell array fabricated by selectively etching the cores of the individual fibers composing the imaging fiber. (b) The sensing microspheres are distributed in the microwell. (c) Fluorescence image of a DNA sensor array with ~13 000 DNA probe microspheres. (d) Small region of the array showing the different fluorescence responses obtained from the different sensing microspheres (Walt, 2000). Reprinted with permission from the American Association for the Advancement of Science (see Figure 1.18, p. 45).

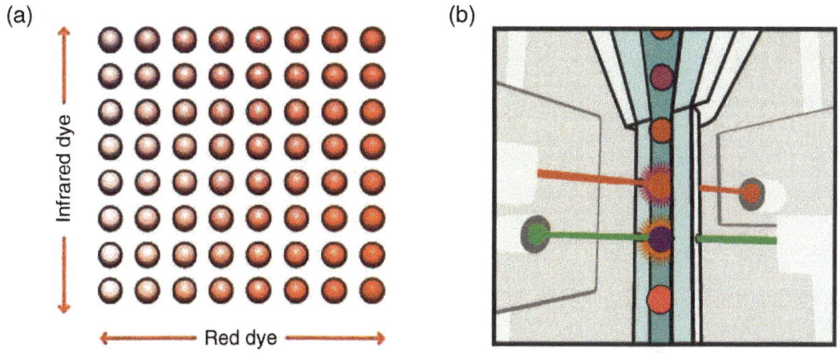

Plate 3 (a) Image represents bead regions dyed with 10 concentrations of two different classification fluorophores – red and near infrared – to produce 100 distinct bead "codes." Color coding enables each microsphere set to be classified individually and to be multiplexed with other microsphere sets. (b) Once inside the instrument, fluidics cause the microspheres to line up in single file as they pass by two lasers – a red laser to classify each color-coded microsphere for determining which assay is being carried out on that particle, and the green laser to measure the assay result on its surface. The presence and abundance of the reporter tag quantifies the occurring reactions precisely. (see Figure 11.1, p. 458).

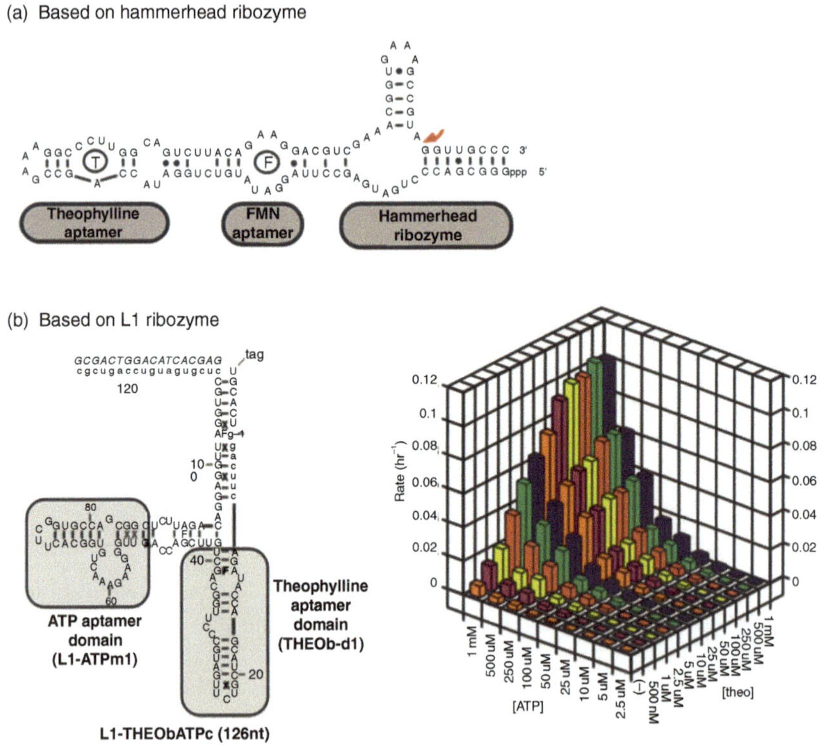

Plate 4 Different "dual effector" aptazymes. (a) A dual effector hammerhead aptazyme designed by Breaker and co-workers. The two aptamers are joined in series. Stabilization of one ligand-binding domain results in stabilization of the adjacent domain, and overall stabilization of this structure results in stabilization of the catalytic core. (b) A dual effector ligase aptazyme designed by Michael Robertson and co-workers. The two aptamers are mounted in parallel on different stems of the L1 ligase. The responsivity profile as a function of effector concentration is shown to the right. Maximal activity of the nucleic acid "and gate" is only seen when both effectors are present at high concentrations. In the presence of single effectors, activation is minimal (see Figure 13.7, p. 521).

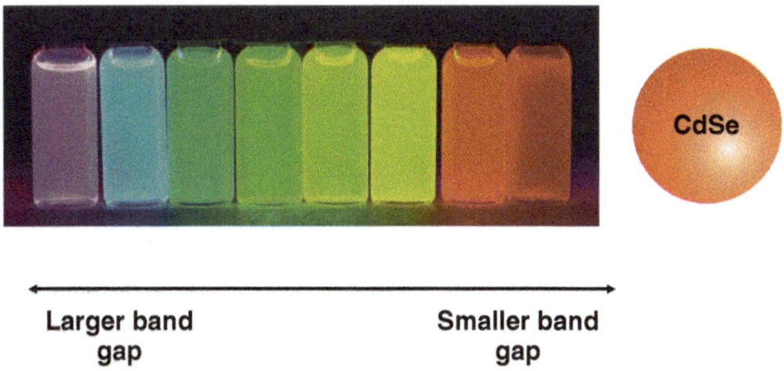

Plate 5 Optical properties of semiconductor nanocrystals. As the nanoparticles' size changes from 2 to 6 nm, there is a significant red shift in the luminescence of the CdSe QDs (see Figure 15.1, p. 586).

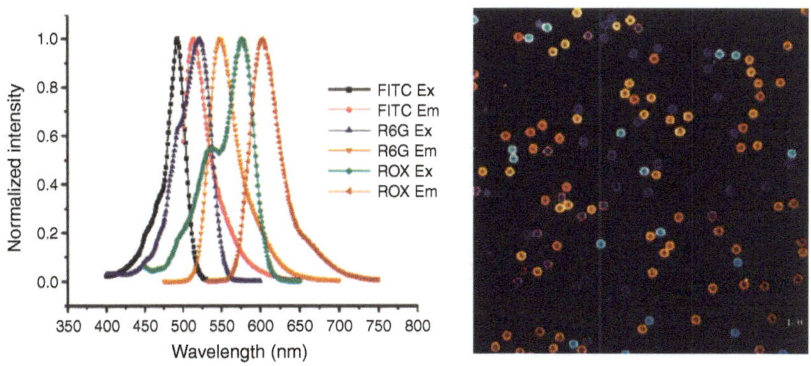

Plate 6 (a) Normalized excitation and emission spectra of FITC, R6G, and ROX dyes in pH 7.4 phosphate buffer (*left*). (b) Confocal fluorescence images of a mixture of five types of microsphere–nanoparticle complexes (*right*). Published with permission from the American Chemical Society (Wang *et al.*, 2006) (see Figure 15.5, p. 596).

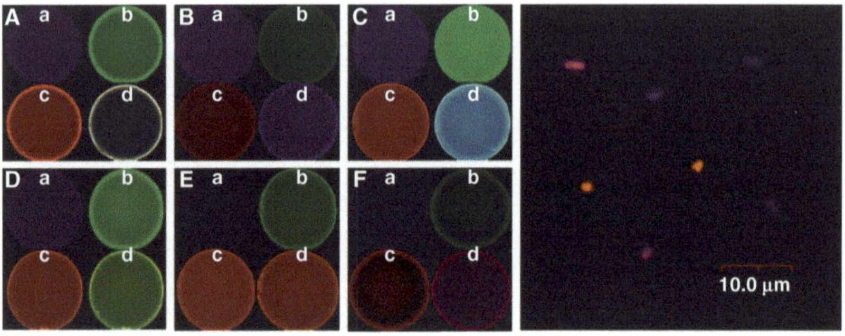

Plate 7 (*Left panel*, A–F) Fluorescence images of FRET nanoparticles taken under a confocal microscope: (a) FITC emission channel; (b) R6G emission channel; (c) ROX emission channel; (d) combinatorial color of the three detection channels. The six types of nanoparticles (dye doping ratios: 1:1:1, 1:0:1, 4:4:1, 0:1:1, 0:1:3, 0.5:0.5:3) exhibit six different resultant colors. (*Right panel*) Confocal image of three bacteria species (*Escherichia coli*, *Salmonella typhimurium*, and *Staphylococcus aureus*) specifically covered with three types of monoclonal antibody-labeled nanoparticles. Published with permission from the American Chemical Society (Wang *et al.*, 2006) (see Figure 15.6, p. 596).

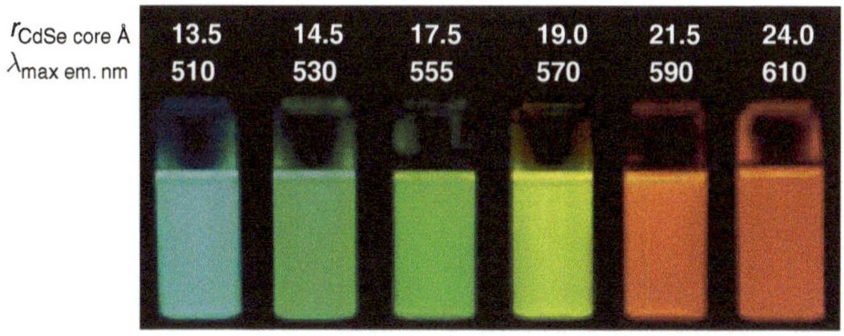

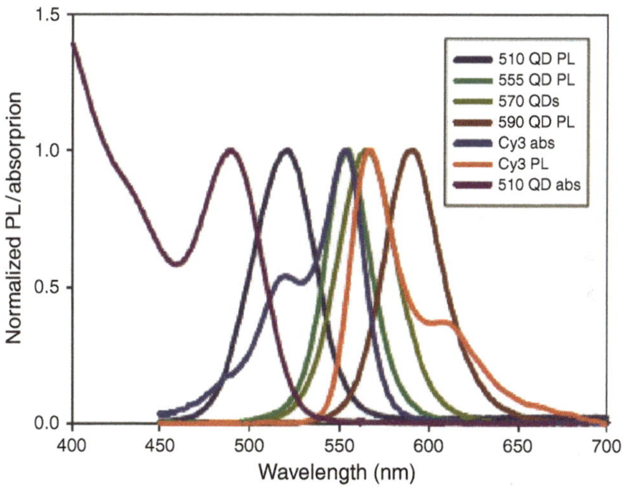

Plate 8 Properties of QDs (Ozkan *et al.*, 2004). (Above) Color photo demonstrating the size-tunable fluorescence properties and spectral range of six CdSe/ZnS core/shell QD dispersions. All samples were excited at 365 nm. For the 610 nm emitting QDs, this translates into an effective Stokes shift of ~250 nm. Figure reprinted with permission of the Nature Publishing Group. (Bottom) Absorption of 510 nm QDs and emission of 510, 555, 570, and 590 nm QDs superimposed over the absorbance and emission spectra of Cy3 dye. Published with permission from Molecular Diversity Preservation International (Sapsford, K.E. *et al.*, 2006) (see Figure 15.10, p. 604).

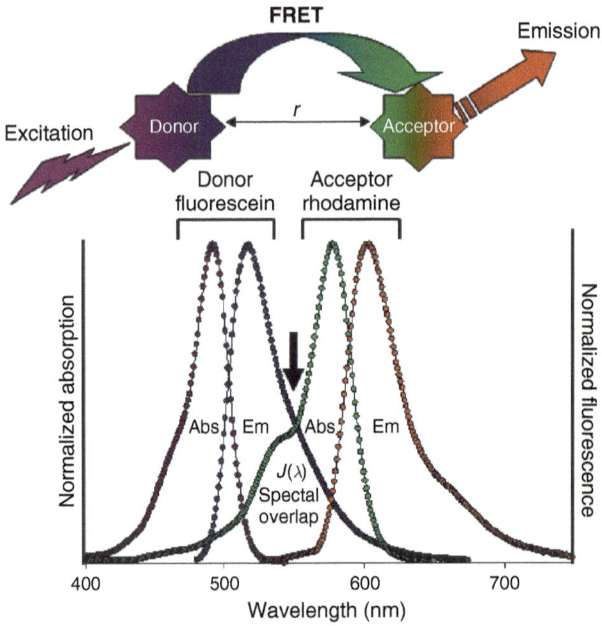

Plate 9 Schematic describing the FRET process. Upon excitation, the excited state donor molecule transfers energy non-radiatively to a proximal acceptor molecule located at distance *r* from the donor. The acceptor may either remit the energy fluorescently or dissipate it through non-radiative channels. The spectra shows the absorption (Abs) and emission (Em) profiles of one of the most commonly used FRET organic dye pairs: fluorescein donor–rhodamine acceptor. Fluorescein can be efficiently excited at 480 nm and emit at ~520 nm. The spectral overlap (between fluorescein emission and rhodamine absorption) occurs from 500 to 600 nm. The Förster distance R_0 for this pair is 55 Å. Thus, in an optimal configuration (donor–acceptor separation distance $r<55$ Å), excitation of fluorescein at <500 nm can result in significant rhodamine emission at >600 nm due to efficient FRET. Figure adapted from Sapsford *et al.* (2006) (see Figure 16.1, p. 626).

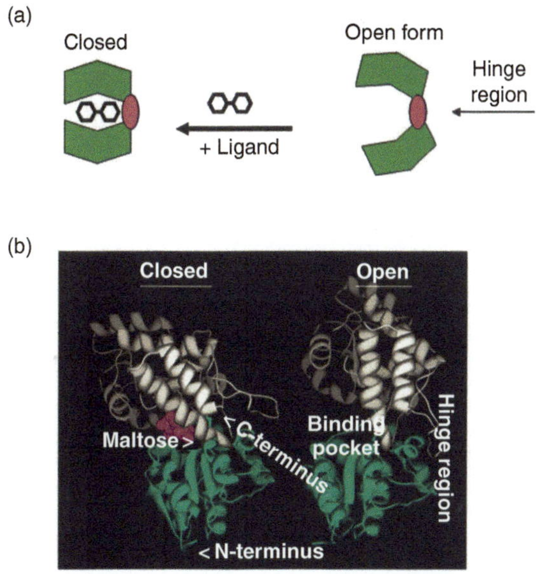

Plate 10 MBP structure and function. (a) A prototypic PBP structure is depicted; this has two lobes (green) that create the ligand-binding pocket and are joined by the hinge-binding region (pink). Upon binding the ligand, the PBP undergoes a transition to the closed structure. (b) Ribbon and string rendering of MBP structure in the open form and ligand-bound, closed form (Spurlino *et al.*, 1991; Sharff *et al.*, 1992, 1993). MBP dimensions are ~30 Å × 40 Å × 65 Å (Sharff *et al.*, 1992; Fehr *et al.*, 2002). The two domains (lobes) are highlighted in green and grey. Upon binding maltose (purple), the lobes rotate ~35° and twist laterally ~8° relative to each other. Overall the amino and carboxy termini move ~7 Å closer to each other after binding. Note the change in conformation of the overall structure upon binding maltose. Figure reprinted from Medintz and Deschamps (2006) with permission from Elsevier (see Figure 16.2, p. 632).

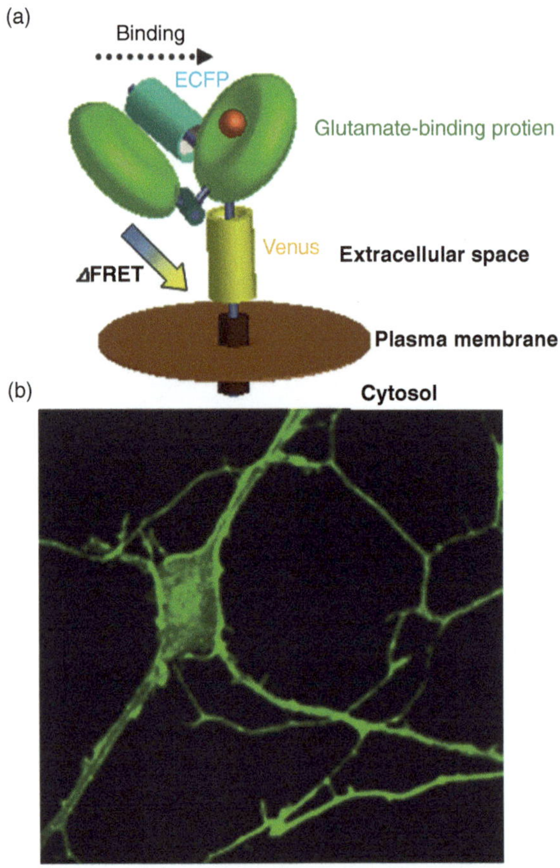

Plate 11 Membrane immobilized glutamate sensor (Okumoto *et al.*, 2005). (a) Model of the glutamate sensor, as displayed on the cell surface. The two lobes of the bacterial glutamate-binding protein YBEJ are shown in green, with glutamate (red) in the central binding pocket. The fluorescent proteins ECFP (blue) and Venus (yellow) are fused to one of the lobes. Upon binding glutamate the two lobes close relative to each other, which alters the FRET efficiency between the two fluorescent proteins and signals that binding has occurred. (b) Confocal image of hippocampal neurons expressing a 600 nM affinity surface-localized glutamate nanosensor. The fluorescence is highest at the plasma membrane. Figure courtesy of W. Frommer and reproduced from Okumoto *et al.* (2005) with permission from the National Academy of Sciences, USA (see Figure 16.4, p. 642).

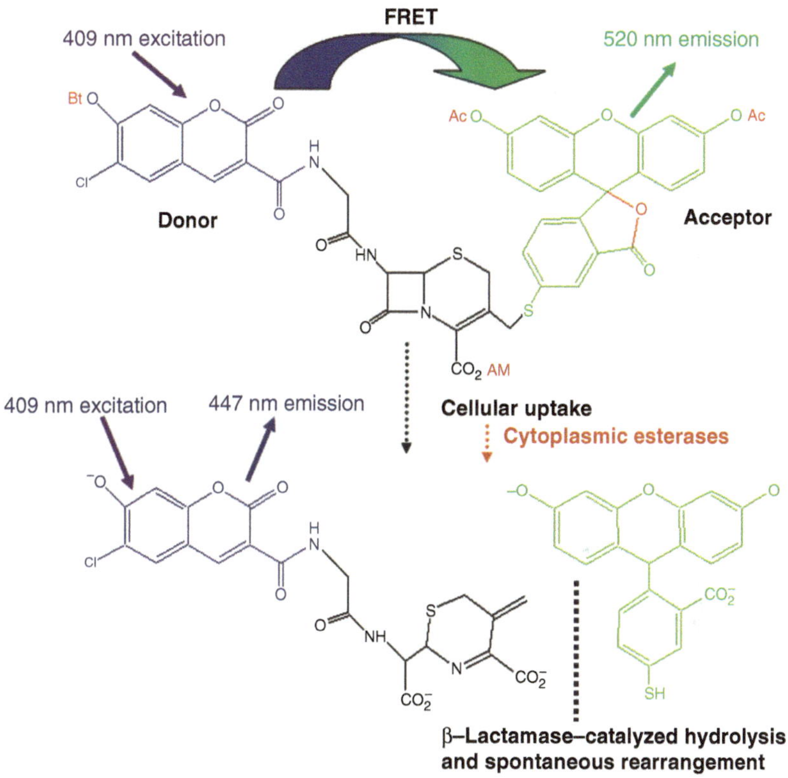

Plate 12 Cell permeable FRET-based substrate for monitoring intracellular β-lactamase activity. The donor coumarin (blue) and acceptor fluorescein (green) are joined by a central cephalosporin (black). Red designates ester functionalities which include butyryl (Bt) and acetoxymethyl (Ac) groups. Upon exposure to cells, the ester groups allow cellular uptake across and are then cleaved by cytoplasmic esterases, trapping the substrate within the cell. β-lactamase expression and activity catalyzes the hydrolysis and structural rearrangement of the cephalosporin, altering FRET efficiency. Structures adapted from Zlokarnik *et al.* (1998) (see Figure 16.5, p. 645).

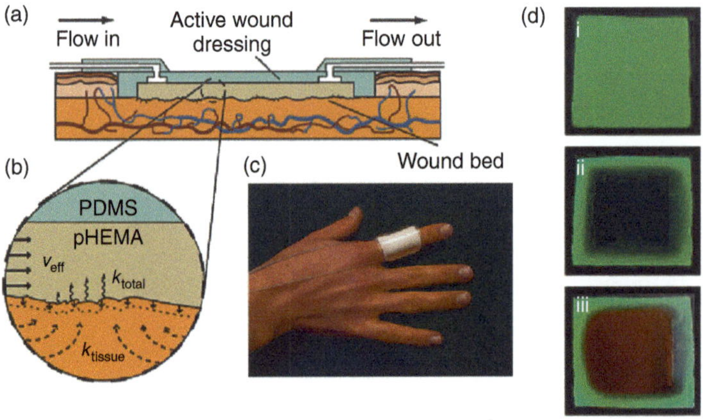

Plate 13 Active wound dressing (AWD). (a) Schematic cross-sectional view of an AWD placed in a wound bed, with input and output flow connections. (b) Expanded view of the interface between the AWD and the wound bed illustrating modes of action of AWD: the wiggly arrows represent convective mass transfer with mass transfer coefficient k_{total} into a flow through the porous pHEMA sponge; the dashed arrows represent flow through the tissue (which has hydraulic permeability κ_{tissue}); the bold arrows pointing into the tissue represent mechanical stress. (c) Image of an AWD adhered to a curved substrate (an uninjured index finger). (d) Images showing a top view of a hydrogel model of a wound bed at various stages of a sequential process of extraction and delivery (scale bar = 1 cm): (i) before extraction ($t = 0$ min); (ii) completed extraction ($t \sim 150$ min); (iii) completed delivery ($t \sim 320$ min). Adapted from Cabodi *et al.* (2007) (see Figure 17.7, p. 677).

FSC
www.fsc.org
MIX
Papier | Fördert gute Waldnutzung
FSC® C083411

Kolibri 360 GmbH
Ferdinand-Jühlke-Straße 7
99095 Erfurt, Deutschland
produktsicherheit@kolibri360.de